Synchrotron Radiation Research

Synchrotron Radiation Research

Edited by
HERMAN WINICK
Stanford Synchrotron Radiation Laboratory
Stanford, California

and
S. DONIACH
Stanford University
Stanford, California

PLENUM PRESS · NEW YORK AND LONDON

Library of Congress Cataloging in Publication Data

Main entry under title:

Synchrotron radiation research.

Includes bibliographies and index.
1. Synchrotron radiation. I. Winick, Herman II. Doniach, S.
QC787.S9S94   539.7′35   80-14378
ISBN 0-306-40363-3

©1980 Plenum Press, New York
A Division of Plenum Publishing Corporation
227 West 17th Street, New York, N.Y. 10011

Printed in the United States of America

Contributors

Frederick C. Brown, Department of Physics, University of Illinois, Urbana, Illinois 61801

George S. Brown, Stanford Synchroton Radiation Laboratory, Stanford Linear Accelerator Center, Stanford, California 94305

R. L. Cohen, Bell Laboratories, Murray Hill, New Jersey 07974

S. Doniach, Department of Applied Physics, Stanford University, Stanford, California 94305

P. Eisenberger, Bell Laboratories, Murray Hill, New Jersey 07974

Keith O. Hodgson, Department of Chemistry, Stanford University, Stanford California 94305

K. C. Holmes, European Molecular Biology Laboratory, and Max Planck Institut fur medizinische Forschung, Heidelberg, Germany

J. Kirz, Department of Physics, State University of New York, Stony Brook, New York 11794

Manfred O. Krause, Transuranium Research Laboratory, Oak Ridge National Laboratory, Oak Ridge, Tennessee 37830

I. Lindau, Stanford Synchroton Radiation Laboratory and Stanford Electronics Laboratories, Stanford University, Stanford, California 94305

F. W. Lytle, The Boeing Co., Seattle, Washington 98124

Ian H. Munro, Daresbury Laboratory, Daresbury, Warrington WA4 4AD, United Kingdom

Andrew R. Neureuther, Department of Electrical Engineering and Computer Sciences, and Electronic Research Laboratory, University of California at Berkeley, Berkeley, California 94720

C. Pellegrini, Brookhaven National Laboratory, Upton, New York 11973

J. F. Petroff, Laboratoire de Mineralogie-Cristallographie, associe au CNRS, Universite P. et M. Curie, 4 Place Jussieu, 75230 Paris Cedex 05, France and LURE (CNRS, Universite Paris-Sud) Bât. 209c, 91405 Orsay, France

James C. Phillips, European Molecular Biology Laboratory, Hamburg Outstation, c/o DESY, 52 Notkestieg 1, 2000 Hamburg, Germany

G. Rosenbaum, European Molecular Biology Laboratory, and Max Planck Institut fur medizinische Forschung, Heidelberg, Germany

Andrew P. Sabersky, Stanford Linear Accelerator Center, Stanford University, Stanford, California 94305

M. Sauvage, Laboratoire de Mineralogie-Cristallographie, associe au CNRS, Universite P. et M. Curie, 4 Place Jussieu, 75230 Paris Cedex 05, France and LURE (CNRS, Universite Paris-Sud) Bât. 209c, 91405 Orsay, France

D. Sayre, Mathematical Sciences Department, IBM Research Center, Yorktown Heights, New York 10598

J. H. Sinfelt, Exxon Research and Engineering Co., Linden, New Jersey 07036

C. J. Sparks, Jr., Metals and Ceramics Division, Oak Ridge National Laboratory, Oak Ridge, Tennessee 37830

James E. Spencer, Stanford Linear Accelerator Center, PEP Group, Stanford University, Stanford, California 94305

W. E. Spicer, Stanford Synchroton Radiation Laboratory and Stanford Electronics Laboratories, Stanford University, Stanford, California 94305

H. B. Stuhrmann, European Molecular Biology Laboratory, Oustation at the Deutsches Elekstronen-Synchroton (DESY), Hamburg, Germany

G. H. Via, Exxon Research and Engineering Co., Linden, New Jersey 07036

Herman Winick, Stanford Linear Accelerator Center, Stanford Synchroton Radiation Laboratory, P.O. Box 4349, Bin 69, Stanford, California 94305

Preface

This book has grown out of our shared experience in the development of the Stanford Synchrotron Radiation Laboratory (SSRL), based on the electron–positron storage ring SPEAR at the Stanford Linear Accelerator Center (SLAC) starting in Summer, 1973. The immense potential of the photon beam from SPEAR became obvious as soon as experiments using the beam started to run in May, 1974. The rapid growth of interest in using the beam since that time and the growth of other facilities using high-energy storage rings (see Chapters 1 and 3) demonstrates how the users of this source of radiation are finding applications in an increasingly wide variety of fields of science and technology.

In assembling the list of authors for this book, we have tried to cover as many of the applications of synchrotron radiation, both realized already or in the process of realization, as we can. Inevitably, there are omissions both through lack of space and because many projects are at an early stage. We thank the authors for their efforts and cooperation in producing what we believe is the most comprehensive treatment of synchrotron radiation research to date.

Synchrotron radiation has been utilized in more than 20 laboratories throughout the world. The fact that many of the applications described in this book have been developed at SSRL is a reflection of the close association of the editors and many of the authors with this laboratory and also due to the fact that it was at SSRL that the first utilization was made of a multi-GeV storage ring for a broad research program.

Consequently, in a real sense, much of the work described in this book would not have been possible without the contributions to and support of many people for the development of SSRL.

Outstanding among these is Bill Spicer, who was the first to conceive of a synchrotron radiation laboratory at Stanford and as Deputy Director participated heavily in its genesis and early development. Among the many others who made contributions we would like to call particular attention to the following and apologize in advance for our inability to list them all: *at SLAC*: Mark Baldwin, Norm Dean, Sid Drell, Gerry Fisher, John Harris, Bruce Humphrey, Fred Johnson, Joe Jurow, Bob Melen, Wolfgang Panofsky, Ewan Paterson, Fred Pindar, Burt Richter; *at Stanford*: Tom Eccles, Keith Hodgson, Brian Kincaid, Bill Massey, Marsh O'Neill, James

Phillips; *SSRL Staff*: Don Baer, Art Bienenstock, George Brown, Katherine Cantwell, John Cerino, Priss Dannemiller, Axel Golde, Ron Gould, Bob Gutcheck, Gail Hamaker, Jerome Hastings, Sally Hunter, Chris Jako, Leif Johansson, Ingolf Lindau, Rino Natoli, Mike Nolan, Paul Phizackerley, Piero Pianetta, Ben Salsburg, Jim Spencer, Jo Stöhr, Bill Warburton; *outside scientists*: Neil Ashcroft, Bob Bachrach, John Baldeschweiler, Fred Brown, Peter Eisenberger, Stig Hagstrom, Farrel Lytle, Vic Rehn, Dale Sayers, Andrew Sessler, David Shirley, Ed Stern, Al Thompson, Ken Trueblood, Nick Webb; *and at the National Science Foundation*: Howard Etzel, Bill Oosterhuis.

One of the problems in producing a volume of this kind is the delay between completion of the manuscripts and actual publication. Most of the manuscripts were completed early in 1979. The reader interested in developments since that time should consult the activity reports of the various laboratories, recent conference proceedings (particularly the National Conference on Synchrotron Radiation Instrumentation, Gaithersberg, Maryland, June 4–6, 1979, proceedings to be published in *Nuclear Instruments and Methods*) and, of course, the various pertinent scientific journals.

We are particularly grateful to Joanne Marchetti, who provided excellent support and coordination among the editors, authors, and the publisher and to Jill Fukuhara for assistance with the index.

Stanford

H. Winick
S. Doniach

Contents

Chapter 4. Inner-Shell Threshold Spectra
Frederick C. Brown

Chapter 5. Electron Spectrometry of Atoms and Molecules
Manfred O. Krause

Chapter 6. Photoemission as a Tool to Study Solids and Surfaces
I. Lindau and W. E. Spicer

Chapter 7. Microlithography with Soft X Rays

Andrew R. Neureuther

Chapter 8. Soft X-Ray Microscopy of Biological Specimens
J. Kirz and D. Sayre

Chapter 9. Synchrotron Radiation as a Modulated Source for Fluorescence Lifetime Measurements and for Time-Resolved Spectroscopy
Ian H. Munro and Andrew P. Sabersky

Chapter 10. The Principles of X-Ray Absorption Spectroscopy

George S. Brown and S. Doniach

Chapter 11. *Extended X-Ray Absorption Fine Structure in Condensed Materials*

George S. Brown

Chapter 12. *X-Ray Absorption Spectroscopy: Catalyst Applications*

F. W. Lytle, G. H. Via, and J. H. Sinfelt

Chapter 13. *X-Ray Absorption Spectroscopy of Biological Molecules*

S. Doniach, P. Eisenberger, and Keith O. Hodgson

Chapter 14. X-Ray Fluorescence Microprobe for Chemical Analysis
C. J. Sparks, Jr.

Chapter 15. Small-Angle X-Ray Scattering of Macromolecules in Solution
H. B. Stuhrmann

Chapter 16. Small-Angle Diffraction of X Rays and the Study of Biological Structures

G. Rosenbaum and K. C. Holmes

Chapter 17. Single-Crystal X-Ray Diffraction and Anomalous Scattering Using Synchrotron Radiation

James C. Phillips and Keith O. Hodgson

Chapter 18. Application of Synchrotron Radiation to X-Ray Topography

M. Sauvage and J. F. Petroff

Chapter 19. *Inelastic Scattering*
P. Eisenberger

Chapter 20. *Nuclear Resonance Experiments Using Synchrotron Radiation Sources*
R. L. Cohen

Chapter 21. *Wiggler Systems as Sources of Electromagnetic Radiation*
James E. Spencer and Herman Winick

Chapter 22. The Free-Electron Laser and Its Possible Developments
C. Pellegrini

1

An Overview of Synchrotron Radiation Research

HERMAN WINICK and S. DONIACH

1. Introduction

The outpouring of research results using synchrotron radiation during the past ten years and the explosive growth of interest in this extraordinary research tool constitutes a major event in the recent history of scientific instrumentation. This is not surprising when one considers that, for many types of experiments, synchrotron radiation provides five orders of magnitude more continuum vacuum ultraviolet (VUV) and x-radiation than conventional sources such as x-ray tubes. Normally a factor of two to ten improvement is considered a major breakthrough.

By using grating or crystal monochromators researchers can select any wavelength from the intense synchrotron radiation continuum. Most other radiation sources in the UV and x-ray regions of the electromagnetic spectrum (e.g., gas discharge lamps and x-ray tubes) provide high intensity at certain characteristic lines but only a very weak continuum. Thus, synchrotron radiation makes it possible for a research worker to select the wavelength most appropriate for the experiment and also to scan the wavelength over a large range. The tunability and intensity provided by synchrotron radiation are having a profound effect on photon physics and its applications. This may be compared to the impact that the development of particle accelerators with MeV energies had on nuclear physics research in the 1930s. In that case also, experimenters were freed from the restrictions of working only with radiation of particular energies (provided by radioactive sources) and this led to a rapid development of the field.

HERMAN WINICK • Stanford Synchrotron Radiation Laboratory, Stanford Linear Accelerator Center, P.O. Box 4349, Bin 69, Stanford, California 94305. **S. DONIACH** • Department of Applied Physics, Stanford University, Stanford, California 94305

The intense x-ray continuum provided by synchrotron radiation has already led directly to many important results. The most dramatic development is the technique known as extended x-ray absorption fine structure (EXAFS). Starting out as an interesting laboratory curiosity, it has become a powerful analytical tool for the determination of local atomic environments in complex systems.

However, synchrotron radiation, particularly as produced by the recently developed multi-GeV electron storage rings, has several other properties that further add to its capabilities as a research tool. These include an extremely broad spectral range (from infrared to x-rays), natural collimation, high polarization, pulsed time structure, high-vacuum environment, small source size, and high stability. The combination of these properties (described in more detail in Chapter 2) have made possible a large variety of previously impossible measurements, opening up new areas of research capability in disciplines such as surface physics, catalytic chemistry, protein crystallography, amorphous materials science, trace element analysis, many areas of biochemistry, microscopy, lithography, and topography, as well as many others. The status of current work in these fields and the prospects for the future are reviewed in this book.

These accomplishments are even more impressive when one realizes that a large fraction of them, particularly in the x-ray region of the spectrum, have been achieved in spite of the severe limitations imposed by parasitic use of the synchrotron radiation emitted during high-energy physics experiments. In the very near future these limitations will be largely vanquished and a major increase in research capability will be available to a much larger community of scientists. This will come about for the following reasons:

1. A number of new major storage ring facilities totally dedicated to synchrotron radiation research are now in construction or design in China, England, Germany, Japan, The Netherlands, the U.S.A., and the U.S.S.R. (See Chapter 3).

2. Several existing storage rings (e.g., DCI, DORIS, SPEAR, VEPP-3) will be increasingly used as dedicated synchrotron radiation sources operated in a high-current, high-energy, single-beam mode. This will provide higher intensities and higher brightness, particularly for x-ray experiments.

3. Expansion of research facilities is now underway on several existing storage rings (e.g., ADONE, DORIS, SPEAR, VEPP-3). Combined with the new facilities mentioned in (1) above this will greatly enlarge the scientific community that has easy access to synchrotron radiation.

4. The further development of insertion devices (e.g., wigglers and undulators; see Chapter 21) is likely to provide two or more orders of magnitude additional improvement in intensity and brightness. The most intense x-ray beam now in use for synchrotron radiation research is produced by a rather modest seven-pole 18-kG wiggler, which began operation at SSRL in March, 1979. More powerful wigglers are clearly possible, and undulators offer the promise of even higher brightness quasi-monochromatic peaks.

5. Further development of specialized research instrumentation (e.g., mirrors, monochromators, detectors; discussed in almost every chapter and reviewed in Chapter 3) will broadly benefit all present synchrotron radiation research and also

will open up new possibilities. For example, new monochromators will provide radiation in the 1–3 keV spectral region not presently available and two-dimensional position-sensitive x-ray detectors are likely to have a profound effect on x-ray diffraction experiments, including protein crystallography and time-dependent biological studies.

2. An Interdisciplinary Tool

It is clear from the foregoing, and from scanning the chapter titles of this book, that synchrotron radiation is an interdisciplinary tool. Although most researchers using synchrotron radiation, as well as most readers of this book, have one primary research interest, there is a common element: the use of electromagnetic radiation. It is this unifying thread that makes it sensible to collect in one place reviews of work in such diverse fields as catalysis and lithography. In fact, a nontrivial by-product of synchrotron radiation research is the fruitful interchange that occurs between workers in different fields who come to the same laboratory, often share the same instruments, and who meet at users' meetings and national and international conferences. An instrument, such as a position-sensitive detector or fluorescence monochromator, that has been developed for one application can and has found other uses, sometimes with even greater effect than in the original application. Techniques, such as EXAFS, developed for one kind of study (e.g., structure of solids, liquids, and gases) can be extended by other scientists to other areas (e.g., surfaces).

3. Some Recent History

In the past five years projections have been made in several parts of the world of the expected need for synchrotron radiation research facilities and this has lead to decisions to greatly expand present facilities and to construct new ones.

In early 1976 a panel was formed to assess the U.S. national needs for facilities dedicated to synchrotron radiation research. The panel report,[1] issued in the summer of 1976, projected that the 200 users of 23 experimental stations in the U.S. in 1976 would grow to at least 1155 users in 1986, needing at least 100 experimental stations. The conservatism of this estimate may be seen from the fact that, at SSRL alone, in January 1979 there were more than 400 users of 10 experimental stations and it is expected that this will increase perhaps by as much as a factor of two by 1982.

The major recommendation of the panel report[1] was that "an immediate commitment be made to construct new dedicated national facilities and expand existing facilities so that optimized XUV and x-ray capabilities are provided ..." (p. 5).

In late 1977 a report[2] was issued by the European Science Foundation that projected the European needs for synchrotron radiation research facilities. This report gives estimates of the numbers of scientists expected to be actively engaged in synchrotron radiation research by the early to mid-1980s in England, Germany, The Netherlands, Denmark, France, and Italy. The report projects that a total of about 1000 scientists in all these countries will be involved with synchrotron radiation research by 1982. The report[2] concluded that "there will be a large discrepancy between the number of scientists who wish to use synchrotron radiation and the number of stations available on existing or proposed machines" (p. 1). It also states that "a large effort to build a dedicated hard x-ray storage ring and appropriate advanced instrumentation with a design which goes beyond that of present day projects is recommended. This machine should be operational by 1985" (p. 1).

The widespread acceptance of synchrotron radiation as a research tool and the above-mentioned panel reports plus assessments made in other countries (e.g., England[3]) have lead to the authorization for major expansion of existing facilities and construction of new sources and facilities in many countries including England, France, Germany, Japan, the Soviet Union, and the United States. Proposals for synchrotron radiation sources and research facilities have recently been made in several other countries including Canada, The Netherlands, and the People's Republic of China. More details on these efforts are given in Chapter 3.

The developing interest in synchrotron radiation as a research tool may be clearly seen from reviews[4–8] and proceedings of conferences[9–11] and workshops,[12–14] plus the aforementioned panel reports.[1–3] See also the proceedings of The National Conference on Synchrotron Radiation Instrumentation held at the National Bureau of Standards, Washington, D.C., June 4–6, 1979 (to be published in *Nuclear Instruments and Methods*).

4. Earlier History

Although what is now called synchrotron radiation was first observed (accidentally) at the General Electric 70-MeV synchrotron in 1947, the theoretical consideration of the radiation by charges in circular motion goes back to 1898 to the work of Liénard.[15] Further theoretical work was done by Schott,[16] Jassinsky,[17] Kerst,[18] Ivanenko and Pomeranchuk,[19] Arzimovitch and Pomeranchuk,[20] and others through 1946. Blewett[21] was one of the first to be concerned with the radiation effects on the operation of electron accelerators and observed its effects on the electron orbit in 1945. A more detailed history of these developments is given by Lea.[6]

With the observation of the radiation in 1947 and with the construction of electron accelerators in the post-war period there was renewed interest in synchrotron radiation. Comprehensive theoretical treatments were presented by Sokolov and co-workers[22] and by Schwinger[23] starting in the late 1940s. With these works the theory was fully developed so that accurate predictions could be made regarding the intensity, spectral and angular distributions, polarization, etc. Quantum mechanical effects were shown to be negligible in practical cases. A review of the theory is given in the textbook by Jackson.[24]

The first experimental investigations of the properties of the radiation were carried out in the late 1940s by Pollack and co-workers[25] using the General

Electric 70-MeV synchrotron. In the 1950s these were extended by several groups[26] using the 250-MeV synchrotron at the Lebedev Institute in Moscow; by Corson[27] and Tomboulian and co-workers[28] using the Cornell 300-MeV synchrotron; by Codling and Madden[29] using the 180-MeV synchrotron at the National Bureau of Standards (NBS) in Washington, D.C. In the mid-1960s Haensel and co-workers[30] were the first to utilize radiation from a multi-GeV electron accelerator, the 6-GeV electron synchrotron in Hamburg. These investigations verified the basic theoretical predictions and provided much useful data and experience in the use of the radiation. For example, the systematic study at NBS on the physics of photoabsorption in gases in the far UV led to a considerable increase in understanding the physics of autoionization. It also aided understanding of the shape of spectral lines corresponding to many-electron excited states degenerate with the photoionization continuum.

By the mid-1960s synchrotron radiation research programs had begun on the Frascati, NBS, Tokyo, and Hamburg synchrotrons. By the early 1970s programs were underway at several other synchrotrons including Bonn, Glasgow, Lund, Moscow, and Yerevan. In most cases the primary function of the synchrotron was high-energy physics research.

In 1968 the first synchrotron radiation program was begun on a storage ring; the 240-MeV ring, Tantalus, at the University of Wisconsin.[31] Although limited to the VUV region of the spectrum, it was quickly evident that storage rings offered research opportunities far superior to those available through synchrotrons because of their constant spectral distribution, stability, and relatively low radiation hazards. Furthermore, the Wisconsin machine was available as a dedicated source of synchrotron radiation—a major advantage. The pioneering efforts of Ed Rowe and collaborators in providing this fine source are certainly responsible for much of the impetus to provide additional synchrotron radiation research facilities.

As new storage rings were built for high-energy physics research the desirable characteristics of their radiation extended to higher photon energies, and several parasitic programs began. In 1971 a synchrotron radiation program in the VUV was started on ACO, the 540-MeV ring[32] in Orsay, France. The x-ray part of the spectrum was opened in 1972 at the Cambridge Electron Accelerator (CEA) in Massachusetts, operating as a storage ring at 3 GeV.[33] At about the same time a program was also started on VEPP-2M (670 MeV) in Novosibirsk.[5]

The Wisconsin machine continues as a dedicated source, although it is soon to be replaced by the 0.8–1.0-GeV dedicated ring, Aladdin. ACO is now a dedicated synchrotron radiation source and VEPP-2M is increasingly used for synchrotron radiation research. The CEA ceased operation in 1973 for high-energy physics and synchrotron radiation research.

At Stanford, the SPEAR storage ring started colliding-beam operation in 1972 with a maximum energy of 2.5 GeV. A group of Stanford faculty, headed by S. Doniach, proposed the development of a synchrotron radiation facility at SPEAR in late 1972. The proposal was funded by the National Science Foundation and the Stanford Synchrotron Radiation Project[34–36] was formed in May 1973 with Doniach as Director and W. Spicer as Deputy Director. H. Winick joined the project in July 1973 and construction was underway in late autumn of that year. Research operation started in May 1974 on five experimental stations, each equipped with a monochromator. This was the first time that synchrotron radiation from a multi-GeV storage ring became available to a large community of users. The spectral range from 150 eV to 40 keV was now accessible with the intensity and stability provided by a storage ring, rather than a synchrotron. This spectral region proved to be even richer in application than the region below 150 eV that had been already explored on smaller storage rings and synchrotrons. The most important early results were in the development of EXAFS as a broadly applicable structural tool and in the use of photoemission for surface studies in the region above 100 eV. By late 1974 SPEAR had been extended to 4 GeV. Also at about this time parasitic synchrotron radiation programs began on two other multi-GeV storage rings; the 3.5-GeV storage ring DORIS[37] (extended to 5 GeV in 1978) in Hamburg and the 2.2-GeV ring VEPP-3[5] in Novosibirsk. In 1977 a parasitic program on the 1.8-GeV ring DCI was started.

5. Photon Physics

The interactions of photons with matter is, of course, one of the basic physical processes underlying human perception of the physical world. By extending the spectral range and available intensity of electromagnetic radiation, synchrotron radiation sources are starting to make possible, or considerably improve, the study of photon–matter interactions in increasingly sophisticated ways.

To explore the possible range of photon experiments, it is convenient to classify photon–matter interactions in terms of the input and output radiation observed (as is done in nuclear physics). Using hv to denote the incoming photon beam, the following are the simpler types of reaction possible:

  a. Total absorption (hv)
  b. Photon scattering and/or fluorescence (hv, hv')
  c. Photoelectric excitation (hv, e)
  d. Photoionization or photodissociation (hv, I) (where I denotes an ionized or dissociated atomic species)
  e. Compton-induced photoeffect $(hv; hv', e)$

It is also useful to characterize the possible reactions in terms of energy and momentum transfer to the sample: $\Delta E = hv - hv'$, $\Delta p = \hbar c\,|\mathbf{K} - \mathbf{K}'|$, where $\mathbf{K}$ and $\mathbf{K}'$ are the ingoing and outgoing photon wave vectors. Thus, photon scattering may either be elastic (as in x-ray diffraction) so that the relevant cross section is a function only of the momentum transfer, or inelastic (as in visible or UV fluorescence), where the cross section depends both on the incoming photon energy and on the energy absorbed as in the Raman effect. For x-ray inelastic processes, the dependence of the cross section on the momentum transfer as well as energy transfer becomes important.

Before the advent of synchrotron radiation sources, most experiments in the UV and x-ray regions of the spectrum relied on fixed photon energy input, so that the variable used to characterize the scattering was that of the outgoing radiation (i.e., $\mathbf{K}'$ for diffraction experiments and hv' or $\varepsilon_{\text{photoelectron}}$ for inelastic experiments). Now that tunable photon beams are available, the additional dependence of cross sections on incoming photon energy is starting to be explored. This is leading to a new dimension for many conventional photon physics experiments. Thus, in x-ray diffraction, the incoming photon beam can be tuned to the vicinity of the absorption edge of a constituent atom in a sample, leading to a considerable increase in the power of anomalous diffraction as a crystallographic tool. In photoemission, the ability to tune the incoming photons to the energy of an internal core → valence transition is leading to some very interesting new resonant photoemission studies. The dependence of the photoemission cross section on incoming photon energy also makes possible the study of "constant final state" spectra, which allows for unravelling of initial state effects (valence orbitals, etc.) from final state effects in the cross section.

6. The Future

Research with synchrotron radiation, particularly in the x-ray part of the spectrum, is still in its infancy. In the VUV region radiation is available from several dedicated storage rings (Wisconsin, NBS, INS-SOR, ACO). VEPP-2M is used as a partly dedicated VUV and soft x-ray source. See Chapter 3 for a listing and discussion of synchrotron radiation sources and research facilities. Radiation in the x-ray region is available now from only a few multi-GeV storage rings (DCI, VEPP-3, DORIS, and SPEAR) and, for the most part, still severely limited by parasitic operation. Yet, in spite of these limitations, far-reaching results, based on new or revitalized techniques (e.g., EXAFS, photoemission), have been achieved.

It is apparent to those of us close to the development of synchrotron radiation research that many more important breakthroughs are on the horizon. In many cases preliminary work has clearly indicated the great potential of synchrotron radiation. Examples include:

1. Replication of microstructures by soft x-ray lithography (see Chapter 7). The possibility exists, for example, for integrated circuits to be produced with 2–3 orders-of-magnitude higher density of elements than is now possible.

2. Soft x-ray microscopy (see Chapter 8). Synchrotron radiation is opening up the possibility of operating high-resolution (~ 100 Å) microscopes that can be used to study live specimens in an atmospheric environment and with chemical element discrimination.

3. Time-resolved spectroscopy (see Chapter 9). The sharply pulsed time structure and tunability of synchrotron radiation offer unique possibilities for studying conformational changes in biological materials on a subnanosecond time scale.

4. Trace element analysis by x-ray fluorescence (see Chapter 14). The limits of sensitivity have already been improved by more than one order of magnitude. Further improvements are clearly possible.

5. Anomalous dispersion as a tool for protein structure determination (see Chapter 17). Promising initial studies indicate that the way may now be open to

solve many protein structures for which present techniques (e.g., multiple isomorphous replacement) are not feasible.

6. Crystal defect determination by x-ray diffraction topography (see Chapter 18). The intensity and collimation of synchrotron radiation are making possible major extensions in the sensitivity and range of application of this technique.

In other cases refinements of already proven synchrotron radiation research techniques have further extended their range of application. These include:

1. Application of EXAFS to surface analysis (see Chapter 11).
2. Angle-resolved photoemission to determine, for example, the orientation of adsorbates on surfaces (see Chapter 6).
3. Use of tuneable U.V. to vary the escape depth of photoelectrons emitted from a solid surface and hence to distinguish surface from bulk contributions to the electronic structure of the sample. One promising application is to the study of chemical bonding and electronic structure at semiconductor-metal interfaces, thus providing information of considerable interest for device applications. (See Chapter 6.)
4. Time-dependent x-ray diffraction as a tool to observe structural changes in live, biological samples (see Chapter 16).

With the increased intensities and higher brightness that will be available with new sources and when present multi-GeV machines are operated routinely as dedicated sources, and with the vastly increased scientific community that will have access to synchrotron radiation in the next few years, an enormous extension of scientific capability and activity will occur. Even further extensions will come from the development of insertion devices as radiation sources (wigglers and undulators—see Chapter 21) in existing rings and in newly constructed rings. In addition, the development of specialized instrumentation for synchrotron radiation research (mirrors, monochromators, detectors) has already had a significant effect and is likely to have even more impact in the future.

The developments of synchrotron radiation sources and instrumentation will certainly result in the maturation of some of the promising applications listed above and the possibility of developing others (e.g., nuclear resonance scattering, holography) for which present intensities and/or facilities are marginal. Although it is not possible to say with certainty which of the presently promising applications will prove successful and broadly applicable, their number is large enough that, based on past experience, it is almost a statistical certainty that several will make it. It is, therefore, highly probable that additional large scientific communities (e.g., protein crystallographers, metallurgists) will be seeking access to synchrotron radiation in the way that surface physicists, catalytic chemists, and bioinorganic chemists are already doing.

It is fortunate that research facilities are being expanded around the world so that access to synchrotron radiation will no longer be so severely limited. The frustration of parasitic operation should also become a thing of the past. Synchrotron radiation researchers will finally be able to enjoy the status of prime users of very powerful radiation sources and very important and exciting new scientific progress is likely to follow.

References

1. R. Morse (chairman), An Assessment of the National Need for Facilities Dedicated to the Production of Synchrotron Radiation, Solid State Sciences Committee, National Research Council, 2101 Constitution Ave., Washington, D.C. 20418, August 1976.
2. H. Maier-Leibnitz (chairman), "Synchrotron Radiation—A Perspective View for Europe," European Science Foundation, 35–37 Rue du Fossé-des-Trieze, F6700, Strasbourg, France; December 1977.
3. The Scientific Case for Research with Synchrotron Radiation, Daresbury Report DL/SRF/R3, 1975.
4. C. Kunz (ed.), Topics in Current Physics—Synchrotron Radiation; Techniques and Applications, Springer-Verlag (1979).
5. G. N. Kulipanov and A. N. Skrinskii, *Usp. Fiz. Nauk.* **122**, 369–418 (1977); English translation: *Sov. Phys. Usp.* **20**, 559–86 (1977).
6. K. R. Lea, *Phys. Rep.* (Section C of Physics Letters) **43**, 337–75 (1978).
7. H. Winick and A. Bienenstock, *Annu. Rev. Nucl. Part. Sci.* **28**, 33–113 (1978). See also A. Bienenstock, *IEEE Trans. Nucl. Sci.* **26**, 3780–3784 (1979).
8. N. G. Basov (ed.), Synchrotron radiation, *Proc. P. N. Lebedev Phys. Inst.* [*Acad. Sci. USSR*] **80**. Translated from the Russian as *Synchrotron Radiation*, Plenum Press (Consultants Bureau), New York and London (1976).
9. R. E. Watson and M. L. Perlman (eds.), Research Applications of Synchrotron Radiation, BNL Report 50381, (1973).
10. E. E. Koch, R. Haensel, and C. Kunz (eds.), *Vacuum Ultraviolet Radiation Physics*, Vieweg-Pergamon, Braunschweig (1974).
11. F. Wuilleumier and Y. Farge (eds.), *Proceedings of the Internation Conference on Synchrotron Radiation Instrumentation and New Developments*, North-Holland Publishing Co., Amsterdam (1978). Also published as *Nucl. Instrum. Methods* **152** (1978).
12. J. W. McGowan and E. M. Rowe (eds.), Quebec Summer Workshop on Synchrotron Radiation Facilities, June 1976. Distributed by the Center for Interdisciplinary Studies in Chem. Phys., Univ. W. Ontario, London, Ontario, Canada N6A 3KF.
13. H. Winick and T. Knight (eds.), "Wiggler Magnets" SSRP Report 77/05 (1977).
14. H. Winick and G. Brown (eds.), "Proceedings of Workshop on X-ray Instrumentation for Synchrotron Radiation Research" SSRL Report 78/04 (1978).
15. A. Liénard, *L'Eclairage Elect.* **16**, 5 (1898).
16. G. A. Schott, *Ann. Phys.* (*Leipzig*) **24**, 635 (1907); G. A. Schott, *Electromagnetic Radiation*, Cambridge University Press, London (1912), Chapters 7 and 8.
17. W. W. Jassinsky, *J. Exp. Theor. Phys.* (*USSR*) **5**, 983 (1935); W. W. Jassinsky, *Arch. Electrotech.* (Berlin) **30**, 590 (1936).
18. D. Kerst, *Phys. Rev.* **60**, 47 (1941).
19. D. Ivanenko and I. Pomeranchuk, *Dokl. Akad. Nauk. SSSR* **44**, 315 (1944); D. Ivanenko and I. Pomeranchuk, *Phys. Rev.* **65**, 343 (1944).
20. L. Arzimovitch and I. Pomeranchuk, *J. Phys.* (*Moscow*) **9**, 267 (1945); L. Arzimovitch and I. Pomeranchuk, *J. Exp. Theor. Phys.* (*USSR*) **16**, 379 (1946).
21. J. P. Blewett, *Phys. Rev.* **69**, 87 (1946).
22. D. Ivanenko and A. A. Sokolov, *Dokl. Akad. Nauk SSSR* **59**, 1551 (1948); A. A. Sokolov, N. P. Klepikov, and I. M. Ternov, *Dokl. Akad. Nauk SSSR* **89**, 665 (1953); A. A. Sokolov and I. M. Ternov, *Sov. Phys. JETP* **1**, 227–30 (1955); *Sov. Phys. JETP* **4**, 396–400 (1957); *Sov. Phys. Dokl.* **8**, 1203–5 (1964); *Synchrotron Radiation*, Pergamon Press, New York (1968).
23. J. Schwinger, *Phys. Rev.* **70**, 798 (1946); *Phys. Rev.* **75**, 1912–25 (1949); *Proc. Natl. Acad. Sci. U.S.A.* **40**, 132 (1954).
24. J. D. Jackson, *Classical Electrodynamics*, Chapter 14, Wiley, New York (1975), p. 848.
25. F. R. Elder, A. M. Gurewitsch, R. V. Langmuir, and H. D. Pollack, *Phys. Rev.* **71**, 829–30 (1947) and *J. Appl. Phys.* **18**, 810 (1947); F. R. Elder, R. V. Langmuir, and H. C. Pollack, *Phys. Rev.* **74**, 52 (1948).

26. I. M. Ado and P. A. Cherenkov, *Sov. Phys. Dokl.* **1**, 517–19 (1956); F. A. Korolev, V. S. Markov, E. M. Akimov, and O. F. Kulikov, *Sov. Phys. Dokl.* **1**, 568 (1956); F. A. Korolev and O. F. Kulikov, *Opt. Spectrosc. (USSR)* **8**, 1–3 (1960); F. A. Korolev, A. G. Ershov, and O. F. Kulikov, *Sov. Phys. Dokl.* **5**, 1011 (1961); F. A. Korolev, O. F. Kulikov, and A. S. Yarov, *Sov. Phys. JETP* **43**, 1653 (1962).

27. D. A. Corson, *Phys. Rev.* **86**, 1052–53 (1952) and *Phys. Rev.* **90**, 748–52 (1953).

28. P. L. Hartman and D. H. Tomboulian, *Phys. Rev.* **87**, 233 (1952); D. H. Tomboulian, U.S.A.E.C. NP-5803 (1955); D. H. Tomboulian and P. L. Hartman, *Phys. Rev.* **102**, 1423–47 (1956); D. E. Bedo and D. H. Tomboulian, *J. Appl. Phys.* **29**, 804–9 (1958).

29. K. Codling and R. P. Madden, *Phys. Rev. Lett.* **10**, 516–18 (1963); *Phys. Rev. Lett.* **12**, 106–8 (1964); *J. Opt. Soc. Am.* **54**, 268 (1964); *J. Appl. Phys.* **36**, 830–37 (1965).

30. G. Bathov, E. Freytag, and R. Haensel, *J. Appl. Phys.* **37**, 3449 (1966).

31. E. M. Rowe and F. E. Mills, *Part. Accel.* **4**, 211–27 (1973).

32. P. Dagneaux, C. Depautex, P. Dhez, J. Durup, Y. Farge, R. Fourme, P. -M. Guyon, P. Jaeglé, S. Leach, R. Lopez-Delgado, G. Morel, R. Pinchaux, P. Thiry, C. Vermeil, and F. Wuilleumier, *Ann. Phys. (N.Y.)* **9**, 9–65 (1975); P. -M. Goyon, C. Depautex, and G. Morel, *Rev. Sci. Instrum.* **47**, 1347 (1976).

33. H. Winick, *IEEE Trans. Nucl. Sci.* **20**, 984–88 (1973).

34. H. Winick, *Proceedings of the 9th International Conference on High Energy Accelerators*, Stanford, California, pp. 685–8 (1974).

35. A. Baer, R. Gaxiola, A. Golde, F. Johnson, B. Salsburg, H. Winick, M. Baldwin, N. Dean, J. Harris, E. Hoyt, B. Humphrey, and J. Jurow, *IEEE Trans. Nucl. Sci.* **22**, 1794–7 (1975).

36. J. Cerino, A. Golde, J. Hastings, I. Lindau, B. Salsburg, H. Winick, M. Lee, P. Morton, and A. Garren, *IEEE Trans. Nucl. Sci.* **24**, 1003–5 (1977).

37. E. E. Koch, C. Kunz, and E. W. Weiner, *Optik (Stuttgart)* **45**, 395–410 (1976).

2

Properties of Synchrotron Radiation

HERMAN WINICK

1. Introduction

Synchrotron radiation has several important properties including:

1. High intensity
2. Broad spectral range
3. High polarization
4. Pulsed time structure
5. Natural collimation

In addition, synchrotron radiation produced by storage rings (rather than synchrotrons) offers:

6. High-vacuum environment
7. Small-source-spot size
8. Stability

Any one of these properties would make synchrotron radiation an important experimental tool. The combination of all of them makes it a unique and rather extraordinary source for a wide variety of scientific and technological applications as described in this book.

The basic equations describing the radiation have been given by Schwinger,[1] Sokolov and Ternov,[2] Tomboulian and Hartman,[3] Godwin,[4] and Jackson.[5]

HERMAN WINICK • Stanford Synchrotron Radiation Laboratory, Stanford Linear Accelerator Center, P.O. Box 4349, Bin 69, Stanford, California 94305.

Many others have contributed to the development of the theory. Other references and brief reviews of the properties of the radiation are given in recent reviews by Winick and Bienenstock,[6] Kulipanov and Skrinsky,[7] and Lea.[8] Detailed tables and graphs describing the spectrum, angular distribution, and polarization functions are given by Green,[9] who also gives an approximate treatment of the optical properties of the source. Elegant field plots showing the directionality of the radiation are given by Tsien.[10]

In our treatment we will assume that the radiation is incoherent between electrons so that the intensity for N electrons is simply N times that of a single electron. In general, electrons can radiate coherently only if the electron density is high enough and if a density variation with short wavelength components is present in the electron bunch. This has been analyzed by Csonka.[11] Bénard and Rousseau[12] have analyzed the statistical properties of synchrotron radiation and they have concluded that the coherence length (maximum distance between correlated emission points) is about 200 λ, where λ is the wavelength of the emitted radiation and is assumed to be smaller than the length of the electron bunch. In the free-electron laser, (see Chapter 22), electrons radiate coherently. In fact, a primary requirement for a free-electron laser to operate in a storage ring at high power is for the bunch density modulations produced in traversing the laser to persist as the electron bunch goes around the ring so that it reenters the laser with some bunch density modulation still present.

2. Radiated Power

As a starting point we may take the Larmor formula for the power radiated by a single nonrelativistic accelerating particle with charge e,

$$P = \frac{2}{3} \frac{e^2}{c^3} \left| \frac{d\mathbf{V}}{dt} \right|^2 \tag{1}$$

For a particle of mass m in circular motion with velocity $\beta = v/c$, energy E, and a radius of curvature ρ, the relativistic form of this equation is

$$P = \frac{2}{3} \frac{e^2 c}{\rho^2} \beta^4 \left(\frac{E}{mc^2} \right)^4 \tag{2}$$

and the energy lost per turn by one particle is

$$\Delta E = \frac{4\pi}{3} \frac{e^2}{\rho} \beta^3 \left(\frac{E}{mc^2} \right)^4 \tag{3}$$

From these equations it is clear that heavy particles such as protons produce negligible synchrotron radiation power compared to electrons.

In practical units (E in GeV, ρ in meters, B in kilogauss, I in amperes) for a highly relativistic electron we obtain

$$\Delta E(\text{keV}) = 88.47E^4/\rho$$

$$P(\text{kW}) = 88.47E^4 I/\rho = 2.654 B E^3 I \tag{4}$$

For example, an intermediate energy storage ring with $E = 1$ GeV, $B = 10$ kG, and $I = 0.5$ A would radiate 13.3 kW.

It is often useful to know the total power radiated in traversing a given length L of magnetic field. A general expression for this is

$$P(\text{kW}) = 1.267 \times 10^{-2} E^2(\text{GeV})\langle B^2(\text{kG})\rangle I(\text{A})L(\text{M}) \tag{5}$$

where $\langle B^2 \rangle$ is the average over the length L. This equation can be used for a constant field or for a field that varies along the trajectory such as in a wiggler or undulator.

3. Spectral and Angular Distribution

At nonrelativistic energies, electrons in circular motion radiate in a dipole pattern as shown schematically in Figure 1, Case I. At relativistic energies this pattern becomes sharply peaked in the direction of motion of the electron, as shown in Figure 1, Case II. For example, light emitted at an angle θ relative to the electron direction of motion in the rest frame is viewed at an angle θ' in the laboratory frame. The transformation is given by[5]

$$\tan \theta' = \frac{\sin \theta}{\gamma(\beta + \cos \theta)}, \qquad \text{where } \gamma = \frac{E}{mc^2}$$

At $\theta = 90°$, $\tan \theta' \approx \theta' \approx \gamma^{-1}$. Thus γ^{-1} is a typical opening half-angle of the radiation in the laboratory system. The above arguments are qualitative and are approximately valid near the critical energy [see equation (8)].

Knowing that the opening half-angle of the radiation is approximately given by γ^{-1}, it follows that the radiation spectrum will extend to photon energies that have a value proportional to γ^3. This may be seen by calculating the duration, as seen by a stationary observer, of the pulse of radiation produced by an electron in circular motion with radius ρ. The observer receives radiation produced by the electron as it traverses an arc of angle $2\gamma^{-1}$. The duration of the radiation pulse is just the difference between the time for the electron to traverse the arc and the time

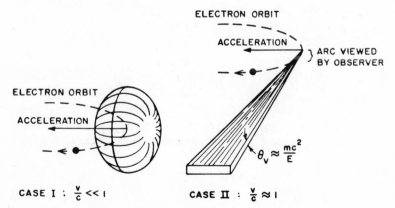

Figure 1. Radiation emission pattern of electrons in circular motion: Case I, nonrelativistic electrons. Case II, relativistic electrons.

for the light to travel along the chord corresponding to the arc. It can be easily shown that this time interval is

$$\tau = \frac{\rho}{c}\left[\frac{1}{\gamma\beta} - 2\sin\left(\frac{1}{2\gamma}\right)\right] \approx \frac{\rho}{c\gamma^3}$$

A light pulse of this duration has frequency components up to about $\omega \approx \tau^{-1} = c\gamma^3/\rho$ corresponding to photon energies of $\varepsilon = \hbar\omega = \hbar c\gamma^3/\rho$.

This qualitative analysis exhibits the γ^3 dependence of the spectral cutoff and gives a photon energy close to the critical energy [equation (8)] that characterizes the spectrum. This approach applies only when the particle is bent by an angle greater than about $2\gamma^{-1}$. When the particle is bent through a smaller angle the spectrum can extend to higher energies as has been pointed out by Coïsson,[13] who has suggested that this should produce visible light in high-energy proton synchrotrons and storage rings. This has recently been verified at the 400-GeV proton synchrotron at CERN in Geneva, Switzerland, by observing visible light from the fringing fields of the ring bending magnets.[14]

The natural collimation becomes very dramatic for electron energies above about 1 GeV. At this energy most of the radiation is emitted in a cone with a full vertical-opening angle of about 1 mrad. This makes possible an extremely high flux on targets as small as 1 mm in height even at distances of 10 to 40 m from the storage ring. This characteristic can also lead to difficulties in alignment of synchrotron radiation beams and experiments, particularly from multi-GeV machines. For these large machines the small vertical angular deviations of the electron beam (of the order of a few tenths of a milliradian) that frequently occur can result in deviations of several millimeters at the target, causing most of the radiation to miss the target or defining slits, which are often only 1 mm high. This has led to the development of synchrotron radiation beam position monitors and feedback systems that control the electron orbit by introducing local orbit distortions to keep the synchrotron radiation locked onto the target.[15] (See also Chapter 3 for a more detailed discussion.)

The results of Schwinger,[1] expressed in terms of photon wavelength[3, 4] rather than frequency, give for the instantaneous power (c.g.s. units) radiated per unit wavelength and per radian (in vertical angle ψ) by a monoenergetic electron in circular orbit:

$$P(\lambda, \psi, t) = \frac{27}{32\pi^3}\frac{e^2c}{\rho^3}\left(\frac{\lambda_c}{\lambda}\right)^4\gamma^8(1 + X^2)^2$$

$$\times \left[K_{2/3}^2(\xi) + \frac{X^2}{1+X^2}K_{1/3}^2(\xi)\right]\frac{\text{erg}}{\text{sec rad cm}} \qquad (6)$$

where $X = \gamma\psi$, $\xi = \lambda_c[1 + X^2]^{3/2}/(2\lambda)$, and $K_{1/3}$ and $K_{2/3}$ are modified Bessel functions of the second kind, ψ is the angle between the direction of photon emission and the instantaneous orbital plane, $\gamma = E/mc^2$, and λ_c is the critical wavelength given by

$$\lambda_c = \frac{4\pi\rho}{3\gamma^3} \qquad (7)$$

[Note: Some authors, including Jackson,[5] define the critical wavelength to be twice that given by equation (7).]

Alternatively we can use a critical energy, ε_c, given by

$$\varepsilon_c = \frac{3\hbar c\gamma^3}{2\rho} \tag{8}$$

In practical units

$$\lambda_c(\text{Å}) = 5.59\rho/E^3 = 186.4/(BE^2)$$

$$\varepsilon_c(\text{keV}) = 2.218E^3/\rho = 0.06651BE^2 \tag{9}$$

Thus, our earlier example of a 1-GeV storage ring and 10-kG field corresponds to a critical energy of 0.665 keV.

Half the total power is radiated above the critical energy and half below. In most storage rings the magnetic field in the bending magnets is about 12 kG or less. However, in straight sections between bending magnets special high-field insertions with higher magnetic field can be used. (See Chapter 21 on wiggler magnets.)

The angular distribution of radiated power integrated over all wavelengths is

$$P(\psi, t) = \frac{7}{16}\frac{e^2c}{\rho^2}\gamma^5[1 + X^2]^{-5/2}\left[1 + \frac{5}{7}\frac{X^2}{1 + X^2}\right]\frac{\text{erg}}{\text{sec rad}} \tag{10}$$

Integrated over all emission angles the spectral power distribution is

$$P(\lambda, t) = \frac{3^{5/2}}{16\pi^2}\frac{e^2c}{\rho^3}\gamma^7y^3\int_y^\infty K_{5/3}(\eta)\,d\eta\,\frac{\text{erg}}{\text{sec cm}} \tag{11}$$

where

$$y = \lambda_c/\lambda = \varepsilon/\varepsilon_c$$

Green[9] has provided an excellent summary of a variety of useful distribution functions. Since his results are not generally available we reproduce some of them here. The Appendix gives the values of the various Bessel functions and integrals as a function of y. Green uses the following notation:

$$G_0(y) = \int_y^\infty K_{5/3}(\eta)\,d\eta \qquad G_i(y) = y^iG_0(y)$$

$$H_0(y, 0) = K_{2/3}^2(y/2) \qquad H_i(y, 0) = y^iH_0(y)$$

$$\gamma = E/mc^2 = 1.957 \times 10^3E(\text{GeV})$$

$$B\rho = 1704\gamma\ \text{G cm} = 33.35E(\text{GeV})\ \text{kG m}$$

$$\varepsilon = 12.398/\lambda(\text{Å})\ \text{keV} \qquad k = \Delta\lambda/\lambda = \Delta\varepsilon/\varepsilon$$

Using this notation, Green presents many power, spectral, and angular distribution functions including the following:

$$P(\lambda) = 1.421 \times 10^{-19} \frac{\gamma^7}{\rho^2} G_3(y) \frac{\text{erg}}{\text{Å sec mA (mrad } \theta)}, \qquad \text{for all } \psi \quad (12)$$

$$N(\lambda) = 2.998 \times 10^{-1} \frac{\gamma^4}{\rho} G_2(y) \frac{\text{photons}}{\text{Å sec mA (mrad } \theta)}, \qquad \text{for all } \psi \quad (13)$$

$$N_k(\lambda) = 1.256 \times 10^{10} k\gamma G_1(y) \frac{\text{photons}}{(k\lambda) \text{ sec mA (mrad } \theta)}, \qquad \text{for all } \psi \quad (14)$$

$$N_{\Delta\varepsilon}(\lambda) = 4.24 \times 10^{16} \frac{\rho}{\gamma^2} G_0(y)$$

$$= 1.013 \times 10^6 \gamma\lambda_c G_0(y) \frac{\text{photons}}{\text{eV sec mA (mrad } \theta)}, \qquad \text{for all } \psi \quad (15)$$

$$\text{At } \psi = 0$$

$$P(\lambda, 0) = 3.918 \times 10^{-23} \frac{\gamma^8}{\rho^2} H_4(y, 0) \frac{\text{erg}}{\text{Å sec mA (mrad } \theta) \text{ (mrad } \psi)} \quad (16)$$

$$N(\lambda, 0) = 8.263 \times 10^{-5} \frac{\gamma^5}{\rho} H_3(y, 0) \frac{\text{photons}}{\text{Å sec mA (mrad } \theta) \text{ (mrad } \psi)} \quad (17)$$

$$N_k(\lambda, 0) = 3.461 \times 10^6 k\gamma^2 H_2(y, 0) \frac{\text{photons}}{(k\lambda) \text{ sec mA (mrad } \theta) \text{ (mrad } \psi)} \quad (18)$$

$$N_{\Delta\varepsilon}(\lambda, 0) = 1.169 \times 10^{13} \frac{\rho}{\gamma} H_1(y, 0) \frac{\text{photons}}{\text{eV sec mA (mrad } \theta) \text{ (mrad } \psi)} \quad (19)$$

$$N_{\Delta\varepsilon}(\varepsilon, 0) = 3.951 \times 10^{19} \frac{\varepsilon\rho^2}{\gamma^4} H_0(y, 0) \frac{\text{photons}}{\text{eV sec mA (mrad } \theta) \text{ (mrad } \psi)} \quad (20)$$

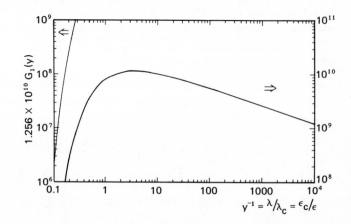

Figure 2. $1.256 \times 10^{10} G_1(y)$ versus y^{-1}. When the ordinate is multiplied by $k\gamma$ one obtains photons $(k\lambda)$ sec mA (mrad θ), for all ψ.

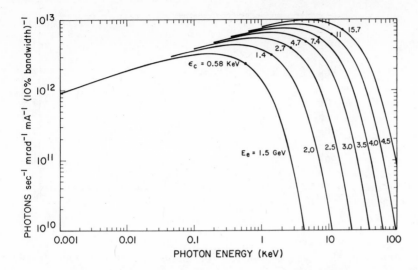

Figure 3. Spectral distribution of synchrotron radiation from SPEAR, ($\rho = 12.7$ m).

The foregoing equations can be used to determine many quantities of interest to the user of sychrotron radiation. For example, using the Appendix we find that at the critical wavelength ($y = 1$), $G_0 = 0.6514$. Thus, $G_1 = yG_0 = 0.6514$. Then using equation (14) we can show that the flux at the critical wavelength within a 10% bandwidth ($k = 0.1$) is

$$N_{0.1}(\lambda_c) = 8.812 \times 10^9 \gamma \text{ photons/sec mA (mrad } \theta), \quad \text{for all } \psi \quad (21)$$

and as a function of electron energy E in GeV

$$N_{0.1}(\lambda_c) = 1.601 \times 10^{12} E \text{ photons/sec mA (mrad } \theta), \quad \text{for all } \psi \quad (22)$$

These are particularly useful equations since it is often desirable to know the flux at the critical wavelength.

The quantity most often needed to determine the flux for a given experiment is the number of photons at a given wavelength and within a given bandwidth, integrated over all vertical-opening angles. This is given by equation (14), which is plotted in Figure 2. Figure 3 is a particular plot for the SPEAR storage ring of this

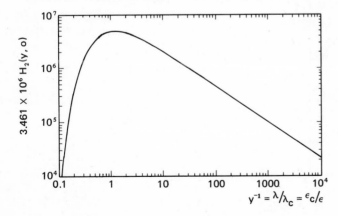

Figure 4. $3.461 \times 10^6 H_2(y, 0)$ versus y^{-1}. When the ordinate is multiplied by $k\gamma^2$ one obtains photons/$(k\lambda)$ sec mA (mrad θ) (mrad ψ).

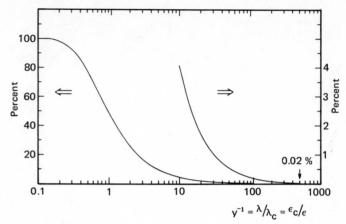

Figure 5. Percent of power at all wavelengths greater than λ versus λ/λ_c. Note that half the power is radiated at wavelengths longer than λ_c.

vertical-angle-integrated flux as a function of photon energy rather than wavelength. Because of integration over all vertical-opening angles, equation (14), Figure 2, and Figure 3 are independent of the electron beam divergence.

If the vertical angular acceptance of the experiment is small compared to the vertical-opening angle of the radiation, then the flux for a finite vertical-acceptance angle $\Delta\psi$ near $\psi = 0$ can be obtained from equation (18), which is plotted as Figure 4. Equation (18) and Figure 4 are valid only when the vertical angular divergence of the electron beam is small compared to the wavelength-dependent vertical-opening angle of the radiation from a single electron.

Figure 5 [also from Green[9]] is a plot of the percent of power at all wavelengths greater than λ versus λ/λ_c.

Figure 6 [Green[9]] is a plot of the percent of photons at all wavelengths greater than λ versus λ/λ_c.

The spectrum of equation (14) has a broad peak at about three times the critical wavelength. It drops as the cube root of the wavelength for much longer wavelengths and it drops rapidly as $(\lambda_c/\lambda)^{1/2}e^{-\lambda_c/\lambda}$ for much shorter wavelengths.

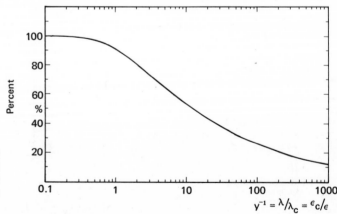

Figure 6. Percent of photons at all wavelengths greater than λ versus λ/λ_c. Note that only about 9% of the photons have wavelengths less than λ_c, although they contain half the radiation power.

Useful approximations for the photon flux at photon wavelengths much greater and much less than the critical wavelength are given respectively by:

$$N(\lambda) \approx 9.35 \times 10^{16} I(A)[\rho(m)/\lambda_c(\text{Å})]^{1/3}(\Delta\lambda/\lambda) \text{ photons sec}^{-1} \text{ mrad}^{-1} \qquad (23)$$

for $\lambda \gg \lambda_c$ and

$$N(\lambda) \approx 3.08 \times 10^{16} I(A)E(\text{GeV})$$
$$\times (\lambda_c/\lambda)^{1/2} e^{-\lambda_c/\lambda}(\Delta\lambda/\lambda) \text{ photons sec}^{-1} \text{ mrad}^{-1} \qquad (24)$$

for $\lambda \ll \lambda_c$.

4. Polarization

The radiation is predominantly polarized with the electric vector parallel to the acceleration vector. The intensity of this polarization is given by the first term in the last bracket of equation (6). The intensity of the perpendicular component of polarization is given by the second term. In the electron's direction of motion the radiation is 100% polarized with the electric vector in the instantaneous orbital plane. Integration over all angles and all wavelengths yields about 75% polarization in the orbital plane. However, when many electrons are present, the incoherent vertical and radial betatron oscillations result in a range of angles for the electron beam at each point in the orbit. This angular divergence reduces the polarization. Elliptical polarization is present off the median plane. One of the axes of the polarization ellipse is always in the orbital plane because the phase difference between the parallel and perpendicular polarization components is always 90°.

Figure 7 summarizes the intensities of parallel and perpendicular polarization for a single electron as a function of emission angle for different wavelengths. As can be seen from Figure 7, the opening angle is approximately γ^{-1} only for $\lambda \approx \lambda_c$. At shorter wavelengths it becomes smaller and at longer wavelengths it becomes much larger, reaching about $4\gamma^{-1}$ at $\lambda \approx 100\lambda_c$.

It is possible to produce synchrotron radiation polarized predominantly in the vertical direction by using a vertically deflecting wiggler magnet (see Chapter 21).

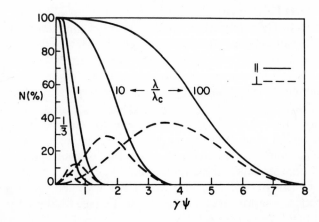

Figure 7. Vertical angular distribution of parallel and perpendicular polarization components.

Helical wigglers cause electrons to execute helical trajectories and emit circularly polarized radiation.

5. Pulsed Time Structure

Electrons in synchrotrons and storage rings are grouped in bunches or buckets whose length is determined by the rf system that replenishes the radiated energy. Thus, the radiation is pulsed with each pulse typically about 10% of the rf period. The orbital period must be an integral multiple of the rf period. This integer is called the harmonic number of the ring. Often this is a very large number (greater than 100) so that a large number of discrete bunches of electrons can be used. In colliding-beam storage rings, usually only a small number of buckets are filled, sometimes only one.

Most machines use an rf frequency in the range 50–500 MHz. As an example, SPEAR operates at an rf frequency of 358 MHz, the 280th harmonic of the 1.28-MHz orbital frequency. Thus, the pulse length of the radiation is typically 0.2 to 0.4 nsec and up to 280 buckets can be filled. Under special conditions (low energy, low current, high r.f. voltage) pulse lengths as short as 50 psec have been observed in SPEAR. Depending on the fill pattern a range of pulse intervals from 2.8 to 780 nsec may be obtained. Only one bunch is used in the colliding-beam operation of SPEAR. The resulting time structure (0.3-nsec pulses repeating at 1.28 MHz) and constant pulse amplitude make this an ideal source for a variety of lifetime studies as explained in Chapter 9.

The amount of current that can be stored in a single bucket is, however, limited. Higher currents can be obtained by filling more buckets. One of the limits on single-bunch currents is the so-called higher-mode or parasitic loss limit.[16] A brief current pulse excites electrical oscillations in parts of the vacuum chamber, particularly in places where there is a change in cross section of the chamber. This can result in significant power dissipation, overheating, and failure of components such as vacuum flanges and bellows. The power deposited varies linearly with the number of bunches and with the square of the current per bunch. Also, the power deposited increases with decreasing bunch length and is minimized when the vacuum chamber is smooth.

For SPEAR, serious heating has been observed when currents in excess of about 60 mA are stored in a single bunch. The total current in many bunches can increase as the square root of the number of bunches without increasing the heating problem. Thus, the same heating effects as that produced by 60 mA in one bunch would result from a total current of 120 mA in four bunches or a total current of 600 mA in 100 bunches. Of course, other effects may also limit the total current, particularly total synchrotron radiation power.

In newly designed storage rings attempts are being made to reduce these higher-mode losses by suitable design of vacuum components. A lower rf frequency, resulting in a longer bunch, also minimizes higher-mode losses but, of course, this results in a less sharply pulsed time structure of the radiation.

For synchrotrons there is also a time structure associated with the cyclic nature of the injection and acceleration process; typically 50–60 Hz. (See Chapter 9 for additional discussion of pulsed time structure.)

6. Brightness and Emittance

In many experiments the source brightness is an important consideration. This is defined as the flux emitted by the source per unit source area and per unit of solid angle. This is related to the size and angular spread of the electron beam (the electron beam emittance) and the angular distribution of synchrotron radiation emission as discussed earlier in this chapter.

In a storage ring or synchrotron, the position of an electron and its angle (both measured relative to the equilibrium orbit) are correlated.[17] Phase-space plots of angle in the median plane versus radial electron position summarize this correlation in a so-called radial electron emittance ellipse. A similar plot of vertical angle versus vertical position gives the vertical electron emittance ellipse. The precise shape and orientation of these phase-space ellipses change from point to point on the orbit, but their area, called the electron emittance, is invariant—an example of Liouville's theorem.[18]

In many applications (e.g., protein crystallography using x-ray diffraction, surface physics studies by VUV photoemission), low emittance, or high source brightness, is desirable to maximize the flux on small sample areas (often $\lesssim 1 \text{ mm}^2$) with minimum angular divergence. Optical systems, generally employing reflection from the curved surfaces of mirrors, gratings, or crystals, image the source onto an experimental sample, usually monochromatizing the radiation at the same time. In such optical systems demagnification of the source can be achieved, but only at the

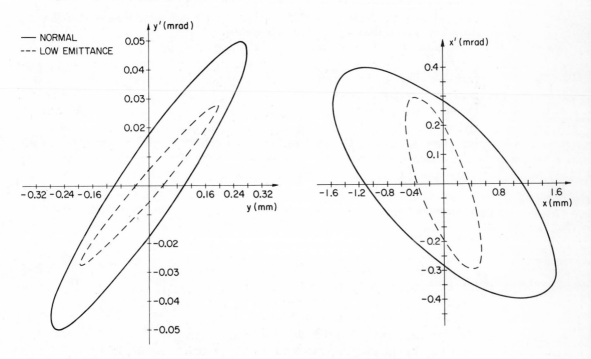

Figure 8. SPEAR electron beam emittance ellipses at 3 GeV at the bending magnet source point: (left) vertical; (right) radial. The low emittance configuration has not yet been implemented.

expense of increased angular divergence, i.e., the brightness is not increased. This is another example of Liouville's theorem. Of course, collimating slits and pinholes can be used to reduce image size and divergence at the expense of total flux.

In other applications, where sample sizes are large (e.g., x-ray lithography), high emittance or low brightness is desirable. To obtain the actual emittance or brightness of the synchrotron radiation, the distribution of emission angles must be convoluted with the electron beam emittance. Examples of electron emittance ellipses are given in Figure 8.

In most storage rings the electron beam emittance can be varied. Increased emittance can be achieved in the vertical plane, for example, by using skew quadrupole magnets to couple the large horizontal betatron oscillation amplitudes into the vertical plane. Smaller emittance can be achieved by increasing the focusing strength of the ring quadrupoles, i.e., raising the betatron tune.[19] It is important to realize that emittance does not vary around the orbit; i.e., a low-emittance configuration has the same low emittance all around the ring. Thus, in dedicated synchrotron radiation facilities serving users interested in high brightness and other users interested in low brightness, it is likely that some operation would have to be scheduled in each mode.

Once the source has been characterized by an emittance ellipse it is possible to determine the effects of optical systems (slits, mirrors, drift space, monochromators, etc.) on the intensity expected in a particular experiment. The use of phase-space technique has been discussed further by Green[9] and others.[20–22] It is a powerful technique for calculating the expected flux and energy resolution of monochromators and beam-line optical systems and increasingly it is being used to optimize the design of synchrotron radiation instrumentation.[22, 23]

Matsushita[24] has pointed out that the phase-space technique does not explicitly include the photon wavelength whereas the Dumond diagram does. The Dumond diagram,[25] a plot of x-ray wavelength versus angle of incidence, is a convenient way to display the angular width of Bragg diffraction by a crystal and has been used, for example, to analyze several kinds of two crystal monochromator systems for use with synchrotron radiation.[26] Matsushita[24] proposes combining the phase space and Dumond diagram methods in a position-angle-wavelength three-dimensional space as a more powerful method for designing and analyzing x-ray optical systems and monochromators.

ACKNOWLEDGMENT

In this summary of the properties of synchrotron radiation I have borrowed heavily from the work of the late K. Green, who made major contributions to synchrotron radiation research including the systematic presentation of the properties of the radiation in a most usable form. Comments on the manuscript by Robert Gutchek and Harry Yakel are gratefully acknowledged. The work at SSRL was supported by the National Science Foundation under contract DMR 77-27489 in cooperation with SLAC and the Basic Energy Division of the Department of Energy.

Appendix: Table of Various Bessel Functions and Integrals[a, b]

y	$K_{1/3}(y)$	$K_{2/3}(y)$	$K_{5/3}(y)$	$H_0(y, 0)$	$G_0(y)$
0.0001	36.284	498.86	6.652 + 6	6.271 + 5	973
0.001	16.715	107.46	1.4330 + 5	2.910 + 4	213.6
0.002	13.192	67.686	4.514 + 4	1.155 + 4	133.6
0.004	10.376	42.621	1.422 + 4	4.581 + 3	83.49
0.006	8.995	32.509	7.233 + 3	2.677 + 3	63.29
0.008	8.116	26.820	4.478 + 3	1.817 + 3	51.92
0.010	7.486	23.098	3.087 + 3	1.348 + 3	44.50
0.020	5.781	14.498	9.723 + 2	5.335 + 2	27.36
0.030	4.932	11.017	4.946 + 2	3.096 + 2	20.45
0.040	4.386	9.052	3.061 + 2	2.102 + 2	16.57
0.050	3.991	7.762	2.110 + 2	1.555 + 2	14.03
0.060	3.685	6.837	1.556 + 2	1.214 + 2	12.22
0.070	3.437	6.136	1.203 + 2	98.37	10.85
0.080	3.231	5.581	96.25	81.94	9.777
0.090	3.054	5.130	79.05	69.69	8.905
0.100	2.900	4.753	66.27	60.25	8.182
0.150	2.343	3.513	33.57	34.15	5.832
0.200	1.979	2.802	20.66	22.59	4.517
0.250	1.714	2.329	14.14	16.26	3.663
0.300	1.509	1.987	10.34	12.34	3.059
0.350	1.343	1.725	7.915	9.713	2.607
0.400	1.206	1.517	6.263	7.850	2.225
0.450	1.809	1.347	5.082	6.474	1.973
0.500	9.890 − 1	1.206	4.205	5.424	1.742
0.550	9.018 − 1	1.086	3.534	4.602	1.549
0.600	8.251 − 1	9.828 − 1	3.009	3.947	1.386
0.650	7.571 − 1	8.933 − 1	2.589	3.414	1.246
0.700	6.965 − 1	8.148 − 1	2.249	2.975	1.126
0.750	6.422 − 1	7.455 − 1	1.967	2.610	1.020
0.800	5.932 − 1	6.839 − 1	1.733	2.302	9.280 − 1
0.850	5.489 − 1	6.288 − 1	1.535	2.040	8.465 − 1
0.900	5.086 − 1	5.794 − 1	1.367	1.816	7.740 − 1
1.00	4.384 − 1	4.945 − 1	1.098	1.454	6.514 − 1
1.25	3.079 − 1	3.406 − 1	6.712 − 1	8.771 − 1	4.359 − 1
1.50	2.202 − 1	2.402 − 1	4.337 − 1	5.557 − 1	3.004 − 1
1.75	1.594 − 1	1.722 − 1	2.906 − 1	3.642 − 1	2.113 − 1
2.00	1.165 − 1	1.248 − 1	1.998 − 1	2.445 − 1	1.508 − 1
2.25	8.581 − 2	9.132 − 2	1.399 − 1	1.672 − 1	1.089 − 1
2.50	6.354 − 2	6.726 − 2	9.941 − 2	1.160 − 1	7.926 − 2
2.75	4.727 − 2	4.981 − 2	7.142 − 2	8.145 − 2	5.811 − 2
3.00	3.531 − 2	3.706 − 2	5.178 − 2	5.772 − 2	4.286 − 2
3.25	2.645 − 2	2.767 − 2	3.781 − 2	4.123 − 2	3.175 − 2
3.50	1.988 − 2	2.073 − 2	2.778 − 2	2.964 − 2	2.362 − 2
3.75	1.497 − 2	1.558 − 2	2.051 − 2	2.144 − 2	1.764 − 2
4.00	1.130 − 2	1.173 − 2	1.521 − 2	1.558 − 2	1.321 − 2
4.25	8.545 − 3	8.853 − 3	1.132 − 2	1.138 − 2	9.915 − 3
4.50	6.472 − 3	6.693 − 3	8.455 − 3	8.338 − 3	7.461 − 3
4.75	4.909 − 4	5.069 − 3	6.332 − 3	6.132 − 3	5.626 − 3
5.00	3.729 − 3	3.844 − 3	4.754 − 3	4.523 − 3	4.250 − 3
5.50	2.159 − 3	2.220 − 3	2.697 − 3	2.481 − 3	2.436 − 3

Appendix (continued)

y	$K_{1/3}(y)$	$K_{2/3}(y)$	$K_{5/3}(y)$	$H_0(y, 0)$	$G_0(y)$
6.00	1.255 − 3	1.287 − 3	1.541 − 3	1.373 − 3	1.404 − 3
6.50	7.317 − 4	7.495 − 4	8.855 − 4	7.659 − 4	8.131 − 4
7.00	4.280 − 4	4.376 − 4	5.113 − 4	4.299 − 4	3.842 − 4
7.50	2.509 − 4	2.562 − 4	2.965 − 4	2.426 − 4	2.755 − 4
8.00	1.474 − 4	1.504 − 4	1.725 − 4	1.376 − 4	1.611 − 4
8.50	8.679 − 5	8.842 − 5	1.007 − 4	7.837 − 5	9.439 − 5
9.00	5.118 − 5	5.209 − 5	5.890 − 5	4.480 − 5	5.543 − 5
9.50	3.023 − 5	3.073 − 5	3.454 − 5	2.569 − 5	3.262 − 5
10.00	1.787 − 5	1.816 − 5	2.030 − 5	1.478 − 5	1.922 − 5

[a] From K. Green.[9]

[b] The positive or negative number following the numerical values represents powers of ten (e.g. 6.652 + 6 represents 6.652×10^6).

References

1. J. Schwinger, *Phys. Rev.* **75**, 1912–25 (1949).
2. A. A. Sokolov and I. M. Ternov, *Synchrotron Radiation*, Pergamon, New York (1968).
3. D. H. Tomboulian and P. L. Hartman, *Phys. Rev.* **102**, 1423–47 (1956).
4. R. P. Godwin, *Springer Tracts Mod. Phys.* **51**, 1–73 (1968).
5. J. D. Jackson, *Classical Electrodynamics*, Chapter 14, p. 848, Wiley, New York (1975).
6. H. Winick and A. Bienenstock, *Annu. Rev. Nucl. Part. Sci.* **28**, 33–113 (1978).
7. G. N. Kulipanov and A. N. Skrinsky, *Usp. Fiz. Nauk* **122**, 369–418 (1977); English translation: *Sov. Phys. Usp.* **20**, 559–86 (1977).
8. K. Lea, *Phys. Rep.* **43**, 337–375 (1978).
9. G. K. Green, BNL Report 50522, 90 pp. (1977); BNL Report 50595, Vol. II (1977).
10. R. Y. Tsien, *Am. J. Phys.* **40**, 46–56 (1972).
11. P. Csonka, *Part. Accel.* **8**, 161–65 (1978).
12. C. Bénard and M. Rousseau, *J. Opt. Soc. Amer.* **64**, 1433–44 (1974).
13. R. Coïsson, *Opt. Commun.* **22**, 135–37 (1977).
14. A. Hofmann (private communication).
15. A. D. Baer, R. Gaxiola, A. Golde, F. Johnson, B. Salsburg, H. Winick, M. Baldwin, N. Dean, J. Harris, E. Hoyt, B. Humphrey, J. Jurow, R. Melen, J. Miljan, and G. Warren, *IEEE Trans. Nucl. Sci.* **22**, 1794–97 (1975).
16. P. B. Wilson, J. B. Styles, and K. L. F. Bane, *IEEE Trans. Nucl. Sci.* **24**, 1496–98 (1977).
17. M. Sands, *Proc. Int. Sch. Phys. "Enrico Fermi"* **46**, 257–411 (1970). Also SLAC Report 121, Nov. 1970.
18. M. S. Livingston and J. P. Blewett, *Particle Accelerators*, McGraw-Hill, New York (1962), p. 117.
19. J. Cerino, A. Golde, J. Hastings, I. Lindau, B. Salsburg, H. Winick, M. Lee, P. Morton and A. Garren, *IEEE Trans. Nucl. Sci.* **24**, 1003–5 (1977).
20. A. P. Sabersky, *Part. Accel.* **5**, 199–206 (1973).
21. I. Lindau and P. Pianetta, in *Proceedings of a Course on Synchrotron Radiation Research*, A. N. Mancini and I. F. Quercia (eds.) Alghero, Italy (1976), 372–387. Sponsored by the International College of Applied Physics and the I.N.F.N. Also *Nucl. Instrum. Methods* **152**, 155–160 (1978).
22. J. Hastings, *J. Appl. Phys.* **48**, 1576–84 (1977).

23. J. B. Hastings, B. Kincaid, and P. Eisenberger, *BNL Report 23353* (1977). Also *Nucl. Instrum. Method* **152**, 167–171 (1978).
24. T. Matsushita, in *Workshop for X-ray Instrumentation for Synchrotron Radiation Research*, H. Winick and G. Brown (eds.), SSRL Report 78/04 (May 1978), p. III-17 to III-25.
25. J. W. M. Dumond, *Phys. Rev.* **52**, 872–883 (1937).
26. K. Kohra, M. Ando, and T. Matsushita, *Nucl. Instrum. Methods* **152**, 161–166 (1978).

3

Synchrotron Radiation Sources, Research Facilities, and Instrumentation

HERMAN WINICK

1. Introduction

In this chapter we discuss the hardware associated with synchrotron radiation research. Of course, the major piece of equipment is the source itself—the synchrotron or storage ring. We discuss the characteristics of these two types of source and list those in operation and in construction throughout the world. The research facilities attached to the source are also discussed, particularly the beam channels and associated equipment to conduct the radiation from the ring to the experimental station. Lastly, we discuss the specialized instrumentation of synchrotron radiation research; particularly the equipment that has been developed to collect, monochromatize, and focus the radiation onto an experimental sample and then to detect the transmitted, diffracted, or re-emitted photons or secondary electrons produced.

2. Sources of Synchrotron Radiation

2.1. General

Synchrotron radiation research is truly an outgrowth of high-energy physics research. More than 20 storage rings and synchrotrons that were originally intended only for high-energy physics research have also been used as sources of synchrotron radiation. In fact, at this time only one operating machine (the

---

HERMAN WINICK • Stanford Synchrotron Radiation Laboratory, Stanford Linear Accelerator Center, P.O. Box 4349, Bin 69, Stanford, California 94305.

Table 1. Synchrotron Radiation Sources

Machine	Location	Completion Date	E (GeV)	I (mA)	R (meters)	ε_c (KeV)	Remarks
Storage Rings in Operation as of December 1979							
PETRA	Hamburg, Germany		18	18	192	67.4	Possible future use for synchrotron
			15	80		39.0	radiation research
			10	50		11.6	
CESR (Cornell)	Ithaca, USA		8	100	32.5	35.0	Used parasitically
			4	50		4.4	
VEPP-4	Novosibirsk, USSR		7	10	16.5	46.1	Initial operation at 4.5 GeV
			4.5			12.2	
			4.5		(18.6)	(10.9)	(from 8-kG wiggler)
DORIS	Hamburg, Germany		5	50	12.1	22.9	Partly dedicated
			2.5	300		2.9	
SPEAR	Stanford, USA		4.0	100 [35]	12.7	11.1	[ ], colliding beams 50% dedicated
			2.5	300 [20]		2.7	
			2.5		(4.6)	(7.5)	(from 18-kG wiggler)
VEPP-3	Novosibirsk, USSR		2.25	100	6.15	4.2	Partly dedicated
					(2.14)	(11.8)	(from 35-kG wiggler)
DCI	Orsay, France		1.8	500	4.0	3.63	Partly dedicated
ADONE	Frascati, Italy		1.5	60	5.0	1.5	Partly dedicated
					(2.8)	(2.7)	(from 18-kG wiggler)
VEPP-2M	Novosibirsk, USSR		0.67	100	1.22	0.54	Partly dedicated
ACO	Orsay, France		0.54	100	1.1	0.32	Dedicated
SOR Ring	Tokyo, Japan		0.40	250	1.1	0.13	Dedicated
SURF II	Washington, DC, USA		0.25	25	0.84	0.041	Dedicated
TANTALUS I	Wisconsin, USA		0.24	200	0.64	0.048	Dedicated
PTB	Braunschweig, Germany		0.14	150	0.46	0.013	Dedicated
N-100	Karkhov, USSR		0.10	25	0.50	0.004	
Storage Rings Under Construction as of December 1979							
PEP	Stanford, USA	1979	18	10	165.5	78	Synchrotron radiation facility
			15	55		45.2	planned
			10	35		13.4	
			12	45	(23.6)	(163)	(from 17-kG wiggler—part of PEP lattice)
PHOTON FACTORY	Tsukuba, Japan	1982	2.5	500	8.33	4.16	Dedicated
					(1.67)	(20.5)	(for a 50-kG wiggler)
NSLS	Brookhaven National Lab USA	1981	2.5	500	8.17	4.2	Dedicated
					(1.67)	(20.5)	(for a 50-kG wiggler)
SRS	Daresbury, UK	1980	2.0	500	5.55	3.2	Dedicated
					(1.33)	(13.3)	(for a 50-kG wiggler)
ALADDIN	Wisconsin, USA	1980	1.0	500	2.08	1.07	Dedicated
BESSY	West Berlin, Germany	1982	0.80	500	1.83	0.62	Dedicated; industrial use planned
NSLS	Brookhaven National Lab USA	1980	0.70	500	1.90	0.40	Dedicated
Synchrotrons in Operation as of December 1979							
DESY	Hamburg, Germany		7.5	10–30	31.7	29.5	
ARUS	Erevan, USSR		4.5	1.5	24.6	8.22	
BONN I	Germany		2.5	30	7.6	4.6	
SIRIUS	Tomsk, USSR		1.36	15	4.23	1.32	
INS–ES	Tokyo, Japan		1.3	30	4.0	1.22	
PAKHRA	Moscow, USSR		1.3	300	4.0	1.22	
LUSY	Lund, Sweden		1.2	40	3.6	1.06	
FIAN, C-60	Moscow, USSR		0.68	10	1.6	0.44	
BONN II	Germany		0.5	30	1.7	0.16	

400-MeV storage ring at the Institute of Nuclear Studies in Tokyo) was conceived and built exclusively for synchrotron radiation studies.

At present most synchrotron radiation research programs operate primarily as parasitic or secondary efforts, using the radiation produced during high-energy physics operation of storage rings and synchrotrons. A few small storage rings are now fully dedicated to synchrotron radiation research (see Table 1). Some dedicated operation for synchrotron radiation research is usually available on larger machines. In the past this has been a very small amount. For example, the SPEAR ring operated for a total of about 2000 eight-hour shifts in the four-year period 1975 to 1978. Of these only about 1.5% were used for dedicated synchrotron radiation research operation. This pattern of usage of large machines is now changing, with dedicated operation for synchrotron radiation research increasing, as will be discussed in more detail later.

During dedicated operation there is the considerable advantage that the parameters of operation (primarily electron energy and current) are determined by synchrotron radiation requirements. For example, in $e^- - e^+$ colliding-beam operation of storage rings such as SPEAR and DORIS the stored current cannot exceed the beam–beam interaction limit, which is often only 10–30 mA. In dedicated synchrotron radiation research operation of these machines, hundreds of milliamperes can be stored in an electrons-only mode. Also, when stored beams collide, the natural dimensions and divergences are increased, thus reducing source brightness. Furthermore, the high-energy physics programs often call for operation well below the high-energy limit of these machines, resulting in synchrotron radiation spectra that have much lower critical energy than that which is available in a dedicated mode. Reduced electron energy is particularly serious because the critical energy and the high-energy cutoff of the synchrotron radiation spectrum vary as the third power of the electron energy [see equation (9) in Chapter 2]. Since the high-energy physics program all too often requires electron energies of half, or less than half, of the peak energy, the highest usable photon energy is often reduced by a factor of eight or more.

With the increasing interest in synchrotron radiation research, its former, largely parasitic, role is now changing. New storage rings dedicated only to synchrotron radiation research are now in construction in England, Germany, Japan, and the U.S. (see Table 1). Furthermore, on many existing colliding-beam storage rings a consistent pattern of increased utilization for synchrotron radiation research is developing. The parasites are consuming the hosts. This is often due to the availability of new, larger colliding-beam machines in the same laboratory.

For example, the ACO ring at Orsay is now used exclusively as a light source. The larger DCI ring at the same laboratory is used for both high-energy physics and synchrotron radiation research. At Stanford, the SPEAR ring will be available as a dedicated synchrotron radiation source for 50% of its operations time starting in 1980 when the larger PEP colliding-beam ring becomes operational. Similarly, in Hamburg, the DORIS ring is used as a part-time dedicated synchrotron radiation source now that PETRA is operational for high-energy physics. In Novosibirsk, VEPP-3 is increasingly used as a dedicated synchrotron radiation source with the high-energy physics work shifted to VEPP-4. Furthermore, on several existing storage rings (ACO, SPEAR, DORIS, VEPP-3) expanded synchrotron radiation research facilities are now being completed. In addition, proposals have been made for dedicated storage rings in Amsterdam (1.5–2 GeV), the Electrotechnical Laboratory, Tokyo (0.6 GeV), the Institute for Molecular Science, Okasaki, Japan (0.4–0.5 GeV), Canada (1.2 GeV), Moscow (0.6–2 GeV), Erevan, USSR (2.5 GeV), the China University of Science and Technology, Hofei, Anhwei, The People's Republic of China (0.6–0.8 GeV), and Lund, Sweden (0.4 GeV). Of particular interest is the study[1] being made by the European Science Foundation for a state-of-the-art, high-energy (~ 5 GeV) dedicated source of synchrotron radiation as a joint European venture.

The newest colliding-beam storage rings (CESR, PEP, PETRA, and VEPP-4) are opening up an entirely new spectral region for synchrotron radiation research. Intense fluxes of radiation with photon energies extending to the 500-keV region are produced by these machines. However, unlike the situation in the spectral region below about 30 keV, there does not appear to be a large scientific community with applications for such high-energy radiation. Thus, it will probably take a few years before the scientific potential of synchrotron radiation in this high-energy region is established. A synchrotron radiation facility is now in construction at Cornell using (largely parasitically) the 8-GeV colliding-beam storage ring CESR. In Novosibirsk there are plans to utilize the 7-GeV storage ring VEPP-4 for synchrotron radiation research as well as high-energy physics. It is likely that proposals will soon be made to add parasitic synchrotron radiation research facilities to PEP and possibly PETRA (both 15–20 GeV).

The use of undulators (see Chapter 21) in these high-energy machines would offer the possibility of producing extremely high-brightness, quasi-monochromatic radiation in the 5–30-keV range, a spectral region of well-established scientific interest. This could be the most important synchrotron radiation impact of these high-energy machines.

It is clear that the cumulative effect of expanded facilities on existing sources and increased dedicated use of these sources, plus new dedicated sources and parasitic use of the newest colliding-beam rings, will result in a vast increase in synchrotron radiation research capability. The rapid and universal recognition of the extraordinary properties of this remarkable radiation is a phenomenon in itself.

2.2. Storage Rings

The most capable source of synchrotron radiation is the storage ring. The basic elements of a storage ring designed as a dedicated synchrotron radiation source are shown in Figure 1. A closed, continuous high-vacuum chamber threads through the various ring elements including (a) bending magnets that make the electrons travel in circular arcs producing synchrotron radiation (only one magnet is shown); (b) special insertions (optional), such as the wiggler magnets (see Chapter 21) shown in the drawing, to produce modified or enhanced synchrotron radiation; (c) an rf cavity and associated power supply, which replenishes the energy lost by the electron beam into synchrotron radiation; (d) vacuum pumps to evacuate the chamber; and (e) an inflector that permits electrons from a separate accelerator (not shown) to be injected.

Of course, there are many other elements required to make an operational storage ring, such as quadrupole magnets to provide focusing forces that keep the electron-beam transverse dimensions small, sextuple magnets that compensate for

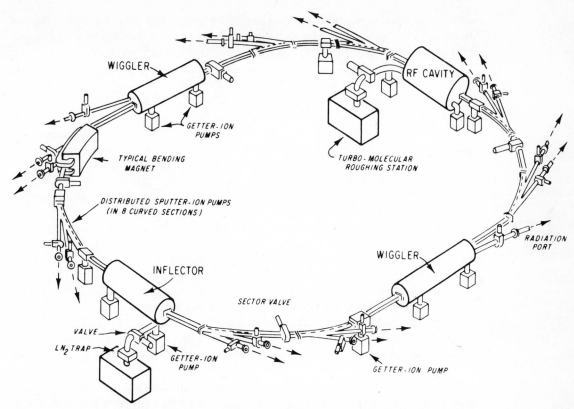

Figure 1. Schematic of an intermediate-energy (0.5–1.0 GeV) electron storage ring designed as a source of synchrotron radiation. Only one bending magnet is shown. Notice the several tangent ports for radiation (courtesy of J. Godell, Brookhaven National Laboratory).

certain effects due to the energy spread in the electron beam, beam position monitors and steering coils used to correct deviations from the design orbit, and often a computer control system. For more information the interested reader is referred to the rich literature of accelerators and storage rings such as the journal *Particle Accelerators* and the proceedings of international accelerator conferences and national accelerator conferences in Japan, the Soviet Union, and the United States.

An excellent treatment of the physics of electron storage rings is given by Sands.[2] The design of storage rings as synchrotron radiation sources is covered in several reports in the Proceedings of the 1979 Accelerator Conference in San Francisco[3–8] and in detailed design reports published by individual laboratories.[9] Storage ring design concepts are also treated by Rowe.[10]

In a properly designed storage ring currents of the order of 0.5 A can be accumulated within several minutes and adjusted within several additional minutes to any desired energy within range of the ring. Injection can be below the operating energy, with the storage ring then used briefly as an accelerator to increase the energy. Injection at the operating energy is quicker, provides the highest stored current, and maximizes experimental running time, but, of course, requires a more powerful injector accelerator. Once a beam is stored it decays slowly with a time constant of several hours, depending largely on the average pressure in the ring (see below). Thus, the synchrotron radiation intensity, as well as the spectral distribution and other source properties, is quite stable over long periods of time.

A storage ring is a high-vacuum device. In order for the stored-beam decay time to be of the order of many hours, the average pressure must be in the 10^{-9} torr range to sufficiently minimize encounters between stored electrons and the residual gas. With present high-vacuum technology it is reasonably straightforward to achieve a pressure in the 10^{-9}–10^{-10} torr range even in such large systems as storage rings. However, in large electron storage rings this pressure must be maintained in the presence of 100 kW or more of synchrotron radiation.

Gas desorption due to synchrotron radiation[11–13] (primarily due to the two-step process of photoemission followed by electrodesorption) is large, and determines the average pressure and lifetime of a stored beam. For example, in a "clean" system with a base pressure in the low 10^{-9} or 10^{-10} torr range, synchrotron radiation can cause a pressure rise of more than two orders of magnitude. Generally there is a gradual improvement with days or weeks of "scrubbing" of the vacuum chamber walls by the radiation.

To minimize gas loads due to synchrotron radiation, special care must be taken in the fabrication, assembly, cleaning, baking, and installation of storage ring vacuum components. In particular, organic materials must be minimized or avoided completely. This concern extends to synchrotron radiation beam channels attached to the storage ring and to experimental equipment installed in the vacuum of these beam channels. In the x-ray region (above about 3 keV) thin beryllium windows[14] are used to separate experiments from the ring vacuum. At longer wavelengths, windows capable of sustaining atmospheric pressure are unavailable. Other techniques are used such as isolation with conductance-limiting slits or differential pumping. Ultrathin windows that can support a small differential pressure and still provide adequate transmission are also used. In this way experiments using gas cells[15] or potentially contaminating samples (e.g., biological materials) are performed.

2.3. Synchrotrons

A synchrotron is quite similar to a storage ring in some ways, i.e., it consists of a roughly circular ring of magnets, a vacuum chamber, and other components similar to the storage ring. The similarity is so close that a single ring could be made to operate as a synchrotron and as a storage ring, as was done at the Cambridge Electron Accelerator.[16] However, a synchrotron is designed to accelerate rapidly (in about 8–10 msec) groups of about 10^{10}–10^{11} electrons from a low injection energy to a peak energy some 10–100 times higher. When used for high-energy physics purposes, at the top energy of the machine the electrons are extracted or made to strike an internal target. The injection and acceleration of a new group of electrons then begins with the process repeating, usually at a 50–60-Hz rate.

Thus, in synchrotrons the electron energy, and consequently the synchrotron radiation spectrum, are not constant. Also, the electron-beam current, position, and cross-sectional area vary within an acceleration cycle and from one cycle to the next. Furthermore, the large amount of high-energy radiation produced by synchrotrons requires that synchrotron radiation experiments be enclosed in shielded areas, remotely controlled and, therefore, less accessible than those performed on storage rings.

The pressure in a synchrotron is typically 10^{-7} torr or higher. Experiments requiring a much better vacuum (e.g., surface physics experiments, which often require a pressure of 10^{-11} torr in the sample chamber) must, therefore, be isolated from the synchrotron. However, poor-vacuum experiments can be directly coupled to the synchrotron with less concern about contamination of the ring than would be the case for a storage ring.

Table 1 gives a list of synchrotrons now in operation.

3. Synchrotron Radiation Research Facilities

3.1. General

Facilities for synchrotron radiation research exist or are planned on almost all operating storage rings and synchrotrons. Descriptions of several of these facilities, in many cases including details of beam lines, monochromators, and other instrumentation, are available, including, for example, the facilities at Hamburg,[17–19] Orsay,[20] and Stanford.[21–25] In addition, much additional information on these subjects is available in activity reports distributed by several laboratories including LURE at Orsay and SSRL at Stanford.

Figures 2–10 give examples of the overall layout of several research facilities.

3.2. Beam Channels

Tangential beam channels or ports attached to the ring permit the synchrotron radiation to enter the experimental hall. In some cases a mirror can be installed directly into a straight section of the ring, deflecting long-wavelength radiation at a large angle radially outward or upward.[21] However, because of their limited spectral range such beam lines are not common.

One beam channel can accept sufficient radiation (10–50 mrad of arc) to serve several simultaneous experiments. Figures 11–14 give examples of the manner in which a single-beam channel may be split and mirrors, gratings, and crystals used to provide monochromatic or white radiation to several experimental stations. Since many beam channels (10–30) can be installed on most rings and each channel can be split to serve several simultaneous users, it is clear that a large number of simultaneous experiments are possible on a single ring. Thus, a storage ring and associated injector, which may cost several million dollars, is really a very cost-effective instrument.

Figure 2. Aerial photograph of SSRL and SPEAR: (1) north arc experimental hall; (2) north arc office extension; (3) south arc experimental hall; (4) lifetimes port shed; (5) trailers (office and storage); (6) colliding-beam interaction regions; (7) SPEAR control room; (8) SLAC end station A (photo by J. Faust, SLAC).

The design of beam channels, particularly on the larger, more powerful storage rings, is a very complex undertaking, involving considerations of high vacuum, cooling, alignment, beryllium windows, radiation shielding, control systems, and other matters. Some details can be found in the Proceedings of the 1979 Particle Accelerator Conference in San Francisco[14, 26–29] and in a review.[30]

3.2.1. Vacuum Considerations

Because beam channels attach directly to the storage ring high-vacuum system, these channels must be designed to vacuum standards used in designing the storage ring itself. In general, this means ion-pumped, all-metal, bakable systems with little or no organic materials such as elastomer "O" ring valve seals.

Special procedures have been developed[27, 31] at storage ring and synchrotron radiation laboratories for designing, fabricating, and assembling these high-vacuum systems. In general, these involve careful selection of materials, specified machining

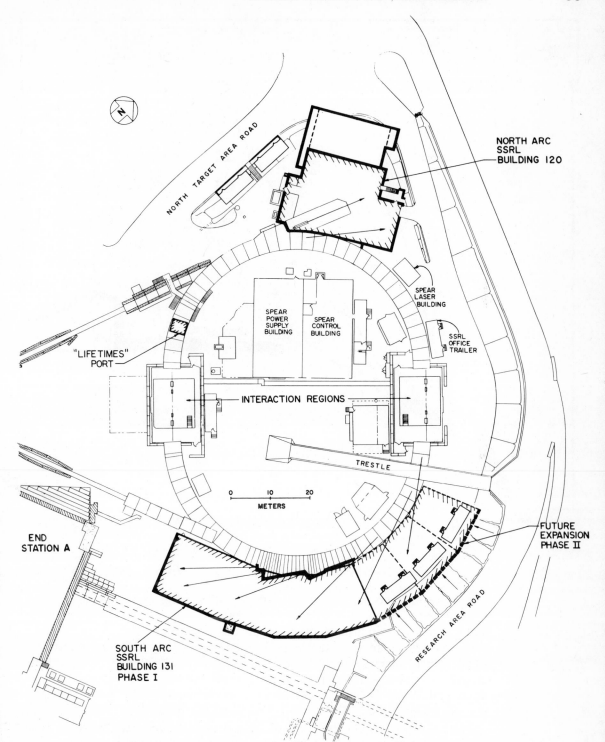

Figure 3. Layout of SPEAR showing synchrotron radiation experimental halls and beam line.

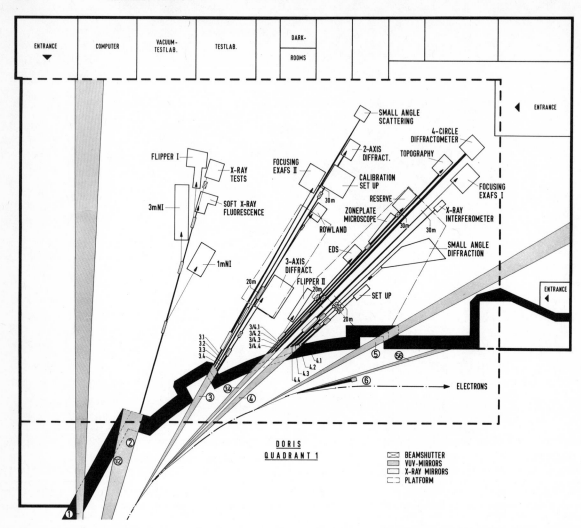

Figure 4. Layout of new research facilities in construction at DORIS (courtesy of C. Kunz).

techniques (e.g., no sulfur-bearing cutting oils), welding techniques (e.g., inside welds whenever possible, avoidance of water to vacuum joints), chemical cleaning, clean assembly, clean installation procedures, and bakeouts.

Alignment is an important consideration, particularly for x-ray beams with small vertical opening angles ($\lesssim 0.2$ mrad). For machines that produce large amounts of power, care must be taken to avoid radiation striking uncooled surfaces, particularly bellows and welds. This is accomplished by installing water-cooled collimators at intervals of 2–5 m along the beam line. These define the extreme rays so that, in the event of vertical or horizontal electron-beam orbit displacements in the ring, radiation strikes only these water-cooled surfaces.

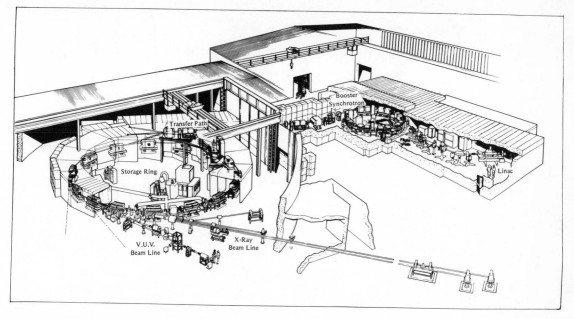

Figure 5. The 2.0-GeV *synchrotron radiation source* (SRS) under construction at Daresbury, England (courtesy of D. J. Thompson, Daresbury).

Although the vacuum system is designed to be rugged and reliable, provision must be made to minimize the adverse effects of accidental venting due to equipment failure (e.g., a leak or a rupture in a thin beryllium window) or experimenter error. Thus, the main beam line and each vacuum branch line are generally equipped with an automatic gate valve that can be closed to isolate experiments from each other and from the storage ring. Since most all-metal high-vacuum gate valves take about two seconds to close and seal, beam lines are often also equipped with fast-acting valves or vanes that close in less than 0.1 sec in order to minimize the gas flow in the event of a catastrophic failure. Acoustic delay lines[32] may also be used to provide a 200–300 msec delay in the arrival time of a sonic wave that may be initiated by an abrupt vacuum failure.

To insure that valves close in the event of an accident and that they can only be opened under proper vacuum conditions, a control system[25] is essential. This system permits the storage ring operator or the experimenter to open gate valves only when the pressure, generally measured by a Bayard–Alpert-type ionization gauge, is below a preset level. Since such gauges have a slow response time ($\gtrsim 1$ sec), faster-responding sensors are often also used to protect against a rapidly rising pressure. One such unit, developed at SLAC, consists of a fine stainless steel wire (0.015″ diameter and 7.5″ long) in a 1.8″ diameter cylinder. The wire is kept at a potential of 1000 V. When the pressure is less than 10^{-4} torr, essentially no current flows. In the event the pressure rises into the 10^{-3} torr range, a discharge will occur in less than 1 msec and this can be used to generate signals that result in valve closure.

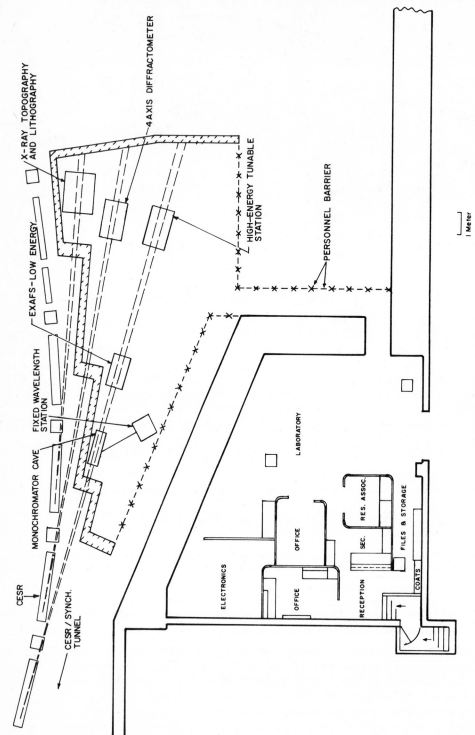

Figure 6. Layout of beam lines at the Cornell *high-energy* synchrotron source (CHESS), which uses (parasitically) the 8-GeV Cornell electron storage ring (CESR) as a source of synchrotron radiation (courtesy of B. Batterman, Cornell University).

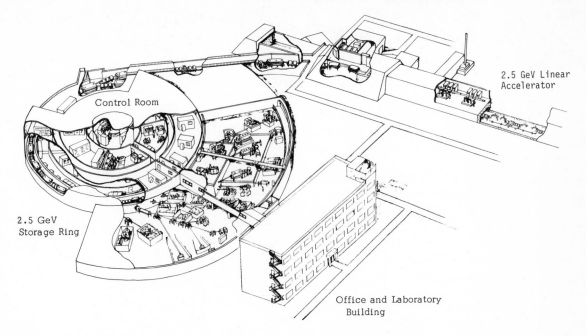

Figure 7. The 2.5-GeV photon factory under construction at KEK, Tsukuba, Japan (courtesy of T. Ohta, KEK).

3.2.2. Thermal Problems

The high radiation power densities produced by multi-GeV machines pose serious thermal problems. Machines such as SPEAR and DORIS have produced more than 100 kW of synchrotron radiation into 2π radians horizontally but into only about 0.3 mrad of vertical opening angle. Even higher power can be radiated during dedicated high-current operation.

Handling this high power is, of course, most difficult when the absorbing surface is close to the orbiting beam and when the radiation strikes at normal incidence. This is the case for the special exit chambers, which often have the tangential spout joined to the main vacuum chamber at a point that may be only two meters from the source point in orbit. The septum, or crotch, between the tangential spout and the main vacuum chamber is subject to a power density in excess of 1 kW cm^{-2}. High-field wigglers (see Chapter 21) and dedicated operation will increase this several fold. This problem has been analyzed at several laboratories and designs have been developed to handle the high power density on exit chambers.[27, 28]

3.2.3. Beryllium Windows

Much effort has been devoted to the development of reliable, vacuum-tight, thin beryllium windows for synchrotron radiation beam lines.[14, 30, 34] Such windows are used to separate x-ray experimental areas from the ring vacuum. Depend-

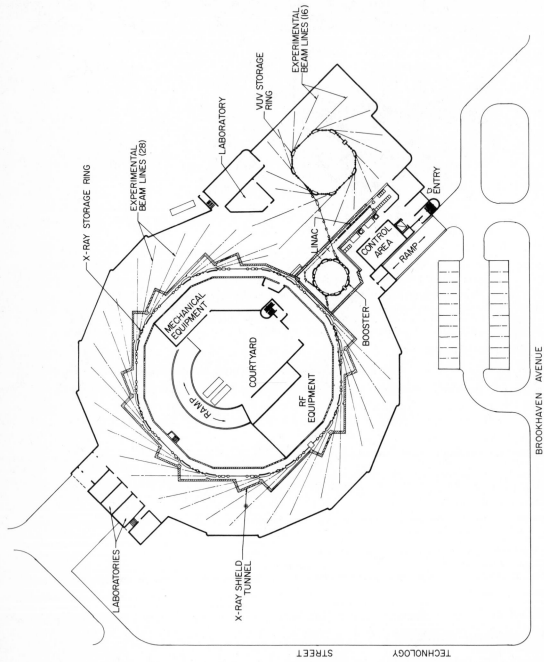

Figure 8. The 2.5-GeV x-ray and 0.7-GeV VUV storage rings under construction at the national synchrotron light source (NSLS), Brookhaven National Laboratory, USA (Courtesy of A. van Steenbergen, NSLS).

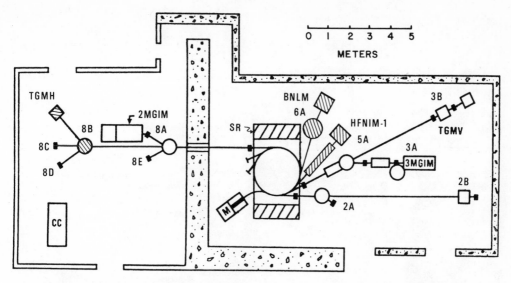

Figure 9. The 250-MeV storage ring and Synchrotron Ultraviolet Radiation Facility (SURF II) at the National Bureau of Standards in Washington, D.C. SR = storage ring; M = injector; CC = control console; TGMH = *toroidal-g*rating *m*onochromator (*horizontal* dispersion); TGMV = *toroidal-g*rating *m*onochromator (*vertical* dispersion); 2MGIM = *grazing-incidence m*onochromator (2.2-m radius); 3MGIM = *grazing-incidence m*onochromator (3-m radius); BNLM = Brookhaven National Laboratory Monochromator; HFNIM = *h*igh-*f*lux *n*ormal-*i*ncidence *m*onochromator; cross-hatching represents new monochromators. (Courtesy of R. Madden, NBS.)

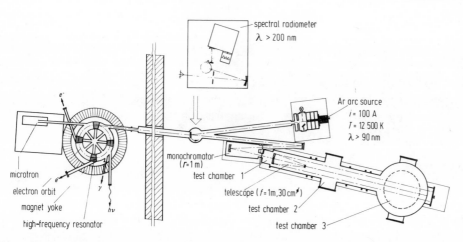

Figure 10. The 140-MeV storage ring and ultraviolet radiation facility at PTB, Braunschweig, West Germany. Parameters: particle energy, $E = 140$ MeV; magnet radius, $R = 0.46$ m; maximum current, $I = 150$ mA; critical wavelength, $\lambda_c = 94$ nm. (Courtesy of H. Kaase.)

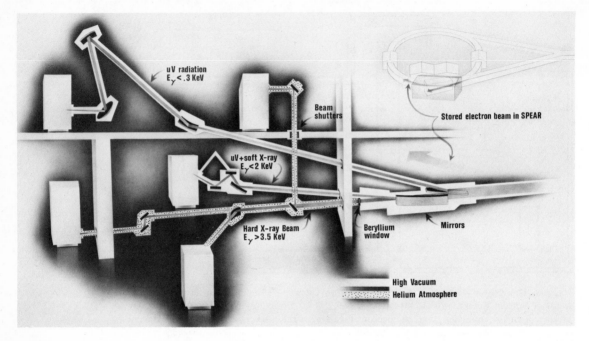

Figure 11. Artist conception of the first beam line for both UV and x-rays, implemented in 1974 at SSRL.

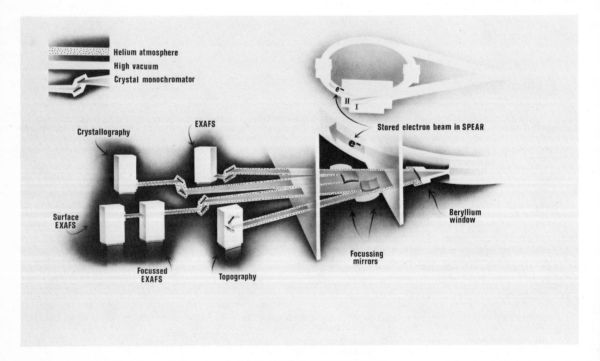

Figure 12. Artist conception of beam line II for x-rays at SSRL.

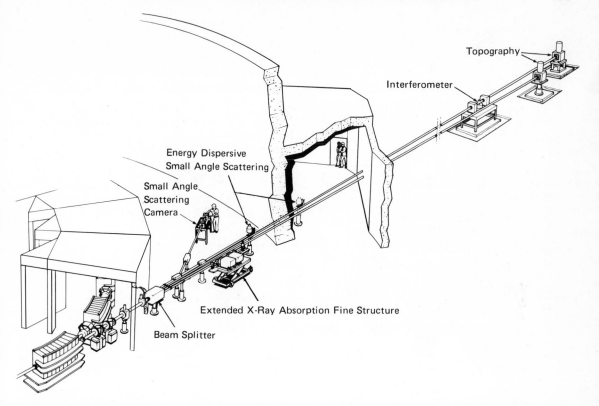

Figure 13. Plans for the first x-ray beam line at the Daresbury SRS (courtesy of D. J. Thompson, Daresbury).

ing on the location of the window and the angular acceptance of the beam line, these windows range from 10–20 cm in width and 0.5–2 cm in height. Smaller windows (e.g., 0.5-cm diameter) can be used in focused beam lines. Usually helium systems are used to transport the radiation from the beryllium window to the experiment since air causes too much attenuation and scattering (see Table 2 for the attenuation lengths of beryllium, helium, and other elements of interest in beam-line design). Because of the reactivity of beryllium, care must be taken to exclude other gases from the helium system, such as gases that may be liberated from plastic components when struck by x-rays and ozone that may be formed by the x-ray beam if there is air in the system.

In most laboratories a window thickness of 100–250 μm is used, often in pairs to provide protection even if one window fails. In many cases the windows must be cooled to remove the absorbed power. The design must take into account thermal stress as well as atmospheric loading and still not exceed conservative stress limits on the beryllium. It is possible to reduce the absorbed power in the beryllium by installing thin (5–10 μm) pyrolitic graphite absorbers in the vacuum system upstream of the beryllium. These absorb long-wavelength radiation, are not subjected to atmospheric loading, and can withstand quite high temperatures.

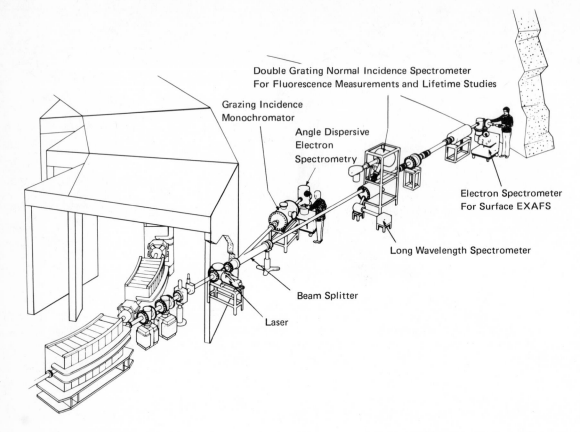

Figure 14. Plans for the first VUV beam line at Daresbury, SRS (courtesy of D. J. Thompson, Daresbury).

Table 2. X-ray Mass Absorption Coefficients and Absorption Lengths[a]

Element		E_γ (keV)						
		1.0	1.5	2.0	3.0	5.0	10.0	20
Hydrogen	cm²/gm	7.32	2.16	1.06	0.561	0.419	0.385	0.370
$\rho = 9.0 \times 10^{-5}$ g/cm²	l_0 (m)	1.52	5.14	10.48	19.80	26.52	28.86	30.03
Helium	cm²/gm	65.2	17.2	6.68	1.89	0.536	0.244	0.196
$\rho = 1.79 \times 10^{-4}$ g/cm³	l_0 (m)	0.85	3.25	8.36	29.56	104.3	229	286
Beryllium	cm²/gm	567	174	72.8	20.5	4.07	0.589	0.218
$\rho = 1.85$ g/cm³	l_0 (mm)	0.0095	0.031	0.074	0.264	1.33	9.18	24.8
Carbon	cm²/gm	2200	710	307	90.6	18.6	2.19	0.414
$\rho = 1.58$ g/cm³	l_0 (mm)	0.0028	0.0089	0.021	0.070	0.340	2.89	15.3
Nitrogen	cm²/gm	3395	1122	494	149	31.3	3.67	0.580
$\rho = 1.25 \times 10^{-3}$ g/cm³	l_0 (cm)	0.236	0.714	1.62	5.38	25.6	218.3	1379
Aluminum	cm²/gm	1178	394	2306	807	198	26.4	3.39
$\rho = 2.70$ g/cm³	l_0 (mm)	0.0031	0.0094	0.00157	0.0046	0.0187	0.140	1.092

[a] W. H. McMaster, N. Kerr Del Grande, J. H. Mollett, and J. H. Hubbell, UCRL 50174, Sec. II, Rev. 1 (May 1969).

Although high-purity (99.67%) beryllium sheet and foil is commercially available, it is difficult to obtain pinhole-free material in areas of about 10–20 cm² with a thickness below 50–100 μm. Several techniques are used to make thin foils, with rolling the most commonly used.

A major problem in utilizing thin foils concerns joining the beryllium to other materials so that it can be made part of the vacuum system. Cements, such as epoxy resins, are unacceptable in high-vacuum applications. Several other techniques have so far been tried, including brazing or soldering beryllium to copper or aluminum in an inert atmosphere, diffusion bonding to copper or Monel, glass frit bonding to Monel, electron beam bonding to aluminum, and mechanical sealing with indium wire as a gasket. Only marginal success has so far been achieved in fabricating beryllium windows with thicknesses below about 100 μm that will still meet the exacting requirements and safety factors required for use in an ultrahigh-vacuum storage ring.

3.3. Radiation Shielding and Personnel Protection Interlock Systems

3.3.1. General

Although the seriousness of a potential radiation exposure increases greatly with machine energy, consideration to personnel protection must be given for all machines. The following factors affect the approach taken to radiation protection:

1. Machine energy.
2. Is synchrotron radiation research a primary or secondary program?
3. Number of simultaneous users.
4. Operations schedule.

Consideration of these and other factors may lead to different solutions as discussed below with specific examples.

3.3.2. Low-Energy Storage Rings

For low-energy storage rings (generally less than about 800 MeV) the major radiation problem occurs during injection because this is a repetitive (1–60 Hz) and frequently lossy process. Once the beam is stored and injection is stopped, the relatively small number of electrons stored at relatively low energy presents little hazard even if all of the particles were suddenly to strike a part of the vacuum chamber. Thus, if personnel are evacuated during injection, only minimal shielding is required (sometimes the vacuum chamber wall itself is an adequate shield) and experimenters are allowed free access to their equipment under stored beam conditions. This is the procedure followed at the Wisconsin 240-MeV storage ring and at the Tokyo 400-MeV storage ring. There is still danger to the eye from intense visible and UV radiation and care must be taken to avoid observation of the direct beam or reflected beams, which are often focused to small spot sizes.

Free access close to synchrotrons is not possible because much higher radiation levels are present due to the constant injection.

3.3.3. High-Energy Storage Rings

For storage rings above about 800 MeV and particularly for multi-GeV storage rings, more shielding is required and access close to the ring is generally not possible even under stored beam conditions. A notable exception is the procedure used on the VEPP-3 storage ring (2.2 GeV), where personnel were permitted up close to the ring, but below the median plane and only when the stored current was very low. This permitted experimenters to manually adjust experimental equipment situated close to the ring and was facilitated by the fact that the storage ring is suspended from the ceiling and the median plane is more than two meters above the floor. Such an access procedure was particularly useful when synchrotron radiation experiments on VEPP-3 were located in the storage ring tunnel. Now a separate shielded room is available.

At SSRL, to maximize accessibility to experimental equipment, shielding and a personnel protection interlock system were designed[21, 23, 24, 30] to permit experimenters to be close to their equipment during all phases of SPEAR operation (filling, storing, and dumping of the beam). Concrete, lead, and wax (for neutron shielding) are used in sufficient thickness to guarantee that the highest possible radiation levels in occupied areas under worst case accident conditions are less than 25 rad/hr. Radiation monitors are set to dump stored beams and stop injection when levels in occupied areas exceed 100 mrad/hr. Regular radiation surveys in occupied areas typically show no measurable radiation. Sometimes levels of 2 mrad/hr are observed in certain areas during injection tuning.

For some beams at SSRL a permanent magnet is installed in the beam line close to SPEAR (about 5.5 m from the source point in orbit). This significantly reduces possible radiation hazards from charged particles by deflecting them vertically so that they do not pass through small vertical apertures (about 1-cm high) located in transverse shield walls 4–5 m downstream of the permanent magnet. This approach is possible for direct x-ray beams with small vertical opening angles and has the advantage that beam lines can remain open even during injection.

VUV and soft-ray beams have larger vertical opening angles requiring larger apertures in the transverse wall. Also, when mirrors are used to deflect beams, including harder x-ray beams, it is often necessary to have larger apertures in the transverse walls. Then it is more difficult to use magnets to provide the larger deflections required. Instead, in these cases at SSRL, shutters that block these apertures are closed automatically during injection so that personnel need not be evacuated from the experimental hall during injection.

For VUV and soft x-ray beams that employ a beam splitter mirror, the angular deviation from the direct line of sight into the storage ring significantly reduces worst case radiation levels. When this deflection is $\gtrsim 4°$ it is usually possible to work freely around the experiment during stored beam conditions. In these cases the synchrotron radiation is also very soft (the hard component is absorbed in the mirror) requiring that the experiments be performed in vacuum. Usually the walls of the vacuum chambers are all the radiation shielding required. In some cases shutters are closed automatically during injection. When the angular deflection of the beam is large ($\gtrsim 8°$) it is not necessary to close shutters during injection.

Figure 15. An enclosure (or hutch) for x-ray experiments at SSRL.[24] An experimenter-controlled interlock system permits access (as shown) to the experiment (through the grill) only when shutters are closed, blocking radiation to that station. A solid sliding door must be closed and locked before shutters can be opened. During installation of large experimental equipment, shutters may be locked closed, and the door and the grill may be temporarily removed.

For hard x-ray beams more shielding is required. Maximum advantage is taken of the small vertical opening angle of the harder x-rays by placing small vertical apertures, often made of lead, along the beam line to reduce worst case radiation levels. Whenever possible the radiation is deflected by a mirror or crystal to avoid a direct line of sight into the storage ring. By using such deflections and small apertures it is possible to allow the radiation to enter individual shielded interlocked experimental enclosures and still provide convenient access for each user. These enclosures, called hutches[24] at SSRL (see Figure 15), are made of steel (thickness ≈ 3 mm) and are lined with 3 mm of lead when white radiation is used. A system of shutters and interlocks[21, 23, 24, 30] permit safe access to each hutch by the experimenter without permission of an operator and independently of the condition of the storage ring or other hutches. To prevent inadvertent or unauthorized entry into monochromator tanks (which are separate from the hutches), these are also interlocked. X-ray beams generally enter the hutch in a helium system terminated in a thin (25 μm) beryllium window inside the hutch. Ozone is produced when x-rays pass through air. This is particularly severe with high-intensity (e.g., polychromatic or white) beams and then the hutch must be equipped with an exhaust system to ventilate the ozone from the building.

A separate hutch, with associated shutters and an interlock system, is provided for each x-ray and some VUV experimental end stations at SSRL. The independent operation of each experiment has proved to be vital at SSRL where up to ten simultaneous experiments have been in operation and more stations are now being added.

3.4. Synchrotron Radiation Beam Position Monitoring and Control

Monitoring and control of the vertical position of synchrotron radiation beams is an important concern in high-energy storage rings. This is due to the fact that the opening angle of the radiation becomes exceedingly small as the electron energy increases. At the critical energy the opening angle of the radiation is

$$\theta_v \approx mc^2/E_e$$

where mc^2 is the electron rest energy (0.51 MeV) and E_e is the electron energy. Thus $\theta_v = 2 \times 10^{-4}$ at $E_e = 2.5$ GeV. This is increased slightly by the range of angles present in the electron beam but still remains of this order.

Because of the coupling between vertical electron-beam displacement and the vertical angle, very small variations in vertical electron orbit (< 0.1 mm) can produce unacceptable changes in the position of the synchrotron radiation beam at the experiment.

Experience at SPEAR has shown that the small changes in orbit from run to run and even the small drifts that occur during a run (both of which are usually acceptable for colliding beams) must be compensated at the synchrotron radiation source point.

This problem was anticipated in the design of the SSRL facility. A simple, accurate, and reliable system was developed,[23, 24, 30] in cooperation with the storage ring physicists and engineers, for reproducing the vertical position of the synchrotron radiation beam. The system consists of a pair of coils (actually trim windings on quadrupoles) that produce horizontal dipole fields. The coils straddle the synchrotron radiation source point and are spaced by about 180° (about 15 m) in vertical betatron phase. Because of the long distance from source point to experiment, the height of the synchrotron radiation beam at the experiment is more sensitive to the angle of the electron beam than to its position. Therefore, one of the steering coils should be close to the source point. It is also possible for the coils to be 360° apart in vertical betatron phase if the source point is close to 180° from each coil. Powered in series, these coils produce a local orbit distortion in the vicinity of the source point with a small residual ($< 5\%$) elsewhere in the ring. A position monitor located in the synchrotron radiation beam run produces an error signal when the beam is off center in the vertical direction. This signal is amplified and fed back to the power supply for the orbit correcting coils. Thus, minute variations in the position of the synchrotron radiation beam are automatically compensated. The beam is held constant to within ± 0.25 mm at 22 m. All experiments are initially aligned to accept a beam in the SPEAR median plane. The need for realignment of experiments is eliminated.

The position monitor[23, 30] consists of a pair of copper photoemitting surfaces that are placed in the beam-line vacuum system so that their leading edges are a few mm above and below the beam center line. The gap between the plates is adequate to pass the x-ray beam. The UV radiation has a larger vertical opening angle and some of this will strike the copper plates, causing photoelectron emission from the plates.

The photocurrent from the copper plates is differentially amplified. The bipolar output voltage is used to control the current in the orbit-correcting power supplies, nulling the position monitor signal and hence correcting the beam back to the median plane.

Compensation for differences in the characteristics of the two plates is provided. The system is calibrated by positioning the beam on a fluorescent screen viewed by a TV camera. It should be possible to use a pair of horizontal wires, rather than plates, for the monitor. This would have the advantage of simplicity and independence of small upstream-masking misalignments, which result in unequal illumination of extended plates.

The problem of orbit control becomes more complex with multiple synchrotron radiation beams on one machine. If the source points are far enough apart ($\gtrsim$ one half-wavelength of the vertical betatron oscillation) then independent orbit control is, in principle, possible. For closer-spaced source points it is difficult to provide independent control. When there are many beam lines it is likely that a more complex system of adjusting the vertical orbit everywhere in the ring, based on signals from position monitors in each beam line, plus electron-beam position monitors in the storage ring, will be required. This will require computer control.

3.5. Experimental Support Facilities

Much equipment is needed at a synchrotron radiation research facility to support the ongoing experiments. This is particularly true at the major installations where a large number of experiments in many disciplines are performed. In some cases at smaller facilities operated for only a few local users, individual users provide almost all the equipment needed for their experiments, and they may also make use of general laboratory facilities situated apart from the synchrotron radiation laboratory, e.g., in other departments of a university or laboratory. Such a mode of operation is not practical for larger facilities operated to serve many simultaneous users and where beam time is more limited and is scheduled for each user. In these cases experimental support equipment and facilities located right at the synchrotron radiation laboratory can significantly increase the effective utilization of the radiation.

A basic necessity for experimental support is a machine shop equipped with the usual hand tools, power tools, and machines including a drill press, band saw, lathe, and milling machine.

The electronics shop and general electronics equipment requirements are considerably more extensive and include oscilloscopes, power supplies, Camac and NIM equipment, general test equipment, analogue to digital and digital to analogue converters, electrometers, voltage to frequency converters, pulse height analyzers, signal averagers, time to pulse height converters, ratemeters, amplifiers, fluorescent detectors, solid state detectors, position-sensitive detectors, and stepping motors.

Computers are ubiquitous at synchrotron radiation research facilities and are used to position mirrors, operate monochromators, monitor incident-beam intensity, and record experimental data. Usually some on-line data analysis is also done, but most analysis is done off-line with computers that have more speed and capacity than the minicomputers generally used on each experimental station. A flexible data acquisition system for control of several independent experiments is described by Clout and Ridley.[33]

At most synchrotron radiation research facilities a vacuum shop is also a necessity because of the large amount of surface physics and other research performed at ultrahigh vacuum. A clean room for assembly and preparation of beamline equipment, monochromators, and sample chambers is essential. The usual range of high-vacuum pumps, gauges, power supplies, residual gas analyzers, laminar flow hoods, and bakeout equipment must be provided. Also, much specialized equipment for surface preparation and characterization is used such as coating systems (evaporation and sputtering), cleavers, LEED, and Auger systems. Chemical cleaning facilities are particularly important in the fabrication of high-vacuum beam lines and instrumentation. Since a high-vacuum shop is also a necessity for servicing the storage ring vacuum system, these facilities can serve both purposes.

A variety of gases are used by many experiments including helium, nitrogen, argon, and helium–neon mixtures. These, plus compressed air and liquid nitrogen, should be conveniently accessible to experimenters.

Biochemistry experiments very often require special facilities to prepare and characterize the samples used. Often it is also necessary to monitor radiation damage to the sample during the x-ray exposure. A well-equipped biochemistry laboratory would include a spectrophotometer, pH meter, analytic balance, ultracentrifuge, microscope, water still, ultrafiltration cells, stirrer, and a rotary vacuum pump. This equipment should be located in a separate room within the synchrotron radiation lab.

Experiments in soft x-ray lithography and microscopy require specialized equipment to prepare, develop, and coat resists and then to enlarge and inspect the images formed. Darkroom facilities are useful for these experiments and also for a large variety of experiments where film is used as a detector (e.g., protein crystallography, topography).

Storage and handling facilities for live animals are required for many biological experiments. Dissection facilities are also needed.

A conventional x-ray generator is a very useful tool for preparing and orienting samples before an experiment is put on line. This can save valuable beam time and also permits this work to be done when the storage ring is not in operation.

Alignment equipment such as lasers and a variety of optical levels and theodolites are needed to install and align beam lines and experiments.

A technical staff is required to operate and maintain all the foregoing equipment and to assist experimenters in using this equipment to set up and perform their experiments.

4. Instrumentation for Synchrotron Radiation Research

4.1. General

With the increasing availability of synchrotron radiation, much effort is being applied to the development of experimental equipment well matched to the properties of the radiation. Instruments that have been developed to meet the specific requirements of particular research areas or techniques are described in the appropriate chapter of this book. In this section we present an introduction and a general survey that may be useful to those seeking an overview of this important subject. For further information the interested reader is referred to the many reports on instrumentation presented at the Stanford[34] and Orsay[35] instrumentation meetings, and at the National Conference on Synchrotron Radiation Instrumentation [at the U.S. National Bureau of Standards (NBS), June 4–6, 1979, proceedings to be published]. See also a recent review by Gudat and Kunz.[36]

4.2. Mirrors

Mirrors perform the following functions in synchrotron radiation beam lines:

1. Deflection: Beam splitting mirrors can be used to deflect part of a single tangential synchrotron radiation beam so that several experimental stations can share a single beam with space between them for equipment (for example, see Figure 10).

2. Focusing: Curved mirrors (generally cylindrical or toroidal) are optical elements that can be used to image the electron-beam source point at some distant location (for example, see Figure 12).

3. Filtration: Mirrors can be made to have a sharp cutoff in reflectivity above a certain photon energy. Thus, they can act as low-pass filters, absorbing unwanted x-rays in VUV beam lines and offering control over harmonics in VUV and x-ray lines.

These and other topics are discussed in considerable detail by several authors[36–39] and in several contributions to the X-ray Instrumentation Workshop.[34] See also Chapters 14 and 16 of this book. Here we offer a brief review of some of the important points.

The reflectivity of a mirror depends on several factors including the photon energy, the angle of incidence, the mirror surface material, and the mirror smoothness. For example, at a 10-mrad grazing angle of incidence on a platinum mirror surface the reflectivity is high up to photon energies of about 8 keV. The cutoff energy varies inversely with grazing angle so that harmonics can be controlled by varying the grazing angle of incidence.

Several different materials have been used as mirror substrates, including Be, Cu, SiC, fused SiO_2, cervit, zerodur, and float glass. The most frequently used coatings are Au and Pt, although uncoated mirrors are also used. Polishing techniques are being perfected with microroughness as low as 5 Å rms reported on 3″ × 3″ blanks of SiC (See Rehn[34]), and as low as 3 Å rms on small spherical fused SiO_2 or Si mirrors.

The surface smoothness achievable in larger and more complex mirrors does not currently approach these values. It should be noted that the auto-covariance[34] distribution of the surface microroughness becomes more important at x-ray wavelengths than the microroughness height. Since mirror manufacturers generally have no way of measuring the microscopic roughness distribution, it is incumbent on the mirror user to develop ways to determine acceptability of mirrors for use in low-grazing-incidence x-ray applications.

Grazing-incidence mirrors used in highly collimated or finely focused synchrotron radiation beam lines have a high sensitivity to optical figure error (e.g., the surface slope error over macroscopic surface elements). Although polishing of planar or spherical surfaces on standard optical glasses is a highly developed and accurate art, this is not the case for toroidal or other nonspherical surfaces or for surfaces polished on nonstandard optical materials.

For mirrors located in the direct x-ray beam in multi-GeV storage rings, thermal loading is a severe problem. In general, cooled metal mirrors must be used in these applications. At extreme grazing angles of incidence ($\lesssim$ 10–20 mrad) nonmetal mirrors can sometimes be used because much of the beam power is reflected and the absorbed power is distributed over a large area. Segmented mirrors are sometimes used, particularly at extreme grazing angles of incidence, because of the length of mirror required (often more than 1 m).

At photon energies above the K absorption edge of carbon (280 eV), and up to about 2 keV, contamination of mirror surfaces with carbon deposits can be a serious problem causing severe reduction in mirror reflectivity. The problem has been analyzed by Shirley[34] who suggests that carbonaceous overlayers are the result of radiation striking catalytically active surfaces such as Pt (which is often used as a coating on mirrors) in the presence of low concentrations of certain residual gases (such as CO) commonly found in high-vacuum systems. It is clearly desirable to avoid sources of carbon in beam lines (e.g., epoxy-based replica gratings) and to develop techniques of rapid replacement and *in situ* monitoring and cleaning of surfaces prone to carbon build up.

Curved mirrors may be used to image the electron-beam source point at VUV and x-ray wavelengths.[34, 38, 40] Depending on the curvatures of the mirror surface, the grazing angle of incidence, and the mirror location in the beam line, it is possible to achieve magnifications of greater than 1, less than 1 (demagnifications), or equal to 1. Demagnifying optics are useful to produce high-flux density on samples as small as 100 μm in diameter. Magnifying optics may be useful to illuminate several square centimeter areas as is required for pattern replication by soft x-ray lithography.

Developments in the fabrication of surface coatings layered on an atomic scale offer the possibility of achieving high values of reflectivity at near normal incidence in the soft x-ray region. Layered structures on a silicon substrate synthesized by vapor deposition with individual layered thicknesses of 7 Å to thousands of angstroms appear achievable according to Barbee.[34, 41] Alternating layers of pairs of materials such as tungsten and carbon or copper and niobium are used. Hundreds of layer pairs can be deposited on areas up to about 100 cm^2. The resulting structure is high-vacuum compatible, stable over a period of years, and capable of withstanding high-temperature and high-radiation levels.

4.3. Monochromators

Because most experiments with synchrotron radiation select a particular wavelength from the continuum and often scan over a range of wavelengths, a tunable monochromator is vital. Many specialized monochromator systems have been developed for use in the vacuum ultraviolet, soft x-ray, and hard x-ray regions of the spectrum and some of these are described in the various chapters of this book. Often we must speak of a mirror–monochromator system because the optical properties of both elements must be considered in designing a beam line to achieve certain performance specifications.

The literature on the subject is quite large. Many reports on grating monochromators and crystal monochromator designs were presented at the Stanford[34] and Orsay[35] instrumentation meetings. More recent developments will be included in the proceedings of the National Conference on Synchrotron Radiation Instrumentation to be held at the National Bureau of Standards, June 4–6, 1979 (to be published in *Nucl. Instrum. Methods*). Here we offer a brief description of a few selected devices.

Monochromators using gratings as dispersive elements are used in the UV region of the spectrum and up to photon energies of 1000–1500 eV. The "grasshopper" monochromator of Brown, Bachrach, and Lien[42] covers a particularly broad spectral range (32–1500 eV) and is described in Chapters 4 and 6. Monochromators using holographically made toroidal reflection gratings[43] appear to offer extremely high efficiency up to about 150 eV. Monochromators based on holographically made transmission gratings[44] have been suggested for the vacuum ultraviolet and soft x-ray regions of the spectrum (see Chapter 4).

A good example of a specialized grating monochromator is the device called "flipper"[45] developed at DESY. This grazing-incidence monochromator (see Figure 16) operates over the photon energy range of 20–300 eV. By selecting one of six plane mirrors with different grazing angles of incidence and different coatings, the cutoff energy may be controlled by the experimenter, thus offering excellent suppression of harmonics. The resolution ($\Delta\lambda/\lambda$) ranges from 10^{-3} to 2×10^{-4}.

The monochromator is built to ultrahigh-vacuum standards to be compatible with storage ring vacuum requirements and to permit direct connection to ultrahigh-vacuum surface physics sample chambers. The overall layout of the device as used to provide monochromatic radiation to be photoemission experiment is shown in Figure 17.

In the x-ray part of the spectrum Bragg diffraction from crystals (e.g., Si, Ge, graphite) is the basis for monochromator designs. Many designs exist in the harder

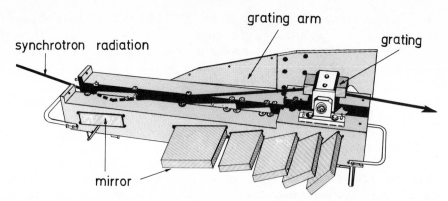

Figure 16. Internal construction of the plane grating "Flipper" monochromator, showing one of the six plane mirrors inserted. By selecting a mirror with the appropriate grazing angle of incidence, harmonics may be suppressed (courtesy of the authors).[45]

x-ray region ($h\nu > 3$ keV) and these have been summarized by several authors.[46, 47] One of the simplest yet versatile devices is the channel-cut crystal monochromator. It provides rapid tunability over a broad spectral range (e.g., 4–30 keV), high transmission, narrow bandwidth ($\Delta E/E \approx 10^{-4}$ with perfect Si crystals), and almost constant exit-beam position and direction. By using two closely spaced crystals in a parallel configuration with a common axis of rotation, the same properties as a channel-cut crystal may be realized once the two crystals have been aligned. However, with independent alignment of the crystals it is also possible to produce slight misalignments that can be used to suppress harmonics that have narrower Darwin widths than the fundamental. Also, relative adjustment of the crystals can be used to compensate for thermal effects due to the higher power density on the first crystal, which must be located in the strong "white"-radiation beam. A two-crystal monochromator incorporating these features has been used with good results on high-current dedicated runs at SSRL. See also Chapter 10 for a more detailed discussion of two crystal monochromators.

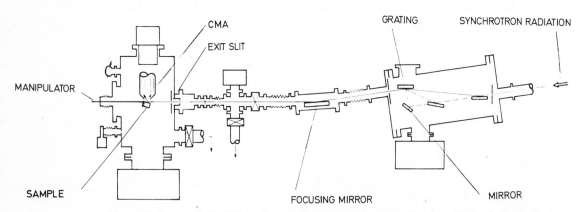

Figure 17. Set up of the "Flipper" monochromator and sample chamber. CMA = cylindrical mirror analyzer—a photoelectron spectrometer (courtesy of the authors).[45]

By using crystals with large lattice spacings, monochromators operating down to photon energies of 500 eV have been constructed.[48]

Many organic crystals (e.g. KAP; $2d = 26.2$ Å) have the required large lattice constants but they suffer radiation damage in the intense x-ray beam. More radiation-resistant crystals are being considered for a long-wavelength crystal monochromator at SSRL,[49] including β-alumina ($NaAl_{11}O_{17}$; $2d = 22.29$ Å), α-quartz (SiO_2; $2d = 8.512$ Å), and indium antimonide ($InSb$; $2d = 7.4866$ Å). Synthetic layered surface structures, as described in the previous section, may also find application here.

By combining a toroidal double-focusing mirror (accepting about 6 mrad horizontally and about 0.3 mrad vertically) with a rapidly tunable two-crystal monochromator, Hastings, Kincaid, and Eisenberger[50] have designed a very powerful system well matched to the characteristics of synchrotron radiation. In use at SSRL for a wide variety of x-ray absorption, diffraction, and other studies since 1976, the system (see Figure 18) has produced a monochromatic flux (at 7.1 keV $\pm$ 2.5 eV) of 10^{12} photons/sec at a focal point (2×4 mm^2) located 20 m from the source point in SPEAR operating at 3.7 GeV and 40 mA.[50]

In an unfocused channel-cut monochromator system at SSRL accepting 1 mrad horizontally and 0.05 mrad vertically, a flux of 2×10^{10} photons/sec was measured in a 1×20 mm^2 spot at 7.1 keV $\pm$ 0.5 eV with SPEAR operating at 3.7 GeV and 40 mA. Thus, the focusing system produces about 50 times the flux and more than 100 times the flux density as the unfocused system, due partly to the five times larger bandwidth.

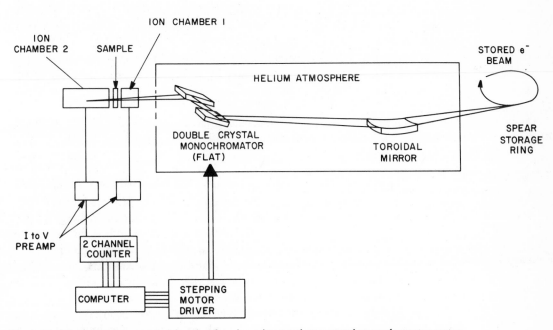

Figure 18. Schematic of a separated function focusing mirror and two-crystal monochromator system by Hastings, Kincaid, and Eisenberger.[50]

At the 10 mrad grazing angle of incidence on the platinum-coated fused quartz mirror,[40] the reflectivity drops sharply for photons with energies above about 8 keV. Although this limits the spectral range well below the 30–40 keV that is possible without mirrors, it has the major advantage of providing excellent control of harmonics. With more grazing angles of incidence it should be possible to extend the spectral cutoff to about 20 keV. Longer mirrors (which could be segmented) are, of course, required.

Specialized monochromators are described in other chapters of this book, including a broad bandpass, curved graphite, crystal monochromator (Chapter 14) and a fluorescence monochromator made up of a mosaic of individual crystals mounted inside a curved surface (Chapter 13).

4.4 Detectors

Many different types of detectors are routinely used in synchrotron radiation research, including scintillation counters, photographic film, proportional counters, channeltrons, image intensifier systems, solid state detectors, ionization chambers, and a variety of position-sensitive detectors. A discussion of several detectors is given in Chapters 10 and 16. Here we present a brief review of position-sensitive detectors.

Development of one- and two-dimensional x-ray position-sensitive detectors is proceeding at an extremely rapid pace, with dozens of groups all over the world active in this field. This was evident at the Workshop on X-ray Instrumentation for Synchrotron Radiation Research[34] held at Stanford on April 3–5, 1978, where, in addition to two invited talks, there were 24 contributed talks on detectors, mostly on position-sensitive devices. The interested reader will find details on the design and performance of a wide variety of detectors in the workshop proceedings[34] and an overview of the subject in the summary by P. Phizackerley in the proceedings. Six presentations on x-ray position-sensitive detectors were made at the Orsay Conference on Synchrotron Radiation Instrumentation and New Developments.[35] Several groups following different approaches have constructed a variety of devices that offer a wide range of performance specifications (spatial resolution, counting rate, linearity, efficiency).

Perhaps the most popular area detector is the multiwire proportional chamber (MWPC), due largely to the extensive development of these devices for high-energy physics research. However, as an x-ray detector in synchrotron radiation experiments the detector must meet different requirements than for high-energy physics experiments. For example, in most high-energy physics applications, counting rates in a MWPC are low and only one coordinate of the track of a charged particle is recorded in one chamber. For synchrotron radiation use the counting rates are high and both coordinates of a photon must be recorded in a single chamber. A review of MWPC systems for x-ray detection has been given by Faruqi.[51]

MWPC detectors have recently been built or are presently in construction for use at synchrotron radiation facilities at Hamburg, Orsay, Novosibirsk, and Stanford. More extensive experience with a MWPC has been accumulated by Xuong and

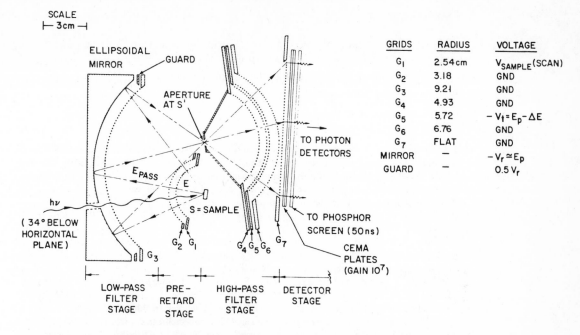

GRIDS	RADIUS	VOLTAGE
G_1	2.54 cm	V_{SAMPLE}(SCAN)
G_2	3.18	GND
G_3	9.21	GND
G_4	4.93	GND
G_5	5.72	$-V_f = E_p - \Delta E$
G_6	6.76	GND
G_7	FLAT	GND
MIRROR	—	$-V_r \simeq E_p$
GUARD	—	$0.5 V_r$

Figure 19. Schematic diagram (top view) of an ellipsoidal mirror display analyzer system for electron energy and angular measurements.[58, 59] Electrons undergo a nearly parabolic trajectory in the retarding field region and have focusing properties similar to those of photons reflected from a virtual mirror. The *channel electron multiplier array* (CEMA) is an area detector. The entire system will be described in the proceedings of the National Conference on Synchrotron Radiation Instrumentation, held at the U.S. National Bureau of Standards on June 4–6, 1979 to be published by *Nucl. Instrum. Methods* (courtesy of D. Eastman, IBM).

co-workers[52, 53] who have used a chamber developed by Perez-Mendez to solve protein structures by x-ray diffraction with a conventional x-ray generator.

The factors controlling the spatial resolution of MWPC systems as two-dimensional x-ray detectors have been analyzed by Charpak and co-workers,[54] who conclude that resolutions of about 100 μm should be attainable. In most applications 1–2 mm spatial resolution is adequate and chambers with such a resolution and counting rates of 10^5–10^6 Hz seem imminent.

Other approaches to x-ray area detectors include backgammon detectors,[55] microchannel multiplier plates,[56] charge-coupled devices, and a variety of electro-optical systems[57] using phosphor screens, fiber optics, image intensifiers, and TV cameras.

A *channel electron multiplier array* (CEMA) has been used[58, 59] as an area detector for photoelectrons in a two-dimensional display-type electron spectrometer for photoemission, LEED, and Auger spectroscopy. Electrons are amplified by a factor of 10^7 and converted to visible light with a phosphor screen. They are then counted with an optical detector. Figure 19 is a schematic of the system.

Acknowledgment

Much of the material and figures describing other laboratories was kindly provided by colleagues at these laboratories. The work at SSRL was supported by the National Science Foundation under Contract DMR 77-27489 in cooperation with SLAC and the Basic Energy Division of the Department of Energy.

References

1. D. J. Thompson, An x-ray synchrotron radiation source for Europe, *Proceedings of the 1979 Particle Accelerator Conference—San Francisco*, in *IEEE Trans. Nucl. Sci.* **26**, 3809–3811 (1979); *European Science Foundation; Reports on a European Synchrotron Radiation Facility, May 1979*: Y. Farges, The feasibility study; Y. Farges and P. J. Duke (eds.), The scientific case; D. J. Thompson and M. W. Poole (eds.), The machine; B. Buras and G. V. Marr (eds.), instrumentation.
2. M. Sands, *Proc. Int. Sch. Phys. "Enrico Fermi"* **46**, 257–411 (1970). Also, SLAC Report 121 (Nov. 1970).
3. S. Krinsky, L. Blumberg, J. Bittner, J. Galayda, R. Heese, and A. van Steenbergen, Design status of the 2.5-GeV National Synchrotron Radiation Light Source x-ray ring, *Proceedings of the 1979 Particle Accelerator Conference—San Francisco*, in *IEEE Trans. Nucl. Sci.* **26**, 3806–3808 (1979).
4. L. Blumberg, J. Bittner, J. Galayda, R. Heese, S. Krinsky, J. Schuchman, and A. van Steenbergen, National Synchrotron Light Source VUV storage ring lattice, *Proceedings of the 1979 Particle Accelerator Conference—San Francisco*, in *IEEE Trans. Nucl. Sci.* **26**, 3842–3845 (1979).
5. D. J. Thompson, The SRS: Progress report on the dedicated synchrotron radiation source at Daresbury and operation of a 600-MeV booster, *Proceedings of the 1979 Particle Accelerator Conference—San Francisco*, in *IEEE Trans. Nucl. Sci.* **26**, 3803–3805 (1979).
6. S. Kamada, Y. Kamiya, and M. Kihara, Lattice of photon factory storage ring, *Proceedings of the 1979 Particle Accelerator Conference—San Francisco*, in *IEEE Trans. Nucl. Sci.* **26**, 3848–3850 (1979).
7. A. van Steenbergen, Synchrotron radiation sources, *Proceedings of the 1979 Particle Accelerator Conference—San Francisco*, in *IEEE Trans. Nucl. Sci.* **26**, 3785–3790 (1979).
8. D. Einfeld, W. D. Klotz, G. Mülhaupt, Th. Mueller, and R. Richter, BESSY, an 800-MeV electron storage ring dedicated to VUV-synchrotron radiation, *Proceedings of the 1979 Particle Accelerator Conference—San Francisco*, in *IEEE Trans. Nucl. Sci.* **26**, 3801–3802 (1979).
9. J. P. Blewett (ed.), Proposal for a National Synchrotron Radiation Light Source. BNL Report 50595 (2 volumes) (1977); Proposal for a 1.5-GeV Electron Storage Ring as a Dedicated Synchrotron Source, FOM Institute Report 39760, Univ. of Tech. Report NK-235 (Oct. 1976).
10. E. M. Rowe, The synchrotron radiation source, in *Topics in Current Physics—Synchrotron Radiation*, C. Kunz (ed.), Springer–Verlag, Heidelberg (1979).
11. E. L. Garwin, 3-GeV Colliding-Beam Vacuum System, Memorandum, SLAC (1963).
12. M. Bernardini and L. Malter, *J. Vac. Sci. Technol.* **2**, 130 (1965).
13. G. E. Fischer and R. A. Mack, *J. Vac. Sci. Technol.* **2**, 123 (1965).
14. J. Cerino and R. Cronin, Beryllium windows for synchrotron radiation beam lines, *Proceedings of the 1979 Particle Accelerator Conference—San Francisco*, in *IEEE Trans. Nucl. Sci.* **26**, 3816–3818 (1979).
15. R. Z. Bachrach, A. Bianconi, and F. Brown, *Nucl. Instrum. Methods* **152**, 53–56 (1978).
16. R. J. Averill, W. F. Colby, T. S. Dickenson, A. Hofmann, R. Little, B. J. Maddox, H. Mieras, J. M. Paterson, K. Strauch, G. A. Voss, and H. Winick, *IEEE Trans. Nucl. Sci.* **20**, 813–15 (1973).
17. E. E. Koch, C. Kunz, and E. W. Weiner, *Optik (Stuttgart)* **45**, 394–410 (1976).
18. E. E. Koch and C. Kunz, *Synchrotronstrahlung bei DESY; Ein Handbuch für Benutzer*, 420 pp., (1977).
19. H. J. Behrend, E. E. Koch, C. Kunz, and G. Mülhaupt, *Nucl. Instrum. Methods* **152**, 37–41 (1978).
20. P. Dagneaux, C. Depautex, P. Dhez, J. Durop, Y. Farge, R. Fourme, P. M. Goyon, P. Jaegle, S. Leach, R. Lopez-Delgado, G. Morel, R. Pinchaux, P. Thiry, C. Vermeil, and F. Wuilleumier, *Ann. Phys. (N.Y.)* **9**, 9–65 (1975).

21. K. O. Hodgson, G. Chu, and H. Winick (eds.), *SSRP Report* **100** (1976).
22. H. Winick, *Proceedings of the 9th International Conference on High Energy Accelerators*, pp. 685–688, Stanford, California (1974).
23. A. D. Baer, R. Gaxiola, A. Golde, F. Johnson, B. Salsburg, H. Winick, M. Baldwin, N. Dean, J. Harris, E. Hoyt, B. Humphrey, J. Jurow, R. Melen, J. Miljan, and G. Warren, *IEEE Trans. Nucl. Sci.* **22**, 1794–97 (1975).
24. H. Winick, SSRL: Past experience, present development, future plans, *Proceedings of the 1979 Particle Accelerator Conference—San Francisco, IEEE Trans. Nucl. Sci.* **26**, 3798–3800 (1979).
25. J. Cerino, A. Golde, J. Hastings, I. Lindau, B. Salsburg, H. Winick, M. Lee, P. Morton, and A. Garren, *IEEE Trans. Nucl. Sci.* **24**, 1003–5 (1977).
26. R. Melen, B. Salsburg, and J. Yang, Vacuum control system for synchrotron radiation beam lines, *Proceedings of the 1979 Particle Accelerator Conference—San Francisco, IEEE Trans. Nucl. Sci.* **26**, 3819–3820 (1979).
27. C. Jako, N. Hower, and T. Simons, Installation and thermal design of synchrotron radiation beam ports at SPEAR, *Proceedings of the 1979 Particle Accelerator Conference—San Francisco, IEEE Trans. Nucl. Sci.* **26**, 3851–3853 (1979).
28. D. Mills, D. Bilderback, and B. W. Batterman, Analysis and design of synchrotron radiation exit ports at CESR, *Proceedings of the 1979 Particle Accelerator Conference—San Francisco, IEEE Trans. Nucl. Sci.* **26**, 3854–3856 (1979).
29. G. Rakowsky and L. R. Hughey, SURF's up at NBS: A progress report, *IEEE Trans. Nucl. Sci.* **26**, 3845–3847 (1979).
30. H. Winick, Considerations for the design of synchrotron radiation research facilities, pp. 43–62 of the *Proceedings of the Course on Synchrotron Radiation Research, Alghero, Italy, Sept. 1976*; A. N. Mancini and I. F. Quercia, (eds.), Int. Colloq. Appl. Phys. INFN.
31. M. Baldwin and J. Pope, SLAC TN-73-13, Stanford Linear Accelerator Center (Oct. 1973).
32. R. Jean and J. Rauss, *Vide* **111**, 123–127 (1964). Also available in English translation as SLAC Translation No. 159. H. Betz, P. Hofbauer, and A. Heuberger, *J. Vac. Sci. Tech.* (to be published).
33. P. N. Clout and P. A. Ridley, *Nucl. Instrum. Methods* **152**, 145–49 (1978).
34. H. Winick and G. Brown (eds.), *Workshop on X-ray Instrumentation for Synchrotron Radiation Research, April 3–5, 1978*, SSRL Report 78/04 (1978).
35. F. Wuilleumier and Y. Farge (eds.), *Proceedings of the International Conference on Synchrotron Radiation Instrumentation and New Developments, Orsay, France, Sept. 12–14, 1977*, North-Holland Publishing Co., Amsterdam (1978). Also published as *Nucl. Instrum. Methods* **152**, (1978).
36. W. Gudat and C. Kunz, Instrumentation for spectroscopy and other applications, in *Topics in Current Physics—Synchrotron Radiation*, C. Kunz (ed.), Springer–Verlag, Heidelberg (1979).
37. A. Franks, X-ray optics, *Sci. Prog.* (*London*) **64**, 371–422 (1977).
38. Y. Sakayanagi and S. Aoki, Soft x-ray imaging with toroidal mirrors, *Appl. Opt.* **17**, 601–603 (1978).
39. V. Rehn, J. L. Stanford, A. D. Baer, V. O. Jones, and W. J. Choyke, VUV scattering by polished surfaces of CVD SiC, *Appl. Opt.* **16**, 1111 (1978); V. Rehn and V. O. Jones, *Opt. Eng.* **17**, 504–11 (1978). Also, *Opt. Eng.* **17**, 504 (1978). [Also available as SSRL Report 77/13].
40. J. A. Howell and P. Horowitz, *Nucl. Instrum. Methods* **125**, 225–230 (1975).
41. T. Barbee, a talk given on layered synthetic microstructures at the Lithography/Microscopy Beam Line Design Workshop held at SSRL on February 21, 1979. See pp. 185–194 of SSRL Report 79/02 (1979).
42. F. C. Brown, R. Z. Bachrach, and N. Lien, *Nucl. Instrum. Methods* **152**, 73–80 (1978).
43. C. Depautex, P. Thiry, R. Pinchaux, Y. Petroff, D. Lepére, G. Passereau, and J. Flamand, *Nucl. Instrum. Methods* **152**, 101–2 (1978).
44. E. Källne, H. W. Schnopper, J. P. Delvaille, L. P. Van Speybroeck, and R. Z. Bachrach, *Nucl. Instrum. Methods* **152**, 101–107 (1978).
45. W. Eberhardt, G. Kalkoffen, and C. Kunz, *Nucl. Instrum. Methods* **152**, 81–83 (1978).
46. J. H. Beaumont and M. Hart, *J. Phys. E* **7**, 823–9 (1974).
47. K. Kohra, M. Ando, T. Matsushita, and H. Hashizume, *Nucl. Instrum. Methods* **152**, 161–166 (1978).
48. M. Lemonnier, O. Collet, C. Depautex, J. Esteva, and D. Raoux, *Nucl. Instrum. Methods* **152**, 109–111 (1978).
49. J. Stöhr, V. Rehn, I. Lindau, and R. Z. Bachrach, *Nucl. Instrum. Methods* **152**, 43–51 (1978).

50. J. B. Hastings, B. M. Kincaid, and P. Eisenberger, *Nucl. Instrum. Methods* **152**, 167–172 (1978).
51. A. R. Faruqi, in *The Rotation Method in Crystallography*, U. W. Arndt and A. J. Wonacott (eds.), Chapter 16, pp. 227–243, North-Holland, Amsterdam (1977).
52. C. Cork, D. Fehr, R. Hamlin, W. Vernon, Ng. H. Xuong, and V. Perez-Mendez, *J. Appl. Crystallogr.* **7**, 319–23 (1974).
53. Ng. H. Xuong, S. Freer, R. Hamlin, C. Nielson, and W. Vernon, *Acta Crystallogr. Sect. A* **34**, 284–9 (1978).
54. G. Charpak, F. Sauli, and R. Kahn, *Nucl. Instrum Methods* **152**, 185–90 (1978).
55. R. Allemand and G. Thomas, *Nucl. Instrum. Methods* **137**, 141 (1976).
56. A. W. Woodhead and G. Eschard; *Acta Electron.* **14**, 181 (1971).
57. U. W. Arndt, in *The Rotation Method in Crystallography*, U. W. Arndt and A. J. Wonacott (eds.), Chapter 17, pp. 245–261, North-Holland, Amsterdam (1977).
58. D. E. Eastman, F. J. Himpsel, and J. J. Donelon, *Bull. Am. Phys. Soc.* **23**, 363 (1978).
59. F. J. Himpsel and D. E. Eastman; *Phys. Rev. B* **18**, 5236–9 (1978).

4

Inner-Shell Threshold Spectra

FREDERICK C. BROWN

1. Introduction

1.1. Kossel–Kronig Structure

It can be argued that lithium is the lightest element with an *inner*-shell spectrum. In this case the spectrum begins at the threshold for excitation of one of the two $1s$ electrons in the K shell of lithium (binding energy = 54.75 eV).[1] These energies progressively increase throughout the periodic system to beyond 150 keV for the K shells of the transuranic elements. Thus photons from the vacuum ultraviolet to hard X-rays are effective in exciting the various inner shells. We will emphasize the low-energy part of this range, and our main concern will be for the details within ten or fifteen volts of threshold, the so-called Kossel structure. This threshold part of an element's inner-shell spectrum may contain Rydberg or exciton effects, and valence orbital, ionization, and chemical shift information. It is most sensitive to the state of aggregation, whether it be atomic, molecular, or solid. Extended x-ray absorption fine structure (EXAFS)[2] (see also Chapters 10–13) or Kronig structure is likewise influenced by neighboring atoms, but in this case data must be taken several hundred or even a thousand electron volts above threshold. EXAFS is a kind of internal diffraction effect dependent upon interference between outgoing electron waves and waves back scattered from surrounding atoms.

1.2. Lifetime Effects

Figure 1 shows the K-edge absorption spectrum for bromine gas (Br_2) taken with high enough resolution (~ 1 eV) to reveal both threshold structure and the EXAFS.[3] The Kronig structure or EXAFS is evident as a slight oscillation super-

---

FREDERICK C. BROWN • Department of Physics, University of Illinois, Urbana, Illinois 61801.

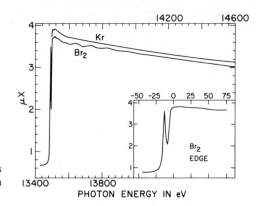

Figure 1. The *K* threshold spectra of bromine (Br$_2$) and krypton gas (Kr) recorded with about 1-eV resolution using intense synchrotron radiation (from B. M. Kincaid[3]).

imposed upon smoothly decreasing absorption far above threshold. A different behavior is observed within 25 eV of threshold. In fact a distinct resonance appears at the Br$_2$ edge.

The *K*-edge spectrum for atomic krypton has been displaced and super-imposed on the Br$_2$ spectrum in Figure 1 for comparison. Notice that EXAFS oscillations are absent. The threshold region is similar to that for Br$_2$ except a resolved line is not seen. This is surprising since we expect a Rydberg-like bound state 5 to 10 eV below the ionization limit due to the core hole potential. Evidently this is not resolved because of lifetime broadening. Figure 2 shows intrinsic linewidths for *K*-

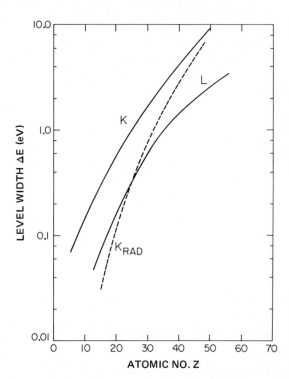

Figure 2. *K*- and *L*-level width as a function of atomic number *Z*. The dashed line shows the contribution of luminescence to *K* width (from estimates of L. G. Parratt[4]).

and L-shell excitation as a function of atomic number Z. These linewidths correspond to estimates made by L. G. Parratt[4] from available experimental data. It can be seen that the K-shell lifetime broadening for krypton, $Z = 36$, is as large as ∼4 eV, comparable to Rydberg binding energies. In the case of Br_2 an intense geometrical resonance rather than a Rydberg line apparently occurs. The intensity and position of this resonance relative to the continuum makes it just discernable in the lifetime-broadened spectrum. An analogous case, that of Cl_2, will be discussed below in Section 3.2. See Chapter 10 for further discussion of near-edge structure.

1.3. Early Work on Excitons

Let us not leave the impression that structure near threshold is unique to gas-phase spectra. The excitonic nature of x-ray threshold spectra in the alkali halides was demonstrated long ago by Parratt and Jossem.[4] Figure 3 shows the K emission and absorption edges for potassium and chlorine in KCl as recorded by these workers. In this early work the energy zero was chosen more or less arbitrarily, as shown in the figure, to represent transitions to the bottom of the conduction band or core ionization energy. The A peaks were then explained qualitatively as x-ray excitons with observed binding energies of 3.2 eV and 4.1 eV for potassium and chlorine, respectively. The apparent linewidths ∼0.4 eV are reasonably explained in terms of the lifetime of core holes in this energy range, determined mainly by Auger processes and to a small extent by x-ray emission processes.

Quite a satisfactory theory of core excitations in KCl was given by Muto and Okuno in 1956.[5] In this theory the localized exciton was described in terms of Wannier functions, an appropriate potential was set up, and then the energies were calculated with effective nuclear charge as a variational parameter. Carrier mass

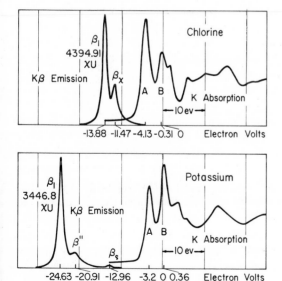

Figure 3. K-emission and absorption spectra for potassium and chlorine in KCl (after Parratt and Jossem[4]).

and effective dielectric constant were also regarded as adjustable parameters. For what we now know are very reasonable values of these parameters, Muto and Okuno obtained exciton binding energies (peaks A in Figure 3) of 3.4 eV and 2.5 eV for potassium and chlorine, respectively. In this work the B bands and higher-lying structures of Figure 3 were not properly explained. Nevertheless, these calculations appear remarkably successful, even though detailed work along the same lines has not been attempted during the intervening years.

1.4. Atomic Effects

It is apparent that spectral detail near threshold is better studied toward lower energies because of lifetime effects. The soft x-ray region includes the outer L and M shells of intermediate elements and the K shells of the lightest elements. There are numerous characteristic features of vacuum-ultraviolet or soft x-ray spectra that have their origin in atomic effects and it is found that these atomic effects are relatively insensitive to the state of aggregation. They include both electron correlation effects as well as processes describable in terms of one-electron models such as (1) a wide distribution of oscillator strength far above threshold, (2) the dominance of $l \rightarrow l + 1$ over $l \rightarrow l - 1$ transitions, (3) a delayed onset of absorption due to centrifugal barrier effects, (4) a resonance near threshold for $nl \rightarrow n'$, $l - 1$ transitions, and (5) Cooper minima when the initial wave functions have nodes. These effects are described in the papers by Cooper and Manson[6] and also in the review by Fano and Cooper.[7] In recent years, synchroton radiation spectroscopy has played a dominant role in the exploration of inner-shell ionization processes. See reviews by Brown,[8] by Haensel,[9] and recently by J. P. Connerade,[10] who has emphasized atomic processes. A general theoretical treatment of inner-level spectroscopy has been given by Kotani and Toyozawa.[11]

A useful experimental approach is to compare gas-phase spectra with the spectra of solids. In this way the effect of neighboring atoms on the final state, as well as mean-free-path effects for the escaping photoelectron, can be found. When this comparison is made a close relation between excitation of the atom and excitation of the solid is frequently obtained as shown in several instances by B. Sonntag and by Koch and Sonntag.[12] In some cases one expects a close correspondence between excited atomic states and excitation in the solid. For example, Dietz et al.[13] have used the interference of atomic excited states with continuum states to successfully explain the $3p$ threshold (65 eV) of metallic nickel observed both by synchrotron radiation and by electron-loss spectroscopy. These authors point out the grounds for assuming that the atomic approach is a reasonable assumption, namely a small $3d$ bandwidth and multiplet splittings small compared to the core excitation energy. In Ni the unoccupied d-band density of states just above the Fermi level is almost singular—it resembles a narrow line. An atomic approach was also used to interpret the $3p$ core spectrum for Ni^{2+} ions in NiO.[14]

There is of course no question about the local atomic character of the inner-shell wave function used for initial states in core excitations. Because of this local initial state a core excitation involves a relatively small region of wave-function overlap in evaluating the transition matrix elements. Actually, the matrix elements depend upon proper knowledge of the final state, its symmetry and extent in the crystal or molecule, the effect of occupied state density on the atom in question and of neighboring atoms (the final state must be orthogonal to occupied state density in the crystal), and finally upon many-electron effects such as screening or adjustment of the electron density to the potential suddenly created by producing a hole in an inner shell. Certainly a correct description of a core excitation in condensed matter goes beyond the atomic picture, although the approximations used may closely resemble atomic or molecular models.[15, 16]

Let us turn in the next section to a brief description of experimental methods before discussing these matters further. Some hydrogenlike absorption spectra will then be shown in Section 3, followed by comments on separation of the discrete part of the spectrum from the continuum part in Section 4. Core excitons as presently understood are covered in Sections 5 and 6. We then turn to some recent work on polyatomic gases in Section 7.

2. Experimental Techniques

2.1. Use of Synchrotron Radiation

Until quite recently x-ray spectroscopy was carried out exclusively with the use of x-ray tubes. However, this has changed with the advent of synchrotron radiation. Tubes are still used in the laboratory when a strong line of characteristic radiation is desired. Nevertheless, the synchrotron continuum is especially suited to high-resolution work near threshold on small samples because of its intensity and collimation, polarization properties, and the high stability of storage rings as photometric sources. (See Chapter 2 for a more detailed discussion of the properties of synchrotron radiation.) A synchrotron or storage ring is a broadband source so the radiation must be dispersed in such a way that a monochromatic band can be tuned and selected. Furthermore, it is desirable to collect light from a large segment of orbit and focus it so as to pass through the slits of the monochromator onto the sample. Ideally, a fixed exit beam of high intensity with variable wavelength and passband is required.

2.2. Synchrotron Radiation Monochromators

Figure 4 shows the outline of three different monochromators that have been used successfully at storage ring sources. They all efficiently collect and transmit radiation. The first instrument[17] is designed for short wavelengths. As shown in Figure 4a, it utilizes a toroidal mirror with a 0.5-degree grazing angle of incidence. This mirror is made of quartz and focuses the point in orbit at or just beyond the exit slit S. Nearly parallel radiation is diffracted by a channel-cut silicon crystal with bandwidth $E \lesssim 1.0 \text{ eV}$ at 10 keV. The optical path (a crude vacuum) is

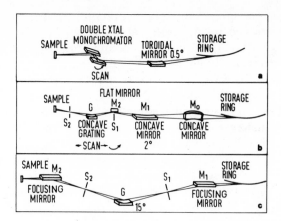

Figure 4. Schematic of monochromator optics used to tune three different parts of the synchrotron continuum. (a) Focusing x-ray monochromator (3000–12000) eV; (b) grasshopper monochromator (25–1000 eV); (c) toroidal grating monochromator (10–150 eV). The elements M are figured mirrors focusing at low or at grazing angles, G's are gratings, and S's are slits.

isolated from the ultrahigh vacuum of the storage ring by thin beryllium windows. These windows and ultimately the d-spacing of the crystal determine the low-frequency cutoff, which does not quite overlap the range covered by the instrument shown in Figure 4b.

The so-called "grasshopper" monochromator[18] shown in Figure 4b uses focusing optics with a two-degree grazing angle of incidence and a high-resolution grating in an ultrahigh vacuum (2×10^{-10} torr), commonly found in storage rings. A Rowland circle mounting is used, but M_1, S_1, and the pivot for the grating arm $\overline{S_1G}$ move in the plane of the synchrotron orbit in such a way as to provide an exit beam fixed in space. By using a 2400-line/mm grating, a relative bandwidth $\Delta\lambda/\lambda$ better than 10^{-3} at 40 Å can be achieved. The resolution is much better at long wavelengths. The instrument has been employed with moderate resolution out to 800 eV.

The toroidal grating monochromator shown in Figure 4c has been developed for highest intensity throughout the low-energy range.[19] It is also an ultrahigh-vacuum instrument so that carbon contamination of the mirrors and of the grating is minimized. Two interchangeable gratings are used to cover the range 10–150 eV. (See Chapters 3 and 6 for more on monochromators.)

2.3. Various Spectroscopic Techniques

A variety of spectroscopic observations can be made beyond the exit slits of the monochromators at a synchrotron source. These observations are sometimes carried out on samples prepared *in situ*. For example, groups at DESY in Hamburg have studied exciton dynamics in the solid rare gases using the apparatus outlined in Figure 5.[20] The measurement of transmission, reflection, and photoemission was possible on samples condensed onto substrates at liquid-helium temperatures. Photoemission as a technique can be further divided into *density of states* (DOS) spectroscopy (in which electron analysis is performed at a fixed photoenergy), *constant final states* (CFS) spectroscopy (in which the photoelectron energy is kept constant but $h\nu$ is varied), and *constant initial state* (CIS) spectroscopy (in which

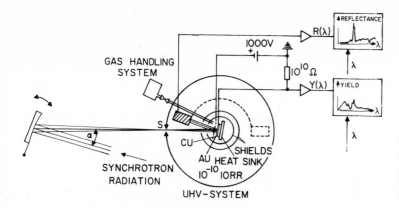

Figure 5. Arrangement for simultaneous measurement of transmission, reflectivity, and photoemission yield on rare-gas samples condensed at low temperature (after Pudewill *et al.*[20]).

both electron energy and *hv* are scanned synchronously). Photoemission can be observed angle-resolved or angle-integrated and there are various forms of partial yield and total yield spectroscopy. These different experimental techniques go well beyond observations of simple photon attenuation or total cross section. At the same time, a knowledge of total cross section is often important in understanding a partial-cross-section or emission experiment. In this review we will emphasize the importance of total-cross-section measurements. For solids, thin films from 500 to 5000 Å thick are often required in a transmission experiment.

Figure 6 shows the apparatus used by Scheifley[21] to evaporate alkali halides onto a variety of substrates (Au, Al, Formvar, etc.) for transmission measurements in the extreme ultraviolet. Two liquid-nitrogen-cooled sample holders could be rotated into either the evaporation or beam-transmission position. A liquid-nitrogen-cooled baffle surrounds the sample region. (See Chapter 6 for more details on experimental sample chambers.)

2.4. Absorption Measurement on Solids and Gases

From a thin-film transmission measurement at frequency ω_1, one obtains the absorption coefficient $\mu(\omega)$ in units of inverse centimeters. Consider a beam of intensity I propagating within a medium in the x direction. Because the decrease in intensity for an infinitesimal thickness dx can be written as $dI = -\mu I\,dx$, the absorption coefficient is just the energy removed per second per unit volume from a beam of unit intensity:

$$\mu = -(1/I)\,dI/dx \qquad (1)$$

For a slab of thickness x, the intensity decreases exponentially,

$$I = I_0 e^{-\mu x} \qquad (2)$$

and one determines μ from the transmission I/I_0, neglecting effects due to reflection.

If we now take into account the ratio of reflected to incident intensity R for both front and back faces of the sample, it can be shown that the transmitted

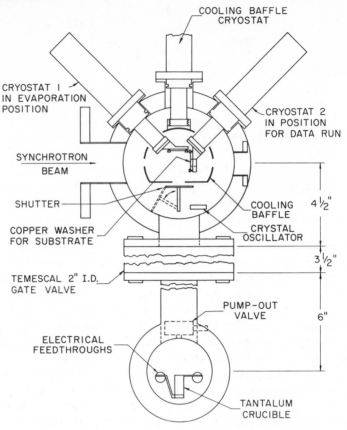

Figure 6. Chamber for evaporation and measurement of thin films. Thickness is monitored by mass changes in an oscillator crystal (after Scheifley[21]).

intensity is given by

$$I = I_0(1 - R)^2 e^{-\mu x}/(1 - R^2 e^{-2\mu x}) \tag{3}$$

This expression neglects interference of multiply reflected radiation, which must be considered under certain circumstances at the longer wavelengths. Throughout the extreme ultraviolet the reflectivity is very small, so that equation (3) reduces to the simple form, equation (2). Then one simply has to make a photometric determination of I/I_0 and carefully measure the sample thickness x. Better yet, measurements are made on several samples of different thickness so as to improve accuracy and estimate stray light. In analyzing the data the coefficient μ can be computed from the optical density D, defined as

$$D = \log_{10}(I_0/I) = \mu x/2.303 \tag{4}$$

Sometimes a mass absorption coefficient $\mu_m = \mu/\rho$ is defined where ρ is the density in grams per cubic centimeter. In such a case, μ_m obviously has units of square centimeters per gram and sample thickness would be given in grams per square centimeter.

In the case of absorption by gases, it is customary to quote an absorption cross section σ per atom or per molecule. In cgs units $\mu = N_A \sigma$ where N_A is the number of molecules per cubic centimeter. It is therefore convenient to reduce an observed absorption coefficient for a gas at a certain pressure and temperature to standard temperature and pressure. The cross section in megabarns $(1 \text{ Mb} = 10^{-18} \text{ cm}^2)$ can then be found by dividing by $10^{-18} L$ where L is Lochschmidt's number, $L = 2.69 \times 10^{19}$ molecules/cm^3. If μ refers to STP, the result is $\sigma(\text{Mb}) = \mu(\text{cm}^{-1})/26.9$. Hudson[22] and also Henke[23] have discussed the optimal sample thickness for accurate values of cross section, as well as the accumulation of errors in such a determination.

Figure 7 shows a gas cell and protective system[24] developed for transmission measurements beyond the exit slits of the grazing-incidence monochromator shown in Figure 4b. Measurements were made on polyatomic gases at the carbon and nitrogen K edges, 285 and 400 eV, respectively. The gases were admitted into a cell containing thin (1000 Å) titanium windows that cut off at about 460 eV due to absorption above the Ti L edge. Further details appear in Section 7 below.

Again, as for solids, the total cross section for a gas can be thought of as being made up of various partial-cross-sections: photoionization, fluorescence, ion yields, etc. A general review and survey of photoionization methods for gases has been given by G. L. Weissler.[25] See also the recent review by Koch and Sonntag.[12]

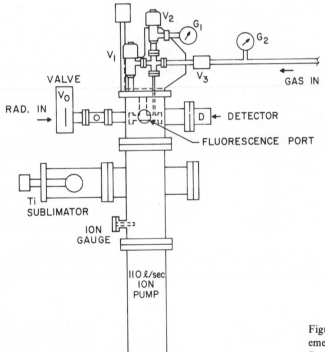

Figure 7. Outline of a gas cell absorption apparatus with emergency pumping and protective apparatus (after Bachrach *et al.*[24]).

3. Hydrogenlike Photoabsorption Spectra

3.1. The Hydrogen Model for X-Ray Absorption; K Edge of Argon

It is well known that in the x-ray region, above about 1 keV, the overall shape of *K* absorption edges can be largely described by means of a hydrogenlike model.[7] The potential in the problem due to the nucleus and inner-shell hole is approximately a Coulomb-like central potential with inner and outer screening corrections. Consequently it can be shown that absorption decreases monotonically from threshold toward higher photon energies, each inner shell contributing a sawtooth pattern to absorption over a very wide range. A detailed quantum calculation shows that the *K* absorption coefficient decreases approximately as $\lambda^{8/3}$ in the immediate vicinity of threshold and more like λ^3 well above the edge.[26]

Figure 8 shows the *K* edge of argon gas plotted against wavelength in angstroms. Notice the characteristic sawtooth shape. When the absorption is recorded with high resolution close to threshold (3.8 Å or 3203 eV for argon gas), Rydberg structure is resolved as shown in Figure 9a. As pointed out by Parratt[4] this structure can be resolved into a series of lines merging into the continuum.

3.2. K Edge of Chlorine in Cl₂ Gas

Figure 9b shows the *K* absorption edge of chlorine gas (Cl_2)[27] showing a somewhat more complicated spectrum than argon because of the diatomic molecular field. In this case it is not correct to ascribe the large peak to a Rydberg-like

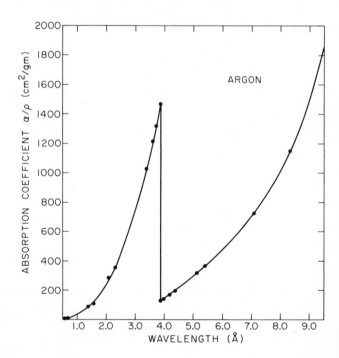

Figure 8. *K* edge of argon gas. [From A. H. Compton and S. K. Allison, *X-rays in Theory and Experiment*, Van Nostrand-Reinhold, Princeton, New Jersey (1949).]

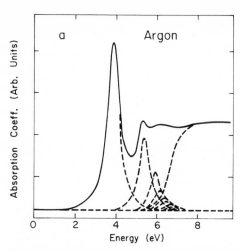

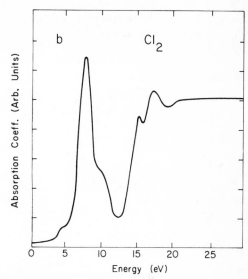

Figure 9. (a) High-resolution scan of the K edge of argon, showing resolution into a series of lines merging into the continuum (after Parratt[4]). (b) The K absorption edge of chlorine gas Cl_2 (after Stephenson *et al.*[27]).

bound state. Again a kind of shape resonance is probably involved as mentioned in Section 1.2 for Br_2. The $L_{2,3}$ absorption edge of the Cl_2 molecule at 200 eV has only recently been resolved.[28] It shows quite an intricate spectrum consisting of excitation to antibonding valence-type molecular orbitals plus higher-lying Rydberg states. Before moving on to these more complicated cases it would be well to review the theory for attenuation of a beam of radiation, especially in the hydrogenlike case.

Oscillator Strength for Rydberg and Continuum Transitions

4.1. Atomic Absorption in the Dipole Approximation

Let us first take a look at the spectral distribution for hydrogen and then for an alkali atom. As mentioned above these have some bearing upon inner K-shell absorption in the x-ray region. We have at least two kinds of absorption: (1) transitions to the continuum and (2) transitions to discrete states usually, but not always, below the ionization limit. By proper normalization the discrete spectrum can be regarded as an appendage of the continuum. We begin by considering the nomenclature describing the absorption of electromagnetic radiation traversing matter.

The loss that gives rise to absorption of radiation is most directly related to the imaginary part $\varepsilon_2(\omega)$ of the dielectric response function,

$$\varepsilon(\omega) = \varepsilon_1(\omega) + i\varepsilon_2(\omega) \tag{5}$$

Sometimes it is convenient to write ε_2 in terms of a frequency-dependent conductivity,

$$\varepsilon_2 = \frac{4\pi\sigma(\omega)}{\omega} \tag{6}$$

In terms of the conductivity the average rate of energy lost per cubic centimeter dW/dt is reasonably given by

$$\frac{dW}{dt} = \frac{1}{2}\sigma EE^* = \frac{1}{2}\sigma E_{0y}^2 \tag{7}$$

where we assume for simplicity an electromagnetic wave $E_y = E_{0y}e^{-i\omega t}$ polarized in the y direction. This energy loss can be written in terms of the transition rate between initial states i and final states j,

$$\frac{dW}{dt} = \sum_{ij} (E_j - E_i)W_{ij} \tag{8}$$

The transition rate W_{ij} is given by the golden rule,

$$W_{ij} = \frac{2\pi}{\hbar} |\langle j|H'|i\rangle|^2 \, \delta(E_j - E_i - \hbar\omega) \tag{9}$$

where because of the energy-conserving delta function, $\hbar\omega = E_j - E_i$. Combining equations (6), (7), and (8), it can be seen that $\varepsilon_2(\omega)$ is given by

$$\varepsilon_2(\omega) = \frac{4\pi\sigma}{\omega} = \frac{8\pi}{\omega|E_{0y}|^2} \sum_{ij} (E_j - E_i)W_{ij} \tag{10}$$

By choosing a proper gauge, the perturbation H' can be written in terms of a vector potential. This vector potential is simply related to field amplitude, which cancels with E_{0y} in the denominator of equation (10). The result is[29]

$$\varepsilon_2(\omega) = \frac{4\pi^2 e^2}{m^2\omega^2} \sum_{ij} |M_{ij}|^2 \, \delta(E_j - E_i - \hbar\omega) \tag{11}$$

where the matrix element $M_{ij} = \langle j|e^{i\mathbf{k}\cdot\mathbf{r}}\hat{\boldsymbol{\eta}}\cdot\mathbf{p}|i\rangle$. The quantity $\hat{\boldsymbol{\eta}}$ is a unit vector in the direction of light polarization and $\mathbf{p}$ is the momentum operator. In all but the deep x-ray region we expect the dipole approximation to be valid, i.e., $e^{i\mathbf{k}\cdot\mathbf{r}} = (1 + i\mathbf{k}\cdot\mathbf{r} + \cdots) \approx 1$. Using this approximation the matrix elements can be rewritten as follows:

$$M_{ij} \approx \langle j|\hat{\boldsymbol{\eta}}\cdot\mathbf{p}|i\rangle = -i\omega m\langle j|r_A|i\rangle \tag{12}$$

where r_A is the component of the radius vector $\mathbf{r}$ in the direction of polarization $\hat{\boldsymbol{\eta}}$.

The absorption coefficient μ and ε_2 are related by means of the expression

$$\mu = \frac{\omega\varepsilon_2}{nc} \tag{13}$$

where n is an index of refraction for the medium, a quantity very close to 1.0 throughout the x-ray region. Let us now combine equations (11)–(13) in order to obtain an expression for the measured quantity, the absorption coefficient. The result is

$$\mu(\omega) = \frac{4\pi^2}{n}\left(\frac{e^2}{\hbar c}\right)\hbar\omega \sum_{ij} |M'_{ij}|^2 \; \delta(E_j - E_i - \hbar\omega) \tag{14}$$

Here we define $M'_{ij} = \langle j|r_A|i\rangle$ as the dipole matrix element.

In general the absorption coefficient for a dilute gas sample is just the product of the atomic cross section and the number of atoms per unit volume as mentioned in Section 2. The cross section for a multielectron atom involves the coordinates of all N electrons in the atom. In the usual formulation,[7] the matrix elements are evaluated by integrating over the entire electron configuration and summing over the number of electrons in the atom. This results in an expression for M'_{ij} of the form

$$M'_{ij} = \int \Psi_j^*(\mathbf{r}_1, \mathbf{r}_2, \dots, \mathbf{r}_N) \sum_n \mathbf{r}_n \Psi_i(\mathbf{r}_1, \mathbf{r}_2, \dots, \mathbf{r}_N)\, d\tau_n \tag{15}$$

Here Ψ_i and Ψ_j are the initial and excited wave functions for the atom written as properly antisymmetrized products of one-electron functions. These many-electron wave functions are then approximated by Slater determinants in which $N - 1$ one-electron functions ψ_i remain unchanged over the transition. In this way the matrix elements can be evaluated in the one-electron approximation,

$$M'_{ij} = \int \psi_j(\mathbf{r})\mathbf{r}\psi_i(\mathbf{r})\, d\tau_r = \int_0^\infty u_{nl}(r)ru_{n,\,l\pm1}(r)\, dr \tag{16}$$

In the last part of equation (16) we have introduced a radial wave function $u_{nl}(r)$. The functions $\psi = \psi_{nlm} = u_{nl}(r)Y_{lm}(\theta, \varphi)/r$ satisfy the Schrödinger equation,

$$\left[-\frac{\hbar^2}{2m}\nabla^2 + V(r)\right]\psi_{nlm} = E_{nlm}\psi_{nlm} \tag{17}$$

where $V(r)$ is an effective central potential. Separation of variables gives the radial Schrödinger equation, which $u_{nl}(r)$ satisfies,

$$\frac{d^2 u_{nl}}{dr^2} + \frac{2m}{\hbar^2}\left[E_{nl} - V(r) - \frac{l(l + 1)\hbar^2}{2mr^2}\right]u_{nl} = 0 \tag{18}$$

4.2. Oscillator Strength and Spectral Density

We now define oscillator strength, a dimensionless quantity containing the important factors in equation (14),

$$f_{ij} = \frac{2m\hbar\omega}{\hbar^2} |M'_{ij}|^2 \tag{19}$$

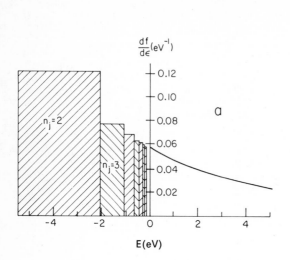

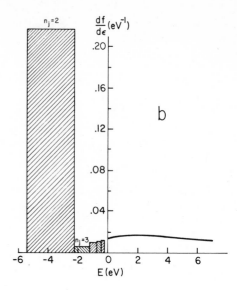

Figure 10. (a) Oscillator strength (theoretical for atomic hydrogen). (b) Measured oscillator strength for lithium (after Fano and Cooper[7]).

In the case of discrete atomic transitions below ionization, this quantity is a measure of the strength of individual lines in a Rydberg series. An atom in a given initial state i obeys a sum rule,

$$\sum_j f_{ij} = N \tag{20}$$

where N is the total number of electrons in the atom. On the other hand, much of the oscillator strength is concentrated in the continuum above the ionization threshold, especially for core transitions in the high-energy part of the spectrum. It is therefore appropriate to replace f_{ij} by a spectral density df/dE in the continuum and to add an integral to the sum in equation (20),

$$\sum_j f_{ij} + \int_{E_I}^{\infty} (df/dE)\, dE \tag{21}$$

where E_I is the ionization energy or beginning of the continuum. Spectral density and absorption coefficients are closely related, i.e.,

$$\mu(\omega) = \frac{\pi e^2 h}{mc} N_a \frac{df}{dE} \tag{22}$$

where N_a is the number of atoms per unit volume in the sample. In this approach the continuum wave functions for final states must be properly normalized per unit energy interval.[26]

Absorption strength in the discrete part of the spectrum merges into the continuum. It is therefore useful to treat the two parts in a similar way as discussed by Fano and Cooper.[7] For this purpose these authors construct a histogram of line intensity with each block centered upon the corresponding line position given by the usual Rydberg formula,

$$E_j = E_I - R_H/(n_j - \delta)^2 \qquad (23)$$

where δ is a quantum defect. Figure 10a shows the theoretical oscillator strength for atomic hydrogen (for which $\delta = 0$ and the ionization threshold $E_I = R_H$). The top of each block is of height $f_j\, dn_j/dE$. The width is dE/dn_j, or the slope of a plot of energy versus quantum number nj. In this way the area under the discrete lines forms a staircase merging into the continuum.

4.3. Comparison of Hydrogen and Lithium Valence Transitions

It is well known that the intensity of atomic Rydberg lines need not decrease exactly as shown in Figure 10a for hydrogen. For example, Figure 10b indicates schematically how much more intensity is concentrated in the first line for the alkali atom lithium—a minimum actually occurs in the discrete part of the spectrum. This is an example of a resonance near threshold. An understanding of the strength of the discrete, as well as the continuum, part of the spectrum comes about through evaluation of the transition matrix elements in each specific case. The aalkali metals are somewhat different than hydrogen and the transition matrix elements must be evaluated in order to precisely understand the observed spectrum. Very few detailed comparisons between measured oscillator strengths and theory are to be found in the literature, especially for Rydberg structure near threshold at high photon energies.

4.4 K Edge of Neon Gas

Figure 11 shows the K absorption spectrum for neon gas near 865 eV.[30] Notice that several lines appear, merging into a fairly strong continuum. Although this is a *core* transition, unlike the prototype case of hydrogen discussed above,

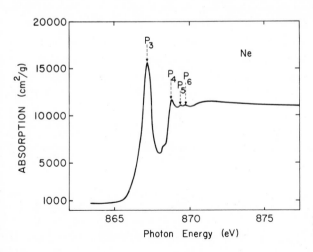

Figure 11. K absorption spectrum for neon gas (after F. Wuillemier[30]).

similarities to Figure 10a are apparent. The transition for example is $^1S_0$ to $^1P_1$, corresponding to a change in electronic configuration, $1s^2 2s^2 2p^6 \rightarrow 1s^1 2s^2 2p^6 np$, with $n \geq 3$. The first line is most intense and subsequent lines decrease in intensity as the ionization energy is approached.

Neon of course has also been studied at lower energies in the valence bond region. It has a large first ionization energy (~ 18 eV), so high-resolution absorption measurements in the laboratory are difficult. Synchrotron radiation has been utilized to study the spin–orbit split ($\Delta E \sim 0.1$ eV) $1s^2 2s^2 2p^6 \rightarrow 1s^2 2s^2 2p^5 ns$ threshold spectra. Higher-lying excitions are also very interesting due to interference between different configurations.[31] Excitations in solid neon have also been extensively studied,[20] but this brings us to the subject of excitons, which we will discuss briefly in the next section.

5. Core Excitations in Insulators

5.1. Valence Excitons in Solid Neon

Exitons in the solid rare gases bear a close relationship to the excitation states of rare-gas atoms. There are, however, distinct differences and in most cases the band or at least the continuum properties of the solid must be taken into account in the theory. This is also true for semiconductors due to the screening of electron-hole Coulomb interactions by sizeable dielectric polarizabilities.

Consider the valence region for solid neon[20] where a well-developed exciton series is resolved as shown in Figure 12. This is a good example of a Wannier exciton in which at least the higher resonances $n \geq 2$ fit an effective- or reduced-mass formula,

$$E_n = E_I - \frac{\mu}{m_0} \frac{1}{\varepsilon^2} R \frac{1}{n^2} \tag{24}$$

where the ionization energy $E_1 = 21.69$ eV and the effective Rydberg $(\mu/m_0) \times (1/\varepsilon^2)R = 5.24$ eV. This is consistent with reasonable values for the dielectric constant $\varepsilon = 1.25$ and electron and hole effective masses 0.8 and 10, respectively. The

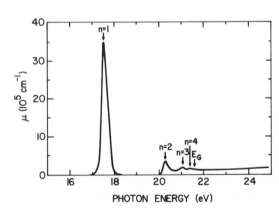

Figure 12. Exciton series in solid neon (from Pudewill *et al.*[20]).

first line of the exciton series is at a smaller energy than predicted by equation (24), therefore a sizeable central-cell correction is required. Also, this excitation corresponds to producing a hole in a p shell, e.g., a $1s^2 2s^2 2p^6 \rightarrow 1s^2 2s^2 2p^5 ns$, nd transition. The spin-orbit splitting is very small ($\sim$0.1 eV) and not resolved in the solid. It does however contribute to the multiplicity of lines observed in the gas spectra.[31]

The theory of valence excitons in solids has a substantial conceptual framework. Certainly a good part of this theory is applicable to the core exciton problem. In concept an exciton involves an electron and a hole coupled through Coulomb interactions. In principle, direct and exchange interactions should be taken into account and very often spin-orbit effects are important. The exciton Hamiltonian can be set up and an integral equation for exciton states written down.[32] In general, solutions fall into two categories, the Wannier, or extended exciton model, and the more localized Frenkel–Peierls model. Recently Bassani and others have shown the formal equivalence between a simple two-particle electron-hole Hamiltonian and the full many-electron Hamiltonian in the problem. This approach can be used to derive band-to-band as well as exciton transitions on an equal footing, at least in the extended- or effective-mass approximation.

5.2. *L* Edge of Solid Argon; Altarelli–Bassani Theory

The solid rare gases have always been a kind of proving ground for exciton theory. Now, this statement can also be made for *core excitons* in the soft x-ray region. Figure 13 shows the L threshold spectra of gaseous and solid argon.[33] A very close correspondence occurs between excitations in the gas and in the solid. Here we are exciting an inner $2p$ shell. Spin–orbit splitting is dominant in that the hole can be created with inner quantum number $j = 3/2$ or $j = 1/2$. The final states are s- or d-like, eg., $2p^6 3s^2 3p^6 \rightarrow 2p^5 3s^2 3p^6\ ns$ or nd, with a spin–orbit splitting of close to 2.0 eV for all lines. Using the integral equation for excitons in an *ab initio* calculation, Altarelli *et al.*[34] obtain a binding energy of about 3.4 eV for the

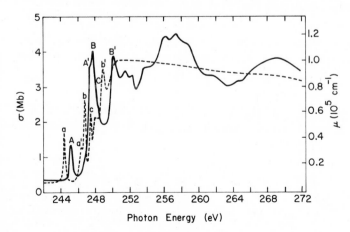

Figure 13. Absorption of solid (solid line) and gaseous (dashed line) argon in the vicinity of the $L_{2,3}$ edge. Corresponding peaks in the two spectra are labeled by upper and lower case letters, respectively (from Haensel *et al.*[33]).

two lowest peaks in solid argon. These lowest peaks, labeled A and A' in Figure 13, are associated with the s-like conduction band for solid argon ($2p_{3/2, 1/2} \to 4s$ transitions on the atom) and the intense peaks labeled B and B' are associated with the higher conduction band of d-like symmetry. The weak feature C is a higher exciton ($n = 2$ in the effective-mass notation) lying closer to the continuum transition (at ~ 248.5 eV). This comparison of theory and experiment allows a proper identification of the continuum edge and a classification of the excitation spectrum of the crystal based upon band structure. At the same time it explains the obvious correspondence with atomic transitions. There are very few instances of this kind of comparison in the literature. Let us turn to the alkali halides next, materials where the comparison is not quite so satisfactory.

Taking spin–orbit interaction into account, Onodera and Toyozawa[35] have applied Wannier exciton theory to the alkali halides with some success. Their approach begins with a band theory and is thought to be appropriate in those cases where the Coulomb interaction between electron and hole is smaller than the bandwidths. They do not deal specifically with core excitons in their original paper, although their results on the relative importance of spin–orbit as opposed to electron-hole exchange energy probably can be carried over directly to the extreme ultraviolet.[36]

5.3. $N_{2,3}$ Edge of Rubidium in RbCl; Satoko–Sugano Theory

The Frenkel–Peierls model of an exciton is perhaps more appropriate to localized excitons as they occur in the alkali halides and often in the core region. An early paper by Dexter[37] shows how a Frenkel exciton in the alkali halides can be viewed as an excited halogen atom within a Madelung well connected with a conduction electron wave function and modified by an envelope function outside the well. This is known as the excitation model as opposed to the transfer model of a Frenkel exciton. Satoko and Sugano[16] have used a somewhat different excitation model in treating core excitons in the rubidium halides around 20 eV. Here transitions corresponding to $4p^6 \to 4p^5\, ns$ and $4p^6 \to 4p^5\, nd$ excitations of Rb$^+$ ions take place. In Figure 14 we show the data of Watanabe et al.[38] on the optical density of

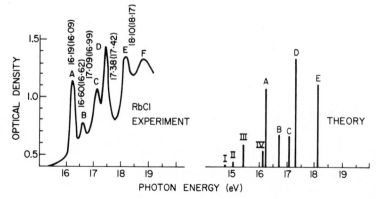

Figure 14. Rb$^+$ $4p$ core excitons in RbCl. The left part of the figure shows experimental data of Watanabe et al.[38], and the right part indicates theoretical estimates by Satoko and Watanabe[16]. Peak energies of both theory and experiment are indicated in upper part of the figure.

thin films of RbCl in the vacuum ultraviolet. Line positions and intensities calculated by Satoko and Sugano[16] are indicated in the lower part of Figure 14. In general, the five peaks labeled A through E are observed in all four rubidium halides with equally good agreement between experiment and theory. It should be emphasized that one does not need the energy band structure in this approach but it is necessary to know the point symmetry around the hole and the one-electron orbitals relevant to the excitation. It should be possible to use a similar detailed analysis on the Rb $3d^{10}$ and possibly on the Cl $2p^6$ cores that occur at higher energy in the extreme-ultraviolet spectrum.

5.4. Deeper Core Structure in RbCl

Figure 15 shows absorption data on a thin evaporated film of RbCl taken with the use of synchrotron radiation by W. Scheifley.[21] Transitions from the Rb^+ $3d^{10}$ core begin at 112 eV where the extinction due to the outer p shells is quite low (actually close to a p to d minimum). Near the d threshold the transitions are mainly d to p due to the atomic effects mentioned earlier. In any case, the d threshold spectrum is interesting because transitions to the conduction-band minimum, which is s-like, are symmetry-unallowed. Furthermore, the $M_{4,5}$ splitting is such that the two spin–orbit components can be readily separated. Transitions from the Rb d core to f-like final states are also possible, but these are very weak at

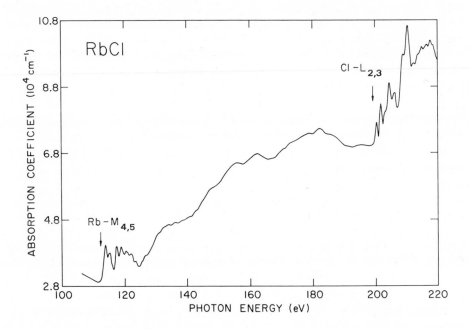

Figure 15. Photoabsorption spectrum of evaporated RbCl showing the Rb $3d^{10}$ and Cl $2p^6$ thresholds; Spectral bandwidth = 0.05 Å (from W. Scheifley[21]).

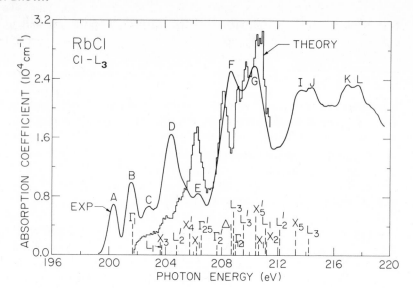

Figure 16. $Cl^-2p^6L_3$ core exciton spectrum in RbCl. The spin–orbit-split L_2 spectrum has been subtracted.

threshold. They do account for the broad rise and gradual decrease near 200 eV (where the Cl-L threshold appears). This prominent d to f maximum is one of the important atomic effects seen in both solids and rare gases.[12]

The fine structure that appears at 200 eV in Figure 15 arises from excitations of the Cl^- $2p^6$ inner shell. The final states are undoubtedly fairly localized excitonic states. These appear either below the band edge, in which case they are pure excitons, or within the band continuum, in which case they are metastable resonances. Excitons below the band edge would bear resemblance to the excitons present in the ultraviolet spectrum arising from excitations of the Cl^- $3p$ valence bands. This is not like the cation core exciton. However, an isolated Cl^- ion has no excited bound state. We must take into account the crystal potential. It is appropriate to speak of band wave functions and band density of states, at least in the framework of Dexter's excitation model.[37] A density-of-states calculation should allow one to distinguish between excitons and band-to-band processes. The positioning of such a density-of-states curve or histogram on the energy axis is, however, crucial and can best be chosen by using accurate high-energy photoemission data.[39]

In order to obtain the state density we rely upon the published OPW conduction-band results for RbCl.[40] A combined pseudopotential tight-binding fit is made to this band structure. The fitting procedure is essentially an interpolation scheme that allows one to solve the band problem at a sufficient number of points throughout the Brillouin zone so as to produce an accurate and nearly continuous density-of-states curve. The resulting histogram representing a mesh of nearly 50 000 points within the first Brillouin zone is superimposed upon the L_3 absorption (separated numerically from the observed $L_{2,3}$ spectrum) in Figure 16. The Γ_1 minimum of the density of states has been aligned so as to agree (±0.5 eV) with electron emission data on core and valence band energies plus the known ultraviolet energy gap. It can be seen that strong resonances appear below the Γ_1 minimum. These are core excitons (binding energy ~ 1.5 eV) as suggested by M. Watanabe and co-workers.[41] The structure at higher energies is also probably due to more or less local excitations associated with different regions of the Brillouin zone.

5.5. Recent Reflectivity Data on the Potassium Halides

Sugano's ligand field approach to cation core excitons has been applied to the potassium[16] as well as the rubidium halides. Figure 17 shows high-resolution reflectivity data on KF, KCl, KBr, and KI taken by Skibowski and co-workers.[42] The region of the K $3p$ edge is shown for single crystals cleaved in ultrahigh vacuum. The first group of peaks marked A_1, A_2 corresponds to final states which are s-like, whereas the second group B, C relates to d-like final states on the potassium ion. Notice that a different temperature shift occurs in these two cases. The first group of peaks shifts to high energy with decreasing temperature, whereas the second group shifts to low energy or does not shift at all. There is of course a considerable thermal broadening at high temperature. It can be seen that the two groups of peaks do not behave in the same way with temperature. The ligand

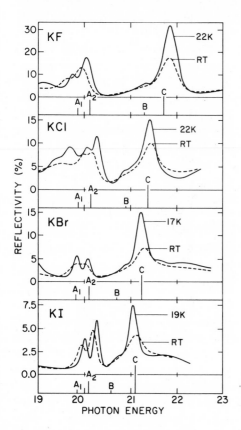

Figure 17. The potassium $3p$ threshold structure for the potassium halides taken by reflectively measurements at two different temperatures (after G. Sprussel, V. Saile, and M. Skibowski[42]).

field model as originally introduced does not explain these charges. On the other hand, a theoretical description of the exciton–phonon interaction has been given[43] and it might be appropriately extended to the core exciton problem.

5.6. *K Edge of Lithium in the Lithium Halides; Zunger–Freeman Theory*

A very detailed theory of optical excitations in LiF has recently been given by Zunger and Freeman.[44] The lithium K threshold at about 60 eV is shown for LiF, LiCl, and LiBr in Figure 18.[45] Lithium fluoride is perhaps a prototype for cation excitation in the alkali halides. Here we are concerned with core excitation of a small Li^+ ion surrounded by six highly electronegative fluorine ions. An understanding of the strong first peak at 61.9 eV as well as higher-lying structure is the goal. To begin with, one suspects that the strong first peak is a rather localized core exciton. The spectrum might bear some resemblence to the highly excited states of a Li^+ ion—except for the surrounding F^- ions. Actually the detailed calculations of Zunger demonstrate that this first line is excitonic and also that solid state, and not just atomic, effects are present.

Zunger and Freeman evaluate the difference in energy between ground and excited states for a small periodic cluster of 16 atoms. In the final state the locally excited ion or defect is at the center of this cluster. Their starting point is a self-consistent band calculation using free-electron exchange in the local density approximation plus an electron correlation potential. Besides Coulomb exchange

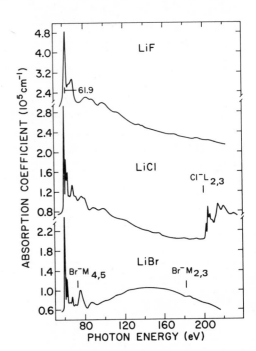

Figure 18. Absorption coefficients in the 50–230 eV range for LiF, LiCl, and LiBr, determined from thin-film transmission measurements (from F. C. Brown *et al.*[45]).

Table 1. Core Excitation Energies in LiF Calculated in the Small Periodic Cluster Model of Zunger and Freeman[44]

Transition	Type	Theory (eV)	Experiment (eV)	Exciton binding (eV)
Li $1s \rightarrow \Gamma_1(2s)$	Interband	63.3	64.4	
	Exciton	61.3	60.8	2.0
Li $1s \rightarrow L_3(2p)$	Interband	71.4	64–71	
	Exciton	62.2	61.9	9.2

[a] From References 45, 46, and 48.

and correlation, solid state effects such as electronic relaxation (for a given state, ground or excited) are fully included in a self-consistent fashion. The important conduction-band states in the problem are the Γ_1 minimum, which has 97% Li $2s$ character, and a level extending outward from L_3 (leading to a pronounced maximum density of states), which has 93% Li $2p$ character. Calculated interband and also exciton energies for these two transitions are given in Table 1. Experimental data from References 44 and 45 are also given. It can be seen that the prominent line at 61.9 eV for LiF in Figure 18 corresponds to a $1s$-to-$2p$-like transition. In this case the exciton binding energy is nearly 9.2 eV. Emission due to radiative decay of this exciton has been found at about 61 eV.[47]

Table 1 indicates a lower-lying ($\sim$61 eV) exciton state corresponding to a $1s \rightarrow 2s$ transition of the ion. Such an atomic transition is unallowed and even in the solid only weak absorption should occur. In fact a shoulder is found on the lower side of the main 62-eV band,[48] and recent electron energy-loss experiments indicate an increase in intensity of this shoulder with increasing momentum transfer, suggesting a forbidden transition. This is all in quite good agreement with the Zunger–Freeman theory, to which the reader is referred for further discussion.

Most likely the strong lines that also appear in Figure 18 for LiCl and LiBr are of similar local exciton origin, namely, closely related to $1s \rightarrow 2p$ transitions on the Li^+ ion. This is contrary to suggestions made by Brown et al.[45] where a fortuitous agreement between observed spectra and calculated density of states (freely shifted) was noted for LiCl. The observed spectra give little direct information about the density of states.

6. Threshold Resonances in Solids

6.1. White Lines, the K Edge of Germanium, and the L Edge of Tantalum

In years past x-ray absorption spectra were almost always taken with the use of photographic plates. Frequently, absorption edges appeared as unexposed bands on the plate, *rayon blanc* or "white lines." These white lines or threshold spikes

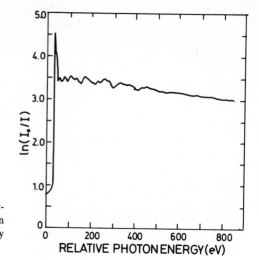

Figure 19. *K*-edge x-ray photoabsorption spectrum of crystalline germanium at 77 K, taken using synchrotron radiation with a resolution of about 1 eV. An arbitrary energy zero is used—the edge actually occurs near 11.3 keV (from B. M. Kincaid[3]).

appear at the *L* edges of the transition metals and at the *K* edges of certain other elements. Figure 19 shows an example, the *K* edge of crystalline germanium.[3] A fairly prominent spike with two or more components occurs at threshold, followed by extended x-ray absorption fine structure. Similar threshold spikes occur at the *K* edges of arsenic and selenium. No such spike is found at the *K* edge of copper. The *K* edges of Zr, Mo, Pd, and Ag are nowhere near as singular, although some structure is evident.[49]

Figure 20 shows another example, the L_3 edge of tantalum metal.[50] Here a very intense line appears over ten volts wide apparently unaccompanied by additional members of a series. The origin of these lines is quite important, especially in regards to the overall interpretation of threshold spectra.

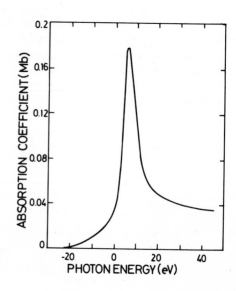

Figure 20. L_3 edge of Ta metal at about 11.1 keV, illustrating the prominent white line. The background below the edge has been subtracted (from M. Brown, R. E. Peirels, and E. A. Stern[49]).

These spikes are certainly not excitons in the usual sense. In germanium, for example, the high polarizability of the medium should effectively screen Coulomb interaction (at least in the soft x-ray region[51]), so that Rydberg binding energies are far less than instrumental resolution. In the hard x-ray region, broadening will obscure exciton lines lying below an ionization limit. In metals one expects the conduction electrons to respond to the core hole potential in such a way as to replace the exciton resonance by an x-ray "singularity".[52, 53] These many-body effects should lie very close to threshold, however, and are probably not the origin of the white lines under consideration.

White lines were explained qualitatively long ago by Cauchois and Mott[54, 55] in terms of a high density of final states to which transitions are made in the presence of an appropriate core potential. In the case of a K shell, the dipole selection rules for transitions from a $1s$ level single out p-like final states. On the other hand, spikes at the L edges involve $2p$ initial states; consequently they depend upon a high density of unoccupied d-like final states just above the Fermi level. The transition metals fall in this last category. Brown, Peierls, and Stern[50] have recently reviewed white lines in x-ray absorption. In particular, these authors evaluate approximately the oscillator strength for the L_3 edge of platinum, a case with very large spin–orbit splitting leading to a different intensity at the L_3 as compared to the L_2 edge. They conclude that the $2p_{3/2} \rightarrow 5d_{5/2}$ transition is much stronger than the $2p_{1/2} \rightarrow 5d_{3/2}$ and that the unoccupied levels retain their $d_{5/2}$ or $d_{3/2}$ character even in the solid. Mott[54, 55] speculated that this was the case for platinum many years ago.

6.2. *K Edge of Arsenic; Recent Theory*

The resonant character of white line spectra extending directly from threshold suggests more than just a singular density of states. For example, the Ge K edge appears more singular than a projected p-like density throughout the lowest ten volts of the conduction band. The adjacent element arsenic shows a similar resonance behavior at the K threshold. Holland, Pendry, Pettifer, and Bordas[56] suggest the importance of scattering of the outgoing electron wave in the periphery of the excited atom. That is, they calculate the atomic cross section for K-shell excitation with matrix elements as in equation (11). Bound exciton states below the ionization limit are excluded. The final continuum wave function involves the separated $l = 1$ solution of the radial wave equation with a muffin-tin potential, and they separate out a normalizing factor sensitive to the muffin-tin radius. This factor is shown to be closely related to the p-wave phase shift induced by the potential of the emitting atom (corrected using a $Z + 1$ analogy for the core hole). They carry out the calculation using a simple octahedral model for arsenic. A strongly resonant behavior occurs at threshold as seen in Figure 21, which compares experiment and theory. In no way does this relate to the Mahon singularity problem.[52] Solid arsenic has valence orbitals that are p-like, partly occupied and partly unoccupied. The excited core electron wave function must be properly orthonormal in the crystal, and a strong k dependence at the periphery of the atom affects overlap with

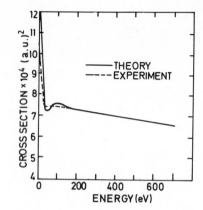

Figure 21. Theoretical (full line) and experimental (circles) cross sections for K absorption of arsenic (after B. W. Holland *et al.*[56]).

the compact initial-core wave function through the normalization factor. The result is a prominent resonance just above the ionization limit and a modified oscillation strength far above threshold.

6.3. L Edge of Silicon and the Elliot Exciton

Like germanium, silicon probably also has some resonant behavior at the K edge ($\sim$1840 eV). However, there has been little high-resolution work in this part of the spectrum. On the other hand, the L edge of silicon at 100 eV has been thoroughly studied.[52, 58] Figure 22 shows this threshold for both crystalline and amorphous silicon. The dotted curve in Figure 22a indicates the joint density of states allowing for the $2p^{3/2}$–$2p^{1/2}$ spin–orbit splitting of the initial state (0.61 eV). Notice that the observed spectrum reveals some of the detail of the crystalline density of state. The peaks around 103 eV are not seen for amorphous silicon because of the absence of long-range order.

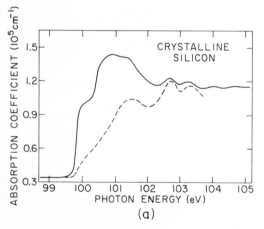

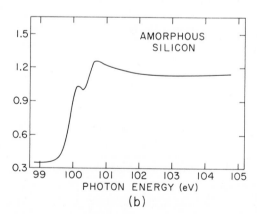

Figure 22. Absorption coefficient at the $L_{2,3}$ threshold for crystalline and amorphous silicon (from F. C. Brown *et al.*[57]).

The enhancement just above threshold in crystalline silicon (from 100 to 102 eV) has been explained by Altarelli and Dexter[59] in terms of the Elliot[60] effective-mass exciton theory. No bound resonances are resolved but the exciton envelope function above the continuum causes considerable enhancement for many Rydbergs. Altarelli and Dexter actually calculate absorption strength, but it is necessary to include a central-cell connection for good agreement with experiment. The agreement is good enough, however, that one does not look for strong enhancement of the kind observed at K edges and calculated for As.[56] In general, when spikes or white lines occur at the s-core thresholds, they do not also appear at the $2p$ thresholds and vice versa.[50]

Debate has occurred as to just what the exciton Rydberg is in silicon. That is, do we really have a case here of a core exciton with a localized hole but an electron wave function which is spread out in the crystal? Is the envelope function extended because of dielectric polarizability? If so the modified Eliot theory is expected to hold. A recent careful comparison of photoabsorption with photoemission data[61] indicates an exciton-binding or L-threshold correction of 0.18 ± 0.2 eV. This is in agreement with early estimates,[58] but not with the surface studies of Bauer *et al.*[62] and Margaritondo *et al.*[63] It is crucial that no exciton line appears below the silicon L edge, although in high-resolution measurements the edge rises within almost 0.1 eV. Resolved lines should be evident if the exciton binding were as large as suggested,[63] namely 0.6–0.7 eV. The effect of different broadening constants on an exciton edge is shown schematically in Figure 23. These matters and the difficult question of carrier screening in doped Si crystals are discussed further by Brown *et al.*[57] and in a recent review by Altarelli.[51]

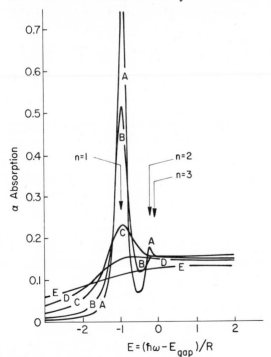

Figure 23. Illustrating the effect of damping on resolution for Elliot excitons. Broadening parameters $\Gamma/R = 0.1(A)$, 0.2 (*B*), 0.5(*C*), 1.0(*D*), and 2.0(*E*) [from D. Blossey, *Phys. Rev. B* **3**, 1382 (1971)].

6.4. K Edge of Titanium and Some Transition-Metal Compounds, $TiSe_2$ and MnO_2

Let us now turn to two cases of transition-element K absorption in solids. The first case involves Ti octahedrally coordinated in the well-studied layer crystal $TiSe_2$ (which has CdI_2 structure). The energy bands[64] and density of states[65] are quite well known for this material. Do we really have a chance of directly revealing the band structure in photoabsorption?

Figure 24 shows the Ti K absorption spectrum taken with about 1 eV resolution using synchrotron radiation.[66] The conduction-band density of states[64, 65] is indicated schematically in the Figure. It would appear that the soft x-ray absorption data can be utilized as a check and for scaling purposes on the theoretical energy-band results. This was demonstrated at lower energies in the case of $NbSe_2$ by Sonntag and Brown.[67] Notice that matrix element selection rules appear to be operative, at least to some extent. The lower band just above the Fermi level involves mainly d-like conduction bands, although s/p–d mixing occurs in the solid. Transitions to the second intense band mainly involve the higher-lying s/p conduction levels. Similar observations about these transition-metal compounds were made some time ago using a molecular orbital approach.[68] In that language the lower weaker band involves the t_{2g}–e_g d-like orbitals, the upper resonance the σ^* antibonding orbitals.

Again there may not be much need for atomic resonance effects to explain the threshold for a transition metal like Ti. Here, because of a projected d-like unoccupied state density just above the Fermi level, we expect white lines at the L edge and not the K edge. The L edge of Ti has not yet been studied in these crystals and

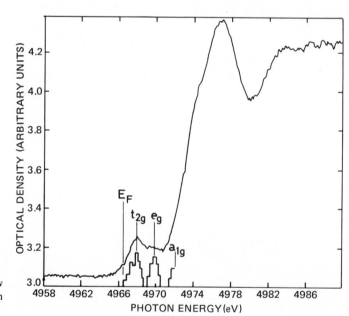

Figure 24. The K edge of $TiSe_2$ recorded at low temperature with about 1-eV resolution (from B. M. Davies and F. C. Brown[66]).

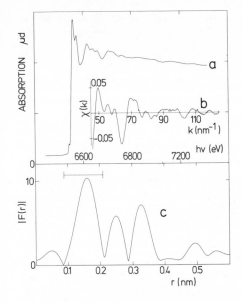

Figure 25. K-edge absorption spectrum for Mn in MnO_2, together with the separated EXAFS spectrum (after Rabe et al.[70]).

no strong singularity has been reported at this edge in Ti metal.[68] Recently, however, Denley et al.[69] have observed a strong spin–orbit-split line at the L threshold for Ti metal. This is presumably an atomic enhancement associated with the projected d-state density as discussed earlier.

Finally, we present in Figure 25 the K edge absorption of MnO_2. This was also obtained with the use of synchrotron radiation by Rabe, Tolkiehn, and Werner.[70] The separated EXAFS spectrum is also shown in the figure. This compound has the rutile structure but the Mn is sixfold coordinated rather like the Ti in $TiSe_2$. Notice the low-lying t_{2g}–e_g peak followed by a stronger band just preceding the EXAFS structure. Similar features appear in other transition metal compounds,[68] including those of vanadium. It would appear that high-resolution core-level spectroscopy is at least a guide to the conduction-band structure in some materials.

7. Simple Polyatomic Gases

7.1. Inner–Outer Well Potential

The x-ray thresholds of a number of polyatomic gases have been systematically studied in recent years. A number of interesting features have been found such as resonances both below and above the ionization limit, the importance of symmetry on transitions, and the sensitivity of Rydberg-type final states to condensation in the solid state. A number of important theoretical principles have evolved, and we will discuss these with a few selected examples.

Inner-shell molecular spectra are in some ways much simpler than the ultra-violet valence-shell spectra of molecules. Levels and series tend to be more widely separated. Nonbonding electrons are involved so one expects discrete lines without as strong vibronic coupling. At the same time, information about the bonding orbitals can be obtained through excitation to antibonding states and by means of the $Z + 1$ analogy.[71] Of course, core spectra have the great disadvantage of greater lifetime and instrumental broadening.

The resonance behavior observed in some molecules has been explained in terms of inner-well resonances. A prototype case is that of the K-shell excitation of sulphur in SF_6.[72] As first proposed by Nefedov,[73] an excited electron experiences a strong repulsive interaction in the vicinity of the electronegative fluorine atoms of the molecule. The effective potential therefore must include a potential barrier separating inner and outer wells. Rydberg structure is absent or extremely weak in the case of SF_6 and this is explained in terms of the diffuse Rydberg states lying in the outer region beyond the molecule with little overlap of the core. Strong overlap occurs however for final states within the F barrier. The spectrum of SF_6 begins with a strong inner-well resonance at 2486 eV, indicating almost ten times the initial–final state overlap for sulfur K-shell excitation in SF_6 compared to the same initial state in H_2S. Dehmer[74] points out that three types of interaction repel the photoelectron in the case of a surrounding shell of electronegative ions: (1) direct Coulomb interaction, (2) exchange interaction, and (3) forces associated with the Pauli exclusion principle that prohibit the electron from moving freely in the space containing occupied molecular orbitals, ie., the excited-state wave function must be orthogonal to all occupied orbitals. This last effect is probably the most important in cases like SF_6.

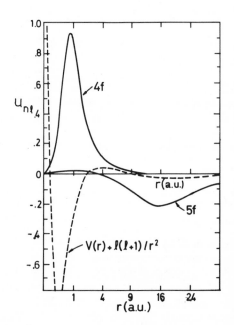

Figure 26. Showing the 4f and 5f wave functions for cesium. The dotted line indicates the $l = 3$ effective potential with a barrier region between 4 and 8 a.u.

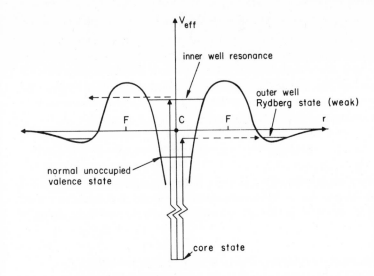

Figure 27. Schematic showing core excitation to inner- and outer-well states. The inner-well transitions tend to be strong, the outer-well transitions weak (because of lifetime and small overlap).

A barrier separating inner and outer regions is well known in the case of certain atoms. For example, Figure 26 shows the inner $4f$ and outer $5f$ wave functions for cerium (Herman–Skillman potential).[75] Because of the importance of the centrifugal term in the radial Schrödinger equation [see equation (18)] for $1 = 3$, the $5f$ function is greatly reduced for r less than 4 a.u. relative to the $4f$ (which scarcely penetrates the barrier). Here the barrier is a centrifugal barrier—for an excited electron in a molecule a pseudopotential barrier may arise from the orthogonality constraints mentioned above, and also sometimes from centrifugal effects. A schematic of core, valence, inner-well resonance, and Rydberg levels is shown in Figure 27, as described by Schwarz.[71]

7.2. *K Edge of Nitrogen in N₂ Gas*

In the case of light elements and sufficiently small molecules it has been possible to compare observation with calculated oscillator strength using Hartree–Fock theory for the ground and highly-excited states of the molecule. In Figure 28

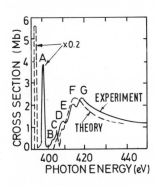

Figure 28. K edge of N_2 gas is shown in comparison with the theory of Rescigno and Langhoff.[77] Notice that the first intense line has been reduced by the factor 0.2 in order to plot on the same scale as the continuum resonance G (from Bianconi *et al.*[76]).

we illustrate this with data on the nitrogen K spectrum for N_2 gas, using the grazing-incidence monochromator shown in Figure 4b.[76] The spectrum is very unlike the Rydberg series of hydrogen. Notice that the first intense line has been multiplied by the factor 0.2 in order to plot it on the same scale as the continuum resonance G. The dotted line shows the results of a calculation by Rescigno and Langhoff.[77] Reasonable agreement is found between theory and experiment both as to shape and magnitude. Peak A corresponds to the first member of the π_g final-state Rydberg series. A partial wave expansion of this state shows it to have a dominant d-symmetric term. Here the low symmetry of the molecule is important because it permits access to this final state from photoabsorption by a K-shell electron. Because of its predominantly d nature, the π_g final state is concentrated within the centrifugal barrier for d-symmetric wave functions. This leads to a high overlap of initial and final states and thus to the observed strong oscillator strength.

7.3. K Edge of Carbon in Methane and the Fluoromethanes

Let us turn to the K edge of carbon in a series of light molecules, methane and the fluromethanes.[78] The K edge of carbon in CH_4 is shown in the upper part of Figure 29. This is a molecule with T_d symmetry and little electronic charge outside the periphery of the central atom. Notice that the spectrum is approximately Rydberg-like except for the small first peak and some asymmetry in the main line,

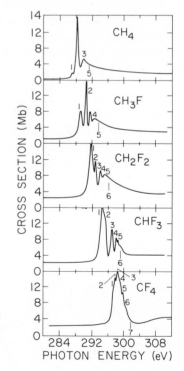

Figure 29. Carbon K-edge spectra in methane and the fluoromethanes taken with the grazing-incidence monochromator in Figure 4b (spectral bandpass about 0.3 eV at 300 eV). The vertical lines marked 5, 6, and 7 refer to ionization energies determined by x-ray photemission (from F. C. Brown et al.[78]).

which is due to transitions to a $3t_2$ molecular orbital with $3p$ parentage on the carbon. Higher-lying absorption below the ionization limit (291.0 eV) is due to nt_2, $n \geq 4$ transitions. The weak first line is a vibronic transition to the $3a_1(3s)$ final state. These assignments have been checked by experiments on the deuterated analogue of $CH_4^{(78, 79)}$ and also by *ab initio* calculations.[80, 81]

The next three molecules CH_3F, CH_2F_2, and CHF_3 have lowered C_{3v} symmetry. It can be seen that the first line becomes increasingly allowed. Also, a considerable chemical shift accompanies electron transfer to the electronegative fluorine. Finally, in CF_4 with T_d symmetry the lines have merged into an intense, somewhat broader band. It is probably no longer appropriate to talk about an np Rydberg series but rather about the accessible molecular orbitals (a_1, t_2) with overlap affected by a developing potential barrier due to the electronegative fluorine shell.

In Reference 78 the absorption of CF_4 was observed with a good signal-to-noise ratio out to about 460 eV, well beyond the region shown in Figure 29. Definite maxima and minima were observed that can be explained as EXAFS due to the interference of outgoing waves with waves backscattered from the fluorines. This is the first reported EXAFS structure above the carbon edge. EXAFS is not seen in the case of CH_4 because of the small scattering cross section of hydrogen.

Other types of carbon K-edge spectroscopy have been carried out on simple organic compounds. For example, Eberhardt *et al.*[82] report high-resolution electron yield measurements for photon energies from 280 to 300 eV on methane and ethane. Structure like that shown in the upper part of Figure 29 is ascribed to C $1s \rightarrow$ Rydberg excitations. These workers also recorded the C $1s$ threshold for ethylene C_2H_4, benzene C_6H_6, and acetylene C_2H_2, compounds that contain π electrons. In these cases strong resonances occur below the ionization threshold that are interpreted as due to excitations involving the unoccupied π antibonding states.

7.4. Second-Row Hydrides and Fluorides; Effect of Condensation on the SiH_4 and SiF_4 Spectra

The difference between valence and Rydberg-like final states is illustrated by the $L_{2,3}$-threshold spectra for a sequence of hydrides. Hayes and Brown[83] reported on the sequence HCl, H_2S, PH_3, and SiH_4—all molecules that have argon as a united atom. Figure 30 shows these $2p$ core spectra compared with the argon gas spectrum of Figure 13. The abscissa is divided into two-volt intervals in each case and the thresholds, which occur in the range 100–244 eV, have been displaced so that the spectra can be superimposed. Schwarz[84] and Robin[85] discuss these spectra and argue that features can be classified as transitions to either valence (antibonding) or Rydberg molecular orbitals. The former lie lowest and account for the first broad bands as opposed to the higher-lying Rydberg states. The distinction is especially evident in HCl but also can be seen in the detailed SiH_4 spectrum.

It is interesting to compare molecular gas spectra with spectra in the condensed phase. Figure 31 shows the results of Sonntag[86] on the $L_{2,3}$ spectra for silicon in both solid and gaseous SiH_4 and SiF_4. A wider range of energy is covered than in Figure 30. First, consider the SiH_4 gas spectra in the lower part of the figure. The intense valence transitions occur at threshold followed by Rydberg structure and a broad p to d maximum around 135 eV (an atomic effect). Upon condensation the Rydberg lines (~ 106 eV) are grossly modified, in fact entirely washed out. This is to be expected for states on the periphery or outside of the molecule. The valence states are less affected by condensation.

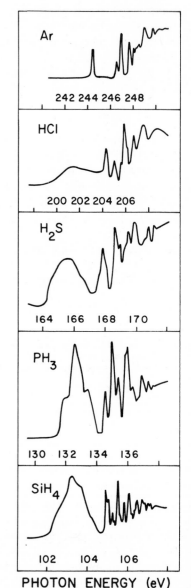

Figure 30. The $L_{2,3}$ edge of Cl, S, P, and Si in second-row hydrides (gases) compared with the $L_{2,3}$ edge of argon (the "united atom" for these molecules). The first broad band involves transition to unoccupied valence orbitals, the higher structure to Rydberg-like states (after W. Hayes and F. C. Brown[83]).

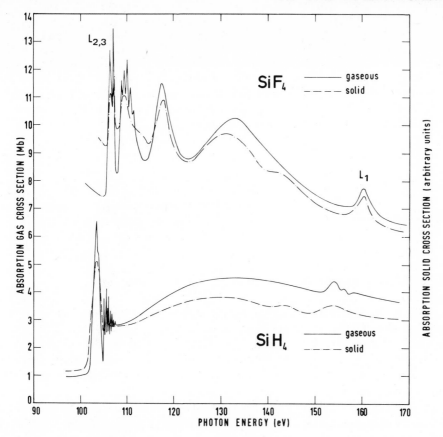

Figure 31. The $L_{2,3}$ edge of Si in SiH_4 and SiF_4 in both gaseous and solid states (from B. Sonntag[86]).

The upper curve in Figure 31 indicates what happens when hydrogen is replaced by fluorine. We now have a sequence of four strong (inner-well) resonances. These correspond to the molecular orbital states a_1, $t_2(p)$, and $t_2(d)$, all accessible from the L core as argued by Dehmer.[74] Upon the first two of these bands are superimposed remnants of the Rydberg transitions. Notice that these are very sensitive to condensation, whereas the four inner-well resonances are rather insensitive. Compare the solid and gaseous curves in Figure 31.

The K edge of silicon (~ 1840 eV) should appear different than the L edge because of selection rules and also because of a greater lifetime broadening. Unfortunately, very little high-resolution work has been done in this part of the spectrum. Both the K and the L spectra of silicon in $SiCl_4$ gas have been reported, however. According to D. L. Mott,[87] the Si K edge shows a strong first band followed by two lower-lying peaks. The Si L spectrum of $SiCl_4$ obtained by Zimkina et al.[88] looks rather like Figure 31 for SiF_4 except that the first two peaks a_1 and $t_2(p)$ are somewhat closer together. These two states may be unresolved in the K spectrum, which seems to have three rather than four identifiable features.

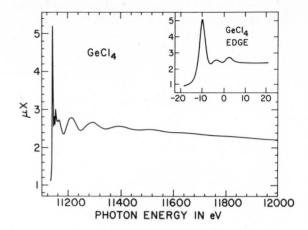

Figure 32. *K* edge absorption of germanium in GeCl₄ gas (from Kincaid and Eisenberger[3]).

7.5. *K* Edge of Germanium in GeCl₄, GeBr₄, and GeH₄

The K edge of germanium in simple polyatomic molecules has been well studied. For example, the x-ray spectra of germane, $GeCl_4$, and $GeBr_4$ were recorded years ago with high resolution.[89] Figure 32 shows recent measurements[3] of the Ge K edge in $GeCl_4$ using synchrotron radiation. Extended fine structure is well developed and the edge shows a strong resonance, the $a_1-t_2(p)$ band followed by two low intensity e, t_2 peaks. (See Figure 33 where these suggested assignments are given.) The near-edge structure and the EXAFS are absent in the case of germane (GeH_4), whose threshold is also shown in Figure 33. Clearly the hydrogens scatter too weakly for EXAFS, and the inner-well resonance behavior is also modified. This illustrates the close connection between inner-well resonance phenomena and EXAFS.

The K-edge features are very similar for $GeBr_4$ as compared to $GeCl_4$. Early measurements on both compounds were made by Glaser,[89] whose results have

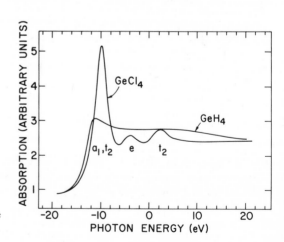

Figure 33. Comparing the K edge spectrum of $GeCl_4$ with the spectrum of germane GeH_4 (after Glaser[89] and Kincaid[3]).

sufficient resolution to be compared directly with Figure 33. Again in $GeBr_4$ a strong line followed by two weaker features is obtained.

The $L_{2,3}$ spectrum of germanium in $GeCl_4$ (at ~ 1248 eV) shows three lines within 25 eV of threshold. (See the paper by D. L. Mott.[87]) These three lines or bands are more equally weighted compared to the K spectrum because of parity selection rules for dipole transitions between molecular orbital states.[90]

7.6. Summary Remarks

In conclusion, we have selected only a few examples of molecular gas spectra from many that appear in the literature. These examples show the importance of identifying the appropriate molecular orbitals and applying selection rules between them. Strong effects of molecular symmetry and of effective potential appear in the threshold structure. There also seems to be a violation of strict selection rules through vibronic coupling. Correct assignments in the observed structure can only be made with certainty in those few cases where detailed theoretical calculations of the highly-excited states of the molecule can be made. It is hoped that these calculations will be extended to more cases in the coming years. Meanwhile a general description of molecular-orbital states for the purpose of understanding inner-level spectroscopy is appropriate.

In molecules it has long been known that molecular orbitals can be roughly classified into three types[71]: (1) *core* orbitals, which are nonbonding orbitals concentrated near atomic nuclei, (2) *valence* orbitals, either of the bonding or antibonding variety, which reflect the structure of the molecule, and (3) *Rydberg* orbitals, which are bound orbitals of the excited state located in the periphery or outside the molecule.

The experimental and theoretical work of recent years has shown that at least two additional categories of molecular-orbital states must be added to the three listed above. These are (4) *shape resonances* or continuum wave functions largely concentrated within the molecule by the effective potential (inner-well resonances belong in this category) and (5) *continuum wave functions* for states whose energy is far above threshold. These states are little affected by the molecular potential and are approximated by atomic continuum wave functions. A synchrotron continuum source allows selective access to these various states and the study of transitions between them. The reader is referred to the recent review by Koch and Sonntag[12] and to the important theoretical paper by Kotani and Toyozawa[11] for further discussion.

Molecular spectroscopy in the higher-energy region involves a wide range of energy and good resolution. A synchrotron continuum properly monochromatized is an ideal probe. The inelastic scattering of electrons can also be used, however, with the additional advantage that momentum transfer can be exploited for scattering out of the forward direction. As an example, data at the carbon K edge in methane as well as in chloromethane has been obtained using kilovolt electron beams by Brion and co-workers.[79] Their resolution equals or exceeds that of Figure 29. In fact these methods are so promising that an increasing amount of core-level spectroscopy may be done with electron rather than photon beams in the future. On the other hand, it may be difficult to employ electron beams for losses much above 2 keV, at least in very-high-resolution threshold work.

There is also a recent trend to employ energy-loss techniques in analytical electron microscopy. In this case the energy loss is measured in transmission for say 100-keV electrons in selected areas of the sample < 100 Å in diameter. The preparation of thin samples is therefore simplified and the microsctructure can be used to explore, for example, crystalline as opposed to amorphous areas of the sample. Whatever the fate of these techniques, the complimentary methods of synchrotron radiation spectroscopy will probably continue to be used. Although momentum transfer studies are not possible in the low-energy range, the polarization characteristics of synchrotron radiation are just beginning to be used. They come into their own in photoemission work, in the study of highly anisotropic crystals, and in the study of oriented molecules on surfaces or interfaces. The short duration and precise repetition of intense synchrotron pulses is also an advantage in lifetime studies and in experiments that involve time-of-flight synchronization of coincidence. One concludes that the versatility of the synchrotron source assures its usefulness for the new spectroscopic studies of molecules and solids that will certainly be pursued in the coming years.

ACKNOWLEDGMENTS

The author would like to express his appreciation for comments and discussion with many colleagues, especially M. Skibowski, R. Haensel, P. Rabe, E. Koch, and B. Sonntag. He is especially grateful for the hospitality extended to him at the University of Kiel during sabbatical leave from the University of Illinois. An Alexander von Humboldt award is also much appreciated. Early support by the National Science Foundation under Grants no. DMR 76-20644 and 77-23999 is acknowledged.

References

1. J. A. Bearden and A. F. Burr, *Rev. Mod. Phys.* **39**, 125 (1967).
2. F. W. Lytle, D. E. Sayers, and E. A. Stern, *Phys. Rev. B* **11**, 4825 (1975).
3. B. M. Kincaid, Ph.D. thesis, Stanford University, 1975 (unpublished); B. M. Kincaid and P. Eisenberger, *Phys. Rev. Lett.* **34**, 1361 (1975).
4. L. G. Parratt, *Rev. Mod. Phys.* **31**, 616 (1959); L. G. Parratt and E. L. Jossem, *Phys. Rev.* **97**, 916 (1955).
5. T. Muto and H. Okuno, *J. Phys. Soc. Jpn* **34**, 761 (1973).
6. J. W. Cooper, *Phys. Rev.* **128**, 681 (1962); S. T. Manson and J. W. Cooper, *Phys. Rev.* **165**, 126 (1968).
7. V. Fano and J. W. Cooper, *Rev. Mod. Phys.* **40**, 441 (1968).
8. F. C. Brown, *Solid State Phys.* **29**, 1 (1974).
9. R. Haensel, *Advances in Solid State Physics*, H. J. Queisser (ed.), Pergamon-Vieweg, New York (1975), p. 203.
10. J. P. Connerade, *Contemp. Phys.* **19**, 415 (1978).
11. A. Kotani and Y. Toyozawa, in *Synchrotron Radiation*, C. Kunz (ed.), *Topics in Current Physics*, Vol. 10, p. 169, Springer-Verlag, Heidelberg (1978).

12. B. Sonntag, in *Rare Gas Solids*, Vol. 2, p. 1021, M. L. Klein and J. A. Venables (eds.), Academic Press, London (1978); see also E. E. Koch and B. Sonntag, in *Synchrotron Radiation*, C. Kunz (ed.), *Topics in Current Physics*, Springer-Verlag, Heidelberg (1978).
13. R. E. Dietz, E. G. McRae, Y. Yafet, and C. Caldwell, *Phys. Rev. Lett.* **33**, 1372 (1974).
14. F. C. Brown, C. Gähwiller, and A. B. Kunz, *Solid State Comm.* **9**, 481 (1971).
15. T. Aberg and J. Dehmer, *J. Phys. C.* **6**, 1450 (1973).
16. C. Satoko and S. Sugano, *J. Phys. Soc. Jpn* **34**, 701 (1973).
17. T. B. Hastings, B. M. Kincaid, and P. Eisenberger, *Nucl. Instrum. Methods* **152**, 167 (1978).
18. F. C. Brown, R. Z. Bachrach, and N. Lien, *Nucl. Instrum. Methods* **152**, 73, 1978).
19. C. Depautex, T. Thiry, R. Pinchaux, Y. Petroff, D. Lepere, G. Passereau, and J. Flamard, *Nucl. Instrum. Methods* **152**, 101 (1978).
20. D. Pudewill, F. J. Himpsel, V. Saille, N. Schwentner, M. Skibowski, E. E. Koch, and J. Jortner, *J. Chem. Phys.* **65**, 5226 (1976).
21. W. Scheifley, *Extreme Ultraviolet Response of Ionic Crystals*, Ph.D. thesis, University of Illinois, 1973 (unpublished).
22. R. D. Hudson, *Rev. Geophys. Space Phys.* **9**, 305 (1971).
23. B. L. Henke, R. L. Elgin, R. E. Lent, and R. B. Ledingham, *Norelco Rep.* **14**, 112 (1967).
24. R. Z. Bachrach, A. Bianconi, and F. C. Brown, *Nucl. Instrum. Methods* **152**, 53 (1978).
25. G. L. Weissler, *Sci. Light (Tokyo)* **21**, 5226 (1976).
26. H. Bethe and E. E. Salpeter, *Quantum Mechanics of One- and Two-Electron Atoms*, Plenum/Rosetta Press, New York (1977).
27. S. T. Stephenson, R. Krogstad, and W. Nelson, *Phys. Rev.* **84**, 806 (1951).
28. E. S. Gluskin, A. A. Krasnoperova, and L. N. Mayalov, Extended abstracts, Fifth International Conference on Vacuum Ultraviolet Radiation Physics, Montpellier, Sept. 1977.
29. L. D. Landau and E. M. Lifshitz, *Electrodynamics of Continuous Media*, Pergamon, Oxford (1960).
30. F. Wuillemier, *J. Phys. (Paris), Colloq.* C4, Supp. 10, **32**, 88 (1971).
31. K. Codling, R. P. Madden, and D. L. Ederer, *Phys. Rev.* **155**, 26 (1967).
32. F. Bassani and Y. Pastori-Parravicini, *Electronic States and Optical Transitions in Solids*, Academic Press, New York (1976); J. J. Forney, A. Quattopani, and F. Bassani, *Nuovo Cimento B*, **22**, 153 (1974).
33. R. Haensel, G. Keitel, N. Kosuch, U. Nielsen, and P. Schreiber, *J. Phys. (Paris) Colloq.* **32**, C4, 236 (1971).
34. M. Altarelli, W. Andreoni, and F. Bassani, *Solid State Comm.* **16**, 143 (1975).
35. Y. Onodera and Y. Toyozawa, *J. Phys. Soc. Jpn* **22**, 833 (1967).
36. P. Rabe, B. Sonntag, T. Sagawa, and R. Haensel, *Phys. Status Solidi B* **50**, 559 (1972).
37. D. L. Dexter, *Phys. Rev.* **108**, 707 (1957).
38. M. Watanabe, A. Ejirei, H. Yamashita, H. Saito, S. Sato, T. Shibaguchi, and H. Nishida, *J. Phys. Soc. Jpn* **31**, 1085 (1971).
39. S. T. Pantelides and F. C. Brown, *Phys. Rev. Lett.* **33**, 298 (1974); S. T. Pantelides, *Phys. Rev. B* **11**, 2391 (1975).
40. A. B. Kunz, *Phys. Status Solidi* **29**, 115 (1968).
41. M. Watanabe *et al.*, *J. Phys. Soc. Jpn* **34**, 755 (1973).
42. G. Sprüssel, V. Saile, and M. Skibowski, Extended abstracts, Fifth International Conference on Vacuum Ultraviolet Radiation Physics, Montepellier, France, Sept. 1977.
43. K. Cho and Y. Toyozawa, *J. Phys. Soc. Jpn* **30**, 1555 (1971); S. Sakoda, *J. Phys. Soc. Jpn* **34**, 1254 (1973).
44. A. Zunger and A. J. Freeman, *Phys. Lett. A* **60**, 456 (1977); *Phys. Rev. B* **16**, 2901 (1977).
45. F. C. Brown, C. Gähwiller, A. B. Kunz, and N. O. Lipari, *Phys. Rev. Lett.* **25**, 927 (1970).
46. R. Haensel, C. Kunz, and B. Sonntag, *Phys. Rev. Lett.* **20**, 262 (1968).
47. E. T. Arakawa and M. W. Williama, *Phys. Rev. Lett.* **36**, 333 (1976).
48. A. A. Maiste, A. M. Saar, and M. A. Elargo, *Fiz. Tverd. Tela (Moscow)* **16**, 1720 (1977); *Sov. Phys. Solid State* **16**, 1118 (1974); see also J. R. Fields, P. C. Gibbons, and S. E. Schnatterly, *Phys. Rev. Lett.* **38**, 430 (1977).
49. J. E. Muller, O. Jepson, O. K. Anderson, and J. W. Wilkins, *Phys. Rev. Lett.* **40**, 120 (1978).
50. M. Brown, R. E. Peierls, and E. A. Stern, *Phys. Rev. B* **15**, 738 (1977).

51. M. Altarelli, Extended abstracts, Fifth International Conference on Vacuum Ultraviolet Radiation Physics, Montpellier, France, Sept. 1977; *J. Phys. (Paris)* **39**, *C*4, 95 (1978).
52. G. D. Mahan, *Solid State Phys.* **29**, 74 (1974).
53. P. Nozieres and C. T. de Dominicus, *Phys. Rev.* **178**, 1097 (1969).
54. N. F. Mott, *Proc. R. Soc. (London)* **62**, 416 (1949).
55. Y. Cauchois and N. F. Mott, *Philos. Mag.* **40**, 1260 (1949).
56. B. W. Holland, J. B. Pendry, R. F. Pettifer, and J. Bordas, *J. Phys. C* **11**, 633 (1978).
57. F. C. Brown, R. Z. Bachrach, and M. Skibowski, *Phys. Rev. B* **15**, 4781 (1977).
58. F. C. Brown and O. P. Rustgi, *Phys. Rev. Lett.* **28**, 497 (1972).
59. M. Altarelli and D. L. Dexter, *Phys. Rev. Lett.* **29**, 1100 (1972).
60. R. J. Elliott, *Phys. Rev.* **108**, 1384 (1957).
61. W. Eberhardt, G. Kalkoffen, C. Kunz, D. Aspnes, and M. Cardona, *Phys. Status Solidi* **88**, 135 (1978).
62. R. S. Bauer, R. Z. Bachrach, J. C. McMenamin, and D. E. Aspnes, *Nuovo Cimento B* **39**, 409 (1977).
63. G. Margaritondo and J. E. Rowe, *Phys. Lett. A* **59**, 464 (1977).
64. A. Zunger and A. J. Freeman, *Phys. Rev. B* **17**, 1839 (1978).
65. H. W. Myron and A. J. Freeman, *Phys. Rev. B* **9**, 481 (1974).
66. B. M. Davies and F. C. Brown (to be published).
67. B. Sonntag and F. C. Brown, *Phys. Rev. B* **10**, 2300 (1974).
68. D. W. Fischer and W. L. Baun, *J. Appl. Phys.* **39**, 478 (1968); D. W. Fischer, *J. Appl. Phys.* **41**, 3561 (1970); D. W. Fischer, *Phys. Rev. B* **8**, 3576 (1973).
69. I. Stöhr, D. Denley, and P. Perfetti, *Phys. Rev. Lett. B* **18**, 4132 (1978).
70. P. Rabe, G. Tolkiehn, and A. Werner, *J. Phys. C.* **12**, 899 (1979); **12**, 1173 (1979).
71. W. H. Eugen Schwarz, *Angew. Chem.* **13**, 454 (1974).
72. R. E. La Villa and R. D. Deslattes, *J. Chem. Phys.* **44**, 4399 (1966).
73. V. I. Nefedov and V. A. Fornichev, *Zh. Strukt. Khim.* **11**, 299 (1970); *J. Struct. Chem. (USSR)* **11**, 277 (1970).
74. J. L. Dehmer, *J. Chem. Phys.* **56**, 4496 (1972); J. L. Dehmer and D. Dill, *J. Chem. Phys.* **65**, 5327 (1976).
75. A. R. P. Rau and U. Fano, *Phys. Rev.* **167**, 7 (1968).
76. A Bianconi, H. Peterson, F. C. Brown, and R. Z. Bachrach, *Phys. Rev. A* **17**, 1907 (1978).
77. T. N. Rescigno and P. W. Langhoff, *Chem. Phys. Lett.* **51**, 65 (1977).
78. F. C. Brown, R. Z. Bachrach, and A. Bianconi, *Chem. Phys. Lett.* **54**, 425 (1978).
79. H. P. Hitchcock, M. Pocock, and C. E. Brion, *Chem. Phys. Lett.* **49**, 125 (1977).
80. L. A. Curtis and P. W. Deutsch, *J. Electron Spectrosc. Relat. Phenom.* **10**, 193 (1970).
81. P. S. Bagus, M. Krauss, and R. E. La Villa, *Chem. Phys. Lett.* **23**, 13 (1973).
82. W. Eberhardt, R. P. Haelbich, M. Iwan, E. E. Koch, and C. Kunz, *Chem. Phys. Lett.* **40**, 180 (1976).
83. W. Hayes and F. C. Brown, *Phys. Rev. A* **6**, 21 (1972).
84. W. H. Eugen Schwarz, *Chem. Phys.* **9**, 157 (1975); **11**, 217 (1975).
85. M. B. Robin, *Chem. Phys. Lett.* **31**, 140 (1975).
86. B. Sonntag, Extended abstracts, Fifth International Conference on Vacuum Ultraviolet Radiation Physics, Montpellier, France, Sept. 1977. *J. Phys. (Paris)* **39**, C4, 9 (1978).
87. D. L. Mott, *Phys. Rev.* **144**, 94 (1966).
88. T. M. Zimkina and A. C. Vinogradov, *J. Phys. (Paris)* **32**, 3 (1971).
89. H. Glaser, *Phys. Rev.* **82**, 616 (1951).
90. G. Herzberg, *Molecular Spectra and Molecular Structure*, Vol. III, Electronic Spectra and Electronic Structure of Polyatomic Molecules, Van Nostrand, Princeton (1966), p. 132.

5

Electron Spectrometry of Atoms and Molecules

MANFRED O. KRAUSE

1. Introduction

Electron spectrometry using synchrotron radiation (ESSR) is a new endeavor, having had its first trials in 1972[1, 2] and its real beginning shortly thereafter.[3] Of course, neither electron spectrometry nor synchrotron radiation are very old. Electron spectrometry matured in the late sixties relying on conventional excitation sources,[4–9] and synchrotron radiation was widely applied to photoabsorption studies in the gas phase during the sixties.[10–12] The combination of electron spectrometry and synchrotron radiation has already led to results of great specificity that would be difficult or impossible to obtain by either continuous absorption measurements or by electron spectrometry that uses discrete photon sources. The ESSR technique allows the x-ray physicist to decompose a photoabsorption spectrum into its constituent components, and it allows the electron spectroscopist to study the features of a photoelectron spectrum as they vary with the photon energy. Thus, structure and dynamics of atomic and molecular systems can be delineated at a highly differentiated level. Specifically, ESSR has the potential of probing many-electron effects in bound and in continuum states, interactions between continuum and bound states, de-excitation pathways, and relativistic effects.

The increased specificity now possible in experimentation has its counterpart in the increased sophistication of theory, in which many-body aspects are now combined with relativistic aspects of atomic structure and processes. Viewing the past, we find that into the fifties atomic theory had remained basically at the level of the single-particle hydrogenic approximation, while in the sixties and seventies

MANFRED O. KRAUSE ● Transuranium Research Laboratory, Oak Ridge National Laboratory, Oak Ridge, Tennessee 37830.

photoabsorption measurements with synchrotron radiation and photoelectron spectrometry measurements with discrete sources have each played important roles in the stepwise improvement of theory.

Much of the ESSR work done so far on free atoms and molecules has been carried out at the synchrotron in Daresbury, England and the storage ring in Orsay, France. In both installations, the photon fluxes have been sufficiently intense at the exits of the monochromators to perform electron spectrometry in the photon energy range from about 15 to 130 eV. The future will see more existing and planned synchrotron radiation sources applied to the electron spectrometry of atoms and molecules, and the range of photon energies will be extended beyond 130 eV to some 300–400 eV and ultimately to 1 keV and beyond. Then the entire UV, ultrasoft, and soft x-ray ranges will be covered continuously by a single excitation source, synchrotron radiation, and a single analyzing instrument, the electron spectrometer. However, reality enters into this seemingly ideal constellation by way of the photon dispersive element, which must be changed in different energy ranges.

This chapter contains an outline of the experimental and theoretical background and describes the ESSR experiments that have been performed. The ramifications of these experiments and their implications for future experiments will be discussed. Results from related experiments, mostly traditional electron spectrometric measurements and mass spectrometric studies, will be interwoven with ESSR data throughout the text. By doing so, the potential and value of ESSR will become especially apparent.

Extensive literature exists on the various facets of both electron spectrometry and synchrotron radiation research; perhaps the most relevant reviews in the present context are those with emphasis on ESSR by Wuilleumier,[13] Marr,[14] and Codling.[15] The latest developments in synchrotron radiation instrumentation are reported in Wuilleumier and Farge;[16] ultravoilet spectroscopic techniques are summarized by Samson,[17] and electron energy analyzers are reviewed by Kuyatt[18] and Sevier.[19] Books by Siegbahn et al.,[4, 5] Turner et al.,[6] and Carlson,[20] as well as the articles edited by Brundle and Baker,[21] offer good accounts of electron spectrometry, especially photoelectron spectrometry applied to chemical systems, while the books edited by Crasemann,[22] and Wuilleumier[23] contain much useful information on electron spectrometry in physics and on the theory of atomic structure and interactions. Studies with high-temperature vapors are summarized by Berkowitz,[24] and photodissociation is treated by Chupka.[25] Finally, the articles by Krause,[26, 27] Manson,[28, 29] Samson,[30] Mehlhorn,[31] and Burhop and Asaad[32] are of particular interest within the scope of the present chapter.

2. The Domain of Synchrotron Radiation

The domain of synchrotron radiation in electron spectrometry encompasses the entire range from about 10 eV to 1 keV. In this range, few conventional photon sources are available with desirable characteristics such as high intensity, narrow linewidth, and absence of satellite lines. In this low-energy range, the manifestations

of a realistic atomic potential and the various electron–electron and channel interactions become important, erasing the sawtooth shapes that we have customarily
associated with the behavior of the photoabsorption cross section on the basis of
measurements at higher energies. Photons from gas discharge lamps, especially the
He I(21.22 eV) and He II(40.8 eV) radiation sources, and from x-ray tubes, providing the $M\zeta$ lines of Y to Mo (132–192 eV) and the $K\alpha$ lines of F (676 eV), Mg
(1.25 keV) and Al (1.49 keV), are too widely spaced in energy to fully map out total
and partial photoionization cross sections and to clearly reveal the dynamic
properties of atoms and molecules. It is then desirable and often crucially important to have a finely tunable continuous photon source such as that provided by
synchrotron radiation.

Properties most frequently determined in the domain of ESSR are the partial
photoionization cross sections, angular distributions of photoelectrons, and energies of photo- and Auger electrons. Energy determinations are of special interest in
regard to (a) correlation satellites, (b) molecular dissociation, and (c) post-collision
interactions. In the energy measurements, emphasis is placed on the low-energy
region, $10 \lesssim h\nu \lesssim 80$ eV, and on threshold regions of electronic core levels. Determinations of cross sections and angular distributions are important throughout the
ESSR regime, but are of particular significance in regions where Cooper minima,
delayed maxima, autoionization, and resonances occur and following the onset of
new excitation and deexcitation processes. In a special case of ESSR, threshold
electron spectrometry, energy deposition in a molecule can be controlled and, in
combination with mass spectrometric and fluorescence techniques, pathways of
molecular deexcitation can be delineated.

As an illustrative example of the significance of ESSR, the photoabsorption cross
section* of xenon is displayed in Figure 1 for photon energies up to 1 keV. Among
the various peculiarities seen in the behavior of the cross section, the absence of
abrupt absorption jumps at the energies of the less tightly bound subshells is
particularly evident. Also evident is the failure of the central field model, which in
the Herman–Skillman (HS) formulation[33] has proven quite adequate at higher
energies[28, 29, 34] to account for the experimental total photoionization cross
section.[35, 36] By contrast, the Hartree–Fock (HF) model, even if applied only to
the final state, gives a fair overall agreement with experiment over the entire energy
range.[37] However, discrepancies (somewhat ameliorated by the logarithmic
display in Figure 1) do occur, and length [HS–HF(L)] predictions (not shown)
differ markedly from velocity [HS–HF(V)] predictions.[37] For example, the experimental data of the total cross section near $h\nu = 100$ eV exceed the HS–HF(V)
prediction by 50%. The question then arises, which of the HS–HF(V) partial cross
sections plotted in Figure 1 are underestimated: the $4d$ cross section by 50%, or the
$5p$ and $5s$ cross sections by a factor of 100, or each of these by some amount within
these limits? Furthermore, the question of the contributions from multiple-electron
transitions remains open since, by their very nature, neither the experimental data

---

* Because the photoeffect is dominant at $h\nu \lesssim 1$ keV, photoabsorption and total photoionization cross
  sections need not be distinguished in the continuum regions.

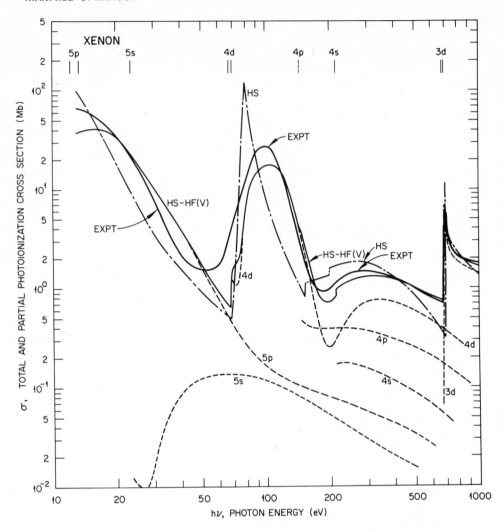

Figure 1. Photoionization cross section of xenon. Experimental data[35, 36] are compared with single-particle model results HS[33] and HS-HF(V).[37] Partial cross sections are HS-HF(V) predictions.

of the total cross section nor the HS or HS–HF theoretical results explicitly refer to these types of processes. However, ESSR has the capability of delineating the individual contributions from the various subshells and combinations of subshells continuously over a wide energy range. To indicate the type and amount of detailed information contained in a photoelectron spectrum, the spectrum of xenon representative of photoeffect at a photon energy of 1.5 keV[5, 38] is displayed in Figure 2. In the following, we shall return repeatedly to the xenon case which, because of its central position in the periodic table, seems to have become a *cause célèbre* and an important testing ground for theoretical developments.

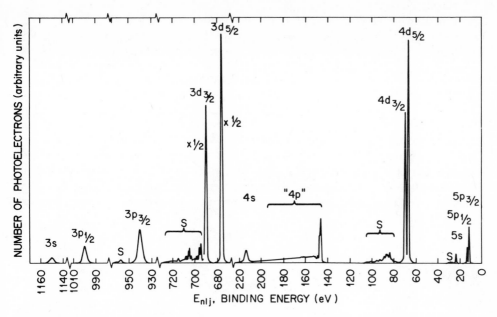

Figure 2. Photoelectron spectrum of xenon excited by 1.5 keV photons. Spectrum is a schematic composite of several spectra.[5, 38]

3. Basic Relations and Background

In electron spectrometry the kinetic energies of electrons ejected from an atom or molecule by photons, charged particles, an Auger process, or some other excitation mode are measured with an energy-dispersive focusing device such as an electrostatic hemispherical or cylindrical mirror analyzer. At the same time, the number of electrons within a given energy interval are measured in relation to a fixed direction, which is usually that of the photon beam. An electron spectrum is then obtained that yields data on (a) energies of levels and transitions within the electronic, vibrational, and rotational structure, (b) widths of levels and transitions, (c) angular distributions of emitted electrons, and (d) transition probabilities for the processes under study. These properties are measured at certain predetermined excitation energies when using conventional discrete photon sources, or at the energies chosen by the experimenter to gain maximum information when using the continuous synchrotron radiation source.

Because of the central role of photoionization in ESSR, basic relations presented in this section will emphasize the photoeffect, while relations pertaining to decay processes, foremost among them being the Auger process, will be discussed in the appropriate sections. The Einstein relation of the photoeffect,

$$hv = E_{e,\text{kin}} + E_B \qquad (1)$$

forms the basis of photoelectron spectrometry by interrelating the energy hv of the photon with the kinetic energy $E_{e,\text{kin}}$ of the photoelectron ejected from the atomic or molecular level with energy E_B. For a more detailed description[28] of the interaction of a photon γ (with energy $h\nu$, angular momentum j, and parity π) with an atom $A°$ (characterized by its total ground-state energy E_0, total angular momentum J_0, and parity π_0), equation (1) is presented in the form of a reaction:

$$\gamma(h\nu, j = 1, \pi_\gamma = -1) + A°(E_0, J_0, \pi_0)$$
$$\rightarrow A^+(E_f, J_f, \pi_f) + e^-(E_{e,\text{kin}}, lsj, \pi_e = (-1)^l) \quad (2)$$

where the ionized atom is characterized by the final-state parameters E_f, J_f, π_f, and the outgoing photoelectron by $E_{e,\text{kin}}$, lsj, and π_e. Reaction (2) has been written for the case of single-electron emission from an atom in an electric dipole interaction; for more complex systems, additional terms and parameters must be added. Energy, angular momentum, and parity are conserved and the following balances hold*:

$$E_{e,\text{kin}} = h\nu - (E_0 - E_f) \quad (3)$$

which is a slight variation of the basic relation (1),

$$\mathbf{J}_0 + \mathbf{j}_\gamma = \mathbf{J}_f + \mathbf{s} + \mathbf{l} \quad (4)$$

and

$$\pi_0 \cdot \pi_\gamma = \pi_f \cdot \pi_e = -\pi_0 = (-1)^l \pi_f \quad (5)$$

While the energy balance, by itself, has provided adequate guidance in the interpretation of much of the work done in the past, the angular momentum balance in conjunction with parity conservation is needed, as demonstrated in more recent work,[39–47] to gain a more thorough understanding of the photon interaction with atoms, especially open-shell atoms, and in particular with molecules. Doubtless, future work will increasingly profit from relations (4) and (5).

3.1. Energies

According to equation (3), the photoelectron energies and hence the features of a spectrum, such as those of Figure 2, correlate uniquely with the various final states that can be reached in the photoeffect. Each E_f corresponds to a photoionization channel from the ground state E_0.

We have for single photoionization from an atomic level nlj,

$$E_{nlj} = E_0 - E_f = E(A°) - E(A^+, nlj, SLJ) \quad (6)$$

where the hole state of the ion A^+ is designated $nlj(SLJ)$; for simultaneous excitation and ionization of two electrons

$$E_{nlj, n'l'j' \rightarrow n''l''j''} = E_0 - E_f = E(A°) - E(A^{+*}, nlj, n'l'j' \rightarrow n''l''j'', S''L''J'') \quad (7)$$

and for double ionization

$$E_{nlj, n'l'j'} = E_0 - E_f = E(A°) - E(A^{++}, nlj, n'l'j', S'L'J') \quad (8)$$

---

* Note that the subscript f refers to the final *ionic* state only.

Calculations of the energies $E_{nlj, \ldots}$ follow standard texts.[48] The wave functions used to calculate the energies for initial and final energies E_0 and E_f may be obtained in a crude approximation from the single-particle central potential model using the Hartree–Slater or Herman–Skillman (HS) approach,[49] or may be obtained more accurately from a many-body model using, for example, the multiconfiguration Hartree-Fock (MCHF)[50] or Dirac–Fock (MCDF)[51] model, many-body perturbation theory (MBPT),[52] and the Green's Function (GF) formalism.[53–55] For deep core levels, the inclusion of quantum-electrodynamic contributions is essential.[56–59] But for the shallow levels, which are important in ESSR, consideration of many-body aspects are crucial; this holds true not only for the two-electron process, equations (7) and (8), but also for the single-electron process, equation (6). Many-body inter-actions lead to a shift in the single-particle energy; hence, the accurate measure-ment of energies can reveal the presence of many-body or electron-correlation (EC) effects when compared to the theoretical single-particle predictions. If the energy E_{nlj} of equation (6) is to be interpreted as the binding energy of subshell nlj, it should be noted that finding the lowest nlj (SLJ) state may not be sufficient because the final ionic state of lowest energy may be a state where one or more of the quantum numbers nlj have changed from those of the ground state $E(A°)$. Such flips between close-lying nlj states are expected to occur in transition-element regions in which the energies of nd and $(n + 1)s$ or nf and $(n + 1)d$ levels depend critically on the atomic potential, and thereby on the state of ionization.[60, 61]

An interesting showcase of a drastic change in configurations when going from $A°$ to A^+ is the $4p$ level of xenon shown in Figure 2. There, the peak of lowest energy, and of highest intensity, does not correspond to a $4p_{3/2}$ hole but a $4d^2$ double hole with one of the $4d$ electrons promoted to a $4f$ level, due to the near energy degeneracy and the strong interaction between the alternative configurations.[55] Since in molecules a propensity for many closely spaced levels is the rule rather than the exception, spectral features similar to those for Xe $4p$ will be a common occurrence. This was clearly demonstrated by Cederbaum and co-workers[62] in a theoretical treatment of the inner valence electrons of molecules. Under these circumstances the seemingly strict division between one- and two-electron processes expressed in equations (6)–(8) is no longer justified.[62]

3.2. Photoionization Cross Sections

Following Fermi's rule[63] and the developments by Bethe[64] and Bethe and Salpeter,[65] the photoionization cross sections can be written as[28, 66]

$$\sigma_{0f} = \frac{4\pi^2\alpha^2}{k_v} \left| \langle \Psi_0^* | \sum_\mu \exp(-i\mathbf{k}_v \cdot \mathbf{r}_\mu)\varepsilon\nabla_\mu | \Psi_f \rangle \right|^2 \tag{9}$$

where α is the fine-structure constant, $\mathbf{k}_v$ and ε the momentum and polarisation vectors of the incident photon, $\mathbf{r}_v$ and ∇_μ the position and gradient operators of the μth electron, and Ψ_0 and Ψ_f represent the initial (ground) state and final state with their wave functions properly normalized. Because the continuum electron, which is part of the final state Ψ_f, can be specified by its direction, equation (9) also represents the formula for differential cross section $d\sigma_{0f}/d\Omega$.[66] Equation (9) shows that the cross section depends on the multipolarity of the photon field through $\exp(-i\mathbf{k}_v \cdot \mathbf{r}_\mu)$, on the polarization, and, most importantly, on the initial- and final-state wave functions of the system through Ψ_0 and Ψ_f. Only if Ψ_0 and Ψ_f are *exact* can the photoionization cross sections be expected to be in agreement with the *perfect* experimental observation. However, a precautionary note is indicated at this point. If comparison of theoretical predictions with experimental data is made at the subshell level, it is important that in the summation over μ the pertinent configurations are included that correspond to a given observed spectral feature. This requires an examination of whether a partial cross section calculated for a subshell nlj pertains only to single-electron transitions or also to many-electron transitions.

Since the photoionization cross section depends on Ψ_0 and Ψ_f according to equation (9), the photoeffect represents a very direct probe of the atomic and molecular ground-state properties and final-state properties. Thus, theory has concentrated on calculating the best Ψ_0 and Ψ_f with the fewest approximations in the solution of the Schrödinger or Dirac equation. The frequently used dipole approximation, or neglect of retardation viz. $\exp(-i\mathbf{k}_v \cdot \mathbf{r}_\mu) = 1$ because $|\mathbf{k}_v \cdot \mathbf{r}_\mu| \ll 1$, is generally permissible for the *total*-cross-section calculation in the domain of ESSR. General accounts and reviews of theory can be found in Bethe and Salpeter,[65] Heitler,[67] Fano and Cooper,[68] Pratt *et al.*,[69] Cooper,[70] Manson,[28] and Amusia and Cherepkov.[71]

3.3. Angular Distribution of Photoelectrons

Based on general arguments[72, 73] or on equation (9), the photoelectron angular distribution is characterized by coefficients B_n in a Legendre expansion[74]:

$$\frac{d\sigma_{0f}}{d\Omega} = \frac{\sigma_{0f}}{4\pi} \sum_{n=0}^{\infty} B_n P_n(\cos \theta) \tag{10}$$

We note at this point that in the frequently used nonrelativistic dipole approximation all coefficients B_n are zero except for $B_2 (B_0 \equiv 1$ always). Then B_2 is the single parameter that determines the angular distribution and assuming polarized radiation and a randomly oriented target, equation (10) becomes

$$\frac{d\sigma_{0f}}{d\Omega} = \frac{\sigma_{0f}}{4\pi} \left[1 + \frac{\beta_{0f}}{2} (3 \cos^2 \theta - 1) \right] \tag{11}$$

where θ is the angle between the photoelectron direction and the polarization vector $\mathbf{E}$, $P_2(\cos \theta) = (3 \cos^2 \theta - 1)/2$, and $B_2 = \beta_{0f}$, where β_{0f} is the asymmetry parameter, which for a given level nl is a function of the radial matrix elements $R_{l\pm1}$ and the phase shift δ for the outgoing waves.[28, 75] Again in the dipole approximation as for equation (11), the angular distribution for partially polarized photons is given as

$$\frac{d\sigma_{0f}}{d\Omega} = \frac{I_x \sigma_{0f}}{I_0 \, 4\pi} \left[1 + \frac{\beta_{0f}}{2} (3 \cos^2 \theta_x - 1) \right] + \frac{I_y \, \sigma_{0f}}{I_0 \, 4\pi} \left[1 + \frac{\beta_{0f}}{2} (3 \cos^2 \theta_y - 1) \right] \tag{12}$$

where the angle θ of equation (11) is indexed by x and y of a coordinate system in which the photons travel along the z axis, and the two orthogonal and incoherently superposing contributions to σ_{0f} are weighted by the components I_x and I_y of the total photon intensity $I_0 = I_x + I_y$. Introducing the polarization $p = (I_y - I_x)/I_0$ in its usual definition, equation (12) becomes

$$\frac{d\sigma_{0f}}{d\Omega} = \frac{\sigma_{0f}}{4\pi} \left\{ 1 - \frac{\beta_{0f}}{4} \left[(3 \cos^2 \theta_z - 1) - 3p(\cos^2 \theta_x - \cos^2 \theta_y) \right] \right\} \qquad (13)$$

using the geometric relation $\cos^2 \theta_x + \cos^2 \theta_y + \cos^2 \theta_z = 1$, or for the special case $\theta_z = 90°$:

$$\frac{d\sigma_{0f}}{d\Omega} = \frac{\sigma_{0f}}{4\pi} \left[1 + \frac{\beta_{0f}}{4} (1 + 3p \cos 2\theta) \right] \qquad (14)$$

where θ is the angle in the xy plane between the photoelectron and the electric vector. For unpolarized photons, $p = 0$ or $I_y = I_x = I_0/2$, we obtain from equation (12) or (13),

$$\frac{d\sigma_{0f}}{d\Omega} = \frac{\sigma_{0f}}{4\pi} \left[1 - \frac{\beta_{0f}}{4} (3 \cos^2 \phi - 1) \right] = X + Y \sin^2 \phi \qquad (15)$$

where, it is to be noted, $\phi = \theta_z$ is now the angle between the photoelectron and the photon propagation directions and $\beta = 4Y/(3X + 2Y)$.

The presence of circular polarization, which turns partially polarized light into elliptically polarized light, does not change the angular distributions, from those given in equations (12) to (14).[73, 76, 77] This circumstance is fortunate for the experimenter who uses synchrotron radiation because this radiation exhibits a degree of ellipticity that increases rapidly outside the plane of the circulating electron current.[78]

According to equation (10) different multipoles, including monopoles, can contribute to the angular distribution. As elaborated by Tseng *et al.*[74] the coefficients up to B_4 are important for our purposes. Calculated coefficients are given in Figures 3 and 4 for unpolarized photons as a function of energy for three cases: photoelectrons from C $2p_{1/2}$, Hg $6s$, and U $6s$. It can be seen that even in the domain of ESSR, $h\nu \lesssim 1$ keV, coefficients B_1, B_3, and B_4 are not negligible. Of course, B_2 of C $2p_{1/2}$ shows the expected energy dependence,[37] and B_2 of the s levels of the heavy elements Hg and U depart markedly from the nonrelativistic dipole value $B_2 = -1$, or $\beta_s = +2$, in this relativistic calculation.[74] Figure 5 shows the actual manifestation of these nonzero B_n ($n = 1, 3, 4$) factors for C $2p_{1/2}$, namely a skewing of $d\sigma/d\Omega$ in the forward direction discernible at photon energies as low as 400 eV. In heavier atoms, B_1, B_3, B_4, etc., may have finite values right at threshold[74] even for shallow levels, a possibility to be remembered by the experimenter who wishes to determine partial cross sections from the measured differential cross sections.

In the theory of angular distributions of photoelectrons, the j_t angular momentum transfer formalism developed by Dill and collaborators[39–47, 79, 80] has proven a powerful vehicle in the interpretation and prediction of directional photoemission

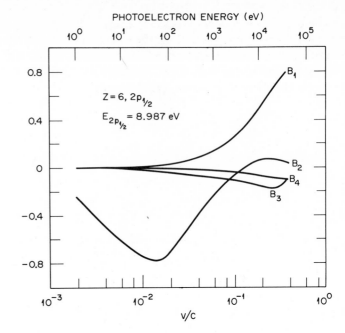

Figure 3. Coefficients B_n of the Legendre polynomials in equation (10) for photoionization of the $2p_{1/2}$ level of carbon (from Tseng et al.[74]).

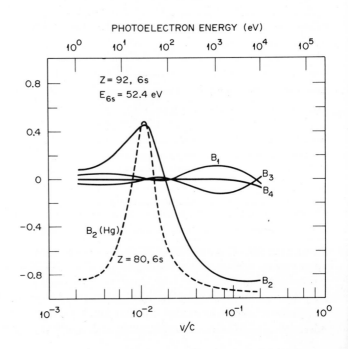

Figure 4. Coefficients B_n of the Legendre polynomials in equation (10) for photoionization of the $6s$ levels of uranium and mercury (after Tseng et al.[74]).

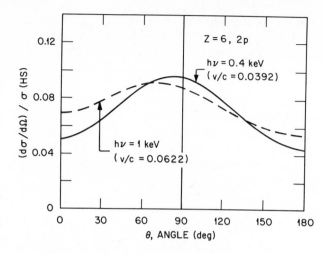

Figure 5. Theoretical angular distributions of photoelectrons ejected from the 2p level of carbon by soft x-rays (after Tseng et al.[74]).

from atoms and molecules. The various possible values of j_t for a photoionization process, equation (2), are given by

$$\mathbf{j}_t = \mathbf{j}_\gamma - \mathbf{l} = \mathbf{J}_f + \mathbf{s} - \mathbf{J}_0 \tag{16}$$

contingent on momentum and parity balances. This formalism offers a convenient way of calculating the photoelectron angular distribution from the various incoherently superposing exit channels. The ion–photoelectron interaction in the final state becomes especially apparent.[42] This interaction is generally anisotopic for molecules and open-shell atoms, leading often to a dramatic deviation of the β value from that predicted by the independent-particle model. For example, the parameter β_{3s} for Cl, an open-shell atom, deviates from $\beta = 2$ of the Cooper–Zare model[75] at all photoelectron energies, but especially at low energies where the interactions between outgoing electron and residual ion are particularly strong.[47] The subject of angular distributions has been reviewed in recent years by Dill,[39] Manson,[28] Dehmer,[81] and Tseng et al.[74]

3.4. Two-Electron Processes

The energetics of two-electron processes in photoionization are given by equations (7) and (8) and can readily be generalized to any number of electrons going either into discrete or continuum channels. However, we shall consider here only two-electron processes, which are the most common and the most amenable to theoretical treatment at the present state of development.

The cross sections of two-electron processes are contained in the general formula equation (9). They may be contained in σ_{0f} implicitly or explicitly depending upon the choices made for Ψ_0 and Ψ_f. For example, in a central field frozen-structure calculation the so-called subshell photoionization cross section σ_{nl} was shown to contain both single- and two-electron transitions.[82] This is seemingly a paradox analogous to that of the Koopmans energy of a level nl, which in the single-particle frozen-structure model is the weighted mean of the energies of all states, single and multiple, involving the level nl.[83] In order to explicitly calculate two-electron cross sections the many-body aspects of the atom or molecule must be included in both the initial and final states Ψ_0 and Ψ_f. Amplitudes for the various allowed transitions between the states Ψ_0 and Ψ_f, each of which represents a superposition of all pertinent configurations within a given symmetry, then correspond separately to the cross sections for single- and two-electron processes. Unfortunately, no single comprehensive calculation has been performed to date giving a complete account of the photoeffect in terms of *all* single and double photoionization events involving the various subshells of an atom. However, individual calculations have been made for either the two-electron processes or for supposedly single-electron processes using wave functions that contain the important electron-electron interactions or correlations in initial and final states.

Various approaches to the many-body problem can be taken to calculate two-electron photoionization cross sections (and at the same time improve on the single-electron cross section). These include the configuration or channel interaction (CI) models,[28, 48, 66, 68] applicable most profitably to the final state, the multiconfiguration Hartree–Fock[84, 85] or Dirac–Fock[51] model, the many-body perturbation theory,[52, 86, 87] the R matrix,[88, 89] the close-coupling method,[90–92] the random-phase approximation in its nonrelativistic[71, 93, 94] (RPA) and relativistic[95, 96] (RRPA) versions, and the Green's Function formalism.[54, 62] So far, CI, MBPT, R matrix, close-coupling, and GF models have been used in calculations of two-electron photoionization cross sections. In the case of GF, the spectral function of the photoeffect in a particular shell is calculated giving simultaneously cross sections for both single and multiple processes.[62, 97]

Although the various many-body formulations have in common that in going beyond the single-particle model they all improve on Ψ_0 and Ψ_f, the relationship between them and their relative levels of sophistication has not always been evident. However, cross links are being established between the different approaches.[62, 85, 98]

It is profitable to characterize two-electron transitions, which manifest themselves as correlation satellites in a photoelectron spectrum, according to whether the electron–electron interactions are predominantly in the initial or final state.[27, 99, 100] We speak of (a) the initial-state configuration interaction (ISCI), or ground-state correlation, and (b) the final-state configuration interaction (FSCI), or final-state correlation. To emphasize the two-part nature of the final state and the differences between the discrete states of the ion and the continuum states (channels) of the photoelectron(s), a subdivision of FSCI into the final *ionic*-state configuration interaction (FISCI) and final *continuum*-state channel interaction (FCSCI) is helpful.[100] The historically important shakeoff process or core rearrangement (CR) pathway remains a viable description of processes involving a deep core level *and* an outer level.[101] While CR, being an outgrowth of perturbation theory treatment, evolves naturally as an individual term or diagram in the MBPT approach,[86] it is composed of certain ISCI and FISCI terms in the CI approach.[102]

As an example of an FSCI process, a two-electron process in argon is depicted in Figure 6 in the MBPT diagrammatic representation.[86, 99] Arrows indicate time

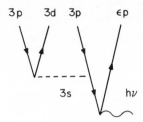

Figure 6. Diagrammatic representation of photoionization in the $3s$ level of argon leading to a correlation satellite (after Chang and Poe.[86]).

evolution; a photon ejects a $3s$ electron into an εp channel, but the $3s$ hole is filled by a $3p$ electron with the simultaneous promotion of another $3p$ electron into a $3d$ state by Coulomb interaction, symbolized by a dash line. The equivalent description of the process in the CI model is given by the interaction of two configurations of the same symmetry, namely $3s3p^6(^2S) \leftrightarrow 3s^23p^43d(^2S)$.

Recent overviews of two-electron processes have been given in Wuilleumier,[23] by Manson,[28] and Shirley[103]; molecular aspects have been treated by Cederbaum and co-workers,[54, 62, 104] and Bagus and Viinikka.[105]

3.5. Level Widths

An important property contained in a photoelectron spectrum is the width of a level. In the absence of interfering multiplets, changes in internuclear distance, and post-collision effects, these widths are *natural* level widths and hence are correlated with the lifetime τ of the hole or, alternatively, the total decay rate S of the hole by radiative (R), Auger (A), and Coster–Kronig (C) transitions.[26, 106] According to the uncertainty principle we have

$$\Gamma_{nlj} = \hbar/\tau = \hbar(S_R + S_A + S_C) \qquad (17)$$

where Γ_{nlj} is the natural level width, which is convoluted in the observed photoelectron line. In molecules determination of core level widths is generally possible even if the internuclear distance changes with ionization,[107] but that of inner valence levels is complicated by strong correlations and the presence of autoionizing and dissociative channels.

4. Experimental Apparatus and Procedures

The principle of the ESSR arrangement is presented in Figure 7, and a sketch of an actual setup, that used at ACO by Wuilleumier and collaborators,[13, 108, 109] is shown in Figure 8. Photons from the continuous spectral distribution of the synchrotron radiation source are energy selected by a grating monochromator, operated in grazing incidence, and directed into a source volume. Electrons emitted from the electron source are analyzed in an electron spectrometer and counted individually. Physically, the apparatus is arranged in the plane of the electron orbit of the storage ring, and by special design of the exit slit[110] the photon beam is fixed in

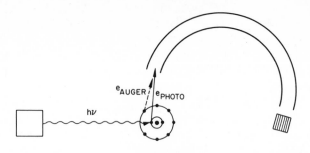

Figure 7. Principle of experimental setup.

PHOTON SOURCE
INCLUDING MONOCHROMATOR

ELECTRON SOURCE
ATOM OR MOLECULE

ENERGY ANALYZER
AND DETECTOR

space upon exiting the monochromator. The electron spectrometer is a cylindrical mirror analyzer (CMA), which accepts electrons from the source volume in a hollow cone at an angle of $\theta_z = 54.7°$ around its symmetry axis, which coincides with the photon propagation direction.

In the range of validity of the dipole approximation, relative partial photoionization cross sections can be obtained with a CMA in a setup of Figure 8 without the need for auxiliary measurements of either the asymmetry parameter β, or the polarization of the photon beam, or the orientation of the electric vector **E** in the xy plane. This follows from equation (12) or (13) or an equivalent expression derived by Schmidt.[76] With the photon beam and spectrometer axis in the z coordinate, the CMA, due to its 2π acceptance angle, performs in effect an integration over θ_x and θ_y or the azimuthal angle $\rho = \rho(\theta_x, \theta_z)$,[76, 77] so that with $\theta_z = 54.7°$ the measurement is independent of β. If in an alternative arrangement the CMA axis is to be aligned with one of the components of the electric vector, for example the major component $\mathbf{E}_y$ of the elliptically polarized light leaving the monochromator, both the orientation of the ellipse and the ratio I_y/I_x of the electric vector magnitudes in the axes of the ellipse need to be ascertained (equation 12).[77] In a third arrangement, electrons are accepted within a small solid cone at $\theta_y = \theta_x = 54.7°$, and hence $\theta_z = 54.7°$, by a suitable spectrometer, preferably a hemispherical or

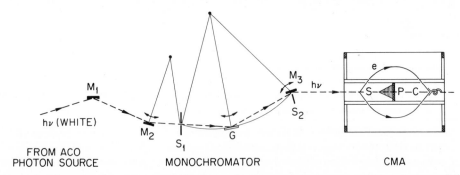

hν (WHITE)

M₁

M₃   hν

S₂

M₂   G

S₁

e

S—◁▦◁—P—C—▷∞—

FROM ACO
PHOTON SOURCE

MONOCHROMATOR

CMA

Figure 8. Schematic of ESSR setup at ACO, Orsay, showing monochromator consisting of mirrors (M), slits (S), and grating (G), electron spectrometer (CMA) with electron source (S), photon (P) and electron (C) detectors (after Wuilleumier.[13]).

spherical-sector analyzer. For this position, as for the setup of Figure 8, equation (12) simplifies to

$$\left.\frac{d\sigma_{0f}}{d\Omega}\right|_{54.7°} = \frac{\sigma_{0f}}{4\pi} \qquad (18)$$

regardless of the ratio I_y/I_x, but with the constraint that major and minor axes of the ellipse coincide with the y and x coordinates. In yet another arrangement, the partial cross section σ_{0f} can be measured directly according to equation (14), if the polarization p has been determined and the spectrometer, for example a spherical sector analyzer, is located in a plane perpendicular to the photon beam at an angle θ such that $1 + 3p \cos 2\theta = 0$.

Under the usual operating conditions, the component I_y is much greater than I_x, and the $\mathbf{E}_y$ vector remains very nearly in the plane of the circulating electron current even after passing through the monochromator. However, changes in I_y and $\mathbf{E}_y$, and correspondingly in I_x and $\mathbf{E}_x$, will occur during operation due to gradual contamination and degradation of the optical components and, possibly, spatial fluctuations of the synchrotron radiation.[108, 109, 111] It is then that the economy of the arrangement of a CMA along the photon direction becomes especially apparent.

If the dipole approximation no longer holds, the angular distribution of the photoelectrons must be determined, and an analyzer should be mounted in the xy plane, viewing the photon beam within a small solid angle around $\theta_z = 90°$. For this a spherical-type analyzer as proposed above for the last two arrangements is particularly well suited.

4.1. The Monochromator

Physical and operational characteristics of the monochromators in use and under development are described in several proceedings[16, 112] and books.[17, 113] The monochromator, including mirrors, is one of the more important components in making optimum use of the photons from a synchrotron radiation source. This has been demonstrated recently[108, 109] at the ACO facility where the change from a flat mirror (Figure 8) to a toroidal mirror and the use of a grating in pristine condition resulted in a photon flux increase of two orders of magnitude producing 10^{11} photons per second in the source volume for $30 \leq h\nu \leq 110$ eV in a 1% bandwidth. With this photon flux, obtainable under ideal conditions, processes with cross sections as low as 0.01 Mb are amenable to investigation.

A recurring problem in diffraction of photons from a continuous energy distribution has been the occurrence of higher orders and the lack of quantitative procedures for minimizing their intensities. However, Wuilleumier and co-workers[108, 109] have overcome the problem by the use of the PAX technique in which Photoelectrons are used for the Analysis of X rays under well-defined conversion conditions.[26, 114, 115] The resulting spectra, which were obtained with a helium converter, are shown in Figure 9. The dependence of the second-order contribution on the angle of incidence is clearly and quantitatively

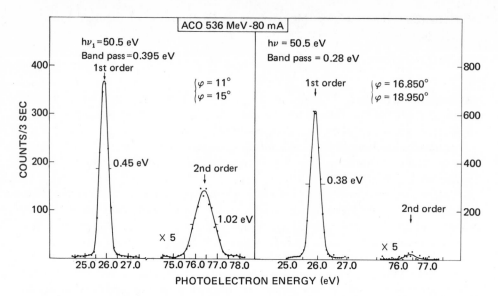

Figure 9. Dependence of the second-order diffraction intensity on the angle of incidence (after Wuilleu-mier *et al.*[108]).

demonstrated. At the greater angle the second-order photon intensity is seen to be drastically reduced, namely to about 5% of the first-order intensity.

Magnitudes and directions of the two electric vector components of the photon beam can also be measured with the PAX technique. In this case the angular distribution of the He 1s photoelectrons is recorded,[116] yielding in the $\mathbf{E}_y$ and $\mathbf{E}_x$ directions maxima whose heights are proportional to I_y and I_x to a very high accuracy, since the asymmetry parameter β_{1s} of helium deviates very little from the nonrelativistic value of two at low photon energies. At $h\nu \lesssim 40$ eV, the deviation was calculated[96] to be about 10^{-6}.

4.2. The Electron Spectrometer

The cylindrical mirror and the hemispherical analyzer, or its spherical-sector version, are well adapted for use in ESSR. Both exhibit outstanding optical characteristics, which have been discussed by many authors.[4, 18, 19, 26] Two other types of analyzers, the time-of-flight (TOF) spectrometer and the zero-energy filter, capitalize on the continuous tunability and the time structure of synchrotron radiation. The TOF analyzer was introduced into ESSR by Bachrach *et al.*,[117] and the zero-energy filter described first in 1969[118] is now being used extensively in the ESSR of molecules. Both analyzers may be used alone,[117–120] but form particularly powerful tools in tandem with a mass spectrometer[121, 122] or a fluorescence detector.[123, 124] A sketch of a typical setup is shown in Figure 10. In this apparatus, ions are selected by TOF and electrons are detected either in the TOF mode or the zero-energy mode.[122] In both instances, the photon pulse provides the time

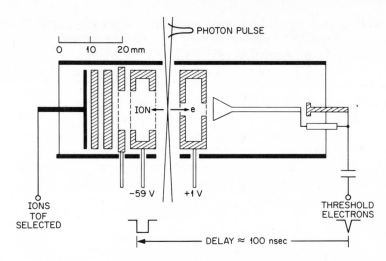

Figure 10. Electron–ion coincidence apparatus combining a steradiancy zero-energy electron spectrometer and a time-of-flight ion spectrometer (after T. Baer et al.[122]).

marker, and electrons and ions may be placed into coincidence. For zero-energy analysis, which is based on discrimination against fast electrons by way of angular dispersion,[118, 121] temporal gating by the photon pulse eliminates signals from faster electrons traveling directly toward the detector.[122] In a hybrid design, a deflection voltage is applied so that only zero-energy electrons can reach the off-line detector.[119] Since this arrangement in effect incorporates a tunable bandpass, electrons with energies up to about 100 meV can be selected at a resolution of 2 to 3 meV.[120] In principle, TOF offers this capability, too, but with an energy-dependent resolution. Collection efficiency of zero-energy electrons is constant by definition and for a given design can be greater than 50%; however, to be able to compare intensities of threshold electrons from different electronic and vibrational levels, the photon flux must be measured as a function of energy.

4.2.1. Energy Calibration

For electrostatic deflection spectrometers, the relationship between the electron kinetic energy and the voltage V applied to the deflection plates is given relativistically by

$$E_{e,\,kin} = fV\left(1 + \frac{E_{e,\,kin}}{E_{e,\,kin} + 2m_0 c^2}\right) \tag{19}$$

In practice, the difference between the energy E of the electron under study and the known energy E_0 of the electron from a standard, $\Delta E = E - E_0$, is determined from a measurement of the corresponding plate voltages V and V_0. For a spherical-sector plate analyzer the formula

$$\Delta E = f(V - V_0)\left[1 + f\frac{(V + V_0)}{2m_0 c^2}\right] \tag{20}$$

was derived,[125] where f is the spectrometer constant, and the last member in the equation represents the relativistic correction in terms of the applied voltages.

Similar expressions have been derived for other analyzers and the relativistic correction for the CMA was found to be smaller than for the spherical instruments and to depend slightly on the focusing conditions.[126, 127]

According to equation (20), energy calibration of an analyzer requires the determination of the factor f by creating electrons whose energy *differences* are precisely known and, because of the possibility of a zero offset, the use of an absolute energy standard. Procedural details can be found elsewhere.[125, 127–129] We present in Table 1 a collection of level and transition energies suitable for calibration purposes. After the electron spectrometer is calibrated, the analyzer may also be used in the PAX mode to determine the energies of the photons entering the source volume. To this end, implementing the setup with a helium discharge lamp can be useful.[130]

Table 1. Level and Transition Energies Suitable for Calibration

Level or line	Energy[a] (eV)	Reference	Level or line	Energy[a] (eV)	Reference
He $1s$	24.5876(1)	b	Xe $4d_{5/2}$	67.541(9)	g, j, k, l
Ne $2p_{3/2}$	21.5647	c	$4d_{3/2}$	69.525(10)	g, j, k, l
$2p_{1/2}$	21.6615	c	He Iα	21.2182(1)	b
Ar $3p_{3/2}$	15.7597	c	He Iβ	23.0872(1)	b
$3p_{1/2}$	15.9372	c	He IIα	40.8135(1)	b
Kr $4p_{3/2}$	13.9997	c	He IIβ	48.3718(2)	b
$4p_{1/2}$	14.6656	c	He $2s2p\,{}^1P$	35.537(15)	b
Xe $5p_{3/2}$	12.1300	c	Zr $M\zeta$	151.65(3)	l, m
$5p_{1/2}$	13.4364	c	Mg $K\alpha$	1253.64(3)	n
Ne $2s$	48.47	d	Ne $KL_2L_3\,{}^1D_2$	804.50(3)	c, e, f
Ne $1s$	870.33(3)	e, f	Ar $L_3M_{2,3}^2\,{}^1S_0$	201.10(2)	c, e, f, g
Ar $2p_{3/2}$	248.616(15)	f–h	Kr $M_5N_1N_{2,3}\,{}^1P_1$	37.87(3)	o
$2p_{1/2}$	250.763(15)	f–i	Xe $N_5O_{2,3}^2\,{}^1D_2$	32.28(3)	o
Ar $\Delta 2p$	2.147(5)	e–i	N $1s(N_2)$	409.93(4)	e, f, p
Kr $3d_{5/2}$	93.795(18)	e, g, j			
$3d_{3/2}$	95.038(20)	e, g, j			

[a] Values are weighted averages if multiple references are given; where applicable, the conversion factor is 1.2398520(32)$nm \times$ eV. Uncertainties are indicated in parentheses.

[b] W. C. Martin, *J. Phys. Chem. Ref. Data* **2**, 257 (1973).

[c] C. E. Moore, U.S. Department of Commerce NSRDS-NBS 34 (1970).

[d] C. E. Moore, Atomic energy levels, *Natl. Bur. Stand. (U.S.) Circ.* **467** (1949).

[e] G. Johansson et al., *J. Electron. Spectrosc. Relat. Phenom.* **2**, 295 (1973).

[f] T. D. Thomas and R. W. Shaw, Jr., *J. Electron Spectrosc. Relat. Phenom.* **5**, 1081 (1974).

[g] G. C. King et al., *J. Phys. B* **10**, 2479 (1977).

[h] M. Nakamura et al., *Phys. Rev. Lett.* **21**, 1303 (1968).

[i] J. Nordgren et al., UUIP-944 (1976), University of Uppsala, Institute of Physics (unpublished).

[j] K. Codling and R. P. Madden, *Phys. Rev. Lett.* **12**, 106 (1964).

[k] U. Gelius, *J. Electron Spectrosc. Relat. Phenom.* **5**, 985 (1974).

[l] M. O. Krause and F. Wuilleumier, *Phys. Lett. A* **35**, 341 (1971); Values were revised using present values of Xe $4d$ and Xe $N_5O_{2,3}O_{2,3}\,{}^1D_2$ as references; M. O. Krause, *Phys. Lett. A* **74**, 303 (1979).

[m] G. Dannhäuser and G. Wiech, *Phys. Lett. A* **35**, 208 (1971).

[n] J. A. Bearden, *Rev. Mod. Phys.* **39**, 78 (1967).

[o] S. Ohtani et al., *Phys. Rev. Lett.* **36**, 863 (1976).

[p] For vibrationally resolved spectra, see G. C. King et al., *Chem. Phys. Lett.* **52**, 50 (1977).

4.2.2. Spectrometer Function

The spectrometer function can be traced by a beam of monokinetic electrons having a sufficiently small energy spread. At the lowest energies, photoelectrons ejected by HeI radiation from Ar, Kr, Xe, or light molecules are excellent candidates, and at medium energies Auger electrons from penultimate shells of Ar, Kr, and Xe are well suited. The widths of a number of recommended levels and lines are listed in Table 2.

As a rule, the spectrometer function is reproduced well by a Gaussian function.[26, 131] The full width at half maximum of this function, Γ_G, is obtained from the Voigt width Γ_V of the observed line using the relation[131]

$$\frac{\Gamma_L}{\Gamma_V} = 1 - \left(\frac{\Gamma_G}{\Gamma_V}\right)^2 - 0.114\left(1 - \frac{\Gamma_G}{\Gamma_V}\right)\left(\frac{\Gamma_G}{\Gamma_V}\right)^2 \tag{21}$$

where Γ_L is the Lorentzian width of the tracing line. After Γ_G is established, equation (21) can be used in turn to determine the widths of levels and lines under study, provided their widths are natural and hence Lorentzian.

4.2.3. Transmission Function and Intensity Measurements

One of the more salient features of ESSR is the capability to determine cross sections as a function of energy. Specifically, the number of pulses at the output of the detector is related to the partial photoionization cross section σ_{0f} of a process

Table 2. Suggested Values of Level Widths (FWHM) Suitable for Calibration Together with X-Ray and Auger Lines Having the Same Widths

Level	Line	Width (meV)[a]	Reference
Ar 2p	2p → 3s	127(10)	b, d
	2p → 3p3p		
Ar 2p3p ³D₃	2p3p ³D₃ → 3s3p ³P₂	40	c
	2p3p ³D₃ → 3p³ ²P		
Ar 2p3p ¹P₁	2p3p ¹P₁ → 3s3p ¹P	<40	c
Kr 3d	3d → 4p	90(10)	d
	3d → 4p4p		
Xe 4d	4d → 5p	110(15)	d, e
	4d → 5p5p		
Photolines Ar 3p, Kr 4p, and Xe 5p excited by He I radiation		1–10	f

[a] Values apply to either component of a spin–orbit doublet. Probable errors are given in parentheses.
[b] M. O. Krause and J. H. Oliver, *J. Phys. Chem. Ref. Data* **8**, 329 (1979).
[c] J. Nordgren et al., *Phys. Scr.* **16**, 280 (1977).
[d] G. C. King, et al., *J. Phys. B* **10**, 2479 (1977).
[e] D. L. Ederer et al., in *Electron and Photon Interactions with Atoms*, H. Kleinpoppen and M. R. C. McDowell (eds.), pp. 69–81, Plenum Press, New York (1976).
[f] J. A. R. Samson, *Rev. Sci. Instrum.* **40**, 1174 (1969); actual value depends on lamp design and source temperature.

defined by initial and final states [see equations (2) and (9)]. Assuming that an electrostatic deflection analyzer with equal entrance and exit slits and with no preacceleration is used, the number of photoelectrons $N(e)$ counted in the energy interval dE is given by

$$N(e)\, dE = GN\eta(1 - \alpha)f(E_{e,\,\text{kin}})N(h\nu)\left.\frac{d\sigma_{0f}}{d\Omega}\right|_{\theta} d\Omega\, dE \qquad (22)$$

where G is a geometry factor, a constant of the system that includes the volume of the electron source, slit widths, solid angles, and dispersion of apparatus; N is the number of atoms in the electron source, $\eta = \eta(E_{e,\,\text{kin}})$ the efficiency of the detector; $\alpha = \alpha(E_{e,\,\text{kin}})$ the fraction of electrons scattered from their trajectory; $f(E_{e,\,\text{kin}})$ the transmission function of the analyzer; $d\sigma_{0f}/d\Omega|_{\theta}$ the differential photoionization cross section at angle θ, and $N(h\nu)$ the number of photons traversing the source assuming negligible photon attenuation in the source.

The factor G is a constant for a given setup and can in principle be calculated, although this is a formidable task.[9, 19, 132] Assuming the size of the electron source has been determined and uniformity of gas pressure in the source ascertained, N can be obtained from an absolute pressure measurement, made for example with a capacitance manometer. Factors η and $N(h\nu)$ can be measured; for example, $N(h\nu)$ can be measured with an ion chamber,[17, 133] an Al_2O_3 photodiode,[134] or a sodium salicylate converter.[17] Inelastic and elastic scattering cross sections that enter into α have previously been tabulated for many species[135] but are being

Table 3. Photoionization Cross Section of
Helium below the Excitation of Autoionization
Features[a,b]

$h\nu$ (eV)	σ (Mb)	$h\nu$ (eV)	σ (Mb)	$h\nu$ (eV)	σ (Mb)
24.6	7.56	32.0	4.70	42.0	2.80
25.0	7.35	32.5	4.56	43.0	2.67
25.5	7.10	33.0	4.43	44.0	2.56
26.0	6.86	33.5	4.31	45.0	2.44
26.5	6.63	34.0	4.19	46.0	2.34
27.0	6.40	34.5	4.08	47.0	2.25
27.5	6.20	35.0	3.97	48.0	2.15
28.0	6.00	35.5	3.86	49.0	2.06
28.5	5.81	36.0	3.76	50.0	1.98
29.0	5.63	36.5	3.67	51.0	1.90
29.5	5.45	37.0	3.57	52.0	1.83
30.0	5.29	38.0	3.39	53.0	1.76
30.5	5.13	39.0	3.23	54.0	1.70
31.0	4.98	40.0	3.08	55.0	1.63
31.5	4.84	41.0	2.94	56.0	1.57

[a] From Marr and West.[143]
[b] The uncertainty is less than 3%.

Table 4. Total Photoabsorption Cross Section of Neon, Partial Photoionization Cross Section, and Asymmetry Parameter for the 2p Subshell [a, b, c]

$h\nu$ (eV)	σ_{tot} (Mb)	σ_{2p} (Mb)	β_{2p}	$h\nu$ (eV)	σ_{tot} (Mb)	σ_{2p} (Mb)	β_{2p}
30	8.86	8.86	0.42	100	3.98	3.30(7)	1.34
32	8.95	8.95	0.51	105	3.71	3.04	1.35
34	8.96	8.96	0.60	110	3.45	2.80(6)	1.36
36	8.91	8.91	0.67	115	3.21	2.56	1.37
38	8.82	8.82	0.73	120	2.96	2.34(5)	1.38
40	8.69	8.69	0.79	125	2.72	2.13(4)	1.39
42	8.54	8.54	0.84	130	2.53	1.96	1.41
44	8.37	8.37	0.90	135	2.36	1.81	1.43
46	8.18	8.18	0.94	140	2.20	1.67	1.45
48	7.96	7.96	0.99	145	2.06	1.55	1.47
50	7.92	7.72(4)	1.03	150	1.93	1.44	1.49
52	7.66	7.44(4)	1.06	155	1.81	1.34	1.49
54	7.42	7.17(5)	1.09	160	1.69	1.24	1.48
56	7.30	6.86(8)	1.12	165	1.57	1.14(3)	1.48
58	7.07	6.62	1.15	170	1.47	1.06	1.47
60	6.85	6.38	1.18	175	1.40	1.00	1.47
62	6.63	6.16	1.20	180	1.33	0.93	1.46
64	6.47	5.97	1.22	185	1.24	0.87	1.46
66	6.28	5.76(9)	1.24	190	1.16	0.81	1.45
68	6.10	5.58	1.25	195	1.09	0.75(2)	1.45
70	5.95	5.40	1.27	200	1.03	0.70	1.44
72	5.81	5.23	1.28	210	0.93	0.62	1.44
74	5.66	5.06	1.29	220	0.82	0.54	1.43
76	5.51	4.90	1.30	230	0.75	0.48(1)	1.43
78	5.35	4.73	1.31	240	0.69	0.43	1.42
80	5.20	4.57	1.31	250	0.61	0.38	1.42
82	5.06	4.41	1.31	260	0.54	0.33	1.41
84	4.92	4.26	1.32	270	0.48	0.29	1.41
86	4.78	4.12	1.32	280	0.44	0.26	1.40
88	4.65	3.98(8)	1.32	290	0.40	0.23	1.40
90	4.53	3.85	1.33	300	0.36	0.21	1.39
92	4.41	3.73	1.33				
94	4.30	3.62	1.33				
96	4.19	3.52	1.34				
98	4.08	3.41	1.34				

[a] The uncertainty in σ_{tot} is less than 6% and in β_{2p} less than 3%. In the autoionization region $41 \lesssim h\nu \lesssim 49$ eV, values are averaged.
[b] From Wuilleumier and Krause.[144]
[c] Error of partition is indicated in parentheses and only when a change occurs in its value.

constantly improved.[136–138] The function $f(E_{e, \text{kin}})$ is simply $E_{e, \text{kin}}$ provided the width of the initial electron distribution is much less than the width of the spectrometer function and effects from residual magnetic fields and surface charges are negligible. These conditions are generally met only at energies above 100 eV, but at the lowest energies encountered in ESSR the respective widths are usually comparable or even reversed. It is therefore necessary to ascertain the function $f(E_{e, \text{kin}})$ for a given setup. Kemeny *et al.*[139] calculated $f(E_{e, \text{kin}})$ for different conditions, including variable preacceleration of the electrons, and others[140–142] measured the function emphasizing the low energies where the ideal conditions for calculation no longer exist.

Up to now, neither the geometry nor the density of an electron source has been determined and, as a result, measurements of $d\sigma_{0f}/d\Omega$ have been made on a relative basis. However, several of these data have been converted to absolute cross sections σ_{0f} and $d\sigma_{0f}/d\Omega$ (see Section 17). These values, of which the most reliable ones[142–144] are collected in Tables 3 to 6, can now be used to determine accurately the transmission function of the apparatus from about 0.1 to 300 eV in electron energy. They may also be used as reference standards for the direct determination of absolute partial cross sections, provided the densities of target and reference gases are known. As references at higher energies, $2s$ and $1s$ partial cross sections of neon have previously been tabulated.[145] At the lowest energies, the

Table 5. Total and Partial Photoionization Cross Sections, Photoelectron Branching Ratio, and Asymmetry Parameters of Xenon below Excitation of Autoionization Features

$h\nu$ (eV)	$\sigma_{\text{tot}}{}^a$ (Mb)	$\dfrac{\sigma_{3/2}}{\sigma_{1/2}}{}^b$	$\sigma_{3/2}$ (Mb)	$\sigma_{1/2}$ (Mb)	$\beta_{3/2}{}^c$	$\beta_{1/2}{}^c$
13.5	66.1	1.63	41.0	25.1	0.85	0.70
14.0	64.4	1.65	40.1	24.3	1.00	0.85
14.5	62.5	1.68	39.2	23.3	1.10	0.95
15.0	60.1	1.71	37.9	22.2	1.20	1.05
15.5	58.0	1.73	36.8	21.2	1.25	1.00
16.0	55.4	1.75	35.3	20.1	1.30	1.15
16.5	52.7	1.76	33.6	19.1	1.34	1.20
17.0	49.9	1.77	31.9	18.0	1.38	1.25
17.5	47.2	1.76	30.1	17.1	1.42	1.30
18.0	44.5	1.75	28.3	16.2	1.46	1.34
18.5	41.7	1.74	26.5	15.2	1.50	1.38
19.0	39.1	1.73	24.8	14.3	1.54	1.42
19.5	36.6	1.72	23.1	13.5	1.58	1.46
20.0	34.2	1.71	21.6	12.6	1.62	1.50

[a] From West and Morton[35]; probable error 2–3%.

[b] On the basis of data reported by Adam *et al.*[109]; error is estimated to be ± 0.03 units.

[c] From J. L. Dehmer *et al., Phys. Rev. A* **12**, 1966 (1975); weighted average of values quoted therein; estimated error 0.05–0.1 units.

Table 6. Relative Intensities in the He I Photoelectron Spectra of O_2 and N_2 Suitable for Calibration at Energies below 9 eV [a,b]

Level	v	Photoelectron energy (eV)	Relative intensity	Asymmetry[c] parameter β
		O_2^+		
$X^2\Pi_g$	0	9.14	46.7(1.8)	
	1	8.90	100.0(5.0)	-0.25
	2	8.67	91.6(3.2)	
	3	8.45	46.1(2.1)	
	4	8.23	15.4(0.5)	
$b^4\Sigma_g^-$	0	3.05	79.3(4.4)	
	1	2.90	68.5(1.3)	
	2	2.76	44.6(1.2)	0.65
	3	2.63	27.3(0.5)	
	4	2.50	15.2(0.2)	
$B^2\Sigma_g^-$	0	0.92	32.1(1.8)	
	1	0.78	38.3(2.3)	
	2	0.65	31.9(2.9)	1.15
	3	0.53	21.4(1.9)	
	4	0.40	12.5(0.5)	
		N_2^+		
$X^2\Sigma_g^+$	0	5.64	100.0(0.3)	0.65
$A^2\Pi_u$	0	4.51	39.9(0.2)	
	1	4.28	48.4(0.2)	
	2	4.05	34.5(0.2)	0.37
	3	3.83	21.5(0.2)	
	4	3.61	8.0(0.1)	
$B^2\Sigma_u^+$	0	2.47	25.8(0.3)	1.28

[a] After Gardner and Samson.[142]

[b] The uncertainties are given in parentheses.

[c] These suggested values are averaged from the values quoted in Reference 146.

convenience of using vibrational progressions for calibration should be particularly noted (see Table 6).[142, 146]

Factor α in equation (22) deserves special attention. Ideally, operating conditions should be chosen to make $\alpha \ll 1$; this requires minimizing the integral $\int_0^l p(x)\,dx$ that enters into α, where $p(x)$ is the gas density along the entire electron trajectory of length l. At energies up to about 100 eV the scattering cross sections change rapidly and, especially in molecules, are subject to large excursions in resonance regions.[147–149] At resonances for negative-ion formation,[149–151] cross sections are often high, making the target gas an effective electron scavenger.[152] In practical terms, the target gas pressure can be relatively high, 1 Pa $\approx 10^{-2}$ torr, if confined to a small volume by the use of a gas jet or a small gas cell with provision for strong, effective pumping outside the source region. However, in contrast to the spherical analyzer, the CMA, because of its physical characteristics, scarcely meets these requirements, so that operation at a much lower pressure, 10^{-2} to 10^{-3} Pa, is mandatory. In fact, experiments utilizing a CMA have been carried out at these low pressures at the ACO and Daresbury machines.[108, 109, 153]

4.3. Source for Circularly Polarized Light

Although synchrotron radiation is elliptically polarized, the component of the electric vector lying in the orbit plane is by far the dominant one. Thus synchrotron radiation offers itself as a primary source for the production of circularly polarized light by the use of quarter-wavelength plates following the monochromator. Light up to 10 eV can be obtained with MgF_2 plates. This could provide a strong source for the study and analysis of autoionization resonances[154, 155] and the Fano effect.[156] As Cherepkov pointed out,[157] spin-polarized electrons could be produced by photoionization of Tl in the $6\,^2P_{1/2}$ autoionization resonance with a greater efficiency than by other methods.

At energies greater than 10 eV, a useful flux of light that has a high circular polarization can be obtained out of the plane of the electron orbit using a small-diameter diaphragm.

4.4. Comparison with Discrete Sources

While comparisons of *calculated* photon intensities have been presented in the past for various photon sources (Chapters 2 and 14), comparisons of *measured* intensities have only recently been made at one installation.[108] The results are shown in Figure 11 by way of the xenon $5p$ photoelectron spectrum. At the given resolution, the flux of the Ne II radiation from an efficient discharge lamp is seen to be about an order of magnitude higher than that from the ACO source. However, while the Ne II photon spectrum contains other strong lines, the monochromatized synchrotron radiation contains only weak higher-order components. Thus, upon monochromatization, the intensity advantage of the Ne II lamp would all but vanish. By contrast, the He II and of course the He I gas discharge lamps remain superior, even if dispersed, to present synchrotron radiation sources at the isolated energies of 40.8 and 21.2 eV, especially if the inherently narrow linewidth of the resonance radiation of a discharge source is utilized. At higher energies, an indirect comparison of 102-eV synchrotron radiation with 109-eV Be K photons[158] shows the ACO source to be considerably more intense.[159]

The photon intensity in the Hopfield continuum of a helium discharge lamp has recently been compared with the intensity available from the Tantalus I storage ring.[160] The laboratory source was shown to produce photon intensities comparable to this synchrotron radiation source only in a very small energy range. However, the rare-gas continuum sources have the advantage of not introducing problems from higher-order radiation.

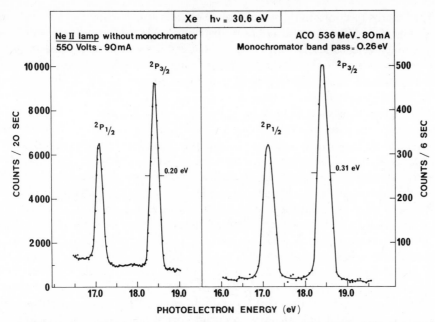

Figure 11. Comparison of photon intensities available from a nonmonochromatized Ne II lamp and the ACO radiation source. The xenon 5p levels are used as the photon–electron converter (from Wuilleumier et al.[108]).

5. Restricted Photoelectron Spectrometry – Energies

For chemical analysis and identification of levels and processes, the measurement of energies is the most important facet of photoelectron spectrometry. In principle, only a few discrete photon energies are needed and are available for this purpose, and the tunability inherent in a synchrotron radiation source seems superfluous. But given the photon intensities delivered by a modern storage ring and given the narrow bandpass obtainable with an efficient high-resolution monochromator, synchrotron radiation will doubtless take its place in restricted photoelectron spectrometry. On general grounds, the possibility of optimizing sensitivity and minimizing potential interferences is a legitimate motivation for using ESSR. More specifically, the following possibilities offer additional motivation: (a) the study of inner valence levels of molecules, (b) the study of core levels of molecules, (c) determination of energies and shifts of *deep* core levels, and (d) the wide use of zero-energy electron spectrometry. We shall briefly elaborate on these points in the paragraphs that follow.

Inner valence levels of molecules, even diatomic molecules, have been shown to be strongly perturbed by correlation effects.[62, 104, 105] Hence, as demonstrated for H_2S and PH_3,[104, 161] photoionization produces numerous lines that may be resolved if a single, narrow photon line is used for excitation. However, the best suited photon line, the He IIα line at 40.8 eV, is accompanied by other lines of the He II and the He I series. Although this problem can be alleviated by monochromatization, a serious problem remains, namely the possible occurrence of minima in the partial cross sections in the vicinity of 40 eV, affecting both the sensitivity of the measurement and the spectral intensity distribution. In addition, many molecules possess inner valence levels with energies greater than 40 eV. As a consequence, it appears that ESSR with photons in the energy range 60–80 eV and with a bandpass of about 50 meV could provide ideal conditions for the study of inner valence levels. In the past, information has been obtained with Al $K\alpha$,[5, 38] Y $M\zeta$,[162] and He II[104] photons, as well as with the dipole $(e, 2e)$ technique,[163, 164] in which a continuously variable photon source is simulated by energetic electrons that undergo energy losses with near zero momentum transfer in inelastic collisions.

Fine structure of K levels in compounds of second-row elements, L levels of third-row elements, and so on, can be studied with narrow photon lines somewhat above the respective level energies. Multiplet and vibrational structure of inner-shell ionized molecules can then be revealed clearly, and questions can be resolved about localization of a core hole, hopping times between orbitals (e.g., $1s\sigma_g$ and $1s\sigma_u$), internuclear distances, and potential curves. These prospects of ESSR may be inferred from an earlier study of CH_4 with 1.5-keV photons,[38] and in particular from excitation of the K electron of N_2 by electrons.[165] The high-resolution electron-energy-loss (EEL) spectrum, displayed in Figure 12, shows clearly the vibrational structure for the $N_2(1s, \pi2p)$ $^1\Pi$ state. In contrast to the EEL spectrum and a photoabsorption spectrum, an ESSR spectrum corresponds to a simpler electronic configuration as the core electron is removed into the continuum. Thus, interpretation is facilitated. However, the experimental requirement of a bandpass of 50 to 100 meV places high demands on the dispersive system.

In advocating photoionization of deep core levels, we extend the regime of ESSR into the keV range. The major advantage of including deep core levels lies in the applicability of the simple potential model to chemical shifts of levels,[4] as demonstrated previously for the K level (2.5 keV) of sulphur compounds.[166]

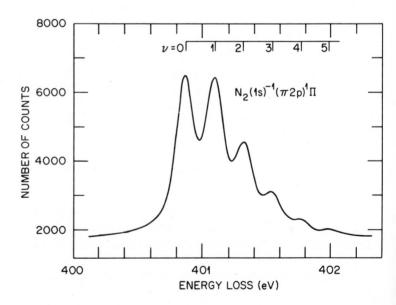

Figure 12. Vibrational structure of N_2 with a hole in the K level and an electron in the $(\pi2p)^1\Pi$ state measured by electron-energy-loss spectrometry (after King *et al.*[165]).

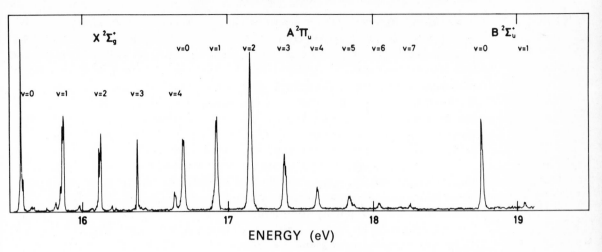

Figure 13. Nitrogen threshold electron spectrum (after Peatman *et al.*[120]).

To determine level energies we may either keep the photon energy constant and scan the electron kinetic energy or keep the electron energy constant and scan the photon energy. In the latter mode it is advantageous to set the electron energy window at zero because of the attending high collection efficiency. Under these conditions, a scan of the photon energy yields the energies of electronic, vibrational, and rotational levels in the very sense of the definition of the ionization energy: that the electron from a given level be removed to infinity *at rest*. The potential of this branch of photoelectron spectrometry, variously called "threshold," "zero-energy," or "photoionization resonance"(PIR) electron spectrometry, has become apparent within the past year with the use of synchrotron radiation as a photon source.[120–122] As an example of the data that have been obtained at DESY, Hamburg and ACO, Orsay, the nitrogen threshold electron spectrum[120] is shown in Figure 13. Since, independent of the thermal motion of the target, a resolution of 2–3 meV is attainable,[167, 168] the method provides a detailed view of the rovibronic structure of molecules and the consequences of ionization. An abundance of PIR studies may be expected in the future for the valence region of molecules. Whether the method holds promise for spectrometry of inner levels that decay rapidly by the Auger effect depends largely on the probability of post-collision interactions (Section 10). Ion charge and (e, 2e) experiments of the Xe $4d$[169] and Ar $2p$[170] levels indicate that this probability is high, so that the production of zero-energy electrons at threshold is reduced.

6. Level Widths and Line Widths

Thus far ESSR has not been applied to the measurement of level widths and linewidths. However, synchrotron radiation was used for this purpose at a very early time in photoabsorption measurements[171] and recently in x-ray fluorescence studies.[172] While recent width determinations by high-resolution EEL spectrometry have yielded accurate data for atoms[173] and molecules[165] having an inner-shell hole *and* an electron in a Rydberg level, ESSR is capable of yielding equally accurate data for the singly ionized species. To use a specific case, the widths of the ground and vibrationally excited states of the N_2 molecule in which a K electron is promoted to a $\pi 2p$ level can be obtained by EEL as seen in Figure 12, but those of the K-shell-ionized N_2 molecule must be obtained by ESSR. Provided contributions can be excluded from nonlifetime effects, such as from a competing close-lying series of states, widths data give the total decay rate for the inner-shell hole and its dependence on vibrational excitation and chemical environment.

As a rule, the widths of photoelectron lines and Auger lines correspond to the singly ionized species, since multiple-hole satellites are separated in energy.[174] However, in cases in which overlap occurs between single- and multiple-hole lines, interference can be eliminated or minimized by proper tuning of the photon energy.

A prerequisite for reliable widths measurements is a bandpass of 20 to 50 meV at photon energies up to about 1 keV and 0.1 to 1 eV above 1 keV, so that the contribution to the measured width from the exciting radiation remains a fraction of the natural width of the level or line under study.

7. Partial Photoionization Cross Sections

Knowledge of partial photoionization cross sections is critical for a deeper understanding of atomic and molecular dynamics. A measurement of these cross sections σ_{0f} provides a stringent test of theory, and in the case of molecules aids in the identification of molecular orbitals and gateways to dissociation or fragmentation. Furthermore, analytical methods based on inner-shell photoionization depend on reliable values of σ_{0f}. Cross sections, it must be emphasized, need to be determined continuously as a function of energy to be of value in fully exploring wave functions [equation (9)] and correlation effects of various types. This is then the area in which the merits of synchrotron radiation become particularly apparent.

Determination of absolute partial photoionization cross sections σ_{0f} is based on equation (22). If the dipole approximation applies, the measurement may be carried out at $\theta = 54.7°$ so that the differential cross section can be replaced by the integrated cross section [equation (18)]. Measurement of a given σ_{0f} may be made on an absolute basis, if referred to a standard cross section such as one of those tabulated in Tables 3–6, or measurement may be made on a relative basis and subsequently normalized to the total, absolute photoionization cross section. The latter method requires that *all* photoionization processes are identified and their relative cross sections are measured.[175] Specifically, this includes a measurement of (a) single-electron processes leaving the ion with a hole in any of the energetically accessible levels, and (b) two-electron processes in which the ion is left with two holes in any combination of the energetically accessible levels. In case (b) both electrons may go into continuum channels (shakeoff) or one electron into the continuum and the other into a discrete, bound level of the ion (shakeup), where the terms shakeoff and shakeup are used in a general sense without implying a particular mechanism. Evidently single electron processes and shakeup processes manifest themselves as discrete peaks (see Figure 2), but shakeoff processes give rise to continuous distributions, and are investigated preferably by ion charge and by Auger spectrometry.[175, 176]

The number of events counted for each of the photoionization channels may be referred to the strongest peak in a spectrum or to the total number of events.[177] In the latter case, we speak of branching ratios. Care must then be exercised that all events have been included.

In open-shell atoms or in molecules, a photoionization channel corresponding to a given total angular momentum or a given electronic state can be subdivided into its multiplet, vibrational, and sometimes rotational components. The intensity distribution of these components can reveal specific electron correlations, such as the strong configuration interactions for the ^5S component of Mn^{3+} $3s3p^63d^5$,[178] and autoionization channels,[120–124] as well as vibration–vibration interactions (Fermi resonances).[179]

7.1. Atoms

Relative partial cross sections were first measured for krypton,[180] and absolute cross sections σ_{0f} were reported for neon in 1973 in the first complete partition of the photoabsorption cross section into its components over a wide energy range.[175] The partition of neon is depicted in Figure 14, showing the contributions

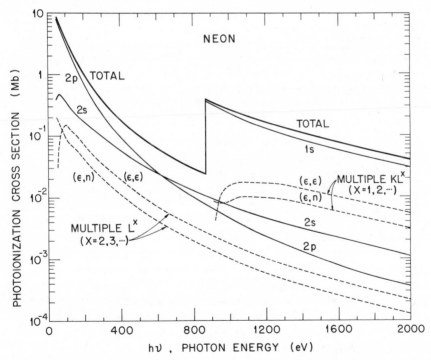

Figure 14. Partition of the photoionization cross section of neon. Shakeoff processes are denoted by $(\varepsilon, \varepsilon)$ and shakeup processes by (ε, η) (from Wuilleumier and Krause[175]).

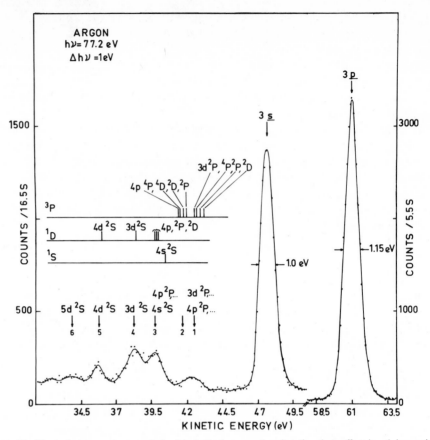

Figure 15. Photoelectron spectrum of the *M* shell of argon, showing the photoeffect involving a single electron (3s or 3p) and two electrons (structure below 44 eV) (from Adam *et al.*[199]).

of the various single and multiple processes. Due to the simple energy dependences of the partial and total cross sections of neon, the limited number of available conventional discrete photon sources posed no problems but would cause serious limitations in a complex situation such as that presented in Figure 1 for xenon.

Recently, helium,[181, 183] argon,[3, 116, 182, 183] and xenon[182–185] have been studied from threshold to about 130 eV. In addition, the earlier neon partition[145, 175] was completed from threshold to 110 eV.[144] In these studies, ESSR provided the data for single ionization and shakeup events, and ion charge spectrometry provided the data for double ionization. Parallel to the ESSR work, determinations of σ_{0f} were made by the dipole (*e, 2e*) technique[186] over about the same energy range, and by photoelectron spectrometry with discharge lamps[187] at energies up to about 40 eV. Unfortunately, partitions did not always take into account all types of multiple processes, although their occurrence had been demonstrated before.[26, 175, 188, 189]

In Figure 15 the photoelectron spectrum of the *M* shell of argon is presented. From this type of spectrum observed at 54.7° [equation (18)], and from the

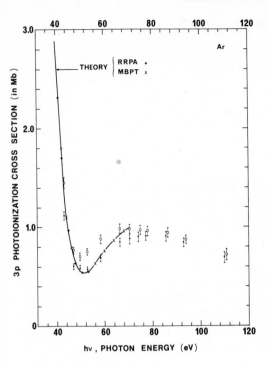

Figure 16. Partial photoionization cross sections of the $3p$ level of argon. MBPT[192] and RRPA[193] results are compared with experimental data[182] normalized to the total cross sections reported by Samson (open squares)[191] and Marr and West (closed circles)[143] (from Adam et al.[182]).

measured probability for double ionization,[190] the partial cross section σ_{3p} corresponding to the peak labeled $3p$ was derived by normalization to the total cross section σ_{tot}. Two sets of σ_{tot} values were used for normalization, and the discrepancy noted between the two sets[143, 191] below 70 eV reappears in the partial cross section σ_{3p}, plotted in Figure 16. On the other hand, two calculations[192, 193] using different many-electron models agree excellently with each other and generally with the experiment. It appears that we are faced with the paradox that the accuracies of the experimental results are limited by the uncertainty in the *total* photoionization cross section σ_{tot} and not in the *relative partial* cross sections σ_{0f}. A similar observation has been made for neon[144] and xenon.[182] Part of the problem can be ascribed to the presence of a poorly known amount of multiple-order radiation in absorption measurements. However, with the help of photoelectron spectrometry, multiple-order contributions can now be quantitatively accounted for (see Section 4), so that, in turn, improved cross-section values are obtainable.

Figure 17 shows a demanding test case of both experimental[3, 116, 182, 186] and theoretical[37, 192–196] results. The cross section σ_{3s} of argon is small at low energies and, because of a node in the ground-state wave function, goes through a Cooper minimum, which is deeper than that for σ_{3p} because only the $l + 1$ channel is available. In the central field approximation, σ_{3s} goes through zero exactly, but in the relativistic case a zero does not occur because the two spin–orbit channels $\varepsilon p_{3/2}$ and $\varepsilon p_{1/2}$ have their respective zeros at slightly different energies. Also, zero in σ_{3s} does not occur in nonrelativistic many-body calculations because of interchannel

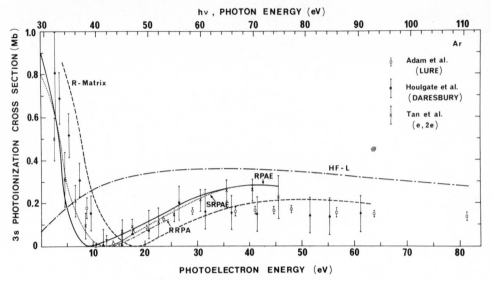

Figure 17. Partial photoionization cross section of the 3s level of argon in the vicinity of the Cooper minimum. Theoretical results, HF-L,[37] *R* matrix,[194] RPAE,[195] SRPAE,[196] and RRPA,[193] are compared with ESSR results from LURE[109, 182] and Daresbury,[116] and (*e*, 2*e*) data by Tan *et al.*[186] (from Adam *et al.*[182]).

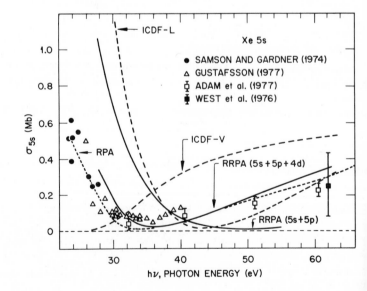

Figure 18. Partial cross section of the Xe 5s level near the Cooper minimum. Theoretical results[197, 198] are compared with experimental data by Samson and Gardner,[187] Gustafsson (90° data),[185] Adam *et al.*,[182] and West *et al.*[184]

coupling, and this behavior is of course maintained in the relativistic many-body model.[197] Thus, it is easily understood that the region of a Cooper minimum constitutes a most sensitive testing range for theoretical models as relativistic and many-electron effects appear amplified in σ_{0f} and also in β_{0f}, as seen below. Positions of minima respond strongly to the model used; for example, the HF model moves the minimum below threshold, while the R-matrix model moves it too far above threshold. Not unexpectedly, the RRPA calculation exhibits the best agreement with what appears to be the most accurate data. This conclusion can also be drawn from the comparison of σ_{5s} of xenon presented in Figure 18. There the agreement between various experimental results[182–184, 195, 198] is slightly better than in the case of argon, so that the findings are placed on more solid grounds. Figure 18 demonstrates how critical the inclusion of the proper number of electron–electron interactions is in the calculation of σ_{0f} near a Cooper minimum.

In spite of the excellent agreement at the Cooper minimum between experiment and the random-phase (with exchange) calculation, either relativistic (RRPA) or nonrelativistic (RPA), discrepancies occur at higher energies, beyond the onset of two-electron processes. According to the comparison in Figure 19, the RPA model,[198] done with frozen and relaxed cores, overestimates the experimental σ_{5s} value considerably, but overestimates the sum $\sigma_{5s} + \sigma_{\text{sat}}$ to a lesser degree. Whether this might indicate that the RPA results implicitly contain this type of two-electron interaction is an open question at present. From experiments on argon,[199, 200] krypton,[200] and xenon,[38, 200–202] and from explicit calculations of two-electron transition rates,[86, 203] it is known that the $ns\,np^6(^2S)$ configuration mixes strongly with the $ns^2np^4nd(^2S)$ configuration. This case was presented in Figures 6 and 15.

While the rare gases have been, and continue to be, favorites in ESSR studies, the research groups at Daresbury and Orsay have recently embarked on ESSR studies of metal vapors. Partial cross sections for the outer levels have been obtained for Cd,[204] and Hg[204] up to $h\nu = 43$ eV, and for Pb[130, 205] up to

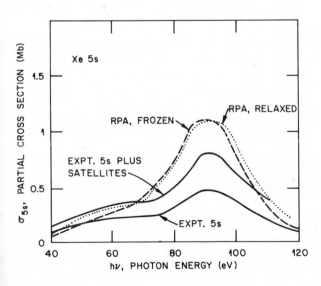

Figure 19. Photoeffect in the 5s shell of xenon. Theoretical RPA predictions[198] are compared with experimental data[183] for the single-electron process and for the sum of single and double processes, but excluding double *ionization*.

$h\nu = 106$ eV. Earlier measurements[206] made with line sources from discharge lamps are thus augmented in a systematic fashion. Photon flux was high enough that a Cooper minimum could be observed in Hg for the 6s level.

7.1.1. *Spin–Orbit Photoelectron Intensity Ratios*

In principle, intensity ratios of spin–orbit components contain less information than the individual partial cross sections. Factors that affect both components equally or similarly may easily be concealed in the ratios. Thus, tests of theoretical models by way of these ratios lack the definitiveness of tests concerning the partial cross sections. However, measurements of intensity ratios are valuable, for they can be made readily and accurately, and can probe interactions that affect the components dissimilarly. To be sure, statistical $(l + 1)/l$ ratios are not to be expected. At low energies, the ratio is generally greater for d and f levels, because cross sections are rising above the threshold, and lower for p levels, because cross sections are falling.[37] This is mostly an energy effect, but it reflects also the different spatial extent of the wave functions of the doublet components.[207] As photon energies increase, generally beyond the domain of ESSR, all spin–orbit intensity ratios fall increasingly below the $(l + 1)/l$ ratio, since the interaction takes place in the inner regions of the atom where the low-spin wave function penetrates deeper than the high-spin wave function. This effect is accentuated in the heavy atoms in which relativistic effects are pronounced, as evidenced in the increased spin–orbit splitting.

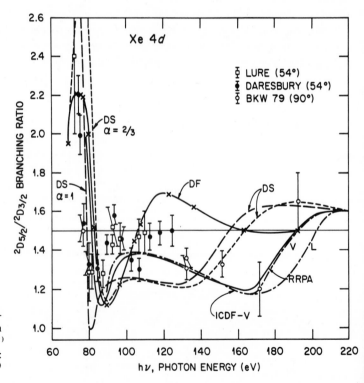

Figure 20. Spin–orbit photoelectron intensity ratio for Xe 4d. Experimental data from LURE,[109,209] Daresbury,[183] and BKW[210] are compared with theoretical results: DS,[209] DF,[212] ICDF-V,[211] and RRPA.[211]

The DS calculation by Scofield[34] places these qualitative statements on a quantitative basis.

Many data on spin–orbit intensity ratios have been reported. For example, the xenon $5p$[208] and $4d$ doublets have been studied over a wider range. Figure 20 shows the $4d$ data obtained with synchrotron[109, 183, 209] and x-ray[210] sources and the theoretical results[209, 211, 212] that range from the simple DS model[209] to the sophisticated RRPA model.[211] The RRPA and ICDF calculations are in good agreement with the experimental data. However, only the RRPA predictions are in good accord with the individual partial cross sections.

7.1.2. Two-Electron Transitions

Shakeup or correlation satellites have been the subject of intense studies with conventional photon sources[27, 38, 200, 213] in the past, but have been investigated by ESSR only recently and only at Orsay. Satellite spectra have been reported for the outer levels of helium,[181, 214] argon,[199] xenon,[109, 183, 201] and lead.[130, 205] The argon spectrum is shown in Figure 15. Since in these cases, FSCI is a dominant route via the $ns \, np^6(^2S) \leftrightarrow ns^2 np^4 nd(^2S)$ configuration interaction as indicated in Figures 6 and 15, intense satellites are produced. Dependence of satellite intensities on the photon energy was observed for both argon and xenon between $hv \approx 60$ eV and $hv \approx 90$ eV, and differences between spectra obtained at low (60–150 eV) and high (1–2 keV) excitation energies were noted. This permitted assignments of the final states and most probable pathways to be made for a number of peaks that on energetic grounds alone resist decomposition. For example, peaks 1 and 2 in Figure 15 could be associated with photoionization in the $3p$ level rather than the $3s$ level of argon.[199] These pilot studies of two-electron transitions by ESSR indicate that shakeup satellites can now be investigated as a function of photon energy, thus affording a better insight into the nature of electron correlation.

7.2. Threshold Laws

The behavior of cross sections just above threshold is of fundamental interest; the cross sections at threshold can be derived rigorously from theory if two particles are created in the collision of two particles.[215] An experimental test of threshold law has been made for photodetachment[216] in which a neutral and a charged particle are produced. In that experiment laser excitation and high-resolution apparatus provided a resolution of about 1 meV over a sufficiently wide range starting right at threshold. It now appears feasible to probe the close vicinity of thresholds of outer levels to determine the range of validity of threshold law for photoionization. In this case two particles of opposite charge are formed and the cross section has a finite value at threshold. A threshold electron-energy analyzer in combination with a synchrotron radiation source has been shown to possess good sensitivity and a resolution of 2 to 3 meV over the energy range of interest.[120] So far, the method has been applied to the Ar $3p$ level, but only for the purpose of studying autoionization features (see Section 9).

7.3. Quasi-Atomic Systems

Atomic features in the photo*absorption* spectra of solids were demonstrated some time ago,[68, 217] most notably for the $3d$ transition-series metals and lanthanides in which $3d$ and $4f$ levels, respectively, display pronounced atomic characteristics. Corresponding photo*ionization* studies in which the partial cross sections are measured by ESSR have recently been reported. These data[218] highlighted the atomic behavior of σ_{4d} of In and Sb from threshold to 350 eV. Within the accuracy of experimental and theoretical[37] results, the single-particle HF calculation was found to give a satisfactory description of the quasi-atomic system. On this basis, one is inclined to state, *cum grano salis*, that many-body effects once thought to be all important in solids in the ultrasoft x-ray range are quite small, while many-body effects once thought to be insignificant in free atoms are found to be substantial at low energies.

In Chapter 6, such quasi-atomic systems are treated in full; here, another interrelation between atomic and solid state systems is pointed out. Electrons in their conduction and valence bands can be characterized in terms of their quantum numbers s, p, d, etc., and their distributions can be obtained by recording the band structure as a function of photon energy and matching the photoelectron spectra with spectra calculated from the density (the variable) and the partial cross sections of the components with different l.[111] But these cross sections, to be reliable, must be measured first in atoms where the levels with different l are separated in energy. In 1971, for the first time, advantage was taken of quasi-atomic properties in solids to locate f electrons within the bands of EuS, GdS, and US.[219]

7.4. Molecules

Traditionally, electron spectrometry of molecules has restricted itself to energy measurements. Progress in partial cross-section measurements of valence levels has been slow for lack of adequate theory and suitable photon sources over a wide range. At the same time, corresponding studies of core levels have shown that molecular cross sections at higher energies are the sums of the atomic cross sections of the constituents.[220] The idea that molecular orbitals can be described as linear combinations of atomic orbitals has been put to good use in determining the parentage of molecular orbitals.[5, 38] This idea implies that the cross section of a molecular orbital will vary with photon energy approximately like the sum of the individual atomic cross sections. The spectra of N_2, Figure 21, illustrate that this approximation is of great value in the identification of orbitals and their composition.[221] For example, the intensities of the $\sigma_u 2s$ state and the $\sigma_u 2p$ state of N_2 are reversed when going from 132 eV to 1.5 keV, just as in the atomic case in which s and p levels show the same trend (see Figure 14). By necessity, the approximation is crude as it neglects such basic properties as the noncentral field of the molecule and the inevitability of strong correlations among the closely spaced levels. A more quantitative experimental approach depends on mapping out partial cross sections with a tunable photon source.

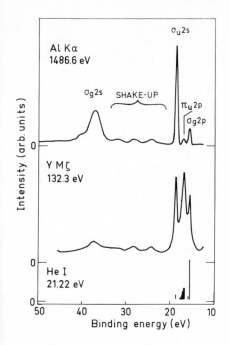

Figure 21. Photoelectron spectra of valence levels of nitrogen excited by photons of greatly different energies (from Nyholm *et al.*[221]).

Within a year's span, partial cross sections of N_2 and CO were measured by ESSR,[153, 222] by the dipole $(e, 2e)$ technique,[223] and with gas discharge lamps,[177] and were calculated[224] by a realistic method such as the scattered wave $X\alpha$ method. The theoretical result and one of the experimental results are compared in Figure 22. Note that those data that are not plotted are in essential agreement with those shown. Theory agrees remarkably well with experiment, if we consider the single-particle nature of the model. The success achieved here might be likened to that seen earlier for atoms when the single-particle HS model was introduced.[33, 68, 225] The major discrepancies seen in Figure 22 are presumably due to the presence of shape resonances, to be discussed in Section 9.

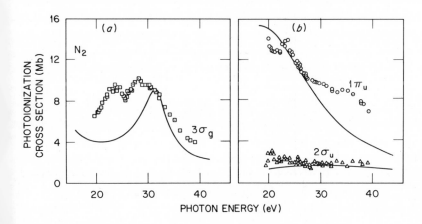

Figure 22. Partial photoionization cross sections of the three electronic states of nitrogen. [Data points are from Plummer *et al.*[222] and theoretical results are from Davenport[224] (solid lines).]

At sufficiently high photon energies, the vibrational intensity distribution within an electronic state is independent of energy and is described by the Franck–Condon principle.[226] However, at low photon energies the distribution is subject to variations because of the competition between autoionization and direct ionization processes. At the thresholds proper, the Franck–Condon principle presumably becomes a poor approximation, so that the intensity distribution could look quite different even in the absence of autoionization. Recordings of the intensities of the vibrational levels have been made for N_2 at low energies.[120, 227] In Figure 23, a survey is presented, showing the two extreme cases of threshold excitation and "high-energy" excitation. While the data at $h\nu = 21.2$ eV agree well with the Franck–Condon prediction, the threshold spectrum looks quite different and emphasizes the higher vibrational levels. The similarity of the threshold spectrum with

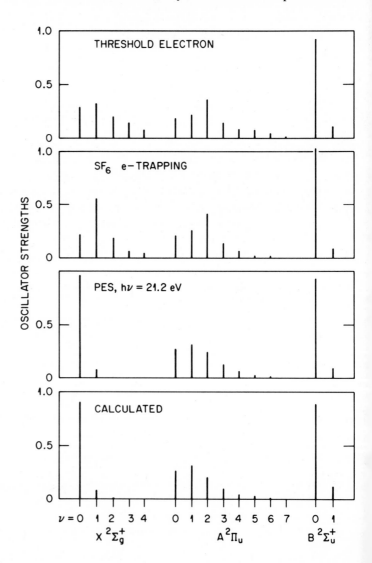

Figure 23. Vibrational intensity distribution in nitrogen excited at threshold, 0.46 eV above threshold, and by 21.2 eV photons. Also shown are the Franck–Condon factors (from Peatman *et al.*[120]).

that obtained at 0.46 eV above threshold by the SF_6 scavenger technique is probably accidental, as indicated by the ESSR study[153] that used a relatively fine mesh from threshold to above 20 eV.

8. Angular Distributions of Photoelectrons

In many ways, photoelectron angular distributions are more specific probes of photoionization dynamics than are the partial photoionization cross sections. They respond sensitively to the relative phases of the outgoing waves, whereas σ_{0f} depends only on the magnitudes of the matrix elements or amplitudes. Thus, the final states of the system, including autoionization and shape resonances, are examined in detail as described especially clearly in the j_t momentum transfer formalism (see Section 3). Furthermore, relativistic[197] and multipole[74] effects are emphasized in angular distributions, and the degeneracy of levels and bands with different angular momentum constituents can be resolved.[158, 228] But even in cases where no more information is gained from the differential cross section than from the integrated cross section, a measurement of both is valuable to assure internal consistency.

8.1. Closed-Shell Atoms

To date, angular distribution measurements by ESSR have been made only at Daresbury. This work has covered many of the levels whose wave functions have nodes and hence exhibit Cooper minima. Data were obtained for photoelectrons of energies up to about 100 eV from the outer p levels of the rare gases and Ar $3s$, Xe $4d$, and Xe $5s$. Thus the body of work accumulated during the past ten years with conventional photon sources could be extended considerably. Since in all of the cases covered so far by ESSR the dipole approximation is valid, the angular distribution is characterized by a single parameter, B_2, or the asymmetry parameter β (see Section 3.3).

For Ne $2p$[229] the parameter β varies slowly with energy, rising from somewhat above threshold to about 150 eV and then dropping off slowly. Only very close the threshold does β undergo a rapid variation. This sharp change was shown to be a general feature, for all Z and all l, due to a rapid variation in the Coulomb phase shift.[28, 29] Experimental results agree well with single-particle model calculations, such as HF, indicating that correlation effects are small for Ne $2p$. Nevertheless, the RPA prediction,[195] which includes correlations, results in the best accord with the most accurate determinations of β.

For Ar $3p$,[3, 116] Kr $4p$,[230] Xe $5p$,[231, 232] and Xe $5s$,[233] the data display dips, characteristic of the occurrence of a Cooper minimum. Since at the minimum $\beta = 0$ in the central field approximation and $\beta \approx 0$ in the many-body formulation, the recording of β as a function of energy offers a sensitive approach to pinpointing the locus of the minimum, and by that an excellent test of the various predictions. The RPA results[195] for Ar $3p$ agree well with the data, but for Xe $5p$, plotted in

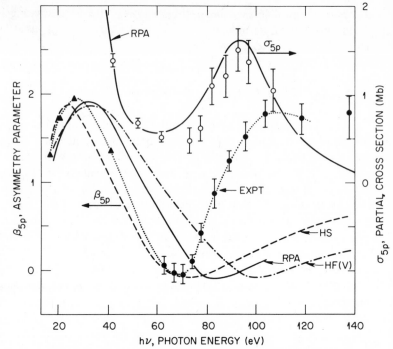

Figure 24. Asymmetry parameter β and partial cross section of Xe $5p$. RPA results are from Amusia *et al.*,[195, 198] HS from Manson and Cooper,[33] and HF(V) from Kennedy and Manson[37]; open circles measured by Adam *et al.*,[182] triangles by Dehmer *et al.*,[232] and solid circles by Torop *et al.*[231]

Figure 24, this is no longer the case. Evidently the particular correlations included in these calculations are sufficient for Ar $3p$, but not for Xe $5p$ or for Kr $4p$. Discrepancies are greatest in the region of d-electron excitation. Although the intershell correlations between $5p$, $5s$, and $4d$ are included in the calculation and lead to a good account of the partial cross section σ_{5p}, they fail to reproduce β correctly. Presumably the details of the interactions and perhaps the failure to include the virtual $5d$ states are responsible for the discrepancies. In comparing the β and σ curves of Figure 24, we note that (a) the minima occur at the same energy for the experimental data but not for the RPA results, and (b) the experimental data of β and σ do not track near the maximum between 90 and 100 eV. Evidently there is need for more extensive measurements and for further improvements in theory.

In Figure 25, the point is made that a single datum,[233] and perhaps two data, can accommodate without difficulty theoretical predictions from different models. This makes a definitive choice difficult, unless we select the most sophisticated model on general grounds, or unless we consult results of analogous properties. In the present case, a comparison between Figure 25 and Figure 18 is advisable.

Data on Xe $5s$ give evidence of relativistic effects that allow for the two continuum channels, $\varepsilon p_{1/2}$ and $\varepsilon p_{3/2}$. If A_1 and A_2 denote the amplitudes for these channels, then the partial cross section is given by[197, 208]

$$\sigma_{ns} = \tfrac{1}{3}|A_1|^2 + \tfrac{2}{3}|A_2|^2 \qquad (22)$$

and the asymmetry parameter β by

$$\beta_{ns} = 2 - 2\{|A_1 - A_2|^2/(|A_1|^2 + 2|A_2|^2)\} \qquad (23)$$

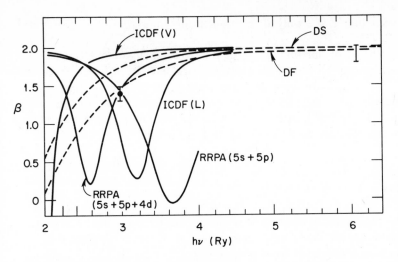

Figure 25. Asymmetry parameter β of Xe 5s in various theoretical models: DS,[208] DF,[233] and ICDF and RRPA[197]; datum at 40.8 eV by Dehmer and Dill,[233] and at 85 eV by Torop et al.[231]

This shows that, as stated in Section 7.1, σ_{ns} does not vanish at a Cooper minimum because $A_1 = 0$ and $A_2 = 0$ at different energies. It also shows that β_{ns} is generally smaller than the nonrelativistic value of 2 where $A_1 = A_2 = A$.

The peculiar ways by which multipole contributions enter angular distributions was demonstrated previously[74, 180] for the Kr-*M* shell at relatively high photon energies. Although the photoelectrons from the three subshells have practically the same energy, distributions of 4s and 4p photoelectrons were found to be symmetric, but that of 4d electrons skewed in the forward direction. The presence[74] of multipole effects of similar magnitudes for other elements at energies of a few hundred eV awaits experimental verification.

8.2. Open-Shell Atoms

Angular distributions of photoelectrons from open-shell atoms are governed by anisotropic interactions between the outgoing electrons and the residual ion.[46] The photoelectron and the ion can exchange angular momentum [equations (2) and (16)], leading to several interfering channels. Hence, each channel corresponding to a possible j_t has its own β value, and most importantly none of these β values are identical with the central field (Cooper–Zare) prediction, in which no allowance is made for momentum transfer. This has the important consequence that $\beta \neq 2$ for s electrons[47] even in the absence of relativistic effects,[197, 208, 234] which provide additional interactions between the spin–orbit states in both the discrete and continuum space.

Consider photoionization of an *ns* level in an atom with an open *np* subshell; according to equation (16), j_t can assume the values of 0, 1 and 2, since $j_y = 1$ and $l = 1$. If the matrix elements for these three possible exit channels are denoted $S(0)$, $S(1)$, and $S(2)$ respectively, the asymmetry parameter is given by[47]

$$\beta = \frac{2|S(0)|^2 - 3|S(1)|^2 + |S(2)|^2}{S|(0)|^2 + 3|S(1)|^2 + 5|S(2)|^2} \tag{24}$$

and it can be seen that $\beta = 2$ only if no momentum transfer occurs: $j_t = 0$ and hence $S(1) = S(2) = 0$.

So far, virtually all of our knowledge of β in open-shell systems comes from theoretical considerations.[81] Only two measurements have been reported for the oxygen atom,[235, 236] and both were obtained with discharge line sources. Agreement between measured and calculated values was found satisfactory for the transition $O[2p^4(^3P)] \rightarrow O^+[2p^3(^4S^0)]$ at $hv = 21.2$ and 40.8 eV. Doubtless, more data at narrower energy intervals are needed to confirm this finding.

8.3. Molecules

Much of the discussion of open-shell atoms applies to molecules as well. Anisotropic interactions are the rule since dynamic coupling between the photoelectron and residual ion is not restricted to electronic states, but can include vibrational and rotational substates.[44, 81, 237, 238] In addition, the inherently noncentral field of a molecule causes anisotropic effects.[81, 239] These will be enhanced upon photoionization if removal of an electron leads to a lowering of the molecular symmetry.[240]

Few measurements of β have been carried out at energies other than those of the strong He I and Ne I resonance lines. Theoretical treatments have been more comprehensive in their approach than experiments, but less reliable because of the relatively crude approximations used so far. First ESSR measurements[241] of the β parameters of the various ionic states of N_2, O_2 and CO have been made in the energy range $20 < hv < 45$ eV. The asymmetry parameter was found to vary significantly over this energy range but seldom to exceed $\beta = 1$. Theory was shown to be generally at variance with these experimental data.

In studies with laboratory sources, especially the He I lamp, the β parameter has been measured for the various vibrational levels of a given ionic state of a molecule.[242] In many cases, a small and monotonic change in β was observed, indicating a small energy dependence of β over the 1-eV range over which an ionic state may extend. However, in the cases of the first ionization states of N_2 and CO, the asymmetry parameter showed a strong dependence on the vibrational level.[243] Although autoionization, shape resonances, or a breakdown of the Born–Oppenheimer approximation have been invoked as the cause for this phenomenon (see Section 9), a definitive interpretation is still lacking and may have to await the results of measurements with a finely tunable photon source.

9. Resonances and Autoionization

In a simplified description, resonances take place when a new channel opens up for the outgoing electron in addition to the already open channel. This can be a discrete state embedded in the continuum leading to autoionization,[79, 244] or the rapid penetration of an l component of the continuum wave function through a centrifugal barrier into a potential well leading to a shape resonance.[43, 240] Resonances are confined to narrow photon energy intervals, which can be extremely narrow for autoionization events. Another class of resonances involving degenerate bound states is independent of the photon excitation energy, but leads to an extended photoelectron energy distribution. Evidence for such a resonance is seen in Figure 2, where the Xe $4p$ photoline is spread by a strong interaction between the $4p$ and $4d^2$ states, permitting the electron to escape into $4dnf$ and $4d\varepsilon f$ channels.[55]

Resonances occur both in atoms and molecules.* They are prevalent in molecules and can be induced electronically, vibrationally, and rotationally. In resonance regions the cross section and asymmetry parameter β are subject to rapid variations. Variations in the total cross sections have been known for a long time from absorption measurements[68]; variations in vibrational intensity distributions and β have been delineated more recently.

9.1. Atoms

Evidence for changes in the β value was first presented for Hg[245] and then for Xe[246] in experiments that used line sources. Experimental and theoretical results for Xe are shown in Figure 26 in the region of autoionization between the $5p$ spin–orbit components. Fluctuations in β, and in σ, are caused by the interaction of the $5p_{3/2}$ continuum channel with the discrete ns and nd states, which converge toward the $5p_{1/2}$ level. Good agreement between theory and experiment can be seen to exist, except that the sharp spikes in β predicted by theory are not reproduced by experiment because of limited resolution.

The analogous resonance in Ar was studied with a laboratory source,[168] and the resonance due to the inner $5s5p^66p(^1P_1)$ state in Xe[247] was measured by ESSR. In these studies, measurements were made only of the $\sigma_{3/2}/\sigma_{1/2}$ ratios of the outer p levels through the resonance region. Rapid and pronounced fluctuations were observed, which in the case of argon are indicative of changes in $\sigma_{3/2}$. The work on argon is remarkable in that an extremely narrow resonance structure could be scanned due to an instrumental resolution of 2–3 meV.

A strong resonance was reported to occur in barium excited by 21.22 eV photons,[248, 249] and was tentatively interpreted as being due to the autoionization of the excited $5p^5(^2P_{1/2})6s^2nd$ state into the $5p^5(^2P_{3/2})6s5d$ state. However, a recent ESSR study[249a] of barium points to a more complex intermediate state ($5p^55d6s11s$ in simplified notation) that autoionizes into the simple [Xe core]$11s$ state via the Coster–Kronig transition $5p^55d6s \rightarrow 5p^6 + e^-$. In the same study, resonances at 19.94 and 21.48 eV, corresponding to the $[\{5p^55d(^3D)6s(^2D_{3/2})\}nd]_1^0$, $n = 5$, 6, states, were shown to decay into the [Xe core]nd states by the same Coster–Kronig transition. Interestingly, the observed electron lines are very symmetric, which implies that *direct* two-electron transitions to these final states from the ground state are either weak, or weakly coupled with the indirect channel, or both.

---

* It is worthwhile to note a difference in language used in describing autoionization structure: in atoms, we speak of continuum features (the photoionization cross section) being perturbed by the presence of discrete states; in molecules, we speak of discrete features (the vibrational intensity distribution) being perturbed by the presence of continua.

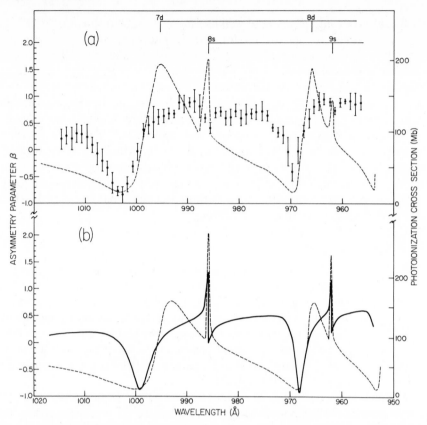

Figure 26. Variation of the asymmetric parameter β in the region of autoionization between the Xe $5p_{3/2}$ and Xe $5p_{1/2}$ levels. Data for β are from Samson and Gardner[246] (closed circles) and predictions of β are from Dill[79] (solid line). For comparison, experimental (a) and theoretical (b) cross sections are shown; dotted lines from Samson and Gardner.[246]

9.2. Molecules

Vibrational spectra of electronic states of molecules have been recorded at different excitation energies using discharge lamps.[227] Anomalies in the intensity distributions, that is, deviations from the Franck–Condon envelope, were found in many molecules, most notably NO and O_2, and attributed to autoionization events.[227, 250] Only for N_2 has a continuous scan of cross sections for individual vibrational states by ESSR been made.[153] Autoionization processes that release zero-energy electrons have been studied for a number of molecules by the PIR technique using both discharge continuum and synchrotron radiation sources.[120–123, 179]

Resonances are not restricted to the region of valence levels; they can also occur near threshold in inner-shell photoionization. Resonances associated with K-shell ionization in N_2 and CO have been investigated theoretically[43, 240] and experimentally by the dipole $(e, 2e)$ technique.[251] Due to the anisotropic nature of

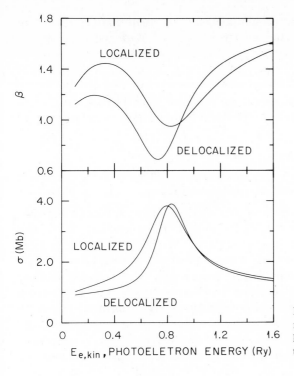

Figure 27. Effect of shape resonance above the K shell of nitrogen on the asymmetry parameter and cross section. Hole localization or delocalization is accentuated in β according to theory (from Dill et al.[240]).

the molecular field, the escaping electron can assume different angular momenta, and if a given l wave function is strongly localized at a certain photoelectron energy, a wide exit channel is provided for the photoelectron resulting in an enhancement of σ_{0f}. Figure 27 shows the theoretical σ_{0f} results, which are in satisfactory agreement with experiment. The variation of σ_{0f} has its counterpart in the variation of β. However, the energy dependence of the asymmetry parameter β responds more sensitively than σ_{0f} to whether or not the K hole is localized, because localization and delocalization can lead to different molecular symmetries and hence to a difference in the anisotropic interactions. Thus the asymmetry parameter β of inner levels becomes an important avenue to basic molecular properties. The variation of β for K ionization of oxygen in CO was predicted to be similar to that for the K shell of N_2.[240]

10. Post-Collision Interactions

In a special class of final-state interactions, two or more charged particles coexist in the continuum for a short time and exchange energy through their Coulomb field. It has become customary to denote these processes as post-collision interactions (PCI). PCI is a long range interaction, up to about 100 Å, and as such is described rather well by classical and semiclassical models. Post-collision interactions manifest themselves in line shifts, asymmetric line broadening, back scattering of electrons into the residual ion, and interference structure.[252]

Evidence for PCI was first seen in autoionization of neutral atoms excited by ion[253] and electron[254] impact and in Auger spectra[255] excited by electrons. Later studies involved the ion charge spectrum created in the ionization of the argon $2p$ level[170] and the Auger spectrum following ionization of the xenon $4d$ level.[256, 257] In the latter cases, which were studied respectively by the $(e, 2e)$ technique and by ESSR, well defined conditions prevailed since the excitation energy was known exactly and could be varied continuously. In the ESSR study, represented schematically by

$$hv + \text{Xe} \rightarrow \text{Xe}^+(N_{4,5}) \quad\quad + e_{\text{photo (slow)}}$$
$$\downarrow \quad\quad\quad \updownarrow \quad \text{PCI} \quad\quad\quad\quad (25)$$
$$\text{Xe}^{2+}(O_{2,3}O_{2,3}) + e_{\text{Auger (fast)}}$$

the $N_{4,5}O_{2,3}O_{2,3}$ Auger electrons were recorded at different photon energies, ranging from 0.8 to 43 eV above the N_5 threshold. Spectra obtained at the two extreme excitation energies are displayed in Figure 28, and the N_5 spectral lines, especially the $N_5O_{2,3}O_{2,3}\,{}^1S_0$ line, can be seen shifted and skewed toward higher energies for near-threshold excitation. The energy gain of the Auger electron, amounting to 0.14 eV at a photon energy of 0.8 eV above the N_5 threshold, occurs at the expense of the energy of the slow photoelectron. Energy exchange between the two free electrons, equation (25), is of course continuous and can range from 0 to 0.8 eV, but as seen from Figure 28 an energy transfer of 0.14 eV is the most probable. The probability function of the energy exchange, which was derived in a semiclassical model, was found to predict satisfactorily the shift of the peak position and the shape of the Auger electron line.[252] The same model also accounts well for the argon ion spectra altered by PCI. This is interesting to note, since the study of ion charges gives a direct measure of the probability of capture of the slow photoelectron into a discrete, bound state while the study of Auger energies relates primarily to the most probable energy exchange. However, a more detailed theory

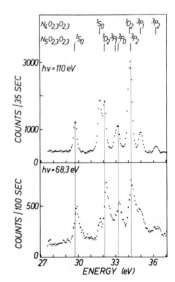

Figure 28. The xenon N_5 Auger spectrum excited by photons far above threshold (top) and 0.8 eV above threshold (bottom) (from Schmidt et al.[257]).

of post-collision interactions, especially those involving the capture of the slow electron, requires a quantum-mechanical approach and the inclusion of intershell correlations.[258] The possibility that angular momentum is exchanged in PCI in an analogous manner as that discussed in Section 3.3 for photoelectron–ion interactions has been considered theoretically,[252] but has not yet been demonstrated experimentally.

11. Photoexcited Auger Spectra

Auger spectra consist of different categories of Auger lines: the diagram lines characteristic of a single initial inner-shell vacancy and satellite lines characteristic of (a) multiple initial vacancies, (b) multiple final vacancies (double Auger process), and (c) the presence of spectator electrons.[259] As a rule, all categories of Auger lines are produced by the various possible excitation modes, but as demonstrated previously for neon[259] and nitrogen,[260, 261] only photoexcitation can provide those critical conditions that are needed for a complete analysis and interpretation of an Auger spectrum. With a tunable photon source, Auger spectra can be selectively excited so that certain categories of lines can be isolated and others can be suppressed. Accordingly, four energy regions may be distinguished, each possessing its own excitation and deexcitation characteristics: (1) the region below threshold, in which a core electron is promoted to an empty bound state giving rise primarily to satellites of the spectator type; (2) the region immediately above threshold, in which post-collision interactions are prevalent; (3) the region just below the onset of the two-electron processes, where diagram Auger lines and double Auger satellite lines are isolated; and (4) the region above the thresholds of two-electron processes, where the probabilities of initial two-electron transitions depend on the photon energy until reaching a constant high-energy value.

Most recently, the xenon $4d$ Auger spectrum has been studied at ACO in the first two regions,[130, 257] and both the Kr $3d$ and the Xe $4d$ spectra have been investigated at DESY in the first region.[262] In each case, a high-energy spectrum (region 4) was obtained for the purpose of reference and comparison. The Xe spectrum that was excited just above threshold is displayed in Figure 28, showing the PCI effects discussed in Section 10, and the reduction of the N_4-series line intensities at $hv = 68.3$ eV. However, the N_4 series could not be completely eliminated because of the presence of higher-order photons in the exciting beam. As an example of Auger spectra excited below threshold, the Kr $M_5 N_{2,3} N_{2,3}$ spectrum in the presence of a $5p$ spectator electron is shown in Figure 29. Another group of lines, peaks 1 to 4, is seen in Figure 29. This group appears offset from the main group by about 3.3 eV and is due to a double Auger event in which the spectator electron is promoted from the $5p$ level to the $6p$ level simultaneously with the emission of an Auger electron. The large probability of double Auger events indicated in Figure 29 is a manifestation of strong correlation effects in the outer regions of the krypton atom.

A summary of the energy shifts of Auger electrons as a function of photon energy is given schematically in Figure 30. The base line is the high-energy limit, which is reached not too far above threshold since PCI quickly subsides.[252] At least

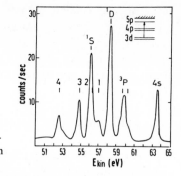

Figure 29. The krypton $M_5 N_{2,3} N_{2,3}$ Auger spectrum in the presence of a $5p$ spectator electron; photon excitation energy = 91.2 eV; the peak labeled $4s$ is a photoelectron
line (from Eberhardt *et al.*[262]).

for shallow levels, lifetimes of core hole states are long enough so that PCI effects are
negligible when the first two-electron excitation (shakeup) threshold is reached.
This implies that the diagram Auger spectrum can be isolated in the absence of
other complications (see Section 9) by photoexcitation in region 3.

Alignment of atoms ionized in an inner shell with $j > \frac{1}{2}$ has been studied by
way of Auger electron angular distributions obtained by electron impact.[263] So
far, measurements of such angular distributions following photoionization have not
been carried out. However, ESSR makes it feasible to study the alignment as a
function of incident photon energy and thus gain information on atomic dynamics,
including correlation effects, that are complementary to studies of partial photoionization cross sections and photoelectron angular distributions.[264]

12. Coincidence Experiments

The selectivity inherent in ESSR can be further increased by coincidence arrangements. The potential of coincidence experiments becomes especially apparent in
studies of complex systems in which a multitude of excitation as well as deexcitation channels is available. In these situations, a given excitation channel can be

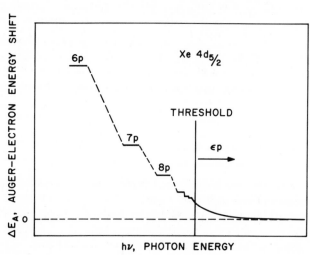

Figure 30. Shift of Auger lines as a function of
photon excitation energy. Only transitions to the
$l' = l - 1$ channel are indicated.

uniquely correlated with a deexcitation channel, thus defining the pathway from the ground state to a selected final state. Naturally, the study of molecules profits most from coincidence experiments. Such experiments may involve photoelectron–ion pairs, Auger or autoionization electron–ion pairs, photoelectron–Auger electron pairs, and photoelectron–fluorescence photon pairs. Ions may be analyzed according to their mass *and* energy, and in the simplest coincidence experiments, particles of either kind are collected over the largest possible solid angle. In more refined experiments, particles are directionally selected and angular correlations are established.

So far, only ions and fluorescence photons have been placed in coincidence with photoelectrons, produced from molecules by either continuous discharge or synchrotron radiation sources. Angular distributions of ions from the fragmentation of H_2^+ have been measured separately,[265] but not in coincidence with the photoelectrons emitted in the initial excitation process. The ESSR coincidence measurements of molecules were made at DESY,[123, 124, 266] ACO,[121, 122] and NBS.[267] In these experiments, ions and fluorescence photons were correlated with threshold electrons. Use of threshold (zero-energy) electrons has the advantage of high collection efficiency (up to 100%) and of specifying the exact amount of energy deposited in the molecule. However, the problem of whether the zero-energy electrons detected are photo- or autoionization electrons is not always easily resolved.

In the threshold electron–ion coincidence work on CO_2, it could be unequivocally shown that the $CO_2^+(\tilde{C}^1\Sigma^+)$ state always dissociates into the $O^+ + CO(\tilde{X}^2\Sigma_g^+)$ fragments.[123] In the work on O_2 the $b^4\Sigma_g^-$ state of O_2^+ was studied[121] in detail by placing threshold electrons from the vibrational excitations $v' = 3$–6 in coincidence with O^+ and O_2^+ ions. It could be demonstrated that the first ion dissociation limit, $O^+(^4S) + O(^3P)$, occurs within the $v' = 4$ vibrational band from the appearance of O^+ ions (which were absent in the $v' = 3$ coincidence spectrum). Because in this study the mass analyzer was of the TOF type, the ion peakwidth could be used as a direct measure of the kinetic energies of the fragment ion. For example, while the $b^4\Sigma_g^-(v' = 4)$ state produced a narrow O^+ peak, the $c^4\Sigma_u^-$ state produced a broad O^+ peak because of the kinetic energy release in that dissociation event.

In threshold electron–photon coincidence measurements of CO_2^+ [124] and N_2O^+,[266] lifetimes and fluorescence quantum yields of selected vibronic states could be determined with good accuracies.

In the work just described the effective electron-energy resolution ranged from 3 to 20 meV, and the coincidence count rate was sufficient to obtain reasonable statistics.

13. Outlook

In the few years since electron spectrometry was combined with synchrotron radiation, the new ESSR technique has made substantial contributions to a better understanding of atomic and molecular structure and dynamics. ESSR studies have also stimulated theoretical developments that culminated in the relativistic generalization of the random-phase approximation, or time-dependent Hartree–Fock model. Interplay between experiment and theory has led to a good accord in the outer regions of the atom for various properties, such as partial photoionization cross sections for single- and two-electron processes, photoelectron angular distributions, resonances, level widths, and level energies. But the agreement between theory and experiment has not always been perfect, especially in regions in which higher-l levels play a dominant role. Experiments have been limited to photon energies below 150 eV and, like theory, have usually been directed toward the determination of only one or two of the atomic properties, although partial cross sections, angular distributions, and electron correlation widths and energies are all closely interrelated.

No doubt, ESSR will soon be extended toward higher energies and measurements will include as many facets of the photoeffect as can possibly be determined at one time and with a particular setup. Most significantly, ESSR studies of closed-shell atoms will be augmented by studies of open-shell atoms. Such work is important, since our knowledge of open-shell atoms is sketchy and based largely on exploratory theoretical work.

Determinations of partial photoionization cross sections in free atoms have been, and will continue to be, important for their fundamental value. In addition, because of the often atomic behavior of outer electrons in solids, extensive mapping of partial cross sections will probably be needed for reference purposes.

Experiments in the vicinity of thresholds and resonances have proven valuable in delineating the nature of electron–electron or configuration interactions. Future improvements in instrumentation should allow the expansion of this type of work.

Selective photoexcitation of Auger electrons offers the opportunity of disentangling the various processes that occur during deexcitation of core vacancies. Work in this area has just begun.

ESSR of molecules is capable of extending and complementing the information that is available from measurements with discharge lamp radiation and other techniques, such as photoabsorption, photoionization mass spectrometry, and (e, $2e$) techniques. Threshold electron or zero-energy spectrometry is particularly well adapted to a continuously tunable synchrotron radiation source, displaying high resolution, high efficiency, and versatility. Direct and detailed studies of molecular excitation and especially deexcitation are now possible by coincidence measurements of threshold electrons with fragment ions and fluorescence photons. The results of the work done so far suggest that a large effort will be devoted to ESSR coincidence studies of molecules in the future.

This outlook, while giving a number of possible and probable directions of future ESSR studies, is by no means exhaustive. In fact, it has given only a small number of the possibilities the future holds.

ACKNOWLEDGMENTS

I am grateful to the many colleagues who provided me so generously with their latest results. Special thanks are due Dr. F. Wuilleumier for helpful discussions and for placing at my disposal many original figures. This work was sponsored by the U.S. Department of Energy, Office of Basic Energy Sciences, under contract W-7405-eng-26 with the Union Carbide Corporation, operator of the Oak Ridge National Laboratory.

References

1. P. Mitchell and K. Codling, *Phys. Lett. A* **38**, 31 (1972).
2. M. J. Lynch, A. B. Gardner, and K. Codling, *Phys. Lett. A* **40**, 349 (1972).
3. R. G. Houlgate, J. B. West, K. Codling, and G. V. Marr, *J. Phys. B* **7**, L470 (1974).
4. K. Siegbahn, C. Nordling, A. Fahlman, R. Nordberg, K. Hamrin, J. Hedman, G. Johansson, T. Bergmark, S. E. Karlson, I. Lindgren, and B. Lindberg, *Nova Acta Regiae Soc. Sci. Ups. Ser. IV* **20**, 1 (1967).
5. K. Siegbahn, C. Nordling, G. Johansson, J. Hedman, P. F. Hedin, K. Hamrin, U. Gelius, T. Bergmark, L. O. Werme, R. Manne, and Y. Baer, *ESCA Applied to Free Molecules*, North-Holland, Amsterdam (1969).
6. D. W. Turner, A. D. Baker, C. Baker, and C. R. Brundle, *Molecular Photoelectron Spectroscopy*, Wiley-Interscience, New York (1970).
7. W. Mehlhorn, D. Stalherm, and H. Verbeek, *Z. Naturforsch. Teil A* **23**, 287 (1968).
8. M. O. Krause, T. A. Carlson, and R. D. Dismukes, *Phys. Rev.* **170**, 37 (1968).
9. C. E. Kuyatt, in *Methods of Experimental Physics*, L. Marton (ed.), Vol. 7A, p. 1, Academic Press, New York (1968).
10. R. P. Madden and K. Codling, in *Autoionization: Astrophysical, Theoretical, and Laboratory Experimental Aspects*, A. Temkin (ed.), p. 129, Mono Book Cp., Baltimore, Maryland (1966).
11. R. Haensel, G. Keitel, and P. Schreiber, *Phys. Rev.* **188**, 1375 (1969).
12. R. P. Godwin, *Springer Tracts in Modern Physics*, Vol. 51, G. Höhler (ed.), p. 1–73, Springer-Verlag, Berlin (1969).
13. F. Wuilleumier, *J. Phys. (Paris) Colloq.* **39**, 1–71 (1978).
14. G. V. Marr, in *Electron and Photon Interactions with Atoms*, H. Kleinpoppen and M. R. C. McDowell (eds.), p. 39–67, Plenum Press, New York (1976).
15. K. Codling, *Rep. Prog. Phys.* **36**, 541 (1973).
16. F. Wuilleumier and Y. Farge (eds.), *Proceedings of International Conference on Synchrotron Radiation Instrumentation and New Developments*, North-Holland, Amsterdam (1978).
17. J. A. R. Samson, *Techniques of Vacuum Ultraviolet Spectroscopy*, Wiley, New York (1967).
18. C. E. Kuyatt, in *Electron Spectroscopy: Theory, Techniques, and Applications*. Vol. IV, C. R. Brundle and A. D. Baker (eds.), Academic Press, London (in preparation).
19. K. D. Sevier, *Low Energy Electron Spectrometry*, Wiley-Interscience, New York (1972).
20. T. A. Carlson, *Photoelectron and Auger Spectroscopy*, Plenum Press, New York (1975).
21. C. R. Brundle and A. D. Baker (eds.), *Electron Spectroscopy: Theory, Techniques, and Applications*, Vols. I and II, Academic Press, London (1977/79).
22. B. Crasemann (ed.), *Atomic Inner-Shell Processes*, Vols. I and II, Academic Press, London (1975).
23. F. J. Wuilleumier (ed.), *Photoionization and Other Probes of Many-Electron Interactions*, Plenum Press, New York (1976).
24. J. Berkowitz, in *Electron Spectroscopy: Theory, Techniques, and Applications*, C. R. Brundle and A. D. Baker (eds.), Vol. I, Chapter 7, Academic Press, London (1977).
25. W. A. Chupka, in *Chemical Spectroscopy and Photochemistry in the Vacuum Ultraviolet*, C. Sandorfy, P. J. Ausloos, and M. B. Robin (eds.), pp. 433–463, D. Reidel Publishing Co., Boston (1974).
26. M. O. Krause, in *Atomic Inner-Shell Processes*, B. Crasemann (ed.), Vol. II, pp. 33–81, Academic Press, London (1975).
27. M. O. Krause, in *Photoionization and Other Probes of Many-Electron Interactions*, F. Wuilleumier (ed.), pp. 133–163, Plenum Press, New York (1976).
28. S. T. Manson, *Adv. Electron. Electron Phys.* **41**, 73 (1976).
29. S. T. Manson, *Adv. Electron. Electron Phys.* **44**, 1 (1977).

30. J. A. R. Samson, *Phys. Rep.* **28**, 303 (1976).
31. W. Mehlhorn, in *Photoionization and Other Probes of Many-Electron Interactions*, F. Wuilleumier (ed.), pp. 309–330, Plenum Press, New York (1976).
32. E. H. S. Burhop and W. N. Asaad, in *Advances in Atomic and Molecular Physics*, A. Marton (ed.), Vol. 8, pp. 163–284, Academic Press, London (1972).
33. S. T. Manson and J. W. Cooper, *Phys. Rev.* **165**, 126 (1968).
34. J. H. Scofield, Lawrence Livermore Laboratory Report #UCRL-51326 (1973).
35. J. B. West and J. Morton, *At. Data Nucl. Data Tables* **22**, 103 (1978).
36. R. D. Deslattes, *Phys. Rev. Lett.* **20**, 483 (1969).
37. D. J. Kennedy and S. T. Manson, *Phys. Rev. A* **5**, 227 (1972).
38. U. Gelius, *J. Electron Spectrosc. Relat. Phenom.* **5**, 985 (1974).
39. D. Dill, in *Photoionization and Other Probes of Many-Electron Interactions*, F. Wuilleumier (ed.), pp. 387–394, Plenum Press, New York (1976).
40. D. Dill and U. Fano, *Phys. Rev. Lett.* **29**, 1203 (1972).
41. U. Fano and D. Dill, *Phys. Rev. A* **6**, 185 (1972).
42. D. Dill, A. F. Starace, and S. T. Manson, *Phys. Rev. A* **11**, 1596 (1975).
43. J. L. Dehmer and D. Dill, *J. Chem. Phys.* **65**, 5327 (1976).
44. D. Dill, *Phys. Rev. A* **6**, 160 (1972).
45. A. F. Starace, S. T. Manson, and D. J. Kennedy, *Phys. Rev. A* **9**, 2455 (1974).
46. D. Dill, S. T. Manson, and A. F. Starace, *Phys. Rev. Lett.* **32**, 971 (1974).
47. A. F. Starace, R. H. Rast, and S. T. Manson, *Phys. Rev. Lett.* **38**, 1522 (1977).
48. J. C. Slater, *Quantum Theory of Atomic Structure*, Vol. II, McGraw-Hill, New York (1960).
49. F. Herman and S. Skillman, *Atomic Structure Calculations*, Prentice-Hall, Inglewood Cliffs, New Jersey (1963).
50. C. F. Froese-Fischer, Comp. Phys. Comm. **4**, 107 (1972).
51. J. P. Desclaux, *Comput. Phys. Commun.* **9**, 31 (1975).
52. H. P. Kelly, in *Advances in Theoretical Physics*, Vol. II, p. 75, Academic Press, New York (1968).
53. D. Pines, *The Many-Body Problem*, Benjamin, Reading, Massachusetts (1962).
54. L. S. Cederbaum and W. Domcke, *Adv. Chem. Phys.* **36**, 205 (1977).
55. G. Wendin and M. Ohno, *Phys. Scr.* **14**, 148 (1976).
56. J. B. Mann and W. R. Johnson, *Phys. Rev. A* **4**, 41 (1971).
57. M. S. Freedman, F. T. Porter, and J. B. Mann, *Phys. Rev. Lett.* **28**, 711 (1972).
58. B. Fricke, J. P. Desclaux, and J. T. Waber, *Phys. Rev. Lett.* **28**, 714 (1972).
59. K. N. Huang, M. Aoyasi, M. H. Chen, B. Crasemann, and H. Mark, *At. Data Nucl. Data Tables* **18**, 243 (1976).
60. B. Breuckmann, Ph.D. thesis, University of Freiburg (1978) (unpublished).
61. A. J. Freeman and J. B. Darby, Jr. (eds.), *The Actinides: Electronic Structure and Related Properties*, Academic Press, New York (1974).
62. L. S. Cederbaum, J. Schirmer, W. Domcke, and W. von Niessen, *J. Phys. B.* **10**, L549 (1977).
63. E. Fermi, *Rev. Mod. Phys.* **4**, 87 (1932).
64. H. A. Bethe, in *Handbuch der Physik*, H. Geiger and K. Scheel (eds.), Vol. 24, Part 1, Springer-Verlag, Berlin (1933).
65. H. A. Bethe and E. E. Salpeter, *Quantum Mechanics of One- and Two-Electron Atoms*, Springer-Verlag, Berlin (1957).
66. J. W. Cooper, in *Photoionization and Other Probes of Many-Electron Interactions*, F. J. Wuilleumier (ed.), pp. 31–48, Plenum Press, New York (1976).
67. W. Heitler, *The Quantum Theory of Radiation*, third edition, Oxford University Press, New York (1954).
68. U. Fano and J. W. Cooper, *Rev. Mod. Phys.* **40**, 441 (1968).
69. R. H. Pratt, A. Ron, and H. K. Tseng, *Rev. Mod. Phys.* **45**, 273 (1973).
70. J. W. Cooper, in *Atomic Inner-Shell Processes*, B. Crasemann (ed.), Vol. I, pp. 160–201, Academic Press, London (1975).
71. M. Ya Amusia and N. A. Cherepkov, *Case Stud. At. Phys.* **5**, 47 (1975).
72. C. N. Yang, *Phys. Rev.* **74**, 764 (1948).
73. M. Peshkin, *Adv. Chem. Phys.* **18**, 1 (1970).

74. H. K. Tseng, R. H. Pratt, S. Yu, and A. Ron, *Phys. Rev. A* **17**, 1061 (1978).
75. J. Cooper and R. N. Zare, in *Lectures in Theoretical Physics*, S. Geltman, K. Mahanthajyra, and W. Brittin (eds.), Vol. XI-C, pp. 317–337, Gordon and Breach, New York (1969).
76. V. Schmidt, *Phys. Lett. A* **45**, 63 (1973).
77. J. A. R. Samson and A. F. Starace, *J. Phys. B* **8**, 1806 (1975).
78. K. C. Westfold, *Astrophys. J.* **130**, 241 (1959); F. Wuilleumier, LURE Report 74/03 (1974) (unpublished).
79. D. Dill, *Phys. Rev. A* **7**, 1976 (1973).
80. D. Dill and J. L. Dehmer, *J. Chem. Phys.* **61**, 692 (1974).
81. J. L. Dehmer, *J. Phys. (Paris) Colloq.* **39**, 4–42 (1978).
82. C. S. Fadley, *Chem. Phys. Lett.* **25**, 225 (1974).
83. R. Manne and T. Åberg, *Chem. Phys. Lett.* **7**, 282 (1970).
84. J. R. Swanson and L. Armstrong, Jr., *Phys. Rev. A* **15**, 661 (1977).
85. J. R. Swanson and L. Armstrong, Jr., *Phys. Rev. A* **16**, 1117 (1977).
86. T. N. Chang and R. T. Poe, *Phys. Rev. A* **12**, 1432 (1975).
87. H. P. Kelly, in *Photoionization and Other Probes of Many-Electron Interactions*, F. Wuilleumier, (ed.), pp. 83–109, Plenum Press, New York (1976).
88. P. G. Burke, *Atomic Physics*, Vol. 5, R. Marrus (ed.), Plenum Press, New York (1976).
89. P. G. Burke and W. D. Robb, *Adv. Atom. Molec. Phys.* **11**, 143 (1975).
90. M. J. Seaton, *Philos. Trans. R. Soc. London Ser. A* **245**, 469 (1953).
91. V. L. Jacobs and P. G. Burke, *J. Phys. B* **5**, L67 (1972).
92. V. L. Jacobs, *Phys. Rev. A* **3**, 289 (1971).
93. G. Wendin, *J. Phys. B* **6**, 42 (1973).
94. G. Wendin, in *Photoionization and Other Probes of Many-Electron Interactions*, F. J. Wuilleumier, (ed.), pp. 61–82, Plenum Press, New York (1976).
95. W. R. Johnson, C. D. Lin, and A. Dalgarno, *J. Phys. B* **9**, L303 (1976).
96. W. R. Johnson and C. D. Lin, *J. Phys. B* **10**, L331 (1977).
97. G. Wendin, *Phys. Scr.* **16**, 296 (1977).
98. T. N. Chang and U. Fano, *Phys. Rev. A* **13**, 263 (1976).
99. M. O. Krause, in *Abstracts of the International Conference on the Physics of X-Ray Spectra*, R. D. Deslattes (ed.), pp. 32–34, National Bureau of Standards, Washington (1976).
100. S. T. Manson, *J. Electron Spectrosc. Relat. Phenom.* **9**, 21 (1976).
101. T. Åberg, in *Photoionization and Other Probes of Many-Electron Interactions*, F. J. Wuilleumier (ed.), pp. 49–60, Plenum Press, New York (1976).
102. R. L. Martin and D. A. Shirley, *Phys. Rev. A* **13**, 1475 (1976).
103. D. A. Shirley, *J. Phys. (Paris) Colloq.* **39**, 4–35 (1978).
104. W. Domcke, L. S. Cederbaum, J. Schirmer, W. von Niessen, and J. P. Maier, *J. Electron Spectrosc. Relat. Phenom.* **14**, 59 (1978); J. Schirmer, L. S. Cederbaum, W. Domcke, and W. von Niessen, *Chem. Phys.* **26**, 149 (1977).
105. P. S. Bagus and E.-K. Viinikka, *Phys. Rev. A* **15**, 1486 (1977).
106. M. O. Krause and J. H. Oliver, *J. Phys. Chem. Ref. Data* **8**, 329 (1979).
107. U. Gelius, E. Basilier, S. Svensson, T. Bergmark, and K. Siegbahn, *J. Electron Spectrosc. Relat. Phenom.* **2**, 405 (1974).
108. F. Wuilleumier, M. Y. Adam, P. Dhez, N. Sandner, V. Schmidt, and W. Mehlhorn, *Jpn. J. Appl. Phys.* **17**, 44 (1978).
109. M. Y. Adam, F. Wuilleumier, S. Krummacher, N. Sandner, V. Schmidt, and W. Mehlhorn, *J. Electron Spectrosc. Relat. Phenom.* **15**, 211 (1979).
110. K. Codling and P. Mitchell, *J. Phys. E* **3**, 685 (1970).
111. D. A. Shirley, J. Stöhr, P. S. Wehner, R. S. Williams, and G. Apai, *Phys. Scr.* **16**, 398 (1977).
112. E. Koch, R. Haensel, and C. Kunz (eds.), *Proceedings of the 4th International Conference on Vacuum Ultraviolet Radiation Physics*, Pergamon-Vieweg, Braunschweig (1974).
113. *Diffraction Gratings Ruled and Holographic–Handbook*, Jobin-Yvon Company, Longjumeau, France (1976).
114. M. O. Krause, *Chem. Phys. Lett.* **10**, 65 (1971).
115. M. O. Krause, *Phys. Fenn. Suppl.* **9**, 1–281 (1974).

116. R. G. Houlgate, J. B. West, K. Codling, and G. V. Marr, *J. Electron Spectrosc. Relat. Phenom.* **9**, 205 (1976).

117. R. Z. Bachrach, F. C. Brown and S. B. M. Hagström, *J. Vac. Sci. Technol.* **12**, 309 (1975).

118. W. B. Peatman, T. B. Borne, and E. W. Schlag, *Chem. Phys. Lett.* **3**, 492 (1969).

119. W. B. Peatman, C. B. Kasting, and D. J. Wilson, *J. Electron Spectrosc. Relat. Phenom.* **7**, 233 (1975).

120. W. B. Peatman, B. Gotchev, P. Gürtler, E. E. Koch, and V. Saile, *J. Chem. Phys.* **69**, 2089 (1978).

121. P. M. Guyon, T. Baer, L. F. A. Ferreira, I. Nenner, A. Tabché-Fouhaillé, R. Botter, and T. Govers, *J. Phys. B* **11**, L141 (1978).

122. T. Baer, P. M. Guyon, I. Nenner, A. Tabché-Fouhaillé, R. Botter, L. F. A. Ferreira, and T. R. Govers *J. Chem. Phys.*, **70**, 1585 (1979).

123. R. Frey, B. Gotchev, O. F. Kalman, W. B. Peatman, H. Pollak, and E. W. Schlag, *Chem. Phys.* **21**, 89 (1977).

124. E. W. Schlag, R. Frey, B. Gotchev, W. B. Peatman, and H. Pollak, *Chem. Phys. Lett.* **51**, 406 (1977).

125. O. Keski-Rahkonen and M. O. Krause, *J. Electron Spectrosc. Relat. Phenom.* **13**, 107 (1978).

126. O. Keski-Rahkonen, *J. Electron Spectrosc. Relat. Phenom.* **13**, 113 (1978).

127. T. D. Thomas and R. W. Shaw, *J. Electron. Spectrosc. Relat. Phenom.* **5**, 1081 (1974).

128. G. Johannson, J. Hedman, A. Berndtsson, M. Klasson, and R. Nilsson, *J. Electron Spectrosc. Relat. Phenom.* **2**, 295 (1973).

129. R. T. Poole, R. C. G. Leckey, J. G. Jenkin, and J. Liesegang, *J. Phys. E* **6**, 201 (1973).

130. N. Sandner, Ph.D. thesis, University of Freiburg, West Germany (May 1978), (unpublished).

131. O. Keski-Rahkonen and M. O. Krause, *Phys. Rev. A* **15**, 959 (1977).

132. G. E. Chamberlain, J. R. Mielczarek, and C. E. Kuyatt, *Phys. Rev. A* **2**, 1905 (1970).

133. J. A. R. Samson and G. N. Haddad, *J. Opt. Soc. Am.* **64**, 47 (1974).

134. E. B. Saloman and D. L. Ederer, *Appl. Opt.* **14**, 1029 (1975).

135. L. J. Kieffer, *At. Data* **1**, 19 (1969).

136. N. Chandra and A. Temkin, *J. Chem. Phys.* **65**, 4537 (1976).

137. J. Siegel, D. Dill, and J. L. Dehmer, *Phys. Rev. A* **17**, 2106 (1978).

138. S. K. Srivawtava, A. Chutjian, and S. Trajmar, *J. Chem. Phys.* **64**, 1340 (1976).

139. P. C. Kemeny, A. D. McLachlan, F. L. Battye, R. T. Poole, R. C. G. Leckey, J. Liesegang, and J. G. Jenkin, *Rev. Sci. Instrum.* **44**, 1197 (1973).

140. R. T. Poole, R. C. G. Leckey, J. Liesegang, and J. G. Jenkin, *J. Phys. E* **6**, 226 (1973).

141. J. L. Gardner and J. A. R. Samson, *J. Electron. Spectrosc. Relat. Phenom.* **6**, 53 (1975).

142. J. L. Gardner and J. A. R. Samson, *J. Electron Spectrosc. Relat. Phenom.* **8**, 469 (1976).

143. G. V. Marr and J. B. West, At. Data Nucl. Data Tables **18**, 496 (1976).

144. F. Wuilleumier and M. O. Krause, *J. Electron Spectrosc. Relat. Phenom.* **15**, 15 (1979).

145. F. Wuilleumier, *Adv. X-Ray Anal.* **16**, 64 (1973).

146. W. H. Hancock and J. A. R. Samson, *J. Electron Spectrosc. Relat. Phenom.* **9**, 211 (1976).

147. G. J. Schulz, *Rev. Mod. Phys.* **45**, 378 (1973).

148. G. J. Schulz, *Rev. Mod. Phys.* **45**, 423 (1973).

149. P. G. Burke and J. F. Williams, *Phys. Rep. C* **34**, 325 (1977).

150. H. Hotop and W. C. Lineberger, *J. Phys. Chem. Ref. Data* **4**, 539 (1975).

151. R. N. Compton, P. W. Reinhardt, and C. D. Cooper, *J. Chem. Phys.* **68**, 2023 (1978).

152. H. S. W. Massey, *Negative Ions*, 3rd edition, Cambridge University Press, Cambridge, England (1976).

153. P. R. Woodruff and G. V. Marr, *Proc. R. Soc. London Ser. A* **358**, 87 (1977).

154. U. Heinzmann, H. Heuer, and J. Kessler, *Phys. Rev. Lett.* **34**, 441 (1975).

155. N. A. Cherepkov, *Zh. Eksp. Teor. Fiz* **65**, 933 (1973) [*Sov. Phys. JETP* **38**, 463 (1974)].

156. U. Fano, *Phys. Rev.* **178**, 131 (1969).

157. N. A. Cherepkov, *J. Phys. B* **10**, L653 (1977).

158. M. O. Krause and F. Wuilleumier, *J. Phys. B* **5**, L143 (1972).

159. F. Wuilleumier (private communication).

160. K. Radler and J. Berkowitz, *J. Opt. Soc. Am.* **68**, 1181 (1978).

161. A. Hamnett, S. T. Hood, and C. E. Brion, *J. Electron Spectrosc. Relat. Phenom.* **11**, 263 (1977).

162. M. S. Banna and D. A. Shirley , *J. Electron Spectrosc. Relat. Phenom.* **8**, 255 (1976).

163. C. E. Brion, A. Hamnett, G. R. Wight, and M. H. Van der Wiel, *J. Electron Spectrosc. Relat. Phenom.* **12**, 323 (1977).

164. A. Hamnett, W. Stoll, G. R. Branton, M. J. Van der Wiel, and C. E. Brion, *J. Phys. B* **9**, 945 (1976).

165. G. C. King, F. H. Read, and M. Tronc, *Chem. Phys. Lett.* **52**, 50 (1977).

166. O. Keski-Rahkonen and M. O. Krause, *J. Electron Spectrosc. Relat. Phenom.* **9**, 371 (1976).

167. W. B. Peatman, *Chem. Phys. Lett.* **36**, 495 (1975).

168. W. B. Peatman and D. Dill, *Chem. Phys. Lett.* **53**, 79 (1978).

169. G. R. Wight and M. J. Van der Wiel, *J. Phys. B* **10**, 601 (1977).

170. M. J. Van der Wiel, G. R. Wight, and R. R. Tol. *J. Phys. B* **9**, L5 (1976).

171. K. Codling and R. P. Madden, *Phys. Rev. Lett.* **12**, 106 (1964).

172. J. P. Briand, P. Chevallier, M. Tavernier, A. Chetioui, A. Touati, and V. Kostroun (to be published).

173. G. C. King, M. Tronc, F. H. Read, and R. Bradford, *J. Phys. B* **10**, 2479 (1977).

174. M. O. Krause, in *Proceedings of the International Conference on Inner-Shell Ionization Phenomena*, R. W. Fink, *et al.* (eds.), pp. 1586–1616, USAEC-CONF-720404 (1973).

175. F. Wuilleumier and M. O. Krause, in *VIIIth International Conference on Electron and Atomic Collisions*, Extended Abstracts, B. C. Cović and M. V. Kurepa (eds.), Vol. II, pp. 559–560, University of Beograd, Beograd, Yugoslavia (1973); *Phys. Rev. A* **10**, 242 (1974).

176. M. O. Krause, *J. Phys. (Paris) Colloq.* **32**, 4–67 (1971).

177. J. A. R. Samson, G. N. Haddad, and J. L. Gardner, *J. Phys. B* **10**, 1749 (1977).

178. S. P. Kowalczyk, L. Ley, R. L. Martin, F. R. McFeeley, and D. A. Shirley, *Phys. Rev. B* **7**, 4009 (1973).

179. R. Frey, B. Gotchev, W. B. Peatman, H. Pollak, and E. W. Schlag, *Int. J. Mass Spectro. Ion Phys.* **26**, 137 (1978).

180. M. O. Krause, *Phys. Rev.* **177**, 151 (1969).

181. F. Wuilleumier, M. Y. Adam, N. Sandner, V. Schmidt, and W. Mehlhorn, (private communication, 1978).

182. M. Y. Adam, F. Wuilleumier, N. Sandner, S. Krummacher, V. Schmidt, and W. Mehlhorn, *Jpn. J. Appl. Phys.* **17**, 170 (1978).

183. M. Y. Adam, Ph.D. thesis, University of Orsay, France (Dec. 1978) (unpublished).

184. J. B. West, P. R. Woodruff, K. Codling, and R. G. Houlgate, *J. Phys. B* **9**, 407 (1976); S. P. Shannon, K. Codling, and J. B. West, *J. Phys. B* **10**, 825 (1977).

185. T. Gustafsson, *Chem. Phys. Lett.* **51**, 383 (1977).

186. K. H. Tan and C. E. Brion, *J. Electron Spectrosc. Relat. Phenom.* **13**, 77 (1978).

187. J. A. R. Samson and J. L. Gardner, *Phys. Rev. Lett.* **33**, 671 (1974).

188. F. Wuilleumier and M. O. Krause, in *Electron Spectroscopy*, D. A. Shirley (ed.), pp. 259–267, North-Holland, Amsterdam (1972).

189. M. O. Krause and F. Wuilleumier, in *Electron and Photon Interactions with Atoms*, H. Kleinpoppen and M. R. C. McDowell (eds.), pp. 89–97, Plenum Press, New York (1976).

190. V. Schmidt, N. Sandner, H. Kuntzemüller, P. Dhez, F. Wuilleumier, and E. Källne, *Phys. Rev. A* **13**, 1748 (1976).

191. J. A. R. Samson (private communication).

192. T. N. Chang, *Phys. Rev. A* **18**, 1448 (1978).

193. W. R. Johnson and K. T. Cheng, *Phys. Rev. A* **20**, 978 (1979).

194. P. G. Burke and K. T. Taylor, *J. Phys. B* **8**, 2620 (1975).

195. M. Ya. Amusia, V. Ivanov, N. A. Cherepkov, and L. V. Chernysheva, *Phys. Lett. A* **40**, 361 (1972).

196. C. D. Lin, *Phys. Rev. A* **9**, 171 (1974).

197. W. R. Johnson and K. T. Cheng, *Phys. Rev. Lett.* **40**, 1167 (1978).

198. M. Ya. Amusia, in *Proceedings of the 4th International Conference on Vacuum Ultraviolet Radiation Physics*, E. Koch, R. Haensel, and C. Kunz (eds.), p. 205, Pergamon-Vieweg, Hamburg (1974).

199. M. Y. Adam, F. Wuilleumier, S. Krummacher, V. Schmidt, and W. Mehlhorn, *J. Phys. B* **11**, L413 (1978).

200. D. P. Spears, H. J. Fischbeck, and T. A. Carlson, *Phys. Rev. A* **9**, 1603 (1974).

201. M. Y. Adam, F. Wuilleumier, N. Sandner, V. Schmidt, and G. Wendin, *J. Phys. (Paris)* **39**, 77 (1978).

202. S. Süzer and N. S. Hush, *J. Phys. B* **10**, L705 (1977).
203. A. L. Carter and H. P. Kelly, *J. Phys. B* **9**, L565 (1976).
204. S. P. Shannon and K. Codling, *J. Phys. B* **11**, 1193 (1978).
205. N. Sandner, S. Krummacher, V. Schmidt, W. Mehlhorn, F. Wuilleumier, and M. Y. Adam, *V. International Conference on Vacuum Uultraviolet Radiation Physics*, Extended Abstracts, Vol. I, p. 105, M. C. Castex *et al.* (eds.), CNRS, Meudon, France (1977).
206. J. Berkowitz, J. L. Dehmer, Y.-K. Kim, and J. P. Desclaux, *J. Chem. Phys.* **61**, 2556 (1974); S. Süzer, S. T. Lee, and D. A. Shirley, *Phys. Rev. A* **13**, 1842 (1976).
207. T. E. H. Walker and J. T. Waber, *J. Phys. B* **7**, 674 (1974).
208. F. Wuilleumier, M. Y. Adam, P. Dhez, N. Sandner, V. Schmidt, and W. Mehlhorn, *Phys. Rev. A* **16**, 646 (1977).
209. F. Wuilleumier, M. Y. Adam, N. Sandner, V. Schmidt, W. Mehlhorn, and J. P. Desclaux, *V. International Conference on Vacuum Ultraviolet Radiation Physics*, Extended Abstracts, Vol. I, p. 14, M. C. Castex *et al.* (eds.), CNRS, Meudon, France (1977).
210. M. S. Banna, M. O. Krause, and F. Wuilleumier, *J. Phys. B* **12**, L125 (1979).
211. W. R. Johnson and V. Radojevic, *J. Phys. B* **11**, L773 (1979).
212. W. Ong and S. T. Manson, *Bull. Am. Phys. Soc.* **23**, 564 (1978).
213. T. A. Carlson, M. O. Krause, and W. E. Moddeman, *J. Phys. (Paris) Colloq.* **32**, 4–76 (1971).
214. F. Wuilleumier, M. Y. Adam, N. Sandner, V. Schmidt, and W. Mehlhorn, in *S. ICPEAC Conference*, Extended Abstracts, p. 1171, C.E.A., Paris, France (1977).
215. J. D. Morrison, in *Energy Transfer in Gases*, R. Stoops (ed.), pp. 397–463, Interscience Publishers, New York (1963); E. P. Wigner, *Phys. Rev.* **73**, 1002 (1948).
216. W. C. Lineberger, H. Hotop, and T. A. Patterson, in *Electron and Photon Interactions with Atoms*, H. Kleinpoppen and M. R. C. McDowell (eds.), pp. 125–132, Plenum Press (1976).
217. J. Sugar, *Phys. Rev. B* **5**, 1785 (1972).
218. I. Lindau, D. Pianetta, and W. E. Spicer, *Phys. Lett. A* **57**, 225 (1976).
219. D. E. Eastman and M. Kuznietz, *Phys. Rev. Lett.* **26**, 846 (1971).
220. B. L. Henke and R. L. Elgin, *Adv. X-Ray Anal.* **13**, 639 (1970).
221. R. Nyholm, A. Berndtsson, R. Nilsson, J. Hedman, and C. Nordling, *Phys. Scr.* **16**, 383 (1977).
222. E. W. Plummer, T. Gustafsson, W. Gudat, and D. E. Eastman, *Phys. Rev. A* **15**, 2339 (1977).
223. A. Hamnett, W. Stoll, and C. E. Brion, *J. Electron Spectrosc. Relat. Phenom.* **8**, 367 (1976).
224. J. W. Davenport, *Phys. Rev. Lett.* **36**, 945 (1976).
225. J. W. Cooper and S. T. Manson, *Phys. Rev.* **177**, 157 (1969).
226. D. L. Albritton, A. L. Schmeltekopf, and R. N. Zare, *Diatomic Intensity Factors*, Wiley, New York (1979).
227. J. L. Gardner and J. A. R. Samson, *J. Electron Spectrosc. Relat. Phenom.* **13**, 7 (1978).
228. T. A. Carlson and C. P. Anderson, *Chem. Phys. Lett.* **27**, 561 (1971).
229. K. Codling, R. G. Houlgate, J. B. West, and P. R. Woodruff, *J. Phys. B* **9**, L83 (1976).
230. D. L. Miller, J. D. Dow, R. G. Houlgate, G. V. Marr, and J. B. West, *J. Phys. B* **10**, 3205 (1977).
231. L. Torop, J. Morton, and J. B. West, *J. Phys. B* **9**, 2035 (1976).
232. J. L. Dehmer, W. A. Chupka, J. Berkowitz, and W. T. Jivery, *Phys. Rev. A* **12**, 1966 (1975).
233. J. L. Dehmer and D. Dill, *Phys. Rev. Lett.* **32**, 1049 (1976).
234. W. Ong and S. T. Manson, *J. Phys. B* **11**, L65 (1978).
235. J. L. Dehmer and P. M. Dehmer, *J. Chem. Phys.* **67**, 1782 (1977).
236. J. A. R. Samson and W. H. Hancock, *Phys. Lett. A* **61**, 380 (1977).
237. C. Duzy and R. S. Berry, *J. Chem. Phys.* **64**, 2421 (1976); E. S. Chang, *J. Phys. B* **11**, L293 (1978).
238. B. Ritchie and B. R. Tambe, *J. Chem. Phys.* **68**, 755 (1978); I. L. Thomas, *Phys. Rev. A* **4**, 457 (1971).
239. I. G. Kaplan and A. P. Markin, *Dokl. Akad. Nauk SSSR* **184**, 66 (1969) [*Sov. Phys. Dokl.* **14**, 36 (1969)].
240. D. Dill, S. Wallace, J. Siegel, and J. L. Dehmer, *Phys. Rev. Lett.* **41**, 1230 (1978); *Phys. Rev. Lett.* **42**, 411 (1979).
241. D. G. McCoy, J. M. Morton, and G. V. Marr, *J. Phys. B* **11**, L547 (1978); G. V. Marr, J. M. Morton, R. M. Holmes, and D. G. McCoy (to be published).

242. T. A. Carlson, G. E. McGuire, A. E. Jonas, K. L. Cheng, C. P. Anderson, C. C. Lu, and B. P. Pullen, in *Electron Spectroscopy*, D. A. Shirley (ed.), pp. 207–231, North-Holland, Amsterdam (1972).
243. T. A. Carlson, *Chem. Phys. Lett.* **9**, 23 (1971); T. A. Carlson and A. E. Jonas, *J. Chem. Phys.* **55**, 4913 (1971); D. M. Mintz and A. Kuppermann, *J. Chem. Phys.* **69**, 3953 (1978).
244. U. Fano, *Phys. Rev.* **124**, 1866 (1961).
245. A. Niehaus and M. W. Ruf, *Z. Physik* **252**, 84 (1972).
246. J. A. R. Samson and J. L. Gardner, *Phys. Rev. Lett.* **31**, 1327 (1973).
247. P. C. Kemeny, J. A. R. Samson, and A. F. Starace, *J. Phys. B* **10**, L201 (1977).
248. H. Hotop and D. Mahr, *J. Phys. B* **8**, L301 (1975).
249. S. T. Lee, S. Süzer, E. Matthias, R. A. Rosenberg, and D. A. Shirley, *J. Chem. Phys.* **66**, 2496 (1977).
249a. R. A. Rosenberg, M. G. White, G. Thornton, and D. A. Shirley, *Phys. Rev. Lett.* **43**, 1386 (1979).
250. P. Natalis, J. Delwiche, J. E. Collins, G. Caprace, and M. Th. Praet, *Phys. Scr.* **16**, 242 (1977).
251. R. B. Kay, Ph. E. Van der Leeuw, and M. J. Van der Wiel, *J. Phys. B* **10**, 2513 (1977).
252. A. Niehaus, *J. Phys. B* **10**, 1845 (1977).
253. G. Gerber, R. Morgenstern, and A. Niehaus, *J. Phys. B* **6**, 493 (1973).
254. G. C. King, F. H. Read, and R. C. Bradford, *J. Phys. B* **8**, 2210 (1975).
255. S. Ohtani, H. Nishimura, H. Suzuki, and K. Wakiya, *Phys. Rev. Lett.* **36**, 863 (1976).
256. V. Schmidt, N. Sandner, W. Mehlhorn, M. Y. Adam, and F. Wuilleumier, *Phys. Rev. Lett.* **38**, 63 (1977).
257. V. Schmidt, N. Sandner, W. Mehlhorn, F. Wuilleumier, and M. Y. Adam, *Xth International Conference on Electron and Atomic Collisions*, Extended Abstracts, p. 1062, C.E.A. Paris, France (1977).
258. M. Ya. Amusia, M. Yu. Kuchiev, S. A. Sheinerman, and S. I. Sheftel, *J. Phys. B* **10**, L535 (1977).
259. M. O. Krause, T. A. Carlson, and W. E. Moddeman, *J. Phys.* (Paris) *Colloq.* **32**, 4–139 (1971).
260. F. Wuilleumier and M. O. Krause, in *Proceedings of the International Conference on Inner-Shell Ionization Phenomena*, R. W. Fink et al. (eds.), pp. 773–791, USAEC CONF-720404 (1973).
261. M. O. Krause, in *Proceedings of the 24th Annual Conference on Mass Spectrometry and Allied Topics*, J. L. Margrave (ed.), pp. 68–72, Rice University Press, Houston, Texas (1976).
262. W. Eberhardt, G. Kalkoffen, and C. Kunz, *Phys. Rev. Lett.* **41**, 156 (1978).
263. W. Sandner and W. Schmitt, *J. Phys. B* **10**, 1833 (1978).
264. E. G. Berezhko, V. K. Ivanov, and N. M. Kabachnik, *Phys. Lett. A* **66**, 474 (1978).
265. J. L. Dehmer and D. Dill, *Phys. Rev. A* **18**, 164 (1978).
266. R. Frey, B. Gotchev, W. B. Peatman, H. Pollack, and E. W. Schlag, *Chem. Phys. Lett.* **54**, 411 (1978).
267. R. Stockbauer, in *Vth International Conference on Vacuum Ultraviolet Radiation Physics*, Extended Abstracts, p. 134, McCarter et al. (eds.), C.N.R.S., Meudon, France (1977).

6

Photoemission as a Tool to Study Solids and Surfaces

I. LINDAU and W. E. SPICER

1. General Considerations

During the last 15 years, photoemission has been established as a very powerful tool for studies of the bulk and surface electronic structure of materials. Different aspects of photoemission spectroscopy have been treated in a series of review articles.[1-24] Synchrotron radiation with its continuous spectral distribution from infrared to x-ray radiation has opened up new avenues in studies of solids and surfaces and strengthened the applicability of the photoemission technique tremendously. In this paper, we will try to review the most prominent advances that have been made since the advent of synchrotron radiation as an excitation source.

A wealth of photoemission work has emerged during the last few years. Therefore, we cannot give a comprehensive treatment of all the work in this review paper. Instead, we have tried to illustrate the new and exciting research directions photoemission is taking with selected examples.

1.1. The Physics of the Photoemission Process

In a photoemission experiment, monochromatic light is allowed to hit a sample and the energy distributions of the photoemitted electrons are then measured. Photoemission is thus an excitation process, and the basic quantity from

I. LINDAU and *W. E. SPICER* ● Stanford Synchrotron Radiation Laboratory and Stanford Electronics Laboratories, Stanford University, Stanford, California 94305.

159

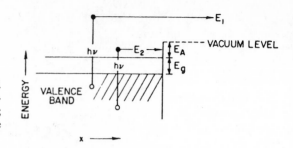

Figure 1. The essence of the photoemission process (from Ref. 12), where $\eta(hv, E)\, dE$ is the probability of hv exciting an electron at energy E to $E + dE$; $\varepsilon_2 \leftrightarrow \int \eta(hv, E)\, dE$; $N(E, hv)\, dE \leftrightarrow \eta(hv, E)\, dE$; and $N(E, hv)$ is equal to the energy distribution of photoemission.

which information about the electronic structure can be extracted is $\eta(hv, E)$, which is the probability for a photon energy hv to excite an electron to a final state with an energy E. This is illustrated in Figure 1. Optical parameters, e.g., ε_2, provide information about the integral of $\eta(E, hv)$, whereas, in photoemission, the integration over final states is removed. Hence, a more unambiguous assignment can be made of the quantum states. This is a great advantage.

The optical excitation extends about 100 Å into the solid,[25] varying with material and photon energy. The escaping electrons must pass through the solid before reaching the surface from which they enter the vacuum. One of the basic problems in interpreting photoemission spectra is to separate the primary unscattered photoelectrons from inelastically scattered ones. Most of the interpretation of photoemission data has so far been based on the three-step model.[26, 27] Within this model, the photoemission process is approximated with three successive events. The first event is the optical excitation, and the second is the transport of the electrons to the surface during which the electrons may suffer from inelastic scattering processes. The third event is the escape of the electrons across the surface potential barrier into vacuum. This is shown schematically in Figure 2. On the left in the figure, a hypothetical $\eta(E, hv)$ has been sketched, illustrating the initial optically excited distribution. Changes in the distribution as the electrons approach the surface and after the escape into vacuum are indicated.

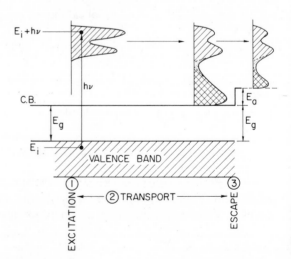

Figure 2. Schematic of the three-step model for the photoemission process (from Ref. 12).

The separation of the photoemission process into three independent steps is, of course, an approximation. The merit of the three-step model lies in its allowance for relatively easy treatment without losing the most relevant physical information in a number of very important applications. More sophisticated models[28-32] have emerged during the last five years, where the photoemission process is treated as a one-step quantum-mechanical event. The first-principle calculations by Liebsch[32] have apparent similarities with LEED (low-energy electron diffraction) formalisms. The wave function is written as $\psi = \psi^0 + \psi^1$, where ψ^0 is the central-site contribution (direct wave) and where ψ^1 takes care of the surrounding atoms. By breaking up the wave function in these two parts, both single and multiple scattering processing are included. The one-step models and their application to different systems have been reviewed recently, and the reader is referred to these references for further details.[33-35]

1.2. The Characteristics of Synchrotron Radiation Important for Photoemission Studies

The tunability of the synchrotron radiation over a large energy region has made it possible to optimize the two most important parameters: (1) photoionization cross sections (matrix elements) and (2) electron escape depths, for the particular problem being studied. Thus, for instance, in order to optimize the surface sensitivity for studies of both valence and core electrons, it is often necessary to work at different photon energies. As an example, we can mention photoemission work on covalent semiconductors. The matrix elements of the valence states fall off very rapidly as the photon energy is increased above 5 eV. Thus, it is usually optimum to study the valence surface electronic structure in the range $20 \leq \hbar\omega \leq 30$ eV. This gives the best compromise between a short escape depth (see Section 1.3) and a reasonably large excitation probability. On the other hand, certain core levels (e.g., $3d$) have matrix elements that are fairly constant over a large energy region (several hundred electron volts). In this case, only the shortest possible escape depth has to be considered in the choice of excitation energy. As will be discussed in more detail in Section 1.4, certain subshells have a very strong energy dependence over a short energy interval due to so-called Cooper minima. We will illustrate in Section 1.3 how this energy dependence can be used in an advantageous way for surface physics studies.

The tunability of synchrotron radiation has also added new versions to the traditional photoemission technique, namely, constant final- (initial-) state spectroscopies. In both these spectroscopies, the photon energy is varied continuously and it is absolutely essential to have a smooth spectral distribution. These techniques will be elaborated on in Section 1.5.

The high degree of polarization of the synchrotron radiation has found somewhat specialized, but very useful, applications for a number of photoemission studies. This is particularly important in connection with angle-resolved photoemission to probe the symmetry of band states and the orientation of gas molecules on surfaces.

Without going into any quantitative comparisons, it should be mentioned that synchrotron radiation provides a more intense flux than the continuum from gas discharges and conventional x-ray tubes. However, the intensity in the discrete lines from conventional sources are typically higher. The details of these comparisons depend on the choice of synchrotron radiation source as well as the specific experimental circumstances.

With one notable exception[36] the pulsed structure of the synchrotron radiation has not been used in connection with photoemission spectroscopy. A time-of-flight spectrometer[36] has certain attractive features but does not presently seem to be competitive with the more conventional energy analyzers discussed in Section 2.

1.3. The Probing Depth in Photoemission

As the photoexcited electrons move toward the crystal surface, they will lose energy due to inelastic scattering. Electron–electron scattering between the excited electron and the valence electrons, and electron–phonon scattering with the crystal lattice are the two important processes that determine the scattering probability. Scattering between the electron and lattice defects (or impurities) is also possible as are plasmons or other collective events. In a first approximation, the scattering probability can be defined in terms of a scattering length $L(E)$ (escape depth, mean free path): $S(E, x) = \exp[-x/L(E)]$, where x is the traveled distance before scattering and E is the electron energy. The mean free path is fairly energy-independent for electron–phonon scattering but extremely dependent on the electron energy for electron–electron scattering. One further important difference is the magnitude of the energy loss. The losses in electron–phonon scattering are small (typically 0.01–0.05 eV) compared to the excitation energy, whereas the losses in electron–electron or electron–plasmon scattering events are typically several electron volts. For the excitation energies discussed in this paper, the electron–phonon scattering will not be of prime importance. In fact, it is only for phonon energies below 6 eV that the electron–electron and electron–phonon scattering lengths are comparable.[12]

Figure 3 gives a summary of experimentally determined electron escape depths versus photon energy for a wide variety of materials.[37] The escape depth (linear scale, in angstroms) has been plotted as a function of electron energy (logarithmic scale, in electron volts). The escape depth shows a strong decrease with increasing electron energy that is characteristic for scattering lengths determined by electron–electron interaction. In fact, the escape depth decreases from 10,000 Å for electron energies a few tenths of an electron volt above the Fermi energy to 20–30 Å for electron energies 5–10 eV above the Fermi level. The solid line to the right in the figure represents results obtained for Au from x-ray photoemission data in the electron energy range 1–3 keV.[38] Here, the escape depth increases with increasing energy. The general behavior of the escape depth as a function of electron energy is thus first a sharp decrease with increasing electron energy, then a fairly flat minimum in the range 50–500 eV with a remarkably small escape depth (≈ 5 Å), and

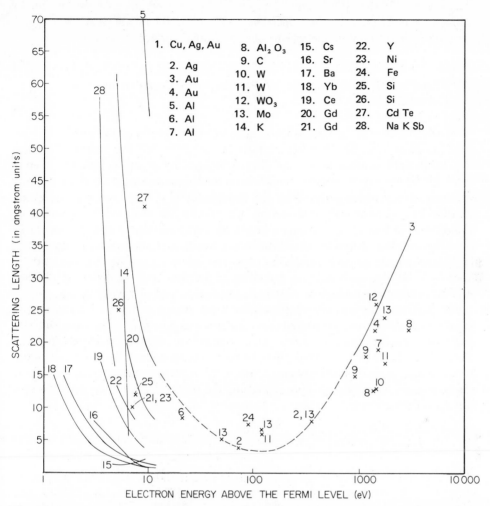

Figure 3. The escape depth, in angstroms, is shown as a function of the electron energy above the Fermi level, in electron volts, for a large number of materials (from Ref. 37).

finally an increase with increasing electron energy. It should also be noted that the details of the energy dependence of the escape depth and the minimum are different for different groups of materials.

We want to emphasize the extreme surface sensitivity the photoemission technique can provide by working around the minima of the electron escape depth, where the probing depth is typically 1 to 2 atomic layers (see Figure 3). At these very small escape depths, the validity of the three-step model becomes more and more questionable.[39] The separation of the photoemission process into excitation, transport, and escape obviously loses its meaning when the escape length turns out to be of the order of the atomic dimensions, which is the case in the energy region 30–200 eV. The plots given in Figure 3 permit an experimenter to estimate the electron energy at which the minimum in the escape depth occurs and therefore to select photon energies to obtain optimal surface sensitivity.

1.4. The Energy Dependence of Partial Photoionization Cross Sections

The cross section for the photoionization process is naturally of utmost importance in determining the emission intensity in the photoemission process. The lack of suitable light sources in the far ultraviolet and soft x-ray regions has been a severe hindrance in obtaining information on partial photoionization cross sections and their energy dependence. With the appearance of synchrotron radiation, this situation has changed and cross-section information has been used in a variety of ways to extract more information on both bulk electronic structure and chemisorption (e.g., bonding geometry of chemisorbed gases). Determination of the energy dependence of partial photoionization cross sections using synchrotron radiation is expected to be of great benefit for a number of different research fields (solid state physics, atmospheric and space physics, astrophysics, and plasma physics).

In this section, we will illustrate the most salient features of some simple one-electron calculations of partial cross sections and how they compare with some recent experimental determinations using synchrotron radiation. Theoretically, the energy dependence of the partial photoionization cross section has been calculated for the noble gases by Kennedy and Manson[40] in the one-electron Hartree–Fock approximation. In the dipole approximation, the partial photoionization cross section σ_{nl} is then given by

$$\sigma_{nl}(E) = \frac{4}{3}\pi^2\alpha a_0^2 \left[N_{nl}(E - E_{nl})\frac{1}{2l+1} \right]\left[lR_{E,l-1}^2 + (l+1)R_{E,l+1}^2 \right]$$

where α is the fine-structure constant, a_0 is the Bohr radius, N_{nl} is the number of electrons in the subshell, E_{nl} is the binding energy, and E is the energy of the ejected electron. The radial dipole matrix elements are

$$R_{E,l\pm1} = \int_0^\infty P_{nl}(r)rP_{E,l\pm1}(r)\,dr$$

where $P_{nl}(r)(1/r)$ is the radial part of the wave function. Of importance for the analysis of our data in the following sections is that σ_{nl} is essentially determined by R_{l+1} since R_{l-1} is, in general, much smaller.

In Figure 4, we present experimental data for the energy dependence of the cross section of the 3d levels for Ga.[41,42] The Ga 3d levels are located 19 eV below the Fermi level, and the energy dependence can thus be followed continuously from threshold to about 350 eV above threshold using one of the monochromators available at SSRL (see Section 2.2). The experimental points in Figure 4 are obtained from the photoemission experiment where the energy distribution of the photoemitted electron is measured and can be plotted as a function of the

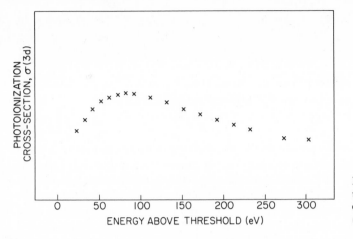

Figure 4. The partial photoionization cross section for the Ga $3d$ core level as a function of electron energy above the excitation threshold (from Ref. 42).

electron binding energy. The energy dependence of the photoionization cross section is then quite simply obtained by measuring the area under the photoelectron peak as a function of the excitation energy, making proper normalizations for the incoming photon flux and the transmission of the energy analyzer. Thus, the energy dependence of the partial cross section can be mapped out in detail from the threshold and up. As seen from Figure 4, the $3d$ photoionization cross section for Ga has a weak energy dependence with a broad maximum 70–80 eV above threshold and a monotomic decrease towards higher energies. The experimental results are in good agreement with Kennedy and Manson's calculation[40] for the $3d$ levels of gaseous krypton (for further discussion, see below).

The $4d$ partial cross section for Sb is plotted in Figure 5 as a function of electron energy above threshold. The experimental data are shown as crosses. The cross section goes through a maximum at 45 eV above threshold and then

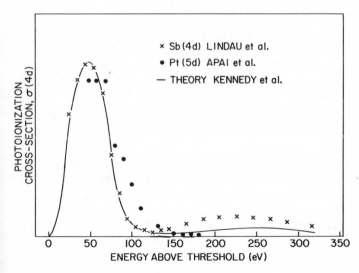

Figure 5. The energy dependence of the Sb $4d$ cross section (crosses) compared to the theoretical calculation (solid line) for Xe $4d$ by Kennedy and Manson[40] (from Ref. 42).

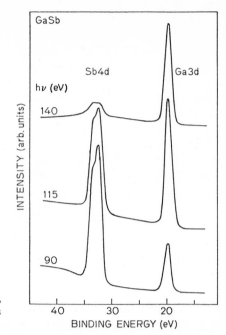

Figure 6. Photoemission spectra of GaSb for three photon energies (90, 115, and 140 eV) demonstrating the strong energy dependence of the 4d cross section.

decreases rapidly before passing through a broad minimum around 130 eV. A flat and extended maximum is observed about 220 eV above threshold before the cross section starts to decrease monotonically towards higher energies. Overall, the 4d partial cross section has a very dramatic energy dependence for the first 200 eV above threshold. The minimum observed at 130 eV is termed the Cooper minimum[43, 44] and arises from the fact that the 4d wave functions have nodes (and thus R_{l+1} passes through zero). In comparison, the 3d wave functions do not have any nodes and the cross section shows a weak energy dependence (for further details, see the paper by Cooper[43]).

The solid line in Figure 5 is the theoretical calculation of the atomic 4d levels in Xe by Kennedy and Manson[40] again using Hartree–Fock wave functions and in the length approximation. The relative amplitudes of the theoretical curve and the experimental data have been fitted at the peak 45 eV above threshold. Overall, there is thus a remarkable agreement between theory and experiment with the energy positions of two maxima and the Cooper minimum reproduced. The cross sections for the 3d and 4d levels for elements (solids) in the third and fourth row of the periodic table are thus in good agreement with Kennedy and Manson's atomic one-electron Hartree–Fock calculations for the corresponding noble gases.[40] However, care should be taken in generalizing these results indiscriminately to other subshells where shakeup,[45] shakeoff,[45] collective resonances,[46, 47] and other effects may considerably complicate the picture, as discussed recently by Hecht, Lindau, and Johansson.[48]

The energy dependence of the photoionization cross section, together with the electron escape depth, are the two important parameters for determining the surface sensitivity of the photoemission technique. This is illustrated in Figure 6.

Energy distributions of photoemitted electrons are plotted here as a function of binding energy with the Fermi level chosen as the reference level. Three distribution curves of the GaSb (110) surface are shown for photon energies 90, 115, and 140 eV. The Ga $3d$ peak is located about 19 eV below the Fermi level. The Sb $4d$ peak is spin–orbit split. The noticeable thing is that the amplitude of the Ga $3d$ peak changes very little with photon energy, whereas the amplitudes of the Sb $4d$ peaks decrease drastically from 90 to 140 eV. This drop occurs in a region of the Cooper minimum, and the observation is in accord with the energy dependence of the $3d$ and $4d$ photoionization cross sections discussed previously in this section. The choice of a suitable excitation energy is thus crucial for obtaining optimal surface sensitivity.

1.5. Different Photoemission Techniques

The classical and still by far the most common photoemission technique is the measurement of an energy distribution curve (EDC) of the photoemitted electrons at a particular photon energy. With the advent of synchrotron radiation, it is possible to obtain a series of EDC's at arbitrary energy spacing for a wide range, 5–300 eV, of excitation energies. As mentioned in the previous two sections, the energy dependence of both the electron escape depth and the partial photoionization cross sections can be utilized to optimize the surface sensitivity. The basic idea behind the EDC was discussed in the introductory sections in conjunction with the description of the photoemission process. An EDC is recorded at a fixed photon energy, but several other parameters can be varied: the angle and polarization of the incident light and the angular distribution of the photoemitted electrons.

In Section 3, where we will describe the photoemission techniques applied to some research problems, most of the experimental data will be presented as EDC's. However, the tunability of synchrotron radiation has made two other techniques possible: constant final-state (CFS) spectroscopy (also named partial yield spectroscopy) and constant initial-state (CIS) spectroscopy.[21] It is harder to illustrate and visualize the CFS and CIS techniques than EDC. In CFS, electrons with a fixed kinetic energy are detected while the photon energy is varied. The CFS technique is illustrated in Figure 7, which is adopted from the review paper by C. Kunz.[49] When

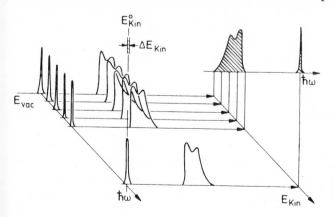

Figure 7. Constant final-state spectra visualized as generated from a series of energy distribution curves at different photon energies $\hbar\omega$. The energy interval ΔE_{kin} at the energy E_{kin} is kept fixed (from Ref. 49).

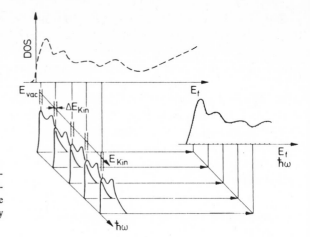

Figure 8. Constant initial-state spectra visualized as generated from a series of energy distribution curves at different photon energies $\hbar\omega$. The interval that accepts the electrons with kinetic energy E_{kin} is shifted synchronously with $\hbar\omega$ (from Ref. 49).

the photon energy is scanned, electrons from occupied states are excited into this fixed final-state energy window. The main difference between the EDC and CFS in probing the occupied density of states is thus that the changes in the final-state density of states is eliminated. And, of course, the energy-dependent matrix elements for the excitation are weighted differently in the two cases. Experience has shown that it is as hard, if not harder, to extract information about the density of occupied states from the CFS as from EDC. In some cases, as will be discussed in Section 3 below, CFS has provided important complementary information. CFS also has the attractive feature that the surface versus bulk contributions can be (de)emphasized by choosing different final-state electron kinetic energies and using the properties of the electron escape depth curve (see Section 1.3).

The CIS technique is also crucially dependent on a continuously tunable light source. In this case, the photon energy and the kinetic energy of the photoemitted electrons are scanned synchronously.[21] Again, we use a figure from the review paper by C. Kunz (see Figure 8) to illustrate this technique. In the CIS method, the density of initial states for a given occupied level is kept fixed while excitations into different empty states are recorded. In a very simplified picture, one should thus be able to obtain information on the density of the unoccupied states above the Fermi level. However, as in the case for CFS, the energy dependence and different weighting of the matrix elements complicate the picture considerably. The usefulness of the CIS method has therefore mostly been as a complementary technique. It has been very successful in many cases.[21]

2. Experimental Details

The photoemission technique was developed with conventional light sources long before synchrotron radiation was available. But, the experimental requirements are, by far, much more demanding when the electron spectrometer has to be interfaced with a synchrotron radiation source. This fact, in combination with the greater experimental possibilities synchrotron radiation offers, have stimulated a rapid advancement and refinement of the technology for photoemission studies. The high photon fluxes required for photoemission have required the development of better optical elements (mirrors, gratings) and novel monochromator designs. New energy analyzers have been developed to handle the larger range of kinetic energies of the photoemitted electrons with sufficient energy resolution.

2.1. Synchrotron Radiation Beam Lines

The complexity of the installations of beam lines for photoemission work utilizing synchrotron radiation depends very much on the size of the storage ring, the main reason being that, at larger storage rings (like DORIS[49] and SPEAR,[50, 51] critical energies ~7–8 keV), radiation shielding is necessary. This means fairly long beam lines are required with limited access to the first optical elements that are necessary to deflect the radiation off the main x-ray beam. On the other hand, experiments at smaller storage rings like TANTALUS I[52] (critical energy 48 eV) or ACO (critical energy 320 eV),[53] short beam lines are used and no shielding is necessary. Although experimenters must leave during injection, they have excellent

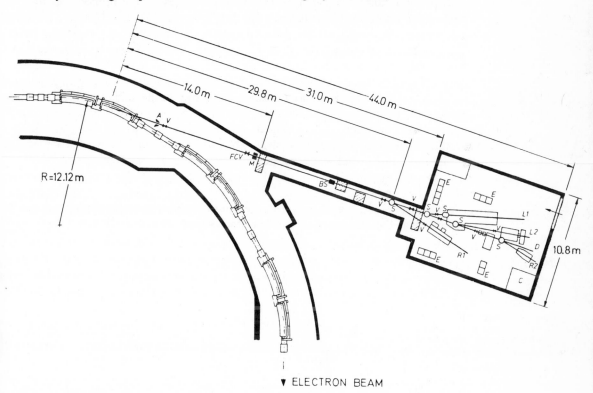

▼ ELECTRON BEAM

Figure 9. Layout of the synchrotron radiation laboratory at DORIS, showing part of the storage ring DORIS, the beam line into the laboratory, and the different branch lines. Note the distance from the source point to the experimental stations (from Ref. 49).

access to the experiment during stored beam operation, facilitating direct adjustment of all optical elements. These considerations will naturally affect the design of
the optical system and the monochromators. To give an idea of what distances are
involved between the source point and the experimental area, we show the layout of
the synchrotron radiation laboratory at DORIS in Figure 9. Radiation safety and
vacuum considerations determine the length of the transport system and the necessity of water-cooled absorbers (A), vacuum isolation valves (V, FCV), beam shutters (BS), and magnet (M) to deflect particles finding their way out into the
transport pipe. A beam of ~4 mrad is split (S) into three to four branch lines
(L_1, L_2, R_1) by grazing-incidence mirrors about 30 m from the source point.
The typical distance between an experimental photoemission station and the source
point at the SPEAR storage ring is similar, 20 m.[50, 51] For both cases, the whole
beam transport system, the optical system, and the monochromators have to be
designed for pressures below 10^{-9} torr, adding to the complexity.[54]

2.2. Monochromators

In this section, we can give only a brief survey of the monochromators of
interest for photoemission spectroscopy in the energy range 5–1200 eV. Different
monochromators are used for different spectral regions within this energy range.
Normal-incidence monochromators [55] are typically used to cover the range up to
40 eV and grazing-incidence instruments for regions beyond.[56, 57] One of the basic
criteria for photoemission measurements is that a fixed exit beam of high photon
flux is required, which is focused onto the focal spot of the analyzer used for energy
analysis of the photoemitted electrons. The required quality of the optical elements
and gratings for such an instrument are extremely severe, and both mirror and
grating technology have been considerably advanced from the experience gained
within synchrotron radiation research.[56, 57]

As an example of what can be accomplished with the present state-of-art
optics, we show in Figure 10 how the grazing incidence monochromator
("grasshopper")[56] has been implemented for photoemission studies on the 4°
beam line at the Stanford Synchrotron Radiation Laboratory.[50]

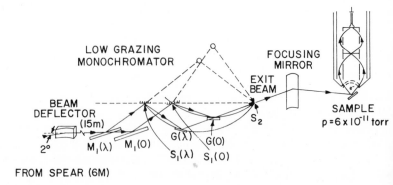

Figure 10. Schematic of the 4° beam
line monochromator at SSRL, showing
the main deflecting mirror (at 2°
grazing incidence), the monochromator, (M_1, S_1, G, S_2), the double
focusing mirror at the exit slit, and
the energy analyzer used for the photoemission experiments (from Ref. 14).

About 2 mrad of the synchrotron radiation is deflected horizontally by a platinum-coated copper mirror (noted beam deflector in Figure 10) onto the enhance slit of the monochromator located 16 m from the source point. The monochromator, which is of grazing-incidence type and operates at 10^{-9} to 10^{-10} torr, was designed and implemented for operation at SSRL by F. C. Brown and collaborators.[56] The exit beam leaves the monochromator at a constant angle from a fixed exit slit. M_1 is a spherical platinum-coated quartz mirror which focuses the source to the entrance slit S_1 (a Codling mirror–slit combination), and G is the grating. The optics in Figure 10 are shown for two positions corresponding to wavelength λ and to zero order. Scanning with a fixed exit beam is accomplished with Vodar geometry and with S_1, G, and S_2 located on a 1-m Rowland circle that rotates about S_2. Further details are given in the paper by Brown *et al.*[56] The monochromatic beam is refocused after the exit slit onto the sample, which is located in the focal spot of the energy analyzer. The monochromator works at a constant resolution of 0.1–0.2 Å and provides monochromatic radiation from 32 to 1200 eV with a 600 lines/mm Bausch and Lomb gold replica grating. It has thus been possible to use this monochromator for certain studies above the carbon *K*-edge, though the resolution gets worse very quickly with increasing photon energy. Further details about the performance of this instrument can be found in the paper by F. C. Brown *et al.*[56]

A plane grating monochromator adopted for ultrahigh vacuum (named FLIPPER)[57] has recently been implemented successfully at the DORIS storage ring for photoemission work up to 280 eV. An important feature of this grazing-incidence monochromator is that one of six different mirrors can be used to deflect the radiation onto the grating. In this manner, higher-order rejection can be accomplished by choosing the appropriate mirror for each photon energy range.

A grazing-incidence monochromator based on a holographic toroidal grating has recently been implemented by Depautex and collaborators.[58] The 3-m instrument is working at a grazing angle of 15°, limiting the high-energy cutoff to 180 eV, with a 600 lines/mm toroidal grating. In its first tests, the instrument showed excellent performance with superior monochromatic flux onto the sample as compared to other instruments in the same energy region.

2.3. Sample Chambers, Energy Analyzers, and Detector Systems

The common feature of sample chambers used for photoemission work is their design for ultrahigh vacuum. More or less, each research group utilizing synchrotron radiation for their photoemission work has sample chambers designed and modified to fit their own specific research purposes. In this section, we will only try to convey to the reader some of the features of a typical setup for photoemission spectroscopy. For this purpose, we give a short description of the apparatus we have been using for part of our research.

In the previous section, we showed schematically the sample and energy analyzer positions behind the exit slit of the monochromator. The sample chamber is a

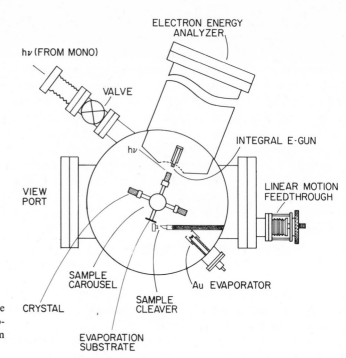

Figure 11. Schematic of the experimental sample chamber with some accessories shown for photoemission studies of surface phenomena (from Ref. 59).

standard ultrahigh-vacuum chamber (base pressure $\leq 4 \times 10^{-11}$ torr) equipped with ion and Ti-sublimation pumping and accessories for surface physics work using the photoemission technique: energy analyzer, LEED, Auger, sputtering gun, cleaver, evaporation beads, gas inlet system, etc. A schematic of our sample chamber is shown in Figure 11 for the case where we want to have the capability of cleaving several samples.[14, 59] Four samples can be mounted simultaneously and positioned at different places for cleavage, photoemission, overlayer deposition, Auger, LEED, and argon ion sputtering (not shown in Figure 11). The emphasis in the design of this sample chamber has been to provide optimal versatility for the application of other surface physics techniques, in addition to the photoemission measurements using the synchrotron radiation, to one and the same sample.

The electron energy analyzer is, of course, a most important component of the experiment and will now be briefly described.[60, 61, 62] To achieve an energy resolution of a few tenths of an electron volt, for electrons with kinetic energies ≥ 50 eV, it is necessary to use an electrostatic analyzer. A retarding field analyzer, commonly used for low photon energies (≤ 11.6 eV), collecting all electrons emitted over 2π does not allow sufficiently good resolution. A certain angular discrimination will occur for electrostatic analyzers. In our case, we use a double-pass cylindrical mirror analyzer (Physical Electronics 15-225G), which accepts electrons into a cone of approximately $\pm 6°$ over the entire 360°, thus averaging over fairly large angles.[62] The spot size on the sample from which electrons are collected is typically 1–2 mm². A schematic of the double-pass energy analyzer and a simple arrange-

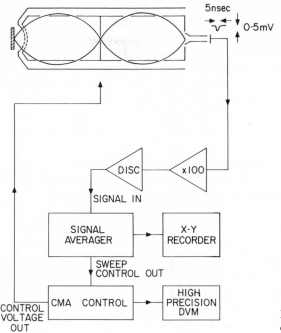

Figure 12. Schematic of the electron energy analyzer and the data acquisition system (from Ref. 14).

ment for the necessary data acquisition are shown in Figure 12. A channeltron (giving a narrow, 5 ns, pulse) is used as the electron detector, and pulse counting is done by standard techniques. The data acquisition is then done by a multichannel analyzer (a Tracor Northern 2048 channel signal averager in our case) or a computer (PDP 11/34 in our case).

Different energy analyzers used for photoemission work in conjunction with synchrotron radiation as the excitation source will not be described here. The reader is referred to the review paper on analyzers by Roy and Carette.[63] However, there is no doubt that the cylindrical mirror analyzer described in the previous section is the most popular.

Recently, there has been some important developments in improving the photoemission detection system beyond what was described above. A two-dimensional displaylike electron spectrometer for angle-resolved photoemission has thus been reported by Eastman *et al.*[64] It combines an electron reflection mirror (elliptical grids) as a low-pass filter with a retarding grid as a high-pass filter to achieve an energy selective analyzer with a large (86°) acceptance cone. A channel plate is used to amplify the energy-selected electrons and convert them to light pulses on a phosphor screen. Angle-resolved as well as angle-integrated detection are possible. A somewhat simpler multiple-angle parallel detection system has also been developed by S. P. Weeks and J. E. Rowe.[65] A series of angle-resolved photoemission spectra can be obtained in a much shorter time with these instruments than with a conventional detection system.

2.4. Concluding Remarks

The next few years should see a very rapid increase in the use of synchrotron radiation for photoemission studies due to a number of different experimental developments. With the increasing activity in synchrotron radiation research (e.g., increased use of existing rings, construction of new dedicated rings), more attention is being directed to optimization of the optics and instrumentation necessary for photoemission studies. The development of better mirrors and gratings will be an important technology for photoemission studies of core levels with high resolution and high intensity in the energy region ≥ 250 eV. From the work on the $4°$ beam line at SSRL, it has been demonstrated that certain types of experiments [e.g., high-resolution (≤ 0.3 eV) absorption and surface EXAFS] can be performed on C 1s (285 eV), N 1s (400 eV), O 1s (530 eV), and other core levels in this energy region.[56] However, so far, photoemission studies of chemisorbed systems have not been possible to perform with sufficient intensity and/or resolution for these important core levels. Presently, there is a strong emphasis in the development of instrumentation for this energy region, and considerable progress is to be expected within the next two years. Furthermore, the new developments of detector systems mentioned in Section 2.3 should have a strong impact, particularly in the area of angle-resolved photoemission.

3. Research Applications

3.1. Introduction

In presenting the research applications of photoemission using synchrotron radiation, we have chosen to select a few problems from different research areas. In Section 3.2, we describe how the energy distributions of the photoemitted electrons can be used to obtain information on the bulk electronic structure of the valence band. Comparisons are made to theoretical band-structure calculations. Section 3.3 deals with probing the electronic states associated with the surface states and surface resonances. The high surface sensitivity makes photoemission a unique tool for these kinds of studies. Section 3.4 illustrates some of the many applications photoemission presently has within surface physics and chemistry: gas chemisorption, oxidation, surface reactions, etc. The studies of interfaces described in Section 3.5 are relatively new but have developed fast, mainly due to the possibility of probing both the valence band for new electronic states and core levels for compositional and chemical changes at the interface. Section 3.6 is devoted to the profiling of the composition in the topmost atomic layers of a binary alloy, CuNi.

3.2. Bulk Electronic Structure

Photoemission measurements have been used extensively to study the valence band for a large number of materials and relating it to theoretical calculations of the bulk band structure. Of course, these studies were started using conventional sources [H$_2$ gas discharge lamps below 11.6 eV, resonance lamps using lines at 16.8, 27.2 (Ne), and 21.2 eV, 40.8 eV (He)], and a wealth of important information became available on the bulk electronic structure before synchrotron radiation was available. In this section, we will illustrate with three examples how the use of synchrotron radiation has resulted in a better understanding of the bulk electronic structure and its relation to theoretical calculations of the band structure.

3.2.1. The Bulk Electronic Structure of the Ge Valence Band

In many ways, Cu has been the metal prototype and Ge the semiconductor prototype for photoemission measurements, and an enormous amount of work has been done on these materials during the last 15 years. In this section, we would like to present part of the extensive work done on Ge by Grobman *et al.*[66-68] The availability of synchrotron radiation opened up the possibility to do (1) a detailed experimental study of the energy band structure for both the valence- and conduction-band states, and (2) a detailed evaluation of the validity of existing one-electron band models used to describe the band structure.

Photoemission energy distribution curves for the valence-band region of Ge (111) are shown in Figure 13 for a large number of photon energies from 6.5 to 23 eV.

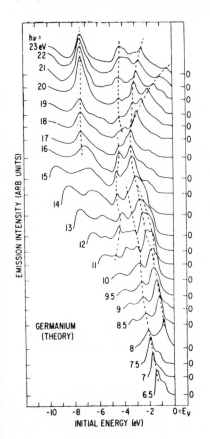
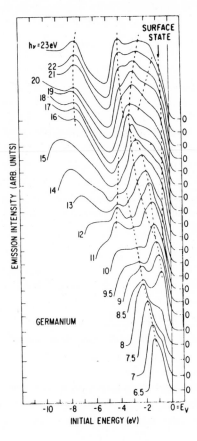

Figure 13. A comparison between experimental and theoretical energy distribution curves for Ge (111) in the photon energy range 6.5–23 eV (from Ref. 68).

The most remarkable feature in this collection of spectra is the strong variations with photon energy in peak locations and relative peak amplitudes. Several of the peaks can be followed over a large photon energy range and show changes in binding energy consistent with direct transitions. Since the movements of these peaks can be followed over a large energy range, it becomes increasingly possible to make a unique identification with theoretically predicted bands and critical points. Such a comparison to theoretically calculated photoemission distribution curves is also shown in Figure 13 as obtained by Grobman *et al.*[68] A nonlocal pseudo-potential due to Phillips and Pandey[69] was used to calculate energy bands, wave functions, and matrix elements. The photoemission process was treated within the three-step model (see Section 1.1) based on direct transitions. It is interesting to note that an improved agreement between theory and experiment was obtained by including an anisotropic surface transmission probability, i.e., a wave vector dependence was included in the transmission function. This would be one indication that the photoemitted electrons are specularly reflected at the surface, which is a necessity if one hopes to extract information from angle-resolved photoemission.

The major peaks in the experimental and theoretical distribution curves are shown as dashed lines. Many of the features show a good correlation and can be identified with direct interband transitions. In particular, it was possible to do a detailed analysis of the energy bands at the critical points L and X in the Brillouin zone for an energy range covering 25 eV (including both valence and conduction bands).

However, there are also important discrepancies between the calculated EDC's and the experimental spectra. It is obvious from Figure 13 that the calculated features do not account for all the emission intensity. One source of this discrepancy are the surface states due to dangling bonds, located at the top of the valence band and not included in the theoretical calculation. A further possible mechanism to explain the additional emission may be surface-associated transitions due to a very short electron escape depth. The direct k-conserving transition would thus be accompanied by an emission resembling the density of states. This contribution should be strongest where the escape depth is shortest, i.e., in the emission from the top part of the valence band (from where the electrons have higher kinetic energies). Some support for this proposal may be found in comparing the spectra in Figure 13. The experimental data appear to have a relatively stronger contribution from the upper part of the valence band.

3.2.2. The Density of States of Some IV, III–V, II–VI, and I–VII Compounds: A Comparison between Theory and Experiment

The band structure (bandwidths, critical-point positions) of group IV, III–V, and II–VI semiconductors have been studied extensively over the years with a number of different techniques, including photoemission. In 1974, several papers appeared at about the same time where new and important experimental information was provided on the valence-band structure of these compounds.[70–72] It had become possible to achieve sufficient resolution and ultrahigh-vacuum conditions using a conventional x-ray source ($hv = 1486.7$ eV) that bandwidths and critical-point assignments could be done with some confidence.[70, 71] A parallel development was the use of synchrotron radiation for this type of study.[72] Since the valence band of these semiconductor compounds is quite wide, typically 10–14 eV, excitation energies well above 25 eV are necessary to separate the valence band from the background of scattered electrons and to get to an energy regime where the spectra are not severely modulated by direct transitions. Below, we illustrate, with a few examples, the significance of these new results in terms of a comparison with existing and improved theoretical calculations.

A comparison between experimental curves (PDS, solid lines) and theoretically calculated density of states [$N(E)$, dashed lines] is shown in Figure 14 for six

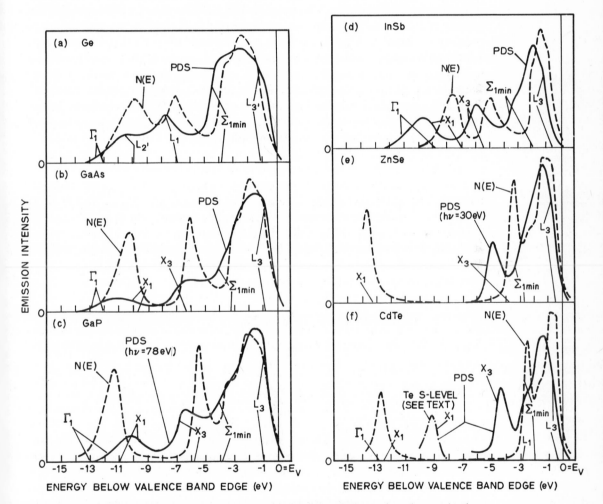

Figure 14. A comparison of energy distribution curves (the background of secondary electrons has been subtracted) for six semiconductors with the calculated density of states according to the empirical pseudopotential method (EPM) (from Ref. 72).

different semiconductors studied by Eastman *et al.*[72] The experimental energy distribution curves are shown with the secondary electron emission subtracted out (PDS = photoemission density of states). The theoretical density of states $N(E)$ are those calculated by Chadi and Cohen (as quoted in Reference 72) using the empirical pseudopotential method (EPM).[73] Energy-band critical points (L_3, Σ_1, X_3, X_1, Γ_1) are indicated by vertical arrows. The EPM results are shown by lines touching the abscissa, and the experimental ones by the lines pointing to the experimental curve. Overall, it can be said that there is good agreement between theory and experiment for the main spectral features: the same number of peaks and edges show up in both spectra at approximately the same energy. However, differences in the energy dependence of the partial photoionization cross sections for the various *s*- and *p*-derived valence bands result in variation in amplitudes between the observed and calculated peaks. The lowest *s*-like band is particularly weak and can typically not be seen at all for photon energies <25 eV. The most noteworthy observation regarding discrepancies in peak positions is that the EPM results (theory) consistently show valence bands that are narrower than those observed experimentally. This discrepancy amounts, for instance, to 0.7 eV for $E_V - X_3$ for GaAs and 1.8 eV for CdTe. It should be emphasized that the original EPM results by Chadi and Cohen were obtained by a fit to optical data[74] giving the band separations accurately but not the absolute energy positions. The discrepancy between theory and experiment led to a reexamination of the EPM method by Grobman *et al.*[68] It was found that the addition of a nonlocal term to the pseudopotential gave considerably better agreement between theory and experiment for Ge.[68] In fact, the theoretical distribution curves presented in Section 3.2.1 had been calculated by including nonlocal terms. Another important observation from Figure 14 is that the gap (X_3–X_1) between the two lowest valence bands increases on going from Ge to the more ionic materials ZnSe and CdTe. A correlation between the ionicity concept and the features X_3 and X_1 in the band structure could thus be established.[70, 72] Another trend is the narrowing of the total bandwidth with increasing ionicity.

3.2.3. The Electronic Structure of the Gold Valence Band

In Figure 15 we show a set of six distribution curves for Au at photon energies between 80 and 180 eV.[75] The binding energy is referred to the Fermi level, and the spectra are recorded at an instrumental resolution of 0.8 eV. Three different classes of features can be observed. Towards lower energies (higher binding energies), there is an increasing background of inelastically scattered electrons on which Auger and one-electron peaks are positioned. The Auger peaks can easily be separated from the photolines when a tunable source is available. The Auger electrons appear with a constant kinetic energy, independent of the excitation energy, and thus move towards higher binding away from the high-energy cutoff of the EDC with the same increment as the change in photon energy, whereas the one-electron peaks have a fixed position relative to the Fermi level. In the EDC taken at a photon energy of 120 eV, we have indicated the positions of the 5*d* levels (valence band), the 5*p* (the $5p_{1/2}$ component is obscured by the 4*f* levels), and the 4*f* levels. We

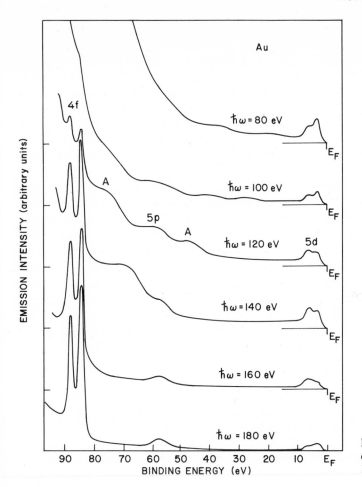

Figure 15. Photoemission spectra of Au for different photon energies between 80 and 180 eV (from Ref. 75).

assign the two Auger peaks (noted A in the figure) as originating from transitions between the valence band and the $5p$ doublet, $5p_{3/2}$ and $5p_{1/2}$.

We want to focus attention on several important experimental observations shown in Figure 15. The first has to do with the relative heights of the two peaks in the valence band with changing photon energy. As was observed by Freeouf *et al.*,[76] the valence band at $hv = 80$ eV is very similar to the x-ray photoemission spectrum at $hv = 1486.7$ eV and the theoretically calculated density of states.[77] Our experimental curve at $hv = 80$ eV is in good agreement with that observed by Freeouf *et al.*[76] In fact, our observations for $hv = 30$–90 eV with an instrumental resolution of 0.3 eV are in excellent agreement with those published by Freeouf *et al.*, shown in Figure 16. However, around $hv = 110$ eV, a reversal in the relative heights of the two peaks occurs, as is shown in some detail in Figure 17. All this energy, the lower peak has the same height as the leading peak, and at $hv = 120$ eV, the lower peak is higher. This trend continues up to 140 eV when the relative heights start to shift back again, and at 180 eV, the shape of the valence band is

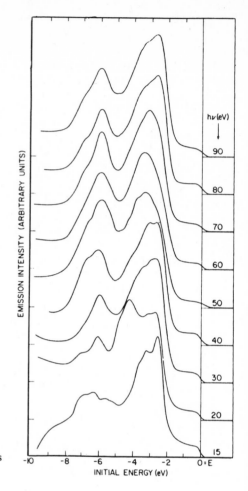

Figure 16. Photoemission spectra of Au for different photon energies between 15 and 90 eV (from Ref. 76).

very close to that at 80 eV. Thus, the relative peak heights are modulated over approximately 100 eV. This is shown in Figure 18, which plots the relative heights of the two peaks as a function of photon energy. It should be pointed out that, for a couple of photon energies, the experiment was repeated at higher resolution (0.3 eV), and the results confirm the data shown in Figure 17. It should also be mentioned that other parameters of the valence band (bandwidth, peak locations relative to the Fermi level, and peak separations) are unchanged over the energy region where the amplitude modulation takes place.

The observed change in the distribution curves of the valence electrons between 100 and 200 eV raises questions concerning the concept of an x-ray limit (above which direct transitions are no longer important), as it was introduced by Freeouf et al.,[76] as well as the experimental determination of this limit.[39] Presently, we have no firm model for the observed modulation but we will discuss several possible options. Modulation of peak amplitudes as a function of electron emission angle[78] has been observed in the x-ray photoemission spectrum taken at

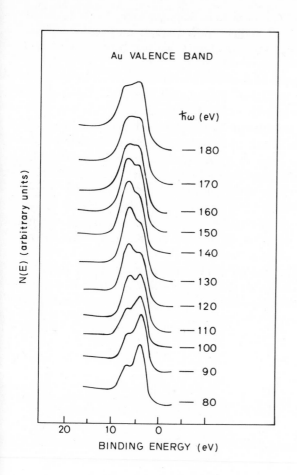

Figure 17. The Au 5d valence band for different photon energies between 80 and 180 eV showing the modulation in the relative peak heights (from Ref. 75).

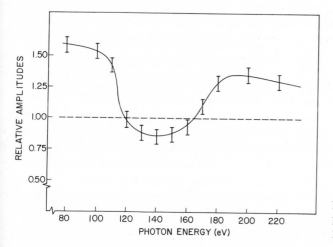

Figure 18. The ratio between the first and second peaks in the Au 5d valence band as a function of photon energy between 80 and 240 eV.

a fixed photon energy, $hv = 1486.7$ eV, as well as in Auger spectra, and it has been ascribed to channeling effects and/or crystallographic diffraction effects.[79, 80] The difference in energy between the two peaks in the Au valence band appears too small (3.5 eV) to explain our experimental results on such grounds. Angular anisotropies in the photoemission from gold at $hv = 21.2$ eV have also been reported by Koyama and Hughey.[81] However, at 21.2 eV, modulations due to joint density-of-state effects are still very important and are probable explanations.

Another possible explanation for changes in the valence band with changing photon energy relates to the escape depth of the photoemitted electrons. As has been discussed in Section 1.3, the escape depth goes through a minimum in the energy range 100–200 eV, with an escape depth of just a few angstroms. [In a separate investigation using synchrotron radiation, we have determined the escape depth for $hv = 30$–80 eV.[82]] However, if a short escape depth would cause changes, it would not primarily show up as an amplitude modulation. Changes in bandwidth would, for instance, be expected, as well as a smaller, more atomiclike, $5d$ spin–orbit splitting.

It is quite conceivable that the photoionization cross sections for the two spin–orbit-split components, $5d_{5/2}$ and $5d_{3/2}$, have a slightly different hv dependence. The cross section for the $5d$ levels go through a Cooper minimum, similar to

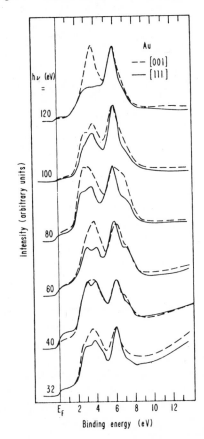

Figure 19. Comparison of photoemission energy distributions along the [001] and [111] directions of Au in the photon energy range 32–120 eV (from Ref. 84).

that shown for the Pt $5d$ valence band in Section 3.4.2. In energy regions where the cross sections, $\sigma(5d_{5/2})$ and $\sigma(5d_{3/2})$, go through this minimum, it may happen that the ratio $\sigma(5d_{5/2}) : \sigma(5d_{3/2})$ shows significant variations. However, from our experimental determination of the energy dependence of the $5d$ cross section, we have concluded that a difference in hv dependence for the two $5d$ components cannot explain the strong modulation effect. It should also be pointed out that the modulation was not observed for the Pb $5d_{5/2}$, $5d_{3/2}$ levels, which are truly atomiclike and do not have any bandstructure contribution.[83].

Finally, we should mention that angle-resolved photoemission measurements along the [001] and [111] directions on single Au crystals by Stöhr *et al.*[84] (See Figure 19) show strong variations up to the highest photon energy used (130 eV). This provides additional evidence that convergence to the x-ray limit has not yet been reached in the case of Au.

In summary, it can be stated that new effects were observed in the valence-band spectra of Au when the photon range was extended well above the limit at which the photoemission curves were thought to reflect the density of states. A satisfactory explanation of the observed phenomena has so far been elusive.

3.3. Surface Electronic Structure

As has been the case with studies of bulk electronic structure, surface states and resonances were first examined by photoemission techniques using conventional light sources. However, the polarization as well as the photon energy dependence of angular-resolved photoemission have, during the last couple of years, contributed significantly to a better understanding of surface states and resonances on both semiconductors and metals and stimulated more thorough theoretical treatments.

3.3.1. Surface States and Resonances on Single Crystals of Metals

The (100) faces of W and Mo have, by tradition, been the most popular for surface states/resonances studies[11, 85-105] on metals. From an extensive investigation, recently completed by Weng *et al.*[85, 86] it was concluded that there are three surface states/resonances on the (100) surface of W and Mo. The experimental photoemission work using synchrotron radiation was combined with a nonrelativistic tight-binding Green's function scheme to calculate the orbital character, the energy position and dispersion, and the penetration depth into the bulk of the surface states/resonances.

In Figure 20 photoemission spectra are shown at different polar angles close to the normal emission direction.[85] The peaks noted with arrows located about 0.3 and 0.65 eV below the Fermi level (E_F) for Mo and about 0.4 and 0.9 eV below E_F for W have been assigned to surface states/resonances. It should be noted that the lower-lying peak is absent in the normal emission spectra and that there is a dramatic change in relative intensities with increasing polar angles. It is also obvious from these spectra that a high angular resolution is required since, in particular, the intensity of the second peak (the low-lying peak) changes rapidly for small incremental increases of the polar angle. Not shown in the figure is a deeper-lying structure, 3.3 eV below E_F for Mo and 4.2 eV below E_F for W, which also has been interpreted as a surface state/resonance. The energy dispersion for all three peaks is small, less than 0.3 eV.

Much more information, in addition to the energy location and dispersion relations, on surface states/resonances can be obtained by utilizing the polarization and tunability of the synchrotron radiation to establish their orbital symmetry and photon energy dependence. The polarization property is particularly helpful to sort out the orbital symmetry. It can be shown that the following symmetry selection rules apply to the initial state. Only those initial states that have the same parity as that of the dipole operator (or the polarization vector $\mathbf{A}$) will be excited. For the (100) face of a bcc crystal, only initial states of symmetry Δ_1 and Δ_5 can be excited. Specifically, this means that initial states with Δ_1 symmetry can only be excited by normal components of the polarization vector, A_z, and those with Δ_5 symmetry only by surface components, A_x and A_y.

For the photoemission spectra shown in Figure 21, the experimental configuration has been arranged to take advantage of these selection rules. For the upper distribution curve, the polarization vector of the incoming p-polarized light is so chosen that only initial states of even parity can be excited. For the lower curves, it has been chosen so that initial states of both even and odd parity can be excited. Figure 21 shows that the second peak at 0.65 eV below E_F is much more pronounced in the lower curve than in the upper one. Therefore, it can be concluded

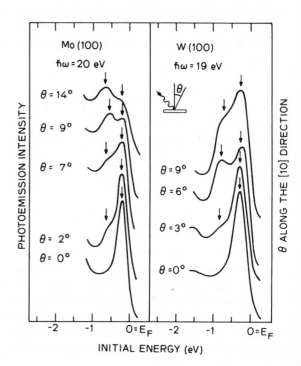

Figure 20. Angle-resolved photoemission spectra of Mo (100) and W (100) at various polar angles θ along the [10] direction (from Ref. 85).

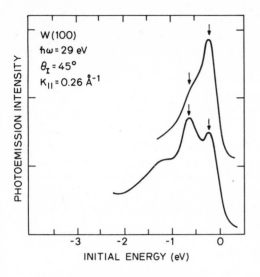

Figure 21. Angle-resolved photoemission spectra of W (100) measured in the (110) mirror plane. For the upper spectrum, the polarization vector **A** is in the same plane as the energy analyzer, and for the lower one, it has been rotated 90° around the surface normal (from Ref 85).

that the structure arising from the peak at 0.65 eV has odd parity, i.e., $d_{zx, xy}$ and $d_{x^2-y^2}$ orbitals. In the same manner, the polarization dependence of the 0.3 eV peak suggests that its parity is even, since the upper curve can only contain initial states with even parity, e.g. d_{z^2} or s.

Theoretical calculations by Weng et al.,[85] as well as others not done self-consistently,[86–94] cannot explain the highest-lying surface state/resonance peak (closest to E_F). However, a self-consistent calculation based on the pseudopotential formalism, recently published by Kerker et al.,[95] predicts the double-peak structure with the correct symmetry as observed experimentally. On the other hand, the second peak is predicted with the correct symmetry orbitals from the theory by Weng et al.[85] Thus, a detailed understanding of the surface states/resonances and comparison with the most relevant theory still does not seem to be settled. The photon energy dependence of the cross section for the different surface state/resonance orbitals, as well as the angular dependence of the electron emission, can be used to extract further information for comparison between theory and experiment. The main idea behind these approaches will be illustrated in Section 3.4.2, where CO chemisorption on Ni is treated.

3.3.2. The Surface Electronic Structure of GaAs (110)

Filled surface states on semiconductors may be easier to detect than on metals if they lie in the absolute band gap separating the valence and conduction bands. Furthermore, the effect of the surface states on band bending can be used for identification. This is the case for Si (111) and Ge (111) where the surface states were first studied with the photoemission technique using conventional light sources concurrently with synchrotron radiation.[106, 107] However, for the (110) cleavage surface of most of the III–V semiconductors, there is a rearrangement of the surface atoms tending to sweep the surface states out of the band gap (see

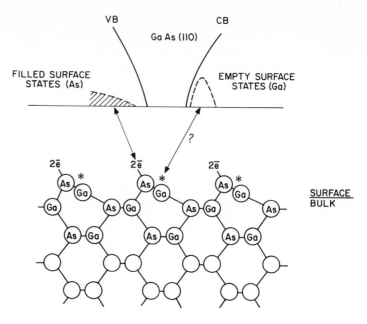

Figure 22. A schematic of the electronic and spatial configurations of the GaAs (110) surface. The As atoms have moved outward and the Ga atoms inward compared to the positions in the bulk of the crystal (from Ref. 16).

below). And, as in the case for the metals discussed in the previous section, there will be an overlap between the emission from the bulk and surface states at the top of the valence band.

The GaAs (110) is, by far, the III–V compound surface that has been studied most systematically both through careful measurements and sophisticated calculations. Therefore, we will concentrate on this system below. Figure 22 gives a schematic indication of our present understanding of the clean GaAs (110) surface.[16, 59] When a covalent bond is broken upon cleavage, each surface atom tends towards its atomic configuration, i.e., p^3s^2 for As and s^2p for Ga. Thus, five electrons will be associated with As and three with Ga. The positions and bond configurations of the surface atoms are determined by achieving the lowest-energy state. This leads to a rearrangement of the atoms within the surface lattice, and the filled and empty surface states are removed from the band gap (calculations for an ideal surface predict the surface states to lie in the band gap[108]). The detailed rearrangement has been the subject for lively and controversial discussions over the last few years, and the issue does not yet seem to be settled. Roughly speaking, there is general agreement that the As atoms will be rotated outwards and the Ga atoms inwards. At issue is the size of these rotations and whether or not they are combined with a relaxation between the surface planes. The surface rearrangement has important consequences for the details of the surface electronic structure and has been discussed extensively from both theoretical and experimental points of view.[108–149] The surface states/resonances are characterized by wave functions that fall off rapidly away from the surface, and their complexity is illustrated by the theoretical calculations by Chadi[143] shown in Figure 23. The density of surface states are shown as the solid line. The notation B_1, B_2, and B_3 is used for filled surface states and B'_1, etc., for empty states. The states noted S_i are strongly model dependent and arises in this case from surface relaxation. Angle resolved photoemission using conventional sources and synchrotron radiation has recently been carried out and an encouraging qualitative agreement between theory and experiment was rendered for the basic features.[123, 127, 128]

The difficulties to be overcome in achieving a more detailed understanding of the surface electronic structure of GaAs (110) are illustrated below. As was mentioned earlier, the electronic structure associated with the last one or two atomic layers can be examined using synchrotron radiation because of the short escape depth that can be attained by selection of the optimal photon energy. Because the valence-band matrix elements decrease bery rapidly with $\hbar\omega$, it is most advantageous to examine the surface valence states of GaAs (110) near 20 eV, i.e., for photon energies lower than those giving the shortest escape depth.

In Figure 24 we illustrate the differences in the surface electronic structure for two different GaAs crystals, given the notations LD1C and MCPB.[117, 118] The effects of a small amount of oxygen is also shown. As can be seen from the figure, there are noticeable differences, particularly in the top 4 eV of the valence band, depending on the crystal surface quality. The oxygen exposures also affect the two surfaces in a very different manner. The curves in the left-hand panel (sample LD1C) show sharp structure in the valence band and an unpinned Fermi level (i.e., the pinning position is right below the conduction minimum) for the clean cleaved surface. Oxygen exposures up to 10^3 L (1 L = 10^{-6} torr sec; note that $\sim 10^8$ L is required for a monolayer) result in a change of pinning position to about midgap, but the surface electronic structure is essentially unchanged. The results for the MCPB sample are drastically different. The structure in the top 4 eV of the valence band is more smeared out, and the Fermi level is pinned at about mid-gap for the clean surface. Sharpening of the structure takes place upon oxygen exposure. It should be noted that the effects of oxygen on the surface electronic structure

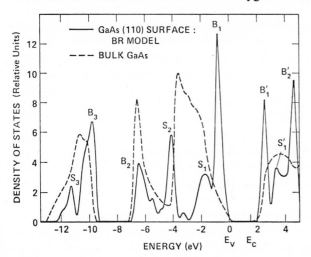

Figure 23. The local density of surface states (solid line) for the bond relaxation model of the GaAs (110) surface. Electronic states localized on the first two layers of Ga and As atoms are shown (from Ref. 143).

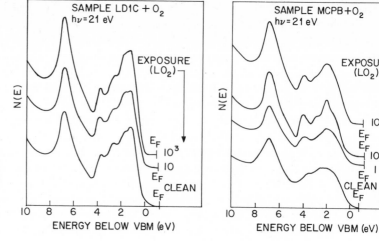

Figure 24. Energy distribution curves from the upper part of the valence band of two different GaAs (110) crystal. The effects of a small oxygen exposure on the Fermi level pinning and valence band structure are also shown (from Ref. 118).

appear long before any emission can be observed from the chemisorbed oxygen itself. The Fermi level position also goes through an interesting evolution. The Fermi level is unpinned for the lowest exposures, 1 to 10 L, but then moves towards a pinned position again as the exposure is increased to 10^3 L.

We ascribe the sharpening of the structure for the MCPB sample upon oxygen exposure to be due to a rearrangement of the surface atoms and relief of strain (probably introduced by the cleavage) within the surface lattice. This may occur for coverages as low as one oxygen per 10^4 to 10^5 surface atoms, indicating a long-range effect of the oxygen on the surface. The sharpening of the valence-band structure occurs at the same time the Fermi level is being unpinned, providing evidence that the surface atom arrangement is closer to that described by the calculation for a relaxed surface with no states within the band gap. Higher oxygen exposures, $\geq 10^3$ L, will introduce a sufficient number of extrinsic states in the band gap to cause pinning anew.

It should be noted that the sharp structure in the MCPB sample produced after 10-L oxygen exposure is considerably different from that of the cleaved LD1C sample. This suggests that the surface atomic rearrangement may be different in the two cases. Since this rearrangement takes place within the unit cell, the low-energy diffraction (LEED) technique has difficulties in distinguishing the two cases. Thus, photoemission may be a more sensitive tool for detecting surface rearrangements. However, it is obvious that a detailed comparison with theoretical calculations presents difficulties since the details of the surface electronic structure are very sensitive to the exact geometrical configuration of the surface atoms. Needless to say, this will also make the interpretation of angle-resolved spectra harder.

3.4. Chemisorption and Oxidation Studies

One very successful use of photoemission in studies of chemisorption has been the "fingerprint" technique in which photoemission spectra from chemisorbed species are compared to known gas-phase spectra for identification of the different molecular levels. Information about dissociation, surface reaction products, etc., can be extracted with this empirical method. In x-ray photoemission, chemical shifts of the inner core level have been used for a number of years to determine the chemical state (e.g., the formation of different oxides) or to follow changes in the chemical environment. But synchrotron radiation made it possible to probe both the molecular levels of the chemisorbed species and the inner core levels of the substrate with high surface sensitivity (1 to 2 molecular layers).

3.4.1. Oxygen Chemisorption and the Initial Oxidation Stages on the GaAs (110) Surface

Since the GaAs (110) surface discussed in Section 3.3.2 is perhaps the best understood of all semiconductor surfaces, both experimentally and theoretically, it is also a natural choice for studies of oxygen chemisorption and the initial stages in the oxidation.[150–162, 113–118, 137] Both the highest-lying core states (Ga $3d$, As $3d$) and the valence-band states were used in the investigations described below. Chemical shifts of the core levels are particularly valuable in following the formation of different chemical states on the surface.[156] The valence-band studies give information on changes in the surface electronic structure (see Section 3.3.2) induced by the oxygen adsorption as well as energy location of oxygen derived electronic levels in the chemisorbed and/or oxidation stages.[118] In all these cases, the excitation energies can be chosen so that optimal surface sensitivity is achieved. In the course of these investigations, we found that the state of excitation of the oxygen could play a key role in determining the sorption process on the GaAs surface. Thus, it was found that the presence of a nude ionization gauge, used to monitor the pressure in the experimental chamber, drastically affected the way the chemisorbed oxygen interacted with the surface. In the following, we will therefore refer to the terms nonexcited molecular oxygen and excited oxygen; the reader is referred to the literature for further details.[118, 156]

In Figure 25, we show spectra for the clean and oxidized GaAs (110) surface at $\hbar\omega = 100$ eV.[156] As the surface is exposed to oxygen, a peak appears 2.9 eV below the arsenic $3d$ peak with a proportionate decrease in the As $3d$ intensity. This is a chemically shifted peak indicating a transfer of charge from the surface arsenic atoms to the adsorbed oxygen. Concurrent with the appearance of the shifted As peak, we see the O $2p$ resonance level at a binding energy of about 5 eV. In contrast to this, the Ga $3d$ peak only shows a slight broadening (0.5 eV) at the highest exposure. When the oxygen exposure is increased, the shifted As $3d$ peak and O $2p$ level grow simultaneously until saturation is reached between 10^9 and 10^{12} L O_2. (10^{12} L corresponds to atmospheric pressure for 20 min). It should be noted that the chemical state characterized by the 2.9-eV-shifted As $3d$ is extremely stable at exposures to nonexcited oxygen.

The results for exposure to excited oxygen exhibit several important differences as shown in Figure 26.[156] First, the oxygen uptake is 2 to 3 orders of magnitude faster. As in the case for exposure to nonexcited molecular oxygen, a 2.9-eV-shifted As $3d$ peak appears. However, increasing oxygen exposure gives a second As $3d$

peak, shifted by 4.5 eV, together with a Ga 3d peak shifted by 1.0 eV. The latter two peaks are dominant for exposures above 10^6 L.

The results presented above have provided new insight into our understanding of the chemisorption and oxidation properties of the GaAs surface, though some ambiguity still exists in the detailed interpretation. It should again be emphasized that about two molecular layers[156] (see Section 1.3) are probed in the data shown in Figures 25 and 26. The oxygen interaction with the GaAs surface can thus be followed with extreme surface sensitivity. The interaction stage where a 1.0-eV-shifted Ga 3d and a 4.5-eV-shifted As 3d peak are observed may be easiest to understand. The Ga 3d shift corresponds to the formation of Ga_2O_3, since the chemical shift is identical to that observed for the bulk oxide. The 4.5-eV-shifted As 3d falls closest to the value observed for the As_2O_5, though there is a slight discrepancy as discussed by Pianetta *et al.*[156] Thus, it appears quite clear that we are forming Ga and As oxides at this stage and that back bonds are being broken between the surface atoms and the bulk. True oxide formation thus occurs. It is more complicated to understand the earlier stage in the chemisorption (exposure to nonexcited molecular O_2), characterized by a 2.9-eV-shifted As 3d peak. The Ga 3d peak did not seem to be affected in this case, except for a slight broadening. Therefore, it is hard to reconcile the Ga 3d core level data with the presence of any large amounts of Ga_2O_3. Even the highest exposure (10^{12} L O_2) did not give evidence for the 1.0-eV-shifted Ga 3d peak, and only one, 2.9-eV-shifted, As 3d peak appeared. This observation led to the conclusion that the oxygen chemisorption in

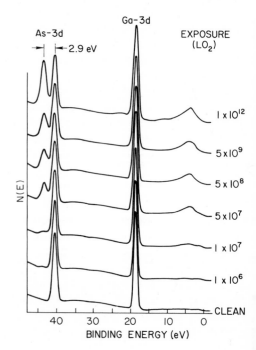

Figure 25. Photoemission spectra at $hv = 100$ eV for GaAs (110) exposed to molecular oxygen (from Ref. 156).

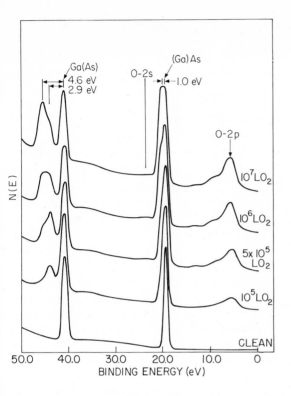

Figure 26. Photoemission spectra at $hv = 100$ eV for GaAs (110) exposed to activated oxygen (from Ref. 156).

this stage takes place on the surface arsenic atoms. This model would be consistent with the current picture of a rearranged (110) surface with the As atoms rotated outwards and the Ga inwards. The surface Ga atoms have three electrons, all involved in back bonding, while the As has five electrons. Three of these electrons take part in the back bonds, and the remaining two form "dangling bond orbitals." Since all the bonding electrons associated with the surface Ga atoms are involved in the back bonds, the oxygen will bond to the Ga only after one or more of the back bonds are broken. Even on exposure to excited oxygen, the 2.9-eV-shifted As 3d peak is present before the appearance of the 4.5-eV peak, supporting the conclusion that oxygen must be chemisorbed to the As before the bonds between the surface atoms and the bulk are broken.

An important question is whether oxygen is chemisorbed as an atom or as a molecule. Mele and Joannopoulos[160] have reconciled a large body of experimental data with a model of molecular chemisorption on the As sites. Goddard *et al.*,[162] also predict chemisorption on the As but favor atomic oxygen. The valence-band photoemission data are presently ambiguous as to whether the oxygen dissociates or not. Finally, it should be remarked that the evolution of different oxides on the GaAs (110) surface has been followed to higher exposures and has demonstrated the usefulness of photoemission for studies of adsorption and reactions on semiconductor surfaces.[156]

3.4.2. Chemisorption of CO on Transition Metal Surfaces— Molecular Levels

The application of the photoemission technique to the chemisorption studies of gases on surfaces emerged as an exciting field in the early 1970s.[18] As will be shown in this section, both the tunability and polarization of the synchrotron radiation have proved very important for further developments in this area. We will illustrate this with two examples: CO chemisorption onto Pt and Ni. CO has been chosen for several reasons. The molecular levels of CO are fairly well understood, and the assignment of the orbital symmetries can be carried over from the gas phase to the chemisorbed system. Furthermore, CO chemisorbs nondissociatively on both Ni and Pt. The reader is referred to the literature for further information on CO chemisorption on transition metal surfaces.[9, 10, 13, 15, 18, 163–179]

For CO in the gas phase, the molecular levels 4σ, 1π, and 5σ have binding energies of 20 eV or less,[180] whereas the 3σ has a binding energy of about 38 eV.[181] The 3σ levels are essentially O 2s and play an important role for the molecular bonding. The 4σ and 5σ levels are the lone-pair electrons on the oxygen atoms (but with some charge on the carbon) and the carbon atoms, respectively. The 1π levels are bonding orbitals with the charge distribution slightly weighted towards the oxygen atoms. When CO is adsorbed on a metal, the 5σ levels (the lone-pair electrons) are involved in bonding to the surface and are moved downward in energy so that they merge in energy with the 1π orbitals, whereas the 4σ levels retain their independent identity. Prior to the work to be described below, the 3σ levels had not been clearly observed for the adsorbed molecule. There were at least two reasons for this. First, the matrix elements for the 3σ transition are weak[182] in the range where most of the previous work was done. Second, the secondary electrons produced by scattering of valence-band electrons tended to obscure the weak 3σ excitation. However, by choosing the proper photon energy from continuously tunable synchrotron radiation, the 3σ levels can be studied.

As discussed in detail elsewhere,[41, 42] the cross section for the $5d$ states decreases from a maximum above threshold and goes through a low and shallow minimum about 150 eV above threshold. This is the Cooper minimum discussed in Section 1.4. This Cooper minimum can be used in chemisorption studies to suppress the emission from the substrate and enhance the emission from the molecularly chemisorbed levels. This is shown in Figure 27 for CO chemisorption onto Pt.[183] The Pt sample was cleaned *in situ* (10^{-11} torr base pressure) by heating to 1800° K in 10^{-7} torr oxygen, followed by argon ion sputtering and annealing according to a standard procedure. The sample cleanliness as well as the CO coverage were checked with Auger electron spectroscopy. The emission from the Pt substrate ($5d$ band) is low due to the Copper minimum, and the background of scattered electrons from the Pt valence band is minimized. The emission from the CO induced molecular levels (3σ, 4σ, 1π–5σ) is correspondingly enhanced.[184] By taking advantage of working around the Cooper minimum, several new and important observations could be made for the CO chemisorption onto Pt. This data shows the first unambiguous identification of the 3σ orbital of chemisorbed CO. The nonbonding molecular orbitals (1π, 4σ) undergo a relaxation shift of typically a

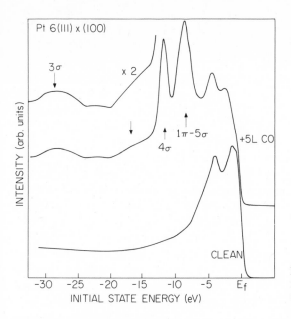

Figure 27. The valence-band spectra at a photon energy of 150 eV for clean Pt and Pt exposed to about a monolayer of CO (from Ref. 183).

few electron volts upon chemisorption as a result of the efficient screening on the metal surface.[18] It has been found experimentally that this relaxation shift is uniform for most of the shallow orbitals (binding energy ≤ 12 eV).[185] The relaxation shift of the 3σ level in CO therefore constitutes an interesting test case for a more tightly bound level, 28.5 eV below E_F, but still a level of molecular character. By comparison to gas-phase data (see Figure 28), we showed that the 3σ electrons suffer a 2.4 eV larger relaxation shift than the 4σ electrons upon chemisorption. The higher degree of localization[187, 188] of the 3σ was interpreted as the main reason for this effect. Another important observation from Figure 28 is that structure is present between the 3σ and 4σ levels, reminiscent of shakeup satellite structure observed earlier for CO in the gas phase.[186] But this appears to be the first time that shakeup structure has been observed for valence orbitals of a chemisorbed molecule. The exact nature of the observed satellite structure is not completely clear. It bears some resemblance to the shakeup structure observed in transition-metal carbonyls, where peaks appear 5–7 eV below each of the CO-related orbitals and are attributed to a mechanism involving charge-transfer excitation from the metal atoms into the $2\pi^*$ CO-derived orbital.[189, 190]

In summary, it should be emphasized that these new observations were made possible only by tuning the photon energy so that optimal contrast and, therefore, optimal surface sensitivity was achieved for the emission from the substrate and the chemisorbed levels, respectively.

In chemisorption studies on single-crystal surfaces, it is also of great interest to know the orientation of the chemisorbed molecule. The unique properties of synchrotron radiation, combined with angle-resolved photoemission (ARPES), have recently been applied to study the CO chemisorption on Ni (100) by Allyn *et al.*[177]

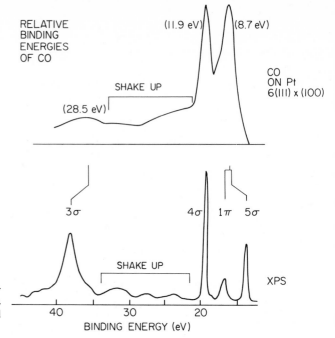

Figure 28. The gas-phase spectrum of CO (Reference 185) is shown in the lower part of the figure. The upper curve gives the difference between Pt and Pt + 5 L CO (from Ref. 183).

Figure 29 shows EDC's taken by ARPES for four different geometries of the polarization vector of the incident light and the emission angle of the photoelectrons. The structure at energies above −5 eV is due to the valence states of the Ni, that below −5 eV to the adsorbed CO (see Figure 29). The peak labeled "P" is due to the 1π and the 5σ orbitals, which overlap in energy. The latter is strongly shifted due to its involvement in the bonding of the CO to the metal as discussed above. The significance of Figure 29 lies in the strong modulation of the two CO peaks as a function of the direction of the polarization of the radiation and the direction of emission of the emitted electrons. By keeping these angles fixed and varying the photon energy, Allyn et al.[177] studied the variation of peak intensities with photon energy in ARPES.

The theoretical work by Dill and Dehmer[191, 192] and by Davenport[193, 194] had suggested that a resonance, i.e., a peak in the absorption cross section, would occur for emission from the 4σ and 5σ (but not 1π) states when the direction of polarization of the radiation **A** and the direction of electron emissions were along the axis of the CO molecule. The resonance is associated with scattering of the excited electrons between the carbon and oxygen atoms, and thus its directional dependence is not surprising. Experimentally, Allyn et al.[177] observed a resonance only when both the direction of emission is close to $\theta = 0°$ and when the **A** vector has a component along the CO axis ($\theta_I = 45°$ but not 0°). Thus, evidence was established that the CO molecule is perpendicular to the surface. Further experiments on the strength of the 4σ peak as a function of polar angle emission (and

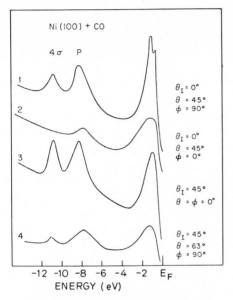

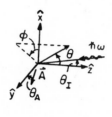

Figure 29. Angle-resolved photoemission spectra at $hv = 32$ eV for CO chemisorbed on Ni (100) (from Ref. 177).

comparison to Davenport's theory) showed that the CO molecule could not be tilted more than 5° from the normal.

One additional experiment was performed by Allyn *et al.*[177] to confirm that the CO molecule is bonded to the Ni surface by the carbon atom. The results are shown in Figure 30. As can be seen, the experiment agrees much better for the case with carbon "down," i.e., next to the surface, than for the other calculated cases.[193]

This example illustrates how ARPES experiments utilizing synchrotron radiation combined with theoretical calculations can be applied to obtain detailed knowledge of the orientation of a gas molecule chemisorbed on a metal surface.

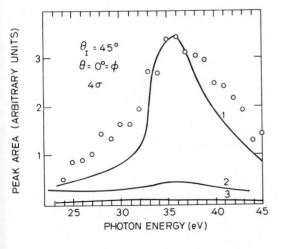

Figure 30. The amplitude of the 4σ peak for electrons emitted in the direction normal to the Ni (100) surface as a function of photon energy (from Reference 177). Solid curves are calculated for the CO axis normal to the surface with carbon (1) and oxygen (2) facing down, respectively, and for the CO axis in the surface plane (3).

3.4.3. The Effect of Chemisorption on the Substrate Core Levels

A wealth of information has been obtained about chemisorption and oxidation by using synchrotron radiation for studies of the atomic and molecular levels of the chemisorbed species. However, as was pointed out in Section 2.4, it has, to date, not been possible to study the C 1s, N 1s, and O 1s core levels for chemisorbed species with synchrotron radiation due to the lack of adequate monochromators in the few-hundred electron volt range. However, as was illustrated in Section 3.4.1, measurements of the substrate core level binding energies and the chemical shifts of these levels can provide important information on the chemical state of the substrate atoms involved in the chemisorption and/or oxidation.

In this section, we will discuss the important question of whether the substrate core levels undergo any chemical shift upon chemisorption of molecular species at (sub)monolayer coverages.[195] For this purpose, photoemission measurements were performed on a clean Pt (100) surface and on a surface exposed to saturation coverage of CO.[196] Synchrotron radiation was used as the excitation source in the photon energy region 30–280 eV. Photoemission spectra of the valence band for clean Pt and Pt exposed to 5 L CO were taken at a photon energy of 150 eV and were discussed briefly in Section 3.4.2. The valence band of clean Pt has a total width of 7.5 to 8.0 eV and exhibits structure in good agreement with results reported in the literature,[197] as shown in Figure 27. Upon adsorption of 5 L CO, which produces saturation coverage, we observe two CO-induced molecular levels at binding energies (E_B) of 11.9 and 8.7 eV. These peaks are assigned to the 4σ levels ($E_B = 11.9$ eV) and to a composite of 5σ–1π levels ($E_B = 8.7$ eV). The intensity of the leading peak (closest to the Fermi level) in the Pt valence band is preferentially decreased on the CO adsorption, analogous to the observation for most transition metals.[198] Overall, there are substantial changes in the emission from the Pt valence band upon CO adsorption.

In Figure 31 we compare the $4f$ core level spectra[196] for tne same sample under the same conditions as discussed in Figure 27: clean Pt and Pt + 5 L CO.

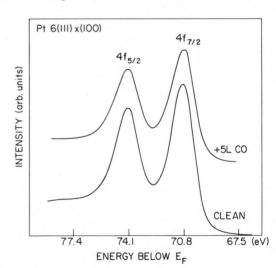

Figure 31. Photoemission spectra of the Pt $4f$ levels for clean Pt and Pt exposed to 5 L of CO taken at a photon energy of 200 eV (from Ref. 196).

The Pt $4f$ doublet has binding energies of 70.8 eV ($4f_{7/2}$) and 74.1 eV ($4f_{5/2}$), respectively. On the chemisorption of CO, there is an attenuation of the intensity of the $4f$ levels but no chemical shift is observed within the experimental accuracy (± 0.15 eV) for coverage up to saturation (monolayer). Furthermore, there is no measurable change in the line shape of the core levels. Similar results for the $4f$ levels are obtained for chemisorption of O_2 into W.[196] These results should be contrasted to the chemisorption of, for instance, oxygen on semiconductor surfaces, as discussed in Section 3.4.1, where pronounced chemical shifts are observed for the substrate core levels well below monolayer coverage. Oxygen chemisorption on Al is still another case, which will be discussed in the next section.

The question of any observable chemical shifts for substrate core levels upon chemisorption on metal surfaces has been quite controversial since earlier x-ray photoemission work[199] using conventional x-ray tubes ($hv = 1486.7$ eV). In most of these cases, the surface sensitivity was enhanced by collecting the photoemitted electrons at a grazing angle to the surface, giving a shorter effective escape depth but introducing the complication of possible distortion in the electron optics. The results shown in Figure 31 are for a photon energy of 200 eV, i.e., the photoemission electrons have an energy of about 130 eV, which falls in the minimum (≈ 5 Å) of the escape depth curve (see Section 1.3), and a high surface sensitivity is achieved without resorting to a shallow exit angle.

We consider that the very efficient screening provided by the metal valence electrons is the most plausible explanation for the lack of a chemical shift of the Pt substrate core levels upon CO adsorption. Any charge transfer between the chemisorbed CO and the substrate may be screened out by the electrons in the Fermi sea. The situation is different for semiconductors and insulators with a band gap.

In earlier work for chemisorption of O_2 on W, a chemical shift of 0.9 eV was claimed.[200] Based on our own experiments showing no such shift,[196] we are inclined to suggest that the 0.9 eV shift can be attributed to an incipient surface oxide of W. Chemical shifts of 0.1 to 0.2 eV for O_2 adsorption on W (110) and CO adsorption on Ru (001) have been reported.[199] These shifts are very small and not necessarily inconsistent with our results for CO on Pt.

3.4.4. The Chemisorption and Oxidation Properties of Al Surfaces

The results discussed in the previous section should be contrasted with the results presented here for the chemisorption of oxygen atoms on an aluminum surface. The oxygen–aluminum system is of particular interest since the relatively simple bulk electronic structure of this system (Al is a free-electron-like metal) has made it suitable as a model system for both theoretical[201–206] and experimental[207–217] work. The recent work by S. A. Flodström et al.[216] has shown that a well-defined chemisorption phase for an ordered oxygen overlayer on the (111) surface of Al is reflected in a chemical shift for the Al $2p$ core level.

In Figure 32 are shown the Al $2p$ core level spectra taken at a photon energy of 130 eV for clean Al (111) surfaces and for surfaces exposed to different amounts of oxygen up to 100 L.[216] Two chemically shifted Al $2p$ levels appear with increasing oxygen exposure. The 1.4-eV-shifted peak is assigned to the interaction of the

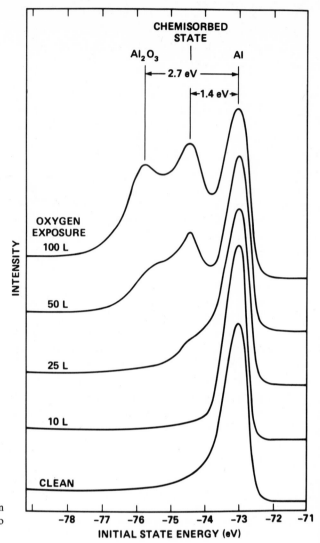

Figure 32. Photoemission spectra of the Al $2p$ region for a clean Al (111) surface and surfaces exposed to different amounts of oxygen (from Ref. 216).

chemisorbed oxygen atoms with the Al substrate. The 2.7-eV-shifted peak has the same binding energy position as the Al $2p$ levels in bulk aluminum oxide Al_2O_3. After 100 L oxygen exposure, the two shifted peaks are of about the same strength. At higher exposures, the 2.7-eV-shifted peak becomes the dominant feature. The chemisorption phase was also studied as a function of temperature. This is illustrated in Figure 33. The 1.4-eV-shifted peak disappears in the temperature range 160–200°C and, at the same time, the intensity of the 2.7-eV-shifted peak increases. This observation unambiguously suggests that oxygen in the chemisorbed phase is transformed into the bulklike oxide Al_2O_3. It is also possible to deduce from this data the required activation energy for this transformation.[216]

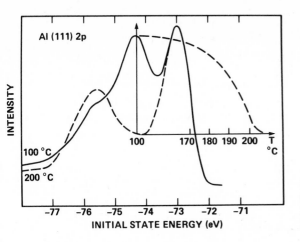

Figure 33. Photoemission spectra of the temperature dependence of the two chemically shifted peaks in the Al 2p region, exposed to 100 L of O_2 at a photon energy of 130 eV (from Ref. 216).

Flodström *et al.*[216] further demonstrated with their photoemission work that the two-step chemisorption/oxidation process is unique to the (111) surface, which is the most closely packed surface of Al. The association of the 1.4-eV-shifted peak with an ordered oxygen overlayer was further corroborated by LEED studies. For the other two simple single-crystal faces—(100) and (110)—the Al_2O_3 oxide forms directly without any intermediate chemisorbed state.[212, 216]

The interaction of oxygen with the different single-crystal surfaces of Al as it is reflected in the binding energy positions of the Al 2p core levels constitutes a beautiful example of how chemical shifts can be used to distinguish between different chemisorption/oxidation stages.

3.5. The Electronic Structure of Interfaces

The electronic properties of interfaces are of great importance in many practical devices and have been studied with several techniques. By using sputter Auger profiling, material is removed gradually in a controlled manner. The Auger technique is used to monitor compositional changes on a microscopic scale as the sputtering proceeds through the interface. Photoemission, on the other hand, can be used to follow the electronic and compositional changes (sensitivity 1 to 2 monolayers) as the interface layer is gradually being built up on the substrate. By using synchrotron radiation, the feasibility of the latter approach has recently been demonstrated and will be illustrated with a few examples in this section.

3.5.1. Oxygen Chemisorption onto Si (111) and the Si–SiO₂ Interface

The $Si–SiO_2$ interface is, without doubt, the most important oxide–semiconductor interface from the point of view of practical applications. Therefore, there have been numerous studies in the past of the bonding of oxygen to the Si surface, the oxidation process, and the electronic structure of the interface.[218–241]

Much controversy still exists regarding the initial formation of a thin oxide layer on the Si surface. One central issue is whether the interface is ideally abrupt (on an atomic scale) or compositionally graded (typically over ~30 Å). Another point of controversy is the bonding configuration between the Si and O atoms at the interface. Photoemission has recently been used to address both these issues.[240] Here, we will mainly present the information that can be extracted from studies of the Si $2p$ core levels. It should be pointed out that these photoemission studies using synchrotron radiation have just been initiated and that further progress should be expected in the near future.

Photoemission spectra taken at an excitation energy of 130 eV are shown in Figures 34 and 35 for binding energies around the position of the Si $2p$ core level (binding energy ≈99 eV). Also shown in Figure 34 (left panel) are the difference curves between the oxygen-exposed and the clean Si surface.[240] Upon oxygen exposure, structure appears on the high binding energy side of the Si $2p$ level. These shifted peaks can be interpreted as due to charge transfer from silicon surface atoms to the oxygen, resulting in a higher binding energy for the silicon atoms involved in the bonding.

Curve c in Figure 34 is for a surface exposed to 10^3 L of nonexcited molecular oxygen. A clear shoulder is observed, and the difference curve, $c - a$, shows that the new structure is shifted from the main line by about 2.0 eV. It should be noted that this occurs at an exposure where the surface states have been depleted[106, 107] (see Section 3.3). As the oxygen exposure is increased up to 10^{12} L, a well-defined peak

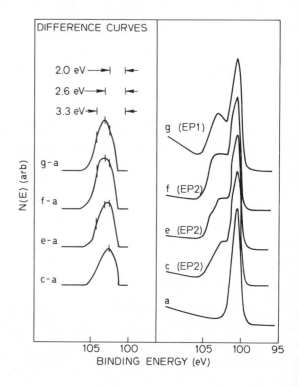

Figure 34. The Si $2p$ core level shifts upon exposure to molecular oxygen at a photon energy of 130 eV: (*a*) clean surface, (*c*) 10^3 L, (*e*) 10^6 L, (*f*) 10^9 L, (*g*) 10^{12} L (from Ref. 240).

shifted by 2.6 eV develops. A Si (111) surface exposed to a large amount of non-excited molecular oxygen can therefore be said to be characterized by this Si $2p$ peak shifted by 2.6 eV. As we will discuss shortly, our present understanding is that the bulk SiO_2 has *not* been formed at this stage. The growth of a SiO_2 layer will thus not be initiated by exposure to nonexcited molecular oxygen at room temperature.

The oxygen interaction proceeds in a different manner when excited oxygen is used in the exposures (Figure 35). Curve a for the clean surface and curve g for the surface saturated by nonexcited oxygen peak shifted by 2.6 eV are shown as a comparison. After exposure to excited oxygen, the shifted Si $2p$ becomes the dominant peak. An exposure of 4×10^6 L O_2 gives a broad (FWHM of 2.3 eV) peak shifted by 3.3 eV (curve h). After another exposure to 4×10^6 L O_2, the shift is 3.8 eV, which is very close to the value observed for bulk silicon dioxide, SiO_2. It should also be noted that the FWHM of this shifted peak is only 1.7 eV. Since the

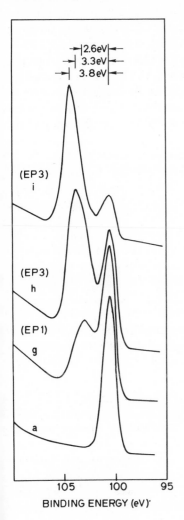

Figure 35. The Si $2p$ core level shifts in the early stages of oxidation at a photon energy of 130 eV: (a) clean surface, (g) 10^{12} L of nonexcited molecular oxygen, (h) 4×10^6 L of excited oxygen, (i) an additional 4×10^6 L of excited oxygen (from Ref. 240).

3.3-eV peak is very broad and asymmetric, it strongly suggests that several states are present at the interface. When the oxygen exposure increases, more and more of these states convert into SiO_2 or a thicker layer of SiO_2 grows on top of the interface region. The present work also showed that the details of the experimental conditions (excited versus nonexcited oxygen, temperature, etc.) are important in determining the relative strengths of the shifted peaks.

We now return to the questions of whether any information can be obtained from the shifted peaks on the bonding configuration at the interface and how wide the interface is. Garner et al.[240] applied ligand shift analysis and argued that, if there are silicon atoms at the interface bonded to one, two, three, and four (dioxide) oxygen atoms, respectively, the 3.3-eV peak should be composed of peaks with chemical shifts of 0.95, 1.9, 2.9, and 3.8 eV respectively. Garner et al.[240] went on to show that a convolution of four Gaussian peaks with these shifts could fit the experimentally observed 3.3-eV peak very well. Furthermore, the peak shifted by 3.8 eV (curve i) could also be fitted very well by the same four peaks with different relative intensity distributions. Garner et al.[240] could also rationalize the peak shifted by 2.6 eV with a modified version of ligand field theory applied to surfaces.[156] This analysis should be considered as very tentative, and no claim is made that the fitting procedure gives a unique answer. An alternative explanation has been offered by Hollinger et al.,[230] who propose that the gradual shift of the Si 2p core level towards higher binding energy with increasing oxygen exposure is the result of changing the relaxation (the efficiency in the electronic screening of the Si 2p hole) with increasing SiO_2 thickness. Thus, it is suggested that only Si and SiO_2 are present at the interface without any intermediate states.[230]

And, finally, we will comment on the width of the interface. Garner et al.[240] have shown that the ratio of the areas under the shifted and unshifted Si 2p peaks can be used to determine the kinetic energy at which the minimum electron escape depth occurs. By further introducing a simple model for the Si surface and combining it with the knowledge of the escape depth, it was possible to estimate the width of the interface layer. The analysis by Garner et al.[240] showed that the multiplicity of states discussed above peak shifted by 3.3 eV exists only over 1 to 2 atomic layers. The analysis for the case of a thicker oxide (curve i in Figure 35) also gave the interface width to about 2 layers. Thus, the important conclusion is that the transition from Si to SiO_2 is very abrupt, ≤ 4 Å.[236–239] By combining the photoemission results with other techniques, the prospects should be good that much more will be learned about the bonding configuration of the Si–SiO_2 interface in the near future.

3.5.2. Metal Overlays on III–V Semiconductor Surfaces

The interaction of a metal overlayer with a semiconductor surface has been studied extensively in the past to gain a better understanding of Schottky barriers[121, 242–285] as well as for other reasons. In this section, we will illustrate how core level spectroscopy can be used to follow the metal–semiconductor interaction on an atomic scale for Au overlayers on III–V surfaces.[277–279] The core level spectra can, in particular, be used to obtain information on (1) chemical

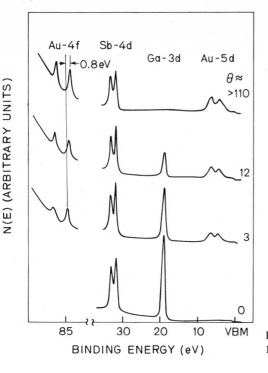

Figure 36. Photoemission spectra taken at a photon energy of 120 eV for GaSb (110) with different Au coverages (from Ref. 279).

bonding at the metal–semiconductor interface via chemical shifts of substrate as well as adatoms, and (2) compositional changes as a function of metal thickness via the relative intensity changes of the core levels.

The overlayers of Au were evaporated onto a (110) surface cleaved *in situ*. The evaporation rate was typically 1 Å/min, and the Au coverage was followed in a controlled manner from a tenth of a nonolayer and up. Figure 36 shows a set of energy distribution curves for *n*-GaSb (110) with increasing Au layer coverages up to at least 110 monolayers.[279] The photon energy, $hv = 120$ eV, has been chosen so that the photoelectrons originating from the Ga 3*d* and Sb 4*d* core levels have kinetic energies that lie within the broad minimum of the electron escape depth curve. Thus, the surface sensitivity is optimal and most of the signal comes from only the last two molecular layers (see Section 1.3). Most noteworthy in Figure 36 is the decrease in Ga 3*d* emission and its complete disappearance as increasing amounts of Au are deposited. On the other hand, the Sb 4*d* emission drops to about two-thirds of that from the clean surface and then remains constant with increasing Au deposition. These results unambiguously show that the GaSb decomposes with the interaction of the deposited Au metal, with Sb moving to the surface, leaving behind a highly nonstoichiometric interface. After an initial drop in intensity of the Sb 4*d* emission, it stays about constant with increasing Au deposition, which suggests a saturation layer of Sb on top of the Au layer. This interpretation is supported by depth profiling using the sputter Auger technique.[279] The significance of the Au 4*f* level shifted by 0.8 eV is discussed elsewhere.[279]

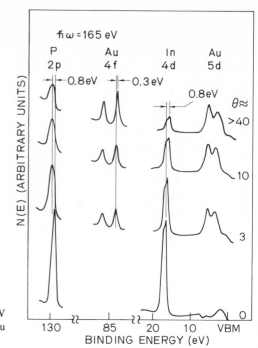

Figure 37. Photoemission spectra taken at photon energies of 120 eV (In 4*d*, Au 5*d*) and 165 eV (P 2*p*, Au 4*f*) for InP (110) with Au coverages (from Ref. 279).

The results for GaSb should be contrasted with those for Au deposited onto InP (110) and GaAs (110) surfaces.[279] The InP results are shown in Figure 37. In the case of InP, a photon energy of 120 eV was used to monitor the In 4*d* core levels and 165 eV for the P 2*p* and Au 4*f* core levels for the same reasons as given for GaSb, namely, to optimize the surface sensitivity. Both the In 4*d* and P 2*p* emission decrease together as the Au coverage is increased. There is thus no measurable preferential removal from the interface of one of the two constituents in InP. However, it should be recognized that we are only sensitive to departures from stoichiometry of about 10% or larger. But, Au is deposited in such amounts that a total depression of the InP emission would have occurred long before the last evaporation shown in Figure 37 if the Au was deposited without any interaction on the semiconductor surface. Instead, there is evidence from the core level intensities that some InP is removed from the interface and transported to the surface. The GaAs (110) surface behaves in a similar manner with the interaction of Au.

The present results give unambiguous evidence that the Au overlayers interact very strongly with the III–V semiconductor surface and that a large amount of intermixing takes place. This is an important observation since most theoretical treatments are based on ideally abrupt interfaces. Such theories are obviously not applicable for Au overlayers on III–V semiconductors. The disruptive effect of the Au may be explained as due to the large heat of condensation of Au as compared to the heat of formation of III–V compounds. A detailed discussion can be found elsewhere.[278, 279] The strong and disruptive interaction between the Au overlayer and the III–V substrate was also an important consideration in establishing a new physical model for the Schottky barrier formation, based on defect states at the interface.[278, 279]

In summary, we can state that it is now possible to study the interaction between the metal and semiconductor on a microscopic scale and follow the evolution of the interface properties from a fraction of a monolayer coverage to thick metallic layers.

3.5.3. Cesium–Oxygen Overlayers on GaAs (110)

In the two previous sections, we have treated the interface properties of an oxide–semiconductor and a metal–semiconductor system, respectively. As a final example of how the photoemission technique can be applied to studies of the electronic structure of interfaces, we will discuss how the oxygen uptake on GaAs (110) surface is affected by the presence of a metal adlayer. The Cs–O system has been chosen since it is of great practical interest in the application to negative-electron-affinity photocathodes.[286–292]

A monolayer of Cs was deposited onto a cleaved GaAs (110) surface, which was then exposed to different amounts of oxygen in a controlled manner.[293, 294] Energy distribution curves at a photon energy of 120 eV are shown in Figures 38

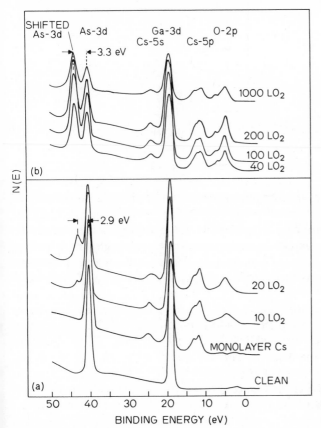

Figure 38. Photoemission spectra ($hv = 120$ eV) of clean GaAs (110) and the same surface exposed to Cs (monolayer) and oxygen (from Ref. 293).

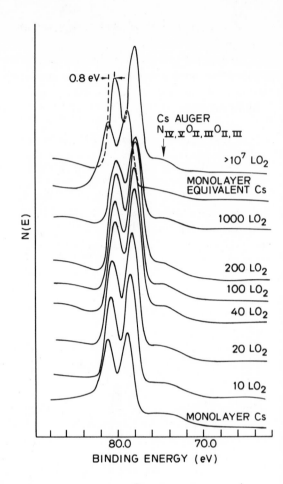

0.8 eV

Cs AUGER
$N_{IV,V}O_{II,III}O_{II,III}$

>10^7 LO_2

MONOLAYER
EQUIVALENT Cs

1000 LO_2

200 LO_2

100 LO_2

40 LO_2

20 LO_2

10 LO_2

MONOLAYER Cs

N(E)

80.0   70.0
BINDING ENERGY (eV)

Figure 39. Photoemission spectra ($hv = 120$ eV) of the Cs $4d$ core levels for the same surface as shown in Figure 38 (from Ref. 293).

and 39 for different oxygen exposures between 10 and 10^3 L O$_2$. The spectra are shown in the binding energy regions for the accessible core levels of GaAs ($3d$) and Cs ($4d$, $5s$, $5p$). A chemically shifted peak appears 2.9 eV below the As $3d$ peak at 10 L O$_2$. The binding energy position of this shifted peak is identical to that observed for oxygen chemisorbed on the clean GaAs (110) surface (Section 3.4.1). However, the rate of oxygen uptake is about seven orders of magnitude larger on the cesiated surface! The presence of a monolayer of a metal thus has a very drastic effect on the oxygen uptake. As the oxygen exposure is increased, the Cs core level binding energies decrease (a total of 0.8 eV), while that of the shifted As $3d$ level increases (a total of 0.4 eV) up to exposures of 40 L O$_2$. Thereafter, the core levels stay constant in energy position to the highest exposure (10^3 L). This correlation indicates interaction between the oxidizing substrate and the Cs–O overlayer Previous models to explain the negative electron affinity have not considered any interaction between the Cs–O layer and the substrate.[286–290] However, from the substrate core level shifts shown in Figure 38 (shifted As $3d$ peaks), we can conclude that the oxygen interacts with the GaAs substrate at a very early stage of the oxygen exposure.

When the surface is exposed to a second dose of Cs, several prominent effects occur (see Figure 39). First, the Cs 4d peaks shift to higher binding energy by 0.8 eV, i.e., back to the original binding energy positions for the Cs core levels on GaAs before oxygen exposure. The same shift is observed in the oxygen related peaks. It was also noted that the shifted as 3d peak (not shown in the figure) is split into two peaks: the original one with a chemical shift of 3.3 eV and a new peak with a shift of 4.5 eV. Furthermore, a small shoulder appears 1.1 eV below the Ga 3d peak. These effects on the Ga and As core levels are very similar to those observed for GaAs exposed to excited oxygen and discussed in Section 3.4.1. Thus, the surface has now reached a stage where true arsenic and gallium oxides are being formed.

From the discussion above, it is clear that the interaction of the Cs–O over-layer with the GaAs substrate is very complex and that much more experimental work needs to be done for a complete understanding. However, this example serves to demonstrate that core level spectroscopy can provide important information for very complex interfaces.

3.6. The Electronic Structure of Cu–Ni Alloy Surfaces

Copper–nickel alloys have been the prototype for studies of surface segregation. A characterization of the surface of a binary alloy must include a description of the composition of the alloy at the surface and as a function of depth below the surface. It is now well established by several experimental techniques that the topmost layer is enriched in copper, which is consistent with theoretical models predicting that the surface region will be enriched in the component with the lowest heat of sublimation.[295–311]

Once it is known that the composition in the surface region differs from the bulk composition, two other questions follow:

1. In how many layers does the composition approach the bulk value?
2. Does the layer composition approach the bulk value in a monotonic fashion?

In order to answer these questions, photoemission measurements were performed on Cu–Ni alloys in the photon energy region 32–240 eV at SSRL.[312] Because the electron escape length is a strong function of energy (see Section 1.3), photoemission will probe to different depths, dependent on the photon energy chosen, and it should be possible to follow compositional changes.

Two surfaces of the same sample was studied: Cu–Ni I with a surface composition of 65% Cu, 35% Ni and Cu–Ni II with 35% Cu, 65% Ni. In Figure 40 we present a partial set of photoemission energy distribution curves (EDC's) of Cu–Ni I taken at different photon energies.[312] The EDC's show two prominent pieces of structure. The first, labeled A, extends from the Fermi level to about 1.8 eV below E_F, the second, labeled B, from 1.8 eV below E_F to ~ 5.6 eV below E_F. As the photon energy is varied, both peaks change somewhat in shape, but the most striking change is in the relative amplitudes of the peaks. It is well known that the

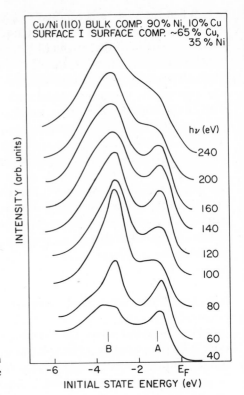

Figure 40. Energy distribution curves for the Cu–Ni alloy in the photon energy region 40–240 eV. Note the changes in the amplitudes of the two-peak structure (from Ref. 312).

EDC's reflect the electronic densities of states. Prior UPS experiments on the Cu–Ni alloy[299–302] and comparison of the experimental data with theoretical calculations using the CPA[296, 297] and ATA[298] methods reveal that peak A is related to emission from the Ni d states, whereas peak B reflects emission from Cu d states. The overall strength of either peak A or B will therefore be predominantly determined by two factors: (1) the amount of Ni or Cu, respectively, present in the region sampled by the experiment, and (2) the 3d cross sections of Ni and Cu. Matrix element effects, which are evident when direct transitions dominate, may still play a role in this disordered alloy at low photon energies,[302] but will be much weaker at the higher photon energies used in this experiment, especially since our experiment is not angularly resolved. Although the peak shapes may vary due to these matrix elements, their overall amplitudes should not be significantly affected.

In Figure 41 we plot the ratio of the Ni emission intensity to Cu emission intensity [hereafter denoted as r(Ni–Cu)] as a function of photon energy. Two sets of data points are shown, one for the Cu–Ni I surface taken from the EDC's shown in Figure 40 and the other set for the Cu–Ni II surface. A simple straight-line background subtraction was performed on the EDC's before the ratios were measured; however, the exact form of the background subtracted has a negligible influence on the overall shape of the plot obtained. In both cases, we obtain a curve that decreases at low photon energies, reaching a minimum at $hv \approx 80$ eV, increases until a maximum at $hv \approx 140$ eV, decreases, and then increases again. At a fixed photon energy $r(Ni/Cu)$ of Cu–Ni II is greater than $r(Ni/Cu)$ of Cu–Ni I, reflecting the higher Cu concentration in the surface region of Cu–Ni I. As the photon energy is varied, the $3d$ cross sections of Cu and Ni will vary; however, their ratio should be a monotonically varying function in the photon energy range of interest (see Section 1.4). It is most likely, then, that the oscillatory behavior of $r(Ni/Cu)$ is due to the amount of Ni and Cu present in the region sampled by the experiment. This region is determined by the escape length of the outgoing electrons. Although the details of this function are not known for the Cu–Ni alloy, it is expected to follow closely the energy dependence discussed in Section 1.3, i.e., it falls to a minimum at 80–100 eV and increases thereafter.

The behavior of $r(Ni/Cu)$ is therefore suggested to be a reflection of the composition of the alloy as a function of depth below the surface. At low photon energies where the escape depth is relatively large, $r(Ni/Cu)$ is large since the bulk is very Ni rich. At first, as the photon energy is increased ($40 < hv < 80$ eV), $r(Ni/Cu)$ decreases since the surface region is enriched in Cu. If we assume, in agreement with theoretical predictions, that the surface layer is the richest in Cu, the minimum in escape depth occurs at $hv = 80$ eV. Thereafter, when the photon energy is increased further ($hv > 80$ eV), the escape depth again increases and, in the photon energy range accessible, $r(Ni/Cu)$ shows a single oscillation. The composition of the alloy below the surface apparently does not approach the bulk value in a monotonic fashion, but instead shows at least one oscillation. A simple model calculation, as discussed elsewhere,[312] indicates that further oscillations of a realistic

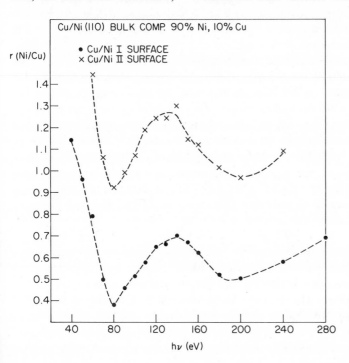

Figure 41. Plot of the ratio of the Ni and Cu emission intensities as a function of photon energy for two Cu–Ni alloys with different surface compositions (from Ref. 312).

nature would be difficult to see in photoemission since the number of electrons collected from a given depth declines exponentially with increasing depth. Note that a similar oscillation in $r(Ni/Cu)$ is expected below 80 eV. However, it is expected that this will be harder to see because (1) the escape depth tends to rise more quickly in this region, and (2) the photoemission EDC's begin to show much more structure, indicating the onset of band-structure-like effects.

In conclusion, the photoemission data presented here confirm the earlier results that the surface of Cu–Ni alloys is enriched in Cu. Furthermore, it is shown that the composition below the topmost surface layer does not approach the bulk value smoothly but has at least one compositional oscillation.

3.7. Concluding Remarks

In this section, we have tried to illustrate, with a few examples, the application of synchrotron radiation to photoemission studies of bulk and surface electronic properties. In doing this, we have, to a large extent, used material from our own research activity at the Stanford Synchrotron Radiation Laboratory. But we have also presented material from other groups in an attempt to achieve a balanced presentation of the current research and possible future directions. We have also attempted to provide sufficient background information for each of the presented areas so that the reader can get the proper perspectives of the research applications. However, within the limited space available for this review article, we make no claims to have covered all the important aspects of photoemission spectroscopy using synchrotron radiation. Realizing these shortcomings, we would like to conclude this section by pointing out a few additional research areas and direct the reader to the existing literature.

As was discussed in Section 3.2, the first, photoemission measurements with synchrotron radiation mainly used the tunability and higher available photon energies to extract information on the bulk electronic structure. This still continues to be an important area of research. Among the materials studied over the years can be mentioned: alkali halides[313–319] (including excitonic effects), solid rare gases,[320, 321] valence mixing in rare-earth compounds,[322] transition metal oxides,[323] tungsten bronzes,[324] thorium hydrides,[325] and layered compounds.[326–343] The work on layered compounds was first done with angle-integrated photoemission but was later extended to the angle-resolved technique to obtain the energy versus momentum dispersion relations. The usefulness of combining the angle-resolved measurements with the polarization for identification of the symmetry character of different levels has been discussed earlier in this section and was applied early to TaS_2.[335] Dispersion relations and the assignment of band symmetries have since been worked out for a large number of layered materials with a very gratifying agreement between theory and experiment.[332–343] The study of the valence band electronic structure of the nobel metals Cu, Ag, and Au has had a similar evolution, as alluded to already in Section 3.2.3.[75, 76, 84, 344–352] The bulk band structure of Ni has been studied,[353] and detailed experimental plots of the dispersion of the energy bands now seem to reveal the magnetic exchange splitting.[354] Excitation energy-dependent lifetimes have been observed for both core levels (outer d levels in Pb, In, and Sn)[355] and valence-band levels (Cu).[356] An additional example of the more and more detailed studies of the noble metals is the strong temperature dependence of the Cu (110) valence-band spectrum observed for normal photoemission.[357] The investigation of surface states and resonances discussed in Section 3.3.1 has recently been extended to Cu,[358] Ni,[359] and Au[360] surfaces. The identification of adsorbate levels and the orientation of molecules on single-crystal metal surfaces[361] has progressed for a number of systems, H_2 on W (001),[362] CO on Ir (100),[363] CO on Pt (111),[364] CO on Cu (110),[365] CO on Rh (111),[366] CO on Pd,[367] and Cl on Si and Ge,[368–370] concurrent with recent theoretical advances.[371–373] The oxidation properties of Bi have been examined in some detail,[374–377] and an increased activity in the study of interfaces (see Section 3.5) [Ga and Al on GaAs,[284, 378] Ge on GaAs,[379] and Si–SiO$_2$,[380]] has taken place during the last year. Photoelectron diffraction has shown some promise as a useful technique for surface structure determinations.[381–383] And, much more work for exploration of this technique should be expected in the near future. Studies of the energy dependence of collective excitations (plasmons)[384, 385] and resonance effects[386, 387] above the ionization threshold of inner-core levels have recently been initiated and have shown important applications, for instance, in the determination of mixed valence states.[387]

The research areas covered in this section have been restricted to photoemission studies of solids and surfaces. A number of other spectroscopies based on synchrotron radiation can provide information relevant to the research fields described here. The reader is referred to other chapters in this book and a number of excellent review papers covering various aspects of electronic states in solids.[388–392]

4. Future Prospects and Developments

The short presentation in the previous section illustrates that there has been a very rapid increase in the application of synchrotron radiation to photoemission studies. The future research directions will be closely tied to the advancement of the experimental technology (better synchrotron radiation sources, monochromators, and electron detector systems). There is presently an intense activity for improvements in all these areas as mentioned briefly in Section 2.4.

The feasibility of studying a number of surface and interface phenomena (chemical bonding, interdiffusion, alloying, decomposition, surface and interface induced electronic states, etc.) has already been demonstrated by a combination of valence-band and core level studies. So far, these studies have been for photon energies below 300 eV, due to lack of suitable monochromators at higher energies. Within the next two years, it is anticipated that new monochromators covering the range 300–2000 eV will be in operation, and the 1s core levels of carbon, nitrogen, oxygen, and fluorine will be reached. Since these elements are present in a large number of overlayers on substrate surfaces, the impact on interface studies should be significant. The same holds true for chemisorption studies of organic and other molecules containing one or several of these elements. The availability of higher photon energies (particularly up to 2000 eV) with sufficient intensity and mono-chromaticity will thus add tremendously to the potential of the photoemission technique.

In Section 3 we discussed several cases where the high degree of polarization of the radiation has been utilized. However, this is, with all certainty, only the very beginning, and the full potential of the polarization properties has not yet been fully explored. A combination of polarization and angle-resolved studies makes it possible to examine in detail (1) the symmetry of energy bands for both bulk and surface states resulting in a plot of their energy dispersion, and (2) the orientation and position of chemisorbed molecules on single-crystal surfaces. Further developments of these approaches and a rapid increase in their application to a number of different problems may be expected with the development of improved electron detection systems for angle-resolved measurements and new monochromators.

Much progress should also be expected in the theoretical treatment of the photoemission process itself and its application to various problems. A more adequate comparison can then be made between theory and experiment to check the accuracy of calculated surface and bulk density of states. The advancement of the theoretical treatment of chemisorbed systems has been progressing at an encouraging pace. New aspects of photoemission, like photoelectron diffraction for determination of the surface structure, will undoubtedly be thoroughly tested in the near future.

ACKNOWLEDGMENT

The support from the U.S. Defense Advance Projects Agency, the Army Office of Research, the National Science Foundation, and the Office of Naval Research is gratefully acknowledged. Part of this work was performed at the Stanford Synchrotron Radiation Laboratory, which is supported by the National Science Foundation in cooperation with the Stanford Linear Accelerator Center and the Department of Energy.

References

1. W. E. Spicer, in *Optical Properties of Solids*, F. Abeles (ed.), p. 757, North-Holland Publishing Co., Amsterdam (1972).
2. D. E. Eastman, in *Techniques of Metals Research VI*, E. Passaglia (ed.), p. 413, Interscience, New York (1972).
3. N. V. Smith, *Crit. Rev. Solid State Sci.* **2**, 45 (1971).
4. D. E. Eastman, in *Electron Spectroscopy, Proceedings of an International Conference at Asilomar*, D. A. Shirley (ed.), p. 487, North-Holland Publishing Co., Amsterdam (1972).
5. W. E. Spicer, *J. Phys. (Paris) Colloq.* **34**, 19 (1973).
6. D. E. Eastman, in *Vacuum Ultraviolet Radiation Physics*, E. E. Koch, R. Haensel, and C. Kunz (eds.), p. 417, Pergamon Press, Vieweg (1974).
7. W. E. Spicer, in *Vacuum Ultraviolet Radiation Physics*, E. E. Koch, R. Haensel, and C. Kunz (eds.), p. 545, Pergamon Press, Vieweg (1974).

8. I. Lindau and W. E. Spicer, in *Charge Transfer/Electronic Structure of Alloys*, L. H. Bennett and R. H. Willens (eds.), p. 159, Metallurgical Society of the American Institute of Mining, Metallurgical, and Petroleum Engineers, New York (1974).

9. E. W. Plummer, in *Interactions on Metal Surfaces*, R. Gomer, (ed.), p. 144, Springer-Verlag, New York (1975).

10. E. W. Plummer, in *The Physical Basis for Heterogeneous Catalysis*, E. Drauglis and R. I. Jaffe (eds.), p. 203, Plenum Press, New York (1975).

11. B. Feuerbacher and R. F. Willis, *J. Phys. C* **9**, 169 (1976).

12. W. E. Spicer, in *Optical Properties of Solids—New Developments*, B. O. Seraphin (ed.), p. 631, North-Holland Publishing Co., Amsterdam (1976).

13. W. E. Spicer, K. Y. Yu, I. Lindau, P. Pianetta, and D. M. Collins, in *Surface and Defect Properties of Solids*, J. M. Thomas and M. W. Roberts (eds.), Vol. V, p. 103, The Chemical Society, London (1976).

14. I. Lindau, in *International College on Applied Physics: Synchrotron Radiation Research*, A. N. Mancini and L. F. Quercia (eds.), p. 372, Alghero, 12–24 September 1976.

15. T. Rhodin and C. Brucker, in *Characterization of Metal and Polymer Surfaces*, Vol. 1, p. 431, Academic Press, New York (1977).

16. W. E. Spicer, I. Lindau, J. N. Miller, D. T. Ling, P. Pianetta, P. W. Chye, and C. M. Garner, *Phys. Scr.* **16**, 388 (1977).

17. D. A. Shirley, J. Stöhr, P. S. Wehner, R. S. Williams, and G. Apai, *Phys. Scr.* **16**, 398 (1977).

18. B. Feuerbacher, B. Fitton, and R. F. Willis (eds.), *Photoemission and the Electronic Properties of Surfaces*, Wiley, New York (1978).

19. W. E. Spicer, in *Electron and Ion Spectroscopy of Solids*, L. Fiermans, J. Vennik, and W. Dekeyser (eds.), p. 54, Plenum Press, New York (1978).

20. A. Liebsch, in *Electron and Ion Spectroscopy of Solids*, L. Fiermans, J. Vennik, and W. Dekeyser (eds.), p. 93, Plenum Press, New York (1978).

21. G. J. Lapeyre, R. J. Smith, J. Knapp, and J. Andersson, *J. Phys. (Paris) Colloq.* **39**, 4–134 (1978).

22. Y. Petroff, *J. Phys. (Paris) Colloq.* **39**, 4–149 (1978).

23. N. V. Smith, *J. Phys. (Paris) Colloq.* **39**, 4–161 (1978).

24. N. V. Smith, in *Photoemission in Solids I*, M. Cardona and L. Ley (eds.), p. 237, Springer-Verlag, Berlin (1978).

25. F. C. Brown, in *Solid State Physics*, H. Ehrenreich, F. Seitz, and D. Turnball (eds.), Vol. 29, p. 32, Academic Press, New York (1974).

26. W. E. Spicer, *Phys. Rev.* **112**, 114 (1958).

27. C. N. Berglund and W. E. Spicer, *Phys. Rev. A* **136**, 1030 (1964).

28. W. L. Schaich and N. W. Ashcroft, *Phys. Rev. B* **3**, 2452 (1971).

29. C. Caroli, D. Lederer-Rozenblatt, B. Roulet, and D. Saint-James, *Phys. Rev. B* **8**, 4552 (1973).

30. N. W. Ashcroft, in *Vacuum Ultraviolet Radiation Physics*, E. E. Koch, R. Haensel, and C. Kunz (eds.), p. 533, Pergamon Press, Vieweg (1974).

31. A. Liebsch, *Phys. Rev. B* **13**, 544 (1976).

32. J. B. Pendry, *Surf. Sci.* **57**, 679 (1976).

33. W. L. Schaich, in *Photoemission in Solids I*, M. Cardona and L. Ley (eds.), p. 105, Springer-Verlag, Berlin (1978).

34. G. D. Mahan, in *Electron and Ion Spectroscopy of Solids*, L. Fiermans, J. Vennik, and W. Dekeyser (eds.), p. 1, Plenum Press, New York (1978).

35. J. B. Pendry and J. F. L. Hopkinson, *J. Phys. (Paris) Colloq.* **39**, 4–142 (1978).

36. R. Z. Bachrach, S. B. M. Hagstrom and F. C. Brown, *J. Vac. Sci. Technol.* **12**, 309 (1973).

37. I. Lindau and W. E. Spicer, *J. Electron. Spectrosc. Relat. Phenom.* **3**, 409 (1974).

38. M. Klasson, J. Hedman, A. Berndtsson, R. Nilsson, C. Nordling, and P. Melnik, *Phys. Scr.* **5**, 93 (1972).

39. P. J. Feibelman and D. E. Eastman, *Phys. Rev. B* **10**, 4932 (1974).

40. D. J. Kennedy and S. T. Manson, *Phys. Rev. A* **5**, 227 (1972).

41. I. Lindau, P. Pianetta, and W. E. Spicer, *Phys. Lett. A* **57**, 225 (1976).

42. I. Lindau, P. Pianetta, and W. E. Spicer, *Proceedings of the International Conference on the Physics of X-Ray Spectra*, R. D. Deslattes (ed.), p. 78, Gaithersburg, U.S. Government Printing Office, Washington, D.C., 30 August–2 September 1976.

43. J. W. Cooper, *Phys. Rev.* **128**, 681 (1962).
44. U. Fano and J. W. Cooper, *Rev. Mod. Phys.* **40**, 441 (1968).
45. M. Mehta, C. S. Fadley, and P. S. Bagus, *Chem. Phys. Lett.* **37**, 454 (1976).
46. M. Y. Amusia and N. A. Cherephov, *Case Stud. At. Phys.* **5**, 47 (1975).
47. G. Wendin, in *Vacuum Ultraviolet Physics*, E. E. Koch, R. Haensel, and C. Kunz (eds.), p. 225, Pergamon Press, Vieweg (1974).
48. M. Hecht, I. Lindau, and L. I. Johansson, *Proceedings of the International Conference on X-Ray and XUV Spectroscopy, Jpn. J. Appl. Phys. Suppl.* **17-2**, 249 (1978), Sendai, 28 August–1 September 1978.
49. C. Kunz, in *Photoemission and the Electronic Properties of Surfaces*, B. Feuerbacher, B. Fitton, and R. F. Willis (eds.), p. 501, Wiley, New York (1978).
50. S. Doniach, I. Lindau, W. E. Spicer, and H. Winick, *J. Vac. Sci. Technol.* **12**, 1123 (1975).
51. I. Lindau, in *Proceedings of the ERDA Symposium on X- and Gamma-Ray Sources and Applications*, Conference 760539, Ann Arbor, 19–21 May (1976), H. C. Griffin (ed.), p. 11, The Institute of Electrical and Electronics Engineers, Inc., New York (1976).
52. E. M. Rowe and F. E. Mills, *Part. Accel.* **4**, 211 (1973).
53. P. M. Guyon, C. Depautex, and G. Morel, *Rev. Sci. Instrum.* **47**, 1347 (1976).
54. I. Lindau and R. Z. Bachrach, in *International Workshop on the Development of Synchrotron Radiation Facilities*, Quebec, 15–18 June 1976, J. Wm. McGowan and E. M. Rowe (eds.), p. 9-3, Centre of Interdisciplinary Studies in Chemical Physics; London, Canada (1976).
55. V. Saile, *Nucl. Instrum. Methods* **152**, 59 (1978).
56. F. C. Brown, R. Z. Bachrach, and N. Lien, *Nucl. Instrum. Methods* **152**, 73 (1978).
57. W. Eberhardt, G. Kalkoffen, and C. Kunz, *Nucl. Instrum. Methods* **152**, 81 (1978).
58. C. Depautex, P. Thiry, R. Pinchaux, Y. Petroff, D. Lepere, G. Passereau, and J. Flamand, *Nucl. Instrum. Methods* **152**, 101 (1978).
59. P. Pianetta, I. Lindau, and W. E. Spicer, in *Quantitative Surface Analysis of Materials, A Symposium Sponsored by ASTM Committee E-42 on Surface Analysis*, N. S. McIntyre (ed.), p. 105, ASTM Special Technical Publication 643, Cleveland, Ohio, 2–3 March 1977.
60. G. Derbenwick, D. Pierce, and W. E. Spicer, in *Methods of Experimental Physics*, Collman (ed.), Vol. 10, p. 67, Academic Press, New York (1974).
61. I. Lindau, J. C. Helmer, and J. Uebbing, *Rev. Sci. Instrum.* **44**, 265 (1973).
62. P. W. Palmberg, *J. Electron Spectrosc. Relat. Phenom.* **5**, 691 (1974).
63. D. Roy and J. D. Carette, in *Electron Spectroscopy for Surface Analysis*, H. Ibach (ed.), p. 13, Springer-Verlag, Berlin (1977).
64. D. E. Eastman, F. J. Himpsel, and J. J. Donelon, *Bull. Am. Phys. Soc.* **23**, 363 (1978).
65. S. P. Weeks and J. E. Rowe, *J. Vac. Sci. Technol.* **15**, 659 (1978).
66. W. D. Grobman and D. E. Eastman, *Phys. Rev. Lett.* **33**, 1034 (1974).
67. W. D. Grobman, D. E. Eastman, J. L. Freeouf, and J. Shaw, *Proceedings of the XII International Conference on the Physics of Semiconductors*, p. 1275, Teubner, Stuttgart (1974).
68. W. D. Grobman, D. E. Eastman, and J. L. Freeouf, *Phys. Rev. B* **12**, 4405 (1975), and references given therein.
69. J. C. Phillips and K. C. Pandey, *Phys. Rev. Lett.* **30**, 787 (1973).
70. L. Ley, R. A. Pollak, F. R. McFeeley, S. P. Kowalczyk, and D. A. Shirley, *Phys. Rev. B* **9**, 600 (1974).
71. N. J. Shevchik, J. Tejeda, and M. Cardona, *Phys. Rev. B* **9**, 2627 (1974).
72. D. E. Eastman, W. D. Grobman, J. L. Freeouf, and M. Erbudak, *Phys. Rev. B* **9**, 3473 (1974).
73. J. Chelikowsky, D. J. Chadi, and M. L. Cohen, *Phys. Rev. B* **8**, 2786 (1973).
74. M. L. Cohen and T. K. Bergstresser, *Phys. Rev.* **141**, 789 (1966).
75. I. Lindau, P. Pianetta, K. Y. Yu, and W. E. Spicer, *Phys. Rev. B* **13**, 492 (1976).
76. J. Freeouf, M. Erbudak, and D. E. Eastman, *Solid State Commun.* **13**, 771 (1973).
77. D. A. Shirley, *Phys. Rev. B* **5**, 4709 (1972).
78. C. S. Fadley, R. Baird, W. Siekhaus, T. Novakov, and S. A. L. Bergstrom, *J. Electron Spectrosc. Relat. Phenom.* **4**, 93 (1974).
79. T. Matsudaira, M. Watanabe, and M. Onchi, *Jpn. J. Appl. Phys. Suppl.* **2**, 181 (1974).
80. E. G. McRae, *Surf. Sci.* **44**, 321 (1974).

81. R. Y. Koyama and L. R. Hughey, *Phys. Rev. Lett.* **29**, 1518 (1972).
82. I. Lindau, P. Pianetta, K. Y. Yu, and W. E. Spicer, *J. Electron Spectrosc. Relat. Phenom.* **8**, 487 (1976).
83. L. I. Johansson and I. Lindau, *Solid State Commun.* **29**, 379 (1979).
84. J. Stöhr, G. Apai, P. S. Wehner, F. R. McFeely, R. S. Williams, and D. A. Shirley, *Phys. Rev. B* **14**, 5144 (1976).
85. S.-L. Weng, E. W. Plummer, and T. Gustafsson, *Phys. Rev. B* **18**, 1718 (1978), and references given therein.
86. S.-L. Weng, T. Gustafsson, and E. W. Plummer, *Phys. Rev. Lett.* **39**, 822 (1977).
87. N. Nicolaou and A. Modinos, *Phys. Rev. B* **11**, 3687 (1975).
88. N. Nicolaou and A. Modinos, *Surf. Sci.* **60**, 527 (1976).
89. A. Modinos and N. Nicolaou, *Phys. Rev. B* **13**, 1536 (1975).
90. M. C. Desjonqueres and F. Cyrot-Lackmann, *J. Phys. F* **5**, 1368 (1975).
91. N. Kar and Paul Soven, *Solid State Commun.* **20**, 977 (1976).
92. N. V. Smith and L. F. Mattheiss, *Phys. Rev. Lett.* **37**, 1494 (1976).
93. S.-L. Weng, *Phys. Rev. Lett.* **38**, 434 (1977).
94. R. V. Kasowski, *Solid State Commun.* **17**, 179 (1975).
95. G. P. Kerker, K. M. Ho, and M. L. Cohen, *Phys. Rev. Lett.* **40**, 1593 (1978).
96. S.-L. Weng and E. W. Plummer, *Solid State Commun.* **23**, 515 (1977).
97. C. M. Bertoni, C. Calandra, and F. Manghi, *Solid State Commun.* **23**, 255 (1977).
98. B. Laks and C. E. T. Gonçalves da Silva, *Solid State Commun.* **25**, 401 (1978).
99. O. Bisi, C. Calandra, P. Flarioni, and F. Manghi, *Solid State Commun.* **25**, 401 (1978).
100. B. Feuerbacher and R. F. Willis, *Phys. Rev. Lett.* **37**, 446 (1976).
101. R. F. Willis, B. Feuerbacher, and B. Fitton, *Solid State Commun.* **18**, 1315 (1976).
102. W. F. Egelhoff, J. W. Linnett, and D. L. Perry, *Phys. Rev. Lett.* **36**, 98 (1976).
103. C. Noguera, D. Spanjaard, D. Jepsen, Y. Ballu, C. Guillot, J. Lecante, J. Paigne, Y. Petroff, R. Pinchaux, P. Thiry, and R. Conti, *Phys. Rev. Lett.* **38**, 1171 (1977).
104. G. J. Lapeyre, R. J. Smith, and J. Anderson, *J. Vac. Sci. Technol.* **14**, 384 (1977).
105. R. C. Cinti, E. Al Khoury, B. K. Chakraverty, and N. E. Christensen, *Phys. Rev. B* **14**, 3296 (1976).
106. L. F. Wagner and W. E. Spicer, *Phys. Rev. Lett.* **28**, 1381 (1972).
107. D. E. Eastman and W. D. Grobman, *Phys. Rev. Lett.* **28**, 1378 (1972).
108. S. G. Louie, J. R. Chelikowsky, and M. L. Cohen, *J. Vac. Sci. Technol.* **13**, 790 (1976).
109. J. van Laar and J. J. Scheer, *Surf. Sci.* **8**, 342 (1967).
110. A. Huijser and J. van Laar, *Surf. Sci.* **52**, 202 (1975).
111. J. van Laar and A. Huijser, *J. Vac. Sci. Technol.* **13**, 769 (1976).
112. J. van Laar, A. Huijser, and T. van Rooy, *J. Vac. Sci. Technol.* **14**, 894 (1977).
113. W. E. Spicer, I. Lindau, P. E. Gregory, C. M. Garner, P. Pianetta, and P. W. Chye, *J. Vac. Sci. Technol.* **13**, 780 (1976).
114. W. E. Spicer, P. Pianetta, I. Lindau, and P. W. Chye, *J. Vac. Sci. Technol.* **14**, 885 (1977).
115. P. W. Chye, P. Pianetta, I. Lindau, and W. E. Spicer, *J. Vac. Sci. Technol.* **14**, 917 (1977).
116. I. Lindau, P. Pianetta, C. M. Garner, P. W. Chye, P. E. Gregory, and W. E. Spicer, *Surf. Sci.* **63**, 45 (1977).
117. I. Lindau, P. Pianetta, W. E. Spicer, P. E. Gregory, C. M. Garner, and P. W. Chye, *J. Electron Spectrosc. Relat. Phenom.* **13**, 155 (1978).
118. P. Pianetta, I. Lindau, P. E. Gregory, C. M. Garner, and W. E. Spicer, *Surf. Sci.* **72**, 298 (1978).
119. W. Gudat and D. E. Eastman, *J. Vac. Sci. Technol.* **13**, 831 (1976).
120. D. E. Eastman and J. L. Freeouf, *Phys. Rev. Lett.* **33**, 1601 (1974).
121. D. E. Eastman and J. L. Freeouf, *Phys. Rev. Lett.* **34**, 1624 (1975).
122. K. C. Pandey, J. L. Freeouf, and D. E. Eastman, *J. Vac. Sci. Technol.* **14**, 904 (1977).
123. J. A. Knapp, D. E. Eastman, K. C. Pandey, and F. Patella, *J. Vac. Sci. Technol.* **15**, 1252 (1978).
124. G. J. Lapeyre and J. Anderson, *Phys. Rev. Lett.* **35**, 117 (1975).
125. G. M. Guichar, C. A. Sebenne, and G. A. Garry, *Phys. Rev. Lett.* **37**, 1158 (1976).
126. J. A. Knapp and G. J. Lapeyre, *J. Vac. Sci. Technol.* **13**, 757 (1976).
127. G. P. Williams, R. J. Smith, and G. J. Lapeyre, *J. Vac. Sci. Technol.* **15**, 1249 (1978).
128. A. Huijser, J. van Laar, and T. L. van Rooy, *Phys. Lett. A* **65**, 335 (1978).

129. R. Ludeke and A. Koma, *J. Vac. Sci. Technol.* **13**, 241 (1976).
130. H. Froitzheim and H. Ibach, *Surf. Sci.* **47**, 713 (1975).
131. J. E. Rowe, S. B. Christman, and G. Margaritondo, *Phys. Rev. Lett.* **35**, 1471 (1975).
132. R. S. Bauer, *J. Vac. Sci. Technol.* **14**, 899 (1977).
133. J. C. McMenamin and R. S. Bauer, *J. Vac. Sci. Technol.* **15**, 1262 (1978).
134. J. D. Joannopoulos and M. L. Cohen, *Phys. Rev. B* **10**, 5075 (1974).
135. J. R. Chelikowsky and M. L. Cohen, *Phys. Rev. B* **13**, 826 (1976).
136. J. R. Chelikowsky, S. G. Louie, and M. L. Cohen, *Phys. Rev. B* **14**, 4724 (1976).
137. P. E. Gregory, W. E. Spicer, S. Ciraci, and W. Harrison, *Appl. Phys. Lett.* **25**, 511 (1974).
138. C. Calandra and G. Santoro, *J. Phys. C.* **8**, L86 (1975).
139. C. Calandra and G. Santoro, *J. Vac. Sci. Technol.* **13**, 773 (1976).
140. E. J. Mele and J. D. Joannopoulos, *Surf. Sci.* **66**, 38 (1977).
141. E. J. Mele and J. D. Joannopoulos, *Phys. Rev. B* **17**, 1816 (1978).
142. D. J. Chadi, *J. Vac. Sci. Technol.* **15**, 631 (1978).
143. D. J. Chadi, *Phys. Rev.* **18**, 1800 (1978).
144. C. Calandra, F. Manghi, and C. M. Bertoni, *J. Phys. C.* **10**, 1911 (1977).
145. A. R. Lubinsky, C. B. Duke, B. W. Lee, and P. Mark, *Phys. Rev. Lett.* **36**, 1058 (1976).
146. S. Y. Tong, A. R. Lubinsky, B. J. Mrstik, and M. A. Van Hove, *Phys. Rev. B* **17**, 3303 (1978).
147. A. Kahn, G. Cisneros, M. Bonn, P. Mark, and C. B. Duke, *Surf. Sci.* **71**, 387 (1978).
148. A. Kahn, E. So, P. Mark, C. B. Duke, and R. Meyer, *J. Vac. Sci. Technol.* **15**, 580 (1978).
149. D. J. Chadi, *Phys. Rev. Lett.* **41**, 1062 (1978).
150. R. Dorn, H. Lüth, and G. J. Russell, *Phys. Rev. B* **10**, 5049 (1974).
151. H. Lüth and G. J. Russell, *Surf. Sci.* **45**, 329 (1974).
152. P. E. Gregory and W. E. Spicer, *Surf. Sci.* **54**, 229 (1976).
153. H. Lüth, M. Büchel, R. Dorn, M. Liehr, and R. Matz, *Phys. Rev. B* **15**, 865 (1977).
154. P. Pianetta, I. Lindau, C. M. Garner, and W. E. Spicer, *Phys. Rev. Lett.* **35**, 1356 (1975).
155. P. Pianetta, I. Lindau, C. M. Garner, and W. E. Spicer, *Phys. Rev. Lett.* **37**, 1166 (1976).
156. P. Pianetta, I. Lindau, C. M. Garner, and W. E. Spicer, *Phys. Rev. B* **18**, 2792 (1978).
157. R. Ludeke, *Phys. Rev. B* **16**, 5598 (1977).
158. P. Pianetta, I. Lindau, C. M. Garner, and W. E. Spicer, *Phys. Rev. B* **16**, 5600 (1977).
159. R. Ludeke, *Solid State Commun.* **21**, 815 (1977).
160. E. J. Mele and J. D. Joannopoulos, *Phys. Rev. Lett.* **40**, 341 (1978).
161. J. D. Joannopoulos and E. J. Mele, *J. Vac. Sci. Technol.* **15**, 1287 (1978).
162. W. A. Goddard III, J. J. Barton, A. Redondo, and T. C. McGill, *J. Vac. Sci. Technol.* **15**, 1274 (1978).
163. G. Blyholder, *J. Phys. Chem.* **68**, 2772 (1964).
164. J. T. Waber, H. Adachi, F. W. Averill, and D. E. Ellis, *Jpn. J. Appl. Phys. Suppl.* **2**, 695 (1974).
165. G. Doyen and G. Ertl, *Surf. Sci.* **43**, 197 (1974).
166. I. P. Batra and O. Robaux, *J. Vac. Sci. Technol.* **12**, 242 (1975).
167. I. P. Batra and P. S. Bagus, *Solid State Commun.* **16**, 1097 (1975).
168. R. V. Kasowski, *Phys. Rev. Lett.* **37**, 219 (1976).
169. D. E. Eastman and J. E. Demuth, *Jpn. J. Appl. Phys. Suppl.* **2**, 827 (1974).
170. J. Küppers, H. Conrad, G. Ertl, and E. E. Latta, *Jpn. J. Appl. Phys. Suppl.* **2**, 225 (1974).
171. T. Gustafsson, E. W. Plummer, D. E. Eastman, and J. L. Freeouf, *Solid State Commun.* **17**, 391 (1975).
172. G. Apai, P. S. Wehner, R. S. Williams, J. Stöhr, and D. A. Shirley, *Phys. Rev. Lett.* **37**, 1497 (1976).
173. P. M. Williams, P. Butcher, J. Wood, and K. Jacobi, *Phys. Rev.* **14**, 3215 (1976).
174. D. R. Lloyd, C. M. Quinn, and N. V. Richardson, *Solid State Commun.* **20**, 409 (1976).
175. R. J. Smith, J. Anderson, and G. J. Lapeyre, *Phys. Rev. Lett.* **37**, 1081 (1976).
176. G. Apai, P. S. Wehner, J. Stöhr, R. S. Williams, and D. A. Shirley, *Solid State Commun.* **20**, 1141 (1976).
177. C. L. Allyn, T. Gustafsson, and E. W. Plummer, *Chem. Phys. Lett.* **47**, 127 (1977).
178. S. P. Weeks and E. W. Plummer, *Solid State Commun.* **21**, 695 (1977).
179. G. W. Rubloff and J. L. Freeouf, *Phys. Rev. B* **17**, 4680 (1978).

180. D. W. Turner, C. Baker, A. D. Baker, and C. R. Brundle, *Molecular Photoelectron Spectroscopy*, Interscience, New York (1970).
181. K. Siegbahn, C. Nordling, G. Johansson, J. Hedman, P. F. Heden, K. Hamrin, U. Gelius, T. Bergmark, L. O. Werme, R. Manne, and Y. Baer, *ESCA Applied to Free Molecules*, North-Holland Publishing Co., Amsterdam (1969).
182. M. S. Banna and D. A. Shirley, *J. Electron. Spectrosc. Relat. Phenom.* **8**, 255 (1976).
183. J. N. Miller, D. T. Ling, I. Lindau, P. M. Stefan, and W. E. Spicer, *Phys. Rev. Lett.* **38**, 1419 (1977).
184. I. Lindau, P. Pianetta, and W. E. Spicer, in *Proceedings of the Vth International Conference on Vacuum Ultraviolet Radiation Physics*, Montpellier, France, 5–9 September 1977, M. C. Castex, M. Pouey and N. Pouey (eds.), Vol. II, p. 244, Centre National de la Recherche Scientifique, Meudon, France (1978).
185. G. W. Rubloff, W. D. Grobman, and H. Lüth, *Phys. Rev. B* **14**, 1450 (1976).
186. U. Gelius, E. Basilier, S. Svensson, T. Bergmark, and K. Siegbahn, *J. Electron. Spectrosc. Relat. Phenom.* **2**, 40 (1973).
187. J. W. Gadzuk, *Phys. Rev. B* **14**, 2267 (1976).
188. P. S. Bagus and K. Herman, *Solid State Commun.* **20**, 5 (1976).
189. D. S. Rajoria, L. Kovnat, E. W. Plummer, and W. R. Salanek, *Chem. Phys. Lett.* **49**, 64 (1977).
190. E. W. Plummer, W. R. Salanek, and J. S. Miller, *Phys. Rev. B* **18**, 1683 (1978).
191. D. Dill and J. L. Dehmer, *J. Chem. Phys.* **61**, 692 (1974).
192. J. L. Dehmer and D. Dill, *Phys. Rev. Lett.* **35**, 213 (1975).
193. J. W. Davenport, *Phys. Rev. Lett.* **36**, 945 (1976).
194. J. W. Davenport, *J. Vac. Sci. Technol.* **15**, 433 (1978).
195. D. Menzel, in *Photoemission and the Electronic Properties of Surfaces*, B. Feuerbacher, B. Fitton, and R. F. Willis (eds.), p. 381, Wiley, New York (1978).
196. J. N. Miller, I. Lindau, D. T. Ling, P. Pianetta, and W. E. Spicer, in *Proceedings of the Vth International Conference on Vacuum Ultraviolet Radiation Physics*, Montpellier, France, 5–9 September 1977, M. C. Castex, M. Pouey, and N. Pouey (eds.), Vol. II, p. 247, Centre National de la Recherche Scientifique, Meudon, France (1978).
197. H. Höchst, L. Hüfner, and A. Goldman, *Phys. Lett. A* **57**, 265 (1976).
198. K. Y. Yu, W. E. Spicer, I. Lindau, P. Pianetta, and S. F. Lin, *Surf. Sci.* **57**, 157 (1976).
199. J. C. Fuggle and D. Menzel, *Surf. Sci.* **53**, 21 (1975).
200. A. Barrie and A. M. Bradshaw, *Phys. Lett. A* **55**, 306 (1976).
201. N. D. Lang and A. R. Williams, *Phys. Rev. Lett.* **34**, 531 (1975).
202. J. Harris and G. S. Painter, *Phys. Rev. Lett.* **36**, 151 (1976).
203. I. P. Batra and S. Ciraci, *Phys. Rev. Lett.* **39**, 774 (1977).
204. R. P. Messmer and D. R. Salahub, *Chem. Phys. Lett.* **49**, 59 (1977).
205. R. P. Messmer and D. R. Salahub, *Phys. Rev. B* **16**, 3415 (1977).
206. G. S. Painter, *Phys. Rev.* **17**, 662 (1978).
207. J. C. Fuggle, L. M. Watson, D. J. Fabian, and S. Affrossman, *Surf. Sci.* **49**, 61 (1975).
208. A. Barrie, *Chem. Phys. Lett.* **19**, 109 (1973).
209. A. M. Bradshaw, P. Hofman, and W. Wyrobisch, *Surf. Sci.* **68**, 269 (1977).
210. K. Y. Yu, J. N. Miller, P. Chye, W. E. Spicer, N. D. Lang, and A. R. Williams, *Phys. Rev. B* **14**, 1446 (1976).
211. S. A. Flodström, L. G. Petersson, and S. B. M. Hagström, *Solid State Commun.* **19**, 257 (1976).
212. S. A. Flodström, R. Z. Bachrach, R. S. Bauer, and S. B. M. Hagström, *Phys. Rev. Lett.* **37**, 1282 (1976).
213. P. O. Gartland, *Surf. Sci.* **62**, 183 (1977).
214. A. Bianconi, R. Z. Bachrach, and S. A. Flodström, *Solid State Commun.* **24**, 539 (1977).
215. R. Z. Bachrach, S. A. Flodström, R. S. Bauer, S. B. M. Hagström, and D. J. Chadi, *J. Vac. Sci. Technol.* **15**, 488 (1978).
216. S. A. Flodström, C. W. B. Martinsson, R. Z. Bachrach, S. B. M. Hagström, and R. S. Bauer, *Phys. Rev. Lett.* **40**, 907 (1978).
217. W. Eberhardt and C. Kunz, *Surf. Sci.* **75**, 709 (1978).
218. J. T. Law, *J. Phys. Chem. Solids* **4**, 91 (1958).
219. M. Green and K. H. Maxwell, *J. Phys. Chem. Solids* **13**, 145 (1960).

220. M. Green and A. Liberman, *J. Phys. Chem. Solids* **23**, 1407 (1962).
221. H. Ibach, K. Horn, R. Dorn, and H. Lüth, *Surf. Sci.* **38**, 433 (1973).
222. F. M. Meyer and J. J. Vrakking, *Surf. Sci.* **38**, 275 (1973); *Surf. Sci.* **33**, 271 (1972).
223. H. Ibach and J. E. Rowe, *Phys. Rev. B* **10**, 710 (1974).
224. R. Ludeke and A. Koma, *Phys. Rev. Lett.* **34**, 1170 (1975).
225. J. E. Rowe, G. Margaritondo, H. Ibach, and H. Froitzheim, *Solid State Commun.* **20**, 277 (1976).
226. W. A. Goddard III, A. Redondo, and T. C. McGill, *Solid State Commun.* **18**, 981 (1976).
227. W. A. Goddard III, J. J. Barton, A. Redondo, and T. C. McGill, *J. Vac. Sci. Technol.* **15**, 1274 (1978).
228. J. M. Hill, D. G. Royce, C. S. Fadley, L. F. Wagner, and J. F. Grunthaner, *Chem. Phys. Lett.* **44**, 225 (1976).
229. S. I. Raider and R. Flitsch, *J. Vac. Sci. Technol.* **14**, 69 (1977).
230. G. Hollinger, J. Jugnet, P. Pertosa, and Tran Mihn Duc, *Chem. Phys. Lett.* **36**, 441 (1975).
231. B. Carriere, J. P. Deville, D. Brion, and J. Escard, *J. Electron Spectrosc. Relat. Phenom.* **10**, 85 (1977).
232. B. A. Jouce and J. H. Neave, *Surf. Sci.* **27**, 499 (1971).
233. A. H. Boonstra, *Philips Res. Rep. Suppl.* **3**, 1 (1968).
234. S. Pantelides and W. A. Harrison, *Phys. Rev. B* **13**, 2667 (1976).
235. S. Pantelides, B. Fischer, R. A. Pollak, T. H. DiStefano, *Solid State Commun.* **11**, 1003 (1977).
236. J. S. Johannessen, W. E. Spicer, and Y. E. Strausser, *J. Appl. Phys.* **47**, 3028 (1976).
237. T. H. DiStefano, *J. Vac. Sci. Technol.* **13**, 856 (1976).
238. C. R. Helms, W. E. Spicer, and N. M. Johnson, *Solid State Commun.* **25**, 673 (1978).
239. C. R. Helms, Y. E. Strausser, and W. E. Spicer, *Appl. Phys. Lett.* **33**, 767 (1978).
240. C. M. Garner, I. Lindau, C. Y. Su, P. Pianetta, and W. E. Spicer, *Phys. Rev. B* **19**, 3944 (1979).
241. C. M. Garner, I. Lindau, C. Y. Su, J. N. Miller, P. Pianetta, and W. E. Spicer, *Phys. Rev. Lett.* **40**, 403 (1978).
242. C. A. Mead and W. G. Spitzer, *Phys. Rev. A* **134**, 713 (1964).
243. C. A. Mead, *Solid State Electron.* **9**, 1023 (1966).
244. T. C. McGill, *J. Vac. Sci. Technol.* **11**, 935 (1974).
245. S. Kurtin, T. C. McGill, and C. A. Mead, *Phys. Rev. Lett.* **22**, 1433 (1969).
246. J. O. McCaldin, T. C. McGill, and C. A. Mead, *Phys. Rev. Lett.* **36**, 56 (1976).
247. J. O. McCaldin, T. C. McGill, and C. A. Mead, *J. Vac. Sci. Technol.* **13**, 802 (1976).
248. J. Bardeen, *Phys. Rev.* **71**, 717 (1947).
249. V. Heine, *Phys. Rev. A* **138**, 1689 (1965).
250. F. Yndurain, *J. Phys. C* **4**, 2849 (1971).
251. F. Flores, E. Louis, and F. Yndurain, *J. Phys. C* **6**, L465 (1973).
252. E. Louis, F. Yndurain, and F. Flores, *Phys. Rev. B* **13**, 4408 (1976).
253. C. Tejedor, F. Flores, and E. Louis, *J. Phys. C* **10**, 2163 (1977).
254. J. C. Inkson, *J. Phys. C* **5**, 2599 (1972).
255. J. C. Inkson, *J. Phys. C* **6**, 1350 (1973).
256. J. C. Inkson, *J. Vac. Sci. Technol.* **11**, 943 (1974).
257. J. C. Phillips, *Solid State Commun.* **12**, 861 (1973).
258. J. C. Phillips, *J. Vac. Sci. Technol.* **11**, 947 (1974).
259. J. M. Andrews and J. C. Phillips, *Crit. Rev. Solid State Sci.* **5**, 405 (1975).
260. J. M. Andrews and J. C. Phillips, *Phys. Rev. Lett.* **35**, 56 (1975).
261. S. G. Louie and M. L. Cohen, *Phys. Rev. Lett.* **35**, 866 (1975).
262. S. G. Louie and M. L. Cohen, *Phys. Rev. B* **13**, 2461 (1976).
263. S. G. Louie, J. R. Chelikowsky, and M. L. Cohen, *J. Vac. Sci. Technol.* **13**, 790 (1976).
264. S. G. Louie, J. R. Chelikowsky, and M. L. Cohen, *Phys. Rev. B* **15**, 2154 (1977).
265. J. R. Chelikowsky, *Phys. Rev. B* **16**, 3618 (1977).
266. E. J. Mele and J. D. Joannopoulos, *J. Vac. Sci. Technol.* **15**, 1370 (1978).
267. M. Schlüter, *J. Vac. Sci. Technol.* **15**, 1374 (1978).
268. H. I. Zhang and M. Schlüter, *J. Vac. Sci. Technol.* **15**, 1384 (1978).
269. J. E. Rowe, S. B. Christman, and G. Margaritondo, *Phys. Rev. Lett.* **35**, 1471 (1975).
270. G. Margaritondo, S. B. Christman, and J. E. Rowe, *J. Vac. Sci. Technol.* **13**, 329 (1976).

271. G. Margaritondo, S. B. Christman, and J. E. Rowe, *Phys. Rev. B* **14**, 5396 (1976).

272. J. E. Rowe, *J. Vac. Sci. Technol.* **13**, 798 (1976).

273. J. E. Rowe, G. Margaritondo, and S. B. Christman, *Phys. Rev. B* **15**, 2195 (1977).

274. J. L. Freeouf and D. E. Eastman, *Crit. Rev. Solid State Sci.* **5**, 245 (1975).

275. W. E. Spicer, P. E. Gregory, P. W. Chye, I. A. Babalola, and T. Sukegawa, *Appl. Phys. Lett.* **27**, 617 (1975).

276. P. W. Chye, I. A. Babalola, T. Sukegawa, and W. E. Spicer, *Phys. Rev. Lett.* **35**, 1602 (1975).

277. P. W. Chye, I. Lindau, P. Pianetta, C. M. Garner, and W. E. Spicer, *Phys. Rev. B* **17**, 2682 (1978).

278. I. Lindau, P. W. Chye, C. M. Garner, P. Pianetta, C. Y. Su, and W. E. Spicer, *J. Vac. Sci. Technol.* **15**, 1332 (1978).

279. P. W. Chye, I. Lindau, P. Pianetta, C. M. Garner, and W. E. Spicer, *Phys. Rev. B* **18**, 5545 (1978).

280. R. R. Varma, A. McKinley, R. H. Williams, and I. G. Higgenbotham, *J. Phys. D* **10**, L171 (1977).

281. R. H. Williams, R. R. Varma, and A. McKinley, *J. Phys. C* **10**, 4545 (1977).

282. L. J. Brillson, *Phys. Rev. Lett.* **40**, 260 (1978).

283. L. J. Brillson, *J. Vac. Sci. Technol.* **15**, 1378 (1978).

284. R. Z. Bachrach, *J. Vac. Sci. Technol.* **15**, 1340 (1978).

285. A. Amith and P. Mark, *J. Vac. Sci. Technol.* **15**, 1344 (1978).

286. R. L. Bell, *Negative Electron Affinity Devices*, Clarendon Press, Oxford (1973).

287. W. E. Spicer, *Appl. Phys.* **12**, 115 (1977).

288. L. W. James and J. J. Uebbing, *Appl. Phys. Lett.* **16**, 370 (1970).

289. L. W. James and J. J. Uebbing, *J. Appl. Phys.* **41**, 4505 (1970).

290. B. Goldstein, *Surf. Sci.* **47**, 143 (1975).

291. G. Ebbinghaus, W. Braun, and A. Simon, *Phys. Rev. Lett.* **37**, 1770 (1976).

292. M. G. Burt and V. Heine, *J. Phys. C* **11**, 961 (1978).

293. W. E. Spicer, I. Lindau, C. Y. Su, P. W. Chye, and P. Pianetta, *Appl. Phys. Lett.* **33**, 934 (1978).

294. C. Y. Su, P. W. Chye, P. Pianetta, I. Lindau, and W. E. Spicer, *Surf. Sci.* **86**, 894 (1979).

295. W. M. H. Sachtler and P. van der Plank, *Surf. Sci.* **18**, 62 (1969).

296. G. M. Stocks, R. W. Williams, and J. S. Faulkner, *Phys. Rev. B* **4**, 4390 (1971).

297. N. F. Berk, *Surf. Sci.* **48**, 289 (1975).

298. A. Bansil, L. Schwartz, and H. Ehrenreich, *Phys. Rev. B* **12**, 2893 (1975).

299. D. H. Seib and W. E. Spicer, *Phys. Rev. B* **2**, 1676 (1970).

300. D. H. Seib and W. E. Spicer, *Phys. Rev. B* **2**, 1694 (1970).

301. K. Y. Yu, C. R. Helms, and W. E. Spicer, *Solid State Commun.* **18**, 1365 (1976).

302. K. Y. Yu, C. R. Helms, and W. E. Spicer, *Phys. Rev. B* **15**, 1629 (1977).

303. C. R. Helms, *J. Catal.* **36**, 114 (1975).

304. C. R. Helms and K. Y. Yu, *J. Vac. Sci. Technol.* **12**, 276 (1975).

305. M. L. Tarng and G. K. Wehner, *J. Appl. Phys.* **42**, 2449 (1971).

306. K. Nakayama, M. Ono, and H. Shimizu, *J. Vac. Sci. Technol.* **9**, 749 (1972).

307. F. L. Williams and D. Nason, *Surface Sci.* **45**, 377 (1974).

308. J. J. Burton, E. Hyman, and D. G. Fedak, *J. Catal.* **37**, 106 (1975).

309. R. A. Van Santen and M. A. M. Boersma, *J. Catal.* **34**, 13 (1974).

310. C. R. Helms, K. Y. Yu, and W. E. Spicer, *Surf. Sci.* **52**, 217 (1975).

311. K. Y. Yu, D. T. Ling, and W. E. Spicer, *J. Catal.* **44**, 373 (1976).

312. D. T. Ling, J. N. Miller, I. Lindau, W. E. Spicer, and P. M. Stefan, *Surf. Sci.* **74**, 612 (1978).

313. A. D. Baer and G. J. Lapeyre, *Phys. Rev. Lett.* **31**, 304 (1973).

314. G. J. Lapeyre, A. D. Baer, J. Hermanson, J. Anderson, J. A. Knapp, and P. L. Gobby, *Solid State Commun.* **15**, 1601 (1974).

315. G. J. Lapeyre, J. Anderson, P. L. Gobby, and J. A. Knapp, *Phys. Rev. Lett.* **33**, 1290 (1974).

316. F. J. Himpsel and W. Steinmann, *Phys. Rev. Lett.* **35**, 1025 (1975).

317. F. J. Himpsel and W. Steinmann, *Phys. Rev. B* **17**, 2537 (1978).

318. M. Iwan and C. Kunz, *Phys. Lett. A* **60**, 345 (1977).

319. M. Iwan and C. Kunz, *J. Phys. C* **11**, 905 (1978).

320. N. Schwentner, M. Skibowski, and W. Steinmann, *Phys. Rev. B* **8**, 2965 (1973).

321. N. Schwentner, F. J. Himpsel, V. Saile, M. Skibowski, W. Steinmann, and E. E. Koch, *Phys. Rev. Lett.* **34**, 528 (1975).

322. J. L. Freeouf, D. E. Eastman, W. D. Grobman, F. Holtzberg, and J. B. Torrance, *Phys. Rev. Lett.* **33**, 161 (1974).

323. D. E. Eastman and J. L. Freeouf, *Phys. Rev. Lett.* **34**, 395 (1975).

324. R. L. Benbow and Z. Hurych, *Phys. Rev. B* **17**, 4527 (1978).

325. J. H. Weaver, J. A. Knapp, D. E. Eastman, D. T. Peterson, C. B. Satterthwaite, *Phys. Rev. Lett.* **39**, 639 (1977).

326. Z. Hurych, D. Buczek, C. Wood, G. J. Lapeyre, and A. D. Baer, *Solid State Commun.* **13**, 823 (1973).

327. Z. Hurych, J. C. Schaffer, D. L. Davis, T. A. Knecht, G. J. Lapeyre, P. L. Gobby, J. A. Knapp, and C. G. Olson, *Phys. Rev. Lett.* **33**, 830 (1974).

328. Z. Hurych, D. Davis, D. Buczek, C. Wood, G. J. Lapeyre, and A. D. Baer, *Phys. Rev. B* **9**, 4392 (1974).

329. I. T. McGovern and R. H. Williams, *J. Phys. C* **9**, L337 (1976).

330. I. T. McGovern, A. Parke, and R. H. Williams, *J. Phys. C* **9**, L511 (1976).

331. R. H. Williams, I. T. McGovern, R. B. Murray, and M. Howells, *Phys. Status Solidi B* **73**, 307 (1976).

332. R. Z. Bachrach, M. Skibowski, and F. C. Brown, *Phys. Rev. Lett.* **37**, 40 (1976).

333. P. Thiry, Y. Petroff, R. Pinchaux, C. Guillot, Y. Ballu, J. Lecante, J. Paigné, and F. Levy, *Solid State Commun.* **22**, 685 (1977).

334. P. Thiry, R. Pinchaux, G. Martinez, Y. Petroff, J. Lecante, J. Paigné, Y. Ballu, C. Guillot, and D. Spanjaard, *Solid State Commun.* **27**, 99 (1978).

335. N. V. Smith, N. M. Traum, J. A. Knapp. J. Andersson, and G. J. Lapeyre, *Phys. Rev. B* **13**, 4462 (1976).

336. P. K. Larsen, G. Margaritondo, J. E. Rowe, M. Schlüter, and N. V. Smith, *Phys. Lett. A* **58**, 423 (1976).

337. P. K. Larsen, M. Schlüter, and N. V. Smith, *Solid State Commun.* **21**, 775 (1977).

338. P. K. Larsen, S. Chiang, and N. V. Smith, *Phys. Rev. B* **15**, 3200 (1977).

339. G. Margaritondo, J. E. Rowe, M. Schlüter, F. Levy, and E. Mooser, *Phys. Rev. B* **16**, 2938 (1977).

340. G. Margaritondo, J. E. Rowe, G. K. Wertheim, F. Levy, and E. Mooser, *Phys. Rev. B* **16**, 2934 (1977).

341. G. Margaritondo, J. E. Rowe, and S. B. Christman, *Phys. Rev. B* **15**, 3844 (1977).

342. G. Margaritondo, J. E. Rowe, M. Schlüter, and H. Kasper, *Solid State Commun.* **22**, 753 (1977).

343. J. E. Rowe, G. Margaritondo, H. Kasper, and A. Baldereschi, *Solid State Commun.* **20**, 921 (1976).

344. D. E. Eastman and W. D. Grobman, *Phys. Rev. Lett.* **28**, 1327 (1972).

345. J. Hermanson, J. Anderson, and G. J. Lapeyre, *Phys. Rev. B* **12**, 5410 (1975).

346. C. Norris and G. P. Williams, *J. Phys. F* **6**, L167 (1976).

347. G. P. Williams, C. Norris, and M. R. Howells, *J. Phys. F* **7**, 2247 (1977).

348. P. S. Wehner, J. Stöhr, G. Apai, F. R. McFeely, R. S. Williams, and D. A. Shirley, *Phys. Rev. B* **14**, 2411 (1976).

349. L. F. Wagner, Z. Hussain, and C. S. Fadley, *Solid State Commun.* **21**, 257 (1977).

350. J. Stöhr, F. R. McFeely, G. Apai, P. S. Wehner, and D. A. Shirley, *Phys. Rev. B* **14**, 4431 (1976).

351. J. Stöhr, P. S. Wehner, R. S. Williams, G. Apai, and D. A. Shirley, *Phys. Rev. B* **17**, 587 (1978).

352. R. S. Williams, P. S. Wehner, J. Stöhr, and D. A. Shirley, *Surf. Sci.* **75**, 215 (1978).

353. R. J. Smith, J. Anderson, J. Hermanson, and G. J. Lapeyre, *Solid State Commun.* **21**, 459 (1977).

354. D. E. Eastman, F. J. Himpsel, and J. A. Knapp, *Phys. Rev. Lett.* **40**, 1514 (1978).

355. G. M. Bancroft, W. Gudat, and D. E. Eastman, *J. Electron Spectrosc. Relat. Phenom.* **10**, 407 (1977).

356. D. E. Eastman, J. A. Knapp, and F. J. Himpsel, *Phys. Rev. Lett.* **41**, 825 (1978).

357. R. S. Williams, P. S. Wehner, J. Stöhr, and D. A. Shirley, *Phys. Rev. Lett.* **39**, 302 (1977).

358. R. S. Williams, P. S. Wehner, S. D. Kevan, R. F. Davis, and D. A. Shirley, *Phys. Rev. Lett.* **41**, 323 (1978).

359. F. J. Himpsel and D. E. Eastman, *Phys. Rev. Lett.* **41**, 507 (1978).

360. Z. Hussain and N. V. Smith, *Phys. Lett. A* **66**, 492 (1978).

361. E. W. Plummer and T. Gustafsson, *Science* **198**, 165 (1977).

362. J. Anderson and G. J. Lapeyre, *Phys. Rev. Lett.* **36**, 376 (1976).

363. G. Brodén, T. N. Rhodin, C. Brucker, R. Benbow, and Z. Hurych, *Surf. Sci.* **59**, 593 (1976).

364. G. Apai, P. S. Wehner, J. Stöhr, R. S. Williams, and D. A. Shirley, *Solid State Commun.* **20**, 1141 (1976).
365. C. L. Allyn, T. Gustafsson, and E. W. Plummer, *Solid State Commun.* **24**, 531 (1977).
366. W. Braun, M. Neumann, M. Iwan, and E. E. Koch, *Solid State Commun.* **27**, 155 (1978).
367. P. S. Wehner, S. D. Kevan, R. S. Williams, R. F. Davis, and D. A. Shirley, *Chem. Phys. Lett.* **57**, 334 (1978).
368. M. Schlüter, J. E. Rowe, G. Margaritondo, K. M. Ho, and M. L. Cohen, *Phys. Rev. Lett.* **37**, 1632 (1976).
369. J. E. Rowe, G. Margaritondo, and S. B. Christman, *Phys. Rev. B* **16**, 1581 (1977).
370. P. K. Larsen, N. V. Smith, M. Schlüter, H. H. Farrell, K. M. Ho, and M. L. Cohen, *Phys. Rev.* **17**, 2612 (1978).
371. S. Y. Tong, C. H. Li, and A. R. Lubinsky, *Phys. Rev. Lett.* **39**, 498 (1977).
372. A. Liebsch, *Phys. Rev. Lett.* **38**, 248 (1977).
373. C. H. Li and S. Y. Tong, *Phys. Rev. Lett.* **40**, 46 (1978).
374. R. L. Benbow and Z. Hurych, *Phys. Rev. B* **14**, 4295 (1976).
375. Z. Hurych and R. L. Benbow, *Phys. Rev. Lett.* **38**, 1094 (1977).
376. Z. Hurych, R. L. Benbow, and J. C. Shaffer, *J. Vac. Sci. Technol.* **14**, 391 (1977).
377. Z. Hurych and R. L. Benbow, *Phys. Rev. B* **16**, 3707 (1977).
378. R. Z. Bachrach and A. Bianconi, *J. Vac. Sci. Technol.* **15**, 525 (1978).
379. P. Perfetti, D. Denley, K. A. Mills, and D. A. Shirley, *Appl. Phys. Lett.* **33**, 667 (1978).
380. R. S. Bauer, J. C. McMenamin, H. Petersen, and A. Bianconi, in *The Physics of SiO₂ and Its Interfaces*, S. T. Pantelides (ed.), p. 401, Pergamon Press, New York (1978).
381. D. P. Woodruff, D. Norman, B. W. Holland, N. V. Smith, H. H. Farrell, and M. M. Traum, *Phys. Rev. Lett.* **41**, 1130 (1978).
382. D. Norman, D. P. Woodruff, C. Norris, and G. P. Williams, *J. Electron Spectrosc. Relat. Phenom.* **14**, 231 (1978).
383. N. V. Smith, P. K. Larsen, and S. Chiang, *Phys. Rev. B* **16**, 2699 (1977).
384. S. A. Flodström, R. Z. Bachrach, R. S. Bauer, J. C. McMenamin, and S. B. M. Hagström, *J. Vac. Sci. Technol.* **14**, 303 (1977).
385. R. S. Williams, P. S. Wehner, G. Apai, J. Stöhr, D. A. Shirley, and S. P. Kowalczyk, *J. Electron Spectrosc. Relat. Phenom.* **12**, 477 (1977).
386. L. I. Johansson, J. W. Allen, T. Gustafsson, I. Linday, and S. B. M. Hagström, *Solid State Commun.* **28**, 53 (1978).
387. J. W. Allen, L. I. Johansson, R. S. Bauer, I. Lindau, and S. B. M. Hagström, *Phys. Rev. Lett.* **41**, 1499 (1978).
388. R. P. Godwin, *Springer Tracts Mod. Phys.* **51**, 1 (1969).
389. K. Codling, *Rep. Prog. Phys.* **36**, 541 (1973).
390. F. C. Brown, *Solid State Phys.* **29**, 1 (1974).
391. R. Haensel, in *Festkörperprobleme*, H. J. Quesser (ed.), Vol. XV, p. 203, Pergamon/Vieweg, Braunschweig (1975).
392. E. E. Koch, C. Kunz, and B. Sonntag, *Phys. Rep.* **29**, 155 (1977).

7

Microlithography with Soft X Rays

ANDREW R. NEUREUTHER

1. Introduction

The use of soft x rays as a practical means of replicating patterns in the fabrication of electronic and optical microdevices was suggested by Spears and Smith[1,2] in 1972. The approach is similar to contact x-ray microscopy, which has been used for several decades (see Chapter 8). Typically, x-ray wavelengths from 0.4 to 8.0 nm are used to proximity print Au mask patterns supported by thin transparent substrates with 0.1-μm resolution. X-ray lithography is an important alternative to optical lithography because it overcomes the fundamental limitations of diffraction and of shallow depth of field. Although x-ray replication is itself dependent on electron beam lithography for generating masks, it has inherently higher resolution for making device features. More importantly, it is a parallel rather than a serial exposure process, which tends to make it much more cost effective than direct electron beam wafer writing. In fact from an economic point of view, x-ray lithography is potentially competitive with optical lithography for fabricating electronic devices with 1.0-μm features. At 0.5-μm feature sizes x-ray lithography may be the only viable approach that has a throughput on the order of one wafer per minute. X-ray lithography systems intended to meet these micron and submicron volume production goals are being developed commercially.

Synchrotron radiation is profoundly influencing x-ray lithography, both as a flexible, quantitative tool for research and as a potential source for the volume production of electronic devices. Patterns with 50-nm features have been replicated in poly(methyl methacrylate), PMMA, and exposure times as small as 1 s have been obtained.[3] The intensity of the flux is several orders of magnitude larger than that

ANDREW R. NEUREUTHER ● Department of Electrical Engineering and Computer Sciences and Electronics Research Laboratory, University of California at Berkeley, Berkeley, California.

of conventional electron bombardment x-ray sources. In fact, the flux is sufficiently intense that the construction of small (0.72 GeV, 12 kOe, 100 mA) dedicated storage rings may be more cost-effective than conventional x-ray sources for volume electronic device production. The divergence of the synchrotron radiation is also small, and it essentially eliminates the problem of geometrical distortion, which imposes severe constraints on mask to wafer positioning with conventional x-ray sources. In addition, the inherent problem of hard x-ray radiation from the bremsstrahlung process that occurs with electron bombardment can be eliminated. This can be accomplished either by the choice of the storage ring parameters or by the use of a grazing-incidence mirror. The small flux divergence, coupled with the broad synchrotron spectrum, is well suited for making absorption measurements with a tuneable monochrometer. This allows the investigation of phenomena such as extended x-ray absorption fine structure (EXAFS), which may be useful in developing and understanding resist materials. Facilities for absorption measurements and resist exposure are currently available at a number of storage ring and synchrotron installations throughout the world. See Chapter 10 for a discussion of EXAFS.

The development of high-resolution lithography, including x-ray lithography, has been stimulated by the technological needs of the electronics industry, and in particular by the economics of the integrated circuit (IC) segment. Since the first integrated circuit was produced in 1959, the cost per function for an electronic device in a circuit has decreased by four orders of magnitude. Further cost reduction, improved circuit performance, and higher levels of complexity in very large-scale integrated (VLSI) circuits are possible and will likely both depend on and provide economic support for the further development of high-resolution lithography.[4]

X-ray lithography has been suggested for a broad range of electrical, optical, and x-ray devices. The most immediate volume application will likely be for bubble memories, which may require only one or at most a few mask steps. Small features on the order of the size of the bubbles are needed, which for the next generation of devices will likely be less than 1.0-μm gaps. X-ray lithography is a possible alternative to optical lithography for short-channel metal oxide semiconductor (MOS) integrated circuits. X-ray lithography is also being considered for high-performance semiconductor devices such as complementary metal oxide simiconductor (CMOS) devices in silicon on sapphire (SOS). Activity is also high in smaller volume special purpose applications that take unique advantage of the very high resolution. These applications include high-resolution diffraction gratings and zone plates for use throughout the electromagnetic spectrum. They also include integrated optics devices, as well as the initial application, surface acoustic wave devices.[2]

Historically, the feasibility of the use of x-ray lithography for microfabrication was first discussed in 1972 by Spears and Smith.[1,2] They also produced the first devices and demonstrated the high-resolution capability. The potential of x-ray lithography was also appreciated by Feder.[5] Shortly after its inception, exploratory studies of the potential of x-ray lithography were carried out in several laboratories. Favorable results stimulated the undertaking of the systematic development of x-ray lithography as a practical technology. Early examples of x-ray lithography systems designed for device production are those developed at Bell Telephone

Laboratories[6] and Hughes Research Laboratories.[7] Today x-ray systems capable of making multilayer device structures exist in many research laboratories.[8-15] A variety of commercial x-ray lithography systems are also being developed.[16,17] They range from modified contact printers with x-ray sources to major new step and repeat systems with automatic alignment.

Synchrotron radiation as a source for x-ray lithography was first explored experimentally in 1976. Experiments were carried out by Spiller et al.[18] using the German electron synchrotron ring DESY in Hamburg, by Fay and Trotel[19,20] with radiation from the ACO storage ring at Orsay in France, and by Aritome et al.[21,22] and Nishimura et al.[23] with the electron synchrotron INS-ES at the University of Tokyo. The SPEAR storage ring at Stanford University has also been used for wavelength tuneable resist absorption measurements[24-25] as well as replication.[24,26,27]

The possibility of x-ray lithography has come about through specific research on technological problems in the areas of x-ray sources, masks and windows, resists, alignment techniques, distortion, and device damage. The technology is relatively new and significant advances are still occurring. Recent advances include a new 25-kW Si source[17] capable of exposing special x-ray resists doped with absorbing material in a few seconds. Thin SiC mask support materials have been developed[10] that should greatly reduce the distortion problems with organic mask materials. The distortion of Si wafers during processing has been shown to be both smaller and more predictable than previously suspected when low-temperature processing is used.[28] X-ray and optical alignment techniques have also been developed. Automatic optical alignment techniques appear to be capable of 0.1-μm accuracy[29] and further improvements may be possible with the use of interference techniques.[30] Most of the deleterious effects introduced during x-ray exposure have also been shown to be removable by a simple annealing process.[31,32] Finally, for high aspect ratio features at micron sizes and below, where the throughput of conventional sources is limited, synchrotron radiation from a dedicated storage ring has been shown to be both cost effective and technically advantageous.

This chapter begins with a discussion of the basic x-ray lithography technique and presents a number of results that demonstrate the unique capability and potential of synchrotron radiation for x-ray lithography. Section 3 characterizes the fundamentals involved in x-ray lithography. This includes summaries of the physics of the generation and absorption of x rays, the generation of photoelectrons, and the resulting effect on resist materials. Section 4 explores the status of developments in the technological areas of sources, masks, and resists. This includes the critical problems of alignment, mask and wafer distortion, and device damage. Finally, various approaches for production x-ray lithography systems are discussed in Section 5.

2. X-Ray Replication and Results

Optical lithography is used predominantly in microfabrication today. The minimum linewidths or critical dimensions of present devices are about 2.0 to 3.0 μm and are produced by optical lithography at UV wavelengths ($\lambda = 0.4$ μm). Two

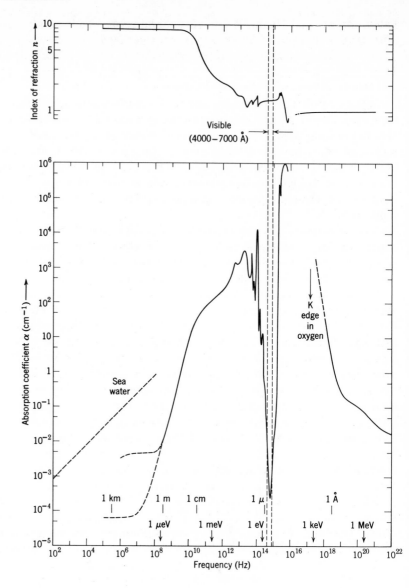

Figure 1. The index of refraction (top) and absorption coefficient (bottom) for liquid water as a function of frequency, energy and wavelength. Note that between the visible region and the K edge in oxygen the absorption constant has a high and broad peak (Jackson[35]).

forms, proximity printing and projection printing, are used. They avoid mask damage caused by direct contact between the mask and the device. However, both approaches are pushing fundamental limits due to diffraction. For proximity printing at a mask to wafer spacing z, the practically producible linewidth is approximately $(z\lambda)^{1/2}$.[33] In projection printing with a diffraction-limited lens of numerical aperture na the resolution useable in practice is about $0.75\lambda/na$ on Si and $1.1\lambda/na$ on Al.[34] New optical machines are being introduced to further reduce linewidths. Some include the use of deep-UV sources with wavelengths down to 0.2 μm. However, as linewidths approach 1.0 μm the technical feasibility of the optical approach over a 1-cm^2 field in a production environment becomes very questionable and the capital equipment cost increases rapidly.

X-ray lithography is a natural extension of optical lithography. Electromagnetic radiation is again used but at a much shorter wavelength. The choice of wavelength is determined by the absorption behavior of materials as a function of wavelength. Figure 1 shows, for example, the absorption constant of water as a function of wavelength.[35] To penetrate a typical thin layer thickness of 1.0 μm an absorption constant of less than 10^4 cm^{-1} is needed. Note that at wavelengths just shorter than the visible region the absorption constant rises extremely rapidly and quickly exceeds this value. Only at the soft x-ray wavelengths near the K edge in oxygen (23.3 Å) does the absorption constant drop below 10^4 cm^{-1}. Here, the absorption constant decreases roughly as the third power of the wavelength. Below 1 Å it is difficult to deposit significant amounts of energy per unit volume in material. Thus from the entire electromagnetic spectrum the only viable alternative wavelengths for lithography are in the soft x-ray region. In particular, only the region from 0.4 to 8.0 nm (4 to 80 Å) is of interest in replication and wavelengths from 0.4 to 2.0 nm will likely be most useful for IC fabrication.

A schematic diagram of a soft x-ray lithography system is shown in Figure 2. This system was designed at Bell Telephone Laboratories.[6, 36] In this system a 25-keV, 4.5-kW electron beam excites the 0.437-nm characteristic radiation from a rotating, water-cooled Pd anode. A 50-μm-thick Be window separates the vacuum

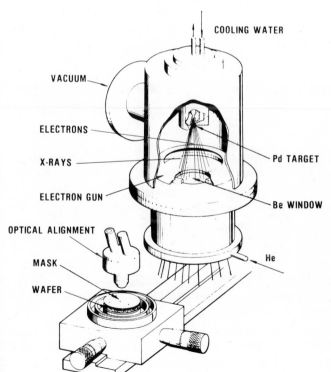

Figure 2. X-ray exposure station with alignment built at Bell Telephone Laboratories (Maydan *et al.*[6]).

source from the exposure area and transmits 80% of the radiation. The exposure takes place in a chamber filled with He to prevent absorption of the characteristic radiation by air. The x-ray mask and substrate are separated by a spacing of 40 μm during exposure. The mask and wafer are located 50 cm away from the source to minimize geometrical distortion. X-ray masks are fabricated with a 25-μm-thick Kapton film stretched and bonded over a Pyrex supporting ring. The absorption pattern is generated by an electron beam exposure system (EBES) and is delineated by a combination of both plasma and ion etching techniques in 0.7-μm-thick Au metalization. A 320× microscope is used to reregister the wafer geometry to the mask pattern. After reregistration the stage containing a vacuum chuck is moved into the exposure position under the He column. The x rays passing through the transparent portion of the mask expose a chlorine-based x-ray resist causing cross-linking of the molecules of the resist. The exposed areas will thus tend to be less soluble and will therefore remain after development (negative-type resist). At present the system has a throughput capability of about 15 wafers per hour and with future developments a throughput of one wafer per minute will be likely. The working resolution as a lithography tool is about 1.0 μm although 0.5 μm may be possible.

The ultimate resolution possible with x-ray lithography far exceeds the wavelength limit of optical lithography and is comparable to that of electron beam lithography. With electron beams, under special circumstances of extremely thin substrates, 8-nm lines have been written by deposition of gaseous contamination in a vacuum system.[37] Arrays of 25-nm lines have been etched in metal[38] from resist openings created by using a positive-type resist, which is made more soluble by

Figure 3. X ray replicated patterns showing high aspect ratios of a positive pattern (a) of 15 : 1, and a negative pattern (b) of 5 : 1 (Feder et al.[56]).

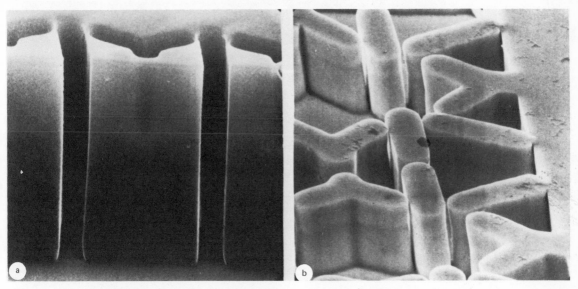

Figure 4. (a) X ray replicated 1 μm linewidth bubble pattern in 15 μm thick resist. (b) Gold pattern obtained by electroplating 5 μm of gold with 5 μm resist thickness (Feder et al.[9]).

exposure. On the other hand, work in x-ray microscopy (see Chapter 8) has demonstrated that x-ray lithography has a resolution capability of about 5 nm and that a resist like poly(methyl methacrylate), PMMA, has sufficient contrast to replicate such structures.[39] Arrays of 17.5-nm lines have been replicated from thin shadow depositions on vertical sidewalls.[40]

For electronic circuits resolution less than 10 nm is not as important as the capability of x-ray lithography to produce high-quality line-edge profiles in the context of device applications. With x rays the profiles are relatively insensitive to underlying materials and topography of the device surface. The nonuniformity in optical exposures due to the interference of the incident light with its own reflection from the substrate does not occur with x rays. The diffraction spreading is also reduced by several orders of magnitude. The x-ray exposure does not spread with depth due to electron scattering because the x-ray absorption process produces much lower-energy electrons than those used in electron beam lithography and these secondary electrons have short ranges. The deposited energy can also be very uniform throughout the resist.

An example of the high aspect ratio (height/width) submicron profiles possible with x-ray lithography is shown in Figure 3.[9] Here the resist is 15.0 μm thick and the opening is 1.0 μm wide. Bubble memory device patterns 5.0 μm high and 1.0 μm wide that have been electroplated are shown in Figure 4.[9] In contrast, the electron beam exposed profiles in Figure 5[41] are more nonuniform due to electron scattering effects. Electron beam profiles with nearly vertical walls can be obtained by compensating the exposure broadening with development effects, as is shown in

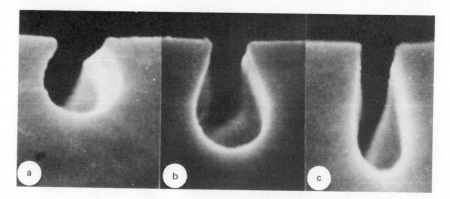

Figure 5. Line edge profiles of PMMA resist exposed at a dose of 10^{-4} C/cm^2 at, from left to right, 10, 15, and 25 kV (Hatzakis[41]).

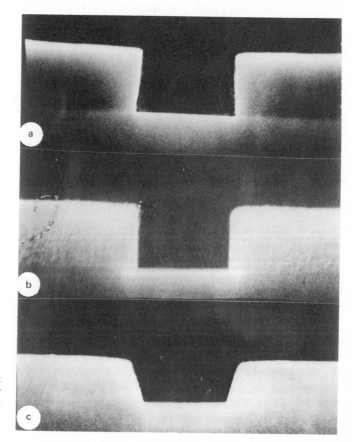

Figure 6. PMMA resist line edge profiles at various exposure charge densities: (a) 10^{-4} C/cm^2; (b) 8×10^{-5} C/cm^2; (c) 5×10^{-5} C/cm^2 (Hatzakis[42]).

1600 Å

Figure 7. Line edge profile of a 0.16 μm linewidth grating pattern exposed in 0.85 μm PMMA resist on a SiO_2/Si substrate using Cu L radiation at 1.33 μm (Flanders *et al.*[43]).

Figure 6.[42] However, the compensation required is sensitive to variations in operating parameters such as dose, resist thickness, substrate material, and the written pattern. A 0.16-μm-linewidth grating exposed in a 0.85-μm film of PMMA using Cu L radiation is shown in Figure 7.[43] Note that the height of the resist lines is greater than the linewidth and that an entire array of lines has been produced over a large area of exposure. This structure as well as the previous bubble structures could probably not be fabricated by electron beam lithography due to the proximity effects produced by electron scattering.

Since the x-ray exposure is only slightly affected by the content or thickness of the underlying material, it is not difficult to fabricate narrow lines across changes in the underlying material. In particular, step coverage is not a problem for x-ray lithography. Figure 8[32] shows a 1.0-μm polysilicon step running up a silicon island on a sapphire substrate. The narrow pattern does not "neck down" as it runs up the step. For both electron beam and optical lithography, the pattern would narrow as it runs up the step.

As device features are reduced to 1.0 μm and below, the throughput with conventional x-ray sources decreases. This is due to the increased source-to-wafer distance necessary to compensate for geometrical distortion for fixed mask-to-wafer spacing and spacing tolerance. The throughput is also decreased by the need for larger x-ray doses to produce narrow lines and high aspect ratio resist profiles. These limitations of the conventional source approach can be overcome by the use of synchrotron radiation from an electron storage ring. Several experiments have been carried out on existing synchrotron and storage ring facilities to assess the feasibility of using synchrotron radiation. The radiation available and the associated optics vary considerably and generally have not been optimized for x-ray lithography. However, the experimental results clearly demonstrate the potential of using synchrotron radiation from a dedicated storage ring for x-ray lithography.

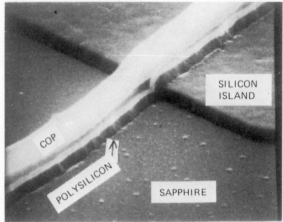

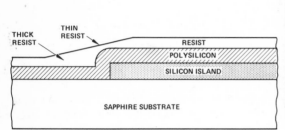

Figure 8. SEM micrographs of 1 μm polysilicon gates made by x ray exposure of COP resist followed by plasma etching (Stover[32]). (left) Illustration of resist coverage of polysilicon step. (right) Uniform resist exposure: no "necking down" of polysilicon gate due to proximity effects or light scattering in resist.

A fairly comprehensive set of experiments to explore the use of synchrotron radiation was carried out by Spiller et al.[18] using the DESY synchrotron ring at 3.5 GeV and 8 mA. The low geometrical distortion with synchrotron radiation is demonstrated in Figure 9, which shows replication in the direct beam of a mask spaced 0.54 mm and 1.04 mm from the wafer. Slight diffraction effects are noticeable at 0.54 mm. This spacing is however very large compared to the few μm tolerance required with conventional sources. Results of replicating 1.0-μm lines in the direct beam for three different resist and filter combinations are shown in Figure 10. In all cases the effective mask contrast was low due to the relatively hard radiation in the direct beam. Case *c* is particularly interesting in that it demon-

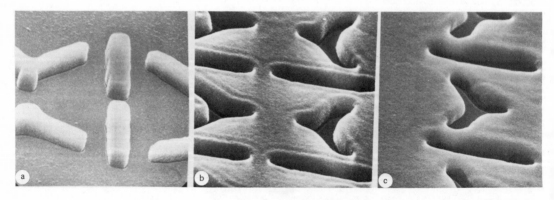

Figure 9. Mask replication with synchrotron radiation from the DESY storage ring at 3.5 GeV and 8 mA. The 1 μm linewidth mask was placed (a) 0.14 mm, (b) 0.54 mm, and (c) 1.04 mm from the wafer. The exposure time of 3 min with a 12 μm Be plus 6 μm Mylar filter produced an effective exposure of 250 J/cm^3 in PMMA (Spiller et al.[18]).

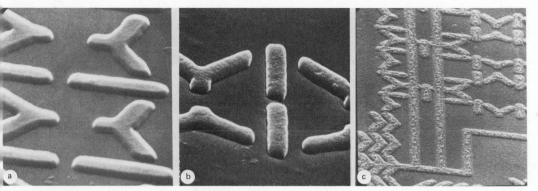

Figure 10. Synchrotron mask replication similar to Figure 9 but with low contrast masks: (a) 6 : 1 contrast and 76/24 copolymer with 30 sec exposure through 6 μm Mylar (79 J/cm^3); (b) 3 : 1 contrast and PMMA with 3 min. exposure through 6 μm Mylar (450 J/cm^3); (c) 2.3 : 1 contrast and 170 min exposure of 76/24 copolymer through a 75 μm Si wafer (80 J/cm^3) (Spiller *et al.*[18]).

strates that it is possible to copy a pattern directly through a Si wafer and thus perform front to back registration.

To achieve high resolution, a grazing-incidence mirror that deflected the beam 4° was introduced to eliminate the hard radiation below 1.0 nm. Patterns replicated with this soft radiation are shown in Figure 11. Linewidths as small as 70 nm were replicated from Au patterns as thin as 0.1 μm. Thus it is possible to use masks with thin absorber films ($t < 100$ nm), which can easily be produced using electron beam techniques. Replicated features on the order of 50 nm were observed in related experiments using synchrotron exposure of biological specimens.[3]

Mask distortion due to heating and damage was also explored. Heating of the Mylar mask and wafer did not have any deleterious effect on resolution. A Mylar mask exposed to the full beam for the equivalent of 10^4 replications in PMMA showed radiation damage as a yellow color but no change in dimensions could be detected.

1 μm

Figure 11. A Fresnel zone plate replica with individual lines as narrow as 0.07 μm. Synchrotron radiation from the DESY storage ring was used with a 4° deflecting Au mirror to eliminate the hard x-ray flux (Spiller *et al.*[18]).

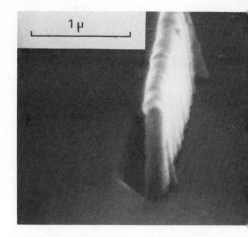

Figure 12. PMMA resist showing a 0.2 μm resist line 0.8 μm high replicated from a 0.1 μm thick Au mask at 540 MeV and 100 mA on the ACO storage ring with a few minute exposure (Trotel *et al.*[19]).

Mask replication with synchrotron radiation has also been explored by Trotel and Fay[19] and Fay *et al.*[20] using the ACO storage ring at Orsay, France. Experiments were performed at 50 mA at beam energies of 540 and 400 MeV. When operated at this high beam current and at 540 MeV the ACO storage ring emits about 10 times as much soft radiation as was available in the DESY experiment. Some distortion due to heating may have occurred, although this could be eliminated by using a thin organic filter such as a Mylar film. At 540 MeV, 0.1 μm of Au on a 4.0-μm Si mask was used. This resulted in an exposure time for PMMA of a few minutes. A resist line less than 0.2 μm wide (shown in Figure 12) was replicated in 0.8 μm of resist. At 400 MeV a 2.5-μm Mylar mask support was used and the exposure time was about 20 min. A typical resist profile in 1.6 μm of PMMA for this beam energy is shown in Figure 13. The effect of a spacing between the mask and wafer was also explored. Figure 14 illustrates the effect of a 250-μm spacing at a beam energy of 540 MeV. The edge of the semi-infinite region is similar to the edges of the narrow lines. The location of edge dip agrees with diffraction theory when

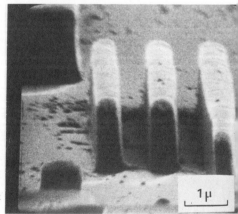

Figure 13. PMMA resist features 1.6 μm high replicated in 20 min at 400 MeV and 100 mA on the ACO storage ring (Trotel *et al.*[19]).

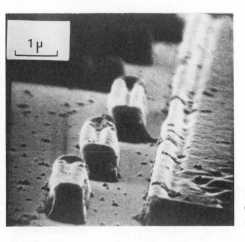

Figure 14. PMMA resist showing diffraction effects at a spacing of 0.25 mm between mask and wafer for 540 MeV operation of the ACO storage ring (Trotel et al.[19]).

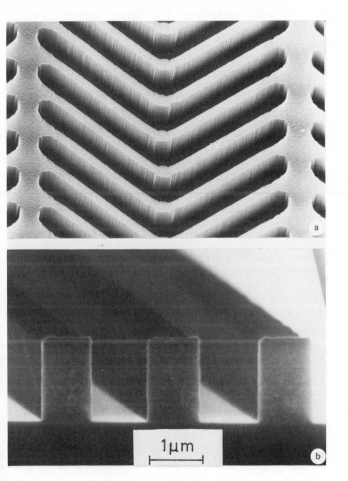

Figure 15. SEM photographs of masks replicated on the 1.3 GeV electron-synchrotron INS-ES at the University of Tokyo. Contact print at 1.16 GeV and 20 mA in 1.7 μm PMMA with 20 minute exposure: (a) 1.5 μm linewidth bubble pattern; (b) 1.0 μm μm linewidth pattern (Nishimura et al.[23]).

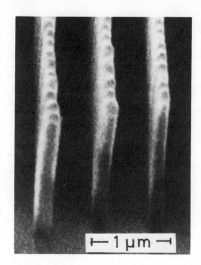

Figure 16. SEM photograph of a 692 nm period grating pattern in 2.2 μm of PMMA replicated from a 0.4 μm mask using INS-ES (Nishimura *et al.*[23]).

the peak photon flux wavelength of 1.5 nm is used. The photon flux for 400-MeV operation peaks at about 5.0 nm and would produce larger diffraction effects.

Synchrotron radiation from the 1.3-GeV electron synchrotron at the University of Tokyo has been used for x-ray lithography by Nishimura *et al.*[23] and Aritome *et al.*[21, 22] An electron energy of 1.1 GeV and a current of 24 mA were used to replicate 0.4-μm Au patterns on 3.0-μm Si substrates. The exposure time for 1.5 μm of PMMA was 20 min. Figure 15 shows the result of replicating a 1.5-μm bubble pattern. A PMMA grating pattern is shown in Figure 16 that has a very large height to width ratio (2.2 μm/0.2 μm). It was obtained by using a 692-nm-period grating pattern fabricated by ion milling an optical interference pattern in photoresist into 0.4 μm of Au. Diffraction was studied by separating the mask and wafer by 400 μm and observing the profiles of lines of various widths. Diffraction noticeably degraded the 1.0-μm lines as can be seen in Figure 17. When the synchrotron was operated at 850 MeV and 120 mA the exposure time for PMMA was again about 20 min. A 380-MeV storage ring (SOR–RING) at the University of Tokyo was also used as a source. With a 2-μm parylene-*N* mask substrate, the exposure time at 100 mA and 300 MeV for PMMA was about 1 min. Figure 18 shows 0.5-μm lines replicated on the storage ring (5-μm substrate, 4 min, PMMA).

At the Stanford Synchrotron Radiation Laboratory (SSRL), both resist replications and resist absorption measurements have been made. Burg *et al.*[27] used a 0.15-μm Al window on a beam line with a 4° deflection mirror to make exposures with both white and monochromatic soft x-ray flux. With SPEAR operating at 2 GeV and 17 mA, the exposure time was 4 min for PMMA and 1 s for PBS. With monochromatic radiation the exposure time for PBS was 30 min. Figure 19 shows a PBS print with monochromatic radiation of a biological specimen inside the window of a Cu grid. The minimum feature size is about 0.5 μm. It illustrates that PBS can form steep, narrow isolated structures about 0.5 μm wide.

Relatively hard x-ray flux from a direct beam line with two 254-μm Be windows has been used by Neureuther *et al.*,[24] Hsu,[25] and Sud[26] to study line-edge

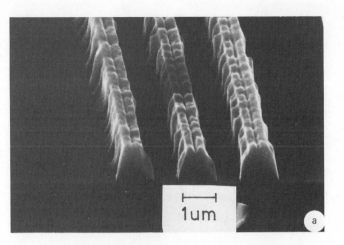

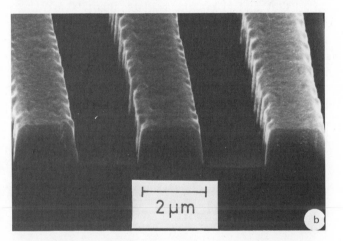

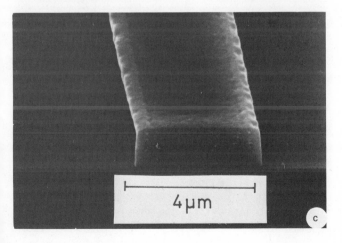

Figure 17. Mask replication of various linewidth patterns at a mask to wafer distance of 0.4 mm: (a) 1 μm line; (b) 2 μm line; (c) 4 μm line (Nishimura et al.[23]).

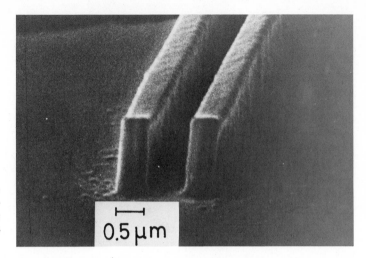

Figure 18. PMMA resist profile after exposure on the SOR storage ring. A 5 μm Paralylene mask substrate was used and the exposure time at 100 mA and 300 MeV was 4 minutes (Aritome *et al.*[22]).

profiles of various resists. At 2.1 GeV and 11 mA the available flux (700 mW/cm^2) was significantly harder (0.2–0.4 nm) than the 0.8-nm flux typically used in x-ray lithography. The exposure time for PBS resist was one minute and PMMA resist required about two hours. Figure 20 shows a PMMA resist line-edge profile. Note that the bottom of the profile is severely undercut due to photoelectrons generated in the Si substrate scattering upwards as much as 0.8 μm into the resist. This substrate photoelectron effect can virtually be eliminated in the case of Si substrates by using only x rays softer than the absorption edge of Si (0.675 nm).

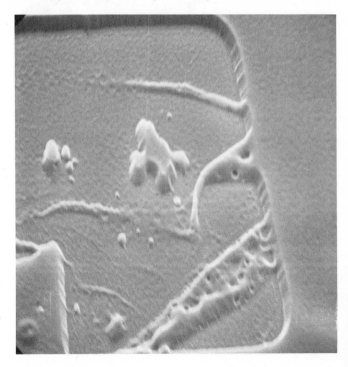

Figure 19. PBS resist profile after 30 min exposure with 500 eV monochromatic radiation at 2.16 GeV and 15 mA on the 4 degree line of the SPEAR storage ring. The small features inside the rectangular window cast by a Cu grid were produced by biological material. They demonstrate that the resolution of the resist is about 0.5 μm (Burg *et al.*[27]).

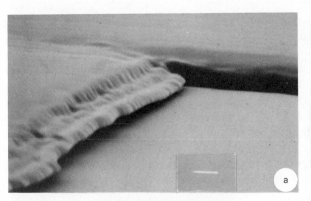

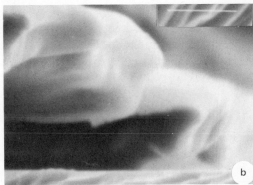

Figure 20. PMMA resist exposed through a Ni grid on the SPEAR topography line. Two 10 mil windows were present and the exposure time was 2 hours at 2.1 GeV and 8 mA. The relatively hard radiation produced significant undercutting. (a) Grid corner showing more rippled edges associated with the horizontal mask wires; (b) line edge profile from a horizontal mask wire; (c) line edge profile of a vertical mask wire (Neureuther *et al.*[24]).

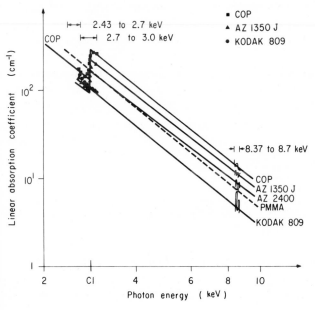

Figure 21. Linear absorption coefficients of several electron beam resists and photoresists (Neureuther *et al.*[24]).

The absorption coefficients of several electron beam and photoresists as a function of wavelength down to energies as low as 2.6 keV (0.47 nm) were measured[24, 25] using a narrow-band tuneable station at SSRL. Figure 21 shows results for the absorption constant of various resists as a function of energy. The behavior of the absorption constant of the resists is similar to that of PMMA. However, abrupt changes in absorption may be present at the Cl and S absorption band edges (2.82 and 2.47 keV, respectively). Although the data shown has been corrected for the third harmonic transmission of the monochrometer, the band edge absorption jump is broader than expected due to instrumentation problems. With proper instrumentation it should be possible to use extended x-ray fine structure (EXAFS) measurements to determine the relative content of absorbing elements in resists and also investigate the nature of the changes in the bonding structure induced by x-ray, electron beam, or optical exposure.

3. Fundamentals

This section describes and quantitatively characterizes the fundamental physical phenomena involved in x-ray lithography. The relationships given here will later be used to study the feasibility and trade-offs in various lithography approaches. Of principal interest is the characterization of the synchrotron radiation and its interaction with material.

3.1. X-Ray Sources

3.1.1. Synchrotron Radiation

Storage rings and synchrotron radiation are described in Chapters 2 and 3. From the point of view of x-ray lithography, the important parameters are the current I, the critical wavelength λ_c, and the relativistic energy ratio $\gamma = E/mc^2$. These parameters completely describe the synchrotron radiation from a storage ring as is shown in Chapters 2 and 3. For x-ray lithography the critical wavelength of interest is between 1 and 6 nm. Thus the shape of the normalized photon flux curve in Figure 3 of Chapter 2 for a critical energy of 0.58 keV, corresponding to a critical wavelength of 2.14 nm, is typical of the spectral distribution for x-ray lithography. The vertical angular spread of the beam is approximately Gaussian and at soft x-ray wavelengths it is dominated by the radiation pattern for a single electron. The standard deviation σ of the radiation angle ψ in radians that approximately fits the data given in Reference [44] is

$$\gamma\sigma = (0.2 + 0.4\lambda/\lambda_c) \tag{1}$$

The angular spread is typically less than 1 mrad.

The secondary effects of the polarization and nonuniformity of the flux are additional considerations in designing a lithography station. The radiation is horizontally polarized in the orbital plane. Out of the orbital plane and at long wavelengths, the radiation is elliptically polarized. The amount of elliptical polarization depends on the ratio of the wavelength to the critical wavelength. For $\lambda/\lambda_c = 1/3$, corresponding to radiation at 10 nm for $\lambda_c = 3.2$ nm, the magnitude of the vertically polarized flux compared to the horizontally polarized flux on axis is at most 6% at its peak at $\gamma\psi = 0.5$. Thus the beam is predominately horizontally polarized. The fact that the beam is strongly polarized may introduce slight differences in the diffraction pattern for the vertical and horizontal mask features. In addition, directional effects may occur in the generation and subsequent exposure by photoelectrons.

The nonuniformity of the beam in intensity and spectral content is a more important consideration. The intensity uniformity requirement will likely be similar to the 3 to 10% specified in optical lithography. For a given x-ray wavelength the flux can be made fairly uniform over the net emission angle in the vertical direction by selecting the central portion of the Gaussian beam. For $\pm10\%$ uniformity, $\pm0.67\sigma$ could be used and this portion would contain 50% of the total flux. The nonuniformity due to changes in polarization would also be small over this range. If the spectral content of the flux is very broad, a nonuniformity may be introduced by the fact that the emission angle described above is a function of wavelength. This variation makes the spectral content of the beam become softer with distance from the orbital plane. If necessary, any nonuniformities could be eliminated and essentially 100% of the flux could be used by scanning the sample vertically through the beam. It may also be possible in a dedicated storage ring to address the beam uniformity problem in the design of the electron orbit and x-ray optics.

3.1.2. Electron Bombardment X-Ray Sources

X-ray radiation can also be produced by bombardment of material with charged particles. Electrons are the most efficient choice as their mass is the same as that of the electrons in the material with which they are interacting. Two types of radiation are produced, continuum and characteristic. The continuum radiation is primarily from interactions with the nucleus. Although the electron does not lose energy to the nucleus because it is so massive that it does not recoil appreciably, the electron does sometimes emit energy by electromagnetic radiation due to its acceleration in the strong electric field near the nucleus (bremsstrahlung). When the incident electron energy is high enough it can cause excitation of atomic electrons or ionization followed by the emission of characteristic lines when vacancies are filled. A tabular listing of the energy of these fluorescence lines for most elements can be found in many handbooks.[45] The corresponding wavelength λ in nm is given by

$$\lambda = 1.24/E \qquad (2)$$

where E is the excitation potential in keV.

The x-ray intensity of both of these processes has been characterized by empirical relationships. In both cases, the angular distribution of radiated energy is rather broad and tends to be largest normal to the surface. The x-ray continuum is described approximately by Kramer's law[46]

$$I(E) = P_0 K Z^n (E_0 - E)/E_0 \qquad (3)$$

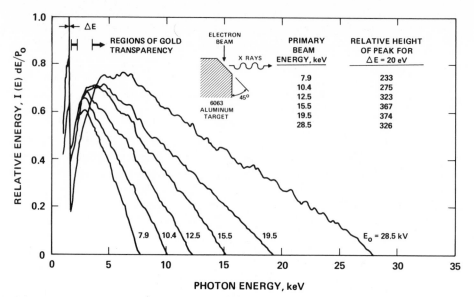

Figure 22. X ray emission spectra from electron bombardment of an aluminum target normalized to constant excitation power (Sullivan et al.[47]).

where $I(E)$ is the total intensity at energy E radiated in all directions and is in units of W/eV energy interval, P_0 is the incident beam power in W, Z is the atomic number of the target, n is about 1, and E_0 is the incident electron energy in keV. An example of experimental measurements of the spectral distribution of the continuum for Al is shown in Figure 22.[47] The value of the constant K for the data shown is 7×10^{-9} eV^{-1}.

In general, the intensity of the characteristic radiation at energy E_c is given approximately by[48]

$$I(E) = BN(Z)F(x)(E_0 - E_c)^m \tag{4}$$

where $N(Z)$ is a function of atomic number and $F(x)$ is the target reabsorption factor, which is dependent on the mass absorption coefficient and the emission angle. The intensity of the aluminum characteristic line at 0.834 nm relative to the continuum is listed in Figure 22. For the data of Figure 22 and voltages less than 15 keV, $m = 1.5$ and $BN(Z)F(x) = 1.6 \times 10^{-2}$ mW/sr W.[47] With increasing voltage the x-ray emission per electron reaches a maximum and then decreases as a result of self-absorption in the target.[49] This reabsorption greatly reduces the intensity of the characteristic and continuum radiation at low takeoff angles from the target. In fact the voltage at which the characteristic efficiency peaks can be directly related to the absorption coefficient.[50] For large wavelengths and mid-Z materials, at normal beam incidence, the voltage of the peak E_p in keV and absorption coefficient α in cm^{-1} are related by

$$(\alpha/\rho)\csc \psi = (2660/E_p)^{1.68} \tag{5}$$

where ψ is the angle of radiation measured from the plane of the target and ρ is in cm^2/g.

Table 1. Absorption Edges and Characteristic Radiation of Primary Interest in X Ray Lithography

Element	Absorption edge (nm)	Characteristic radiation (nm)	Efficiency at 20 keV (mW/sr W)
C_K	4.377	4.382	—
N_K	3.105	3.160	—
O_K	2.337	2.371	—
F_K	1.805	1.831	—
Ne_K	1.419	1.462	—
Cu_L	—	1.336	0.019 (8 keV)
Na_K	1.148	1.191	—
Mg_K	0.951	0.989	—
Al_K	0.795	0.834	0.055
Si_K	0.625	0.713	0.060
P_K	0.579	0.616	—
Mo_L	—	0.541	0.053
S_K	0.502	0.538	—
Cl_K	0.440	0.473	—
Rh_L	—	0.460	0.065
Pd_L	—	0.437	0.068
A_K	0.387	0.419	—
K_K	0.344	0.374	—

[a] The efficiency of generation is for bombardment by 20 keV electrons unless otherwise indicated.
[b] From Maydan et al.[6], Kaelble[45], Sullivan et al.[47]

Figure 22 indicates that the total power in the continuum is generally not negligible and that it exceeds the power at the characteristic wavelength for primary beam energies above 15 keV. The characteristic wavelength and generation efficiency measured at 20 keV are listed in Table 1 for a variety of possible source materials. Unfortunately, these values indicate that the net conversion from electrical energy to x-ray energy is only on the order of 0.01%. The efficiency improves with increasing voltage. Eventually excitation of harder radiation from higher-energy characteristic lines occurs, e.g., at 8.89 keV in Cu. Some hard radiation can be tolerated as it is not strongly absorbed in the resist. However, the hard radiation can result in additional exposure near the substrate interface due to the creation of photoelectrons in the substrate. A more complete discussion of the effects of target materials and operating conditions for optimizing resist absorption are given by Greeneich[51] and Sullivan and McCoy.[47]

3.2. Optics

In making exposures over large areas two geometrical factors, penumbral blurring and geometrical distortion, must be considered. They are illustrated in Figure 23.[31] The size of the x-ray source contributes a penumbral blurring of an edge on the mask. Since the mask to wafer spacing s is much smaller than the source to

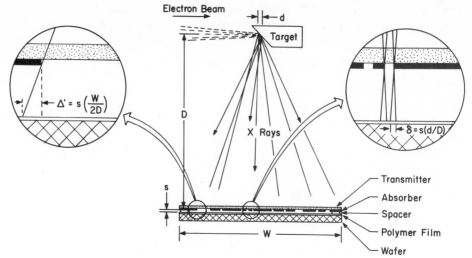

Figure 23. Schematic diagram of an x ray lithography system. (Right inset) Perumbral blurring; (Left inset) Geometric distortion (Bernacki *et al.*[31]).

mask distance D, the penumbral blurring of a source of diameter d, $\delta = (sd)/D$, can easily be made much smaller than the linewidth. The second effect, geometrical distortion, is much more critical in practice. If the mask to wafer spacing s changes, the lateral position of an image at the outer edge of a mask of diameter W will shift by a distance $\Delta' = \Delta's(W/2D)$. With conventional sources the size of the field being replicated may only be about one tenth the source to mask spacing. In this case the lateral translation would be as much as 10% of the vertical displacement. If for example a 4-in. wafer were to be replicated at a source to mask spacing of 50 cm with only a 0.1-μm lateral shift, the vertical displacement would have to be maintained to within 1.0 μm. Geometrical distortion decreases linearly as the source to mask spacing is increased, but the exposure time increases as the square of this distance.

Developing optical components such as mirrors to overcome these problems is complicated by the fact that the reflectivity at normal incidence of most materials at soft x-ray wavelengths is very low. The dielectric constant of most materials is in fact usually slightly less than unity. This property can be used to obtain total external reflection for near grazing angles of incidence. The electrons of the material act like an electron gas and the reflection process is much like reflection of radio waves from the ionosphere. Fortunately, the physical laws of reflectance known as Fresnel laws have been shown to apply both theoretically and experimentally in the soft x-ray range where the wavelength is comparable to the lattice spacing.[52] The critical angle below which total internal reflection occurs is $(2\delta)^{1/2}$. Here δ is given by

$$\delta = (ne^2\lambda^2)/2\pi mc^2 \tag{6}$$

where n is the total number of electrons per unit volume. In general the absorption constant and the effective number of electrons that participate must also be considered.[53] For Au at 0.67 nm the critical angle is about 2°.[54] Thus an Au mirror used at a total deflection angle of 4° could be used to eliminate radiation harder than 0.67 nm. The small critical angle means that it is probably not feasible to design a condensing system for a conventional electron bombardment source. However, mirrors are commonly used with synchrotron radiation from a storage ring both to modify the beam shape and to limit the spectral content (see Chapter 3). For example, for lithography purposes, it appears feasible to focus a horizontal emittance angle of 10 mrad into a 2-cm^2 area at a high-resolution exposure station located about 30 m from the source point in the electron orbit.[55] A dedicated storage ring could support 15 tangential beam lines each of which could be subdivided into five or more of these exposure stations.

3.3. Absorption

For x-ray photon energies less than 1 Mev, the loss of x-ray beam intensity as it traverses matter occurs only by absorption and by scattering.[48] The scattering may be either coherent (Rayleigh scattering including Bragg diffraction), or incoherent, with change of wavelength (Compton scattering). For Rayleigh scattering no energy is transferred to the atoms. For soft x rays with wavelengths longer than 0.1 nm the loss in energy due to Compton scattering is less than 1% of that due to absorption. The exponential decay of the x-ray beam energy with distance into the material is described by the absorption constant, usually measured in cm^{-1} or μm^{-1}.

Absorption constant data of interest for x-ray lithography are given in Figure 24[56] as a function of wavelength. More detailed absorption data is also available.[57] The corresponding energy in keV can be found from equation (2). The absorption constant of x rays of a given wavelength increases as a high power of the atomic number Z and is proportional to the density for material. For a given element the absorption constant increases roughly as the cube of the wavelength, with interruptions at wavelengths corresponding to the characteristic absorption edges. Fortunately it is possible to find highly transparent substrate materials such as SiC or Parylene and opaque mask materials such as Au. The resist material PMMA has an absorption constant similar to Mylar. It is desirable that the resist have an absorption constant in the range 0.05–1.0 μm^{-1} to provide sufficient absorption and allow fairly uniform exposure with depth in the resist. For these reasons x-ray lithography is limited to the wavelength range of 0.4 to 8.0 nm or less. In practice the regions of 0.4 to 2.0 nm and just above the K edge of carbon (4.48 nm) to 8.0 nm are of greatest interest.

3.4. Photoelectrons

When an x-ray photon is absorbed, a photoelectron is generated with kinetic energy E_p where $E_p = E_x - E_b$. Here E_x is the energy of the x-ray photon and E_b is the binding energy required to release an electron from an atom. The vacancy

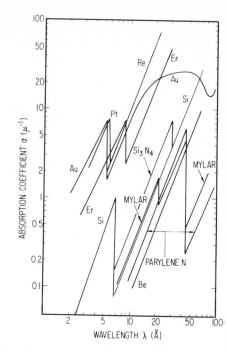

Figure 24. Absorption coefficient of some of the most absorbing and most transparent materials for soft x rays (Feder et al.[56]).

that is created is filled quickly, and the energy E_b is distributed to the surroundings via Auger electrons or fluorescent radiation. Usually the vacancy is filled by an electron from the next higher level and the energy released is a little bit smaller than E_b. An interesting consequence is that the fluorescent radiation occurs at a wavelength slightly longer than the wavelength corresponding to the absorption edge. The ratio of the probabilities for Auger emission to fluorescence for light elements is 9 : 1.[58] Hence for light elements 90% of the binding energy E_b is transferred to the surroundings by Auger electrons. For the main atoms of PMMA, the values of E_b are 0.28 and 0.53 keV, the binding energies of the electrons in the K shells of carbon and oxygen, respectively. In Si $E_b = 1.8$ keV when $E_x > 1.8$ keV (K shell) or $E_b = 0.15$ keV when $E_x < 1.8$ keV (L shell). Figure 25[59] gives the energies of the photo and Auger electrons when the resist and wafer are irradiated with Rh L (2.7 keV) and Al $K\alpha$ (1.49 keV) x rays. In most cases the photoelectrons have the most energy. Only in the case of Rh L radiation do the Auger electrons generated in the silicon have the higher energy.

The photo and Auger electrons lose their energy to the material by excitation and ionization of the neighboring molecules. The angular distribution of the photo and Auger electrons is important in calculating the distribution of energy transferred. The Auger electrons have a spherical or isotropic distribution from their starting point, as shown in Figure 25. The photoelectrons generated by x rays with energies below 10 keV are preferentially emitted perpendicular to the impinging x rays and in the plane of polarization of the x rays.[60] The preferentially

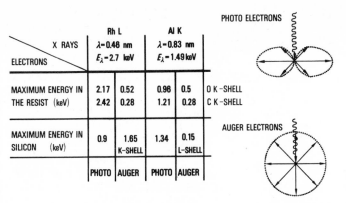

X RAYS / ELECTRONS	Rh L $\lambda=0.46$ nm $E_\lambda=2.7$ keV		Al K $\lambda=0.83$ nm $E_\lambda=1.49$ keV		
MAXIMUM ENERGY IN THE RESIST (keV)	2.17	0.52	0.96	0.5	O K-SHELL
	2.42	0.28	1.21	0.28	C K-SHELL
MAXIMUM ENERGY IN SILICON (keV)	0.9	1.65 K-SHELL	1.34	0.15 L-SHELL	
	PHOTO	AUGER	PHOTO	AUGER	

Figure 25. Maximum energies of photo and Auger electrons in resist and silicon during exposure with Rh L (0.46 nm) and Al K (0.83 nm), and the angular distribution of the photo and Auger electrons (Tisher et al.[59]).

emitted photoelectrons are spread by diffraction and mean-free-path scattering effects. The photoelectron pattern is also shown in Figure 25.

3.5. Energy Deposition

The energy density D in J/cm^3 that is delivered to the material can be found by integrating the energy losses per unit path length of the electrons along their trajectories. Path length loss results from electron beam scattering[61, 62] can be used to characterize this process. The energy loss of the electrons is given under the continuous slowing-down approximation by

$$\delta E/\delta r = E_p \Lambda(f)/r_g \tag{7}$$

where $\Lambda(f)$ is the depth dose function, r_g is the Grun range in μm, and $f = r/r_g$. The values of r_g and $\Lambda(f)$ can be calculated from polynomial approximations,[61]

$$\Lambda(f) = 0.74 + 4.7f - 8.9f^2 + 3.5f^3 \qquad 5 < Z < 12$$
$$\Lambda(f) = 0.6 + 6.21f - 12.4f^2 + 5.69f^3 \qquad 10 < Z < 15 \tag{8}$$
$$r_g = (0.046/\rho)(E_p \text{ in keV})^{1.75}$$

Here ρ is the density in g/cm^3 and is about 1.2 for PMMA. For Rh L and Al $K\alpha$ the Grun range for photoelectrons in PMMA is 0.18 and 0.05 μm, respectively.

Measurements of the maximum penetration depth of electrons from a heavy metal layer into resist are shown in Figure 26.[56] Both the magnitude and dependence on wavelength appear to be somewhat smaller than the predictions from the Grun formula. For Al $K\alpha$ radiation the electron scattering will have little effect on practical resolution at present. However, with radiation harder than the Si absorption edge (0.67 nm), spurious effects have been observed with both negative resist[63] and positive resist[59] at the resist substrate interface. The high-energy bremsstrahlung radiation contributes to these effects by penetrating into the masked areas and by producing electrons with larger scattering ranges. This contribution from harder x rays can be eliminated for a storage ring source by the use

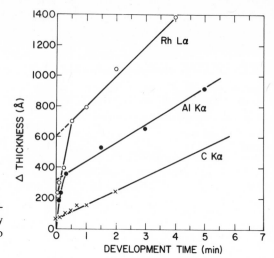

Figure 26. The depth of development of exposed resist as a function of time. The intercept with the vertical axis approximately represents the maximum penetration depth of photoelectrons into resist (Feder et al.[56]).

of a grazing-incidence mirror. In addition, x-ray resists designed for high sensitivity at wavelengths shorter than the Si absorption edge may be less sensitive to these effects.[64] For electron beam lithography the electron scattering effects are a particular problem. The range for 20-keV electrons is about 5 μm in Si and proximity corrections must be made starting at 1-μm linewidths.

3.6. Energy Deposition Effects

The energy deposited in material by x-ray-generated photo and Auger electrons produces several different effects in materials. In semiconductors, for example, hole–electron pairs can be generated. Generally, one hole–electron pair is generated for each $3eV_g$ of energy where V_g is the band gap voltage and e is the charge of an electron. For silicon the generation efficiency is roughly one pair per 3.6 eV.[58] The photo and Auger electrons are not of sufficient energy (< 145 keV[58]) to damage the crystal structure of silicon but they do generate electronic surface states. These states effect the electronic device performance, such as MOS transistor threshold voltage and leakage current. Consequently, it is necessary to anneal the semiconductor device. Typically heating the devices to 400°C for 30 min is sufficient to remove all of the readily measurable effects.

In polymers, the deposited energy affects the molecular bonds. Simultaneous cross-linking (adding) and scission (breaking) of bonds can occur. These processes are described by an efficiency factor G that is proportional to the radiation chemical yield. For PMMA at moderate dosages, scission dominates and its G value for scission is 1.3 events per 100 eV. Scission reduces the average molecular weight and may produce gaseous by-products. Both effects may increase the solubility rate in chemical developers. Cross-linking generally makes the removal of material more difficult.

Positive resists are generally glassy polymers that are made more soluble by scission effects. A simple model[65] is that the solubility S is given in terms of the average molecular weight M_n by

$$S = S_0 + BM_n^{-\alpha} \tag{9}$$

It has been shown that the decrease in the average molecular weight due to radiation induced scission is given by[66]

$$M'_n = M_n/(1 + PM_n) \tag{10}$$

Here M_n is the original number average molecular weight and M'_n is the new number average molecular weight after radiation. The constant P, the probability of scission, is given by

$$P = Dg/e\rho N_a \tag{11}$$

where D is the exposure dose in mJ/cm^2, e is the electronic charge, ρ is the density, and N_a is Avogadro's number. By combining these two formulas it is possible to describe the development of a polymer with an effective surface etch rate versus dose curve.

Using this foundation the parameters affecting the sensitivity of positive resists such as poly(methyl methacrylate), PMMA, have been explored. In studies of solubility rate ratios with electron beam exposure,[67] the weight average molecular weight, before and after exposure, was found to correlate better with the solubility rate ratio than the number average molecular weight. The weight average molecular weight is the average of the square of the molecular weight divided by the average molecular weight. The weight average molecular weight tends to emphasize the influence of the larger molecules, which are likely to play a dominant role in the actual development process. The molecular size of the developer solvent was also shown to have a much greater effect on the solubility than the molecular weight of the resist. The empirical use of a surface etch rate curve by Hatzakis *et al.*[68] and Greeneich[69] for electron beam exposure of PMMA has been quite successful. As a result the measurement of the etch rate versus dose is frequently used to characterize positive resists. Some convenient models are shown in Figure 27.[70] More rigorous models of the dynamics of polymer dissolution[71] include swelling and the formation of a gel layer. Other physical and chemical changes such as the formation of volatile matter (which could result in microporosity in the polymer matrix) may also affect the solubility rate.[72]

Negative resists tend to use material in a glassy or rubbery state that the radiation can gel by cross-linking. This generally requires less dose than breaking bonds in a degrading material to make it soluble. During development a negative resist may swell to three times its normal volume. This may cause high aspect adjacent lines to swell together and may result in bridging upon removal of the solvent by rinsing. The swelling effect makes negative resists much more difficult to characterize with quantitative models. In practice negative resists are described by plotting the thickness remaining versus log dose.[73] To quote sensitivities the linear portion of the curve near the 50%-thickness point is extrapolated as shown in Figure 28. The

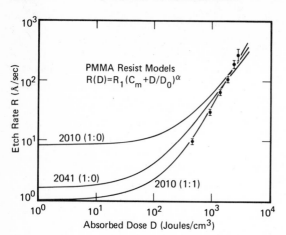

PMMA	Dev	R_1	C_M	D^0	a	Comment
2010	Conc	8.33	1.0	325	1.404	Data fit
2041	Conc	8.33	0.309	325	1.404	M_N scaled
2010	1 : 1	1.0	1.0	199	2.0	Data fit

PMMA ETCH RATE MODEL

Figure 27. Resist etch rate versus absorbed dose for PMMA Elvacite[R] 2010 and 2041 for development in a mixture of MIBK and IPA (Neureuther *et al.*[70]).

zero intercept D^i is the gel initiation dose and the 100%-thickness point is $D^{1.0}$. In making curcuits a dose of $D^{0.5}$ to $D^{0.9}$ might be used. The contrast γ is defined as

$$\gamma = [\log(D^{1.0}/D^i)]^{-1} \tag{12}$$

It is possible to use this same type of curve to characterize positive resists. However, this approach contains less information than the etch rate curve. Furthermore for positive resists it is less meaningful as it does not yield a profile that is open to the substrate and useable for wet etching.

The characterization of the lithography response of negative electron beam resist by a gel-point energy and contrast becomes feature size dependent below

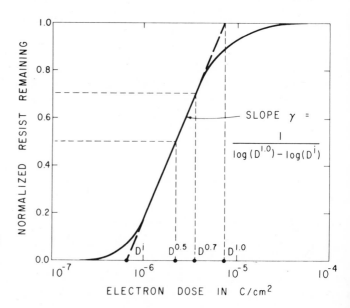

Figure 28. Graph of the normalized remaining thickness after development versus exposure dose used to characterize negative resists for either electron beam or x ray exposure (Thompson *et al.*[73]).

5-μm sizes.[74] As features are progressively reduced, the gel-point advances to higher doses and the contrast increases. These new parameter values correlate poorly with the molecular parameters of the polymer; they depend instead on developer-induced swelling of the polymer and on the gel rupture by forced development.

The parameters that affect the performance of negative resists particularly suited for x-ray lithography have been explored for example by Taylor *et al.*[75] Atoms with high absorption coefficients have been incorporated to enhance the x-ray absorption and hence increase the material sensitivity. The sensitivity also depends on the molecular structure, which can be affected by the inclusion of reactive groups and by molecular weight. However, increasing the molecular weight and content of reactive groups such as glycidal groups adversely affects resist contrast and hence resolution. This is apparently due to the formation of intramolecular cross-links, which are less effective than intermolecular cross-links in providing the solubility differences required to give a developed image of high resolution. The resist molecular structure also affects the resistance to plasma etching with fluorine based gases and gas mixtures. However, the ion milling rate, which is slightly higher than that of glassy positive resists, is almost independent of the molecular structure of the negative resists, which are tough rubbery materials at room temperature.

Figure 29 shows the relationship between the electron beam and x-ray sensitivity for a number of positive and negative resists.[76] The two sensitivities are roughly proportional by a factor of 50 mJ/cm^2 per μC/cm^2 at the Al $K\alpha$ wavelength. Theoretically, a factor of about 25 mJ/cm^2 would be expected from Monte Carlo calculations of electron beam exposures in 0.8 μm of PMMA on Si.[77]

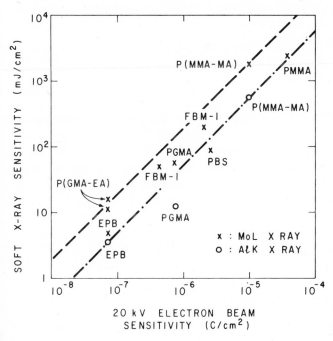

Figure 29. Relationship between electron beam and x-ray sensitivity for various resists (Murase *et al.*[76]).

Doping with absorbing atoms as is done in the case of DCPA[75] should lower this factor of proportionality. This ratio is also wavelength and beam voltage dependent.

4. X-Ray Lithography Technology

The practical use of x-ray lithography, particularly for producing integrated circuit devices, requires technological developments in a number of areas. These include the x-ray source, vacuum window, masks, resists, alignment techniques, compatible post-lithography processing, and removing device damage that occurs during processing. Work in these areas is currently being carried out by many researchers in a number of organizations and the technology will likely change rapidly in the next few years. Several viable approaches have been developed in each area. In general, there is no consensus that one particular approach is best. This is partly because the choice depends to some extent on the lithography requirements and economics of the device family being produced. The discussion here will attempt to summarize developments to date and put these rapidly changing technologies in a general perspective.

4.1. X-Ray Sources

4.1.1. Novel X-Ray Sources

Most x-ray lithography work has been done with either a conventional electron bombardment source or with synchrotron radiation from a storage ring.[78] Two alternative x-ray sources have been reported recently. Both approaches utilize x-ray emission from an extremely high-energy plasma. They differ in the manner in which the energy is pumped into the plasma. The simplest approach uses a novel device based on the interaction of a tightly focused (50 μm diameter) energetic electron beam (5 kA, 90 keV, 120 ns) with a dense plasma produced by a sliding spark in a capillary.[79] Work with sliding sparks in capillaries[80] established the high radiative brightness in the soft x-ray range of such devices. Spectroscopic results for a carbon plasma from 1.0 to 100 nm show numerous characteristic lines.[81] The characteristic lines are shifted in some cases into the more useful high-absorption regions due to the change in binding energies for highly ionized states that occur in the plasma. Apparently little hard x-ray radiation is produced. Although no absolute intensity measurements were undertaken, estimates indicate single-shot line emission of about 0.02 J/sr-line for the more intense lines. This appears to be an efficiency of a few percent. Single-shot direct exposures of about 60 ns duration in PMMA have been made with a resolution of 30 nm. Contamination of the sample by debris ejected from the high-temperature plasma and damage to the mask due to heating are significant problems. The potential of the "sliding spark" as an x-ray lithography source merits further investigation.

The other new source approach uses pulsed laser heating to generate a plasma that emits characteristic x-ray radiation.[82] The advantages of this technique are that theoretically about 10% of the laser energy can be converted to x-rays in the 1–3-keV range without generation of hard x rays. The source can be as small as 1 mm and is extremely bright. PBS was exposed recently[83] by 60 shots of a 40-ns pulse from a 100-J Nd laser at a target to substrate spacing of 5 cm. Experimentally about 0.3% of the laser energy was converted to x rays and 90% of this was in the $K\alpha$ characteristic line of the Al target. The over all efficiency may be significantly lower depending on the electrical efficiency of the laser. With development, single-pulse exposures of sensitive resists might be possible. However, several fundamental problems exist in obtaining the theoretical predictions. To obtain high efficiency and high x-ray intensity the laser must be switched very rapidly. If the target is closer than a few meters, the reflectivity of the target introduces premature interactions with the laser. This affects the laser pulse shape and rise time and makes it difficult to obtain short (10 ns) and uniform pulses. The thermal heating of the glass in the lenses requires a cooling period, which is a fundamental problem. On present systems the cycle time is about 15 min. The high temperature plasma also ejects vapor and debris. High vapor pressure, thin targets may reduce these effects somewhat but low defect densities on devices is of extreme importance. Another major problem is the damage to the mask due to heating during exposure. Finally, the cost effectiveness of this approach has yet to be established.

Another possible x-ray source which is even more speculative is an x-ray laser.[84] At x-ray wavelengths it is difficult to obtain the population inversion needed to sustain stimulated emission. The main problem is the very short time constant for spontaneous decay, which is at most 10 ps. It has not been possible to pump fast enough to obtain a population inversion. Although strong indications of stimulated emission and intense anomalies have been reported in the 35–85-nm range, the evidence for x-ray laser action is still inconclusive.

4.1.2. Electron Bombardment Sources

The dominant factor in the design of electron bombardment sources is the thermal limitation of the target. The temperature rise can result in melting of the target material or can induce stress through thermal expansion that may lead to mechanical failure. For fixed targets melting is the limitation. The power capacity can be determined from a steady-state calculation of heat flow. The capacity depends on the melting point, thermal conductivity, and thickness of the target material.[47] Cu has a high thermal conductivity and is often used as a substrate for a thin layer of target material. The choice of a thin coating material depends on additional factors such as x-ray generation efficiency, wavelength-dependent resist sensitivity, and even effects from photoelectrons generated at the wafer. For simple targets the power is limited to a few hundred watts for a spot diameter of 1 mm.[85] A stationary cone target 8 mm in diameter can handle as much as 4 kW.[86]

Rotating anodes can handle much higher power. The limiting factor in this case is the stress induced by transient thermal heating in passing under the electron beam. This depends on the specific heat capacity, thermal conductivity, thermal expansion factor, Young's modulus, and tensile strength of the material. The stress limitation generally degrades the total power from that which would be expected from thermal analysis alone. An approximate relationship for rotating anode source power capacity can be extracted from the work of Wardly et al.,[87]

$$P_{sl} = K(D\omega)^{1/2}d^{3/2} \tag{13}$$

Here P_{sl} is the stress limited power in W, D is the diameter of the anode, ω is the angular velocity in rpm, and d is the diameter of the electron beam in mm. The constant K is about 5, 10, and 15 for Al, Cu, and W, respectively. It appears feasible to produce an 8000-rpm 25-cm-diameter rotating anode with 6-mm-diameter spot capable of powers up to 25 kW with a deposited target layer.[17, 88] A Si target on such a powerful source operated at 20 keV at a distance of 40 cm would produce a flux of 0.60 mW/cm² at a wavelength of 0.713 nm with a 25-μm Be window. Note that only a few tenths of one percent of the energy supplied to the source is converted to x rays.

4.1.3. Synchrotron Radiation Sources

The synchrotron radiation flux available from a storage ring can easily exceed 1 W/cm². Even for a small storage ring, the effective exposure flux is about three orders of magnitude larger than that of conventional sources. The exact amount depends on the storage ring parameters, the geometry of the lithography port, and the range of the continuous radiation spectrum that can be utilized in the lithography process. The heating of the mask and subsequent distortion is an important consideration that will likely limit the maximum flux that can be utilized. The potential of storage rings as sources for x-ray lithography can be assessed by making several design assumptions about the critical wavelength of the storage ring, the fraction of the total flux that falls on the wafer, the absorption of the window and mask, and the properties of the resist materials. Typical assumptions will be made in this section to assess the suitability of a small dedicated storage ring as a source for x-ray lithography.

The first consideration is the spectral distribution. As shown in Chapter 2, the radiation spectrum is described by the spectral function, which depends only on the critical wavelength λ_c. At about $\lambda_c/3$, the amplitude of this function is about 10% of its maximum. Due to the higher photon energy the energy at this wavelength is 30% of that at λ_c. This wavelength can be roughly considered to be the short-wavelength cutoff. The wavelength range of interest in x-ray lithography may extend down below the K absorption edge of Cl (0.437 nm) when Cl-doped resists are used or may end just above the Si absorption edge (0.67 nm) to avoid problems with photoelectrons from Si substrates. The Cl doping apparently reduces the sensitivity to the effects of photoelectrons.[64] X rays harder than 0.3 nm are generally undesirable as they are more difficult to absorb in the mask and resist. It is possible to tailor the spectrum of flux available from a port on a storage ring by using λ_c to shift the spectral peak and by using a grazing-incidence mirror to abruptly cut off the hard x rays.

The rapid change in absorption with wavelength for materials shown in Figure 24 affects the specific choice of wavelengths for x-ray lithography. Due to the occurrence of absorption edges of Si, O, and C (0.67, 2.34, and 4.38 nm, respectively), it is convenient for analysis purposes to consider the flux in four spectral regions; 0.3–0.67, 0.67–2.34, 2.34–4.38, and 4.38–8.0 nm. The flux in the 0.3–0.67-nm range is usable particularly if resists are doped with absorbers such as Cl and S. However, thick Au masks (0.7 μm) are required, and large numbers of photoelectrons will be emitted from the substrate and will have ranges on the order of 0.2 μm. The number of photoelectrons from the substrate is greatly reduced in the 0.67–2.34-nm range, which is above the absorption edge of Si. In this range thin windows such as 12.5 μm of Be and cooling gases such as He and N are relatively transparent. In the 2.34–4.38-nm range PMMA and other organic resist materials are highly absorbing and it is difficult to expose resist material to a depth of 1.0 μm. The flux in the 4.38–8.0-nm range can penetrate 1.0 μm into the resist but it would only be usable with either a differentially pumped vacuum system or a thin organic window. For wavelengths longer than 8.0 nm the absorption is again too high to penetrate thick resist.

To explore the available flux in a more quantitative manner, the effective power delivered to PMMA resist has been calculated for each of these spectral regions. A small 2-m-bending radius storage ring with a 12 kOe magnetic field operated at 0.72 GeV and 100 mA has been assumed. This corresponds to a critical wavelength λ_c of 3.0 nm. To account for the relative effectiveness of the radiation in exposing resist, the flux at each wavelength has been multiplied by the absorption constant for PMMA. This product is then integrated over the spectral regions of interest. The results are shown in Table 2. The units are W/(100 mA mrad cm of resist). At a source to sample distance of 10 m, 1 cm corresponds to 1 mrad. Since the vertical angular divergence of the radiation is also about 1 mrad, the values in Table 2 can also be interpreted as power deposited in the resist in W/cm^3 at a distance of 10 m. Alternatively and more precisely, the values shown in Table 2 are flux per cm^2 for 1 mrad of horizontal angle. The exposure time for PMMA is calculated on the basis of 1000 J/cm^3. The data can be used to roughly estimate the exposure times for other resists, by comparing the absorption constant of the resist to that of PMMA and adjusting for the required dose. If instead the required resist

Table 2. Power Density Absorbed in PMMA Resist from the Various Spectral Bands[a]

	Wavelength Range (nm)					Exposure Time[b]
Condition	0.4–0.7	0.7–2.33	2.33–4.48	4.48–8.0	Total	(s)
No filter	1.33	380	833	208	1422	0.7
1 μm PMMA	1.26	126	74.7	87.6	290	3.5
1 μm PMMA + 2.5 μm mylar	1.24	90	44.0	65.7	201	5.0
1 μm PMMA + 12.7 μm Be	1.24	90	14.1	0.02	105	9.5

[a] Values are in W/cm^3 or W/(100 mA mrad cm of resist), where $\lambda_c = 3.0$ nm; $I = 100$ mA; and $D = 10$ m. One mrad of horizontal divergence is assumed to uniformly illuminate 1 cm. The flux at each wavelength is weighted by the absorption constant of PMMA at that wavelength and then integrated over the spectral region.

[b] Exposure time = (1000 J/cm^3)/total power density.

dose is given in J/cm^2, the synchrotron flux can be estimated by dividing the power listed in the table by the absorption constant for PMMA. While the table is fairly accurate for PMMA, estimates for other resists will only be approximate if abrupt changes in the absorption constant occur within a spectral region.

The first row of Table 2 gives the power per cm^3 absorbed just inside the surface of a resist layer. The power in the 2.34–4.38-nm range dominates due to the high flux level and very high absorption constant. Also note that very little power is contributed in the 0.4–0.7-nm range. This is primarily because for $\lambda_c = 3.0$ nm little flux is available in the short-wavelength range. The equivalent effective flux at 0.834 nm can be found by dividing the power absorbed by the absorption constant for PMMA at this wavelength. The equivalent effective flux is about 10,000 times larger than the 0.1 mW/cm^2 available from most conventional sources. The corresponding exposure time is less than 1 s. The second row of the table for a $1.0\text{-}\mu\text{m}$ thickness of resist shows how the effective power changes with depth in the resist. The power in the 2.34–4.38-nm range, which is highly absorbed, is attenuated by more than a factor of ten. The total exposure time increases to about 3 s.

If a $2.5\text{-}\mu\text{m}$ Mylar window is added, the total exposure time is increased to about 6 s. Note that in this case both the 0.67–2.34-nm and 4.38–8.0-nm ranges contribute almost equally. If a $12.7\text{-}\mu\text{m}$ Be window is used instead, only the 0.67–2.34-nm range contributes and the exposure time is about 10 s. Still the equivalent effective flux is about 1000 times that of most conventional sources. While the flux in the 4.38–8.0-nm range may be desirable for special lithography or microscopy purposes, its absorption in a membrane other than the mask is advantageous for most IC lithography applications in order to reduce mask heating. Its use would require either very thin or organic windows and masks. It would also contribute diffraction effects at moderate mask to wafer spacing. Using the 0.7–2.3-nm region for resist exposures makes it possible to flow a cooling gas such as N_2 between the mask and wafer. The use of shorter wavelengths also reduces the vulnerability to organic defects.

The synchrotron radiation pattern is fan shaped. Incorporating this beam in a full wafer exposure system is one of the key functions of the lithography port design. The simplest approach is to scan the mask and wafer through the beam at a uniform rate. The required scan speed in this case can be calculated from the PMMA exposure times in Table 2, which are in effect the exposure times normalized to 1 mrad of horizontal angle flux per cm^2. A scanning lithography station could be located a few meters from the source point. A second approach is to use grazing-incidence mirrors to focus the flux on a given area at a high-resolution lithography port about 30 m from the source point.[55] The approach used and horizontal angle employed will depend on the sensitivity of the resist, the size of the area to be replicated simultaneously, and the maximum flux level that can be tolerated during exposure.

The small dedicated storage ring source (0.72 keV, 12 kOe, 2 m, 100 mA) would likely have adequate flux for most purposes. If however even higher flux levels are required they could be obtained by changing the storage ring design parameters.[55] Currents as high as 500 mA are possible in a 0.72-GeV storage ring. However, to reach such high currents may require a full energy injector. The cost of such a high-energy injector may exceed the cost of the storage ring itself. Much higher flux levels in the 0.67–2.34-nm range are possible by decreasing λ_c. For example, increasing the energy while keeping the magnetic field constant (thus increasing the radius) such that $\lambda_c = 1.33$ nm (1.08 GeV, 12 kOe, 3 m, 100 mA) will increase the flux at 1.0 nm by a factor 6. Rather than changing the energy to decrease λ_c it is possible to increase the magnetic field strength. More powerful and even superconducting magnets are a possibility, but the latter are not likely to be used except as local wigglers (see Chapter 21).

4.2. Windows and Masks

To expose a pattern onto a wafer, an absorber pattern must be placed in the beam. At micron resolution it is not feasible to make self-supporting patterns and consequently a thin substrate is used. It is also generally desirable to isolate the high-vacuum source region from the frequently accessed wafer region with a thin x-ray window. The placing of substrates or windows or even He in the x-ray path filters the radiation spectrum. A considerable effort has been made to develop thin windows and substrates that are both relatively transparent and mechanically stable. The mechanical stability is particularly important for masks that must maintain submicron tolerances throughout the exposure process. For high-resolution work, the tolerance of the edge of thick absorber patterns is also important.

An example of the sequence of steps used in the fabrication of thin membrane masks is shown in Figure 30.[89] The approach is related to that used in the first

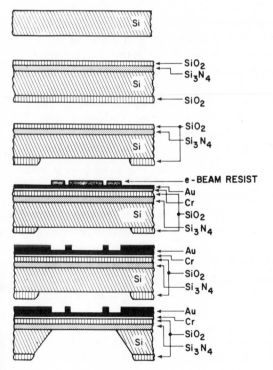

Figure 30. Sequence of steps used in the fabrication of silicon nitride membrane masks (Bassous *et al.*[89]).

masks that were fabricated on thin Si windows.[2] The windows were produced by doping the front side of a wafer with B and then preferentially etching the undoped Si from the back in the areas of the windows. In making circuits the windows are placed in positions corresponding to chip locations on the device wafer.

Similar masks are used today but they are somewhat thinner to reduce the attenuation due to absorption in the mask. In one approach the mask substrate consists of SiO_2 sandwiched between two layers of Si_3N_4 and has about a 1.4-μm total thickness.[90] The multilayer films permit balancing of compressive and tensile stress for maximum strength. Al_2O_3 films have also been used.[47, 91]

Stretched organic films such as mylar and kapton have been used. However spin-on organic films such as polyimide are preferred as they can be made thinner and more uniform. They also avoid the problem of contamination in manufacturing. Flanders and Smith have described the spin-on fabrication process for polyimide films in detail.[92]

Recently thin (2.5 μm) SiC membranes have been developed for mask substrates.[10] Unlike polymer films, SiC is unaffected by humidity changes. It is much stiffer with a Young's modulus two orders of magnitude larger than those of stiff polymers. Thus it is much less susceptible to pattern-induced distortion. SiC is an excellent etch stop so that the Si substrate on which it is deposited can be etched away easily. The SiC–Si interface is free from the striations often seen in Si membranes and the optical quality is generally better. SiC membranes are rather rugged and are easily made with 5-cm diameters without individual chip window structures. The measured burst strength is in the range of $(5-10) \times 10^9$ dyn/cm^2.

The absorber material used in the mask pattern has almost universally been Au. An important factor in mask quality is the mask contrast ratio, which is the inverse of the mask transmission. A contrast ratio of at least 4 and usually 10 is used. The thickness of Au required depends, through the absorption constant, on the wavelength. For a contrast ratio of 10, a minimum of 0.1 μm of Au is needed at wavelengths >2.0 nm. At 1.3 nm, 0.2 μm is needed and at 0.8 and 0.5 nm, as much as 0.7 μm is needed. The original from which the Au is patterned is usually written in a resist coating the Au by means of a high-resolution electron beam system. The Au is then patterned either by an etching (subtractive) process or by an electroplating (additive) process.

Special processing techniques are used to obtain steep edges of the Au absorber. In one approach particularly useful for dark field masks[6, 93, 94] the Au is coated with a thin layer of a metal such as Ti or Cr that can be oxidized. The resist is used to etch only the thin layer. The thin layer is then used as a mask to etch the Au. By allowing O_2 into the sputter etching or ion etching system, the thin layer forms an oxide that etches very slowly.[95] This can result in very steep line-edge profiles in the Au. However, because of lateral erosion effects, the faceting of the mask layer must be considered in choosing its thickness.[96]

Another approach useful for bright field masks is to start with thin Au on Ti layers and electroplate.[93, 97] This accurately reproduces the resist line-edge profiles and allows plating at least as thick as the resist layer. At high resolution it may be advantageous to first make a thin mask and then make a second copy using x rays with a wavelength of 1.3 nm or longer. An example of improving the mask contrast

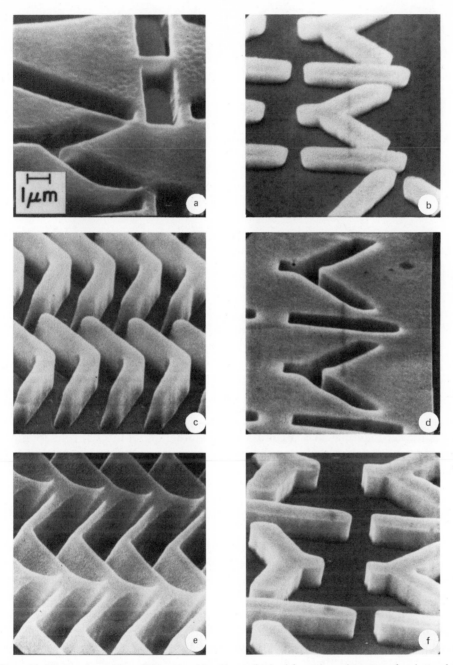

Figure 31. Photoresist patterns (left) and gold patterns obtained from the resist pattern by electroplating (right) for the first three levels of x-ray mask fabrication. Electron beam fabricated mask (top), negative x-ray copy (middle), and positive x-ray copy of the negative (bottom) are shown. All linewidths are 1.0 μm (Spiller[98]).

Figure 32. Photograph of a 61 mm diameter polymide x-ray mask. The mask is patterned with 6 mm square 64k bit bubble circuits (Stein[14]).

ratio by successive copying with x rays is shown in Figure 31.[98] Note that the Au in *f* is thicker than that in *b*. Figure 32 is a photograph of a mask consisting of a 5 × 5 array of 6-mm × 6-mm 64k bubble circuits on polyimide produced by electroplating.[14]

4.3. Resists

The trend of electronic and optical microdevices toward increased packing densities and reduced dimensions has placed new demands on resist materials. Resists developed for photolithography such as AZ1350J can be used directly as they are capable of resolving features smaller than 0.1 μm. Their sensitivity to x-ray or electron beam exposure at 20 kV (4×10^{-5} C/cm^2) is about two times better than that of PMMA (8×10^{-5} C/cm^2).[99] Many of these resists have also under-gone development for process compatibility. This includes uniformity of thickness, step coverage, adhesion to substrate, resistance to etchants and heating, defect density, shelf life, and so on. However, for use in x-ray and electron beam lithography, improvements both in sensitivity and process compatibility, are needed. The writing time for nonoptical lithography is high and an order of magnitude improvement in sensitivity is needed to reduce the cost per wafer. High resolution

Table 3. Characteristics of Typical Resist Materials for X-Ray Exposure with Al K Radiation[a]

Resist	Type	Sensitivity (mJ/cm^2)	Resolution (μm)
PMMA	Positive	1,000	0.1
PBS	Positive (electron beam)	100	0.5
FPM	Positive (electron beam)	100	0.5
PGMA–EA	Negative (electron beam)	50	0.5
DCPA	Negative	10[b]	0.5
TI XR79	Negative	1.5	0.5
Hunt Experimental	Negative	8	0.5

[a] From Hughes *et al.*[100]
[b] When exposed to palladium radiation.

(2.0 to 0.5 μm) also necessitates the replacement of certain post lithography process steps, such as wet etching, with new techniques like plasma etching or reactive ion etching. These new techniques place entirely new requirements on the properties of resist materials.

The development of a high-resolution resist technology is a concern of every major microdevice manufacturer. In general the goals for resist sensitivity are 10^{-6} to 10^{-7} C/cm^2 for electron beam resists and 5 to 20 mJ/cm^2 for x-ray resists. Table 3 summarizes some of the more promising results to date.[100] The sensitivities quoted are at the Al $K\alpha$ wavelength (0.834 nm) unless otherwise specified. Note that the most sensitive x-ray resists such as DCPA have been made more sensitive than the electron beam resists. This has probably been done through the addition of highly absorbing elements. A more detailed listing of resist data through 1977 has been compiled by Spiller and Feder.[101]

While resist sensitivity and contrast can be measured directly and used as an index to compare and optimize resists, the context of the application often constrains the choice of resists. For mask making with planar substrates where the resist need only serve to etch very thin layers, thin resists (<0.5 μm) with very low sidewall angles can be used. Here the most sensitive resists can generally be used. For direct wafer writing, however, topological features of various materials are encountered on the wafer. A typical example of the wafer topology for an NMOS process is shown schematically in Figure 33.[102] To cover step height changes a thick resist coating (about 1.0 μm) is needed. To open small features (1.0 μm) in such a resist layer, resists and exposure levels capable of producing very steep sidewall angles are required. The change in the underlying material with position introduces position-dependent electron scattering and photoelectron exposure effects that may also affect the choice of resists.

Perhaps the most important consideration is the resist strategy to be used in depositing or etching metals. The lift-off approach[103] requires a resist directly

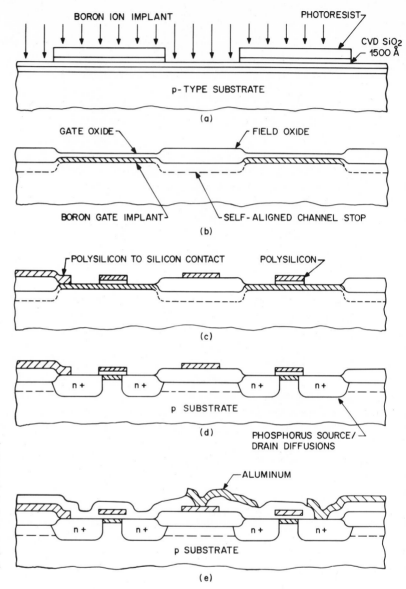

Figure 33. Process sequence for self-aligned polysilicon gate NMOS circuits: (a) After ion implantation of oxide layer in all regions that will not be transistors; (b) after growth of the thick field oxide and the gate oxide; (c) after deposition and etching of the polysilicon layer; (d) after source drain diffusion; (e) after metalization (Glaser et al.[102]).

capable of or modifiable[104] for producing an undercut profile. If the resist is sufficiently robust ion milling can be used.[94] The alternative of first etching a thin Ti or Cr layer that in turn can be used as a mask to etch a second thicker metal layer is more difficult to use on device wafers. Generally, the resists must be thicker to cover device features, which makes swelling more of a problem and increases the required dose. The development of sensitive resists capable of withstanding plasma etching would be very useful.

4.4. Alignment and Distortion

Advantageous use of high-resolution lithography, particularly in the step and repeat mode, requires rapid accurate alignment. Optical alignment is the most common technique used today. The trend toward micron linewidths has put an increasing demand on optical instrument performance, and many factors must be considered in accurate linewidth measurement.[105] Operators typically make 25 to 100 alignments per hour at an accuracy of about 0.5 μm. Careful optical alignment can be done to 0.25-μm accuracy. Automatic alignment can equal this performance in both speed and accuracy, and is the current trend. If the wafer is not moved after alignment, layer to layer accuracy of better than 0.25 μm can be obtained.

Recently a vibrational alignment scheme with an accuracy of 0.1 μm for an x-ray exposure system was reported.[29] Complementary mask and wafer patterns are sinusoidally vibrated with respect to each other. When the patterns are aligned, the signal received by a photodiode contains no fundamental frequency component. The accuracy for each vibrational direction is reported to be 0.05 μm. When combined motions are used for alignment in x, y, and θ the accuracy is 0.1 μm. It is also reported that mask to wafer spacing can be measured to 1.0 μm by monitoring the capacitance between the patterns on the mask and wafer. This approach can be used to insure parallelism between the mask and wafer as well.

Further improvement in optical alignment is likely. Moire pattern techniques have been used to obtain accuracies of 0.2 μm[106] and accuracies of an order of magnitude smaller have been predicted.[107] The most interesting approaches use phase information. For translational tables accuracies of 1/64 of a wavelength (0.01 μm) have been obtained.[108] A new interferometric alignment technique has been suggested for mask to wafer alignment.[30] Two symmetrical gratings are used as shown in Figure 34. When the gratings are aligned, the net reflected signal in the diffracted first orders is symmetrical. Any misalignment of less than half a period results in an imbalance in the first-order beams. Experimental results[109] of 0.02 to 0.05 μm for 1.2-μm-period gratings have been obtained. The fundamental limitations appear to be asymmetry introduced in the individual gratings during processing and sensitivity to mask–wafer spacing.

Finally, direct use of x rays for alignment may be feasible. Three techniques have been reported.[110, 111] One approach is to use hard x rays to transmit through the wafer. This avoids the need for a thin window on the wafer, which is required in a second technique that uses soft x-rays passing through the wafer. A third approach uses $K\alpha$ fluorescence of P and Si mask and wafer alignment marks as shown in Figure 35. All of these approaches are generally limited by the signal-to-noise ratio. Thus the use of synchrotron radiation should afford major improvements in the use of direct x-ray alignment techniques.

A critical problem for replication is the distortion of either the wafer or mask, which would cause their patterns to fail to register throughout the field of view even though they are matched at the alignment marks. The most fundamental concern is the possibility of distortion of the wafer itself during processing. Initial estimates of wafer distortion were as high as a few μm shift across a 3-in. wafer. Careful measurements for processes with temperatures below 900°C show that the distortion is

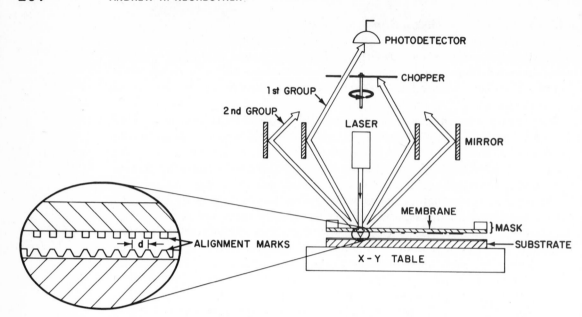

Figure 34. Schematic illustration of an interferometric alignment technique. Each group consists of two or more constituent beams diffracted by one or both of facing alignment mark gratings. The substrate is aligned by moving it laterally until symmetrically diffracted groups are balanced in intensity (Flanders *et al.*[30]).

probably less than 1.0 μm across a 3-in. wafer and that some of this may be due to the chucking of the wafer during the measurement.[112, 113] The nature of the distortion has been explored using an electron beam system for measurements.[28] The growth of an oxide was found to expand the wafer while metalization tended to shrink it. Upon stripping to bare Si the wafer returned to approximately its original size. The largest distortion from processes that included steps at temperatures

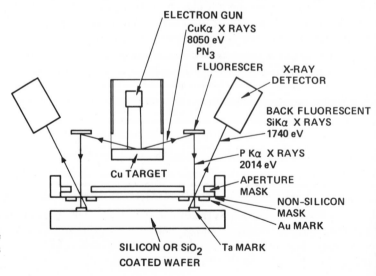

Figure 35. X ray alignment technique using back fluorescent Si K x rays (Sullivan[110]).

exceeding 1000°C was only about 0.75 μm across a 3-in. wafer. Most of the distortion was predictable from measurements of wafer bowing. The residual nonlinear distortion was less than 0.2 μm. This indicates that with mask compensation it may be possible to systematically design a 1.0-μm process using full wafer exposure. For significantly smaller linewidths a step and repeat process will be necessary to expose an entire wafer.

Distortion due to thermal expansion of the mask or wafer during exposure must also be controlled. In addition to gradual shifts over large distances, local distortion due to localized heating of the absorber pattern might also occur. This thermal distortion during exposure will likely limit the x-ray flux density that can be utilized. One approach to control thermal distortion is to use a silicon mask support ring to match thermal expansion coefficients and introduce He or N_2 gas to equalize local mask and wafer temperatures. The x-ray damage to the masks may also lead to distortion but probably only after thousands of exposures.[18] Mask distortion on the order of 0.1 μm can be detected using interferometric techniques.[92]

4.5. Radiation Damage to Devices

Radiation damage during processing is an important issue for all the beam technologies (including plasma etching, ion etching, ion implantation, electron beam lithography, electron beam evaporation, as well as x-ray lithography). The low-energy x-rays produced in the process excite charges in the semiconductor device structure and degrade device performance unless the charges are neutralized by thermal annealing. The typical dose is about 5 Mrad[114] for electron beam metalization and from 10 to 100 Mrad for x-ray lithography.[115] One Mrad is 10^8 ergs per gram of irradiated material or an exposure dose of about 13 mJ/cm² at the Al wavelength. The device degradation is roughly proportional to the dose. For a dose of about 100 Mrad, corresponding to seven exposures of COP resist and one of PMMA resist, the threshold voltage for PMOS and NMOS devices are shifted by 2.5 and 0.25 V, respectively.[32] The leakage current also increases an order of magnitude. With CMOS devices on silicon-on-sapphire (SOS) a positive interfacial charge density of about 10^{11} cm^{-2} is also created at the silicon–sapphire interface.

Fortunately, the damage appears to be annealable at 400°C in about 30 min.[31, 32] In a comparison of C-V and I-V characteristics of annealed and controlled samples, all of the readily measurable x-ray damage was removed. The data to date do not rule out the possibility of some incipient degradation of the MTTF (mean time to failure). A more extensive set of reliability measurements such as thermal stressing and accelerated aging are needed.

5. System Approaches

In developing a new lithography approach such as x-ray lithography the key issues are demonstrating the manufacturing capability, establishing an economic advantage, and overcoming the inertia of existing and alternative approaches. In this section the economics of x-ray lithography with conventional and synchrotron sources will be compared to that of optical and electron beam lithography. The status of the manufacturing capability with x-ray lithography, as reviewed in the previous section, is on the verge of becoming a production reality. However, at this writing there is no example of the manufacturing use of x-ray lithography to which the author can refer. Electron beam lithography, on the other hand, is currently being used in production for certain contact hole mask levels. In fact this particular use of electron beam lithography may have even preceded its use in making optical masks. It was adopted in wafer production in the mid-1970s even at additional expense because it solved the technical problem of quick lithography turn around for customizing computer components.[116] In the long run, for general purpose applications, the favorable economics of x-ray lithography will likely become a dominant factor.

Although storage ring sources are economically feasible their use in production will not occur quickly. X-ray lithography with conventional sources will likely be developed first and used as the primary exposure tool for process development. Once x-ray lithography with conventional sources is in production, the interest in storage ring sources may rise sharply. Meanwhile synchrotron radiation from existing storage rings will continue to be used in research and development to explore fundamental aspects of x-ray lithography. This more conservative two-step development of the use of synchrotron radiation for x-ray lithography is predicated by both manpower and funding limitations.

The goal of this section is to estimate the relative cost per wafer exposure for optical, electron beam, and x-ray lithography systems and to characterize the probable changes in these costs as the resolution is reduced. Published data on existing lithography systems will be used to estimate costs at about 1-μm resolution. The future trends will be estimated on the basis of tradeoffs in fundamental physical phenomena that limit performance. The issues and assumptions that enter into a comparison of this type are worthy of lengthy discussion and debate. The presentation here, however, is of necessity limited to simply illustrate the main point, which is the economic feasibility of x-ray lithography with a storage ring. The choice of the storage ring parameters that have been used have by no means been optimized and rather represent a machine that could be designed immediately with little development effort. The conclusions arrived at here are similar to those found by Broers[117] except that intermediate and large area x-ray step and repeat operation for storage ring sources are considered and found to be cost effective compared to the other lithography approaches.

The assessment is based on the following simplifying assumptions. The wafers are assumed to have a 3-in. diameter and the cost per wafer exposure is calculated as the exposure time for a 3-in. wafer times the cost per production minute for the lithography system. The cost per production minute is assumed to be $0.50 times the estimated capital cost of the machine in units of $100,000. Except for cases where the exposure time is less than 1 min, the throughput estimate is based only on exposure time. When exposure times are on the order of seconds, the time required for wafer interchange and alignment must be added.

The simple machine cost estimate formula is based on the following considerations: amitorization of the lithography tool and associated space over a period of four years at 10 hours per day would contribute about $0.25; a supervised operator, maintenance, and utilities would likely add up to another $0.25 per production minute and be independent of the utilization; the cost of energy to operate the tools is generally negligible even for storage rings when multiple ports are used.

Two fabrication strategies are considered. One strategy assumes it is possible to use a thin layer of highly sensitive resist material. The resist in this case is of necessity thin so that high-resolution patterns can be obtained in the presence of the low sidewall angles or swelling effects characteristic of these sensitive resists. This approach is suitable for patterning a thin underlying layer. Such a process can be used in making masks or in patterning a thin layer that in turn can be used as a mask in a further etching process.[118] The development of sensitive resists and compatible etching processes to make this strategy work is an important area of current research. The minimum dose that can be used in this strategy has not been well established. For comparison purposes, an effective energy density of 50 J/cm^3 deposited in the resist will be used as representative of the sensitive resist strategy.

The second strategy for wafer production is the patterning of a thick robust resist. The advantage of this approach is that the resulting resist profiles can be used directly in further processing of material layers of similar thickness to the resist itself. This is often used in reactive ion etching, metal lift-off, and electroplating. The robust resist strategy is basically similar to the current processes used with optical projection and contact printing. In the robust resist processes the shape of the resist is very important. X-ray lithography will likely make a major contribution to this approach as it produces high aspect ratio resist profiles with good linewidth control. An effective energy density of 1000 J/cm^3 deposited in the resist is assumed to be representative of this robust resist strategy.

The effective energy deposition for the sensitive and robust strategies can be translated into electron dose or x-ray flux requirements. For electron beam exposure at 20 kV the corresponding effective exposure charge densities are roughly 2 and 40 μC/cm^2. The corresponding x-ray flux requirements for Al $K\alpha$ x rays (0.834 nm) are 50 and 1000 mJ/cm^2. By doping with absorbing materials such as Cl and other techniques, dose levels as low as 10 mJ/cm^2 can be used for the sensitive resist strategy.

Electron bombardment sources are currently being developed for x-ray contact printers. The goal is to obtain 1-μm production capability. Of necessity, the sensitive resist strategy is being employed. As pointed out in Section 3, the two basic problems are the low x-ray generation efficiency of about 0.01% and the source spacing limitations due to geometrical distortion. As a point of reference, a 1-kW Al x-ray source at a distance of 20 cm from the wafer would produce a flux of 0.14 mW/cm^2. For the robust resist strategy the resulting exposure time would be about two hours. Sensitive resists could be exposed in 5 min or possibly even 1 min if doped resists are used. In these calculations, the absorption in the window, in the intervening environment, and in the mask substrate have been neglected. For these and other reasons the flux available in actual systems is typically as much as a factor of four lower than the simple estimates from the efficiencies listed in Table 1.

The characteristics of actual systems reported in the literature are shown in Table 4.[100] The corresponding properties of the resists used in these systems are

Table 4. Examples of Existing X-Ray Lithography Systems[a]

Builder	Electron-beam power (kW)	Source-to-wafer distance (cm)	Resist	Throughput (wafers/hr)	Devices reported produced	Source
Bell Laboratories	4.5	50	DCPA	15	CMOS	Palladium
General Instrument Corp.	1.0	20	PGMA	2	MOS EE-PROM	Aluminum
Hughes Research Laboratories	10	24	PGMA	5	CMOS on sapphire	Aluminum
IBM Corp	2.5	10	PMMA	1	Bubble devices	Aluminum
Nippon Telegraph and Telephone	25	50	FPM	4		Silicon (7.1 Å)
Sperry Univac	7	16	PBS	12	64-kilobit bubble circuit	Aluminum
Texas Instrument Inc.	18	43	XR79	70	Bubble memories	Aluminum

[a] From Hughes *et al.*[100]

given in Table 3. Note that high-power sources up to 25 kW have been developed. However, only the most powerful of these can produce a flux greater than 0.100 mW/cm^2 at the wafer position. For robust resists this is still not sufficient flux to make their use practical with conventional x-ray sources. Although simple flux estimates indicate that exposure times of a few seconds are potentially possible for sensitive resists, the lowest exposure times reported to date are slightly less than 1 min per wafer. The resolution for the short exposure time systems is about $1.0 \mu m$. As the linewidth is reduced into the submicron range, the source to wafer spacing in conventional systems will be increased to reduce the geometrical distortion and the exposure time will rapidly increase.

The flux available from a storage ring can be several orders of magnitude higher than that available from conventional sources. Even a small dedicated storage ring with a $12.7-\mu m$ Be window could expose a 1-cm^2 area of PMMA resist in 10 s with just 1 mrad of acceptance angle. This is equivalent to 100 mW/cm^2 of flux at the Al $K\alpha$ characteristic wavelength. Flux levels in excess of 1 W/cm^2 over large areas could be obtained by using collecting optics or higher beam currents or higher energies if necessary. Although high flux levels can be produced, the maximum usable flux level will likely be limited by the difference in the thermal expansion of the mask and wafer during the exposure process. Special techniques such as the flowing of gases between the mask and wafer during exposure might be used to reduce this effect. For purposes of estimating the throughput in the presence of heating effects, it will simply be assumed that thermal distortion during exposure will limit flux levels to 1 W/cm^2 although this may be an optimistic choice.[119]

The area that can be exposed simultaneously with synchrotron radiation from a storage ring will likely be limited by wafer distortion during processing. At $1-\mu m$

linewidths there is a reasonable possibility that distortion-compensated masks could be used either for scanning or for full wafer flood exposure. In either case the actual resist exposure time would be only a few seconds for the most robust resists. The net exposure time per wafer including wafer handling and alignment would probably be about 6 s regardless of the resist used. If wafer distortion limits the simultaneously exposable area to 10 cm^2 the net time to step and repeat a wafer would be about 10 s. Similarly, 4-cm^2 areas could be used to expose a wafer in about 30 s. If it is necessary to step and repeat 1-cm^2 areas, as is done in optical lithography due to lens limitations, the throughput would be about the same as it is optically, 1 min per wafer.

The degree to which a storage ring can be utilized depends on the area being exposed and the number of stations employed. As a rule of thumb, an equivalent effective flux of 1 W/cm^2 at the Al wavelength requires 10 mrad of flux from a small dedicated storage ring with a 12.7-μm Be window. Full wafer scanning could easily be accomplished at a distance of a few meters from the source point. For wafer stepping of 10- or 4-cm^2 areas, 100 or 40 mrad would have to be collected for robust resists. A single tangential beam line could collect this radiation and focus it on to the desired area at a high-resolution exposure station located about 30 m from the source point. Up to 15 such beam lines could be accommodated on a storage ring. The above estimates assume only 100 mA of stored beam, which is a conservative value.

The cost per wafer exposure in the integrated circuits industry is rising rapidly. Table 5 gives estimated capital equipment costs, throughput, and costs per wafer for various lithography approaches. Today the majority of the high-resolution work is slightly larger than 2 μm on projection printers with reflection optics and costs about $0.50 per wafer. With step and repeat cameras the linewidths are

Table 5. Estimated Cost Per Wafer for Various Lithography Approaches

System (type)–(area)	Resolution used in estimate (μm)	Exposure time (min)	Capital cost ($)	Cost per wafer[a] ($)
Optical–wafer scan	2	0.25	200K	0.50
Optical–step and repeat	1	1.0	400K	2.00
Electron–step and repeat	1	1–4	1.6M	8.00–32.00
X ray–conventional[b]				
flood	1	1	200K	1.00
step and repeat	1	1	400K	2.00
X ray–storage ring[c]	1	0.10	0.5–1M	0.25–0.50
100 cm[c]				
10 cm[c]	1	0.16	0.5–1M	0.40–0.80
4 cm[c]	1	0.50	0.5–1M	1.25–2.50
1 cm[c]	1	1.0	0.5–1M	2.50–5.00

[a] Cost per wafer = (exposure time)(capital cost)($0.50)/$100,000.
[b] Sensitive resist strategy.
[c] Robust resist strategy.

approaching 1 μm at a cost of $2.00 per wafer. Direct wafer writing electron beam systems are just becoming commercially available. Advanced machines with aperture projection techniques will probably cost $1.6M and will likely be bandwidth limited to exposure times of from 1 to 4 min per wafer. The cost per wafer for such a system would be from $8 to $32. Full wafer x-ray exposure systems with conventional sources would be cheaper than optical step and repeat systems and for 1-min exposures would cost about $1.00 per wafer. Step and repeat x-ray systems would be similar both in cost and throughput to optical steppers but would presumably provide better linewidth control at 1 μm. These x-ray steppers could be extended well into the submicron range at reduced throughput.

The cost per wafer for a storage ring x-ray source depends on the degree to which it is utilized and on the amount of wafer handling and alignment equipment added to each port. A small dedicated storage ring would likely cost from $4M to $6M. The x-ray optics and alignment equipment might cost as much as $250k per port. Thus the cost for 8 to 15 ports would range from $0.5M to $1M each. (The 16th port location is used for injection.) The cost per wafer as a function of the area that can be simultaneously exposed is given in Table 5. The lower numbers correspond to 15 ports and the higher numbers are for 8 ports.

The cost per wafer for storage ring sources as estimated here compare very favorably with other lithography approaches. The $0.40 to $0.80 per wafer for 10-cm^2-area stepping is similar to the cost with the optical wafer scanning technique in wide use today. Yet it is based on the use of robust resists having 1-μm features with vertical profiles. While conventional x-ray sources could be used at about the same cost per wafer, the sensitive resist strategy requires the additional development of special compatible processing. If wafer distortion during processing limits exposure areas to 4 cm^2, the cost per wafer with a storage ring source increases but is still comparable to that of optical step and repeat cameras. Even in this case the higher quality resist line-edge profiles and better linewidth control with x-ray exposure of robust resists will probably make the use of x-ray lithography advantageous.

As the linewidth is reduced below 1 μm the cost per wafer increases by differing rates for the various lithography approaches. Major problems arise in the optical approach in the areas of lens design, fabrication tolerances, and sources in the deep-UV wavelength region (200 to 300 nm). With conventional x-ray and electron beam lithography the cost per wafer tends to increase inversely as the square of the linewidth. For conventional x-ray sources this is due to geometrical distortion. In electron beam systems it occurs if the throughput is information rate limited. With synchrotron radiation from a properly designed storage ring and port, the cost per wafer is primarily limited by the maximum area permitted by wafer distortion. Here there is at least some possibility that with time the impact of wafer distortion could be reduced through the use of low-temperature processing and careful characterization. Thus the cost per wafer with storage ring sources may not rise as rapidly as it will for the other approaches.

In summary, the cost per wafer of x-ray lithography with storage ring sources appears favorable to that of other lithography approaches at 1-μm resolution. As the resolution decreases to the submicron range the relative economic advantage of the storage ring approach may increase as rapidly as the inverse square of the linewidth. The most important and presently unknown parameter in comparing the economic feasibility of various lithography approaches is the area that can be simultaneously exposed in the presence of mask and wafer distortion. The throughput estimates made here for storage ring sources are based on the use of robust resists. Presumably the resulting higher-quality resist profiles and greater linewidth control would provide additional advantages in device applications. The greatest deterrent to the development of storage ring lithography systems appears to be the large initial investment. When the incremental evolution of other lithography tools can no longer economically meet the demands of device designs for volume products the development of storage systems will be undertaken. Such may be the case for future generations of bubble devices that require high-quality resist profiles and yet involve only a few mask steps and are not subject to significant wafer distortion during processing.

ACKNOWLEDGMENT

The author is grateful to his colleagues and friends who have contributed to the chapter. He would especially like to thank Paul Sullivan, Gary Taylor, and Juan Maldonado for their assistance with the manuscript. He is also grateful to Bill Oldham, Eberhart Spiller, and Hannk Smith for many technical discussions. He would also like to thank Caryn Dombrosky for the preparation of the references.

References

1. D. L. Spears and H. I. Smith, High-resolution pattern replication using soft x-rays, *Electron. Lett.* **8**, 102 (1972).
2. D. L. Spears and H. I. Smith, X-ray lithography—a new high resolution replication process, *Solid State Technol.* **15**, 21–26 (1972).
3. E. Spiller, R. Feder, J. Topalian, D. Eastman, W. Gudat, and D. Sayre, X-ray microscopy of biological objects with carbon K and with synchrotron radiation, *Science* **191**, 1172–1174 (1976).
4. G. Moore, Keynote Address, Kodak Microelectronics Seminar, Proceedings Interface-75, 2–5 (1976).
5. R. Feder, X-ray projection printing of electrical circuit patterns, IBM Report TR22. 1065, August 1970.
6. D. Maydan, G. A. Coquin, J. R. Maldonado, S. Somekh, D. Y. Lou, and G. N. Taylor, High speed replication of submicron features on large areas by x-ray lithography, *IEEE Trans. Electron Devices* **22**, 429–439 (1975).
7. P. A. Sullivan, X-ray lithography system complete with interdigital transducer master, Hughes Research Laboratories Report AFCRL-TR-75-0573, November 1975.
8. D. L. Spears, H. I. Smith, and E. Stern, X-ray replication of scanning electron microscope generated patterns, in *Proceedings of the Fifth International Conference on Electron and Ion Beam Science and Technology*, pp. 80–91, Electrochemical Society, Princeton, New Jersey (1972).
9. R. Feder, E. Spiller, and J. Topalian, Replication of 0.1-μm geometrics with x-ray lithography, *J. Vac. Sci. Technol.* **12**, 1332–1335 (1975).
10. R. K. Watts, K. E. Bean, and T. L. Brewer, X-ray lithography with aluminum radiation and SiC mask, in *Proceedings of the Eighth International Conference on Electron and Ion Beam Science and Technology*, pp. 453–457, Electrochemical Society, Princeton, New Jersey (1978).

11. T. Hayashi, Electron beam and x-ray lithography for very large scale integration devices, in *Proceedings of the Eighth International Conference on Electron and Ion Beam Science and Technology*, pp. 85–97, Electrochemical Society, Princeton, New Jersey (1978).

12. J. Lyman, Lithography chases the incredible shrinking line, *Electronics* **52**(8), 105–116 (1979).

13. E. Hundt and P. Tischer, A simple set-up for making multilayer structures by x-ray lithography, in *Proceedings of the International Conference on Microlithography*, Paris, France, June 1977, pp. 211–215.

14. B. F. Stein and M. J. Casey, Magnetic Bubble Device Fabrication Using X-Ray Lithography, in *Proceedings of the Eighth International Conference on Electron and Ion Beam Science and Technology*, pp. 480–489, Electrochemical Society, Princeton, New Jersey (1978).

15. M. C. Peckerar, C. J. Taylor, and P. D. Blais, Self aligned cross processing with Rh and Ag L line sources, 24th International Electron Device Meeting, 1978 *IEDM Tech. Dig.*, 589–590 (December 1978).

16. G. P. Hughes, X-ray lithography for IC processing, *Solid State Technol.* **20**, 39–42 (1977).

17. X-ray lithography system achieves ultrafine resolution, *Electronics* **51**(16), pp. 69–70, August 3, 1978.

18. E. Spiller, D. E. Eastman, R. Feder, W. D. Grobman, W. Gudat, and J. Topalian, Application of synchrotron radiation to x-ray lithography, *J. Appl. Phys.* **47**, 5450–5459 (1976).

19. J. Trotel and B. Fay, Contrast and exposure time calculations in x-ray lithography: experiments using synchrotron radiation from a storage ring, in *Proceedings of the International Conference on Microlithography*, Paris, France, June, 1977, pp. 201–209.

20. B. Fay, J. Trotel, Y. Petroff, R. Pinchaux, and P. Thiry, X-ray replication of masks using synchrotron radiation produced by the ACO storage ring, *Appl. Phys. Lett.* **29**, 370–372 (1976).

21. H. Aritome, T. Nishimura, H. Kotani, S. Matsui, O. Nakagawa, and S. Namba, X-ray lithography by synchrotron radiation of INS–ES, *J. Vac. Sci. Technol.* **15**, 992–994 (1978).

22. H. Aritome, S. Matsui, K. Moriwaki, S. Hasegawa, and S. Namba, Fabrication of optical devices by x-ray lithography by using synchrotron radiation, in *Proceedings of the Eighth International Conference on Electron and Ion Beam Science and Technology*, pp. 468–479, Electrochemical Society, Princeton, New Jersey (1978).

23. T. Nishimura, H. Kotani, S. Matsui, O. Nakgowa, H. Aritome, and S. Namba, X-ray replication of masks by synchrotron radiation of INS–ES, *Jpn. J. Appl. Phys. Suppl.* **17**, 13–17 (1978).

24. A. R. Neureuther, R. Sud, and Y. T. Hsu, X-ray Lithography, Stanford Synchrotron Radiation Laboratory Activity Report 77/09, pp. VI-116–V-117 (1977); 78-02, pp. VII-59 (1978); 78–10, pp. VII-57 (1978).

25. Y. T. Hsu, Resist Characterization for High Resolution Lithography Using Synchrotron Radiation of SPEAR, M.S. Thesis, University of California, Berkeley, March 1979.

26. R. Sud, Modelling and Characterization of High Resolution Resists for Sub-Micron Integrated Circuits, M.S. Thesis, University of California, Berkeley, December 1977.

27. R. Burg, J. Kirz, H. Rarbach, M. J. Malachowski, and J. Wm. McGowan, X-Ray Microscopy and Microchemical Analysis with the 4 Degree Beam Line, Stanford Synchrotron Radiation Laboratory Report 79/01 (1979).

28. L. D. Yau, Correlation between process-induced in-place distortion and wafer bowing in silicon, *Appl. Phys. Lett.* **33**, 756–758 (1978).

29. S. Yamazaki, S. Nakayama, T. Hayasaka, and S. Ishihara, X-ray exposure system using finely position adjusting apparatus, *J. Vac. Sci. Technol.* **15**, 987–991 (1978).

30. D. C. Flanders, H. I. Smith, and S. Austin, A new interferometric alignment technique, *Appl. Phys. Lett.* **31**, 426–428 (1977).

31. S. E. Bernacki and H. I. Smith, Fabrication of silicon MOS devices using x-ray lithography, *IEEE Trans. Electron Devices* **22**, 421–428 (1975).

32. H. L. Stover, X-ray Lithographic Technique for Fabricating Integrated Circuits, Final Report DELET-TR-77-2669-F. Prepared for U.S. Army Electronics Research and Development Command, Fort Monmouth, New Jersey, February 1979.

33. B. J. Lin, Deep-UV comformable-contact photolithography for bubble circuits, *IBM J. Res. Dev.* **20**, 213–221 (1976).

34. H. Widman, IEEE Workshop on VLSI, Hilton Head, South Carolina, September (1978).

35. J. D. Jackson, *Classical Electrodynamics*, 2nd ed., Wiley, New York (1975).

36. D. Maydan, G. A. Coquin, J. R. Maldonado, J. M. Moran, S. Somekh, and G. N. Taylor, X-ray lithography: one possible solution to VLSI device fabrication, in *Proceedings of the International Conference on Microlithography, Paris, France, June, 1977*, pp. 196–199.

37. A. N. Broers, W. W. Molzen, J. J. Cuomo, and N. D. Wittels, Electron-beam fabrication of 80-Å metal structures, *Appl. Phys. Lett.* **29**, 596–598 (1976).

38. A. N. Broers, J. M. E. Harper, and W. W. Molzen, 250-Å linewidths with PMMA electron resist, *Appl. Phys. Lett.* **33**, 392–394 (1978).

39. R. Feder, E. Spiller, J. Topalian, A. N. Brocks, B. Gudat, B. J. Panessa, Z. A. Zadonaisky, and J. Sedet, High resolution x-ray lithography with carbon *K* radiation, in *Proceedings of the Seventh International Conference on Electron and Ion Beam Science and Technology*, pp. 198–203, Electrochemical Society, Princeton, New Jersey (1976).

40. D. C. Flanders, Replication of 175Å lines and spaces in polymethylmethacrylate using x-ray lithography, *Appl. Phys. Lett.*, **36**(1), 93–96 1980).

41. M. Hatzakis, Electron sensitive ploymers as high resolution resists, *Appl. Polym. Symp.* **23**, 73–86 (1974).

42. M. Hatzakis, Recent developments in electron-resist evaulation techniques, *J. Vac. Sci. Technol.* **12**, 1276–1279 (1975).

43. D. C. Flanders, H. I. Smith, H. W. Lehmann, R. Widmer, and D. C. Shaver, Surface relief structures with linewidths below 2000 Å, *Appl. Phys. Lett.* **32**, 112–114 (1978).

44. Reference documents for the proposal for a National Synchrotron Light Source, BNL Report 50595, Vol. II (1977).

45. *Handbook of X-Rays*, E. F. Kaelble (ed.), McGraw-Hill, New York (1967).

46. H. A. Kramers, On the theory of x-ray absorption and of the continuous x-ray spectrum, *Philos. Mag.* **46**, 836–871 (1923).

47. P. A. Sullivan and J. H. McCoy, Determination of wavelength and excitation voltage for x-ray lithography, *IEEE Trans. Electron Devices*, **23**, 412–418 (1976).

48. N. A. Dyson, *X-Rays in Atomic and Nuclear Physics*, pp. 203–243, Longmans Group Ltd., London (1973).

49. M. Green, Target absorption correction in microanalysis, in *X-Ray Optics and X-Ray Microanalysis*, H. H. Patlee, V. E. Cossleth, and A. Engstrom (eds.), pp. 185–192, Academic Press, New York (1963).

50. D. F. Kyser, Experimental determination of mass absorption coefficients for soft x-rays, in *Proceedings of the Sixth International Conference on X-Ray Optics and Microanalysis*, University of Tokyo Press, Tokyo (1972).

51. J. S. Greeneich, X-ray lithography, I. Design criteria for optimizing resist energy absorption. II. Pattern replication with polymer masks, *IEEE Trans. Electron Devices* **22**, 434–439 (1975).

52. A. H. Compton, *Bull Natl. Res. Counc. U.S.* **20**, 48 (1922).

53. V. Rehn and V. O. Jones, Vacuum ultraviolet (VUV) and soft x-ray mirrors for synchrotron radiation, *Opt. Eng.* **17**, 504–511 (1978).

54. V. Rehn, X-Ray Mirrors, Workshop on X-ray Instrumentation for Synchrotron Radiation Research, SSRL Report 78/04, pp. VII-13-VII-34, (1978).

55. H. Winick, private communication (1978).

56. R. Feder, E. Spiller, and J. Topalian, X-ray lithography, *Polym. Eng. Sci.* **17**, 385–389 (1977).

57. B. L. Bracewell and W. J. Veigele, Tables of x-ray mass attenuation coefficients for 87 elements at selected wavelengths, *Applied Spectroscopy*, Vol. 9, E. L. Grove and A. J. Perkins (eds.), pp. 375–400, Plenum Press, New York (1973).

58. G. F. Knoll, *Radiation Detection and Measurement*, John Wiley and Sons, New York (1979).

59. P. Tischer and E. Hundt, Profiles of structures in PMMA by x-ray lithography, in *Proceedings of the Eighth International Conference on Electron and Ion Beam Science and Technnology*, pp. 444–457, Electrochemical Society, Princeton, New Jersey (1978).

60. C. D. Anderson, *Phys. Res.* **35**, 1139 (1930).

61. T. E. Everhart and P. H. Hoff, Determination of kilovolt electron energy dissipation vs. penetration discharge in solid material, *J. Appl. Phys.* **42**, 5837–5846 (1971).

62. R. D. Heidenreich, L. F. Thompson, E. D. Feit, and C. M. Mellior-Smith, *J. Appl. Phys.* **44**, 4039 (1973).

63. J. R. Maldonado, G. A. Coquin, D. Maydan, and S. Somekh, Spurious effects caused by the continuous radiation and ejected electrons in x-ray lithography, *J. Vac. Sci. Technol.* **12**, 1329–1331 (1975).
64. J. R. Maldonado, private communication (1979).
65. K. Ueberreiter, The solution process, in *Diffusion in Polymers*, J. Crank and G. Park (eds.), Academic Press, New York (1968).
66. A. Charlesby, *Atomic Radiation and Polymers*, Pergamon Press, London (1960).
67. E. Gipstein, A. C. Ouano, D. E. Johnson, and O. U. Need, III, Parameters affecting the electron beam sensitivity of poly(methyl methacrylate), *IBM J. Res. Dev.* **21**(2), 143–153 (1977).
68. M. Hatzakis, C. H. Ting, and N. Viswanathan, Fundamental aspects of electron beam exposure of polymeric resist system, in *Proceedings of the Sixth International Conference on Electron and Ion Beam Science and Technology, Sixth International Conference*, 542–579, Electromechanical Society, Princeton, New Jersey (1974).
69. J. S. Greeneich, Developer characteristics of poly(methyl methacrylate) electron resist, *J. Electrochem. Soc.* **122**, 970–976 (1975).
70. A. R. Neureuther, D. F. Kyser, and C. H. Ting, Electron beam resist edge profile simulation, *IEEE Trans. Electron Devices*, **26**(4), 686–693 (1979).
71. A. C. Ouano, Y. O. Tu, and J. A. Carothers, Dynamics of polymer dissolution, *Structure-Solubility Relationships in Polymers*, 11–20, Academic Press, New York (1977).
72. A. C. Oriano, A study of the dissolution rate of irradiated poly (methyl methacrylate), *Polym. Eng. Sci.*, **18**(4), 306–313 (1978).
73. L. F. Thompson, F. D. Feit, M. H. J. Bowden, and E. G. Spencer, Polymeric resists for x-ray lithography, *J. Electrochem. Soc.* **121**, 1500–1503 (1974).
74. E. F. Feit, M. E. Wurtz, and G. W. Kammlott, Sol-gel behavior and image formation in poly(glycidyl methacrylate) and its copolymers with ethyl acrylate, *J. Vac. Sci. Technol.* **15**, 944–947 (1978).
75. G. N. Taylor, G. A. Coquin, and S. Somekh, Chlorine containing resists for x-ray lithography, *Polym. Eng. Sci.* **17**, 420–429 (1976).
76. K. Murase, M. Kakuchi, and S. Sugawa, Newly developed electron and x-ray resists, in *Proceedings of the International Conference on Microlithography*, Paris, France, 1977, 261–269.
77. D. F. Kyser and K. Murata, Monte Carlo simulation of electron beam scattering and energy ions in thin films on thick substrates, in *Proceedings of the Sixth International Conference on Electron and Ion Beam Science and Technology*, 205–223, Electrochemical Society, Princeton, New Jersey (1974).
78. D. Robman, *Science* **205**, 1239–1241 (1979).
79. R. A. McCorkle and H. J. Vollmer, *Phys. Rev. Lett.* **39**, 1263–1266 (1977).
80. P. Bogen, H. Conrads, G. Gatti, and W. Kohlhaus, *J. Opt. Soc. Am.* **58**, 203–206 (1968).
81. R. A. McCorkle, Soft x-ray emission by an electron beam-sliding spark device, *J. Phys. B* **11**, L407–408 (1978). R. A. McCorkle, *et al.*, Flash x-ray microscopy, *Science* **205**, 401–402 (1979).
82. B. J. Nagel, M. C. Peckerar, J. R. Greig, R. E. Pechacek, and R. R. Whillock, *SPIE Proceedings* **135**, 46–53 (1978).
83. M. C. Peckerar, J. R. Greig, D. J. Nagel, R. E. Pechacek, and R. R. Witlock, High speed x-ray lithography with radiation from laser-produced plasmas, in *Proceedings of the Eighth International Conference on Electron and Ion Beam Science and Technology*, 432–443, Electrochemical Society, Princeton, New Jersey (1978).
84. Evidence of x-ray lasing in Russia and new lasers for enriching uranium, *Laser Focus* **13**(8), 12 (1977).
85. P. A. Sullivan and J. H. McCoy, Optimized source for x-ray lithography of small area devices, *J. Vac. Sci. Technol.* **12**, 1325–28 (1975).
86. J. R. Maldonado, M. E. Poulsen, T. E. Saunders, F. Vratny, and A. Zachrias, X-ray lithography source using a stationary solid Pd target, *J. Vac. Sci. Tech.* **16**(6), 1942–1945 (1979).
87. G. A. Wardly, E. Munro, and R. W. Scott, High brightness ring cathode rotating anode source for x-ray lithography, in *Proceedings of the International Conference on Microlithography*, Paris, France, 217–220 (1977).
88. H. Yoshihara, M. Kiuchi, Y. Saito, and S. Nakayama, X-ray silicon target preparation for x-ray lithographic system, *Jpn. J. Appl. Phys.* **18**, 2021–2022 (1979).
89. E. Bassous, R. Feder, E. Spiller, and J. Topalian, High transmission x-ray masks for lithographic applications, *Solid State Technol.* **1**, 55–58 (1976).

90. K. Suzuki, J. Matsui, T. Kadata, and T. Ono, Preparation of x-ray lithography masks with large area sandwitch structure membranes, *Jpn. J. Appl. Phys.* **17**, 1447–8 (1978).
91. T. Funayama, Y. Takayama, T. Inagaki, and M. Nakamra, New x-ray mask of Al-Al$_2$O$_3$ structure, *J. Vac. Sci. Technol.* **12**, 1324 (1975).
92. D. C. Flanders and H. I. Smith, Polyimide membrane x-ray lithography masks—fabrication and distortion measurements, *J. Vac. Sci. Technol.* **15**, 99–999 (1978).
93. R. K. Watts, D. C. Guterman, and H. M. Darley, Submicron x-ray lithography, *SPIE, J.* **80**, 100–105 (1976).
94. T. Funayama, K. Yanagida, N. Nakayama, K. Komeno, and T. Inagai, Fabrication of micron and submicron-bubble memory devices by a mask transfer technique with subsequent getter-ion etching, *J. Vac. Sci. Technol.* **15**, 998–1000 (1978).
95. M. Cantragrel and M. Marchal, Argon ion etching in a reactive gas, *J. Mater. Sci.* **8**, 1711 (1973).
96. P. G. Gloersen, Ion-beam etching, *J. Vac. Sci. Technol.* **12**, 28–34 (1975).
97. M. C. Blakeslee, L. T. Romankiw, R. E. Acosta, S. Krongelb, and B. Toeber, Electrodeposition process for fabrication of the conductor first, SLM, 2-μm bubble memory, in *Proceedings of the Seventh International Conference on Electron and Ion Beam Science and Technology*, 198–203, Electrochemical Society, Princeton, New Jersey (1976).
98. E. Spiller, R. Feder, J. Topalian, E. Costellani, L. Romankiw, and M. Heritage, X-ray lithography for bubble devices, *Solid State Technol.* **19**, 62–78 (1976).
99. J. M. Shaw and Hatazakis, Performance characteristics of diazo-type photoresists under E-beam and optical exposure, IEEE Trans. Electron Devices **25**, 425–430 (1978).
100. G. P. Hughes and R. C. Fink, X-ray lithography breaks the VLSI cost barrier, *Electronics*, **51**(23), 99–106 (1978).
101. E. Spiller and R. Feder, X-ray lithography, *Topics in Applied Physics*, Vol. 22, H. J. Queisser (ed.), Chapter 3, Springer-Verlag, New York (1977).
102. A. B. Glaser and G. E. Sobak-Sharpe, *Integrated Circuit Engineering*, pp. 279, Addison-Wesley, Reading, Massachusetts (1977).
103. M. Hatzakis, Electron resists for microcircuit and mask replication, *J. Electrochem. Soc.* **116**, 1033–1037 (1969).
104. B. J. Conavello, M. Hatzakis, and J. M. Shaw, Process for obtaining undercutting of a photoresist to facilitate lift-off, *IBM Tech. Discl. Bull.* **19**, 4048 (1977).
105. D. A. Swyt, NBS program in photomask linewidth measurements, *Solid State Technol.* **19**(4), 55–61 (1976).
106. M. C. King and D. H. Berry, *Appl. Opt.* **11**, 2455 (1972).
107. J. Schwider and C. Hiller, *Optica Acta* **23**, 49–61 (1976).
108. J. Pasiecznik and J. W. Reeds, Digitally positioned mechanical stage, *J. Vac. Sci. Technol.* **15**(3), 909–912 (1978).
109. S. Austin, H. I. Smith, and D. C. Flanders, Alignment of x-ray lithography masks using a new interferometric technique—experimental results, *J. Vac. Sci. Technol.* **15**, 984–986 (1978).
110. J. H. McCoy and P. A. Sullivan, Mask alignment for the fabrication of integrated circuits using x-ray lithography, *Solid State Technol.* **19**, 59–64 (1976).
111. H. I. Smith, Fabrication techniques for surface-acoustic-wave and thin-film optical devices, *Proc. IEEE* **62**, 1361–1387 (1974).
112. D. H. Leebrick and D. W. Kisker, an electrical alignment test device and its use in investigating processing parameters, *Kodak Microelectronics Seminar Proceedings Interface* (1977), 66–83.
113. R. E. Gegenwarth and F. P. Laming, Effect of plastic deformation of silicon wafers on overlay, *SPIE J.* **100**, 66–73 (1977).
114. A. M. Goodman and C. E. Weitzel, The effect of electron-beam aluminization on the Si-sapphire interface, *Appl. Phys. Lett.* **31**, 114–117 (1977).
115. K. F. Galloway and S. Mayo. On the compatibility of x-ray lithography and SOS device fabrication, *IEEE Trans. Electron Devices* **25**, 549–550, (1978).
116. H. S. Yourke and E. V. Weber, A high-throughput scanning-electron-beam lithography system, EL 1, for semiconductor manufacture: general description, *1976 IEDM Tech. Digest*, 437 (1976).
117. A. N. Broers, Fine line lithography systems for VLSI, in *1978 IEDM Tech. Digest*, 1–5 (1978).
118. J. M. Moran and D. Maydan, High-resolution, steep-profile, resist patterns, *Bell Syst. Tech. J.*, **58**(5), 1027–1036 (1979).
119. P. A. Sullivan, private communication (1979).

8

Soft X-Ray Microscopy of Biological Specimens

J. KIRZ and D. SAYRE

1. Introduction

X-ray photons in the 1–10 nm wavelength range present several favorable characteristics for the microscopy of biological specimens. These include total reaction cross sections that are appropriate for the examination of whole organelles or in some cases whole cells; the presence of numerous absorption edges (including those of low-Z elements), providing a variety of opportunities for developing contrast even in unstained and undried specimens; scattering cross sections that are small compared with absorption, leading to an effective absence of multiple scattering; and availability of the photons in adequate quantities from sources of synchrotron radiation. Coupled with these characteristics, however, are two factors that complicate the process of image formation: a wavelength of the same order of size as the resolution desired; and a current shortage of technology in optics, detectors, etc., for the radiation involved. At present it is not clear how these issues will ultimately be resolved. We attempt here to present a report on the current status of the subject.*

* In June, 1979, after this review was written, a conference on soft x-ray microscopy and its applications in biological and physical sciences was held in New York under the auspices of the New York Academy of Sciences. This conference produced a large amount of new material on the subject, which may be found in: *Proceedings of the Conference on Ultrasoft X-Ray Microscopy* (D. F. Parsons, ed.), New York Academy of Sciences (in press).

J. KIRZ • Department of Physics, State University of New York, Stony Brook, New York 11794.
D. SAYRE • Mathematical Sciences Department, IBM Research Center, Yorktown Heights, New York 10598.

There exist several excellent reviews[1-3] of x-ray microscopy using x rays from conventional sources. These x rays are of wavelengths mainly in the 0.1–1 nm range, and have reaction cross sections, absorption edges, and scattering characteristics that are less favorable for biological microscopy than those of the longer-wavelength radiation considered here. The problems of producing high-resolution optics and detectors for the harder radiation are also more difficult than for the radiation we shall be considering. The specifically biological aspects of x-ray microscopy using conventional sources have been more recently reviewed by Engström[4] and by Hall *et al.*[5] Since the advent of synchrotron radiation sources, brief reviews have been prepared by Horowitz[6] on scanning microscopy and by Spiller *et al.*[7] and Gudat[8] mainly on contact microradiography. A popular treatment of microscopy and other applications of soft x rays has recently been given by Spiller and Feder.[9]

Historically, much of the potential of the softer photons has been appreciated for many years by workers in x-ray microscopy, even if limitations of sources and technique have made work with them difficult. Experimentation in microradiography of biological materials with photons of $\lambda > 1$ nm goes back at least to 1936 with the work of Lamarque,[10] who built an x-ray generator capable of working at wavelengths to approximately 1.2 nm. During the 1950s the range was extended to approximately 2.5 nm by various workers, notably by Engstrom[4] and his co-workers. Wolter[11] was explicit in observing the potential usefulness of 2.4-nm photons in the examination of wet biological material (these photons fall between the K absorption edges of oxygen on the one hand and nitrogen and carbon on the other, and hence are weakly absorbed by water but strongly absorbed by organic matter), and subsequently Henke[12] stated very clearly the usefulness of the full range of wavelengths to 10 nm in high-resolution microradiographic analysis. In 1972 Horowitz and Howell[13] demonstrated the feasibility of doing microscopy with x rays from a synchrotron source. Other references to the history of the subject will be made at appropriate points throughout the article.

The chapter is organized as follows. Sections 2 and 3 treat the two key topics of contrast mechanisms and imaging. To allow the treatment to go into some detail, and yet not lose sight of the main points, both these sections contain a summary at the end. Section 4 briefly discusses, from the point of view of microscopy, some of the soft x-ray sources available today. With the conclusion of Section 4 the main features of the microscope system have been delineated. Section 5 then discusses the topic of mapping at high resolution the concentration of any given atomic species. Since the topic is somewhat complex, a summary is included in this section as well. Section 6 discusses several additional technical aspects that are of secondary but still high importance if the full potential of the microscopy in biology is to be realized. Section 7 makes some concluding remarks.

2. Contrast Mechanisms

In any form of microscopy, there are two basic elements that must be present. First, for an image to be formable, there must be some respect in which the events occurring in the specimen vary from point to point; this variation supplies the

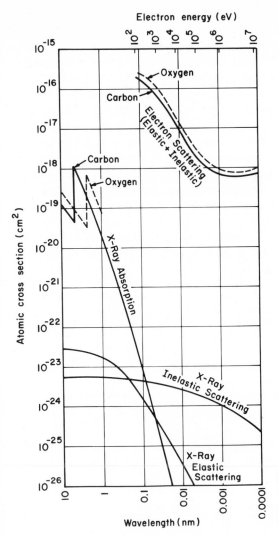

Figure 1. Cross sections for reactions of photons and electrons with carbon, as functions of particle wavelength λ. For comparison, portions of the corresponding curves for oxygen are also shown (dashed curves).

contrast mechanism of the microscopy in question. Second, the arrangement of optical devices and detectors observing the events (or in scanning systems the manner in which the particles inducing the events are delivered to the specimen) must allow a detected event to be accurately referred to the point in the specimen at which it occurred; this arrangement supplies the *imagining system* of the microscopy. In the present section we discuss the contrast mechanisms available in soft x-ray microscopy. Imaging techniques will be discussed in Section 3.

To a good approximation (see Figure 1) soft x-ray photons participate in only one reaction with matter, i.e., absorption of the photon in an atom. This fact considerably simplifies the situation with which we shall be dealing. There are, however, three consequences of an absorption reaction of importance in soft x-ray microscopy, which we shall now discuss in turn.

2.1. Removal of Photons; Transmission X-Ray Microscopy

The first consequence of the absorption of a photon is the removal of the absorbed photon from the x-ray beam. The removal of photons provides the basis for the simplest form of soft x-ray microscopy, *transmission x-ray microscopy* (TXM). The photons are allowed to interact with the specimen; after interaction the photons are counted. Assuming that the various photon paths through the specimen are provided with equal initial numbers of photons, variations in the photon counts will indicate variations in the specimen thickness or composition from path to path. Since scattering is infrequent compared with absorption, all paths are essentially straight lines, greatly simplifying the analysis.* In the most common arrangement, all paths are also parallel. In this case the detector is placed downstream from the specimen and looking back at it; a small detector with a small solid angle of acceptance suffices to collect all photons.

(The above description ignores diffraction effects, which become important when one attempts to operate at resolutions close to the wavelength limit. These effects are discussed in some detail in Sections 3 and 6.)

---

* A further consequence is that there is little harm in having the soft x-ray photons traverse a thin layer of homogeneous material of uniform thickness, such as a short air path. The additional material may weaken the image somewhat, but will not blur it.

Table 1. X-Ray Linear Absorption Coefficients for Substances of Biological interest[a].

λ (nm)	Water	Carbo-hydrate	Protein	Lipid	Nucleic acid	Air
0.13	0.0006	0.0007	0.0006	0.0003	0.0014	0.0005
0.2	0.002	0.003	0.002	0.001	0.005	0.002
0.4	0.018	0.022	0.017	0.008	0.037	0.015
0.8	0.126	0.158	0.113	0.065	0.169	0.108
1.4	0.560	0.719	0.524	0.310	0.768	0.492
2.3	1.93	2.54	1.88	1.13	2.71	1.74
O absorption edge						
2.4	0.112	1.02	1.38	1.00	1.36	1.46
2.7	0.151	1.36	1.84	1.35	1.82	1.95
3.0	0.198	1.77	2.38	1.74	2.34	2.50
N absorption edge						
3.1	0.215	1.92	1.90	1.89	1.74	0.172
3.7	0.336	2.92	2.88	2.87	2.63	0.270
4.2	0.458	3.91	3.85	3.84	3.50	0.369
C absorption edge						
4.4	0.521	0.624	0.522	0.245	1.19	0.415
5.2	0.782	0.941	0.769	0.372	1.69	0.622
7.4	1.84	2.20	1.67	0.872	3.26	1.45
9.0	2.89	3.45	2.38	1.36	4.76	2.28

[a] In units of μm^{-1} except for air, where units are mm^{-1}.

The basic properties of transmission microscopy are easily deduced from this description. The probability that a photon will pass through the specimen is $\exp(-\int \mu \, ds)$, where μ, the linear absorption coefficient, is a function of the photon wavelength and the composition of the material in the neighborhood of ds and is given by $\mu = \sum n_Z \sigma_Z$; here n_Z is the number of atoms of atomic number Z per unit volume of the specimen near ds and σ_Z is the atomic cross section* for the absorption of the photon of wavelength λ by the atom of atomic number Z. μ is listed in Table 1 for several substances of biological interest and for various photon wavelengths. It is seen that, for λ in the range 1–10 nm, μ is of the order of 1 μm^{-1} for these substances, so that specimen thicknesses extending to a few μm are entirely appropriate for transmission microscopy of these substances. This is a range of thicknesses nicely suited to organizational studies of biological cells and organelles in the intact state. In comparison, the thickness range for electron microscopy (roughly 0.01–1 μm) generally requires that these structures be sectioned before study, while that for x-ray microscopy using conventional x-ray sources (roughly 10–1000 μm) produces too much overlap of features for high-resolution work.

Next, a closer look at Table 1 reveals the fact that μ varies significantly with λ and from material to material. An important instance of this is shown in Figure 2,

---

* Atomic cross sections for absorption can be calculated from the tables of soft x-ray mass attenuation coefficients compiled by Henke and his co-workers.[100, 101] (The ranges covered are $\lambda = 0.834$–11.38 nm, $Z = 1$–92. An earlier compilation[102] covers a still larger wavelength range but a smaller range in Z.) See also the tabulation by Viegele.[103] In addition, an extensive table of x-ray absorption edges and emission lines to 10 nm and beyond has been given by Bearden.[104]

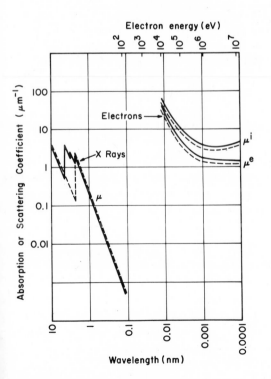

Figure 2. The linear absorption coefficients μ, for protein (solid curve) and water (dashed curve), as a function of photon wavelength λ. For comparison, the linear elastic and inelastic scattering coefficients for electrons, μ^e and μ^i, are also shown for the same two substances. The comparative difficulty of distinguishing between the two substances with electrons is evident.

where for $\lambda > 2.33$ nm (the oxygen K absorption edge) water becomes noticeably transparent, whereas protein (being principally carbon) remains comparatively absorbing until $\lambda > 4.36$ nm. The result, which is highly significant for biological microscopy, is that for wavelengths in the region 2.33–4.36 nm organic features can be seen even in the presence of water. Thus TXM with soft x rays, which was found to be applicable to cells and organelles without sectioning, can also be applied to these materials without dehydration. Indeed we shall see below (Section 2.4) that the contrast between water and organic materials in the 2.33–4.36 nm wavelength region is sufficient to allow staining to be dispensed with as well. In microscopy with soft x rays the way is therefore open for the examination of biological materials in the natural (intact, wet, unstained) or living state.

In electron microscopy, wavelength selection is not a mechanism for contrast enhancement (see Figure 2), and contrast in biological materials must in general be developed by specimen preparation techniques.

2.2. Reaction Products; Fluorescence X-Ray Microscopy and Electron-Emission X-Ray Microscopy

The absorption of a photon produces, initially, a photoelectron and an excited atom. As the photoelectron travels through the absorbing material it undergoes inelastic collisions that excite other atoms; finally the electron loses enough energy to be absorbed itself. The excited atoms undergo fluorescence and Auger deexcitation processes, resulting in the emission of further particles (fluorescence photons, Auger electrons), which may excite yet further atoms. The product of a photon absorption is thus a shower of electrons and photons. The number of particles in the shower is determined principally by the energy of the absorbed photon. The proportion of photons in the shower (fluorescence yield) depends primarily upon the composition of the absorbing material (see Figure 3). The fluorescence photons and Auger electrons have discrete energy spectra that are characteristic of the atomic species in which they originate. The energy of the initial photoelectron is the excess of the photon energy over the photoelectron's binding energy, and is thus

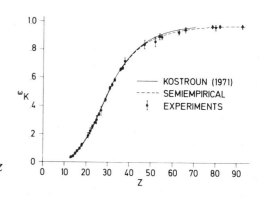

Figure 3. Fluorescence yields for K-shell vacancies as a function of Z (after Bambynek et al.[108]).

determined by both photon energy and Z. Of the two components of the shower (photons and electrons) the photons have considerably the greater range.

It is possible to do x-ray microscopy by counting showers rather than unabsorbed photons. There are two cases to be distinguished, according to whether electrons or photons are detected; the former is *electron-emission x-ray microscopy** (EXM) and the latter is *fluorescence x-ray microscopy* (FXM). Because of the difference in range of the particles, FXM detects showers occurring anywhere within moderately thick specimens, while EXM detects only showers occurring within a few nanometers of a surface of the specimen facing the detector. EXM is thus potentially useful as a method of surface-layer microscopy. In addition, because of the variation of fluorescence yield with Z, FXM is mainly useful for the imaging of medium- or high-Z features. In addition, in both EXM and FXM, the detector design must take into account the fact that the products emerge from the specimen in all directions, and in FXM the fluorescence photons must be distinguished from the scattered or transmitted photons.

We shall see below (Section 2.4) that the counting statistics for the counting of showers are similar to (or in some cases slightly more favorable than) those for the counting of unabsorbed photons; accordingly, the general properties of absorption microscopy noted above carry over into EXM and FXM. In addition, the possibility of identifying the atomic species of an absorbing atom by energy analysis of its emitted Auger electrons or fluorescence photons gives these microscopies a special interest in connection with the mapping of the concentrations of particular atomic species (see Section 5). On the other hand, EXM and FXM require fairly complex methods of image formation (Section 3).

We may note finally that the shower-counting microscopies are dark field (features are bright against a dark background) and that absorption microscopy is bright field (dark features against a bright background).

2.3. Damage to the Absorbing Material

The remaining important effect of the absorption reaction is the production of permanent changes (i.e., damage) in the absorbing material. During the period immediately following an absorption there exist conditions in the material (vacancies in the atomic valence shells, surplus free electrons detached from atoms, electrostatic forces between atoms with net charge due to vacancies) that can easily cause the valence structure finally reestablished in the material to differ from that which originally existed, i.e., can produce structural damage. Such rearrangements are energetically possible in the x-ray region under consideration, where even the 10-nm photon has an energy of approximately 120 eV, or the equivalent of numerous covalent bonds.

An important factor in determining how much damage is done is the dosage,

---

* This microscopy exists also with the incident radiation in the UV, where it is called photoelectron microscopy[32] or photoemission electron microscopy.[33]

i.e., the energy transferred (from particles to absorber) per unit mass of the absorbing material. Dosage is easily calculated in the case of soft x rays and low-Z material, for in this case the only long-range reaction product is the fluorescence photon, which is generated in low yield (Figure 3), so that there effectively exists no mechanism for quick removal of the absorbed energy from the absorbing material. It is easily seen then that in this case $D = n(h\nu)(\mu/\rho)$, where D = dose, n = number of incident photons per unit area, $h\nu$ = energy of the incident photon, and ρ = density of the absorbing material. Figure 4 shows D/n for protein for incident photons of various wavelengths. For comparison the same quantity is also shown for incident electrons of various wavelengths.[14] It is seen that roughly equal dosages are produced by equal fluxes of the two particles.

Isaacson[15] has summarized the available data on the relation between damage and dosage in the case of biological materials. A dose of 10 J/g suffices to kill the most resistant living cell. A dose of 10^4 J/g will cause severe structural rearrangements in most organic materials.

In addition to depending on dosage, the nature and extent of damage is also affected by the composition and structure of the absorbing material. This is a complex topic,[16] which we shall not pursue further here. Damage may also depend upon the type and energy of the particle delivering the dosage. Generally, a particle that delivers dosage in a spatially nonuniform distribution will produce more damage than one that delivers the same dosage in a more uniform spatial distribution.[17] Since the mean energy transfer in one electron scattering event is of the order of 50 eV, as opposed to 1200–120 eV in the absorption of one 1–10 nm photon, this suggests that soft x rays are more damaging, dose for dose, than electrons. Consistent with this, 0.83-nm x rays have been found to be 1 to 3 times

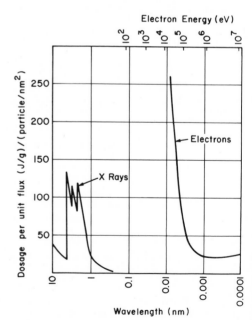

Figure 4. D/n (dosage per unit incident flux) for protein, as a function of wavelength λ, for irradiation by photons (left) and electrons (right). An alternative unit for dosage is the rad (1 rad = 10^{-5} J/g).

more efficient (on an equal-dosage basis) in damaging various organic polymers than are 10-keV electrons.[18] Further information on this point, relating to longer-wavelength x rays and to biological materials, would be of great interest.

It is unlikely that damage to the specimen can be utilized as a contrast mechanism in any practical way. Damage to an x-ray resist, on the other hand, is a highly important process in x-ray microscopy today (see Section 3.1).

2.4. Quantitative Relationships

As noted above, the behavior of the linear absorption coefficient μ of the soft x-ray photon suggests qualitatively that that particle is well adapted to acting as a structural probe for intact (thick, wet, unstained) biological specimens. Sayre *et al.*[19–21] have recently calculated these effects in some detail. We give only a brief summary of these results here, and refer the reader to the original papers for details and for references to earlier studies.

A typical result is shown in Figure 5. Here the specimen under consideration consists of protein features (the features having equal edge lengths in all three dimensions) in a water background. The figure shows (solid contours), for every pair of values of specimen thickness and feature size, the incident flux on the specimen required to provide statistically significant visibility of the features in soft x-ray TXM. In Figure 6 these curves are modified to show specimen dosage instead of incident flux. In either Figure 5 or Figure 6 it is seen that specimen thickness causes little adverse effect on feature visibility until it reaches values of 10 μm or more. That this is so even with very small features indicates that large amounts of water may be present without significantly affecting the visibility of the features.

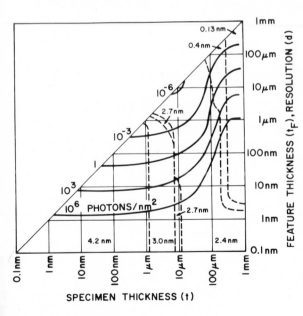

Figure 5. Minimum incident photon flux (solid lines) required to produce an acceptable signal-to-noise ratio ($S/N > 5$) in viewing protein features in a water background. (The photon wavelengths shown are optimal within the regions indicated by the dashed lines, in the sense of minimizing dosage. As expected, the optimal wavelengths lie between the oxygen and carbon absorption edges, except for extremely thick specimens where soft x rays have insufficient penetrating power.) The microscopy is TXM.

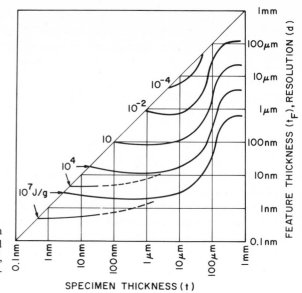

Figure 6. Similar to Figure 5, but showing specimen dosage rather than specimen exposure. The fully solid curves refer to TXM. The solid-dashed curves refer to EXM, and the dashed portion is a reminder that the inner portions of thick specimens will not be seen.

The situation may be contrasted with that shown in Figure 7, which is similar to Figure 6 but for transmission electron microscopy (TEM). With electrons, feature visibility is strongly affected by specimen thickness and the presence of water.

A sense of the scale of the incident flux levels appearing in Figure 5 may be obtained by noting that the monochromator output of currently planned synchrotron radiation sources over small areas is of the order of 10 photons nm^{-2} s^{-1}. A more fundamental scale, however, is obtained by considering the dosages of Figure

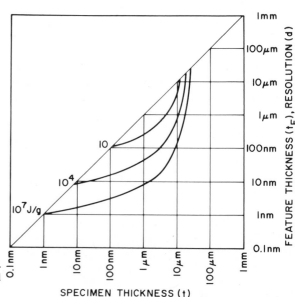

Figure 7. Similar to Figure 6, but for electron microscopy. Dosages have been minimized over a variety of bright- and dark-field CTEM and STEM modes.

6 and recalling (Section 2.3) that most organic materials cannot accept dosages of more than 10^4 J/g without undergoing severe structural damage. We conclude that with wet unstained specimens to thicknesses of a few μm, soft x-ray TXM may reveal features to approximately 10 nm without excessive damage-related distortion. Correspondingly, from Figure 7 it would appear that for specimens of similar thickness in TEM, 100 nm features represent the limit of moderate-damage visibility in the wet unstained case, an estimate which is roughly borne out by experience.*

Comparison of Figures 6 and 7 shows that for sufficiently thin specimens the situation is reversed, i.e., TEM can give slightly better resolution at a given dosage than TXM. However, Figure 6 also shows that for thin specimens the shower-counting x-ray microscopies (EXM in this low-Z case) may have a higher resolution than TEM.

It should be emphasized that the resolutions given in this section are those of the contrast mechanisms. For the resolution of the entire system, the resolution of the imaging technique must also be included. It should also be noted that Sayre *et al.* effectively treated the specimen as a two-dimensional object; in the real case of three-dimensional specimens, doses may be increased very considerably whenever an attempt is made to work at resolutions close to the wavelength limit (see Section 6.3).

2.5. Summary

Soft x-ray photons interact with matter chiefly through absorption. The variation of the number of absorption events from point to point of the specimen provides the contrast mechanisms of soft x-ray microscopy.† In TXM, the variation is detected by counting the transmitted photons. In EXM and FXM the variations are detected by counting the showers of reaction products, in the former case via the electrons contained in the shower, and in the latter case via the photons contained in the shower. EXM is limited to surface-layer microscopy, and FXM is not efficient for low-Z features. TXM is the general-purpose soft x-ray microscopy.

Quantitatively, analysis of these contrast mechanisms suggests that at exposures corresponding to the maximum tolerable dosage to the specimen (10^4 J/g) they will operate at a satisfactory level of signal-to-noise at resolutions of the order of 10 nm with specimens consisting of biological cells and organelles in their natural (i.e., intact, wet, unstained) state. (But see the last paragraph of Section 2.4.) For electrons, corresponding resolutions are probably an order of magnitude less favorable. Physically the high level of performance of the soft x-ray photon is due to its moderate energy, appropriate level of reactivity, and the large variation of its reactivity with λ and Z.

---

* It should be pointed out that at least one group of electron microscopists hope to better this performance with HVEM (high-voltage electron microscopy) (Dr. D. F. Parsons, private communication).

† For completeness, the possibility of phase contrast should be mentioned. See, e.g., the attempt at phase-contrast microscopy by Ando and Hosoya[105] using Mo $K\alpha_1$ radiation ($\lambda = 0.0709$ nm). Such attempts are ruled out at present in the wavelength region under consideration here by the lack of suitable crystals for making the interferometer required.

3. Image Formation

In the preceding section we examined the principal consequences of photon absorption and saw how they may be utilized to obtain contrast mechanisms for x-ray microscopy. The imaging systems of x-ray microscopy must be based upon these same consequences, together with such additional elementary wave properties such as interference and refraction or reflection at an interface. The interface properties are, however, restricted in practice with x rays to the case $n \approx 1$ (here $n =$ ratio of refractive indices across the interface), i.e., to very small angles of refraction, very small ordinary reflecting powers, and near-grazing angles for total reflection. The result is that refraction and reflection, which are at the heart of the imaging systems in use in light and electron microscopy, are of only limited utility in x-ray microscopy, and it has become necessary to devise rather unusual imaging systems for the x-ray case, relying principally upon the properties of interference, removal of photons, generation of reaction products, and damage. These imaging systems are the subject of the present section. They may be divided broadly into two kinds: in the former (Section 3.1) the ability of x rays to damage materials or generate reaction products is the only x-ray property used, and the burden of imaging is thereafter transferred to electron optics; while in the latter (Sections 3.2 and 3.3) an attempt is made to produce x-ray optics of a quality that can meet the imaging requirements directly.

3.1. Magnification by Electron Optics

3.1.1. Contact Microradiography: General

The simplest method of imaging, and the only one to date to achieve resolutions near the theoretical limits of Section 2.4, is shown in Figure 8. The specimen is placed in contact with a thin film of x-ray resist and is exposed to a uniform parallel beam of photons moving roughly normal to the resist surface. The photons that are not absorbed in the specimen enter the resist and damage it, the damage being greatest where the absorption in the specimen was least. A projected image of the specimen is thus recorded as a damage pattern in the resist. After removal of the specimen the resist is developed by dissolving away the damaged (or with some resists the undamaged) material in an appropriate solvent, converting the variations

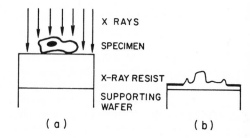

Figure 8. Contact microradiography. (a) Irradiation of the x-ray resist through the specimen. (b) The resist after development. The projected image of the specimen is recorded in the resist profile and can be read out, after light metallization, in an SEM.

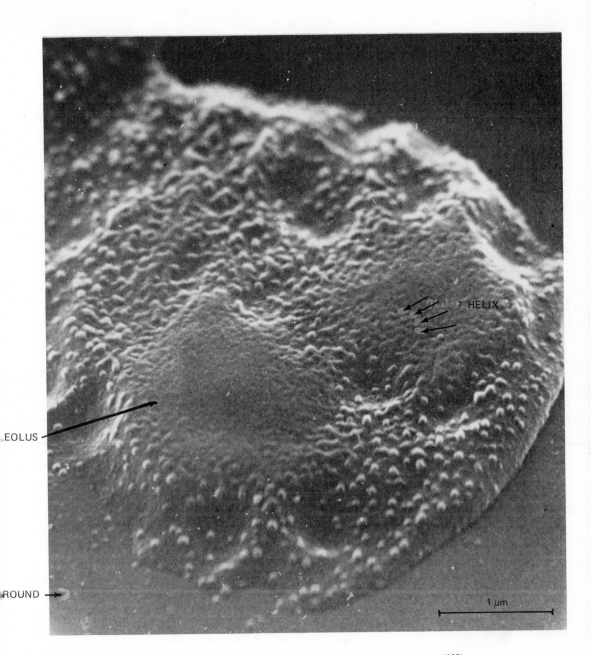

Figure 9. Contact microradiograph of human interphase nucleus from glioblastoma tissue culture[109] in PMMA, made with carbon $K\alpha$ radiation. The x-ray source was of the type shown in Figure 3.1 of Reference (28) and was operated at 5 kV and 40 mA, with a target to specimen distance of 18 cm and exposure time of 40 hr. [Microradiograph courtesy of L. Manuelidis (Yale University), J. Sedat (University of California at San Francisco), and R. Feder (IBM).]

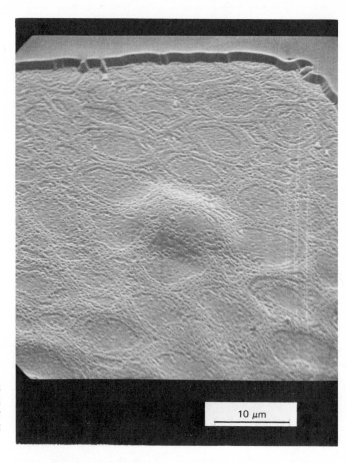

Figure 10. Contact microradiograph of a section of the third nerve of the pigeon in PBS, made with synchrotron radiation at SSRL (from Pianetta et al.[79]). The rise near the center is due to SEM damage of the resist.

in resist damage into variations in the resist profile. The image can now be read out by examining the surface of the resist in a high-resolution SEM (scanning electron microscope). The technique as a whole is CTXM [conventional (i.e., nonscanning) TXM]. Figures 9 and 10 give two examples of the technique. The method was first described by Feder et al.[22]; see also two subsequent papers[23, 24] from the same laboratory. Ladd et al.[25] and Asunmaa[26] had earlier studied a number of techniques, including some similar to the above, for producing EM-visible images with x rays, but none apparently were of sufficient promise to pursue further; their work should remind us, however, that x-ray resists are probably not the only systems useful for high-resolution contact microradiography.* Through the use of resists the technique is closely related to x-ray microlithography, from which it was developed, and the reader will find much useful information on resists and their performances in the article on that topic in the present volume (Chapter 7). The

---

* A novel detector, chalcogenide glass, was used in an experiment[106] at the VEPP-3 synchrotron radiation source in Novosibirsk. The exposure results in a change of the refractive index of the glass. As a result, readout must be optical, and this limits the resolution obtainable.

treatment given here will be correspondingly abbreviated. The technique is also obviously related to the older technique of contact radiography using x rays from conventional sources, silver halide film, and optical readout, but is of much higher resolution.

3.1.2. Contact Microradiography: Resolution and Sensitivity of Resists†

The resist that has been most extensively used to date in x-ray microscopy is high-molecular-weight polymethylmethacrylate (PMMA) with methyl isobutyl ketone (MIBK) as the developer. The resolution of this resist is approximately 5 nm when the resist is used at its best operating point, i.e., with photons of wavelength just below the carbon absorption edge at 4.36 nm and with a dosage to the PMMA of approximately 10^4 J/g. Under these conditions the resolution of the resist is seen to reach the diffraction limit imposed by the photon wavelength. At shorter wavelengths (higher energies) the resolution deteriorates, due to the increased diameter of the shower of damaging particles in the resist, becoming roughly 35 nm at $\lambda = 0.83$ nm. At longer wavelengths the resolution also deteriorates, because of increased diffraction effects.

The above data are of primary relevance in x-ray microlithography, where the specimens employed are masks, i.e., artificial high-contrast structures. In microscopy, where specimen contrast typically is low, another effect becomes important, namely the tendency for small features in the resist to be lost through sideways attack by the solvent (see Figure 11). The problem has been studied by Sayre and Feder,[27] who conclude that a reasonably satisfactory solution can be obtained by restricting the depth of development to, say, five times the minimum resolvable distance d desired.* The price that must be paid for this is that one must be

---

† Data on resists in this section are taken from the review by Spiller and Feder.[28]
* It should be noted that regions of resist with low dissolution rates are affected by nearby regions with high rates, and not vice versa. There may therefore be an advantage in comparing the images of a specimen in positive and negative resists. (The dissolution rate increases with exposure in a positive resist, and decreases with exposure in a negative resist.)

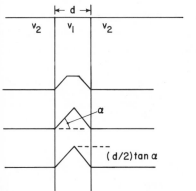

Figure 11. Profile of a feature in the resist after three different depths of development. The angle α is given by $\cos \alpha = v_1/v_2$, where v_1 and v_2 are the dissolution rates of the resist inside and outside the feature, respectively. The feature cannot be made higher than $(d/2)\tan \alpha$, regardless of the amount of development given (d = width of feature).

prepared to detect variations in the resist profile that are only a fraction of d in height, that exposure and development must be closely controlled on a point-by-point basis over the specimen, and that there is a loss in resist sensitivity at high resolutions due to the inability to make use of the photons absorbed at depths greater than $5d$ below the surface of the resist. Figure 12 gives approximate values of the resist exposures and development times required for various values of d. For PMMA, exposures are found experimentally to be in approximate agreement with these predictions, indicating that the sensitivity of PMMA is near optimal when measured by the number of usefully absorbed photons.[28]

Sayre and Feder also briefly consider techniques other than dissolution for the readout of the information stored in the damage pattern in the resist. They note in particular that a perfect method of readout would result if an electron-dense stain could be found that would preferentially stain damaged regions of the detector material. In this case readout without loss of resolution could extend indefinitely into the detector material, improving sensitivity (see Figure 12) and allowing the material to act as a three-dimensional detector of the radiation field. (This capability could be of importance in dealing with the problem of diffraction discussed in Section 3.1.3.) It should be noted that a self-staining detector of adequate resolution might be obtained through the use of sufficiently fine-grained silver halide emulsion.

Several departures from the optimal conditions become possible if a decrease in resist resolution can be accepted. One may remain with PMMA but shift to other wavelengths or to lower exposures; thus lower-resolution images may be formed in PMMA with doses as small as several hundred joules per gram. Or one may shift to other resists that are more sensitive than PMMA. A resist produced by copolymerization of methylmethacrylate and methacrylic acid [CoP(MMA-MAA)] can be used at doses of 50 J/g, and PBS [poly(butene-1 sulfone)] at doses of 14 J/g. On signal-to-noise considerations the resolution of these resists would be expected to decrease at least as rapidly as $D^{1/3}$ (D = dosage of the resist), starting from the 5 nm resolution of PMMA with 10^4 J/g dosage; i.e., a resist operating with 10 J/g dose at $\lambda = 4.3$ nm would be expected to have a resolution not better than 50 nm. In fact the limited

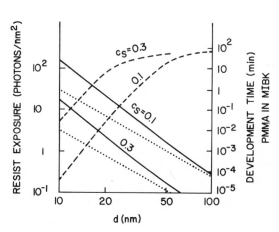

Figure 12. Resist exposures (solid curves) and development times for PMMA (dashed curves) for various values of the desired resolution d. Values are shown for two feature contrasts. The dotted lines show the resist exposures that would be required if readout could extend indefinitely into the detector material. (After Sayre and Feder.[27])

data available seem to indicate that the resolution decreases more rapidly than this; thus one of the very fast resists, DCPA [poly(dichloro propyl acrylate)], which operates at 10 J/g, has a resolution of the order of 500 nm. Thus the development of lower-resolution resists that retain the high absolute sensitivity of PMMA remains an important challenge to resist technology at the present time.

A property of the contact method that is worth noting is that specimen dosage must equal or exceed resist dosage (in the usual case where both resist and specimen are organic materials and thus have similar absorptivities), due to the shielding of the resist by the specimen.

The topic of resist sensitivity is of importance mainly because of its possible correlation with image quality: a low sensitivity implies a damaging, and possibly a long, exposure for the specimen, and both effects increase the likelihood of image degradation due to specimen movement during the exposure.

3.1.3. Contact Microradiography: Resolution of the Technique

It should be noted that the discussion above concerned the resolution of the resist only, and that the microradiographic technique as a whole may show a diminished resolution, due to the effects of finite specimen thickness. These include increased diffraction effects and effects due to finite source size.[2, 29]

Let a and b be the resist-to-feature and resist-to-source distances, S the source diameter, and λ the wavelength. Then the geometric resolving distance in the plane of the feature, r_g, is $(S/b)a$, found by requiring that the penumbras of two features be separated in the resist. The diffraction resolving distance, r_d, is approximately $(\lambda a)^{1/2}$. Here we require that the image of a second feature fall outside the first fringe of the Fresnel diffraction pattern. The former (penumbral) effect is not usually serious. With $a = 1$ μm, $b = 10$ cm, and $S = 0.3$ mm, r_g is only 3 nm.

The diffraction effect is important, however. With $a = 1$ μm and $\lambda = 4$ nm, r_d becomes approximately 63 nm. For r_d to be comparable with the resolution of the contrast mechanism, a must be reduced to 25 nm. Thus the contact technique gives good definition to features in a thin layer near the bottom of the specimen, but blurs features lying higher in the specimen. (Despite the blurring, the definition remains somewhat higher than in the ordinary light microscope.)

This may not be the full story, however. The contact microradiograph is the superposition of the signals diffracted by and transmitted through the specimen, and these may combine coherently if the micrograph is made with suitably coherent x-rays. If so, some form of image reconstruction, including at least some reduction of blurring, should be possible by appropriate optical processing, i.e., contact microradiography would become a special case of Gabor holography, with certain advantages and disadvantages arising from the simplicity of the hologram-generating method used. (In particular, the reference signal, here the transmitted signal, would be nonconstant, due to absorption in the specimen.) To the best of our knowledge, no work on contact microradiography from this point of view has as yet been done. For a brief discussion of other forms of holographic x-ray microscopy, see Section 3.2.4.

3.1.4. Contact Microradiography: Details of the Technique

For those who would like to do contact microradiography with PMMA, the following is a step-by-step description of the process.* The starting material can be obtained from Dupont (Elvacite #2041) or from Esschem, P.O. Box 56, Essington, Pennsylvania. The material as received from the manufacturer is in the form of a powder. The basic working solution is prepared by dissolving the PMMA in chlorobenzene in the proportion 5 g PMMA to 95 ml chlorobenzene, with warming to 30°C and stirring. The solution is stored in clean dropper-type bottles.

When it is desired to prepare some resist surfaces, take a small amount of the stored solution and dilute further by adding an equal volume of chlorobenzene. Place a suitable glass or silicon support wafer (typical diameter ~1 inch) on a spinner, bring to approximately 6000 rpm, drop several drops of the diluted solution on the wafer, and spin for 30–60 seconds. This should produce a resist layer approximately 500 nm thick. The thickness can be controlled by adjusting the amount of chlorobenzene used in preparing the diluted solution, or by using a different spinner speed. The resist surfaces should be baked at 160°C for 1 hour in air. They may be kept stored in clean containers for some months.

The specimen may be placed directly on the resist surface, or it may be placed on a carbon- or formvar-coated EM specimen grid and then (with the grid inverted) brought into contact with the resist. The advantage of the grid procedure is that after the x-ray exposure the specimen can usually be lifted off intact from the resist and is still available for a further x-ray exposure or for examination in the EM. (For the handling of wet specimens, see Section 6.1.)

After exposure of the resist, wash or lift off the specimen, and if there is any doubt about the cleanness of the resist surface, clean it ultrasonically in water. A typical development procedure is as follows. Spray the resist with isopropyl alcohol and blow dry with clean nitrogen or air after 30 seconds. Examine the resist surface under the optical microscope. Proceed with periods of immersion of say 15 seconds in 1:10 or 1:5 mixtures of MIBK in isopropyl alcohol, arresting development with pure isopropyl alcohol, blowing surface dry, and examining under the microscope, until sufficient development has occurred.

The resist surface may be prepared for examination in the SEM by evaporating onto it a thin coating of 40% palladium–60% gold. Examination of the resist by TEM is discussed in Section 6.2.

3.1.5. Contact Microradiography with Direct Photon–Electron Conversion

It is possible to replace the resist by a thin layer of high-Z material that acts as a converter of photons into secondary electrons, and to image the latter in a suitable electron optical system. Such a device was constructed by Mollenstedt and Huang,[30] and has recently been reproposed by La Placa.[31] Potential advantages over the resist-based technique include (1) possibly higher resolution (due to the

* Our thanks are due to Dr. R. Feder, IBM Research Center, Yorktown Heights, New York 10598, for supplying this description.

smaller dimension of the shower of secondary particles in a high-Z material), (2) greater convenience of use, and (3) real-time imaging (ability to follow movements or changes in the specimen). Problems with the technique include (1) loss of electrons in the conversion layer, (2) loss of electrons or resolution in the imaging system, due to emergence of the electrons from the conversion layer with a significant spread in energy and direction, and (3), with wet specimens, need for the conversion layer to support the pressure difference between air and water on the specimen side and vacuum on the imaging side.

3.1.6. Image Formation Using Electrons Emitted from the Specimen

The most direct way to obtain electrons for imaging in an electron optical system would be to remove the conversion layer from the instrument just described and allow the optical system to focus the electrons issuing directly from the specimen. The microscopy would be CEXM (conventional EXM), and the possibility of switching to CEXM would be an additional potential advantage of the instrument described.

An interesting feature of CEXM is that the diffraction limit is set by the wavelength of the emitted electron, and this wavelength can be of the order of 0.1 nm, even when the incident photon is a soft x-ray photon. The result is that imaging methods for CEXM are potentially capable of higher resolutions than those for any other x-ray microscopy mode. This fact could be of considerable interest when taken in combination with the high resolution of the EXM contrast mechanism (Figure 6). (Note however that CEXM is limited to surface-layer microscopy, and requires that the specimen be in a vacuum.)

To the best of our knowledge, the closest approaches to instrumenting CEXM to date are those[32, 33] using UV mentioned in a previous footnote.

3.2. Magnification by X-Ray Optics*

3.2.1. Grazing-Incidence Mirrors

As previously noted, the fact that $n \approx 1$ for soft x rays has made it necessary to develop alternatives to lenses and mirrors in the conventional sense for the formation of magnified images. Among the first techniques to be developed was the use of *grazing-incidence mirrors* based on total external reflection. The work of Kirkpatrick and Baez[34] promised a potential resolution around 7 nm, and demonstrated that magnified x-ray images can be obtained. Wolter[11] called special attention to the advantages of ultrasoft x rays, and recommended improved mirror figures to reduce aberrations. Nevertheless, this approach did not gain popularity in microscopy due to the technical problems in mirror formation and alignment as well as the limited specimen area over which the aberrations could be controlled. Resolution better than 250 nm has not been demonstrated, and the review by Cosslett and

---

* See also the recent review of x-ray optics by Franks.[36]

Nixon[2] remains current. It should be noted, however, that the largest grazing angle for total reflection depends upon $[\text{Re}(1 - n)]^{1/2}$ (where Re denotes the real part of a complex quantity), and that except near absorption edges $\text{Re}(1 - n)$ varies as λ^2 in the wavelength region under consideration. It is therefore considerably easier to build mirrors for the long-wavelength region of interest to us, where $\text{Re}(1 - n)$ is of the order of 10^{-3}, than at the short wavelengths used by Kirkpatrick and Baez. Grazing-incidence mirrors do in fact find extensive use in soft x-ray astronomy and in the instrumentation of synchrotron radiation beam lines.[35-37] Mirrors are likely also to become of importance in scanning microscopy, and we shall return to them in Section 3.3.2. Mirrors are also discussed in Chapters 3, 14, and 16.

3.2.2. Point Projection Microscopy

The next significant advance came with the development of *point projection microscopy* by Cosslett and Nixon[38] after an earlier suggestion by von Ardenne.[39] This is a technique of great simplicity and considerable popularity. A point source of x rays is formed by focusing the electron beam on the exit window of the x-ray tube. The specimen is placed near this source in the path of the diverging x rays, and the transmitted radiation (the shadowgraph) is photographed.* With hard x rays the resolution is limited by the size of the "point source" (typically 0.1–1 μm), and also by the overlapping of images in the thick specimens required. With soft x rays diffraction effects become important. They may be minimized by keeping the specimen close to the source, but in this geometry only a small specimen area is illuminated. Because magnification in projection microscopy makes direct use of the divergence of the x-ray beam, synchrotron radiation with its highly collimated characteristics appears not to have much to offer for this approach. The state of the art is described in several fine reviews.[3-5]

3.2.3. Microscopy with Zone Plate Objectives

While these techniques were being developed, Myers[40] proposed that *Fresnel zone plates* should be useful for image formation with soft x rays in the 1-nm range. The first zone plate for the ultraviolet and soft x-ray range was designed by Baez[41] and fabricated by Buckbee Mears Co., St. Paul, Minnesota. It consists of 19 gold rings held together by struts. Baez demonstrated image formation with visible light and ultraviolet (253.7 nm) radiation, while Pfeifer et al.[42] were the first to use a similar plate with synchrotron radiation at 52.5 nm. The Tübingen group succeeded in making reduced scale copies of the Buckbee–Mears zone plates and using these for imaging at 0.83 nm and 4.48 nm.[43] Zone plates with higher resolution and more zones were made available by Johannes Heidenhain Co., Traunreut, West

---

* A primitive projection microscope was built by Sievert[107] in 1936. He used a pinhole illuminated by an ordinary x-ray tube as the point source, an arrangement severely limited in speed and field of view.

Germany around 1970, and were used[44] to form a micrograph of dried onion membrane at 4.48 nm. Nevertheless, much of the effort in zone plate development around that time was aimed at x-ray astronomy,[44–47] rather than microscopy, at least partly due to the lack of suitable soft x-ray sources.

In a first approximation, zone plates act as highly chromatic lenses, with focal length $f = r_1^2/\lambda$, where r_1 is the radius of the innermost ring and the radii of the other rings are given by $r_n^2 = nr_1^2$. For a plate with N zones the width of the outermost ring $\Delta r = r_N - r_{N-1} \approx r_N/2N$ is the approximate resolution limit, and the f/number is $\Delta r/\lambda$. The depth of field when the zone plate is used as a microscope objective is approximately (f/number) Δr; this is also the depth of focus when the zone plate is used to form a spot for scanning microscopy. Typically λ is chosen to provide good contrast, Δr is the finest linewidth that can be formed, and the third free parameter (f, N, or r_N) is chosen to suit the geometry. To indicate the size of these variables we show in Figure 13 their relationship for $\lambda = 3$ nm. Clearly, these optical elements are small and delicate. To obtain diffraction-limited resolution the x rays must be monochromatic ($\lambda/\Delta\lambda > N$). With alternating opaque and transparent zones, the efficiency with which the plate images photons is ideally $1/\pi^2$, or about 10%. The rest is either absorbed (50%) or is transmitted (40%) in a way that degrades image contrast unless precautions are taken.[46, 48, 49] Phase zone plates[50] or zone lenses[51] should have higher efficiencies and better contrast, but have not been made for x-ray use to date.

The simplest way to eliminate the transmitted background (mainly 0th order) is to use a zone plate with the first several zones opaque. The image can then be

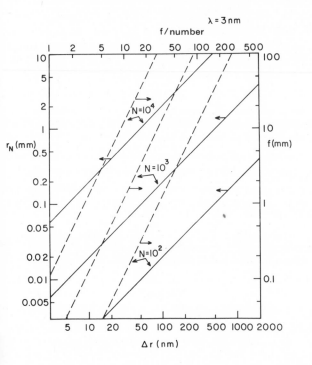

Figure 13. Zone plate characteristics for $\lambda = 3$ nm. Solid lines (left scale) give the radius of the zone plate with N zones as a function of the width of the outermost zone, Δr. Dashed lines (right scale) give the focal length. The top scale relates the f/number to Δr.

formed entirely in the shadow of this opaque disk. Such apodized zone plates (zone rings) must then have large zone numbers (N), and these can be formed most easily by holographic techniques.[46, 52] For large zone numbers the approximate $n^{1/2}$ relationship for the radii is not adequate, and the correct expression[50, 52] depends upon object and image distance as well as the wavelength.

The Microscope of Niemann, Rudolph, and Schmahl.   An arrangement for producing the appropriate zone plate pattern in photoresists is described by Schmahl and Rudolph.[52, 53] Their group also constructed the first zone plate microscope to operate with synchrotron radiation.[54, 55] The diagram of the optical system is shown in Figure 14. The combination of grating and condenser zone plate forms a reduced-scale monochromatic image of the radiation source in the object plane. The micro-zone plate then forms an enlarged image of the object on a photographic plate. All elements of the system are maintained within a vacuum chamber. (Access is accordingly somewhat cumbersome.) The system as a whole is CTXM. Micrographs of cells, fibers, and other biological material have been obtained (see Figure 15) with resolutions as fine as 0.2 μm with exposures of several minutes at $\lambda = 4.6$ nm using the DESY synchrotron. The authors plan to reduce exposure times by using more efficient zone plates and the more intense beam at the DORIS storage ring. Improvement of the resolution obtainable with holographic zone plates appears to be a harder task, since resolution reflects directly the finest interference fringes recorded in the pattern, which are already close to the resolution limit. Schmahl et al.[55] suggest that UV lasers may help record finer patterns. They also demonstrate that holographic zone plates form higher-order images that in principle may have higher resolution[55, 56] but at considerable cost in magnification and efficiency.

To make substantial improvements in resolution, a new approach to zone plate fabrication is necessary. One possibility is to vacuum deposit alternate layers of an "opaque" and a "transparent" material onto a rotating cylinder, in carefully controlled amounts, to build up zone-by-zone a roll, which after slicing on a microtome produces zone plates. Very narrow zones can be achieved this way in principle, but the technical problems are clearly challenging. Another possibility[57] would be to employ electron-beam microfabrication techniques, with which linewidths of less than 10 nm have been achieved.[58]

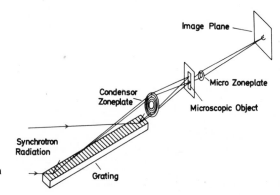

Figure 14. Schematic diagram of the microscope of Niemann et al.[54]

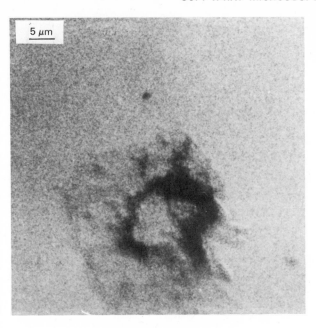

Figure 15. Micrograph of 3T3 mouse cell, made by Schmahl *et al.*[55] with the zone plate microscope. X-ray magnification 15 × (reproduced at 20% reduction), $\lambda = 4.6$ nm.

3.2.4. Holographic Microscopy

Soon after the appearance of Gabor's first papers on holography, Baez[59] examined the possibilities for holographic x-ray microscopy. Gabor had shown[60] that a hologram may be recorded using short-wavelength probes (he used electrons) and reconstructed using visible light, yielding a magnification equal to the ratio of the wavelengths. Baez concluded that, in the x-ray case, source size and film grain limit the resolution attainable. These problems were removed, at least in principle, with the development of Fourier-transform holography and source-effect compensation by Stroke.[61] But further work[62] demonstrated that the difficulties in recording fringes were largely due to the lack of opaqueness at the edges of the objects being imaged and to the lack of monochromaticity in the x-ray beam. In addition, progress in x-ray holography has been slowed by the lack of a laser or other highly coherent x-ray source. However, interest in the technique has persisted,[63] and recognizing that sharp boundaries are more easily realized using soft x rays, Aoki and Kikuta have succeeded in recording x-ray holograms both with synchrotron radiation[64] and with an Al $K\alpha$ source,[65] and in reconstructing a magnified image using a laser. Their best resolution (4 μm) is still rather far from the ultimate prospects. An attempt at higher resolution holography by Reuter and Mahr[66] was only marginally successful. Roughness and graininess of the photographic plates used is cited as the most serious problem by these authors. X-ray resists should overcome this problem, and the use of improved synchrotron radiation sources offering more highly monochromatic soft x-ray beams should also lead to substantial progress in this approach.[67]

Holography is of importance in x-ray microscopy as a technique that is potentially capable of utilizing a large portion of the diffraction pattern of a specimen in the imaging process, and thereby of gaining high resolution. The possibility that this might take the form of a deblurring technique for contact microradiographs has been mentioned earlier.

3.3 Scanning Microscopy

The first design for an x-ray scanning microscope was put forward by Pattee.[68] It places the specimen in contact with the window of an x-ray tube. The electron beam is scanned over the window, in television raster fashion, generating x rays. The transmitted flux is detected electronically, and this signal is used to modulate the brightness of an oscilloscope trace, scanned in synchronism with the beam in the x-ray tube. The microscopy as a whole is STXM (scanning TXM). The device did not live up to expectations in the practical sense,[2] essentially because of the limitations of the x-ray source. The availability of synchrotron radiation brightens the prospects considerably for the future.

3.3.1. The Microscope of Horowitz and Howell

The first—and so far the only—scanning x-ray microscope using synchrotron radiation was built by Horowitz and Howell[13] for operation at the Cambridge Electron Accelerator. They used an ellipsoidal mirror[35] to focus the beam to a 1×2 mm² spot, with a 3.5-keV high-energy cutoff. A 10-μm-thick Be window terminated the vacuum system; the microscope itself was in the open environment (Figures 16 and 17). The beam was collimated with a 2-μm pinhole in a 100-μm-thick gold foil. This spot size sets the resolution of the instrument. The specimen was scanned in raster fashion across this spot. Images formed by fluore-

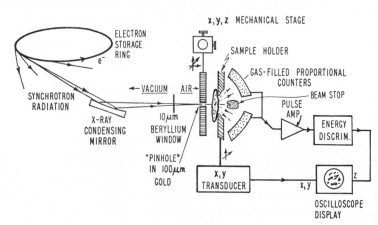

Figure 16. Schematic diagram of the scanning microscope of Horowitz and Howell.

Figure 17. Photograph of the scanning microscope of Horowitz and Howell at the Cambridge Electron Accelerator. Synchrotron radiation enters from the right. The proportional counters to detect fluorescence (cylinders with tubing attached) surround the specimen, which is mounted in the longer cylinder supported on the left. Scanning motion is generated using the electromagnets visible to the left of the micrometers.

scent x rays were obtained, revealing the elemental composition of the specimen, an aspect of the work reviewed elsewhere in this volume (Chapter 14) and also briefly in Section 5 below. This microscopy was scanning FXM (SFXM). In addition, during the last months of the storage ring's existence, they examined specimens, including biological ones, in STXM. For these experiments[69, 70] a smaller (1-μm) pinhole was used, and a miniature proportional counter was placed in front of the "beam stop." Exposure times were around a minute. By shifting the orientation of the specimen, pairs of stereo photographs were also obtained with the apparatus. Examples of images obtained with their instrument are shown in Figure 18.

The authors point out that up to a point ($\sim 0.2\ \mu$m) smaller pinholes improve the resolution almost in direct proportion to size, although counting rates drop quadratically. Pinholes smaller than the limit indicated only hurt, because diffraction spreads the beam before it reaches the specimen.

With the peak of their x-ray spectrum in the vicinity of 0.5 nm, the Horowitz–Howell microscope was suitable for the examination of relatively thick, opaque specimens. The penetrating power of their beam also required that the pinhole be like a long narrow tunnel, which is hard to fabricate and needs careful alignment.

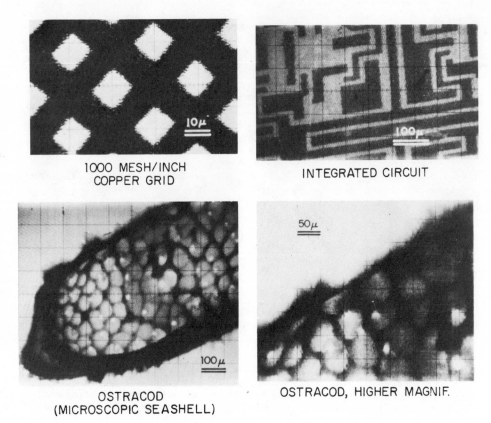

1000 MESH/INCH
COPPER GRID

INTEGRATED CIRCUIT

OSTRACOD
(MICROSCOPIC SEASHELL)

OSTRACOD, HIGHER MAGNIF.

Figure 18. Micrographs obtained by Horowitz and Howell[70] using the scanning transmission setup at the CEA synchrotron radiation source.

3.3.2. Future Prospects for Scanning Microscopy

Scanning is among the highly promising techniques for x-ray microscopy of wet biological specimens. As we shall see, the basic arrangement of the Horowitz–Howell microscope appears to be readily adaptable to the use of x rays in the optimal wet-specimen wavelength range of 2.33–4.36 nm.

Scanning is particularly advantageous for minimizing radiation exposure.[28, 55, 71] The x rays transmitted by the specimen can be detected electronically (e.g., flow proportional counters) with close to 100% efficiency. In the other techniques discussed, significant losses are encountered in either the magnifying optics or detector or both.

The specimen may remain in the open environment if the path between vacuum window and proportional counter is short enough. For $\lambda = 3.10$–4.36 nm, an air path 2.5 mm long will transmit over 30% of the x rays. Below the nitrogen edge at 3.10 nm the corresponding air path is shorter, but is still over 0.5 mm. In a mixture of 20% oxygen and 80% helium the corresponding path is over 1 cm for

$\lambda = 2.33$–3.6 nm and over 5 mm for $\lambda = 3.6$–4.5 nm. Metal foils in the 100–200 nm-thickness range are strong enough to stand 1 atm over areas a few micrometers across, and may therefore serve as vacuum windows in the scanning mode with 40–90% transmission.

Pinholes are easier to make for the less penetrating ultrasoft radiation, and their alignment becomes simple. Beams with adequate flux in the desired wavelength range are now available, or will come on line within the next few years at several synchrotron radiation laboratories. We see no serious obstacles to the construction of a scanning pinhole microscope operating with a resolution of 0.5–1 μm using a broad band beam in the 2.5–4 nm range.

For the microscope to have significant impact, however, it should have considerably higher resolution. To achieve this, three approaches have been suggested for demagnifying the beam spot emerging from the pinhole or from some other small source:

1. Reflection optics with grazing-incidence mirrors, proposed by Horowitz and Howell,[13] appears quite suitable here, since the aberrations of mirror systems are unimportant for the small field of view required. A broad wavelength band can be accepted with good efficiency, and this should give adequate flux in the small image spot. The technical problems of fabrication, alignment, and stability are as yet unsolved at the level of precision necessary. There are recent developments,[72, 73] however, that may lead to suitable toroidal mirrors within the next few years.

2. Reflection optics with multilayer normal-incidence mirrors, proposed by Spiller,[74] will work best at longer wavelengths. These mirrors focus monochromatic radiation, and are under active development. In recent work Haelbich *et al.*[75] report achieving theoretical performance in the 15–20 nm wavelength region and predict the fabrication of useful mirrors for wavelengths as short as 5 nm.

3. Zone plate optics[28, 55, 71] requires a monochromatic source. This may be the exit port of a conventional monochromator or the image of the synchrotron radiation source formed by a grating–zone plate combination, such as the one in the object plane of the microscope of Niemann *et al.* (Figure 14). The single-zone-plate focusing monochromator designed by Spiller[76] could also form an adequate source.

By one or more of these methods a resolution better than 100 nm should be achievable within the next few years.

3.4. Summary

To preserve the resolution of which the soft x-ray contrast mechanisms are capable, the imaging methods in soft x-ray microscopy must work close to the wavelength limit. At $\lambda = 4$ nm, contact microradiography with high-resolution resists reaches this level of performance for a thin (~ 25 nm) layer along the bottom of the specimen, but features lying elsewhere in the specimen are blurred. (Resolution remains better than optical, however, for any but the thickest specimens.) At the present time, imaging methods based on x-ray optics operate very far from the wavelength limit. The situation is accordingly the reverse of that with electrons, where, in the case of biological materials in the natural state, resolutions are limited not by the imaging techniques but by the contrast mechanisms.

Higher-resolution imaging methods will require the development either of wide-angle x-ray optics or possibly wide-angle x-ray holography. The ability currently being acquired in microlithography to fabricate structures and record patterns at resolutions comparable to soft x-ray wavelengths may provide the primary technologies required. Thus the fabrication of zone plates with f/number ≈ 1 in the soft x-ray region is not out of the question today. Numerous additional technical difficulties will undoubtedly have to be solved, however, before success is achieved by either method.

At a more immediate level, contact microradiography is reasonably simple to carry out, and can be added without great difficulty to the activities of any laboratory equipped with a high-resolution electron microscope. [A suitable low-cost x-ray source for this purpose appears to have been found in the high-intensity pulsed plasma source (Section 4.2.3).] The usefulness of contact microradiography could be increased by improving the information readout of high-resolution resists and the quantum efficiency of low-resolution resists.

Some potential advantages of the contact technique with direct photon-to-electron conversion have been noted in Section 3.1.5. A related technique, CEXM, is a special case in terms of imaging, with potentialities for high-resolution imaging not found in the other soft x-ray microscopy modes.

Of the x-ray optical methods, scanning probably places the lowest demand on the quality of the optics. In addition, scanning is of interest because of its versatility; STXM, SFXM, and SEXM should all be realizable on essentially one instrument, with at least the first two of these modes being of interest also in high-resolution microanalysis (Section 5). Scanning also offers the most natural basis for minimizing specimen exposure. Finally, it may simplify work with wet specimens (Section 6.1) and should also offer advantages in signal processing and the formation of three-dimensional images. For a discussion of three-dimensional imaging, see Section 6.3.

4. Sources of Soft X Rays for Microscopy

4.1. Requirements

No one source is ideal for all forms of microscopy. We list here some general criteria for comparison.

Intensity is the first consideration. Long exposures are not only inefficient and time consuming, but also introduce problems in long-term stability both of the specimen and of the microscope. For wet biological specimens flash microscopy would appear to be ideal to avoid movement in the specimen. For dry or frozen specimens a time scale of a minute or less would appear reasonable. Image formation and contrast considerations described in Section 2.4 assume lossless optics and detection, and indicate that 10^{-1}–10^3 photons/nm² are required on a typical biological specimen in the wavelength band chosen to obtain an image with resolution in the 100–10 nm range (Figure 5).

Tunability and control over spectral bandwidth are useful in several respects. It is the possibility of choosing the wavelength of best contrast for a given specimen that allows image formation with minimal radiation damage. In absorption microanalysis, exposures on both sides of an absorption edge must be compared. The basic bandwidth requirement for microscopy is quite modest: $\Delta\lambda/\lambda \approx 0.05$ is typically adequate unless the image forming elements have chromatic aberrations. An example of the latter case is zone plate optics, which requires $\Delta\lambda/\lambda < 1/N$, where N is the number of zones, for diffraction-limited performance.

Source size or beam divergence limits the resolution attainable in a variety of systems. Small sources with small divergence are generally required, though by viewing a source from a distance, or by collimating, one can trade intensity for better beam geometry. X-ray optics can also reshape the beam and trade between size and divergence, but the product of these parameters (volume in " phase space ") can not be reduced without loss of intensity. A discussion of these optical possibilities may be found in a paper by Hastings[77] and in Chapter 3 of this volume.

Stability and uniformity of the beam are desirable to assure uniform illumination of the specimen. In the presence of significant instabilities (temporal or spatial) the beam must be monitored during exposure to make the necessary adjustment to exposure time or location. If the entire specimen is exposed at one time, the beam must either be uniform over the full field, or the microscope should be moved around during exposure in such a way as to even out the nonuniformities. This last alternative appears unattractive except in the case of contact microradiography, where it is easily accomplished.

4.2. Soft X-Ray Sources

4.2.1. Synchrotron Radiation

Because of its intensity, tunability, and small source size and divergence, synchrotron radiation comes close to being the ideal universal source. Aside from flash and projection microscopy, this source is excellent for all the techniques we have considered. Evidently some form of monochromatization or at least filtration is always necessary. The main reason for considering other sources is that we can not all have our own synchrotron radiation source.

Vacuum requirements around storage ring sources require special attention. For electron beams to have lifetimes of many hours it is necessary for the pressure in an electron storage ring to be in the 10^{-9}-torr range or lower. Maintaining this high vacuum in the presence of intense synchrotron radiation bombardment of the chamber walls is feasible only if the walls are clean. In particular, contamination of these surfaces by long-chain hydrocarbon molecules results in large outgassing rates. Also, the cracking of hydrocarbon molecules on beam-line and monochromator optical elements causes a blackening of these surfaces and a degradation in optical performance. It is therefore advisable to introduce biological material and organic resist detectors into specimen enclosures separated from the source by thin windows,[78, 79] or by small apertures with removal of contamination by differential pumping.

The small vertical size of the beam can lead to nonuniformities over large specimens. If no monochromator is used, the wavelength composition also changes with height because of the larger vertical divergence of the longer wavelengths. Beam lines are being designed[80] to eliminate these problems, which also affect x-ray lithography.

4.2.2. Conventional X-Ray Generators

X rays have traditionally been generated by accelerating electrons and slamming them into a solid target. The efficiency of this method is quite low—typically $< 10^{-3}$ of the electron energy is converted to x rays. It is even lower for soft x rays, which tend to be absorbed in the target. An attraction of these sources is that the strong characteristic emission lines can be isolated by simple filters, so that much of the x-ray flux is concentrated into a narrow spectral band. Some of the lines are still too broad, however, for zone plate microscopy.[81] Cosslett and Nixon[2] review the general design criteria and limitations, while sources developed specifically for ultrasoft x-ray generation are discussed by Henke and Tester[82] and Engstrom,[4] See also Spiller and Feder[28] for a discussion of several new sources for Al K (0.83 nm), Cu L (1.33 nm), and C K (4.48 nm) radiation, which have been developed for use in x-ray microlithography. These latter sources, and especially the 4.48-nm source, have proved to be suitable for contact microradiography,[23, 24] where they offer the great advantage of convenience of access. Among their disadvantages are the long exposure times (typically ~ 16 hours) and the fact that the wavelength of 4.48 nm lies outside the most useful range for work with wet biological specimens.

For the more powerful of the conventional devices the source size is typically 1 mm or larger due to problems of space charge (especially at low-voltage operation) and heat load. The x rays are emitted in all directions, and only a small fraction can be collected onto the specimen.

4.2.3. Plasma Sources

Hot plasmas, generated by high-power lasers or electric discharges, produce soft x rays copiously.[83, 84] These sources, although as yet practically untapped, may play a unique role in biological microscopy. The x-ray emission comes in very intense pulses of short (order of nanoseconds) duration from a small (~ 0.1–1 mm) region. A single pulse can provide enough flux to form the image, making "flash" micrography of wet, even live, specimens possible. The time period is probably sufficiently short that neither specimen motion nor radiation damage can blur the image. The effect of such a short pulse on specimens has yet to be determined.* The cost of these sources is moderate (roughly comparable with that of conventional sources).

One such instrument, built by McCorkle and Vollmer,[85] bombards a sliding spark-produced plasma with an electron beam (Figure 19). Their source produces collimated soft x rays with strong lines from highly ionized atomic species,[86] the choice of which is up to the user. The energy levels in these stripped atoms are shifted due to the reduced shielding of the nuclear charge. As a result, the emission lines appear at wavelengths shorter than in conventional sources, and are often

---

* Note that an exposure of 10^2 photons/nm^2 (Figures 5 and 12) corresponds in the 2.33–4.36 nm wavelength region to a dose (Figure 4) of approximately 10^4 J/g (or 2400 cal/g), i.e., to a potential temperature rise in the case of flash microscopy of order 2400°.

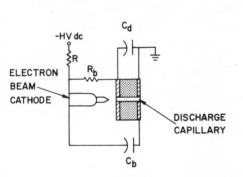

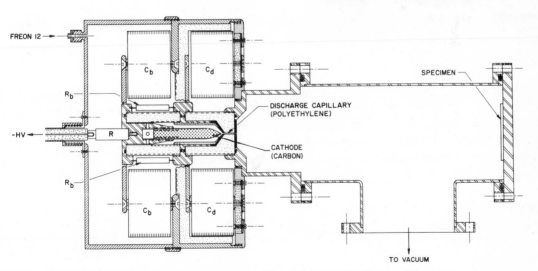

Figure 19. The McCorkle–Vollmer pulsed plasma source. Upper left, schematic diagram of the source. Below, diagram showing some of the construction details. Upper right, photograph of the source. (Courtesy of R. A. McCorkle.)

0.5 μm

Figure 20. Contact microradiograph of 0.5-μm latex spheres using a pulsed plasma source. The resist was PMMA, exposure was from a single flash from the source shown in Figure 19, and the distance from the discharge capillary to the specimen was 22 cm. Development required approximately 3 minutes in a 1:1 mixture of MIBK and isopropyl alcohol. (After McCorkle et al.[110])

shifted past the corresponding absorption edges. This enhances the usefulness of the source, as for instance the carbon K emission is now in the most valuable wavelength range, where it is strongly absorbed by organic material.

A plasma source of somewhat different design is available commercially from Chelsea Instruments Ltd., Epirus Road, London SW6 7UR.

A contact microradiograph taken with the carbon K emission from a McCorkle source is shown in Figure 20.

5. Mapping the Concentration of a Particular Atomic Species

An important possibility in soft x-ray microscopy is that of mapping, at high resolution, the concentration of a given atomic species in a wet intact specimen. This ability acquires considerable significance because the biologist today possesses a number of techniques[87, 88] for carrying a given label to a given structural element in the cell, wherever that element may be found; mapping the concentration of the label then becomes equivalent to mapping the concentration of the structural element.

Two different techniques exist, one based on TXM and the other on FXM.* These are called *absorption microanalysis* and *fluorescence microanalysis*, and are treated in Sections 5.1 and 5.2, respectively. In both sections specimen dosages required for adequate signal-to-noise ratios are calculated by methods somewhat similar to those described in Section 2.4, for intact, wet, biological specimens and for various mapping conditions. Details of some early calculations of this sort are given by Kirz *et al.*[89]

Two electron-based techniques also exist: electron energy-loss analysis, which is roughly analogous to x-ray absorption microanalysis; and electron microprobe analysis, which is analogous to x-ray fluorescence microanalysis. Electron microprobe analysis is widely used today for high-resolution microanalysis of dry specimens. In both sections dosages for the electron technique are also calculated to permit a comparison between the x-ray and electron techniques in the intact, wet specimen case.

5.1. Absorption Microanalysis

In our discussion of the morphological study of the specimen, the source of contrast among the various components of the object under scrutiny was the chemical composition. The contrast is a function of the wavelength, and undergoes dramatic changes at the absorption edges of the constituent elements. Comparison of images formed at two wavelengths straddling an absorption edge then provides information on the spatial distribution of the element in question.

The method was developed as a quantitative tool by Engström,[90] Lindström,[91] Henke,[12] and their co-workers, and has been reviewed in detail.[2, 4] With the development of electron probe and x-ray fluorescence microanalysis, the absorption technique lost some of its appeal, especially due to the lack of intense readily tunable x-ray sources. The availability of synchrotron radiation removes this obstacle, and a revival of the technique may be on the way.

Every method of image formation in TXM that we have described can be used for absorption microanalysis. The contact technique, using photographic plates as detectors, was used with synchrotron radiation by Polack *et al.*[92] to locate iron and titanium in mineralogical specimens. The spatial resolution of their photographs is a few μm, and they suggest using x-ray resists for future experiments. McGowan *et al.*[93] studied chick embryo heart cells using PMMA as the recording medium. They find striking differences (Figure 21) between the image obtained using broad band synchrotron radiation and that using a carbon K source.

---

* A method based on EXM could also exist, which would be suitable for surface-layer microanalysis. However, we do not pursue this case further here.

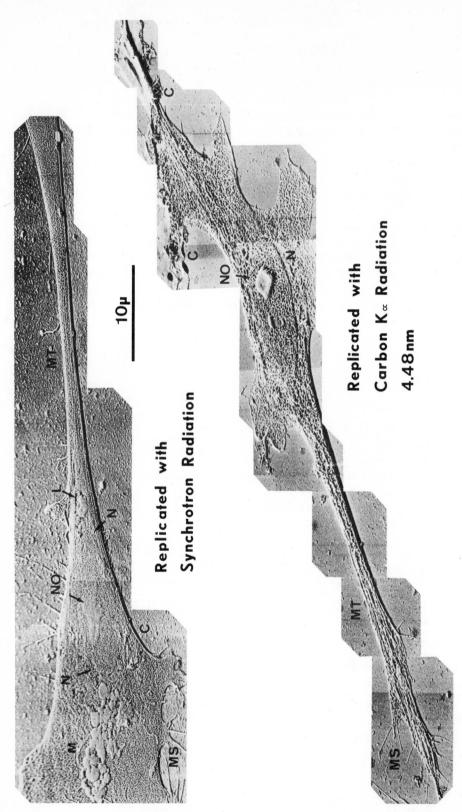

Figure 21. Absorption microanalysis. Contact microradiographs of chick embryo heart cells made (upper left) with broad band synchrotron radiation and (lower right) with carbon Kα radiation. The nucleus (N) and nucleolus (NO) are more prominent in the lower picture, reflecting the presence of oxygen and phosphorus in the nucleic acid. Mitochondria (M) are more readily seen in the upper picture because of their magnesium content. Other structures which are visible in both pictures are microspikes (MS) and microtubules (MT). (From McGowan *et al.*[93])

Synchrotron radiation is the most conveniently tunable source of monochromatic soft x rays today. But with the narrow bandwidth passed by the available monochromators, the intensity on the specimen is too low for use with PMMA, and faster resists must be used. Pianetta *et al.*[79] used polybutyl sulfone (PBS) to image biological materials with monochromatic synchrotron radiation as a feasibility study for higher-resolution micronalysis. Images with a resolution of around 300 nm were obtained in 30-min exposures (see Figure 10).

With the contact method of microanalysis it is necessary to remount the specimen for two or more exposures. The exposures should be as close to identical as possible in all respects (except wavelength). A specimen holder capable of remounting specimens inside a vacuum chamber was built by McGowan.[94] He mounts the specimen on a movable arm that brings it into contact with the resist or lifts it off. The resist is mounted from another port, whose motion allows bringing unexposed resist behind the specimen while it is lifted off.

The quantitative comparison of the micrographs requires that the picture elements be represented in numerical form. For film or resist this requires microdensitometry or some similar processing, while for electronic detection (as in the case of scanning microscopy) the raw information is directly comparable. Methods for the electronic processing and comparison of pictures have been developed for satellite photographs, and these methods should be directly applicable to the analysis of x-ray images.

The exposure necessary in a given situation can be determined using the approach described in Section 2.4, with the comparison in this case being between transmitted counts from the same resolution element at the two sides of an absorption edge. The results indicate that for a constituent to be potentially detectable with 10-nm spatial resolution, at a dose of 10^4 J/g, it must be present with a concentration large enough to absorb at least a few percent of the radiation (as measured on the absorptive side of the edge). For elements with a major absorption edge in the 1–10-nm range (K edges for $4 < Z < 11$, L edges for $15 < Z < 30$, M edges for $38 < Z < 60$) this corresponds to detectable masses in the 10^{-17}–10^{-18} g range, or a concentration of 1–10 % by weight in a 1-μm-thick specimen. For edges in the 0.1–1-nm range the radiation is more penetrating, with the result that more mass (10^{-16}–10^{-17} g) is needed for detection. The required dose varies inversely with the area of the resolution element, and inversely as the square of the concentration for thin specimens.

In essence, absorption microanalysis depends upon measuring the fraction of the incident x-ray probes that create an inner-shell vacancy in the given atomic species, this fraction being proportional to the concentration of the species in the specimen. A similar measurement can be made with incident charged-particle beams as well. In that case, however, inner-shell vacancy formation is recognized as a characteristic energy loss by the incident probe. Electron energy-loss analysis has received considerable attention recently as a tool for the detection of low-Z elements in biological specimens, where for $Z < 11$ it requires lower radiation doses[89] than the fluorescence techniques of Section 5.2. However, when the comparison is made between electron energy-loss analysis and x-ray absorption microanalysis, calculated dosages for this low-Z range are found to be approximately equal.

5.2. Fluorescence Microanalysis

Detection of elemental constituents by x-ray fluorescence is the subject of Chapter 14 of this volume. We shall, therefore, only mention here those aspects related to high-resolution localization of these constituents in biological material.

The sensitivity of the fluorescence technique is a result of the nature of the detected signal: the fluorescent x ray is direct evidence for the presence of the element of its origin. The distribution of the element in the specimen is found by scanning the specimen with a small x-ray probe and recording the fluorescent signal as a function of position. Horowitz and Howell[13] were the first to form images with fluorescent x rays using a synchrotron radiation source (Figure 22). They used

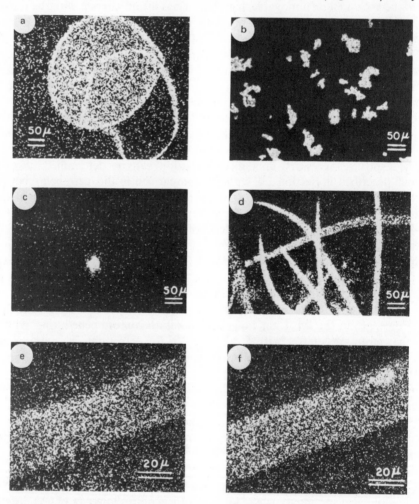

Figure 22. X-ray fluorescence micrographs made by Horowitz and Howell[70] with their scanning microscope. (a) *Coscinodiscus* diatom (silicon fluorescence); (b) sulfur dust (fluorescence); (c) sulfur dust (90° geometry); (d) fruit hairs (SES) (potassium fluorescence); (e) fruit hair (detail) (potassium fluorescence); (f) same fruit hair (calcium fluorescence).

mostly proportional counters as detectors, but also demonstrated that the better energy resolution of solid state detectors can improve the sensitivity considerably.[69]

Radiation damage to the specimen is minimized by efficient detection of the fluorescent signal (using detectors covering a large solid angle), and by using mono-chromatic radiation just energetic enough to excite the fluorescence of the element under investigation. The quantitative relationship between radiation dose and minimum detectable mass[89] shows a strong atomic number dependence. The fluorescence yield for low-Z elements is very low (Figure 3), rendering the method quite inefficient for $Z < 10$. For each successfully detected fluorescent x ray, many damaging events have to take place. The situation improves rapidly with increasing atomic number. For calcium ($Z = 20$) 10^{-17} g (about 1% by weight in a 1-μm-thick specimen in a 30 nm × 30 nm resolution element) should be detectable with less than 10^4 J/g dose to the specimen. In general, for $Z > 12$, fluorescence is comparable to or less damaging than absorption microanalysis for biological specimens. It also lends itself better to the detection of low concentrations than the absorption method, where, as the analysis shows, the comparison of images taken on two sides of an edge involves measuring the small difference of two large numbers, a statistically unfavorable procedure. The dose in the fluorescence case is inversely proportional to the minimum mass to be detected, irrespective of spatial distribution. For elements with $Z < 50$ the L-shell fluorescence rates are quite low; therefore K-shell fluorescence will in principle be preferable.

When x-ray probes are compared with electron, proton, or heavy-ion probes, the x rays are found[89] to impose considerably smaller dosages for most biological specimens, especially for $Z > 12$. In contrast to charged beams, the x-ray wave-length can be chosen to maximize the interaction with a given element, and the overwhelming majority of these interactions will produce the inner-shell vacancies necessary for fluorescence.

5.3. Summary

Methods for imagining a single atomic species in intact wet specimens can be derived from the contrast mechanisms already discussed by specializing them so that they become sensitive to the single atomic species in question. This can be done in TXM by detecting the change in the number of photons transmitted as the incident wavelength is shifted past an absorption edge of the element in question (i.e., absorption microanalysis), and in EXM and FXM by detecting reaction products with energies characteristic of the element in question. In the latter case, we have considered only the microanalysis based on FXM (i.e., fluorescence microanalysis).

Required dosages for adequate signal-to-noise ratios have been calculated for intact wet biological specimens for both methods. Absorption microanalysis is superior for the detection of low-Z elements, fluorescence microanalysis for the detection of medium- and high-Z elements. For dosages of 10^4 J/g, concentrations of the order of 1% by weight should permit mapping at resolutions of the order of 10–30 nm.

Comparison of the soft x-ray methods with their electron analogues have also been made on the basis of calculated dosage for adequate signal-to-noise ratios in intact wet biological specimens. The comparison is complicated by uncertainties concerning instrumentation efficiencies in the soft x-ray cases, but suggest that absorption microanalysis will require dosages comparable with those of electron energy-loss analysis, and that fluorescence microanalysis will require dosages considerably lower than electron microprobe analysis. Fluorescence microanalysis requires adequate soft x-ray scanning optics, however.

6. Additional Technical Aspects

Thus far we have been concerned with the elements of what might be called *basic* soft x-ray microscopy. We now turn to certain additional aspects of the technique that are required if its full potential in biology is to be realized. Work on most of these topics is only at a preliminary stage at present, and the treatment given here is necessarily brief.

6.1. Wet Specimens

We have seen (Section 2.4) that, with properly chosen wavelengths, organic feature visibility can remain high in the presence of water. To take advantage of this potentially important characteristic of the microscopy, some provision must be made for maintaining the important characteristic of the microscopy, some provision must be made for maintaining the environment needed by a wet specimen during exposure. A review of techniques that solve this problem in the case of electron microscopy has been given by Parsons[95, 96]; either a thin window or a series of small, differentially pumped, collinear apertures is placed between the specimen with its local environment and the high vacuum. The problem is simplified somewhat in the x-ray case because of the greater ability of the particles to penetrate windows or air paths, the absence of scattering, and the fact that in the case of contact microradiography the particles do not need to be brought into the vacuum again after interacting with the specimen.

Up to now, no successful wet-specimen soft x-ray picture appears to have been obtained†. McGowan et al.[93] attempted contact radiography with living cells (mouse L cells) in a windowed wet-specimen chamber but failed to obtain an image. The authors believe that failure was due to movement of the cells during the 5–15 min. exposures used.* Feder and his co-workers at the IBM Research Center are currently experimenting with a wet-specimen technique that may provide restriction of specimen movement. The technique is an adaptation of the method described by Basu et al.,[97] in which a thin (~ 10 nm) layer of SiO is evaporated onto a wet specimen to form a specimen replica. Instead of using the layer as a replica, Feder leaves it in place on the specimen, to act as a barrier to water loss. A

---

† **Note added in proof**: Successful wet-specimen pictures have now been obtained by Niemann et al. and by Feder (zone-plate microscope and contact microradiographic technique, respectively).

* The authors suggest cooling the resist surface to approximately 2°C to essentially halt cell movement.

related technique for possible adaptation is that of Nagata and Fukai,[98] in which a wet-specimen chamber is made by sandwiching the specimen between two thin planar films of SiO.

For contact microradiography the wet-specimen chamber defined by apertures has the disadvantage that the apertures limit the field of view to an approximately 100-μm diameter in a typical two-aperture design. Larger apertures might be possible for flash work, where the apertures need be open only briefly.

In scanning microscopy the small cross section of the x-ray beam near the specimen permits a different approach. The vacuum terminates in an exit aperture (< 10 μm diameter), covered by a thin foil. The focus is in the atmospheric environment, and neither specimen nor specimen holder need enter a vacuum chamber.

6.2. Readout of Resist Images by Transmission Electron Microscopy

Examination of the developed resist by transmission electron microscopy (TEM) instead of SEM would avoid the inconveniences arising from the comparative scarcity of high-resolution SEM's. Pictures of the resist taken in TEM should also be more suitable for stereo viewing than those taken with SEM. For this purpose, Feder is currently experimenting with a technique involving the preparation of a platinum–carbon and carbon replica[99] of the resist surface. (Removal of the replica from the resist is easily accomplished by dissolving away the resist in its own developer, the solubility of the resist having been increased by exposing it to ultraviolet radiation prior to the replication.) An alternative technique might involve spinning the resist on a carbon-coated EM specimen grid and then (after exposure and development) viewing the resist directly in TEM after suitable shadowing to enhance contrast.* A stainable detector (see Section 3.1.2) would give an image that would be directly viewable in TEM.

6.3. 3-D Imaging

The ability of soft x-ray microscopy to image a thick specimen at high resolution carries with it the problem that the large amount of information present may render insufficient any simple technique based on the viewing of two-dimensional images. (An example is the obscuring of one feature by another in a projected view.) The result is that soft x-ray microscopy faces in full force both the necessity and the opportunity to do three-dimensional imaging of the specimen.

Three basic methods may be considered: use of an imaging system with a large depth of field, with recording of multiple views of the specimen; use of an imaging system with small depth of field, with focusing on successive layers of the specimen; and holography. Of these, the second (wide-angle optics) and third have already appeared in Section 3 as techniques that are potentially capable of reaching the wavelength limit on resolution everywhere in the specimen. Thus both of the high-resolution imaging techniques possess their own natural 3-D imaging methods.

---

* **Note added in proof**: A technique similar to this has now been brought into successful operation by Feder and his colleagues.

Indeed it appears that a high-resolution imaging technique may not even be separable from its 3-D imaging capabilities, i.e., that the price of wavelength-limited resolution is that a three-dimensional object must be imaged in an essentially three-dimensional way. It would then follow that the first of the methods under consideration does not exist at wavelength-limited resolutions, and in fact we have seen (Section 3.1.3) that contact microradiography, an obvious candidate for the first method, does not succeed in giving large depth of field and wavelength-limited resolution simultaneously. Summarizing, it may not be possible to do 3-D imaging at resolutions close to the wavelength by the multiple-view technique (i.e., stereo tomography), but there appears to be in principle no barrier to doing such imaging by either of the other methods.

Lower-resolution 3-D imaging by the multiple-view method is of course possible, and in Figure 23 we show an example of the stereo pairs obtained by Horowitz and Howell with their scanning technique. In contact microradiography, grid-mounted specimens (Section 3.1.4) may be lifted off the resist after an exposure, shifted to an unexposed portion of resist, lowered again into contact with the

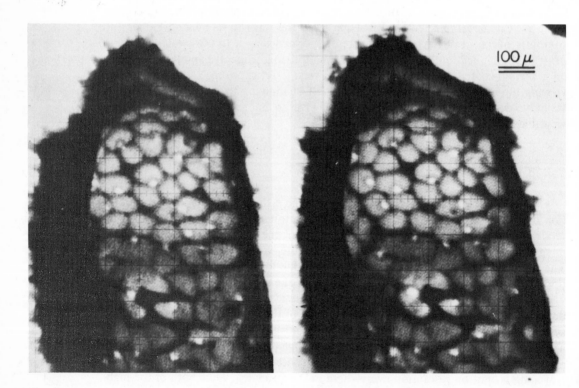

Figure 23. Stereo pair of scanning transmission micrographs of an ostracod shell by Horowitz and Howell.[70]

resist, and exposed with x-rays coming from a different angle. An advantage of the optical techniques for imaging, including scanning, is that the specimen can be viewed from a variety of directions by placing it on a rotating stage.

Figure 24 is an example of the work of Aoki and Kikuta,[65] showing 3-D imaging by holography.

We close this section by pointing out that the calculations of dosage of Section 2.4 assume that it is possible to have high-resolution imaging along two dimensions of the specimen, without having it in the third dimension also. The considerations of the present section cast some doubt on that assumption, in the case that the high resolutions sought are close to the wavelength limit. The meaning of this is that in the high-resolution soft x-ray case (lower portion of Figure 6), one would in fact be required to inflict more dosage on the specimen, and produce more information about the specimen, than is indicated in the figure. (Physically, this would arise, e.g., in the case of wide-angle optics, through the necessity of focusing layer-by-layer

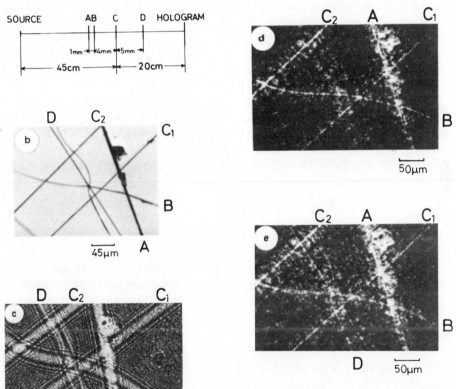

Figure 24. 3-D imaging by holography (Aoki and Kikuta[65]). The specimen consists of five polyethylene-terephthalate fibers with widths of 2–6 μm. (a) and (b) The positions of the fibers. (c) The x-ray hologram, made with aluminum $K\alpha$ radiation ($\lambda = 0.834$ nm). (d) Reconstructed image of fiber B. (e) Reconstructed image of fiber D.

through the specimen, independent of whether layer-by-layer information about the specimen were wanted.) In the electron case, where resolutions are not wavelength limited, one would be free to reject the extra information and the extra dosage. In actual cases this freedom might mean little, however, since the structure of the specimen might not be comprehensible without the extra information.

7. Summary and Conclusions

The soft x-ray photon (wavelength range 1–10 nm) offers favorable contrast mechanisms for the examination of relatively thick biological specimens in a natural state, at resolutions approaching the wavelength of the photon. Imaging techniques are less satisfactory at present, but contact microradiography now images at these resolutions over part of the specimen, and attainment of these resolutions over the entire specimen seems possible, in theory and practice, by several imaging techniques. The question of sources for soft x-ray microscopy is now admirably taken care of by the synchrotron radiation and perhaps the pulsed plasma sources, offering usefully complementary characteristics. An extremely useful property of the contrast mechanisms is that they can be specialized, if desired, to a given atomic species, permitting the concentration of that species in thick biological specimens in a natural state to be mapped at high resolution. Finally, the auxiliary techniques necessary to allow the methods to be applied to biological materials of the type described appear to be feasible.

We conclude that the technique offers valuable characteristics not otherwise available for the imaging of biological material, and hence has a high potential in biology and medicine. Considerable further development, especially of imaging methods, is possible and is needed for full realization of the potential. Enough of the potential should be available, even with the current imaging methods to open up new areas of study of biological material.

ACKNOWLEDGMENT

One of us (J.K.) gratefully acknowledges support for research in x-ray microscopy from the National Science Foundation under Grant # PCM77-16641.

References

1. P. Kirkpatrick and H. H. Pattee, Jr., in *Handbuch der Physik*, S. Flugge (ed.), Vol. 30, pp. 305–336, Springer-Verlag, Berlin (1957).
2. V. E. Cosslett and W. C. Nixon, *X-Ray Microscopy*, Cambridge University Press, Cambridge, England (1960).
3. V. E. Cosslett, X-ray microscopy, *Rep. Prog. Phys.* **28**, 381–407 (1965).
4. A. Engström, in *Physical Techniques in Biological Research*, 2nd edition. A. W. Pollister (ed.), Vol. III, Part A, pp. 87–171, Academic Press, New York (1966).

5. T. A. Hall, H. O. E. Rockert, and R. L. de C. Saunders, *X-Ray Microscopy in Clinical and Experimental Medicine*, Charles C Thomas, Springfield, Illinois (1972).
6. P. Horowitz, in *Principles and Techniques of Scanning Electron Microscopy*, M. A. Hayat (ed.), Vol. 5, pp. 181–192, Van Nostrand Reinhold, New York (1976).
7. E. Spiller, R. Feder, and J. Topalian, Soft x-rays for biological and industrial pattern replication, *J. Phys. Colloq.* **39**, 205–211 (1978).
8. W. Gudat, Soft x-ray microscopy and lithography with synchrotron radiation, *Nucl. Instrum. Methods* **152**, 279–288 (1978).
9. E. Spiller and R. Feder, The optics of long-wavelength x-rays, *Sci. Am.* **239**, 70–78 (1978).
10. P. Lamarque, Historadiographie, *C.R. Acad. Sci.* **202**, 684–685 (1936).
11. H. Wolter, Spiegelsysteme streifenden Einfalls als abbildende Optiken für Röntgenstrahlen, *Ann. Phys.* (Leipzig) **10**, 94–114 (1952).
12. B. L. Henke, in *Advances in X-Ray Analysis*, W. M. Mueller (ed.), Vol. 2, pp. 117–155, Plenum Press, New York (1960).
13. P. Horowitz and J. A. Howell, A scanning x-ray microscope using synchrotron radiation, *Science* **178**, 608–611 (1972).
14. L. Reimer, in *Physical Aspects of Electron Microscopy and Microbeam Analysis*, B. M. Siegel and D. R. Beaman (eds.), pp. 231–246, Wiley, New York (1975).
15. M. Isaacson, in *Principles and Techniques of Electron Microscopy*, M. A. Hayat (ed.), Vol. 7, pp. 1–78, Van Nostrand Reinhold, New York (1976).
16. A. Charlesby, in *Radiation Damage Processes in Materials*, C. H. S. Dupuy (ed.), pp. 231–260, Noordhoff, Leyden (1975).
17. G. W. Barendsen, in *Encyclopedia of X-Rays and Gamma Rays*, G. L. Clark (ed.), pp. 934–936, Reinhold, New York (1963).
18. L. F. Thompson, E. D. Feit, M. J. Bowden, P. V. Lenzo, and E. G. Spencer, Polymeric resists in x-ray lithography, *J. Electrochem. Soc.* **121**, 1500–1503 (1974).
19. D. Sayre, J. Kirz, R. Feder, D. M. Kim, and E. Spiller, Transmission microscopy of unmodified biological materials: comparative radiation dosages with electrons and ultrasoft x-ray photons, *Ultramicroscopy* **2**, 337–349 (1977).
20. D. Sayre, J. Kirz, R. Feder, D. M. Kim, and E. Spiller, Potential operating region for ultrasoft x-ray microscopy of biological materials, *Science* **196**, 1339–1340 (1977).
21. D. Sayre, J. Kirz, R. Feder, D. M. Kim, and E. Spiller, Assessment of the potential of ultrasoft x-ray microscopy, *Ann. N.Y. Acad. Sci.* **306**, 286–290 (1978).
22. R. Feder, D. Sayre, E. Spiller, J. Topalian, and J. Kirz, Specimen replication for electron microscopy using x-rays and x-ray resist, *J. Appl. Phys.* **47**, 1192–1193 (1976).
23. E. Spiller, R. Feder, J. Topalian, D. Eastman, W. Gudat, and D. Sayre, X-ray microscopy of biological objects with carbon $K\alpha$ and with synchrotron radiation, *Science* **191**, 1172–1174 (1976).
24. R. Feder, E. Spiller, J. Topalian, A. N. Broers, W. Gudat, B. J. Panessa, Z. A. Zadunaisky, and J. Sedat, High-resolution soft x-ray microscopy, *Science* **197**, 259–260 (1977).
25. W. A. Ladd, W. M. Hess, and M. W. Ladd, High-resolution microradiography, *Science* **123**, 370–371 (1956).
26. S. K. Asunmaa, in *X-Ray Optics and X-Ray Microanalysis*, H. H. Pattee, V. E. Cosslett, and A. Engström (eds.), pp. 33–61, Academic Press, New York (1963).
27. D. Sayre and R. Feder, Exposure and Development of X-ray Resist in Microscopy, IBM Research Report RC-7498, Yorktown Heights, New York (1979).
28. E. Spiller and R. Feder, in *Topics in Applied Physics*, H.-J. Queisser (ed.), Vol. 22, pp. 35–92 (1977).
29. H. H. Pattee, in *X-Ray Microscopy and X-Ray Microanalysis*, A. Engstrom V. Cosslett, and H. Pattee (eds.), pp. 56–60, Elsevier, Amsterdam (1960).
30. G. Mollenstedt and L. Y. Huang, in *X-Ray Microscopy and Microradiography*, V. E. Cosslett, A. Engström, and H. H. Pattee, Jr. (eds.), pp. 392–396, Academic Press, New York (1957).
31. S. J. La Placa (private communication).
32. O. H. Griffith, G. H. Lesch, G. F. Rempfner, G. B. Birrell, C. A. Burke, D. W. Schlosser, M. H. Mallon, G. B. Lee, R. G. Stafford, P. C. Jost, and T. B. Marriott, Photoelectron microscopy, a new approach to mapping organic and biological surfaces, *Proc. Natl. Acad. Sci. U.S.A.* **69**, 561–565 (1972).

33. G. Pfefferkorn, L. Weber, K. Schur, and H. R. Oswald, in *Scanning Electron Microscopy 1976*, O. Johari (ed.), pp. 130–142, IIT Research Institute, Chicago, Illinois (1976).
34. P. Kirkpatrick and A. V. Baez, Formation of optical images by x-rays, *J. Opt. Soc. Am.* **38**, 766–774 (1948).
35. J. A. Howell and P. Horowitz, Ellipsoidal and bent cylindrical condensing mirrors for synchrotron radiation, *Nucl. Instrum. Methods* **125**, 225–230 (1975).
36. A. Franks, X-ray optics, *Sci. Prog.*, Oxford **64**, 371–422 (1977).
37. V. Rehn, in *Workshop on X-Ray Instrumentation for Synchrotron Radiation Research*, H. Winick and G. Brown (eds.), pp. VII 13–35, Stanford Synchrotron Radiation Laboratory Report SSRL78/04 (1978).
38. V. E. Cosslett and W. C. Nixon, An experimental x-ray shadow microscope, *Proc. R. Soc. London B Ser.* **140**, 422–431 (1952).
39. M. von Ardenne, Zur Leistungsfähigkeit des Elektronen-Schattenmikroskopes und über ein Röntgenstrahlen-Schattenmikroskop, *Naturwissenschaften* **28**, 485–486 (1939).
40. O. E. Myers, Jr., Studies of transmission zone plates, *Am. J. Phys.* **19**, 359–365 (1951).
41. A. V. Baez, Fresnel zone plate for optical image formation using extreme ultraviolet and soft x-radiation, *J. Opt. Soc. Am.* **51**, 405–412 (1961).
42. C. D. Pfeifer, L. D. Ferris, and W. M. Yen, Optical image formation with a Fresnel zone plate using vacuum-ultraviolet radiation, *J. Opt. Soc. Am.* **63**, 91–95 (1973).
43. G. Mollenstedt, H. J. Einighammer, K. H. v. Grote, and U. Mayer, in *X-Ray Optics and Microanalysis*, R. Castaing, P. Deschamps, and J. Philibert (eds.), pp. 15–29, Hermann, Paris (1966).
44. H. Brauninger, H. J. Einighammer, and H. H. Fink, in *X-Ray Optics and Microanalysis*, G. Shinoda, K. Kohra, and T. Ichinokawa (eds.), pp. 17–27, University of Tokyo Press, Tokyo (1972).
45. B. E. Bol Raap, J. B. Le Poole, J. H. Dijkstra, W. de Graaff, and L. J. Lantwaard, in *Small Rocket Instrumentation Techniques*, K.-I. Maeda (ed.), pp. 203–210, North-Holland Publishing Co., Amsterdam (1969).
46. J. H. Dijkstra, W. de Graaf, and L. J. Lantwaard, in *New Techniques for Space Astronomy*, F. Labuhn and R. Lust (eds.), pp. 207–210, Reidel, Dordrecht (1971).
47. D. Rudolph and G. Schmahl, in *New Techniques in Space Astronomy*, F. Labuhn and R. Lust (eds.), pp. 205–206, Reidel, Dordrecht (1971).
48. G. Elwert and J. V. Feitzinger, Zur Verbesserung der Abbildung von Flächenquellen mit Zonenplatten, insbesondere der Sonne im XUV und weichen Röntgengebiet, *Optik (Stuttgart)* **31**, 600–612 (1970).
49. H. H. Fink, Eine Kombination zweier Zonenplatten zur halofreien Abbildung entfernter Objekte, *Optik (Stuttgart)* **31**, 150–153 (1970).
50. J. Kirz, Phase zone plates for x-rays and the extreme ultraviolet, *J. Opt. Soc. Am.* **64**, 301–309 (1974).
51. N. Ceglio, in *X-Ray Optics and Microanalysis, Proceedings of the VIIIth International Conference*, Boston, 1977, Science Press, Princeton, New Jersey (in press).
52. G. Schmahl and D. Rudolph, Lichtstarke Zonenplatten als abbildene Systeme für weiche Röntgenstrahlung, *Optik (Stuttgart)* **29**, 577–585 (1969).
53. G. Schmahl and D. Rudolph, in *Progress in Optics*, E. Wolf (ed.), Vol. 14, pp. 195–244, North-Holland Publishing Co., Amsterdam (1976).
54. B. Niemann, D. Rudolph, and G. Schmahl, X-ray microscopy with synchrotron radiation, *Appl. Opt.* **15**, 1883–1884 (1976).
55. G. Schmahl, D. Rudolph, and B. Niemann, X-ray microscopy of biological specimens, *J. Phys. Colloq.* **39**, 202–204 (1978).
56. G. Schmahl, D. Rudolph, and B. Niemann, in *X-Ray Optics and Microanalysis, Proceedings of the VIIIth International Conference*, Boston, 1977, Science Press, Princeton, New Jersey (in press).
57. D. Sayre, Proposal for the Utilization of Electron Beam Technology in the Fabrication of an Image Forming Device for the Soft X-Ray Region, IBM Research Report RC-3974, Yorktown Heights, New York (1972).
58. A. N. Broers, W. W. Molzen, J. J. Cuomo, and N. P. Wittels, Electron-beam fabrication of 80Å metal structures, *Appl. Phys. Lett.* **29**, 596–598 (1976).
59. A. V. Baez, A study in diffraction microscopy with special reference to x-rays, *J. Opt. Sci. Am.* **42**, 756–762 (1952).

60. D. Gabor, Microscopy by reconstructed wave-fronts, *Proc. R. Soc. London A Ser.* **197**, 454–487 (1949).
61. G. W. Stroke, in *Optique des Rayons X et Microanalyse*, R. Castaing, P. Deschamps, and J. Philibert (eds.), pp. 30–46, Hermann, Paris (1966).
62. A. V. Baez and H. M. A. El-Sum, in *X-Ray Microscopy and Microradiography*, V. E. Cosslett, A. Engström, and H. H. Pattee, Jr. (eds.), pp. 347–366, Academic Press, New York (1957).
63. G. L. Rogers and J. Palmer, The possibilities of x-ray holographic microscopy, *J. Microsc. (Oxford)* **89**, 125–134 (1969).
64. S. Aoki, Y. Ichihara, and S. Kikuta, X-ray hologram obtained by using synchrotron radiation, *Jpn. J. Appl. Phys.* **11**, 1857 (1972).
65. S. Aoki and S. Kikuta, X-ray holographic microscopy, *Jpn. J. Appl. Phys.* **13**, 1385–1392 (1974).
66. B. Reuter and H. Mahr, Experiments with Fourier transform holograms using 4.48 nm x-rays, *J. Phys. E.* **9**, pp. 746–751 (1976).
67. A. M. Kondratenko and A. N. Skrinsky, Use of radiation of electron storage rings in x-ray holography of objects, *Opt. Spectrosc. (USSR)* **42**, 189–192 (1977).
68. H. H. Pattee, Jr., The scanning x-ray microscope, *J. Opt. Soc. Am.* **43**, 61–62 (1953).
69. J. A. Howell, Scanning X-Ray Microscopy Using Synchrotron Radiation, Ph.D. thesis, Harvard University (May, 1974).
70. P. Horowitz, Some experiences with x-ray and proton microscopes, *Ann. N.Y. Acad. Sci.* **306**, 203–222 (1978).
71. J. Kirz, in *X-Ray Optics and Microanalysis*, Proc. VIIIth Int. Conf., Boston, 1977, Science Press, Princeton, New Jersey (in press).
72. Y. Sakayanagi and S. Aoki, Soft x-ray imaging with toroidal mirrors, *Appl. Opt.* **17**, 601–603 (1978).
73. S. Aoki, S. Kawata, and Y. Sakayanagi, Focusing of the synchrotron radiation in the soft x-ray region, *Jpn. J. Appl. Phys.* **17**, 733–734 (1978).
74. E. Spiller, Low-loss reflection coatings using absorbing materials, *Appl. Phys. Lett.* **20**, 365–367 (1972).
75. R.-P. Haelbich, A. Segmüller, and E. Spiller, Smooth multilayer films suitable for x-ray mirrors, *Appl. Phys. Lett.* **34**, 184 (1979).
76. E. Spiller, in Workshop on X-Ray Instrumentation for Synchrotron Radiation Research, H. Winick and G. Brown (eds.), pp. VI 44–49, SSRL Report 78/04 (1978).
77. J. B. Hastings, X-ray optics and monochromators for synchrotron radiation, *J. Appl. Phys.* **48**, 1576–1584 (1977).
78. E. Spiller, D. E. Eastman, R. Feder, W. D. Grobman, W. Gudat, and J. Topalian, Application of synchrotron radiation to x-ray lithography, *J. Appl. Phys.* **47**, 5450–5459 (1976).
79. P. Pianetta, R. Burg, J. Kirz, H. Rarback, J. W. McGowan, and M. Malachowski, X-ray Microscopy with the 4° Beam Line, SSRL Report 79/01 (1979).
80. R. Z. Bachrach, I. Lindau, V. Rehn, and J. Stöhr, Report of the Beam Line III Optics Advisory Committee, SSRL Report 77/14 (1977).
81. B. Niemann, D. Rudolph, and G. Schmahl, Soft x-ray imaging zone plates with large zone numbers for microscopic and spectroscopic applications, *Opt. Commun.* **12**, 160–163 (1974).
82. B. L. Henke and M. A. Tester, in *Advances in X-Ray Analysis*, W. L. Pickles, C. S. Barrett, J. B. Newkirk, and C. O. Ruud (eds.), Vol. 18, pp. 76–106, Plenum Press, New York (1975).
83. P. J. Mallozi, H. M. Epstein, R. C. Jung, D. C. Appelbaum, B. P. Fairand, W. J. Gallagher, R. L. Uecker, and M. C. Muckerheide, Laser-generated plasmas as a source of x-rays for medical applications, *J. Appl. Phys.* **45**, 1891–1895 (1974).
84. D. J. Nagel, in *Advances in X-Ray Analysis*, W. L. Pickles, C. S. Barrett, J. B. Newkirk, and C. O. Ruud (eds.), Vol. 18, pp. 1–25, Plenum Press, New York (1975).
85. R. A. McCorkle and H. J. Vollmer, Physical properties of an electron beam–sliding spark device, *Rev. Sci. Instrum.* **48**, 1055–1063 (1977).
86. R. A. McCorkle, Soft x-ray emission by an electron-beam sliding-spark device, *J. Phys. B* **11**, L407–408 (1978).
87. A. W. Rogers, *Techniques of Autoradiography*, Elsevier, Amsterdam (1973).
88. L. A. Sternberger, *Immunocytochemistry*, Prentice-Hall, New York (1974).
89. J. Kirz, D. Sayre, and J. Dilger, Comparative analysis of x-ray emission microscopies for biological specimens, *Ann. N.Y. Acad. Sci.* **306**, 291–303 (1978).

90. A. Engström, Quantitative micro- and histochemical elementary analysis by Roentgen absorption spectrography, *Acta Radiol. Suppl.* **63** (1946).

91. B. Lindström, Roentgen absorption spectrophotometry in quantitative cytochemistry, *Acta Radiol. Suppl.* **125**, 1–206 (1955).

92. F. Polack, S. Lowenthal, Y. Petroff, and Y. Farge, Chemical microanalysis by x-ray microscopy near absorption edge with synchrotron radiation, *Appl. Phys. Lett.* **31**, 785–787 (1977).

93. J. Wm. McGowan, B. Borwein, J. A. Medeiros, T. Beveridge, J. D. Brown, E. Spiller, R. Feder, J. Topalian, and W. Gudat, High resolution microchemical analysis using soft x-ray lithographic techniques, *J. Cell Biol* **80**, 732–735 (1979).

94. J. Wm. McGowan (private communication).

95. D. F. Parsons, Structure of wet specimens in electron microscopy, *Science* **186**, 407–414 (1974).

96. D. F. Parsons, V. R. Matricardi, R. C. Moretz, and J. N. Turner, in *Advances in Biological and Medical Physics*, Vol. 15, J. H. Lawrence, J. W. Gofman, and T. L. Hayes (eds.), pp. 161–270, Academic Press, New York (1974).

97. S. Basu, G. Hausner, and D. F. Parsons, New wet-replication technique, *J. Appl. Phys.* **47**, 741–761 (1976).

98. F. Nagata and K. Fukai, Effects of electron irradiation on growth of spore, reported at U.S.A.–Japan seminar on new trends in high voltage electron microscopy (1971). Cited by K. Hama, in *Advanced Techniques in Biological Electron Microscopy*, J. K. Koehler (ed.), pp. 275–297, Springer-Verlag, New York (1973).

99. S. Bullivant, in *Advanced Techniques in Biological Electron Microscopy*, J. K. Koehler (ed.), pp. 67–112, Springer-Verlag, New York (1973).

100. B. L. Henke and E. S. Ebisu, in *Advances in X-ray Analysis*, C. L. Grant, C. S. Barrett, J. B. Newkirk, and C. O. Ruud (eds.), Vol. 17, pp. 150–213, Plenum Press, New York (1974).

101. B. L. Henke and M. L. Schattenburg, in *Advances in X-Ray Analysis*, R. W. Gould, C. S. Barrett, J. B. Newkirk, and C. O. Ruud (eds.), Vol. 19, pp. 749–767, Kendall/Hunt, Dubuque, Iowa (1976).

102. B. L. Henke and R. L. Elgin, in *Advances in X-Ray Analysis*, B. L. Henke, J. B. Newkirk, and G. R. Mallett (eds.), Vol. 13, pp. 639–665, Plenum Press, New York (1970).

103. Wm. J. Viegele, Photon cross sections from 0.1 keV to 1 MeV for elements $Z = 1$ to $Z = 94$, *At. Data* **5**, 51–111 (1973).

104. J. A. Bearden, X-ray wavelengths, *Rev. Mod. Phys.* **39**, 78–124 (1967).

105. M. Ando and S. Hosoya, in *Proceedings of the Sixth International Conference on X-Ray Optics and Microanalysis*, G. Shinoda, K. Kohra, and T. Ichinokawa (eds.), pp. 63–68, University of Tokyo Press, Tokyo (1972).

106. V. P. Koronkevich, G. N. Kulipanov, V. I. Nalivaiko, V. F. Pindurin, and A. N. Skrinsky, Contact projection of microobjects by means of x-ray synchrotron radiation, Preprint INP 77-10, Institute of Nuclear Physics, Novosibirsk (1977).

107. R. M. Siebert, Two methods of Roentgen microphotography, *Acta Radiol.* **17**, 299–309 (1936).

108. W. Bambynek, B. Crasemann, R. W. Fink, H.-U. Freund, H. Mark, C. D. Swift, R. E. Price, and P. Venugopala Rao, X-ray fluorescence yields, Auger, and Coster–Kronig transition probabilities, *Rev. Mod. Phys.* **44**, 716–813 (1972).

109. J. Sedat and L. Manuelidis, A direct approach to the structure of eukaryotic chromosomes, *Cold Spring Harbor Symp. Quant. Biol.* **42**, 331–350 (1977).

110. R. A. McCorkle, J. Angilello, G. Coleman, R. Feder, and S. J. La Placa, Flash x-ray microscopy, *Science*, **205**, 401–402 (1979).

9

Synchrotron Radiation as a Modulated Source for Fluorescence Lifetime Measurements and for Time-Resolved Spectroscopy

IAN H. MUNRO and ANDREW P. SABERSKY

1. Introduction: Source Properties

The fundamental properties of synchrotron radiation as a spectroscopic source are enhanced by the additional quality of time modulation of the source intensity over a broad range of modulation frequencies. This time modulation is a result of the dynamics of accelerators and storage rings. It depends mainly on a balance between longitudinal focusing and damping due to the magnetic lattice and the radio frequency drive and to excitation due to the emission of synchrotron radiation. The spectral profile, state of polarization, and angular distribution of emitted radiation as a function of frequency can be calculated explicitly for any accelerator or storage ring source in terms of its known parameters. The time modulation of the emitted intensity is proportional to the electron beam current modulation seen traversing a single azimuth in the accelerator. The circulating beam in the storage ring is preserved by continuously restoring energy lost to synchrotron radiation

IAN H. MUNRO • Daresbury Laboratory, Daresbury, Warrington WA4 4AD, United Kingdom.
ANDREW P. SABERSKY • Stanford Linear Accelerator Center, Stanford University, Stanford, California 94305.

with radio frequency (rf) power, usually in the range of from 10 to 500 MHz. In a working storage ring, although the bunch length is set primarily by the frequency of the accelerating field, the bunch shape is also modified by the magnitude of the circulating current and by the internal geometry of the vacuum chamber that contains the electron beam. The complexity of the interactions between the bunch and the walls of the vacuum chamber are such that the bunch shape is difficult to predict exactly for a new storage ring and is best determined empirically as a function of the various operating parameters of the storage ring.

A storage ring can contain any number of electron bunches from one to a maximum (the harmonic number N) defined by the ratio of ring circulation period to the period of the accelerating rf field. The ring period is set by the size of the accelerator and is the ratio of the ring circumference to the velocity of light. Very short bunches can be stored in an accelerator by using high-frequency, high-voltage rf and by adjusting the machine momentum compaction, α^1. The Berlin storage ring BESSY will be able to work at α close to zero for very short bunches, and 50 ps bunches have been observed at SPEAR, using a streak camera, running at low-energy and high-rf voltage.

2. Time Modulation of Synchrotron Radiation Sources

2.1. Storage Rings

In electron storage rings, the energy loss rate of the stored electrons due to synchrotron radiation is so high that energy must be fed continuously to the beam to keep it circulating for useful times, which are usually of the order of several hours. As in all circular resonance accelerators, the energy is coupled through electromagnetic fields: this implies that the stored electrons have an intermittent or pulsed intensity along their orbital path. In storage rings used as radiation sources, the electrons are sufficiently relativistic for their orbital circulation frequency to be considered constant. The accelerating (or maintaining) rf field can have a frequency equal to any integer multiple of the circulation frequency f_0. The minimum number of circulating bunches is one and the maximum number is N.

The classical model of equilibrium electron bunch length in a storage ring[1] takes into account geometrical properties of the storage ring lattice, intensity of the rf accelerating field, effects on the energy of the electrons due to emission of synchrotron radiation, and the interactions between these factors. This model yields a Gaussian bunch intensity profile independent of bunch current. The electromagnetic interaction between the bunch and its surroundings—usually a metal vacuum vessel and metal rf coupling devices (cavities)—can be described using a model containing a function $\hat{Z}$,[2] the complex coupling impedance of the storage ring. This function can, in principle, be determined from a detailed knowledge of the beam environment, although it has never been calculated for any but simple cases of isolated elements. For some existing machines, this function has been experimentally determined through the study of beam behavior.

A real bunch is not Gaussian and is not symmetric about its first moment. Whereas the classical bunch has a smooth profile that is invariant with time, the real bunch is subject to internal instabilities that distort its profile in a random, unsystematic way. These instabilities also play a part in overall current-dependent bunch lengthening. Some of the internal instabilities can be controlled through feedback techniques, but one tries in general to avoid them by controlling $\tilde{Z}$ appropriately during design and construction of a storage ring.[3] Bunch length can be controlled in an existing ring by adding or subtracting large discrete coupling impedances, or by modifying or adding to the existing rf system.[4] For an ideal rf source with zero internal impedance, the equilibrium phase of the center of a bunch with respect to the rf accelerating voltage is a function of the same factors that determine bunch length and is not affected by bunch current. In existing real storage rings, the bunch phase will change with respect to the rf voltage as a function of current. This change is dependent on the detailed characteristics of the rf system[5] as well as $\tilde{Z}$ and other properties of the storage ring.

2.2. Synchrotrons

Although qualitatively similar to a storage ring, the normal mode of operation of an electron synchrotron is such as to make time studies in the nanosecond region impracticable. An electron synchrotron is injected with low-energy electrons at a rate of 1–60 Hz. The electrons are accelerated to high energies by increasing the magnetic field and are usually extracted during or a little after the time at which the magnetic field has reached its peak value. The magnetic field usually follows an approximately sinusoidal time course and the range of accelerating radio frequencies are much the same as are found in different storage rings. It is probably possible to inject a single bunch in a synchrotron, but usually many bunches are injected over most of the orbit circumference. This produces a train of pulses spaced by the period of the accelerating field with the total number of pulses in the train being rather less than the ring harmonic number. Storage rings can also operate in single-bunch or multibunch modes. The number of electrons in each of the bunches may be dependent on the bunch position in the bunch train. The entire bunch train remains in the synchrotron for a duration corresponding to roughly half the magnet repetition frequency. For example, for the Daresbury synchrotron, NINA, electrons circulated for 18.8 ms. If the accelerating rf field is maintained even while the magnetic field is decreasing, it is possible to retain electrons in the synchrotron for up to three quarters of the magnet cycle and in some cases for the entire cycle. Unfortunately, while electrons are maintained in the synchrotron their energy and hence the synchrotron radiation spectrum they emit and its associated beam divergence are continuously changing. This results in a smaller duty cycle than one might expect, particularly for x radiation.[6] Finally, the number of electrons at each injection may vary by a few percent, yielding intensity fluctuations modulated at the injection frequency. As a consequence of these multifrequency modulations, there are no reports of time-resolved experiments in the nanosecond region using a synchrotron. Results have, however, been reported on time studies in the millisecond time domain in association with time-resolved x-ray diffraction[6] and with the phosphorescence of long-lived states of solid rare gases doped with organic molecules.[7]

2.3. Electron Bunches and Their Behavior in a Storage Ring

For fast time domain measurements such as fluorescence decay studies, where the bunch length is an important factor, it is important to know the detailed bunch shape during an experiment either as a function calculable from measured storage ring working parameters or directly from diagnostic devices. With bunches shorter than 250 ps FWHM one diagnostic successfully used is the combination of a fast coaxial planar photodiode feeding a sampling oscilloscope.[5, 8] This combination has a time resolution of ∼120 ps FWHM.

Once bunch lengths have been measured over a suitable operating range for a particular storage ring, they may be well described as a function of the following storage ring operating parameters: current per bunch; synchrotron tune; and beam energy. Bunch lengthening in SPEAR for some values of the above parameters is shown in Figure 1.[5] The lifetime of a stored beam in a storage ring can vary from 2 to 10 hours and it is important to determine how the bunch length and shape will change during an experimental run.

The instabilities discussed in Section 2.1 can have various instrumental effects and must be identified. In the lowest oscillation mode, the dipole, the shape of the bunch (i.e., the dipole component) is invariant. The whole bunch oscillates in phase with respect to the accelerating rf at a frequency (the synchrotron frequency f_s) in the kilohertz range. These oscillations can have large amplitudes with phase (or time) excursions larger than the bunch length. Their effects may be neglected in timing experiments when the time reference signal is taken from the bunch itself, i.e., a signal from a pickup electrode. These signals vary in amplitude with beam current and are thus subject to discriminator timing walk. The second mode is the

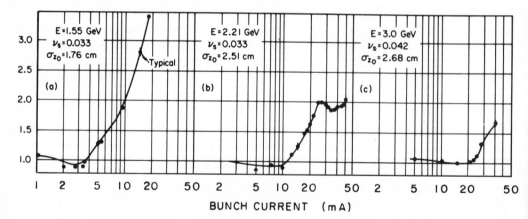

Figure 1. Bunch lengthening in SPEAR II (358-MHz rf). The σ_z values are derived from FWHM measurements: The bunches are not symmetrical in general.

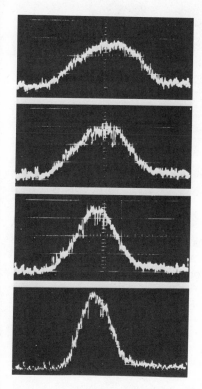

Figure 2. Quadrupole oscillations of the bunch in SPEAR I (50 MHz rf). These oscillations are periodic with a frequency $\simeq 2 \times$ synchrotron frequency. The horizontal scale is 0.5 ns/div.

quadrupole mode (Figure 2). The bunch changes shape cyclically at a frequency close to $2f_s$. The length excursions may extend from $+100\%$ to -50%. This mode is especially troublesome in fast timing work since the changes in the peak amplitude can cause difficulty when using a standard discriminator. In principle a zero-crossing discriminator will provide a solution to this problem since the charge in the pulse is invariant. The quadrupole mode may sometimes be tuned out by varying machine impedances and it has also been successfully damped by active feedback through the rf system at SPEAR. There are even higher modes, up to six being observed on SPEAR. Each mode is close to the appropriate integer multiple of f_s in frequency. Frequency shifts away from the integer multiples are due to distortion of the phase-focusing properties of the beam caused by its interaction with its surroundings. The relative amplitudes of the modes and their effects on the bunch shape as applied to fast timing decrease with higher mode numbers. In storage rings to date, only the dipole and quadrupole modes have caused serious problems.

The effects of the instabilities on the spectrum of beam modulation increase linearly with frequency, since a given time shift Δt of the bunch becomes a phase shift:

$$\Delta \theta = N \, \Delta t (2\pi f_0)$$

where N is the harmonic number of f_0 and $f_0 = c/L_{\text{orbit}}$ is the revolution frequency.

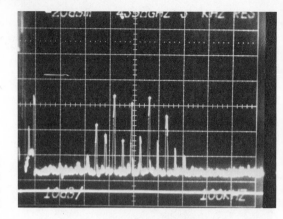

Figure 3. A segment of the power spectrum of the signal picked up from the beam in SPEAR II by an antenna within the vacuum chamber. The line nearest center is a revolution-frequency harmonic. The horizontal scale is 100 kHz/div and the vertical scale is 10 dB/div.

At a sufficiently high frequency the revolution frequency harmonics are surrounded by instability sidebands that may be even larger than the central spectral line (Figure 3). Spectral analysis of pickup electrode signals offers a powerful diagnostic tool for monitoring the strengths of instabilities. Optical beam profile monitors[4] with sufficiently fast scan of beam density distributions can also give useful, quantitative information on instabilities.

2.4. *Optical Properties of the Pulsed Source*

A single, highly relativistic electron moves in a segment of its orbit in a bending magnet. At a distance l from a tangent point (T), we look at the emitted synchrotron radiation with an ideal photodetector (P) of width $2W$ (Figure 4). The photodetector sees an arc of the orbit of angle 2δ.[1] Assuming that the electrons are traveling at the speed of light, it can be shown that length of the electron plus photon flight path L (see Figure 4) is given by

$+\delta$ flight path: $L_+ = z + k = l/\cos \delta + R \tan \delta$

$-\delta$ flight path: $a = R \sin \delta, \qquad \xi = l - a, \qquad m = \xi/\cos \delta$

$$L_- = m = l/\cos \delta - R \tan \delta + 2R\delta$$

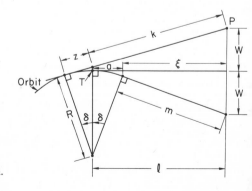

Figure 4. Geometry of the synchrotron light timing problem.

Expanding the trigonometric functions and rejecting terms above third order, we get

$$L_+ = l(1 + \delta^2/2) + R\delta(1 + \delta^2/3)$$
$$L_- = l(1 + \delta^2/2) + R\delta(1 - \delta^2/3)$$

The flight path differences are

$$\Delta L_+ = L_+ - (l + R\delta) = l\delta^2/2 + R\delta(\delta^2/3)$$
$$\Delta L_- = L_- - (l + R\delta) = l\delta^2/2 - R\delta(\delta^2/3)$$

These expressions give estimates of the timing errors that are to be expected while looking at a segment of orbit subtending an angle 2δ. Taking SPEAR as an example:

$$l = 10 \text{ m}, \qquad \delta = 10^{-2}, \qquad R = 12.7 \text{ m}$$
$$\Delta L_+ \cong \Delta L_- = 5.04 \times 10^{-4} \text{ m}$$

Thus,

$$\Delta t = 1.68 \text{ ps}$$

The above assumes that the wave front impinges on a planar surface. The quadratic curvature of the isochronous wave front suggests that we may correct the timing by using a suitable curved photodetector or by using a special optical system.

It is in fact possible to show[9] that there exist optical systems that will bring all the radiation from a large arc or even, in principle, from the full circumference of a storage ring to a point isochronously. The geometrical analysis given here neglects the important effects of the spatial distribution, angular distribution, and finite emission angle of horizontal and vertical radiation in the beam. However, it does give an order of magnitude for the importance of these effects. For the bunch lengths and dimensions of existing and proposed synchrotron light sources, these timing effects are negligible. For observations involving very high-frequency modulation of the bunch, such as in free-electron laser storage rings, geometric effects can be important.

2.5. Source Parameters

Some parameters of synchrotron radiation sources have been tabulated in Chapter 3. Table 1 lists some time and modulation properties associated with each storage ring source. Most storage rings constructed to date have been designed for elementary particle research to operate in the stable colliding-beam single-bunch mode and this has led to the easy application of the single photon coincidence method to time resolved studies at a number of synchrotron radiation facilities. Electron synchrotrons are unsuitable as high-frequency modulated light sources,

Table 1. Parameters of Storage Rings Used as Modulated Light Sources

	ε_c (keV)	E (GeV)	R (m)	I (mA)	f_{rf} (MHz)	f_0 (MHz)	$2\sigma_z$ (cm)
SURF II (Washington, USA)	0.041	0.25	0.83	25	114	57	30
TANTALUS I (Stoughton, USA)	0.048	0.24	0.64	>100	31.8	31.8	30–150
INS-SOR II (Tokyo, Japan)	0.054	0.3	1.1	100	121	17.3	20
ACO (Orsay, France)	0.31	0.54	1.11	100	27.24	13.24	12–40
Brookhaven (USA) VUV Ring[b]	0.40	0.7	1.9	1,000	52.8	5.87	7.5
BESSY (Berlin, West Germany)[b]	0.52	0.75	1.8	100	125	4.8	6
					500		
VEPP 2M (Novosibirsk, USSR)	0.54	0.67	1.22	100	302	16.7	5
ALADDIN (Stoughton, USA)[b]	1.1	1.0	2.08	1,000	50.78	3.39	30
ADONE (Frascati, Italy)	1.5	1.5	5.0	60	8.57	2.86	60
PAMPUS (Amsterdam, Netherlands)[b]	1.8	1.5	4.17	>100	500	4	~6
DCI (Orsay, France)	3.4	1.8	3.82	500	25.3	3.16	30
SRS (Daresbury, UK)[b]	3.9	2.0	5.55	370	500	3.12	6
Photon Factory (Tsukuba, Japan)[b]	4.1	2.5	8.33	500	476	1.67	6
VEPP-3 (Novosibirsk, USSR)	4.2	2.25	6.15	100	76	4.03	20
Brookhaven (USA) X-ray Ring[b]	5.1	2.5	6.8	500	52.8	1.76	~10
DORIS (Hamburg, West Germany)	7.8	3.5	12.1	500	500	1.04	4
SPEAR II (Stanford, USA)	11.1	4.0	12.7	100	358	1.28	6
CESR (Cornell, USA)[b]	12.8	8.0	88	100	500	0.45	~5
VEPP-4 (Novosibirsk, USSR)	26.0	7.0	29	100	180.9	0.8	15

[a] Planned or under construction.
[b] The bunch length, $2\sigma_z$, has been measured directly only for a very few rings. For most rings the only available information is the calculated natural bunch length in the absence of bunch lengthening effects even though such effects will certainly contribute to the real length.

and they are not included in the table. The parameters listed in the table are defined as follows:

ε_c is the critical energy of the synchrotron radiation;
E, the maximum electron energy in GeV;
R, the orbit radius in the dipole bending magnets;
I, the probable maximum circulating current at peak energy;
f_{rf}, the accelerating radio frequency;
f_0, the storage ring fundamental frequency;
$2\sigma_z$, the stored electron bunch length in cm.

The integral harmonic number of the storage ring (N) is given by the ratio f_{rf}/f_0 and is the maximum number of electron bunches that can be accelerated in the storage ring.

2.6. Comparison with Other Sources

A great variety of techniques exist for the high-frequency modulation of visible and ultraviolet light and it is interesting to compare the collective properties of a synchrotron radiation source with its alternatives. Mode-locked lasers can produce pulses with subpicosecond duration, pulse trains with repetition rates of up to several hundred megahertz, and pulsed powers at the operating wavelengths of the laser that are enormously greater than will ever be produced from a synchrotron radiation source.[10] Laser light is collimated, linearly polarized and tunable, albeit over rather narrow frequency ranges in the near ultraviolet, the visible, and infrared regions. At present, tunable laser systems are readily available for wavelengths longer than around 300 nm. For shorter wavelengths, particularly in the vacuum ultraviolet region, only a few relatively low-power laser sources exist, although much research effort is currently directed towards the development of coherent radiation sources in the vacuum ultraviolet and soft x-ray regions. The relative merits of the various source types are summarized in Table 2. In general terms, for all measurements requiring a time resolution not better than 1 ns, the most effective source in terms of cost and simplicity is a normal modulated discharge lamp. For ultrafast timing studies in the subpicosecond domain a laser system is essential but is expensive, rather difficult to set up and use, and seriously restricted in its operating wavelength range by the properties of the laser medium and transmission of the optical system. If access to a storage ring source is available, it can yield around picosecond time resolution and provide excitation at any wavelength over a very broad range.

For the future, when the advantages of undertaking spectroscopic research in the subnanosecond time domain become more widely appreciated and practiced, modifications are likely to be made to storage ring facilities specifically to achieve even shorter bunch lengths than are indicated in Table 2.

Without any changes, SPEAR could be operated now at 1.5 GeV with a total of 4.0 MV applied to the four 358-MHz rf cavities used for beam acceleration to give a calculated natural bunch length, $2\sigma_z$, of 1.6 cm (equivalent to about 55-ps pulse width) (private communication, P. Wilson). Alternatively, the use of higher

Table 2. Comparison of Pulsed Light Sources.

Characteristic	Synchrotron radiation	Lasers	Incoherent sources
Wavelength range	~ 0.1 nm to ~ 1 cm	Tuneable over narrow ranges in ultraviolet and visible. Some lines below 200 nm and many in the infrared.	A wide variety of sources are needed to cover the range ~ 15 nm to radio frequencies.
Intensity (number of photons) per pulse within a 0.1% wavelength band.	$< 10^9$	$\sim 10^{10}$ in 1 ps pulse of width ≤ 10 cm^{-1}	$< 10^6$
Minimum approximate pulse duration	100 ps	0.2 ps	$\gtrsim 1$ ns
Pulse repetition rate	~ 1 to 500 MHz	dc to ~ 100 MHz	dc to < 100 MHz
Source size and divergence	Incoherent ~ 1 mm^2 < 10 mrad	Coherent $\gtrsim 1$ mm^2 $\gtrsim 5$ mrad	Incoherent $\sim$ few mm^2 isotropic

acceleration frequencies will assist in the production of short bunches. The SPEAR "high harmonic cavity" operating at 860 MHz to give control over the bunch shape (in practice, to yield a longer bunch with a squarer profile) could be used along with or instead of the 358-MHz rf power source for SPEAR and this also would result in a reduction in bunch length.

In general, using present day technology and within the constraint of using only a relatively small circulating current (5 mA or so), it is clear that a pulse length of a few tens of picoseconds will become a feature of many storage ring facilities.

3. Experimental Techniques for Time-Resolved Measurements

3.1. Direct Time Measurements

A medium energy (2 GeV) and high current (1 A) storage ring such as the SRS at Daresbury, United Kingdom, will produce a maximum of about 4×10^{-5} J of electromagnetic energy per 100-ps pulse into a horizontal aperture of 1 mrad. Although the energy per pulse is reasonably large, the duration is such that shape detail cannot be resolved by commercially available real time oscilloscopes because of their limited (less than 1 GHz) bandwidth. In addition, the response time (FWHM) of a fast vacuum photodiode is of the same order as the electron pulse width. Real time oscilloscopes are characterized by low sensitivity and low trace brightness at high frequencies, which can be overcome to some extent by using a low-noise, broad-band, high-gain photomultiplier to view the signal. Unfortunately, even at relatively low gains of about 10^3, the tube response, associated with the transit time dispersion of electrons within the electron multiplying structure, will begin to increase significantly. The time response degradation is such that for the fastest photomultiplier tubes commercially available the FWHM increases from ~ 150 to ~ 250 ps as the gain is raised from $\sim 10^3$ to $\sim 10^6$ (Varian tubes, types 148 and 154, respectively). The shape of the circulating electron bunch can (in principle) be studied without the use of a photodetector by measuring the time characteristics of a signal induced on a strip-line antenna or other rf electrode in the orbit chamber of the storage ring.

At present, the highest resolution time studies of bunch shape have been carried out using a sampling oscilloscope and fast photodiode where the extreme stability of the emitted light in terms of intensity and high-repetition frequency were well matched to the relatively high sensitivity ($\sim$ mV/cm) and rise time (~ 25 ps) of the oscilloscope. A photomultiplier used in conjunction with a sampling oscilloscope possesses high sensitivity, good time resolution (limited solely by the photodetector), and signal averaging capability. One of the first fluorescence lifetime measurements using storage ring radiation employed such a method to measure the ultraviolet fluorescence from cerium doped lanthanum trifluoride.[11]

The time resolution associated with windowless electron multipliers, channel electron multipliers, and microchannel plates is being improved, but for all such devices it remains restricted to greater than 100 ps (or 10 Ghz).

The only direct signal measurement device with ultrahigh-frequency reponse is a streak camera, which is normally used in conjunction with single pulse laser excitation to yield a time resolution of a few picoseconds.

As the applications of pulsed synchrotron radiation become increasingly extensive, a detailed knowledge of the electron bunch profile with high time resolution will be needed. Up to the present, electron bunches have been studied either using a fast detector and sampling oscilloscope,[5] or by using a phase-shift technique or power spectrum analyzer to observe longitudinal bunch frequency content as alterations are made in machine parameters such as energy, current, or rf phase.[12]

The latest development for detailed bunch profile studies has been to use a streak camera, which gave a time resolution of about 20 ps when driven synchronously by the storage ring accelerating power source and ∼1-ps resolution when used as a "single-shot" device for individual bunch length measurements.

3.2. Single-Photon Counting Method

The constraints of real time or quasi-real time measurements of short duration pulses from synchrotron radiation sources are associated primarily with low signal intensity. Paradoxically, this constraint can be relaxed by observing only single-photon events and using digital electronic techniques to measure time correlations between pulse pairs. The method is called "single-photon counting" and has been used extensively for the past ten years for measurement of the fluorescence lifetimes of organic materials.[14] Instrumentation has been developed and extended to measure subnanosecond decay times[15] and its limitations in relation to nonsynchrotron radiation sources, such as the wavelength dependence of the apparatus response time, have been carefully considered.[16] Single photon counting has been used to great advantage in measurements of the length and time structure of the electron bunch in the ACO (Orsay) storage ring.[17]

The basis of the technique is to observe for each excitation of the sample the emission of only one fluorescence photon. Using an appropriate zero-time reference signal, and by count accumulating the time distribution of the arrival times of the single-fluorescence photons, one may reconstruct the time-dependent decay curve of the fluorescence emission. Of course, if the photodetector were to produce an output pulse corresponding to a two- or a three-fluorescence photon emission, the accumulated decay curve would no longer represent the true time dependence of the fluorescence emission but would indicate a faster time decay. In addition, should more than one signal photon arrive within the "time window" of the experiment defined by the time scale selected on the time to amplitude converter (TAC), any second or subsequent photons would not be recorded and consequently the measured decay profile would be biased towards short times giving a "pulse-pileup" effect. It has been shown[14-16] that for the measured curve to possess less than 1% distortion relative to the unknown time decay function, the signal collection rate should on average be less than 2% of the sample excitation rate. Higher count rates than this will yield less accurate decay curves, although it is possible to apply electronic rejection of multiphoton events by observing the signal amplitude at the last dynode of the photomultiplier. Another rate enhancement technique is to process the data using the method of modulating functions.[18]

In all photon counting experiments the time difference between the "start" and "stop" pulses is digitized and stored in the appropriate address of a set of counting registers whose location cycle time is restricted at present to longer than one microsecond. Since the repetition frequency of the light pulse from a storage ring is usually from approximately 1 to 20 MHz, there is a good match between the cycle time of the multichannel analyzer (MCA) and the single photon counting rate.

An experimental arrangement that has been used for the measurement of fluorescence lifetimes[19] and to study time-resolved fluorescence anisotropy using the single-photon counting method is shown in Figure 5. Synchrotron radiation from SPEAR was passed through a monochromator tuned to the wavelength required for sample absorption. Fluorescence was detected using a fast photomultiplier such as an RCA Type 8850 with a good single photon pulse height resolution. Single photon counting was ensured by maintaining a maximum fluorescence signal count rate of about 2% of the excitation rate (i.e., 20 kHz in that case). Intensity attenuation can be used to modify the count rate either by altering the slit width setting of the excitation monochromator or by using an iris diaphragm to change the solid angle of collected fluorescence. It is an advantage to use the fluorescence

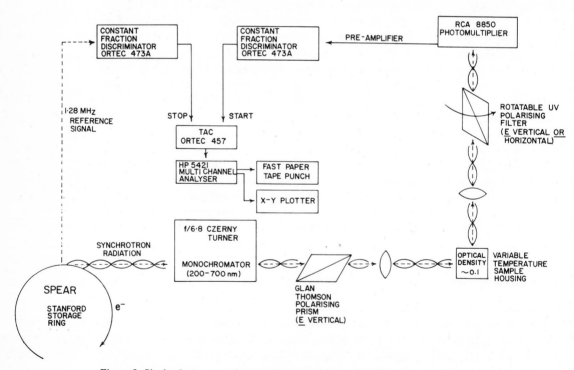

Figure 5. Single-photon counting apparatus used in conjunction with a storage ring for the measurement of flurorescence lifetimes and fluorescence anisotropy.[19]

single-photon signal as the time reference (the "start" pulse) rather than the storage ring "clock" since most TAC's have an operating rate limit of around 1 MHz. A consequence of using a fluorescence pulse as the "start" signal is that time is recorded "backwards" in the MCA and the initial somewhat nonlinear portion of the TAC sweep becomes of diminished significance. The time address of each single-fluorescence photon is thus measured and stored until the statistical quality of the data is sufficient, usually when $\sim 10^5$ counts have accumulated in the peak channel.

An experimentally measured curve consists of the convolution of the instrument response function together with the fluorescence decay function to be expected. The instrument response is determined mainly by the transit-time dispersion of the photodetector and by the time profile of the excitation pulse. Extraction of the unknown decay function is almost an art form in itself, especially when the fluorescence decay is of multiexponential character. Applications of the photon counting method are characterised by the wide variety of mathematical and computing techniques introduced in the extraction of data.[18, 20–22]

The successful operation of a single-photon counting experiment using a synchrotron radiation source is not achieved easily and it is worth noting some of the difficulties. The time resolution is set primarily by the detector and values for a single-electron time spread (FWHM) are about 650 ps for a focused dynode tube, 200 ps for a curved microchannel plate, and around 100 ps or less for a static crossed field photomultiplier tube.[23, 24] Of course, the multiphotoelectron time dispersion is less than that for single photons and also decreases markedly as the intensity per excitation pulse increases. This can, and does, lead to a range of fluorescence lifetime results erroneously biased towards short times. Among many other problems are the shielding of the detector from strong rf and magnetic fields, avoidance of ground loops, careful impedance matching of the output from the detector anode, and the dependence of the photodetector time response on the wavelength of the incident light,[25] which yields a shift of approximately 1 ps per nm in the ultraviolet region.[19]

The source pulse shape is defined by the Poisson process of synchrotron radiation emission and by the electron distribution in the circulating bunch, which is assumed to be Gaussian. This gives an incoherent (for $\lambda \ll$ bunch length) and Gaussian time profile to the synchrotron radiation pulse and also defines the time profile of the emitted light to be totally wavelength independent; an extremely important and advantageous characteristic for lifetime measurements.[26]

The pulse-shaping discriminator, (usually of the "constant fraction" type), time-to-amplitude converter, variable delays, and fast amplifiers all lead to the deterioration of the "instrument function." It is usual and reasonable to assume that the time resolution ($\lesssim 50$ ps) associated with the modular electronics can be neglected by comparison with the time response of the detector and the light pulse profile. The timing of the time reference pulse (the "stop" pulse) is important. The highest amplitude and minimum pulse spread are associated with signals induced directly on rf electrodes placed in the storage ring orbit chamber. Less satisfactory would be fanout signals derived from the storage ring master oscillator, which can be subject to phase shifts with respect to the electron bunches or the use of a second photomultiplier tube looking at the excitation light pulse in the multiphoton mode.

Of critical importance is the rejection of stray light or excitation light scattered directly from the sample into the detector. In fluorescence anisotropy work and in the measurement of very short lifetimes, even small (1%) amounts of scattered light can render the results meaningless. The normal procedure would be to make a measurement, for example in solution, with and without the presence of the solute and under identical conditions. Data can conveniently be accumulated subtractively in the multichannel analyzer to eliminate the effects of sample scattering and of shot noise from the photomultiplier tube.

Very short lifetime measurements are limited by the time resolution of the apparatus but very long lifetimes are modified by the continual reexcitation of the sample at the repetition frequency of the storage ring. The problem is most acute in small storage rings with a ring period of less than 100 ns, but measurement of the decay slope and modulation depth permit the measurement of fluorescence time constants that are many times longer than the interpulse period of the storage ring used for the excitation.[27]

3.3. Phase-Shift Measurements

A well established technique for the measurement of fluorescence lifetimes is that of phase and/or modulation fluorometry. If the intensity of the excitation source is modulated periodically at frequency f (i.e., at an angular frequency of $\omega = 2\pi f$), then the emitted fluorescence signal will be modulated at the same frequency but will appear to have a phase delay (θ) with respect to the excitation light such that

$$\tan \theta = \omega\tau$$

If the modulation amplitude is defined as the ratio of ac to dc components in the signal, then the ratio of the modulation amplitudes of the fluorescence and the excitation light signals (M) are given by

$$M = \left(1 + \omega^2\tau^2\right)^{-1/2}$$

In each case, the expression implies that the fluorescence decay profile is a single exponential function of lifetime τ. A wide variety of techniques have been used to carry out measurements of these types[28] in the majority of which the modulation frequency is restricted to frequencies less than 10 MHz. A storage ring is an interesting example of a light source that possesses an exceedingly high degree of modulation at high frequencies. A storage ring can be filled with as many electron bunches as the rf harmonic number, although it is often operated in the e^+-e^- colliding beam mode with only single bunches. In the case of a stable bunch and repetition period the intensity modulation appears in the storage ring, e.g., SPEAR, as a train of 200-ps-wide, approximately Gaussian pulses spaced at 789-ns intervals. The *frequency* spectrum of the radiation is a train of Dirac delta functions spaced at the 1.28-MHz ring revolution frequency and with an amplitude modulation envelope given by the Fourier transform of the single-pulse shape. For SPEAR this gives a smooth frequency spectrum with a half-peak amplitude width of at least

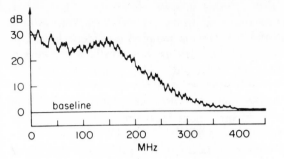

Figure 6. The power spectrum of an RCA Type 8850 photo-multiplier tube measured using synchrotron radiation. The overall frequency response of the tube extends to a maximum usable upper frequency of about 450 MHz.[29]

1.8 GHz, which may extend to as much as 3.5 GHz under operating conditions with a minimum bunch length of about 100 ps.

A description of the first phase-shift measurements using radiation from a storage ring[29] has shown that this broad source frequency spectrum can be used to calibrate the frequency response of the photodetector in a unique way. The power spectrum of the source and photodetector were measured with a Tektronix 7L13 high-frequency spectrum analyzer. Since the phototube time dispersion is several times larger than the storage ring bunch length, the measured power spectrum represents primarily the frequency spectrum of the phototube (Figure 6). The frequency spectrum of the source alone can be measured directly and independently using an antenna to "view" the bunch circulating in the storage ring, by electromagnetic pickup. The response of such antennae are not always well known and bunch information derived from them is therefore not necessarily accurate.

Phase-shift measurements are possible over a frequency range limited at low frequencies by the ring fundamental f_0 (1.28 MHz in the case of SPEAR) and extending at discrete harmonics of the fundamental to an upper frequency where the combined effect of timing dispersion and detector noise still can provide an adequate signal-to-noise ratio of the harmonic lines at the output of the detector. In the case of an RCA 8850 tube operated at 3 kV this limit is approximately 450 MHz which is illustrated very clearly in Figure 6. The actual pulse comparison measurements were made using apparatus shown schematically in Figure 7. A

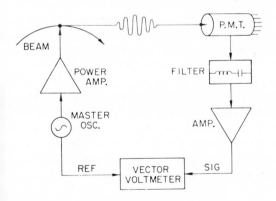

Figure 7. A simple apparatus to make phase-shift measurements using synchrotron radiation from SPEAR.[29]

single operating frequency was selected using a cavity filter ($Q \sim 1.6 \times 10^3$ at 358 MHz) and the phase comparison observed using a Hewlett-Packard Model 8405A Vector Voltmeter. Although the experiments required considerable care in the elimination of geometry induced phase shifts, it was shown to be a straightforward procedure to obtain a time resolution of about $\pm 1°$, which is equivalent to ± 8 ps at 358 MHz. The use of a tunable filter or amplifier would clearly permit measurements to be carried out at a large number of different frequencies, which, in principle at least, would permit any multicomponent decay problem to be analysed using modern methods in multicomponent analysis.[30]

In all phase measurements the selection and use of the proper reference phase source is critical. When working at radio frequencies from the storage ring transmitter and using a signal derived therefrom, systematic phase shifts between the bunch and rf source (see Section 2.1) must be carefully avoided. These can be caused by the internal workings of the rf source and its servo loops, operator interaction, and beam current changes. An antenna or pickup inside the storage ring vacuum chamber will develop a strong electrical signal that is related to the bunch shape. The frequency spectrum of this signal has a structure similar to that of the modulation of the light, but with a modulation envelope and with phase relationships between harmonics that are different from those of the synchrotron radiation modulation. The phase relationship between light signals and pickup signals of the same single harmonic is a constant, regardless of the condition of the bunch. Thus, one may use pairs of matched filters to select any harmonic frequency for the measurement.

The Fourier transform of a symmetric bunch shape is completely real: there are no phase shifts between harmonics of the revolution frequency. Since actual bunches are asymmetric, there is a complex part to the shape transform. There may be systematic phase shifts between harmonics and these shifts can change as the bunch shape changes. It remains to be determined whether it is safe to generate reference phases via harmonic generation or by heterodyning for these types of measurements.

A number of authors have attempted high-frequency phase-shift studies using modulated laser beams[31–33] but the possibility of operation at more than one frequency has not yet been seriously pursued. A new technique has also been described where phase fluorometry can be applied at exceedingly low light levels.[34]

A major limitation in the technique is the difficulty of achieving sufficiently high gain (associated with low noise levels) at the frequency of interest to allow a very accurate phase comparison to be made between the fluorescence and the reference signals. The problem arises because in the single-bunch mode of operation, a pulse amplitude of several volts at the anode of the photomultiplier tube will yield only a few microvolts of signal at the selected frequency. The method of phase comparison is important and it may turn out to be the case that a null detector combined with an optical phase delay will provide the most accurate means of measuring phase delay. Note that in this context 1 ps delay is equivalent to 0.3 mm! In all phase-shift experiments it is critically important to ensure that none of the excitation light is able to leak through into the fluorescence signal since even as little as 1 % of leakage-scattered light will produce a large reduction in the phase shift apparently associated with the fluorescence process under observation.

With these reservations in mind, the possibility of achieving picosecond or perhaps subpicosecond time resolution using a storage ring will be of extreme importance in the study of atoms, small molecules, and solids in the vacuum ultraviolet region of the spectrum. The energetics of solids and the reaction kinetics of ions, radicals, and small molecules are almost unexplored due to the absence of any suitable excitation source and to the problems of carrying out fast-timing studies. Clearly it is a wide area of research where one may expect many fresh developments in the near future.

3.4. Time-Resolved Spectroscopy

The techniques of time-resolved excitation and emission spectroscopy, as distinct from measuring only the time dependence of the total emission, are ideally matched to a synchrotron radiation source that gives continuously tunable, high-repetition-rate, and short-duration excitation light pulses.

The power of the technique is illustrated in Figure 8 where the protolysis of β-naphthol can be observed in stages within a time of a few nanoseconds.[35] The initial excitation of the cation and its fluorescence emission can be observed and the rate of the radiationless protolysis reaction, which competes with cation fluorescence, can be measured. Finally, a new species, the fluorescent anion, is identified by its fluorescence emission spectrum and its fluorescence decay time. The time-resolved emission from fluorescence in acid C_2H_5OH solutions using partially deuterated solvent C_2H_5OD has also been studied.[36] Deuteration of the solvent

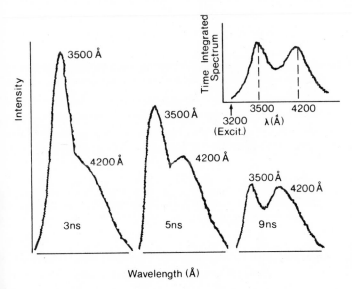

Figure 8. The time-resolved emission spectrum from β-naphthol. The emission spectrum is measured at 3 ns, 5 ns, and 9 ns after the excitation flash. The maximum at 350 nm corresponds to the initially excited neutral species, while the peak at 420 nm is related to the excited anion formed by a protolytic reaction whose emission first grows and then decays with a lifetime of 10.4 ns. The time-integrated spectrum is also shown.[35]

leads to a decrease in the rate of the protolytic reaction and a marked increase in the cation fluorescence intensity. Time-resolved fluorescence emission spectroscopy is a new tool that will greatly assist in the understanding of excited state chemistry of organic molecules[36, 37] and of pure and doped rare gases in the gaseous and solid phases.[38–40]

Apparatus used for time-resolved spectroscopy and for fluorescence decay time studies is shown in Figure 9.[38, 39] Synchrotron radiation from SPEAR was dispersed by an analyzing monochromator and used for excitation of the sample in an evacuated chamber. Pulses from the photomultiplier detector are fed to the "start" input of a time-to-amplitude converter via a fast discriminator. The reference signal is derived from an rf strip-line antenna in the SPEAR orbit chamber. This is used as the "stop" pulse for the TAC and can be continuously delayed in time with respect to the start pulse. The time resolution of the experiment is defined usually by the response time of the photomultiplier tube. The variable time window used for observation of the spectrum extends from 1 ns to approximately 500 ns.

As an alternative to the measurement of the time-resolved emission spectrum it is possible using a synchrotron radiation source to measure a time-resolved excitation spectrum. A striking example of the kind of information that this produces is given in Figure 10. In this study of solid xenon the initial spectrum (0–50 ns) is, in effect, a reflection spectrum from the sample, and it is dominated by scattered light

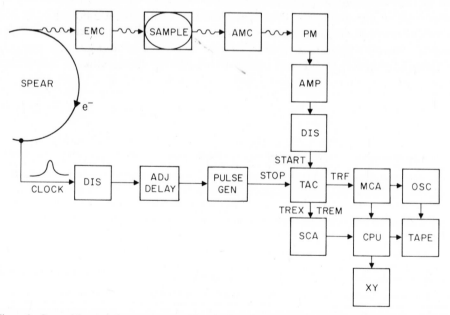

Figure 9. General layout of apparatus used for time-resolved spectroscopy at SSRL (from References 38 and 39): excitation monochromator (EMC), analyzing monochromator (AMC), photomultiplier (PM), amplifier (AMP), discriminator (DIS), time-to-amplitude converter (TAC), multichannel analyzer (MCA), single-channel analyzer (SCA), central processing unit (CPU), oscilloscope (OSC). chart recorder (XY), and magnetic tape drive (TAPE).

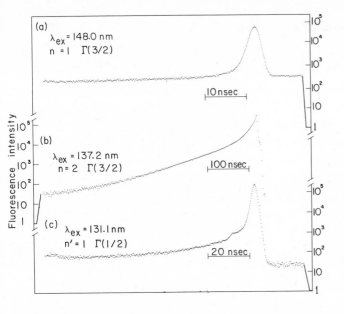

Figure 10. Photoluminescence decay spectrum for pure solid xenon at 80 K taken at excitation wavelengths corresponding to the excitonic absorption peaks for $n(\frac{3}{2}) = 1$, $n(\frac{3}{2}) = 2$, and $n(\frac{1}{2}) = 1$, respectively. The time scale is given backward and is different for each case.[41]

from the excitation pulse. The 5–30-ns excitation spectrum looks similar to the "static" excitation spectrum measured using conventional techniques, but the delayed 150–470-ns spectrum is quite different and contains information about long-lived components in the luminescence from solid xenon.[41]

Another excellent example of time-resolved excitation spectroscopy is given by a comprehensive study of singlet–triplet intersystem crossing and the effects of collisions on this process in the molecule CO. These experiments were carried out using pulsed radiation from the storage ring ACO at Orsay. When a high resolution (.03 nm) excitation bandwidth was used, it was possible to observe changes in the fluorescence decay profile to yield the spectral distortion of rotational levels with substantially singlet or with triplet character defined solely by the delay time used in the experiment.[36] The versatility of the technique is such that the effects of adding an inert gas (argon) at various pressures can be used to show how the spectral distribution of emitting rotational levels for the "slow" photon changes dramatically with the added argon gas pressure. Fluorescence decay curves can readily reveal emitting levels that are not excited directly by photoabsorption and have led, for example, to the suggestion of reversible collision-induced intersystem crossing in the CO molecule. An obvious suggestion at the time of writing is that the technique of phase-shift measurements at high frequency (see Section 3.3) should be incorporated into time-resolved excitation spectroscopy since the fluorescence emission phase shift could be measured accurately and continuously as a function of excitation wavelength. In this way, the reaction kinetics of energy degradation in solids (e.g., exciton diffusion and surface quenching) and also the study of chemical intermediates in solution (such as are involved in retinal photoreception) could be studied with picosecond time resolution.

4. Research Applications

4.1. Fluorescence Lifetime Measurements of Organic Molecules

The fluorescence transition in an organic molecule can be characterized by its lifetime, quantum efficiency, polarization, and emission spectrum. In general, fluorescence measurements include probably the most versatile and sensitive techniques at present available for the study of molecular environment, fast chemical kinetics, and conformational changes. The significant chemical and biological properties of organic molecules are, in the main, associated with the weakly bonded π electrons, which provide the unsaturated linkages between the carbon atoms in the benzene rings. Since the π-electron ionization potentials lie at around 8 eV and we are concerned almost exclusively with bound-state transitions, the wavelength range needed for excitation of organic molecular fluorescence extends from around 160 nm to longer wavelengths. Absorption of a photon by a molecule in its ground state will place the molecule in an excited singlet energy level from which it normally loses excess energy within a time less than 100 ps by falling to the lowest excited singlet state. From this lowest excited singlet level the molecule may change to any vibrational or rotational level of the ground electronic state either by emission of a photon or by some nonradiative process. The various energetic pathways by which the molecule finally reaches its unexcited ground state are selected by molecular geometry, solvent effects, and the possibility of energy transfer by a variety of mechanisms that will modify the fluorescence process in a calculable way. A study of fluorescence lifetimes hence may lead to a fuller understanding of the biophysical processes or chemical reaction kinetics involved.

A number of preliminary measurements of fluorescence lifetimes of organic systems were undertaken using the ACO storage ring.[35–37, 42, 43] These measurements have begun to reveal the wide range of problems in photophysics to which synchrotron radiation is applicable.

Results were obtained on the fluorescence standards quinine sulphate and the fluorescein anion. Using low-pressure (1 torr) vapor samples, it was shown to be possible to measure fluorescence lifetimes associated with absorption into and emission from single vibronic levels in the molecular vapor. The lifetime of analine vapor was studied in the range 33610–37650 cm^{-1}. Even though the quantum yield of pyrazine vapor at 8 torr is only 10^{-3}, a rapid fluorescence decay time of about 500 ps could be measured. The reactivities of β-naphthol and of fluorescein were measured in the excited state. In these compounds, the decay time and time-resolved spectrum yielded the kinetic scheme for the fast prolytic reaction of the excited state.[43]

The qualities of synchrotron radiation have been exploited in a study of the dependence of the fluorescence decay of NO on excitation wavelength. It has been found that all vibrational levels of the $A^2\Sigma^+$ state of NO decay exponentially. However, the $C^2\Pi$ states of NO yield nonexponential decay functions and a detailed analysis of the fluorescence decay of the $C^2\Pi(v=0)$ level revealed three components. The relative intensities from the three components were varied by scanning the excitation wavelength at high resolution through the range of rotational sublevels in the $C^2\Pi(v = 0)$ state.[27]

Excited states of the CN radical have been prepared via the gas-phase photolysis of ICN. Using pulsed synchrotron radiation from SPEAR the radiative decay of $CN^*(B^2\Sigma^+)$ has been measured as a function of ICN pressure and of excitation wavelength from 169.5 to 139.9 nm yielding lifetime values in the range 69.4 to 73.5 ns.[44]

Solid-phase measurements have also been carried out on organic crystals and on molecules trapped in low-temperature glasses. The fluorescence of tetracene single crystals at room temperature gave nonexponential decay curves with the radiative lifetimes of the first singlet exciton level of tetracene being 200 ps. A longer component of lifetime 1.7 ns was also measured.[45] The results differed from data measured using pulsed laser excitation, probably as a result of excitation density effects on the crystal decay function. Such secondary effects are absent when synchrotron radiation is used as the excitation source.

Detailed experimental data were taken using synchrotron radiation to measure the fluorescence decay functions and relative quantum yields of carbazole in the presence of a range of concentrations of potassium iodide, which acted as a quencher in ethanol glass solutions at 77 K. Figure 11 illustrates the data for different solid solutions where the fluorescence decay profiles are measured to show systematic but small changes as the concentration of the KI quencher is altered. The maximum change in lifetime is about 6%.[46]

Measurements have also been made on room temperature silicate-base glasses doped with a range of CeO_2 concentrations from 0.01 to 5%. Strong concentration quenching effects could be observed, giving a decrease in both the fluorescence lifetime and maximum quantum yield with increase of Ce^{3+} concentration.[47]

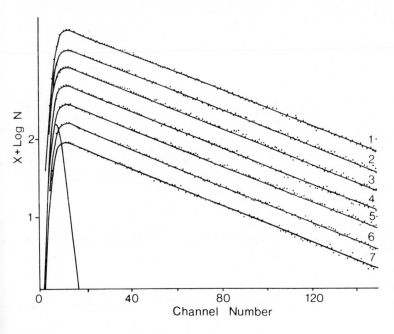

Figure 11. Fluorescence decay functions of carbazole in the presence of KI in ethanol at 77 K. The concentrations of KI are corrected for ethanol contraction. The points are the experimental data and the full lines are calculated fluorescence decay functions. (1) 0.000; (2) 0.025; (3) 0.050; (4) 0.075; (5) 0.100; (6) 0.125; (7) 0.150.[46]

4.2. Time-Resolved Spectroscopy of Rare Gases

The high intensity of synchrotron radiation throughout the vacuum ultraviolet region, combined with its short pulse duration and high repetition rate, far exceeds the present capabilities of laser systems in that region. For these reasons and partly because of attempts to understand and develop tunable laser systems for the vacuum ultraviolet, many studies are presently being carried out on rare-gas atoms.

4.2.1. Pure Rare-Gas Solids

Using pulsed radiation from SPEAR the decay time of the integrated photoluminescence from a solid xenon film at approximately 80 K has been measured as a function of excitation wavelength. The first and second members of the $\Gamma(\frac{3}{2})$ and the first member of the $\Gamma(\frac{1}{2})$ free-exciton series were selected in this way and the resulting liminescence decay profiles used to understand the regions of the two broad Stokes-shifted emission bands near 7.6 and 7.1 eV that are generated by photoabsorption close to the fundamental edge in xenon.[48] The results showed major differences in the emission decay profile and imply that the 7.1 and 7.6 eV bands are associated with the production of a series of self-trapped excitons derived from the $\Gamma(\frac{3}{2})$ free-exciton series. Radiative recombination of a vibrationally relaxed charge center resulting from a lattice-trapped electron–hole pair could result in this characteristic emission.

In emission bands of neon, argon, krypton, and xenon a variety of contributing states have been identified by the observation of a wide range of emission lifetimes. The fine structure in neon emission can be explained in terms of competitive relaxation to either an effective Ne_2^+ charge center or else to a "bubble center" formed by the moving apart of two neighboring atoms from the excited atom, effects which are associated with lifetimes of 2 and 10 ns, respectively.[49]

4.2.2. Solid Rare-Gas Mixtures

A broad range of spectroscopic studies have been carried out on the properties of doped rare gases with the aim of understanding excitonic states, energy transfer, and migration and surface effects in these materials. In particular, xenon and krypton doped with NO have been studied as potential cryogenic laser media. In such mixtures, energy transfer to individual dissociative excitations of the NO can occur. The self-trapped excitons of xenon and krypton that are energetically resonant with the $^1\Delta$ and $^1\Pi$ states of N_2O combine to form $O(^1D)$ atoms, which are normally metastable but which can combine with ground state $Xe(^1S_0)$ atoms to form the $XeO(^1D)$ excimer, which then decays from two excited states yielding emission at 1.66 eV. In an analogous fashion excitation of the $N_2O(^1\Pi)$ yields $O(^1S)$ atoms. Using the time structure of SPEAR, photoluminescence lifetimes of between 110 and 3.6 ns were measured in the $Xe:N_2O$ and $Kr:N_2O$ systems at 30 K and in this way it was shown that dissociative excitation of N_2O takes place via nonradiative energy transfer from self-trapped excitons in the host.[38] Time-resolved emission spectra could be used to separate emission bands that overlap in wavelength but have dissimilar decay times and, for example, to separate the spontaneous and stimulated emission components of the same band.

Time-resolved experiments using synchrotron radiation have been used to study relaxation cascades for radiative and nonradiative transitions for xenon and krypton atoms in argon matrices and for xenon, krypton, and argon atoms in a krypton matrix. For isoelectronic dopants, guest excitons may be observed that have the same binding energies as those of the host and these states help in establishing the link between Frenkel and Wannier excitons and Rydberg states. In addition, experiments have been undertaken to study relaxation mechanisms associated with small molecules such as CO and NO in neon and argon matrices.[50]

4.2.3. Rare-Gas Measurements

Pulsed synchrotron radiation from SPEAR has been used to demonstrate the feasibility of producing oriented atomic states and of observing quantum coherence effects in their decay.[51] Experiments were carried out at pressures of 2–5 μtorr, to avoid resonance radiation trapping, giving count rates of about 15–20 s^{-1}. The initial spatial anisotropy was achieved by photoselection of atoms using the pulsed polarized radiation. In the presence of a variable applied external magnetic field of up to 40 G, Larmor precession about the field axis gave a single periodic modulation of the fluorescence intensity. Figure 12 shows the modulated decay of the $5s(\frac{3}{2})_1$ state of krypton together with a large Rayleigh scattered peak indicating the response function of the detector. The modulation frequency is seen to increase with an increase in the applied magnetic field strength. From data such as this it is possible to extract lifetimes and from the beat frequency information to measure spins, multipolarities, and g factors even when the substates are unresolved in energy.[51] Clearly this technique must become increasingly important in the future.

In other low-pressure rare-gas measurements[52] the influence of the initial state on the branching ratio for emission into the first and second continua have been measured as a function of excitation energy. Emission spectra have been studied as a function of pressure and of excitation energy with the intension of building a full picture of the dynamics of the system and eliminating the radiative lifetimes of the contributing states. The spectrum and time decay of the second continuum fluorescence of xenon at 172 nm has also been studied in detail on the ACO storage ring.[53] The fluorescence decay of the O_u^+ and $1u$ states of Xe_2^* have been measured giving one component of 2.1 ns lifetime at 151.4 nm and two components of 6.9 and 112 ns at 169 nm. The shorter components are assigned to fluorescence from the upper and lower points of the O_u^+ manifold, respectively, and the long lifetime is assigned to fluorescence from lower levels of the $1u$ state.[54] Other attempts have also been made to understand the behavior of this system using synchrotron radiation.

Using wavelengths above the threshold, 52.9-nm emission lifetime measurements have been made for the first time of the Xe II $5s5p^6$ $^2s_{1/2}$ state at pressures of $>10^{-3}$ torr giving a lifetime of 34.4 ns. Other measurements in krypton, argon,[56] and neon[57] are in progress.

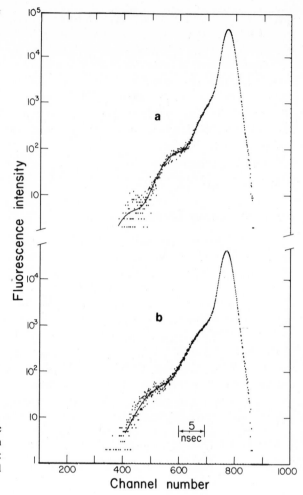

Figure 12. Fluorescence decay of the $5s(\frac{3}{2})_1$ level of atomic krypton (1236 Å, 2×10^{-6} torr) showing quantum beats in the decay curve for high and low external applied fields: (a) 34 G; (b) 27 G. Points represent data with background subtracted and the full curve is the theoretical fit.[51]

4.3. Time-Resolved Fluorescence Spectroscopy of Large Molecules

An excellent illustration of the merits of synchrotron radiation as an excitation source for fluorescence lifetime measurements is its recent application to the study of proteins. There have been a number of studies of protein molecules in which fluorescent chromophore groups were incorporated into the protein in an attempt to establish knowledge of their microenvironment within the protein. Considerable effort has been expended in the development of suitably stable light sources for such work and into the comprehensive computer programs needed to extract the physical parameters from the raw data.[22, 58] Of course, the ideal chromophore for study in a protein is one of its constituent fluorescent amino acids. Unfortunately, the lifetimes of phenylalanine, tyrosine, and tryptophan—the only amino acids that are fluorescent—are rather short (less than 10 ns), the absorption coefficient for the fluorescence transition is often low in dilute aqueous solutions of protein, and the quantum yield may be small.[59] The long-wavelength (red) edge for tryptophan absorption is approximately 300 nm and absorption by tyrosine and phenylalanine occur at 280 and 265 nm, respectively. It is the combination of requirements such as these that can be met using a synchrotron radiation source. At present, although pulsed laser sources should also be suitable, they are subject to the same detector-limited time resolution when the photon counting technique is used. Also lasers may introduce photochemical changes in the sample when used in the single-laser pulse mode along with a streak camera. Since the absorption spectrum of the tryptophan is sensitively dependent on its solvent environment, the ability to tune the wavelength of the excitation source to the region of maximum fluorescence emission anisotropy is important and this is not necessarily a simple procedure with a laser source, particularly for wavelengths in the ultraviolet region below 300 nm.

Using radiation from a storage ring, the lifetime of aqueous *L*-tryptophan solutions have been observed over a range of pH and of excitation wavelengths[60] and measurements have been made on the fluorescence lifetimes of haem proteins excited into the tryptophan absorption band. In a multitryptophan containing protein the intensity decay represents a superposition of the emissions from each of the emitting amino acid subunits. It is usually impossible in practice to differentiate between the emission from a number of different chromophores but nevertheless the integrated information is important since the fluorescence decay curves are in general nonexponential and show significant changes in going, for example, from apohaemoglobin to the intact protein.[61]

An important extension of direct lifetime studies alone of fluorescent amino acids or of other chromophores lies in the measurement of the time dependence of fluorescence polarization, which permits us to infer changes in the position of the chromophore (and hence to measure movements either of the chromophore or of the entire protein) within its excited-state lifetime.

The accessibility of internal tryptophan residues to quenching by oxygen molecules during their excited-state lifetime has shown that proteins undergo structural fluctuations on a nanosecond time scale.[62] A recent theoretical paper studied the dynamics of a globular protein by solving the equations of motion for the atoms with an empirical potential energy function. The results suggested that the interior of the protein was fluidlike and that internal movements of the order of 0.1 nm could occur in a time of the order of a picosecond (10^{-12} s).[63]

Recently, radiation from the SPEAR storage ring has been used to measure the time-resolved fluorescence emission anisotropy of a variety of materials using the apparatus shown schematically in Figure 5.[19] The experiments were carried out on a number of proteins containing only a single tryptophan in the intact material. Photoselection of the tryptophan residues were achieved using linearly (vertically) polarized light tuned to the long-wavelength absorption edge of tryptophan to achieve a maximum in the initial anisotropy at approximately 300 nm. The time dependence of the *y* polarized (vertical) and *x* polarized (horizontal) components of the fluorescence emission were measured as a function of time using a polarizing filter to view the fluorescence emission.

The total fluorescence intensity $F(t)$ and the fluorescence emission anisotropy $A(t)$ are given in the simplest case by

$$F(t) = y(t) + 2x(t) = F_0 e^{-t/\tau}$$

and

$$A(t) = [y(t) - x(t)]/F(t) = A_0 e^{-t/\phi}$$

F_0 is the initial fluorescence intensity and τ is the excited-state lifetime. The initial emission anisotropy (A_0) can have a value between 0.4 and -0.2, depending on the angle between the absorption and emission transition moments in the fixed chromophore. The rotational correlation time ϕ arises from thermal motion of all or part of the molecule and can be related to the shape and dimensions of the molecule and to the microviscosity of its fluid environment. The experimentally observed quantities $y_{exp}(t)$ and $x_{exp}(t)$ are of course modified by the instrument response function $L(t)$ according to the expression

$$y_{exp}(t) = \int_0^\infty L(t')y(t - t') \, dt'$$

and similarly for $x_{exp}(t)$. These modifications have to be taken into account when deriving any information on fluorescence lifetimes or rotational correlation times.

An important characteristic of the synchrotron radiation source for this work was that the full width at half maximum of the light pulse was 650 ps measured with a conventional commercially available electronic detection system. In addition, with a stored current of 10 mA, approximately 10^4 photons per pulse of polarized tunable radiation in a bandwidth of 4 nm was available for excitation of the sample, giving a data collection rate of about 12 kHz in the single-photon detection mode.

The very wide range of rotational correlation times that could be measured with the apparatus were revealed in a study of the emission anisotropy kinetics of the relatively small molecule N-acetyltryptophanamide (nata). The rotational correlation times were measured for the tryptophan chromophore in isotropic media (a range of glycerol-water mixtures) of known viscosity. The measured correlation times of 0.05, 0.38, 3.7, and 16.8 ns correspond satisfactorily with the measured solvent viscosities of 1.4, 17, 304, and 1445 cP at 293 K.

The results shown in Figure 13 are a striking illustration of the range of rotational freedom of the single amino acid in different proteins. The tryptophan residue in nuclease B gave a single rotational correlation time of 9.85 ns, a value that is close to a calculated value of 7.6 ns based on the assumption that the protein acted as a rigid hydrated sphere. Clearly the tryptophan had no measurable rotational freedom separate from the rotational tumbling motion of the entire protein. In contrast, the measured correlation times for basic myelin protein of 0.09 and 1.26 ns were very much shorter than the rotation time (about 7.0 ns) calculated on the assumption that the protein was a rigid sphere. These data are consistent with other experimental evidence showing that basic myelin protein behaves as a random coil. The tryptophan in nuclease B, however, is probably hydrogen bonded and in close contact with the carboxyl-terminal helix.[19]

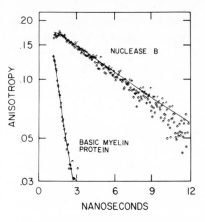

Figure 13. The single tryptophan residue of *Staphylococcus aureus* nuclease has a rotational correlation time of 9.9 ns at 20°C, corresponding to rotation of the whole protein molecule. By contrast, bovine basic myelin protein exhibits much shorter correlation times of 90 ps and 1.3 ns, probably because of the much greater flexibility associated with its random coil configuration.[19]

A further example of rotational freedom in a protein is shown in Figure 14 where, in azurin, two clearly different correlation times are measured both in the holo- and apo- forms of the protein. The very rapid component ($\phi = 0.51$ ns) can be explained as a consequence of the free rotation of the amino acid within the interior of the protein over a large angular range. By removing the prosthetic group, the entire protein becomes less rigid as is evident from the shorter correlation times, and both the correlation times and fluorescence lifetimes can be used to determine the effect of the proximity of the copper atom on the tryptophan residue inside the "fluid" interior of the azurin.

Using the storage ring source for this type of study has extended the observable rates of rotational motion by roughly two orders of magnitude into the 10^{-10} s region and has demonstrated clearly the sensitivity of an amino acid mobility to its environment. It will be interesting to relate protein flexibility to biological activities such as catalysis, allosteric regulation, energy transduction, and signaling since it appears likely that rapid structural fluctuations may be the elementary steps in conformational transitions that are essential for these processes.

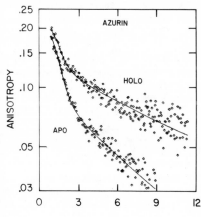

Figure 14. A comparison of the emission anisotropy kinetics of the holo- and apoprotein forms of *Pseudomonas aeruginosa* azurin at 8°C excited at 300 nm. The tryptophan fluorescence yields information about mobility within the protein, about the rotation of the entire protein, and about the increased flexibility resulting from the loss of the prosthetic group. The short correlation time for the apoazurin is 490 ps and the long correlation time for the holoazurin is 11.8 ns.[19]

The same technique has been used to study temperature effects and denaturation of tryptophan containing proteins and, in addition, to attempt to characterize the structural flexibility of immunoglobulins and immonoglubulin fragments.[64] Experiments of this type could be used also in the study of tyrosine-containing proteins, the conformational changes of DNA as a function of solvent environment, of RNA polymerases, and many other systems of biological significance.

The rotational mobility of diphenylhexatriene (DPH) and several analogues of DPH have been measured in various solvents over a range of temperatures.[65] These compounds are "sticklike" polyenes, which have been modified by the attachment of a variety of substituents at the ends of the molecule. The substituents might, for example, modify the solubility and chemical bonding properties of the molecule when used as a probe in, for example, the investigation of membrane structure and properties.

Fluorescence anisotropy studies play an important role in delineating the behavior of all types of polymers, that is, of synthetic as well as biopolymers. Dye-tagged synthetic polymers can be used to derive segmental rotation times as a function of the positional location of the tag along the polymer chain. In addition, a combination of fluorescence energy transfer and fluorescence anisotropy measurements should prove an extremely powerful tool to assist in the understanding of the reaction chemistry of these complex and important substances.

References

1. M. Sands, The Physics of Electron Storage Rings, SLAC-121, Stanford Linear Accelerator Center (1970).
2. P. Germain and H.G. Hereward, Longitudinal Equilibrium Shape for Electron Bunches with Various Self-Fields, CERN-ISR-DI/75-31, CERN (1975).
3. M. Sands, Energy Loss to Parasitic Modes of the Accelerating Cavities, PEP-90, Stanford Linear Accelerator Center (1974).
4. M. A. Allen et al., SPEAR II performance, IEEE Trans. Nucl. Sci. 22, 1366–1369 (1975).
5. P. B. Wilson et al., Bunch lengthening and related effects in SPEAR II, IEEE Trans. Nucl. Sci. 24, 1211–1214, (1977).
6. J. C. Haselgrove, A. R. Faruqi. H. E. Huxley, and U. W. Arndt, J. Phys. E. 10, 1035 (1977).
7. S. S. Hasnain, T. D. S. Hamilton, I. H. Munro, and E. Pantos, DL/SRF/077, Daresbury Laboratory (1977).
8. G. Beck, Photodiode and holder with 60 ps response time, Rev. Sci. Instrum. 47, 849–853 (1976).
9. R. Lopez-Delgado and H. Swarcz, Opt. Commun. 19, 286–291 (1976).
10. C. V. Shank, E. P. Ippen, and S. L. Shapiro (eds.), Picosecond Phenomena, Chemical Physics Series, Vol. 4, Springer-Verlag, Berlin (1978).
11. W. S. Heaps, D. S. Hamilton, and W. M. Yen, Optics Commun. 9, 304–305 (1973).
12. M. A. Allen et al., Beam energy loss to parasitic modes in SPEAR II, IEEE Trans. Nucl. Sci. 22, 1838–1842 (1975).
13. M. C. Adams, D. J. Bradley, and W. Sibbett, in Picosecond Phenomena, Chemical Physics Series, Vol. 4, C. V. Shank, E. P. Ippen, and S. L. Shapiro (eds.), Springer-Verlag, Berlin (1978).
14. A. E. W. Knight and B. K. Selinger, Aust. J. Chem. 26, 1–21 (1973).
15. B. Leskovar, C. C. Lo, P. R. Hartig, and K. Sauer, Rev. Sci. Instrum. 47, 1113–1121 (1976).
16. P. R. Hartig, K. Sauer, C. C. Lo, and B. Leskovar, Rev. Sci. Instrum. 47, 1122–1129 (1976).
17. R. Lopez-Delgado, J. A. Miehe, B. Sipp, and H. Zyngier, Nucl. Instrum. Methods 133, 231–236 (1976).

18. B. Valeur, *Chem. Phys.* **30**, 85–93 (1978).

19. I. H. Munro, I. Pecht, and L. Stryer, *Proc. Natl. Acad. Sci. U.S.A.* **76**, 56–60 (1979).

20. A. Grinwald, *Ann. Biochem. Exp. Med.* **75**, 260–280 (1976).

21. I. Isenberg, R. D. Dyson, and R. Hanson, *Biophys. J.* **13**, 1090–1115 (1973).

22. P. Wahl, *New Tech. Biophys. Cell Biol.* **2**, 233–241 (1975).

23. B. Leskovar and C. C. Lo, Lawrence Berkeley Laboratory Report 6456, University of California, Berkeley (Oct. 1977).

24. J. B. Abshire and H. E. Rowe, N.A.S.A. Tech. Memo 78028, Goddard Space Flight Center, Maryland (December 1977).

25. P. Wahl, J. C. Auchet, and B. Donel, *Rev. Sci. Instrum.* **45**, 28–32 (1974).

26. C. Benard and M. Rousseau, *J. Opt. Soc. Am.* **64**, 1433–1444 (1974).

27. O. Benoist d'Azy, R. Lopez-Delgado, and A. Tramer, *Chem. Phys.* **9**, 327–335 (1975).

28. J. B. Birks and I. H. Munro, *Prog. React. Kinet.* **4**, 239–303 (1967).

29. A. P. Sabersky and I. H. Munro, in *Picosecond Phenomena*, Chemical Physics Series, Vol. 4, C. V. Shank, E. P. Ippen, and S. L. Shapiro (eds.), Springer-Verlag, Berlin (1978).

30. H. J. Stöckmann, *Nucl. Instrum. Methods* **150**, 273–281 (1978).

31. I. B. Salmeen and L. Rimai, *Biophys. J.* **20**, 335–342 (1977).

32. E. R. Menzel and Z. D. Popovic, *Rev. Sci. Instrum.* **49**, 39–44 (1978).

33. H. P. Haar and M. Hauser, *Rev. Sci. Instrum.* **49**, 632–633 (1978).

34. E. W. Schlag, H. L. Selzle, S. Schneider, and J. G. Larsen, *Rev. Sci. Instrum.* **45**, 364–367 (1974).

35. L. Lindquist, R. Lopez-Delgado, M. M. Martin, and A. Tramer, *Opt. Commun.* **10**, 283–287 (1974).

36. R. Lopez-Delgado, in *International Colloquium of Applied Physics*, A. N. Mancini and S. F. Quercia (eds.), Vol. 1., pp. 63–124, Alghero, Italy (1976).

37. R. Lopez-Delgado, *Nucl. Instrum. Methods* **152**, 247–253 (1978).

38. K. M. Monahan and V. Rehn, SSRL Report 77/10 (Oct. 1977).

39. K. M. Monahan and V. Rehn, *Nucl. Instrum. Methods* **152**, 255–259 (1978).

40. V. Hahn, N. Schwentner, and G. Zimmerer, *Nucl. Instrum. Methods* **152**, 261–264 (1978).

41. K. M. Monahan and V. Rehn, *J. Chem. Phys.* **67**, 1784–1785 (1977).

42. L. Lindquist, R. Lopez-Delgado, M. M. Martin, and A. Tramer, in *Proceedings of the International Symposium of Synchrotron Radiation Users*, G. V. Marr and I. H. Munro (eds.), Daresbury Report DNPL/R28, pp. 257–266 (Jan. 1973).

43. R. Lopez-Delgado, A. Tramer, and I. H. Munro, *Chem. Phys.* **5**, 72–83 (1974).

44. E. D. Poliakoff. M. G. White, R. A. Rosenberg, S. Southworth, G. Thornton, and D. A. Shirley, SSRL Annual Users Group Meeting, Stanford (Oct. 1978).

45. R. Lopez-Delgado, J. A. Miehe, and B. Sipp, *Opt. Commun.* **19**, 79–82 (1976).

46. J. Najbar and I. H. Munro, *J. Lumin.* **17**, 135–148 (1978).

47. W. C. Smith, Ph.D. thesis, University of Manchester, United Kingdom (1975).

48. K. M. Monahan, V. Rehn, E. Matthias, and E. Poliakoff, *J. Chem. Phys.* **67**, 1784–1785 (1977).

49. U. Hahn and N. Schwentner, DESY Report SR-78/12 (1978).

50. E. Boursey, U. Hahn, R. Haensel, and N. Schwentner (to be published).

51. E. Matthias, M. G. White, E. D. Poliakoff, R. A. Rosenberg, S. T. Lee, and D. A. Shirley, Lawrence Berkeley Laboratory Report LBL-6651, University of California, Berkeley (July, 1977).

52. G. Zimmerer, DESY Report SR-78/12 (1978).

53. O. Dutuit, R. A. Gutchek, J. Le Calve, and M. C. Castex, SSRL Annual Users Group Meeting, Stanford (Oct. 1978).

54. G. Thornton, E. D. Poliakoff, R. A. Rosenberg, M. G. White, S. Southworth, and D. A. Shirley, SSRL Annual Users Group Meeting, Stanford (Oct. 1978).

55. M. Ghelfenstein, R. Lopez-Delgado, and H. Swarcz, *Chem. Phys. Lett.* **49**, 312 (1977).

56. R. A. Rosenberg, M. G. White, E. D. Poliakoff, G. Thornton, and D. A. Shirley, Lawrence Berkeley Laboratory Report LBL-8061, Preprint, University of California, Berkeley (August, 1978).

57. E. Matthias, R. A. Rosenberg, E. D. Poliakoff, M. G. White, S. T. Lee, and D. A. Shirley, *Chem. Phys. Lett.* **52**, 239 (1977).

58. J. Yguerabide, *Methods Enzymol.* **26**, 498–578 (1972).

59. *Excited States of Proteins and Nucleic Acids*, R. F. Steiner and I. Weinryb (eds.), Plenum Press, New York (1971).

60. B. Alpert, D. M. Jameson, R. Lopez-Delgado, and R. Schooley (to be published).
61. B. Alpert and R. Lopez-Delgado, *Nature (London)* **263**, 445–446 (1976).
62. J. R. Lakowicz and G. Weber, *Biochemistry* **12**, 4171–4179 (1973).
63. J. A. McCommon, B. R. Gelin, and M. Karplus, *Nature (London)* **267**, 585–590 (1977).
64. I. H. Munro, I Pecht, and L. Stryer (to be published).
65. R. B. Cundall, I. Johnson, M. W. Jones, E. W. Thomas, and I. H. Munro, *Chem. Phys. Lett.* **64**(1), 39–42 (1979).

10

The Principles of X-Ray Absorption Spectroscopy

GEORGE S. BROWN and S. DONIACH

1. Introduction–Overview of X-Ray Absorption Spectroscopy and EXAFS Applications

The advent of very broad-band highly collimated intense photon beams from high-energy electron storage rings may be reasonably claimed to have revolutionized the applications of photon absorption spectroscopy in the soft-to-intermediate x-ray energy regime. These applications have been found to have considerable usefulness in a wide variety of fields of science owing to the richness of chemical and structural information which may be obtained on analysis of the spectroscopic data. Using synchrotron radiation sources, these spectra may be taken on a time scale of seconds to hours compared to many days of data taking to produce a single spectrum needed with a conventional x-ray tube source.

Chapters 11, 12, and 13 of this book review the applications of this revitalized spectroscopy to a variety of different fields of science: the study of condensed-matter systems, the study of heterogeneous and homogeneous catalytic systems, and the study of biological systems. For readers who are unfamiliar with the applications of this spectroscopy, we summarize below a few of the salient, qualitative features. The interested reader may then pursue the details in the subsequent chapters.

The text of the present chapter is designed to provide both the theoretical physics underpinning of the interpretations of x-ray spectra used in the subsequent

GEORGE S. BROWN • Stanford Synchrotron Radiation Laboratory, Stanford Linear Accelerator Center, Stanford, California 94305.
S. DONIACH • Department of Applied Physics, Stanford University, Stanford California 94305.

chapters and a brief summary of some of the technical considerations in the design of the experimental set up for the spectroscopic measurements. Finally, in Section 4, we give an overview of the numerical analysis techniques used to extract physical information from the data.

As discussed below, x-ray spectra are distinguished from ordinary visible light or soft-UV spectra by the fact that these photons can excite inner-shell electrons from the absorbing atoms with consequent sharp steps in the absorption cross section as the x-ray energy is increased through an inner-shell ionization threshold.

The resulting spectrum (see Figure 1) can be analyzed in two different regions:

a. the threshold, or pre-edge and near-edge regions, from about 1 rydberg (Ry) below the absorption edge to 3–5 Ry above the absorption edge, and

b. the EXAFS—extended x-ray absorption fine structure—region, from about 3–5 Ry above the edge out to as much as 100–150 Ry above the edge.

The detailed physics of the absorption cross section in these different ranges is discussed in Section 3 of this chapter. We very briefly summarize the important characteristics here for the uninitiated reader.

The Threshold Region. This region of the spectrum contains information about the binding energies, quantum numbers, and multiplicities of low-lying bound electronic excited states of the ionized absorbing atom and of low-lying resonant electronic states in the continuum of the absorbing atom. Although quantitive information about the relation of this region of the spectrum to details of the molecular potential is difficult to obtain (see Section 3.6), it is known that the position of the edge and the qualitative features of the absorption peaks in

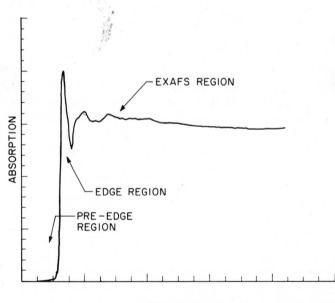

Figure 1. Schematic of x-ray absorption spectrum showing the threshold region (including pre-edge and edge regions) and the EXAFS region.

the near-edge region are sensitive to the chemical valency of the absorbing atom, and to the symmetry of the surrounding near-neighbor atoms.

The EXAFS Region. In this region of the spectrum, the observed series of gentle oscillations in the absorption cross section may be interpreted in a highly quantitative way in terms of the scattering of the excited photoelectrons by the neighboring atoms and the resulting interference of this reflected electron wave with the outgoing photoelectron waves, leading to the observed modulation of the cross section. As shown in Section 3 of this chapter, the analysis of the detailed modulation pattern of the cross section has allowed for the determination of atomic distances from an absorber to a nearby scattering atom with accuracies as good as ± 0.01 Å in a wide variety of absorbing systems. This distance information extends out to about 3.5–4.0 Å from the absorbing species, and in favorable cases one may distinguish up to four different coordination distances in this bond-length range.

Two of the most important general features in the use of EXAFS to determine near-neighbor structural information are the facts that:

a. the absorption spectrum is element specific: the inner-shell absorption steps occur at x-ray energies characteristic of the atomic number of the absorbing element, so that individual element absorption spectra can be picked out in samples containing a complex chemical mixture of different elements simply by changing the x-ray wavelength; and

b. the absorption measurement is entirely independent of the physical form of the sample. It may be applied to gases, liquids, solids (both crystalline and noncrystalline), amorphous materials, living matter, mineral samples, etc. Furthermore, above about 3 keV, the measurement may be done in air or in a helium atmosphere, independently of the ultrahigh-vacuum system of the storage ring, once the x rays have emerged from a beryllium window.

The last two features combine to make EXAFS a new spectroscopic and structural probe of extremely wide applicability. The first EXAFS spectra using a storage ring source were taken by Kincaid and Eisenberger.[22] Since that time, many thousands of spectra have been taken. Many tens of reports have been published in which the resulting analysis has led to new physical and chemical information about a wide variety of materials. In Chapters 11, 12, and 13 a subset of these applications are reviewed. However, the applications are not restricted to the topics mentioned in these chapters, and are proceeding on a wide variety of fronts, including geochemistry, atmospheric pollution studies, metallurgy, polymer science, blood chemistry, etc.

A major reason for the rapid development of the above studies is the fact that the intense photon beams available from high-energy electron storage rings allow for rapid acquisition of absorption cross-section data of high quality. With the further increases in intensity soon to be realized with dedicated storage rings and wiggler magnet assemblies, still further improvements in data rates and in the measurement of spectra for very dilute samples will become possible. These improvements will also open up the possibility for time-dependent studies of x-ray absorption spectra, so that chemical and structural changes associated with kinetic phenomena will be accessible to study by these techniques.

2. Physics of Photoabsorption in Atoms

In this chapter we will be concerned with the physics of photoabsorption from the inner shells of atoms, and, more specifically, with the effects of the near-neighbor environment of an absorbing atom in a molecule or a solid on the absorption spectrum. To discuss the physics of this spectrum, we start with some preliminary remarks on the current understanding of the physics of photoabsorption in an isolated atom.

If the coupling of an atom to the electromagnet field is treated in the dipole approximation, the total photoabsorption cross section may generally be expressed

$$\sigma(\omega) \propto \sum_f \left| \langle \psi_i | \sum_{n=1}^N \boldsymbol{\varepsilon} \cdot \mathbf{x}_n | \psi_f \rangle \right|^2 \delta(\varepsilon_i - \varepsilon_f - \hbar\omega) \tag{1}$$

where $\boldsymbol{\varepsilon}$ is the polarization vector of the electromagnetic field of frequency ω, ψ_i and ψ_f are the initial and final wave functions of the atom, with total energies E_i and E_f, and $\mathbf{x}_n$ are the position variables of each of the N electrons in the atom. Although an individual atom has relatively few electrons, and the Hamiltonian of the system is completely described, at least neglecting relativistic corrections, in terms of Coulomb forces between the electrons and between each electron and the atomic nucleus, the first-principles calculation of equation (1) is nevertheless a very complex problem because of the many-electron excitations that occur during the absorption process.

2.1. Fully Relaxed Transitions

A major part of the total cross section may be calculated by treating the photon-induced transitions in terms of the change of state of a *single electron* accompanied by the *complete relaxation* of the remaining $(N - 1)$ electrons in the atom.[1] Although this type of treatment is not general, because the resulting excited state has finite lifetime due to Auger and radiative decay, it is well defined in terms of a Hartree–Fock or Hartree–Fock–Slater calculation of the wave function for the $(N - 1)$-electron system. Within this framework the inner-shell vacancy (or so-called core hole) may be specified by an appropriate change of configuration (e.g., $1s^2 \to 1s^1$ for K-shell absorption). For this type of transition, which we will call an "elastic" transition, the contribution to the cross section may be written

$$\sigma_{\text{el}}(\omega) \propto \sum_{E_f} \left| \langle \psi_i^N | \boldsymbol{\varepsilon} \cdot \mathbf{x}_f | \psi_{f,R}^{N-1} \varepsilon_f \rangle \right|^2 \delta(E_i^N - E_{f,R}^{N-1} - \varepsilon_f + \hbar\omega) \tag{2}$$

where the subscript el indicates the elastic contribution. In the limit that the final photoelectron energy is far from threshold (see Section 3.5.1), the final-state wave function may be calculated in the sudden approximation:

$$|\psi_{f,R}^{N-1}\varepsilon\rangle = |\psi_{f,R}^{N-1}\rangle|\varepsilon\rangle \tag{3}$$

This is a product wave function consisting of the fully relaxed $(N - 1)$-electron wave function (i.e., lowest-energy configuration in the presence of the core hole) multiplied by the wave function of a valence (bound) or continuum electron calculated by solving the one-electron Schrödinger equation in the self-consistent $(N - 1)$-electron potential generated by $\psi_{f, R}^{N-1}$.

2.2. Shake Up and Shake Off

Still within the Hartree–Fock approximation, other kinds of "inelastic transition" may be considered in which the final state consists of an excited electron with energy ε, but where the remaining $(N - 1)$ electrons are now in an excited state relative to $\psi_{f, R}$, leading to a "partially relaxed" transition: "shake up" in the case of bound excited states or "shake off" in the case of continuum excited states.[2] In this type of treatment, quantum interference or "configuration mixing" between degenerate final states with different quantum numbers, is neglected. This mixing, which leads to effects such as the Fano-effect distortion of resonance lines in the continuum,[3] is also known to have a fairly major effect on the magnitude of the total cross section. (See Chapter 5 for a discussion of these effects in atoms and molecules.)

2.3. Form of the One-Electron Cross Section

The simplest calculation is that of the photoabsorption of the hydrogen atom. Because of the singular nature of the Coulomb wave function,[4] the dipole matrix element between $1s$ and continuum states varies near threshold as $1/k^{1/2}$, where k is the photoelectron wave vector,

$$k = [(2m/\hbar^2)(\hbar\omega - E_I)]^{1/2} \tag{4}$$

(where E_I is the ionization potential for the $1s$ state). This singularity is compensated by the density of final states, varying as k, leading to a step change in the photoabsorption cross section at the K edge. The full cross section above the absorption edge is given by[4, 5]

$$\sigma_{1_H}(\omega) = \frac{8\pi r_0^2}{3} \frac{128\pi}{Z^2\alpha^3} \left(\frac{E_I}{\hbar\omega}\right)^4 \frac{e^{-4 \operatorname{arccot} \xi}}{1 - e^{-2\pi\xi}} \tag{5}$$

giving as a limiting value

$$\sigma_{1_H}(\omega) \cong 1/\omega^{7/2}$$

which is valid well above threshold, where $\xi = E_I/(\hbar\omega - E_I)]^{1/2}$ and r_0 is the Thompson radius, mc^2/e^2.

For $l \neq 0$ core states, the existence of one or more nodes in the radial wave function of the core state leads to the possibility of cancellation between positive and negative parts of the dipole matrix element integral for certain values of the outgoing photoelectron energy, resulting in a zero of the one-electron cross section at some energy above threshold, the so-called Cooper minimum.[6] In practice, many-electron final-state interference effects can substantially alter the energy and amplitude of this minimum relative to the values calculated using one-electron theory.[7, 8]

Table 1. *K-Shell Core-Hole Widths and Radiative Decay Branching Ratios for a Selection of Elements*[a]

Atom	Z	K absorption energy(keV)	Full width (eV)	Fluorescence efficiency
P	15	1.839	0.50	0.058
Ca	20	4.038	0.77	0.16
Fe	26	7.112	1.15	0.34
Cu	28	8.333	1.23	0.45
Ge	32	11.104	1.83	0.55
Kr	35	14.322	2.57	0.66
Mo	42	19.999	4.29	0.77
Ag	47	25.514	6.42	0.84

[a] From Reference 9.

2.4. Effect of Core-Hole Lifetime

For many-electron atoms, the inner-shell electron vacancy is not a stable quantum state, and the system relaxes after some time either by a radiative transition or Auger transition of outer electrons. The radiative linewidth is given (for hydrogen-like wave functions) by

$$\gamma \propto Z^4 \tag{6}$$

For light atoms the main decay mode is by Auger transitions, while for heavy atoms it is radiative. Physical photoabsorption cross sections are therefore obtained by convoluting the core-hole production cross sections with a Lorentzian lineshape function in which the total width is given by the sum of the partial rates for the various radiative and Auger decay channels of the core hole. Because these widths and branching ratios are of importance in interpreting x-ray absorption spectra, their values are given for a selection of atoms in Table 1.

3. Photoabsorption in Molecular and Condensed Systems

3.1. General Features of the Spectrum

By tuning the x rays to the energy characteristic of an inner-shell absorption threshold for a specific chemical element, information may be obtained about the immediate atomic environment of this element in a complex molecular or condensed system (liquid, amorphous, or crystalline). The basic physical process is still given by expression (1), though now, for the inner shells of atoms, the core-hole final state may, to a rather good approximation, be treated in terms of a wave function in which the core hole is localized to a specific atom (of the absorbing chemical species). [This is true even in diatomic molecules of a single chemical element, owing to the rapidity of relaxation of the electrons of a given atom around the core hole.[10, 11] Corrections have recently been discussed by Caroli et al.[12]]

Within the Hartree–Fock or Hartree–Fock–Slater approximations, the elastic part of the cross section may now be written, using the product form of equation (3), as

$$\sigma_{el} \propto \left| M_{el} \right|^2 \sum_{\varepsilon_f} \int d^3x \left| \varphi^*_{i,\,core}(x)\boldsymbol{\varepsilon} \cdot \mathbf{x}\varphi_{\varepsilon_f}(\mathbf{x}) \right|^2 \delta(E_i - E_{f,\,R} - \hbar\omega + \varepsilon_f) \qquad (7)$$

where

$$M_{el} = \langle \psi_i^{N-1} | \psi_{f,\,R}^{N-1} \rangle \qquad (8)$$

is the monopole matrix element of the initial $(N-1)$ electrons (excluding the core electron) in the system with the final, fully relaxed $(N-1)$ electrons, excluding the excited electron with energy ε_f.

The calculation of $\varphi_f(x)$, the wave function of the excited electron moving in the potential of the *fully relaxed* $(N-1)$ remaining electrons in the system, involves the interaction of this electron with atoms neighboring the absorber, in addition to the potential of the absorber itself (modified by the presence of the core hole). Two regions of the absorption spectrum are to be distinguished (see Figure 1); the threshold region, extending from 1–2 Ry below threshold to 4–16 Ry above (50–200 eV, depending on the environment), and the high-energy or "EXAFS" region where, in practice, environmental effects may be seen experimentally up to about 80 Ry ($\sim$1000 eV). The distinction between these two regions is that in the threshold region, multiple scattering or "molecular cage" effects, in which the electron-neighbor scattering cannot be treated in a simple way, are important in determining the form of the spectrum, while the EXAFS region appears to be well treated in a single-scattering approximation[13, 14]:

$$\varphi_\varepsilon(x) \cong \varphi_\varepsilon^{ab}(x) + \sum_{R_j} \varphi_\varepsilon^{sc}(x - R_j) \qquad (9)$$

where the superscripts ab and sc indicate absorber and scattered, respectively.

3.2. Photoabsorption in the EXAFS Region

Using formula (9), the elastic part of the cross section becomes

$$\sigma_{el}(\omega) \propto \sigma_{el}^{ab}(\omega)[1 + \chi(\omega)] \qquad (10)$$

where $\sigma_{el}^{ab}(\omega)$ is the *atomic* elastic cross section (2), while $\chi(\omega)$ is the "EXAFS" modulation of the atomic cross section due to the environment,

$$\chi(\omega) = \frac{1}{\sigma_{el}^{ab}(\omega)} \sum_{\text{neighbors within } R_j} 2R_e \int d^3x \, \varphi_{i, \text{core}}^*(x)\boldsymbol{\varepsilon} \cdot \mathbf{x} \, \varphi_{\varepsilon, l, m}^{ab}(x)$$

$$\times \int d^3x' \, \varphi_{i, \text{core}}^*(x')\boldsymbol{\varepsilon} \cdot \mathbf{x}'\varphi_{\varepsilon, l, m, R_j}^{sc}(x') \quad (11)$$

where $\varphi_{\varepsilon, l, m}^{ab}(x)$ is a continuum outgoing wave with angular momentum quantum numbers l and m, and $\varphi_{\varepsilon, l, m, R_j}^{sc}$ is the resulting scattered wave from a neighboring atom situated at position R_j. The outgoing wave quantum number l is determined by the dipole selection rule to be $l = 1$ for K- and L_1-shell absorption, or a super-position of $l = 0, 2$ for $L_{2,3}$-shell absorption. In (11) a term of second order in $\varphi_{\varepsilon, l, m}^{sc}$ has been neglected. [This term does not produce an oscillating contribution to χ but renormalizes the atomic absorption in (11).]

In the EXAFS region, where the photoelectron wave vector k [equation (4)] is of the order of, or larger than, $(2\pi/\text{atomic dimension})$, it becomes a reasonable approximation to express φ^{sc} in terms of the asymptotic form of the wave function at distances $|x - R_j|$ far from the scattering atom:

$$\varphi_{\varepsilon, l, m}^{sc}(x) = \varphi_{\varepsilon, l, m}^{ab}(R_j)f(\Omega_x) \exp(ik|x|)\exp[-\mu(|R_j| + |x|)]/|x|$$

$$= Y_{lm}(\Omega_{R_j})\exp[i(k|R_j| + \delta_1)]f(\Omega_x)\exp(ik|x|) \quad (12)$$

$$\times \exp[-\mu(|R_j| + |x|)]/k|R_j||x|$$

where Ω_{R_j} is the angle (θ, φ) subtended by the scattering atom relative to the absorber and Ω_x the angle measured relative to the scatterer, $f(\Omega)$ is the scattering length at angle Ω to the absorber–scatterer axis, and δ_l is the phase shift of the outgoing wave due to the potential of the absorber. On calculating the matrix element of the scattered wave with the core hole, however, further account must be taken of the acceleration of the scattered electron in the Coulomb field of the absorbing atom, which leads to an additional factor $\exp(i\delta_l)$. The net result[15-17] for a given distribution $\{R_j\}$ of neighbors is

$$\chi(\omega; \{R_j\}) = \sum_j \frac{|\boldsymbol{\varepsilon} \cdot \mathbf{R}_j|^2}{k|R_j|^2} \text{Im}\{f(\pi)\exp(2ik|R_j|)\exp(2i\delta_1)\}\exp(-2\mu|R_j|) \quad (13)$$

for K-shell absorption, where $f(\pi)$ is the back scattering amplitude,

$$f(\pi) = \sum_{l'} \frac{\exp(2i\delta_{l'}) - 1}{2ik}(2l' + 1)(-1)^{l'} \quad (14)$$

and $\mathbf{R}_j$ is a unit vector along the R_j direction.

In equation (12) an additional factor $\exp[-\mu(|R_j| + |x|)]$ has been inserted to include the fact that the fast photoelectrons have a finite mean free path (which is quite strongly energy dependent) due to electron–electron scattering. After aver-aging over thermal vibrations (see below), equation (13) has been used extensively to analyze K-shell x-ray absorption data in a wide variety of materials. For L_2 and L_3 core-level absorption, equations (10) to (13) require modification since both s

and *d* outgoing electron waves are allowed by the dipole selection rule. However, a recent theoretical observation due to Teo and Lee[18] allows for a practical simplification: they observe that the relative magnitude of the $2p \rightarrow$ (continuum-*s*) transition amplitudes are about a factor five smaller than the amplitudes for the $2p \rightarrow$ (continuum-*d*) transition for elements with atomic numbers ranging from 22 (Ti) to 74 (W) and are rather independent of *k* in the EXAFS range (2 Å^{-1} to 15 Å^{-1}). This leads to an overall factor of about 50 in the relative cross sections for outgoing final *d* waves relative to outgoing *s* waves, so that to a good approximation, the $L_{2,3}$ EXAFS modulation of the elastic absorption cross section may be written in terms of *d*-wave absorption only as

$$\chi_{L_{2,3}} \cong \sum_j \frac{\frac{1}{2}(1 + 3[\mathbf{x} \cdot \hat{\mathbf{R}}_j]^2)}{k |R_j|^2}$$

$$\times \operatorname{Im}\{f(\pi)\exp(2ik|R_j|)\exp(2i\delta_2)\}\exp(-2\mu|R_j|) \tag{15}$$

Teo and Lee observe that the physical reason for the relative smallness of the outgoing *s*-wave amplitude is the fact that this wave must be orthogonalized to the core levels, so that it oscillates faster in the neighborhood of the $2p$ shell than does the outgoing *d*-wave function.

3.3. The Thermal Average of the EXAFS Cross Section

The distances between absorber and scatterers appearing in equation (13) are not unique, owing to zero-point and thermally induced atomic vibrations in molecules or condensed systems. There may also be additional randomness due to molecular diffusion (as, for instance, in solutions) or the intrinsic disorder in amorphous systems and glasses (see discussion in Chapter 11). Since the x-ray absorption process is very fast on the time scale of molecular motion (with a rate set by the lifetime of the core hole $\tau_{\text{core hole}}$, for instance, of the order of $\hbar/1$ eV for the Cu-*K* shell $\sim 10^{-15}$ s—see Table 1 in Section 2.4), only an average over the atomic positions is observed: the absorption measures an instantaneous "snapshot" of the distribution. This leads to

$$\bar{\chi}(\omega) = \sum_j \int d^3R_j \, g(\mathbf{R}_j)\chi(\omega;\{\mathbf{R}_j\})$$

where $g(\mathbf{R}_j)$ is a pair correlation function.

For single crystals this average will depend on the relative direction between the electromagnetic polarization vector ε and the crystal axis since the phonon frequencies determining the spread of atomic positions will be different along different crystalline directions. For polycrystalline (liquid or amorphous) samples, the polarization term averages out and only variations in $|R_j|$ are important. In a general molecular environment, a number of different phonon modes contribute to the spread of $|R_j|$. The simplest approximation is to treat the resulting distribution as a Gaussian with mean-square amplitude σ^2 and to neglect the effects of atomic vibration normal to the absorber–scatter bond direction.

On averaging over equation (13) and replacing the denominator $1/R_j^2$ by its average value, $1/\bar{R}_j^2$, one has

$$\bar{\chi}(\omega) = \sum_j \frac{|f(\pi)|}{k|\bar{R}_j|^2} \sin\left(2k|\bar{R}_j| + \tilde{\alpha}(k)\right)\exp(-k^2\sigma^2)\exp(\mu^2\sigma^2)\exp(-2\mu|\bar{R}_j|) \quad (16)$$

where the phase shift $\tilde{\alpha}(k)$ now acquires a temperature dependent correction,

$$\tilde{\alpha}(k) = (2\delta_l + \arg f(\pi) - 2k\sigma^2/R_j - 2k\mu\sigma^2) \quad (17)$$

The factor $\exp(-k^2\sigma^2)$ in (16) is often referred to as a "Debye–Waller" type of correction. However, it should be noted that the variance σ of the absorber–scatterer distance is quite different from the mean-square atomic displacement entering the usual Debye–Waller correction to Bragg scattering, since the latter is sensitive to motion relative to the crystal center of mass, while in EXAFS only relative absorber–scatterer motions are important.[19, 20]

It should be remarked that part of the temperature-dependent correction $2k\sigma^2(\mu + 1/R)$ to the phase shift is the result of the photoelectron mean-free-path approximation, which is not well founded theoretically, though it seems plausible. For near-neighbor scattering, it may well be an overestimate of inelastic effects, since a large part of these will already be included in complex potential corrections to the absorber and scatterer potentials (see below). At any event, such a correction has not been used in practice in analyzing the data. By omitting this term and the corresponding amplitude correction $\exp(\mu^2\sigma^2)$, equation (16) may be rewritten in the form usually used for the analysis of data as

$$\bar{\chi}(\omega) = \sum_j \frac{|f(\omega)|}{k|\bar{R}_j^2|} \sin\left[2k\bar{R}_j + \alpha(k)\right]\exp(-k^2\sigma^2)(\exp(-2\mu|R_j|)) \quad (18)$$

with

$$\alpha(k) = 2\delta_l + \arg f(\pi) \quad (19)$$

3.4. Determination of Electron–Atom Scattering Phase Shifts and Amplitudes in the EXAFS Region

In order to use equation (18) to obtain new structural information on a given material, it is necessary to have some way of determining the k-dependent phase correction $\alpha(k)$ and amplitude $|f(\pi)|$. In practice, two different approaches have been used to obtain these atomic parameters: (a) semiempirical and (b) *ab initio* (with an adjustable zero of energy). In the semiempirical approach,[14, 21] absorption spectra for model compounds of known structure are used to determine the atomic parameters that are then applied to analyze the spectra from the unknown material whose chemical constituents are similar to the model compound. (See discussion in Chapters 11 and 13.)

The *ab initio* theoretical approach, initiated by Ashley and Doniach[16] and by Lee and Pendry,[17] and applied to a simple molecule (Br_2) by Kincaid and Eisenberger,[22] has been very successfully developed by Lee and Beni[23] to a level where it appears to produce rather reliable results in a wide variety of materials (accuracies in distance determination $|R_j|$ of about 0.01 Å for nearest neighbors[18]). According to the philosophy of the fully relaxed final state, the electron–atom phase shifts needed to determine $\alpha(k)$ and $f(\pi)$ [equations (14), (15), and (19)] should be obtainable by simply integrating the one-electron Schrödinger equation for the outgoing and back scattered photoelectron in the self-consistent Hartree–Fock potential of the fully relaxed absorber atom (with core hole) and scattering atom. However, it is known from LEED and photoemission studies that energy-dependent polarization of the fully relaxed atomic potentials are important.[24] From the shake-up–shake-off point of view, this means that coherent mixing of virtual shake-up excitations with the fully relaxed state occurs in the presence of the moving photoelectron (so-called many-body or configuration-interaction corrections[25]). Lee and Beni[28] have used a relatively simple treatment of these polarization corrections, based on the local electron gas approximation,[25] and shown that the resulting corrections lead to a rather good agreement for the EXAFS phase function $\alpha(k)$ between theory and experiment for a variety of model compounds subject only to one important adjustable parameter, the energy zero for the photoelectron continuum.

In order to understand the significance of the adjustable E_0 parameter [which leads to the definition of the relationship between photon energy and photoelectron wave vector via equation (4)] it is convenient to parametrize the EXAFS phase function $\alpha(k)$ in the form

$$\alpha(k) = a_{-1}k^{-1} + a_0 + a_1 k + a_2 k^2 + \cdots \tag{20}$$

A redefinition of E_0 leads to a new definition of k via

$$k' = \left(k^2 + \frac{2m\Delta E_0}{\hbar^2}\right)^{1/2} \cong \left(k + \frac{m\Delta E_0}{\hbar^2}\frac{1}{k}\right) \tag{21}$$

Hence, the principle effect of an uncertainty in E_0 is to modify the value of the parameter a_{-1}. In a fitting procedure to relate the EXAFS formula (16) to a given spectrum, the value of a_{-1} will be most influenced by the low-k region of the data. This is the region where multiple scattering corrections become most important. Thus, by leaving a_{-1} a free parameter, some of these corrections may be included on an empirical basis. Other contributions to a_{-1} come from the Coulomb potential due to the core hole (and to the total ionic charge) since the Coulomb wave function of the photoelectron at large distances has a distance-dependent correction to the absorber phase shift of the form[4]

$$\delta_l \cong \delta_l^0 + Z_{\text{eff}} \frac{\log(R/a_B)}{a_B k}$$

where a_B is the Bohr radius. This correction will clearly be influenced by the chemical state of an absorbing ion, so it will be expected to vary from compound to compound. Other uncertainties in E_0 result from the difficulty of determining the true continuum threshold experimentally and from uncertainty about the value of the "muffin-tin" potential in the interstitial regions of a condensed system.

In practice, a_{-1} may be determined empirically for model compounds in a chemical state close to that of the unknown, as done by Cramer *et al.*,[21] or determined by a prescription due to Lee and Beni.[23] They point out that in formula (16), if $\alpha(k)$ is known perfectly, the Fourier transform of the empirical EXAFS function $\chi(k)$, should have an imaginary part that peaks at the same R value as the modulus of $F(R)$ for a given single shell of atomic scatterers. The phase shift and amplitude functions are removed using the theoretically calculated forms of these functions,

$$F(R) = \int dk \, \exp(-i2kR)\chi(k) \exp[-i\alpha(k)]k \exp(+k^2\sigma^2)/|f(\pi)| \qquad (22)$$

This fit is achieved by adjusting E_0 (or equivalently a_{-1}). For the molecule Br_2, Lee and Beni use this prescription to find a value of E_0 6 eV below the continuum value estimated by Kincaid and Eisenberger,[22] based on a calculation of the position of the $4p$ pre-edge bound-state absorption. The resulting change of the interatomic distance estimate relative to the experimental value is 0.02 Å [based on the position of the maximum of $|F(r)|$] or 0.013 Å [based on the position of the maximum of Im $\{F(r)\}$]. The values of calculated phase shift and amplitude functions for a variety of elements have been tabulated by Teo *et al.*[18] Some representative values are given in Table 1.

The semiempirical approach—transferability of EXAFS phase shifts. There is a good experimental and theoretical evidence that the phase shifts, $\delta_l(k)$, and arg $f(\pi, k)$ determined by numerical fitting procedures (see Section 4) over a k range with $k_{min} \geq 3$ Å^{-1} are transferable from one chemical environment to another in gases and some solids. In performing an experiment on some model compound with known internuclear separation, one always obtains the sum $\alpha(k)$. Although it is not possible to obtain the central atom and scattering atom phase shifts separately from any finite set of binary measurements, it is possible to obtain their sum. Citrin *et al.*[23a] have shown that $\delta_1(k)$ and arg $f(\pi, k)$ are transferable among vapor phases of Ge and Br compounds, and that arg $f(\pi, k)$ for carbon is similarly transferable, to accuracies of 0.02 Å or better.

3.5. Many-Electron Effects in the EXAFS Spectrum

The treatment of EXAFS as a one-electron process leading to the basic EXAFS formula (18) assumes implicitly that the remaining $(N-1)$ atoms in the system act as "spectators" once they have readjusted into their fully relaxed configuration in the self-consistent potential of the core hole. Two classes of correction to this may be distinguished:

a. final states in which part of the incoming photon energy is left behind in a shake-up or shake-off excitation of the system, in addition to the energy given to the primary photoelectron—these events contribute to the inelastic part of the total absorption cross section;

b. energy-dependent polarization of the valence electrons by the outgoing photoelectron but where the final state is still fully relaxed—these effects contribute to an energy dependent modification of the phase shift and amplitude functions.

3.5.1. Inelastic Events

Neglecting the finite lifetime of the deep hole state, the inelastic part of the total cross section (1) may generally be written as

$$\sigma_{\text{inelastic}}(\omega) = \sum_{\varepsilon_f, s} \left| \langle \omega_i^N | \boldsymbol{\varepsilon} \cdot \mathbf{x} | \psi_s^{N-1}, \varepsilon_f^* \rangle \right|^2 \delta(E_i^N - E_R^{N-1} - \Omega_s - \varepsilon_f^* + \hbar\omega) \quad (23)$$

where $\Omega_s = E_s^{N-1} - E_R^{N-1}$ is the inelastic excitation energy of the spectator electrons measured relative to the fully relaxed final-state energy. In order to make progress on the evaluation of (23), some consideration must be given to the time scale of evolution of the final states $|\psi_s^{N-1}, \varepsilon_f^*\rangle$. (For a recent discussion see an article by Gadzuk.[26]) For final electron energies ε_f^* very close to zero, i.e., for absorption events close to threshold, the interaction of the slowly outgoing electron with the deep hole may not be neglected, so that the final-state wave function can not in general be written as a product of an excited (or fully relaxed) $(N-1)$-electron wave function with the outgoing electron wave function. On the other hand, for final electron energies ε_f^* large compared to typical relaxation times (of the order of plasmon energies) for the $(N-1)$-electron system, the "sudden approximation" may be used in which $|\psi_s^{N-1}, \varepsilon_f^*\rangle$ may be treated as a product wave function $|\psi_s^{N-1}\rangle|\varepsilon_f^*\rangle$, as in the elastic case (2). For the EXAFS region of the spectrum, starting 2–3 Ry above threshold, this is a reasonable approximation. In this region, for a given photon energy, $\hbar\omega$, ε_f^* is now shifted relative to the photoelectron energy in the elastic part of the cross section by the amount of energy loss Ω_s to the $(N-1)$ remaining electrons. Hence, within the sudden approximation, the effective one-electron form of the inelastic cross section may be written[27]

$$\sigma_{\text{inelastic}}(\omega) = \sum_s |M_s^2| \sigma_{\text{inelastic}}^s(\omega)[1 + \chi_s(\omega - \Omega_s)] \quad (24)$$

where M_s is the monopole matrix element $\langle \psi_i^{N-1} | \psi_s^{N-1} \rangle$ between the initial state (with core electron removed) and the final shake-up–shake-off excited state. $\sigma_{\text{inelastic}}^s(\omega)$ is the absorber one-electron cross section calculated in the Hartree–Fock potential corresponding to the shake-up configuration and $\chi_s(\omega)$ is the EXAFS function (18) appropriate to this same excited state potential but in which the photoelectron wave vector is calculated using the down-shifted photoelectric energy

$$k_s = (k^2 - 2m\omega_s/\hbar)^{1/2} \quad (25)$$

It may therefore be seen that the inelastic part of the photoabsorption cross section will contribute some EXAFS-type oscillations to the total cross section, which, however, will have a period which is averaged over a whole spectrum of inelastic shifts of the photoelectron wave vector k, and which depends on the spectrum with respect to Ω_s of the shake-up and shake-off states. These spectra have been studied both theoretically and experimentally (by photoemission spectroscopy), and for nonmetallic samples one or two discrete shake-up levels of relatively low energy ($\varepsilon_s \geq 1$ Ry) have been found, followed by a continuum of shake-off excitations.[28]

It is clear that the continuum part of the cross section will tend to average out the amplitude of the inelastic EXAFS $\chi(\omega - \Omega_s)$ so that this will contribute a relatively smooth background to the total absorption cross section. The discrete levels, on the other hand, will give rise to a definite EXAFS oscillation: for $k^2 \gg 2m\Omega_s/\hbar$, (25) may be expanded to give $k_s \cong k - m\Omega_s/\hbar k$. It may thus be seen that the principal effect of low-lying shake-up states is to contribute a term in the general EXAFS expression (18) in which the phase-shift function $\alpha(k)$ (20) has an altered coefficient of the $1/k$ term in its parametrization (20). As discussed earlier, this will only be important in the low-energy region of the EXAFS spectrum where other corrections are also significant.

3.5.2. Effect of Screening in Metals

The above framework provides an unambiguous way to describe the fact that the conduction electron screening in metals is a time-dependent process. At first sight, it might be supposed that only low-energy EXAFS events will see the fully screened potential, while higher-energy events (several hundred eV above the edge) would tend to see an unscreened potential. In fact, the above description shows this is not the case (see, however, a recent discussion that disagrees with this point of view[29]). In a one-plasmon description of screening, the fully relaxed or screened state of the ion corresponds to the elastic part of the EXAFS spectrum while the unscreened part corresponds to an inelastic contribution in which the final state $|\psi_s^{N-1}\rangle$ contains a one-plasmon excitation. As shown above, this will lead to an EXAFS contribution with a phase shift that is altered principally in the low-k region of the spectrum. At higher-k values, the phase-shift function for the one-plasmon inelastic term $\alpha^s(k)$ would be calculated using an unscreened ion potential, which could lead to a small shift of the term linear in k. However, the relative weight $|M_s^2|$ of this contribution is known to be rather small (of order 10% or so) compared to the fully relaxed matrix element (from studies of plasmon sideband contributions to core lines in x-ray photoemission spectroscopy, XPS[1]) so that even at high photoelectron energies, the principal contribution to the EXAFS comes from the fully relaxed, i.e., fully screened, final state.

The main effect of the inelastic events, therefore, is to contribute a relatively smooth background to the EXAFS spectrum. Lee and Beni have shown empirically that this may be taken into account by renormalizing the magnitude of the EXAFS function calculated using the purely elastic one-electron theory by an energy-independent factor of order 65% to take account of this inelastic background. As discussed by Rehr and Stern,[27] this is reasonably consistent with calculations of the total inelastic contribution in terms of the magnitudes of the matrix elements $|M_s^2|$.

3.5.3. Exchange and Correlation Effects on the One-Electron Potential

The other principal many-electron effect distinguished at the beginning of this section is the dynamic polarization of the electron density in the absorbing and scattering atoms, as a function of the velocity of the primary photoelectron. Since the polarization effects occur as an intermediate state even in the elastic contribution to the total cross section, they correspond to a virtual coherent mixing of shake-up and shake-off states with the fully relaxed ground state.

Lee and Beni[23] have treated this effect, following ideas of Hedin and Lundquist,[25] using an energy-dependent complex potential correction to the Hartree self-consistent potential seen by the photoelectron. This correction (which would reduce to the Kohn–Sham version of the Slater $\rho^{1/3}$ exchange for an electron at the Fermi level of a metal) approximately accounts for the energy-dependent exchange and correlation (i.e., polarization) of the electron gas, treated locally as a free-electron gas of density $\rho(\mathbf{r})$, by the photoelectron. This is done by solving the one-electron Schrödinger equation.

$$\left\{ \frac{\hbar^2}{2m} \nabla^2 + V_{\text{Hartree}}(\mathbf{r}) + \sum \left[p(\mathbf{r}), \frac{1}{2m} p(\mathbf{r}) \right] \right\} \psi = E\psi$$

where $\sum (p, E)$ is an energy-dependent free-electron gas-type self-energy, which is calculated by Lee and Beni using a one-plasmon excitation formula. To calculate the effective local kinetic energy, $\frac{1}{2} p^2(r)$ (and accompanying momentum in a quasi-free-electron approach), Lee and Beni use the ansatz $p(r)^2 \cong k^2 + k_F(r)^2$, where k^2 is the photoelectron wave vector and $k_F(r)$ is the local Fermi wave vector given by $\frac{2}{3}\pi \rho(r)^{1/3}$.

Using this complex energy-dependent potential, Lee and Beni find an appreciable improvement of the calculated EXAFS phase-shift function $\alpha(k)$ for EXAFS Br_2 and $GeCl_4$ molecules. (In terms of the comparison of calculated and observed interatomic distances, the improvement amounts to about 0.1 Å for Br_2.)

3.6. The Near-Edge Region of the X-Ray Absorption Spectrum— Multiple Scattering Effects

The reader is referred to Chapter 4 for a detailed discussion of the phenomenology of photoabsorption in molecules near an inner-shell absorption edge. The purpose of this section is to review recent progress in the theoretical treatment of the near-edge continuum photoabsorption. We will be concerned purely with the "elastic" 1-*electron* contributions to the total photoabsorption, i.e., in which the final state of the $(N-1)$ spectator electrons is fully relaxed. In this framework the outgoing photoelectron moves in a potential $V_{\text{mol}}(r)$ that measures both the direct Coulomb interactions with the atomic nuclei and the electrons composing the molecular cluster and also the exchange and correlation effects measuring the many-electron polarization of the spectator electrons by the moving photoelectron.

In the spectral region immediately above the inner-shell threshold, the outgoing photoelectron is moving with energies comparable to the local variations in molecular potential, and the single-scattering approach appropriate to the EXAFS region breaks down. Ashley and Doniach,[16] and Lee and Pendry[17] have discussed an intermediate-energy region where multiple scattering contributions can be put in by successive approximations, or iterations, of the integral form of the Schrödinger equation

$$\psi(r) = \phi(r) + \int d^3r'\, G_E(r - r')V_{\text{mol}}(r')\psi(r') \tag{26}$$

where G_E is the free-electron Green's function for a photoelectron of kinetic energy E. However, as E approaches zero, (in practice for $0 \le E \lesssim 3$ Ry for low-to-intermediate-Z molecular clusters), a low-order iteration of equation (26) in powers of $V(r)$ is not a good approximation and the equation requires direct numerical solution.

This is done by expanding $\psi(r)$ in a set of spherical wave bases centered on the individual atoms of the cluster. The potential $V_{\text{mol}}(r)$ is treated as spherically symmetric in spheres ("region I") surrounding each atom. Then the entire cluster is surrounded with an outer sphere ("region III" outside the outer sphere) and the potential in the interstitial region between inner and outer spheres ("region II") is treated as constant—the muffin-tin approximation. The amplitudes of the spherical wave projections of ψ on each of the regions are matched by continuity of ψ and $d\psi/dr$ at the sphere boundaries, leading to a homogeneous secular equation for states below threshhold (bound states) and an inhomogeneous matrix equation version of equation (26) for the continuum region. This approach, the so-called multiple scattering method, was introduced by K. Johnson[30] for bound-state calculations in molecules (where it is closely analogous to the Korringa–Kohn–Rostoker method for electronic band calculations in solids) and the continuum version was worked out by Dill and Dehmer[31] and applied to the calculation of photoabsorption and electron scattering in diatomic molecules (see also the work by Davenport[32]).

One of the principal results of the work of Dill and Dehmer is the explanation of resonance structure observed in energy-loss studies for electron–dinitrogen forward scattering (analogous to photoabsorption)—the effect of the molecular potential on an outgoing photoelectron wave is to mix the dipole allowed $1s \to$ continuum p wave with higher-l partial waves. The molecular field also serves to enhance the resulting higher partial wave resonance. This kind of resonance effect could not be calculated in terms of a low-order iteration of the multiple scattering equation, but essentially involves the inversion of the secular matrix representing the continuum states in the molecular potential.

Recently the multiple scattering approach has been extended to polyatomic clusters by Natoli et al.[33] See also work on electron–molecule scattering by Dehmer et al.[34a] and Kennerly et al.[34b] Natoli et al. have performed *ab initio* multiple scattering calculations for K-shell photoabsorption in the tetrahedral cluster $GeCl_4$. One of the features of the calculation was the use of the same energy-dependent Hedin–Lundquist potential used by Lee and Beni for EXAFS cal-

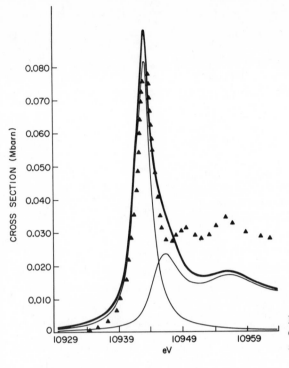

Figure 2. Calculation of K absorption edge structure for $GeCl_4$,[33] and comparison with measurements (▲) of Kincaid and Eisenberger.[22]

culations. The results, shown in Figure 2, are in fair agreement with the x-ray absorption spectroscopy, XAS, measurements of Kincaid and Eisenberger. Of particular note are a series of resonances in the range 5–20 eV above the absorption edge. Projection of the continuum wave function onto partial waves in the outer sphere region shows considerable mixing of higher-l components in the resonance regions, analogous to the mixing found for the diatomic case by Dill and Dehmer. While only a first attempt, these results are an encouraging indication that the richness of observed structure in the near-edge region may be understood in terms of a one-electron theory of the absorption process and hence will provide a useful analytic tool for the study of the details of bonding and the chemical state of absorbing atoms in complex systems.

4. Instrumentation for X-Ray Absorption Spectroscopy

4.1. Introduction

The field of x-ray instrumentation has evolved rapidly over the past ten years, owing to the increased activity in x-ray astronomy, controlled thermonuclear fusion, and synchrotron radiation research. In this section, we will restrict our discussion to the particular problems of x-ray absorption spectroscopy.

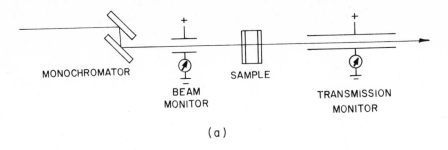

(a)

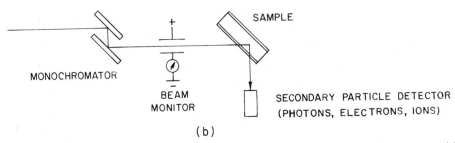

(b)

Figure 3. Schematic for x-ray absorption measurements by transmission and by secondary particle detectors.

The problem can be posed as follows: given a broadband source of x-radiation, what is the best means of measuring the absorption cross section $\sigma_x(E)$ of some atomic species present in the solid, liquid, or vapor phase in some given matrix. Although there are many exotic possibilities, we will confine our discussion to two classes of methods (see Figure 3). In method (a), the radiation is monochromatized by a periodic structure such as a crystal or grating, and the sample absorption is measured directly by monitoring the incident and absorbed flux. In method (b), the radiation is similarly monochromatized, but some signature of the absorption, such as optical or x-ray fluorescence, is detected. Before discussing the details of the hardware, we shall first discuss the relative sensitivities of the two methods.[35]

For the first method, we have, for the measures of incident and transmitted intensity, respectively,

$$M_i = I_0[1 - \exp(-\mu_i L_i)]$$
$$M_t = I_0[\exp(-\mu_i L_i - \mu_m L_x - \mu_x L_x)]$$

where M_i and M_t are the measures of the incident and transmitted intensity, respectively, and the incident monitor is a material of thickness L_i and absorption coefficient μ_i. The sample consists of thickness L_x of species x and thickness L_x of the matrix, with absorption coefficient μ_m. We then have

$$\mu_x = (1/L_x)\{\ln(M_i/M_t) - \ln[1 - \exp(-\mu_i L_i)] - \mu_i L_i - \mu_m L_x\} \qquad (27)$$

For many EXAFS experiments, it is sufficient to know μ_x only up to an additive constant, obviating the need to know the calibration constants of the detectors, as well as absolute sample thickness.

We calculate the sensitivity of the transmission technique by assuming that the statistical uncertainty in the measurement resides in the fluctuations in M_i and M_t, and that these fluctuations are characteristic of a Poisson distribution, i.e., $(\Delta M_i)^2 = M_i$ and $(\Delta M_t)^2 = M_t$. We then have

$$(\Delta \mu_x)^2 = \frac{1}{L_x^2}\left(\frac{1}{\mu_i} + \frac{1}{\mu_t}\right)$$

$$= \frac{1}{I_0 L_x^2}\left(\frac{1}{1 - \exp(-\mu_i L_i)} + \exp(-\mu_i L_i - \mu_m L_x - \mu_x L_x)\right) \quad (28)$$

The optimum sensitivity will be achieved when $\partial(\Delta \mu_x)^2/\partial L_i = 0$ and $\partial(\Delta \mu_x)^2/\partial L_x = 0$; this occurs when

$$L_x = \frac{2.557}{\mu_x}, \qquad L_i = \frac{0.246}{\mu_i}$$

Inserting these values into equation (28) we get

$$(\Delta \mu_x)^2 = (3.223/I_0)(\mu_x + \mu_m)^2 \quad (29)$$

The second method for determining the absorption cross section is to measure the secondary intensity M_s of some physical process that is characteristic of the relaxation of the core hole. If we lump the quantum yield,[36] solid angle, and detector efficiency into one factor $\Omega \leq 1$, we then have (see Figure 3)

$$M_i = I_0[1 - \exp(-\mu_i L_i)]$$

$$M_x = I_0 \exp(-\mu_i L_i)\Omega\{1 - \exp[-L_x(\mu_m + \mu_x + \mu_s)]\} \frac{\mu_x}{\mu_m + \mu_x + \mu_s}$$

$$\approx \begin{cases} I_0 \exp(-\mu_i L_i)\Omega\mu_x L_x & \text{for } (\mu_m + \mu_x + \mu_s)L_x \ll 1 \\ I_0 \exp(-\mu_i L_i)\Omega \dfrac{\mu_x}{\mu_m + \mu_x + \mu_s} & \text{for } (\mu_m + \mu_x + \mu_s)L_x \gg 1 \end{cases}$$

The first approximation applies to thin samples such as films or monolayer adsorbates. The second approximation, which applies to thick samples, is only useful when $\mu_m + \mu_s \gg \mu_x$, i.e., dilute concentrations. For thick samples we shall assume that this inequality is valid.

Performing an analysis similar to that for the transmission case, but assuming that L_x is fixed by the sample preparation technique, we get

$$\left(\frac{\Delta \mu_x}{\mu_x}\right)^2 = \frac{1}{\mu_i} + \frac{1}{\mu_s} = \frac{1}{I_0}\left[\frac{1}{1 - \exp(-\mu_i L_i)} + \frac{1}{\exp(-\mu_i L_i)\Omega\mu_x L^*}\right] \quad (30)$$

where

$$L^* = L_x \qquad\qquad (\mu_m + \mu_x + \mu_s)L_x \ll 1$$

$$L^* = (\mu_m + \mu_x + \mu_s)^{-1} \qquad (\mu_m + \mu_x + \mu_s)L_x \gg 1$$

Minimizing $(\Delta\mu_x)^2$ with respect to L_i gives

$$\mu_i L_i \approx \Omega\mu_x L^*$$

$$(\Delta\mu_x)^2 \approx \frac{1}{I_0}\left(\frac{2\mu_x}{\Omega L^*}\right) \tag{31}$$

The physical content of these expressions is plain: the best performance is achieved in method (b) when the detection rates in the beam monitor and secondary detector are the same. Furthermore, by comparison with equation (29), we see that for $\mu_m \gg \mu_x$, $(\Delta\mu_x)^2$ is independent of μ_x in transmission, but that $(\Delta\mu_x)^2$ is proportional to μ_x in method (b). The techniques have equal sensitivity for thick, dilute samples when

$$\Delta\mu_x = \frac{1.61(\mu_m)^2}{\mu_m + \mu_s} \tag{32}$$

4.2. X-Ray Monochromators

We restrict our discussion of x-ray monochromators to semiquantitative descriptions of existing, proven systems. For details about monochromator design or novel methods for reducing harmonics, improving angular or spatial resolution, etc., the reader is referred to several excellent review articles[37–42] and the discussion in Chapter 3.

The most successful monochromator design realized to date is shown in Figure 4. The first crystal C_1 serves as primary monochromator, and the second crystal C_2, when correctly adjusted, filters out harmonics and unwanted reflections, as well as rendering the output beam parallel to the input beam, displaced by a distance $2D \cos\theta$, where D is the distance between crystal faces.

The bandwidth and efficiency of this parallel crystal monochromator is analyzed by examining a given polychromatic ray originating from point O in Figure 4. The Ewald–von Laue theory of the dynamical diffraction of x-rays gives the probability $P(\theta, E)$ that an x ray of energy E and angle θ is reflected by a perfect crystal. Some reflection profiles for the (220) reflection of silicon are shown in Figure 5. For our purposes it is sufficient to approximate these curves as rectangular functions with height $R < 1$ and width $\delta\theta$, the so-called Darwin width. In this

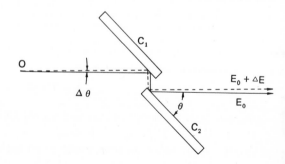

Figure 4. Two-crystal x-ray monochromator in parallel configuration.

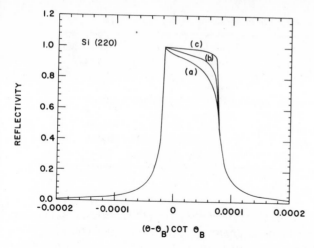

Figure 5. Reflectivity profile for Si(220) Bragg reflection calculated according to the Darwin theory: (a) 5411 eV; (b) 8040 eV; (c) 17 445 eV.

approximation, the energy spread associated with a *given* ray will be approximately $\Delta E = E_0 \cot \theta_0 \, \delta\theta$, and the additional energy spread associated with the angular divergence of the beam will be $\Delta E = E_0 \cot \theta_0 \, \Delta\theta$. Under single-beam conditions at SPEAR, with $E_e = 3.7$ GeV, $\Delta\theta$ is about 1.2×10^{-4}; under colliding-beam conditions, $\Delta\theta$ can be perhaps a factor of two-to-three larger. However, for the (220) reflection at 8040 eV, $\delta\theta$ is about 4×10^{-5}. Thus, silicon monochromators utilizing the full vertical angular spread of the beam have energy resolutions dominated by this divergence. For moderately high resolution spectroscopy one can collimate the beam to an angular spread h/R where h is the effective source height and R is the source-to-slit distance. For example, at SPEAR $h \simeq 0.1$ cm under single-beam conditions, and $R \simeq 2 \times 10^3$ cm, giving $\Delta\theta \simeq 5 \times 10^{-5}$, which still dominates the resolution function.

The second crystal, when adjusted slightly out of parallel with the first crystal, will selectively filter out high-order reflections. This is because the Darwin width is proportional to $F(\sin \theta/\lambda)\tan \theta/(h^2 + k^2 + l^2)$, where $F(\sin \theta/\lambda)$ is the structure factor; therefore, the tolerance on the adjustment of the crystals is much more critical for high-order reflections. Furthermore, if the crystals are randomly oriented about the lattice normal vector, then the $(h' k' l')$ reflections from the first crystal are absorbed by the second one, and only the primary $(h k l)$ beam is transmitted. For the so-called channel-cut monochromator, where the reflecting surfaces are part of a larger, monolithic crystal, one does not have these degrees of freedom, and the harmonics and spurious reflections are efficiently transmitted along with the primary beam.

4.3. X-Ray Detectors

In an EXAFS experiment, the signal of interest may be a 1% modulation of the absorption cross section, and this modulation may have a repetition period of order 100 s in real time. Therefore, it is imperative that the x-ray detectors have a corresponding gain instability. Most pulse counting techniques (scintillation detectors,

Table 2. Summary of Properties of X-Ray Detectors

Detector	Time resolution	Dead time	ΔE(FWHM) at 10 keV
Ion chamber	100 ns	—	—
Conductive wire proportional counter	25 ns	300 ns	1300 eV
Resistive wire proportional counter	100 μs	100 μs	1300 eV
Inorganic scintillation counter	10 ns	500 ns	5000 eV
Organic scintillation counter	1 ns	10 ns	8000 eV
Semiconductor detector	5 ns	10 ns	200 eV

proportional counters, and semiconductor detectors) are sufficiently stable because the method is inherently independent of analog gain. However, for counting rates comparable to, or greater than, the pulse repetition frequency, analog counting techniques must be employed. It is generally true that the low-frequency gain stability of semiconductor amplifiers, whose gain depends upon passive feedback components, is superior to the gain stability of electron-cascade-type amplifiers, such as photomultipliers and proportional counters. Thus, for achieving signal-to-noise figures comparable to $N^{1/2}$, where $10^5 \le N \le 10^{10}$, the ion chamber (or its cousins the dc semiconductor detector and charge-coupled devices) have the best performance. For example, commercially available FET amplifiers have noise figures of 10^{-14} C s^{-1}Hz$^{-1/2}$ or better. This corresponds to an rms quantum noise rate of 234 Hz for 8000-eV photons ionizing a gas with 30-eV/ion-pair mean ionization potential. For signal rates below 5×10^4, the amplifier noise dominates, and pulse counting techniques are therefore preferred. Table 2 summarizes the properties of x-ray detectors. (See also Chapters 3 and 16.)

4.3.1. Ionization Chambers

The simplest gaseous x-ray detector is the ionization chamber, shown in Figure 6.[43] X rays penetrating the active volume are photoelectrically absorbed, resulting in a fast photoelectron and either Auger electrons or fluorescence photons ranging from a few eV to a few keV. The fast electrons produce further electron–ion pairs

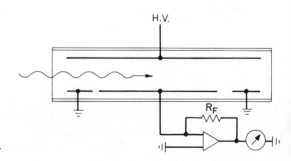

Figure 6. Cross section of a parallel-plate ionization chamber.

by inelastic collisions, and the photons either escape or are photoelectrically absorbed. If the applied electric field exceeds 100 V/cm, the electrons and ions are swept apart before they have the opportunity to recombine. For all common gases, a photon energy of 25–35 eV is required to produce a single ion pair. The detection efficiency of an ion chamber is determined by the active volume, the gas pressure, and gas mixing.

4.3.2. Proportional Counters

The ionization chamber, having no intrinsic gain, is too insensitive for most applications as a secondary particle detector. However, when the electric field in the vicinity of the anode is increased to above 25,000 V/cm, secondary multiplication takes place, and the device exhibits gain. From simple arguments one can show that the gain should follow the law[44]

$$\ln A = kE^{1/2}((E/E_0)^{1/2} - 1) \tag{33}$$

where k and E_0 depend upon the gas mixture, and are typically 0.2 $(cm/V)^{1/2}$ and 25 000 V/cm, respectively. To realize high electric fields at the anode, it is taken to be a fine metallic wire or resistive fiber concentric with the cathode in the single-wire proportional counter, or a plane or wires parallel to a plane cathode.

For counters operating in the region where equation (33) is valid, gains of $A \sim 5000$ can be easily realized without corona, sparking, or discharge taking place.[45] The anode-to-ground capacitance can be reduced to below 20 pF in a carefully designed chamber; a 10-keV x-ray stopping in such a detector will charge the anode to about 10 mV with a rise time of about 10 ns. The energy resolution is given by the formula

$$\langle (E - \bar{E})^2 \rangle^{1/2} = \bar{E}(F/N)^{1/2} \tag{34}$$

where N is the mean number of electrons liberated before multiplication has taken place, and F is the Fano factor.[46, 47] The Fano factor reflects the statistical nature of the division of energy between the charged products (electrons and ions) and neutrals (phonons and photons). If there is strict correlation between these products, the Fano factor is zero and the linewidth is zero; if the division is completely uncorrelated, the factor is one. The latter case obtains for ionization chambers and proportional counters, resulting in an energy resolution of about 1300 eV (FWHM) for 10-keV x rays.

4.3.3. Scintillation Counters

Detectors for which the fluorescence optical radiation resulting from the x-ray absorption process is detected are called scintillation counters. At the present time the most popular scintillation materials can be divided into two classes: the organic scintillators and the single-crystal activated alkali halides. The former materials are inexpensive and machinable in air, and the luminescent lifetime can be as short as 2 ns.[48, 49] The pulse-height resolution is very poor, however; about 400 eV per photoelectron at the photocathode is required, assuming a photocathode efficiency of 0.2. Inorganic alkali halides such as sodium iodide have better energy resolution (350 eV/photoelectron), but the scintillation lifetime is 270 ns, which limits the pulse pair resolution to about 700 ns. The crystals are fragile and hygroscopic, and can be easily fractured by large thermal gradients. In almost every respect they are inferior to proportional counters, except for the fact that they are commercially available in a wide variety of sizes and shapes. For synchrotron radiation experiments, the resolution is good enough to resolve the harmonic frequencies from the fundamental, and the pulse pair resolution is compatible with repetition frequencies of order 1 MHz.

4.3.4. Semiconductor Devices

The solid state analog of the ionization counter is the semiconductor detector, shown in Figure 7. The charge carriers are now electrons and holes rather than electrons and ions, and the ionization potential is the band gap, which is 2.97 eV for germanium and 3.73 eV for silicon at 77 K. At the present time silicon and germanium are the most popular materials used for x-ray detectors. They have high carrier mobility, small band gap, and are available with the required purity.

The principal factors affecting the energy resolution of a semiconductor detector[50] are the presence of crystal imperfections and impurities inhomogenously distributed throughout the crystal. These imperfections may trap the carriers for a period exceeding the pulse-shaping time, which is on the order of 10 μs. Furthermore, to achieve the ideal detector, which is totally depleted, it is necessary to have a net impurity concentration less than 10^9 cm^{-3}. Small germanium crystals with this purity are now available, but large germanium crystals, as well as silicon crystals, must be deliberately compensated. This is accomplished by diffusing lithium into p-type material at elevated temperatures in the present of an electric field of about 5000 V/cm.

The semiconductor detector, like the ionization chamber, has no intrinsic gain, and therefore must be provided with external electronic amplification. The signals are so small that the amplifier noise makes a significant contribution to the pulse-height resolution. Furthermore, there are important applications of semiconductor

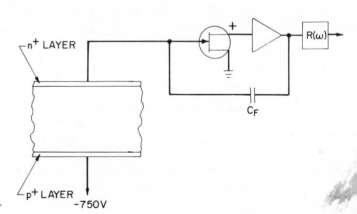

Figure 7. Schematic of semiconductor detectors.

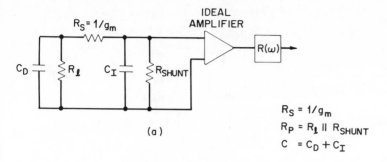

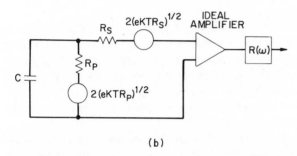

$$R_S = 1/g_m$$
$$R_P = R_\ell \parallel R_{SHUNT}$$
$$C = C_D + C_I$$

Figure 8. Equivalent circuits for the preamplifier stage of a semiconductor detector.

detectors to synchrotron radiation experiments where very high counting rates are desirable, and it is useful to know what the trade-off between averaging time and pulse-height resolution is.

The three principal sources of electronic noise in a typical detector–amplifier circuit are:

a. leakage resistance of the crystal,
b. shunt resistance of the input amplifier, and
c. the channel noise of the input stage of the amplifier.

These sources can be lumped into an equivalent series resistance R_s and parallel resistance R_p, followed by an ideal amplifier and pulse-shaping network (see Figure 8). The series resistance is approximately equal to the reciprocal of the amplifier transconductance g_m, and is roughly 200 Ω for presently available field-effect transistors. The parallel resistance is equal in magnitude to the parallel combination of the leakage resistance of the crystal and the shunt resistance of the input amplifier, and is typically 10^{12} Ω. The series resistance contributes a Johnson noise spectrum $V_s(\omega) = 2(ekTR_s)^{1/2}$, which is flat to infinite frequencies; the parallel resistance, however, being in parallel with the input capacitance C, contributes a noise spectrum $V_p(\omega) = 2(ekTR_p)^{1/2}/(1 + b\omega)$, where $b = R_p C$. A good pulse-shaping network will minimize the squares of the sums of the noise from these two sources. A prototype network might have the step response $(at)\exp(-at)$, which would be represented, in the frequency domain, by the response function $R(\omega) = ia\omega/(1 - ia\omega)^2$. In general, the mean-square noise is given by

$$\langle V^2 \rangle = \int |V(\omega)|^2 |R(\omega)|^2 \, d\omega$$

and, for the parallel and series contributions, with $R(\omega)$ given above,

$$\langle V_s^2 \rangle = 4ekTR_s \int \frac{(a\omega)^2}{(1 + (a\omega)^2)^2} \, d\omega$$

$$\langle V_p^2 \rangle = 4ekTR_p \int \frac{1}{1 + (b\omega)^2} \frac{(a\omega)^2}{(1 + (a\omega)^2)^2} \, d\omega$$

Evaluating the integrals approximately, assuming $a \gg b$, we have

$$\langle V_s^2 \rangle \sim 4ekTR_s/a$$

$$\langle V_p^2 \rangle \sim 4ekTR_p a/b^2$$

and the sum is a minimum when $a^2 = b^2(R_s/R_p)$, or $a^2 = R_p R_s C^2$. Putting $R_p = 10^{12} \, \Omega$, $R_s = 200 \, \Omega$, and $C = 5 \times 10^{-13}$ F, we get $a = 7 \, \mu s$. For averaging times much shorter than this, the mean-square voltage, and therefore the pulse-height resolution, is inversely proportional to the averaging time. However, it must be remembered that the above noise figures must be added to the intrinsic fluctuations in the charge collection process, which contribute about half of the broadening at 6 keV.

In principle one can do relatively fast timing experiments with thin semiconductor detectors, as the typical carrier velocity at 1000 V/cm is about 10^7 cm/sec. The jitter in the pulse arrival time is caused by the R_s Johnson noise and is typically 2 ns for a 1-mm crystal.

The chief disadvantage of semiconductor detectors is the expense. An intrinsic Ge detector, with dewar and preamplifier, costs about \$8,000; a ten-element array of detectors can be built commercially for about \$60,000. However, the techniques for growing intrinsic germanium crystals are improving rapidly, and it is likely that they will be useful in applications where proportional counters and scintillation counters are now used, but where improved energy resolution is important.

4.3.5. Analyzing Detectors

A very promising method for efficiently collecting fluorescensce radiation has been developed by Hastings et al.[51] (see discussion in Chapter 13). A mosaic of 2 mm × 2 mm graphite crystals were arranged to form the surface of a truncated ellipsoid of revolution or "barrel." One can show that such a device condenses all rays originating from one focus O_1 to the other focus O_2, while still satisfying the Bragg condition.

5. Numerical Analysis of X-Ray Absorption Data–Extraction of Physical Parameters

The advent of high-intensity synchrotron radiation sources and the improvement of laboratory x-ray tube sources has led to the generation of very large amounts of high-quality x-ray absorption data for a wide variety of different materials. In order to take advantage of the information about the structure and composition of a material contained in its x-ray absorption spectrum, reliable numerical analysis techniques are needed. However, it turns out that the specific methods needed for extraction of the physical parameters depend to quite a large extent on the nature of the physical system in question. Furthermore, the success of a given technique is sometimes hard to assess (for instance, in the case of amorphous materials or dense metallic systems) so that a variety of somewhat arbitrary prescriptions developed by various practitioners are to be found in the literature, often without readily available objective comparisons. In this section, we attempt to give an overview of current analysis techniques.

5.1. General Considerations in the Analysis of EXAFS Data

All analyses of the EXAFS region of the spectrum starts with the basic back scattering formula derived from (18),

$$\bar{\chi}(k) = \sum_j \frac{|f_j(k, \pi)|}{k} \int_0^\infty dr \, g_{oj}(r) \sin[2kr + \alpha_{oj}(k)] \exp(-k^2\sigma_j^2) \exp[-2\mu(k)r] \quad (35)$$

where $g_{oj}(r)$ is the pair distribution function of an atom of species j measured relative to the absorbing atom at the origin.

In applying equation (35), two general material-dependent factors must be considered:

a. the mean electron density of the absorber environment—which determines the importance of multiple scattering effects; and
b. the degree of coordinational disorder, determining the nature of the distribution function $g_{oj}(r)$.

For crystalline and molecular materials, g_{oj} consists of a series of well-defined coordination shells whose atomic constituents are in general disordered as to orientation (except in single crystal samples) but very well defined as to distance. For amorphous or liquid samples, on the other hand, both angle and iteratomic distances are disordered (though usually less so for the nearest neighbor shell) so that $g(r)$ must be represented by a continuous distribution rather than a series of δ functions as in the crystalline and molecular case.

Point (a) determines a fundamental physical limitation—the lower bound on the range of photoelectron k vectors, k_{min}, below which multiple scattering effects (see Section 2.6) become important and where equation (35) fails. In practice k_{min} is usually determined empirically for a given class of materials—for instance, in organic molecules (where the absorber environment has low Z and low coordination) k_{min} is usually set in the region of 3 Å^{-1} (corresponding to photoelectron energies of about 35 eV). On the other hand, for dense materials, such as metallic Cu, multiple scattering effects may well be important to 300 eV (about 9 Å^{-1}) leading to a much more restricted range of validity for (35).[16] The upper value of the k range, k_{max}, is usually determined by the data itself; since $\chi(k)$ falls off rather rapidly with k, both because of the Debye–Waller-type exponential fall off and because in many materials the back scattering amplitude $f(k, \pi)$ tends to fall off fairly rapidly (particularly for low-Z scatterers such as are found in organic materials), the signal-to-noise ratio of the data rapidly worsens for high k. The value of k_{max} therefore depends strongly on experimental considerations; dilution of absorber species, available photon counts, etc. In practice, values of k_{max} range from about 11 Å^{-1} (for noisy data from very dilute samples) to about 18 Å^{-1} for concentrated materials.

Point (b) determines the nature of the physical information one may hope to extract from EXAFS data. For molecular or crystalline materials, information on shell coordination distances and back scattering amplitudes as well as phase shifts involving as many as four shells in a distance range out to about 3.5 Å have been obtained. For amorphous materials, on the other hand, a lot of information about the form of $g(r)$ is lost owing to the limitation of $k > k_{min}$, so that in practice information on coordination distance and number of atoms for the first coordination shell is the most that may be reliably extracted from the data.[52]

5.2. Subtracting the Background–Setting the k Scale

The general form of an x-ray absorption spectrum is shown in Figure 1. It consists of a decreasing cross section $\sigma_{background}(\omega)$ leading up to the edge (increasing x-ray energy $\hbar\omega$) followed by the edge region, and at higher photon energies by the basic inner-shell contribution to the atomic cross section $\sigma_{inner}(\omega)$, modulated by the EXAFS function $\chi(\omega)$:

$$\sigma_{total} = \sigma_{background}(\omega) + \sigma_{inner}(\omega)\chi(\omega) \tag{36}$$

A numerical procedure to extract $\sigma_{background}(\omega)$ is to fit the 100-eV range below the edge with a low-order polynomial in ω, or a power-law function[53]

$$\sigma_1(\omega) = A/\omega^\beta \tag{37}$$

where β is a free parameter (of order 3.5), or a combination of two powerlaw functions. This is then extrapolated over the range of the data and subtracted, leaving as a residual part, the main inner-shell contribution. It then remains to find a prescription to fit $\sigma_{inner}(\omega)$ and normalize the inner-shell cross section by this fitted function, yielding an empirical evaluation of $\chi(\omega)$. To do this, one relies on the fact that $\sigma_{inner}(\omega)$ is a slowly varying function of ω. A low-order polynomial is fitted to $[\sigma_{total} - \sigma_{background}(\omega)]$ by a least-squares procedure. This fit has the effect of averaging out the EXAFS modulations, so that on subtracting the resultant from the data [i.e., from $\sigma_{inner}(\omega)$] the residual modulations oscillate rather evenly about zero. To ensure the uniform background subtraction, it sometimes turns out to be necessary to least-squares fit to a set of different low-order polynomials over different k ranges in the interval $k_{min} < k < k_{max}$ in such a way that the values of the fitted polynomials and their derivatives match at the boundaries crossing from one fitting range to the next. (A fit of this kind is known as a "spline.") Finally, the resulting background-subtracted data may be normalized by dividing by the background-fit function at each value of ω, yielding an empirical evaluation of the EXAFS modulation function $\chi(\omega)$.

The photoelectron wave-vector scale follows directly from the relation

$$k = [2m(\hbar\omega - E_0)/\hbar^2]^{1/2} = [0.2628\,(\hbar\omega - E_0)]^{1/2}\,\text{Å}^{-1} \tag{38}$$

where the photon energies are measured in eV. However, the value of E_0 (the start of the inner-shell–continuum transition region) is in general very hard to determine on the basis of direct inspection of the data on account of the existence of core level $\rightarrow$ bound state transitions occurring just below the edge, and frequently merging with it as a result of the natural lifetime broadening of the core state. Different approaches have been taken to determination of E_0: it may be set by an arbitrary prescription (e.g., the inflection point of the absorption step, or the position of the main absorption maximum at the edge region); by many workers E_0 is kept as a disposable parameter to be varied in the EXAFS fitting procedure (see discussion in Section 5.4).

5.3. Fourier Filtering

The general form of the EXAFS function (35) represents a series of damped, phase-modulated sine waves, so that a natural way to approach its analysis is to take the numerical Fourier transform of the measured data over the EXAFS k range

$$\tilde{\chi}(k) = \int_{k_{\min}}^{k_{\max}} dk\; e^{ikR}\chi(k) \tag{39}$$

In the case of materials with sharply defined absorber–scatterer distances, $|\tilde{\chi}(k)|$ exhibits a number of peaks characteristic of different shells of scatterers. If the phase function $\alpha(k)$ were purely linear in k, the position of the maxima of $|\tilde{\chi}(k)|$ would represent twice the shell distances shifted by $d\alpha/dk$. In practice, however, the fact that $\alpha(k)$ has appreciable curvature as a function of k leads to nonnegligible skewing of the peaks in $\tilde{\chi}(R)$. Furthermore, it is hard to determine the relative weights (hence coordination numbers) of the different scatterer shells directly from the transform, so that more sophisticated methods of extracting these numbers have been developed.

These methods are based on a technique whereby $\tilde{\chi}(R)$ is multiplied by a filter function $F_{\text{shell}}(R)$ centered on one of the shell peaks (e.g., a flat-topped window with Gaussian edges to remove "ringing" effects) and then retransformed back to k space

$$\chi_{\text{shell}}(k) = \frac{1}{2\pi}\int_0^\infty dR\; F_{\text{shell}}(R)\tilde{\chi}(R)e^{-ikR} \tag{40}$$

where the negative frequency part of $\sin(2kR_s + \alpha)$ has been eliminated as a consequence of the fact that (39) picks up only $R > 0$ contributions to the transform.

As suggested by Kincaid,[54] the amplitude and phase of χ_{shell} are now given directly by

$$A(k) = |\chi_{shell}(k)|$$

$$\Phi(k) = 2ikR_s + \alpha(k) = \arctan\left[\frac{\text{Im } \chi_{shell}(k)}{\text{Re } \chi_{shell}(k)}\right]$$

The value of E_0 for the data set in question may now be determined by reference to a phase-shift function $\alpha_{model}(k)$ determined either theoretically or experimentally by analysis of data on a model compound of known shell distance: if E_0 is varied, leading to a k scale denoted by k', then

$$\Phi(k') - \alpha_{model}(k) = 2kR_s + 2(k' - k)R_s + \alpha(k') - \alpha_{model}(k) \qquad (42)$$

Hence, by varying E_0 till equation (42) becomes a straight line passing through the origin, the unknown E_0 may be adjusted to be consistent with the model E_0. The slope of the resulting line then gives the value of $2R_s$. The amplitude [in equation (41)] may then be compared to the amplitude for a model compound and in this way, the unknown coordination number can be compared to the model coordination number.

5.4. Nonlinear Curve Fitting of EXAFS Data—Application to Model Compounds and Multishell Parameter Determination

An alternative approach to fitting the EXAFS formula to data is the use of a standard nonlinear least-squares-fitting computer routine to fit a parametrized version of equation (35) to single-shell data (in the case of simple model compounds) or to Fourier-filtered data in the case of multishell data.

For this purpose the phase function $\alpha(k)$ is expanded in a series

$$\alpha(k) \sim a_{-1}k^{-1} + a_0 + a_1 k + a_2 k^2 \qquad (43)$$

and the a's together with the unknown shell distance R_s are treated as variables in the fitting routine. Similarly the amplitude function is parametrized in some suitable functional form. For low-Z scatterers, a form chosen is

$$f(k, \pi) \exp(-k^2\sigma^2) = c_0 \exp(-c_1 k^2)/k^{c_2} \qquad (44)$$

where c_0, c_1 and c_2 are treated as fitting parameters. For higher-Z scatterers, some additional resonance structure appears in the back scattering amplitude[18] and a more complicated functional form would be needed.

This approach has two advantages over the more direct (and faster method) outlined in Section 4.3: (a) for single-shell model compounds it avoids some of the numerical artifacts introduced by Fourier filtering (i.e., it can be applied directly to unfiltered data)—in practice, this is not particularly a large effect though; (b) for closely overlapping shells, which are essentially impossible to separate by Fourier filtering, the fitting routine may be used to fit a pair of single-shell EXAFS functions with different sets of parameters defined via (43) and (44). The existence of two overlapping shells in a given compound is often manifested by the appearance of a "beat" in the measured EXAFS function $\chi_{\text{shell}}(k)$ after Fourier filtering the more distinct shells. This beat results from the superposition of two sine waves of very similar frequency.

The disadvantage of the nonlinear technique is that (a) the least-squares deviation may strongly correlate two or more of the variable parameters: typically, the Debye–Waller parameter c_1 and the amplitude fall-off factor c_2 are often strongly correlated (i.e., the least-squares deviation is insensitive to suitably compensating changes in the parameters, so that the resulting fit values are not accurately defined); and (b) fitting routines of this type are often subject to "false minima" (i.e., local solutions that are quite far from the global minimum). In practice, the remedy to (a) is to fix one of the parameters (e.g., the Debye–Waller coefficient) by reference to similar compounds. A remedy to (b) is to try a number of different starting vectors for the parameter set. Usually knowledge of parameters for model compounds is important in the selection of physically reasonable starting vectors and in the discrimination between possible different solutions.

5.5. Numerical Analysis of X-Ray Absorption Edge Data

As discussed in Section 3.6, the interpretation of the structure in the absorption cross section in the region of the absorption edge in terms of physical parameters is a much more complex task than for the EXAFS region of the spectrum. Nevertheless, it has been found useful and practical to perform numerical analysis on the data in terms of the positions, widths, and oscillator strengths of bound-state and resonance peaks, and of an edge position, width, and height for the principle continuum edge.[55]

To do this, a nonlinear fitting routine was applied to fit functions of the form

$$\chi_{\text{edge}} = \sum_i \frac{f_i \gamma_i / \pi}{(\omega - \omega_i)^2 + \gamma_i^2} + \frac{f_0}{\gamma_0} \arctan\left(\frac{\omega - E_0}{\gamma_0}\right) \tag{45}$$

The resulting bound-state and continuum-edge energies and oscillator strengths may be used to compare with the results of multiple scattering calculations described in Section 2.6. The Lorentzian form assumed in (45) describes accurately the natural lifetime broadening of the core state, but would not be expected to fit perfectly to the effects of instrumental resolution, which form, of course, part of the observed data. In principle, more sophisticated line-shape analysis, such as has been applied to XPS data by Citrin et al., could be used.[56] This has not been done to date (to our knowledge) for XAS data.

References

1. See L. Hedin and S. Lundqvist, *Solid State Physics*, H. Ehrenreich, F. Seitz, and D. Turnbull (eds.), Vol. 23, p. 1, Academic Press, New York (1969).
2. See, for instance, *Photoionization and Other Probes of Many-Electron Interactions*, F. J. Wuilleumier (ed.), Plenum Press, New York (1976), for general reviews.

3. U. Fano and J. W. Cooper, *Rev. Mod. Phys.* **40**, 441 (1968).
4. K. Gottfried, *Quantum Mechanics*, W. A. Benjamin, Inc., New York (1966).
5. W. Heitler, *Quantum Theory of Radiation*, 3rd edition, Oxford University Press, New York (1954).
6. J. W. Cooper, *Phys. Rev.* **128**, 681 (1962).
7. H. P. Kelly and R. L. Simmons, *Phys. Rev. Lett.* **30**, 529 (1973).
8. M. Amus'ya, N. A. Cherephov, and L. V. Chernysheva, *Zh. Eksp. Teor. Fiz.* **60**, 160 (1971) [*Sov. Phys. JETP* **33**, 90 (1971)].
9. V. O. Kostroun, M. H. Cehn, and B. Crasemann, *Phys. Rev. A* **3**, 533 (1971).
10. S. Doniach, in *Computational Methods in Band Theory*, P. M. Marcus, J. F. Janak, and A. R. Williams (eds.), p. 500, Plenum Press, New York (1976).
11. P. Bagus, *Phys. Rev. A* **9**, 1090 (1974).
12. S. Abraham-Ibrahim, B. Caroli, C. Caroli, and B. Roulet, *Phys. Rev. B* **18**, 6702 (1978).
13. R. de L. Kronig, *Z. Phys.* **70**, 317 (1931); **75**, 191, 468 (1932).
14. D. E. Sayers, F. W. Lytle, and E. A. Stern, *Advances in X-Ray Analysis*, B. L. Henke (ed.), Vol. 13, p. 248, Plenum Press, New York (1970).
15. E. A. Stern, *Phys. Rev. B* **10**, 3027 (1974).
16. C. A. Ashley and S. Doniach, *Phys. Rev. B* **11**, 1279 (1975).
17. P. A. Lee and J. B. Pendry, *Phys. Rev. B* **11**, 2795 (1975).
18. B. K. Teo and P. A. Lee (to be published).
19. P. Lagarde, *Phys. Rev. B* **14**, 741 (1976).
20. G. Beni and P. M. Platzmann, *Phys. Rev. B* **14**, 1514 (1976).
21. S. P. Cramer, K. O. Hodgson, E. I. Stiefel, and W. E. Newton, *J. Am. Chem. Soc.* **100**, 2748 (1978).
22. B. M. Kincaid and P. Eisenberger, *Phys. Rev. Lett.* **34**, 1361 (1975).
23. P. A. Lee and G. Beni, *Phys. Rev. B* **15**, 2862 (1977).
24. D. W. Jepsen, P. M. Marcus, and F. Jona, *Phys. Rev. B* 5, 3933 (1972).
25. L. Hedin and B. I. Lundquist, *J. Phys. C* 4, 2064 (1971).
26. J. W. Gadzuk, *J. Electron. Spectrosc. Relat. Phenom.* **11**, 355 (1977).
27. J. J. Rehr, E. A. Stern, R. L. Martin, and E. R. Davidson, *Phys. Rev. B* **17**, 560 (1978).
28. T. A. Carlson, M. O. Krause, and W. E. Maddeman, *J. Phys. (Paris) Colloq.* **32**, 4 (1971).
29. C. Noguera, D. Spanjaard, and J. Friedel, *J. Phys. (London) F* **9**, 1189 (1979).
30. K.-H. Johnson, *Advances in Quantum Chemistry*, P.-O. Löwdin (ed.), Vol. 7, p. 143, Academic Press, New York (1973).
31. D. Dill and J. L. Dehmer, *J. Chem. Phys.* **61**, 692 (1974).
32. J. W. Davenport, *Phys. Rev. Lett.* **36**, 945 (1976); J. W. Davenport, W. Ho, and J. R. Schrieffer, *Phys. Rev. B* **17**, 3115 (1978).
33. C. R. Natoli, D. K. Misemer, S. Doniach, and F. C. Kutzler submitted for publication to *Phys. Rev.* (1980).
34a. J. L. Dehmer, J. Siegel, and D. Dill, *J. Chem. Phys.* **69**, 5205 (1978).
34b. R. E. Kennerly, R. A. Bonham, and M. McMillan, *J. Chem. Phys.* (to be published).
35. J. Jaklevic, J. A. Kirby, M. P. Klein, A. S. Robertson, G. S. Brown, and P. Eisenberger, *Solid State Commun.* **23**, 679 (1977).
36. W. Bambynek, B. Crasemann, R. W. Fink, H. U. Freund, H. Mark, C. D. Swift, R. E. Price, and P. V. Rao, *Rev. Mod. Phys.* **44**, 716 (1972).
37. J. Witz, *Acta Crystallogr. A* **25**, 30 (1969).
38. M. Hart, *Rep. Prog. Phys.* **34**, 435 (1971).
39. J. H. Beaumont and M. Hart, *J. Phys. E* 7, 823 (1974).
40. U. Bonse, G. Materlik, and W. Shcroeder, *J. Appl. Crystallogr.* **9**, 223 (1976).
41. J. B. Hastings, *J. Appl. Phys.* **48**, 1576 (1977).
42. W. Bauspiess, U. Bonse, W. Graeffand and H. Raugh, *J. Appl. Crystallogr.* **10**, 338 (1977).
43. H. W. Fulbright, *Handbuch Der Physik, Vol. XLV*, Springer-Verlag, Berlin (1958).
44. S. C. Curran and J. D. Craggs, *Counting Tubes*, p. 42, Butterworth, London (1949).
45. G. Charpak, R. Bouclier, T. Bressani, and J. Favier, *Nucl. Instum. Methods*, **62**, 262 (1968).
46. U. Fano, *Phys. Rev.* **70**, 44 (1946).
47. U. Fano, *Phys. Rev.* **72**, 26 (1947).
48. Z. H. Cho, C. M. Tsai, and L. A. Eriksson, *IEEE Trans. Nucl. Sci.* **22**, 72–88 (1975).

49. L. A. Eriksson, C. M. Tsai, A. H. Cho, and C. R. Hurlbut, *Nucl. Instum. Methods* **122**, 373 (1974).
50. E. S. Goulding and D. A. Landis, *Nuclear Spectroscopy and Reactions, Part A*, pp. 289–343 and 413–481, Academic Press, New York (1974).
51. J. Hastings, P. Eisenberger, and J. Brown (to be published).
52. See review by H. Winick and A. I. Bienenstock, *Ann. Rev. Nucl. Sci.* **28**, 33 (1978), and Chapter 11 of this volume.
53. J. A. Victoreen, *J. Appl. Phys.* **19**, 855 (1948).
54. B. Kincaid (private communication).
55. F. Kutzler and K. O. Hodgson (private communication).
56. P. N. Citrin, G. K. Wertheim, and Y. Baer, *Phys. Lett.* **35**, 885 (1975).

11

Extended X-Ray Absorption Fine Structure in Condensed Materials

GEORGE S. BROWN

1. Introduction

The application of *extended* x-ray *absorption fine* structure, or EXAFS, to a variety of problems in condensed materials is rapidly evolving. To date, the most fully developed applications are toward the fields of catalysis and biology; these subjects are treated in detail in Chapters 12 and 13. In this chapter we review the applications of EXAFS to periodic and quasi-periodic solids; to studies of the structural properties of surfaces; to studies of disordered solids; and, finally, to liquids and solutions.

EXAFS owes its remarkable success partly to the simple physical basis for the phenomenon, and partly to the correspondingly simple mathematical expression for the modulation of the photoabsorption cross section. According to this picture, K-shell photoabsorption results in the emission of an "elastic" photoelectron with probability $\eta \sim 0.5$; in a monatomic gas the wave function would have the symmetry of a p wave, and, above threshold, the photoabsorption cross section would be a smooth function of energy. In the presence of ligands, however, the photoelectron wave may be reflected back toward the origin. Since the photoabsorption cross section is proportional to the square of the matrix element of the dipole operator between the initial- and final-state wave functions,

$$\sigma \propto |\langle f | \mathbf{x} | i \rangle|^2 \tag{1}$$

GEORGE S. BROWN • Stanford Synchrotron Radiation Laboratory, Stanford Linear Accelerator Center, P.O. Box 4349, Bin 69, Stanford, California 94305.

and since the initial-state wave function is highly localized, this matrix element will be modulated with the periodicity of the back scattered wave, viz., $\sin(2kr)$ where k is the photoelectron wave vector and r is the internuclear distance.

This simple model must be refined before being applied to real physical systems. First, the photoelectron may experience a k-dependent phase shift $\phi_c(k)$ upon emerging from the central atom, and it may likewise experience a phase shift $\phi_b(k)$ upon reflection from the ligand. The back scattering probability $f(k, \pi)$ is also k-dependent. Putting these factors together, we get the basic formula for the modulation of the photoabsorption due to a single ligand:

$$\chi(k) = (\eta/kr^2) | f(k, \pi) | \sin[2kr + \phi_c(k) + \phi_b(k)] \qquad (2)$$

To generalize this expression to include atoms with multiple ligands, we make use of the fact that the scattering probability is weak, so that $\chi(k)$ may be regarded as a superposition of the modulations from the various ligands. Also, in such systems the photoelectron may suffer an inelastic collision with probability $\mu(k)$ per unit length; in practice this effect reduces the effective photoelectron range to 4–10 Å. The general expression for the EXAFS modulation then becomes

$$\chi(k) = \eta/k \sum_j (1/r_j^2) | f_j(k, \pi) | \exp[-2\mu(k)r_j] \sin[2kr_j + \phi_c(r) + \phi_{b,j}(k)] \qquad (3)$$

This expression assumes that all photoabsorbing sites have the same structural environment. If this is not true, equation (3) must then be summed and averaged over distinct photoabsorbing sites.

A great deal of experimental and theoretical effort has been devoted to establishing the limits of the validity of equation (3). Although a critical evaluation of these efforts is beyond the scope of this chapter, a few remarks are in order. It is generally accepted that the formalism is invalid for photoelectron wave vectors less than about 3 Å^{-1}. For photoabsorption sufficiently close to threshold, the interatomic potential is comparable to the electron kinetic energy, and the chemical bonding becomes quite significant in determining $\phi_b(k)$. Above 3 Å^{-1}, the central atom phase shift seems to be ligand-independent, and the back scattering phase shift is transferable among ligands even in different chemical environments.[1] Although the most detailed experiments to date have been performed in the vapor phase, theoretical calculations of the phase shifts for solid Ge and Cu yield results accurate to 1% or so. Interestingly enough, the second and fourth shells in Cu metal are so badly distorted, apparently by multiple scattering, that they are uninterpretable.

The preceding discussion assumed that every absorbing site is surrounded by ligands at a fixed distance. In practice, even at zero temperature, the atoms are oscillating about their equilibrium positions with some mean-square relative displacement $\sigma_j^2(T)$. If the oscillations are harmonic, then the expression for the modulation is damped by the factor $\exp[-2\sigma_j^2(T)k^2]$. This effect is readily observable, allowing one to measure mean-square relative displacements of atomic pairs in a straightforward manner. It is important to recognize, however, that for materials with large vibration amplitudes, distance corrections of order $-2\sigma^2(1/r_j + \mu)$ are necessary when a proper average of equation (3) is performed.[2] This correction can be as large as 0.1 Å at room temperature for materials such as metallic zinc.

2. Periodic and Quasi-Periodic Solids

For condensed materials with a high degree of short-range order but with essentially no macroscopic ordering, such as metalloproteins and heterogeneous catalysts, EXAFS is unrivalled as a structural method. However, even for crystalline materials, EXAFS can offer a distinct advantage over x-ray and neutron diffraction when specific information is sought. The precise arrangement of the atoms within the unit cell is usually determined from diffracted intensity measurements, for which the corrections due to primary and secondary extinction are notoriously difficult. By comparison, EXAFS gives nearest-neighbor distances, coordinations, and mean-square relative displacements rather directly, with accuracies approaching 0.01 Å. EXAFS, then, is naturally complementary to diffraction in that it yields local atomic distributions, whereas diffraction naturally reflects the translation symmetry of the material.

In an interesting application of EXAFS to periodic solids, Ingalls, Garcia, and Stern[3] have measured the bond compressibilities in FeF_2 and pyrite (FeS_2). From the measurements at pressures up to 64 kbars, the authors show that the Fe–S bond length contracts at most a factor of ten less than the Fe–Fe bond length. The authors have also observed a dramatic reduction in the Fe–S mean-square displacement with increasing pressure, obtained from a simple analysis of $\chi(k)$. These results are in reasonable accord with estimates based upon the Debye–Gruneisen approximation, and the quality of the data is sufficient to merit a more detailed analysis. This landmark experiment has demonstrated that high-quality absorption spectra can indeed be measured within the constraints imposed by a high-pressure cell. Further research along these avenues is very promising indeed.

The application of EXAFS to the problem of superionic conduction is a classic example of the complementarity of EXAFS to scattering techniques. In a detailed study of crystalline AgI, Boyce et al.[4] were able to resolve competing models for the position and dynamics of the mobile Ag^{+1} ions, which represents an important advance in our understanding of the phenomenon of superionic conduction.

AgI is known to have the hexagonal wurtzite structure below the transition temperature, 147°C, and the I sublattice is known to be body-centered cubic above the transition temperature, where the ionic conductivity increases by about four orders of magnitude. From low-temperature data at the K absorption edge of the Ag ions, the authors were able to extract the phase shift and amplitude functions associated with Ag–I bond. They then attempted to model the Fourier transform of the data with five distinct distributions for the Ag ions: (1) a delta function at the center of the tetrahedral cavity formed by the I sublattice, (2) a delta function displaced toward the center of an edge of the tetrahedron, (3) a delta function displaced toward the center of a tetrahedral face, (4) the complicated 42-site distribution of the Strock[5] model, and (5) two delta functions displaced by the vectors from the tetrahedral center, as proposed by Bührer and Halg.[6] Of these four models, only (2) resulted in a satisfactory fit, apart from an apparent overall reduction in the amplitude of the peak. Since the Ag ions are known to be mobile, the reduction in amplitude can be accounted for by assuming that the Ag ions spend roughly 25% of their time hopping from one site to the next, which is consistent with simple thermodynamic arguments. As emphasized in Section 4, Fourier transform peak amplitudes are very sensitive to mean-square amplitudes of vibration because of the low-k cutoff, and so it might be argued that the 25% jump time is an upper limit; this only reinforces the authors' conclusion that free diffusion is inconsistent with the data. Also, the authors have not ruled out the possibility of other distributions yielding the same or better fits to the data, but have limited the possibilities to physically plausible configurations.

3. Surface EXAFS

The application of EXAFS to the study of the structure of surfaces is developing rapidly, owing to the rapid increase in the sensitivity of detection methods. The effective concentration of a monolayer adsorbate is roughly 10^{15} cm^{-2}, which is six orders of magnitude lower than the effective concentration of bulk species. However, with cross sections of 10^{-19} cm^2, fluxes of 10^{11} s^{-1} and efficiencies of 10^{-4}, counting rates of order 10^3 s^{-1} are feasible with secondary particle detection schemes, such as Auger electron emission,[7–9] secondary ion emission,[10] and secondary electron emission.[11–13] Alternatively, multiple substrates can be stacked normal to the beam, limited only by the attenuation of the primary beam by the substrate.

A good example of the latter method is the work of Stern et al.[14, 15] who measured the EXAFS spectra of Br adsorbed on layers of exfoliated graphite. This material, known by the trade name of Grafoil, consists of about equal parts of randomly oriented carbon crystallites, and quasi-parallel ($\pm 15°$) 200-Å-thick layers of graphite. Since the x-ray absorption cross section for Br is about 10^3 times as great as the carbon cross section, the substrate absorption is nearly negligible. The authors report that the Br–Br EXAFS, at 0.2 monolayer coverage at room temperature, is independent of the orientation of the x-ray polarization vector relative to the basal planes, indicating a random orientation of the Br–Br bond. However, for higher coverages (0.6 and 0.9 monolayer) and for temperatures between 100 and 293 K, the Br–Br EXAFS amplitude is negligible for the polarization vector normal to the basal plane, indicating that the Br_2 molecule is aligned parallel with the basal planes.

In a more recent study, Citrin et al.[9] studied a prototype system, I adsorbed on Ag, in parallel with a careful program to study the phase diagram by conventional techniques. The authors measured the 3300-eV $L_3 M_{4,5} M_{4,5}$ Auger transition with a commercial double-pass cylindrical mirror analyzer. They obtained signal rates of about 10^3 s^{-1} in the presence of a 2×10^4 s^{-1} continuum background, with an incident flux of 10^{11} s^{-1} focussed into a 2 mm × 4 mm spot. Because of limitations in data collection time, the authors restricted their initial studies to the $\frac{1}{3}$ monolayer $(3^{1/2} \times 3^{1/2})R30°$ system. By carefully analyzing the I–Ag amplitude, the authors were able to confirm an earlier LEED result that the I is threefold coordinated, and that the I–Ag distance is 2.87 ± 0.03 Å. This represents a fourfold improvement in our knowledge of the Ag–I bond length.

Although the majority of EXAFS research has been performed with hard $(E > 2800$ eV$)$ x rays, recent experiments have demonstrated the feasibility of studies in the vacuum-ultraviolet region of the spectrum. Stöhr *et al.*[11-13] have demonstrated that the secondary partial-electron yield contains a component that is proportional to the K-shell photoabsorption cross section for both bulk and surface species, but that the surface-to-bulk sensitivity can be enhanced by detecting only electrons with relatively low kinetic energy. Curiously enough, the signal-to-continuum ratio is roughly the same as for the single Auger channel (~ 0.05), but very high effective count rates are feasible by setting the lower limit of the kinetic energy of the accepted electrons to a few electron volts. The effective count rate then approaches $F\sigma n \sim 10^{-4}F$, where F is the incident flux and n is the adsorbate surface density. The method has been applied successfully to the model systems Al_2O_3, S_3N_4, and SiO_2; to the important problem of oxygen chemisorbed on the Si(111) surface; and to a preliminary study of the oxidized Ni(100) surface. On the basis of measured near-neighbor distances and coordinations, the authors found that only two out of seven popular models of the oxidized silicon surface are consistent with the data. Eventually, by making use of the natural polarization of synchrotron radiation, the authors argue that these two models could be resolved. These experiments mark a major advance in the EXAFS technique, by making accessible the important and highly reactive atoms O and N.

It is important to recognize that EXAFS bears the same relation to low-energy electron diffraction (LEED) for surface studies as it does to x-ray diffraction for bulk studies. LEED measurements give the symmetry group of the adsorbate directly, as well as the unit-cell dimensions. However, the arrangement of atoms within the unit cell, as well as the distances normal to the translational symmetry plane, must be obtained from intensity measurements, which are notoriously difficult, owing to the strong multiple scattering amplitude. EXAFS, on the other hand, is ignorant of long-range order, but gives near-neighbor distances and coordinations directly.

4. Disordered Solids

The application of EXAFS to amorphous materials dates to the work of Sayers, Lytle, and Stern[16] who used the method to study the environment of the Ge atom in glassy Ge and GeO_2. Since that time there have been a number of studies of disordered materials in the solid and liquid phase. However, at the present time our ability to extract detailed structural information for disordered materials from the EXAFS spectra is still quite limited, and some of the conclusions presented in the literature may well require revision in the final analysis. It is, therefore, worthwhile to first discuss the current methods of data analysis, with a view toward discriminating between *intrinsic* limitations of the method and limitations based upon our present inability to invert the EXAFS spectrum.

We shall consider, for simplicity, the case of a *monatomic* disordered material that is homogeneous and isotropic. Such a material is characterized by a radial distribution function[17] $RDF(r) = 4\pi r^2 \rho(r)$ where $RDF(r)\ dr$ is the average number of atoms in a spherical shell of radius r and thickness dr. Distributions are sometimes expressed by the so-called pair correlation function, $W(r) = \rho(r)/\rho_0$. Figure 1a shows a representative function $W(r)$ for a collection of hard spheres arranged to satisfy the Percus–Yevick[18, 19] equation. A number of important properties of $W(r)$ are exhibited in this curve: (1) $W(r)$ tends to unity at large distances, expressing the fact that all atom–atom correlations tend to vanish at large distances; (2) $W(r) = 0$ for $r < r_{min}$, which is implicit in the assumption of hard spheres but is nevertheless a fairly good approximation for realistic atomic potentials; and (3) the function consists of a succession of peaks of diminishing amplitude and increasing breadth, representing the most probable positions of neighboring atoms. Such a picture, then, is a complete description of the structure of the material within the constraints of homogeneity and isotropy. Attempts to describe the distribution by neighbor spacings and coordination numbers, by analogy with ordered materials, is somewhat artificial.

The EXAFS modulation for an assembly of identical atoms arranged with a density distribution $\rho(r)$, in analogy with equation (3), is given by

$$\chi(k) = (4\pi/k) f(k, \pi) \int_0^\infty dr \rho(r) \exp[-2\mu(k)r] \sin[2kr + \phi_c(k) + \phi_b(k)] \quad (4)$$

whose inversion presents formidable mathematical problems, on account of the k dependence of the phase-shift functions and the absorption coefficient. For the moment, let us take $\mu(k) = \mu_0$ and $\phi_c(k) + \phi_b(k) \equiv 0$. We can then formally invert equation (4):

$$\rho(r) = (4/\pi) \exp(2\mu_0 r) \int_0^\infty dk \frac{k\chi(k)}{4\pi f(k, \pi)} \sin(2kr) \quad (5)$$

Equation (5) is of little use, however, because the phase functions $\phi_b(k)$ and $\phi_c(k)$ have a significant magnitude and derivative throughout their useful range. For

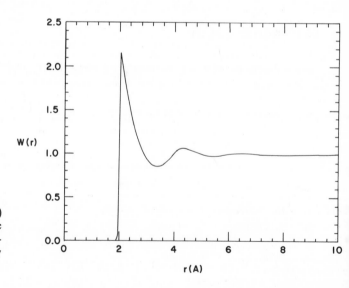

Figure 1a. The pair correlation function $W(r)$ for a collection of hard spheres whose atomic arrangement satisfies the Percus–Yevick equation with a sphere diameter of 2.0 Å and density parameter $n = 0.3$.

example, a good approximation for Ge as both central atom and back-scattering atom is[21]

$$\phi_c(k) + \phi_b(k) = -2.253 - (k/0.845) + (k/6.198)^2 - (4.59/k)^3 \qquad (6)$$

One readily observes that the nonlinear components contribute significantly to the total phase for all k. Unfortunately, the trigonometric functions with nonlinear phase do not form an orthonormal set of basis functions. To bypass this problem, Lee and Beni[1] have devised a prescription for determining neighbor spacings for systems with discrete shells. Define the complex function $F(r)$ by

$$F(r) = (4/\pi) \exp(2\mu_0 r) \int_0^\infty dk \, \frac{k\chi(k)}{4\pi f(k, \pi)} \exp[-i\phi_c(k) - i\phi_b(k)]\exp(-2ikr) \qquad (7)$$

The authors have demonstrated that if $\rho(r)$ is a peaked function, then the imaginary part of $F(r)$ will exhibit peaks and the real part will exhibit nodes at the same values of r. In practice, however, the phase function has nonlinear terms, and Im $F(r)$ represents a distorted image of the true density function $\rho(r)$. To the author's knowledge, no explicit method of extracting $\rho(r)$ has been published that takes into account the nonlinear phase and the k-dependent photoelectron mean free path.

We now turn to the most serious problem of all with disordered solids, namely, that the fundamental formula (2) for $\chi(k)$ is not valid for $k \lesssim 3$ Å^{-1}, as discussed in Chapter 10. But the fundamental theorems of Fourier analysis tell that for $0 \lesssim k \lesssim 3$ Å^{-1}, $\chi(k)$ contains information about the correlations in $\rho(r)$ at "wavelengths" $\Delta r \gtrsim 1$ Å. By truncating the data below $k = 3$ Å^{-1} we have effectively applied a "high-pass" filter to $\rho(r)$; therefore, sharp structures in $\rho(r)$ will be preserved but correlations between $\rho(r)$ and $\rho(r + \Delta r)$ for $\Delta r \gtrsim 1$ Å will be completely lost, including the average value. Therefore, the prescription (5) for inverting $\chi(k)$ invariably leads to a gross distortion of the spectrum. An *ad hoc* procedure employed by many authors is to perform the full complex Fourier transform of

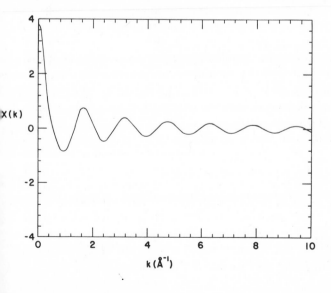

Figure 1b. The corresponding EXAFS spectrum, assuming $f(k, \pi) = k$, $\phi_b(k) + \phi_c(k) = 0.0$, and $\mu(k) = 5.0$ Å^{-1}.

$\chi(r)$, and then take the magnitude of the Fourier transform as a measure of $\rho(r)$:

$$\rho^*(r) = (1/\pi) \exp(2\mu_0 r) \left| \int_0^\infty dk \frac{k\chi(k)}{4\pi f(k, \pi)} e^{2ikr} \right| \tag{8}$$

This prescription, while giving the correct result for hypothetical data sets with zero phase shift and beginning at $k = 0$, again gives a distorted density function for realistic phases and a realistic cutoff, as shown in Figure 2.

The main conclusion of the above analysis is that, within our present understanding of the extended fine structure, it is not possible to reconstruct the radial distribution function from a given data set, as an infinite number of distinct functions can be constructed that match the data set over the allowable range in k. Practically speaking, sharp structure in r space is preserved, which justifies the use of EXAFS in obtaining distance information in ordered materials; in the case of amorphous materials, the sharpest structure in $\rho(r)$ is usually the leading edge of the first peak, corresponding to the hard-core atomic repulsion. In this sense EXAFS is complementary to x-ray scattering,[17] which, because of the experimental limitations, usually cannot be used beyond an effective k of about 5 Å^{-1}.

Despite the interpretational difficulties associated with the EXAFS spectra of disordered materials, much of the earliest work falls into this category. Because the EXAFS technique is atom specific, it is natural to apply it to the problem of dilute impurities in a matrix or solution; it is also tempting to apply it to binary or ternary materials where the complementary techniques (x-ray and neutron scattering) also suffer from interpretational difficulties.

One of the earliest applications of EXAFS to disordered materials is the study of glassy GeO_2 by Sayers et al.[16, 22] From x-ray and neutron scattering measurements, the Ge–O and Ge–Ge distances have been measured to fair accuracy, leading to two models for the structure, viz., the random-network model and the microcrystalline model. According to the random-network model, the structure is describable by a network of GeO_2 tetrahedra, as in the quartz phase, but with some randomness in the Ge–O–Ge bond angle. The microcrystalline model is qualitatively similar, differing only in size of the ordered region, in this case some 15–20 Å. Since this is well beyond the volume sampled by EXAFS, such a material would be indistinguishable from a crystal, whereas the authors observe that the spread in the second-shell Ge–Ge distance is about 0.08 ± 0.01 Å greater than the spread in the crystalline material. This result favors the random network model, and generally all models with this degree of disorder in the second shell. Unfortunately, one cannot rule out from the published data the possibility that the second shell was badly distorted by accompanying noise. However, with the present improved sensitivity of the method, the experiment could be easily repeated with enough accuracy to resolve these doubts.

In a similar study of the glass phase, Hayes et al.[23] have studied the environment about the Ge site in arc-quenched glassy $Pd_{78}Ge_{18}$, sputtered amorphous $Pd_{80}Ge_{20}$, and crystalline PdGe. In a novel analytic approach, the authors show that both amorphous samples can be described by a structure having 8.6 ± 0.5 Pd atoms at a mean distance of 2.49 ± 0.01 Å from the Ge site, and a shell of 4 ± 1 Ge

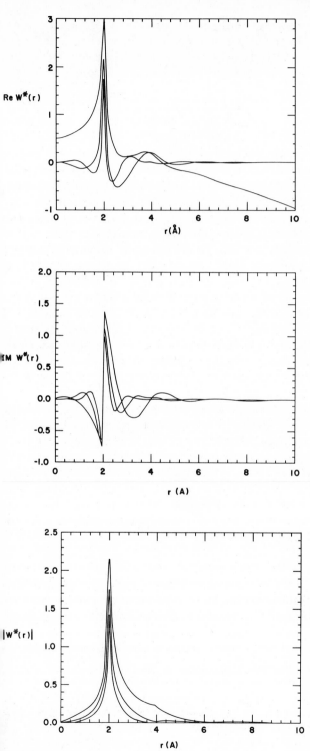

Figure 2. (a) Real part (b) imaginary part, and (c) magnitude of the Fourier transform of the spectra shown in Figure 1b, for various values of the cutoff parameter k_c.

atoms at a mean distance of 3.38 Å. Furthermore, the quality of the fit worsens considerably if it is forced to accommodate one or more Ge atoms within the range of covalent to metallic diameters (2.45 to 2.73 Å). The tacit assumption of this argument is that there is an upper bound to the rms spread in distances of this hypothetical Ge site; for reasons discussed earlier, neighbors with a sufficiently large spread in distance ($\sim$0.3 Å) do not manifest themselves in the spectra above 3 Å^{-1}.

A prototype structural problem involving the study of a dilute impurity in a disordered matrix is the problem of group III and group V doping in amorphous semiconductors. Knights[24] has argued, for example, that the electronic structure of the impurity As in an amorphous Si matrix changes qualitatively from a characteristically substitutional (i.e., fourfold coordinated) structure at low concentrations to a relaxed (i.e., threefold coordinated) structure at high concentrations characteristic of alloys. In a subsequent paper,[25] Knights et al. measured the K absorption spectra of amorphous Si prepared by plasma decomposition of silane-arsine mixtures, with a view toward obtaining accurate As coordination. What is remarkable about this paper is that the authors claim uncertainties in the coordination of $\pm$0.04 Å, based upon goodness of fit criteria. Taking these uncertainties at face value, one definitely observes a decrease in coordination with increasing concentration of As, suggesting that one in five As atoms are fourfold coordinated at 1% concentration, decreasing to none by 10% concentration. The assumption in this model is that the threefold and fourfold distances are the same and sharply defined, i.e., to 0.03 Å or better. Otherwise, a very slight reduction in mean-square relative displacements would appear as an increase in coordination.

In an important set of papers, Hunter and Bienenstock[26] and Hunter, Bienenstock, and Hayes[27, 28] have applied the EXAFS method to the amorphous semiconducting alloy $Cu_x(As_4Se_6)_{1-x}$ with a view toward understanding the relation between the atomic arrangement and the electronic and physical properties of the material. The experimental conclusions can be summarized as follows: for $x = 0.05$ and 0.25, the Cu coordination is 3.75 $\pm$ 0.25 (using the plausible assumption that the Cu coordination of c-$CuAsSe_2$ is 4.0); the Se coordination increases from 2.25 $\pm$ 0.25 to 4.25 $\pm$ 0.25; and the As coordination is approximately 3.4 for $x = 0.05$, and somewhat larger at $x = 0.25$. These data are consistent with the relation proposed by Liang et al.,[29]

$$g_a = 8 - N_e \qquad (9)$$

where g_a is the average coordination and N_e is the number of valence electrons. This model accounts for the fact that the As and Se atoms have higher coordination numbers in the glass than in As_2Se_3, but makes it difficult to understand the low viscosity of the melt, which ordinarily would indicate threefold or fewer coordination among all three species.

In one of the earliest applications of the fluorescence method, Brown et al.[30] studied Nb_3Ge films prepared by argon-ion sputtering. Under appropriate conditions, this material is known to have the highest superconducting transition temperature ($\sim$ 23 K); under these conditions the structure is known to be a single phase with the A15 structure. Under less favorable conditions (e.g., low deposition temperature) the transition temperature drops to 4 K and the material exhibits no long-range periodicity. Nevertheless, EXAFS data taken in the vicinity of the Ge K absorption edge indicates that in the amorphous phase, the Ge remains relatively highly coordinated to Nb (8 ± 2), with a Ge–Nb distance some 0.21 Å shorter than in the A15 phase. Furthermore, the transition temperature versus concentration of the amorphous phase was measured, yielding a correlation subsequently found to be in agreement with a model postulating islands of amorphous material reducing the superconducting coherence length.[31]

5. Liquids

We now turn to the application of EXAFS to metallic ions in solution, a problem formally identical to the problem of impurities in disordered amorphous materials. Such studies have the potential for determining short-range order and the possible formation of complex ions as a function of ion concentration; at the very least, the ion–oxygen distance and coordination can be determined approximately.

In an early study,[32] Eisenberger and Kincaid reported a study of the EXAFS spectra of Cu^{2+} and Br^- ions in aqueous solution at concentrations of about 0.1 at. %. From the spacing of the extrema in $\chi(k)$, and from independent measurements of the Br–O and Cu–O phase shift, the authors estimate that the mean Cu–O and Br–O distances are 1.97 ± 0.08 Å and 3.14 ± 0.10 Å, respectively, which are consistent with the corresponding distances in hydrated salts as measured by x-ray diffraction. In a subsequent study, Fontaine et al.[33] report the Cu and Br EXAFS for highly concentrated $CuBr_2$ (0.5 M to saturation). The Cu–O and Br–O distances are consistent with the earlier results (1.93 ± 0.02 Å and 3.30 ± 0.10 Å, respectively), but the authors report that there is strong evidence for a Cu–Br bond at 2.39 ± 0.02 Å over the entire concentration range studied. The Cu and Br coordination numbers and bond lengths are consistent with half of the Cu^{+2} and Br^- ions organized in either square-planar $(CuBr_4)^{-2}$ ions or $CuBr_2$ molecules. These measurements represent a direct confirmation of a structural model[34] that was proposed to account for results obtained by Raman spectroscopy and transport measurements. More recently, Morrison et al.[35] studied the hydrolytic polymerization of iron (III) in aqueous solution. The authors resolved a 1.89 Å and 2.03 Å Fe–O distance, with relative abundance 1 : 2.5, suggesting the formation of the dimer $(H_2O)_4Fe(OH)_2Fe(OH_2)^{4+}$, which had been previously postulated to exist on the basis of nonstructural methods. By attributing a peak at 2.42 Å in the Fourier transform to the Fe–Fe bond (2.91 Å after correcting for the phase shift) the authors postulate a planar $(FeOH_2)_2$ ring with OH–Fe bond angles of 101° and HO–Fe–OH bond angles of 79°. These results compare favorably with crystalline model compounds studied by x-ray diffraction. It is worth noting that this data was taken in 16 h with a conventional line-focus Mo x-ray tube and curved Ge monochromator crystal with an entirely satisfactory signal-to-noise ratio.

6. X-Ray Sources

The majority of the experimental work discussed thus far has been performed with synchrotron radiation, which is ideally suited for photoabsorption experiments. There are two alternative techniques, however, which deserve mention by virtue of their advantage in certain special circumstances. The first method is to use the continuum bremmstrahlung radiation from a conventional x-ray source, a technique that has been brought to a high state of perfection by Knapp et al.[36] and by Del Cueto and Shevchik.[37] The authors report a flux of 10^6–10^7 photons s^{-1} in an 11-eV bandpass at 10^4 eV, which is roughly five orders of magnitude lower than an EXAFS branch line at SSRL under dedicated operating conditions. However, the limiting signal-to-noise ratio for a concentrated sample measured by the transmission technique is approximately $N^{1/2}$, where N is the number of photons incident upon the sample at a given sampling energy. Experience has shown that it is difficult to achieve signal-to-noise figures much better than 10^4 even when $N > 10^8$ because of other sources of instability. For relatively concentrated samples, the advantage of a dedicated, stable, relatively inexpensive source outweighs the disadvantage of relatively long integration times.

The second method is to use high-energy inelastic electron scattering from thin foils. EXAFS spectra have been reported for the $L_{2,3}$ edge of $Al^{(38)}$ and for the K edge of $C^{(39)}$ on instruments whose design parameters were far from optimum for EXAFS. For bulk materials, electron energy loss is decidedly more sensitive than transmission EXAFS[39] with present synchrotron radiation sources and optical systems. However, for surface studies, secondary yield methods,[11, 13] when used in conjunction with synchrotron radiation sources, are far more sensitive.

7. Future Directions

In the scant four years in which synchrotron radiation has been available for experiments in EXAFS research, the practical sensitivity of the method has increased from some 10^{20} atoms/cm² to below 10^{15} atoms/cm². The electron back scattering theory is firmly established, with published phase shifts now available for routinely determining near-neighbor distances and coordination for relatively simple systems. The experimental method has proved itself adaptable to elevated temperatures and pressures, and to studies of the solid, liquid, and vapor phase. It is therefore worthwhile to consider the possible directions that EXAFS research may take in the future.

From the experimental standpoint, the sensitivity of the EXAFS method will increase in direct proportion to the increase in available intensities arising from increased stored currents and energies, improved wigglers and undulators, and in more efficient x-ray optical elements (mirrors and monochromators). There is also much room for improvement in secondary detector techniques. To date, x-ray and optical fluorescence, Auger electron emission, secondary electron emission, and secondary ion emission have been shown to exhibit EXAFS; the list will doubtlessly grow rapidly with time. Some of these methods are, in principle, capable of resolving elements in distinct chemical states. Should this prospect be born out in practice, one can envisage experiments on systems of inequivalent atoms that would be far too ambiguous to interpret with present methods. Furthermore, since each of these methods yields a "prompt" secondary, improved sensitivity leads naturally to studies of the time evolution of molecular structure. With presently available fluxes (10^{11} s^{-1}) the millisecond time scale is already accessible for concentrated systems; the microsecond time scale will certainly be accessible with the advent of dedicated storage rings.

In addition to the improvements in experimental sensitivity, work remains to be done in the basic physical theory underlying EXAFS and especially in the analytical techniques for extracting physical parameters from the data. The accuracy of *ab initio* phase and amplitude functions needs accurate testing over a wider range of elements and chemical environments. Of the utmost importance to structural studies of disordered systems is the validity of the data at low-electron wave vector. These considerations are also important for systems with large ($\sigma > 0.2$ Å) thermal vibration amplitudes or for ordered systems with a multiplicity of neighbor spacings falling within $\Delta r \sim 0.2$ Å. In order to interpret next-neighbor and higher-lying shells, much work needs to be done in evaluating multiple scattering effects and in correctly evaluating the reduction in intensity due to inelastic losses. At the present time these interpretational questions are the only barriers to reliably understanding the higher-shell contributions that occur in most data sets.

References

1. P. A. Lee and G. Beni, *Phys. Rev. B* **15**, 2862 (1977).
2. P. Eisenberger and G. S. Brown, *Solid State Commun.* **29**, 481 (1979).
3. R. Ingalls, G. A. Garcia, and E. A. Stern, *Phys. Rev. Lett.* **40**, 334 (1978).
4. J. B. Boyce, T. M. Hayes, W. Stutius, and J. C. Mikkelsen, Jr., *Phys. Rev. Lett.* **38**, 1362 (1977).
5. L. W. Strock, *Z. Phys. Chem., Abt. B* **25**, 411 (1934) and **31**, 132 (1936).
6. W. Bührer and W. Halg, *Helv. Phys. Acta* **47**, 27 (1974).
7. Uzi Landman and David L. Adams, *Proc. Natl. Acad. Sci. U.S.A.* **73**, 2550 (1976).
8. P. Lee, *Phys. Rev. B* **13**, 5261 (1976).
9. P. H. Citrin, P. Eisenberger, and R. C. Hewitt, *Phys. Rev. Lett.* **41**, 309 (1978).
10. M. L. Knotele and P. J. Feibelman, *Phys. Rev. Lett*, **40**, 964 (1978).
11. J. Stöhr, D. Denley, and P. Perfetti, *Phys. Rev. B* **18**, 4132 (1978).
12. J. Stöhr, *J. Vac. Sci. Technol.* **16**(1), 37 (1979).
13. J. Stöhr, *Jap. J. Appl. Phys.* **17**, Suppl. 17-2, 217 (1978).
14. E. A. Stern, D. E. Sayers, J. G. Dash, H. Shechter, and B. Bunker, *Phys. Rev. Lett.* **38**, 767 (1977).
15. S. M. Heald and E. A. Stern, *Phys. Rev. B* **17**, 4069 (1978).
16. D. E. Sayers, F. W. Lytle, and E. A. Stern, *J. Non-Cryst. Solids* **8**, 401 (1972).
17. B. E. Warren, *X-Ray Diffraction*, Addison-Wesley Publishing Co., Reading, Massachusetts (1969).
18. J. K. Percus and G. J. Yevick, *Phys. Rev.* **110**, 1 (1958).
19. J. K. Percus, *Phys. Rev. Lett.* **8**, 462 (1962).
20. N. W. Ashcroft and J. Lekner, *Phys. Rev.* **145**, 83 (1966).
21. P. A. Lee, B.-K. Teo, and A. L. Simmons, *J. Am. Chem. Soc.* **99**, 3856 (1971).
22. D. E. Sayers, E. A. Stern, and F. W. Lytle, *Phys. Rev. Lett.* **35**, 584 (1975).
23. T. M. Hayes, J. W. Allen, J. Tauc, B. C. Gressen, and J. J. Hauser, *Phys. Rev. Lett.* **40**, 1282 (1978).
24. J. C. Knights, *Philos. Mag.* **34**, 663 (1976).
25. J. C. Knights, T. M. Hayes, and J. C. Mikkelsen, Jr., *Phys. Rev. Lett.* **39**, 712 (1977).

26. S. H. Hunter and A. Bienenstock, *Proceedings of the VI International Conference on Amorphous and Liquid Semiconductors*, Academy of Sciences of USSR, AF. Ioffe Physical Technical Institute, Leningrad (1975).

27. S. H. Hunter, A. Bienenstock, and T. M. Hayes, *Proceedings of the VII International Conference on Amorphous and Liquid Semiconductors*, Centre for Industrial Consultancy and Liaison, University of Edinburgh, Edinburgh (1977).

28. S. H. Hunter, A. Bienenstock, and T. M. Hayes, *Proceedings of the Symposium on Structure of Non-Crystalline Materials*, Taylor and Francis, Ltd., Cambridge, England (1976).

29. K. S. Liang, A. Bienenstock, and C. W. Bates, *Phys. Rev. B* **10**, 1528.

30. G. S. Brown, L. R. Testardi, J. H. Wernick, A. B. Hallak, and T. H. Geballe, *Solid State Commun.* **23**, 875 (1977).

31. C. S. Pande, *Solid State Commun.* **24**, 241 (1977).

32. P. Eisenberger and B. M. Kincaid, *Chem. Phys. Lett.* **36**, 134 (1975).

33. A. Fontaine, P. Lagarde, D. Raoux, M. P. Fontana, G. Maisano, P. Migliardo, and F. Wanderlingh, *Phys. Rev. Lett.* **41**, 504 (1978).

34. M. P. Fontana, G. Maisano, P. Migliardo, and F. Wanderlingh, *Solid State Common.* **23**, 489 (1977).

35. T. I. Morrison, A. H. Reis, Jr., G. S. Knapp, F. Y. Fradin, H. Chen, and T. E. Klippert, *J. Am. Chem. Soc.* **100**, 2362 (1978).

36. G. S. Knapp, H. Chen, and T. E. Klippert, *Rev. Sci. Instum.* **49**, 1658 (1978).

37. J. A. Del Cueto and N. J. Shevchik, *J. Phys. E* **11**, 616 (1978).

38. J. J. Ritsko, S. E. Schnatterly, and P. C. Gibbons, *Phys. Rev. Lett.* **32**, 671 (1978).

39. B. M. Kincaid, A. E. Meixner, and P. M. Platzman, *Phys. Rev. Lett.* **40**, 1296 (1978).

12

X-Ray Absorption Spectroscopy: Catalyst Applications

F. W. LYTLE, G. H. VIA, and J. H. SINFELT

1. Introduction

Fundamental to the ultimate understanding of catalysts and catalysis are models based upon atomic arrangement. However, the compositional nature of catalysts makes the structure determination difficult. The element specific advantage of x-ray absorption spectroscopy has opened a new dimension in the characterization of catalysts. Determinable parameters involving the catalytic atom include bond distances, number and kind of near neighbors, thermal and static disorder, as well as changes in valence electron states. The penetrating character of x radiation is advantageous for performing *in situ* experiments. Herein we discuss (1) the nature of catalysts, (2) principles of EXAFS used to analyze our data, (3) experimental technique, (4) EXAFS determination of catalyst structure—primarily our own work on supported metal catalysts but also other recent work on metal oxide and homogeneous catalysts, and (5) a correlation of L_3-edge difference spectra with electronic structure.

It is important to note that most of these results were made possible only by the increased x-ray flux available at a synchrotron source. By nature the elements of interest in a catalyst are trace constituents and spectroscopy involving those elements is difficult. For many classes of materials simply increasing scanning time permits the accumulation of satisfactory data; however, for many catalysts this is not a viable option. The desired catalyst conditions cannot be maintained for the time required (days or weeks) with conventional x-ray sources because of their

F. W. LYTLE • The Boeing Co., Seattle, Washington 98124.
G. H. VIA and J. H. SINFELT • Exxon Research and Engineering Co., Linden, New Jersey 07036.

extremely active surfaces. As shown by the results herein, it was possible to prepare *in situ* catalytic surfaces and maintain them for the 0.5–1.0 h necessary for data collection at SSRL.

2. Nature of Catalysts

A catalyst is a substance which increases the rate of a chemical reaction without being consumed in the process. In general, catalytic processes may be classified as homogeneous or heterogeneous. In homogeneous catalysis, the catalyst and reactants are present in a single phase, commonly a liquid solution. In heterogeneous catalysis, the reactants and catalyst are present as separate phases. Frequently the reactants are present as a vapor while the catalyst is a solid phase. However, other combinations of phases such as liquid–solid or liquid–liquid are also possible. Both types of catalytic processes are of considerable scientific interest, and both are important technologically. In the present article, however, the emphasis is mostly on solid catalysts of the type employed in heterogeneous catalytic processes.[1, 2]

Solid catalysts may be classified in two broad categories, metals and nonmetals. The most common metal catalysts contain metals of Group VIII or Group IB of the periodic table as the active components. Nonmetal catalysts are frequently oxides. Solid catalysts of practical interest are generally high-surface-area materials. In some cases the catalyst is a highly porous material. Most of the surface area then resides in the walls of the pores, and is often termed the "internal surface" of the solid. For highly porous solids, the internal surface area of a particle is frequently several orders of magnitude higher than the external surface area. Activated alumina is a highly porous material of this type with surface areas commonly in the range 100–300 m^2/gm. It is employed as a catalyst for dehydration of alcohols or for the skeletal isomerization of olefins. Often the active component of a catalyst is supported on a carrier to achieve high dispersion. The carrier is usually a high-surface-area refractory material such as silica or alumina.

Supported metal catalysts are highly important in industrial catalysis. A method commonly employed in the preparation of such catalysts is called impregnation. The method involves contacting of a carrier with a solution of a salt of the metal of interest. The solute deposits on the carrier, and the resulting material is then dried and often calcined at a higher temperature. The material dispersed on the carrier is then reduced to the metallic form, usually by contact with a stream of hydrogen at elevated temperature. After the reduction step, the material consists of small metal clusters or crystallites dispersed over the carrier surface. For example, in the preparation of a platinum-alumina catalyst, the inpregnation is commonly conducted with a chloroplatinic acid solution. After a drying step, the material is usually calcined in air at a temperature of about 500–550°C, prior to reduction in hydrogen.

Nonmetallic catalysts of interest are frequently oxides such as alumina, chromia, or molybdena. Such catalysts are commonly prepared by procedures involving precipitation or gel formation. For example, an alumina catalyst that is active for alcohol dehydration can be prepared by addition of an aqueous solution of ammonia to an aluminium nitrate solution. A finely divided precipitate of aluminium hydroxide is obtained on contacting the solutions. The precipitate is then dried and calcined at elevated temperatures (400–600°C) to produce the alumina. The use of a carrier in a nonmetallic catalyst, as in the case of metal catalysts, is a common method of improving the degree of dispersion of the active catalytic component. Such catalysts may be prepared by impregnation or coprecipitation procedures. For example, impregnation of alumina with an aqueous solution of chromic acid can be employed in the preparation of a chromia–alumina catalyst. After impregnation, the material is dried and subsequently calcined at temperatures of 350–550°C to complete the preparation. Such a catalyst can also be prepared by coprecipitation of chromia and alumina from a solution of chromic and aluminum nitrates by addition of ammonium hydroxide. The coprecipitated material is then dried and calcined in the same manner as impregnated chromia–alumina.

The degree of dispersion, or specific surface area, is an important property of a catalyst. In the case of catalysts consisting of a single component, a measurement of surface area by the low-temperature physical adsorption of a gas is applicable. This method provides information on the total surface area of a catalytic material, since physical adsorption is nonspecific with regard to the type of surface. For catalysts in which the active catalytic component comprises only a small fraction of the total surface area of the material, as in many supported metal catalysts, the method is no longer useful. In the case of supported metal catalysts, one needs a selective chemisorption method, in which a gas is adsorbed on the metal but not on the carrier. The chemisorption of simple gases such as hydrogen and carbon monoxide at room temperature meets this requirement. From such measurements one can determine the ratio of surface atoms to total atoms in the active metal component of the catalyst. The development of selective chemisorption methods has been extremely important for the characterization of highly dispersed, supported metal catalysts.

3. Analysis of EXAFS Data*

A photoelectron ejected from an atom as a result of x-ray absorption is characterized by a wave vector K, which is given by the equation

$$K = (2mE)^{1/2}/\hbar \tag{1}$$

where m is the mass of the electron, $\hbar$ is Planck's constant, and E is the kinetic energy of the photoelectron, which is equal to the difference between the x-ray energy and a threshold energy associated with the ejection of the electron. The EXAFS function $\chi(K)$ is defined by the equation

$$\chi(K) = (\mu - \mu_0)/\mu_0 \tag{2}$$

where μ and μ_0 are atomic absorption coefficients characteristic of the absorption associated with the ejection of an inner-core electron from an atom. The coefficient

---

* See References 3 and 4 and Chapter 10.

μ refers to absorption by an atom in the material of interest while μ_0 refers to absorption by an atom in the free state, both of which are functions of K.

In the experimental determination of EXAFS, the ratio I_0/I, where I_0 is the intensity of the x rays impinging on the sample and I is the intensity of the transmitted x rays, is determined as a function of x-ray energy on the high-energy side of the absorption edge. In obtaining $\chi(K)$ for a given point corresponding to a particular value of K, the first step is the determination of the difference between the observed value of $\ln I_0/I$ and a background value, the latter being determined from a moving average background line through data points extending over a limited range of K on either side of the data point in question. The "best fit" background line thus varies from one data point to another. The difference between the observed value of $\ln I_0/I$ and the background value, i.e., the fluctuation, is proportional to the difference between μ and μ_0. The increase in $\ln I_0/I$ at the edge, which we call the step height, is proportional to μ_0. The step height is taken as the difference between the minimum value of $\ln I_0/I$ at the x-ray energy corresponding to the onset of the edge and a higher value at the same energy obtained by back extrapolation of a "best fit" background curve through the data on the high-energy side of the edge. We recall that μ_0 is a function of K. To account for this dependence, we determine a hypothetical step height as a function of K by employing the McMaster[5] empirical relations based on x-ray absorption information on pure elements. A value of $\chi(K)$ at a given value of K is then determined by dividing the magnitude of the fluctuation by the step height.

Theories of EXAFS[6–10] based on the single scattering of an ejected photoelectron by atoms in the coordination shells surrounding the central absorbing atom give an expression for $\chi(K)$ of the following form

$$\chi(K) = \sum_j A_j(K) \sin[2KR_j + 2\delta_j(K)] \qquad (3)$$

where the summation extends over j coordination shells. In this expression, R_j is the distance from the central absorbing atom to atoms in the jth coordination shell and $\delta_j(K)$ is the phase shift. The factor $A_j(K)$ is an amplitude function for the jth shell, and is defined by the expression

$$A_j(K) = (N_j/KR_j^2)F_j(K) \exp(-2K^2\sigma_j^2) \qquad (4)$$

where N_j is the number of atoms in the jth shell, σ_j is the root-mean-square deviation of the interatomic distance about R_j, and $F_j(K)$ is a factor accounting for electron back scattering and inelastic scattering. The factor $F_j(K)$ is related to the back-scattering factor $f(K)$ and the mean free path λ for inelastic scattering by the expression

$$F_j(K) = f(K) \exp(-2R_j/\lambda) \qquad (5)$$

The back scattering factor $f(K)$ depends on the kind of atom responsible for the scattering.[6, 7, 10, 11a]

Fourier transformation of EXAFS data yields a function $\phi_n(R)$

$$\phi_n(R) = (1/2\pi)^{1/2} \int_{K_{min}}^{K_{max}} K^n \chi(K) \exp(2iKR) \, dK \qquad (6)$$

where $\chi(K)$ has been multiplied by the weighting factor K^n before the transform is taken.[4, 7] The function $\phi_n(R)$, where R is the distance from the absorber atom, is called a radial structure function.[12] It exhibits a series of peaks at $R = R'_j = R_j - a_j$. The difference a_j from a value of R_j, corresponding to a particular coordination shell, arises from an average over the range of K of the phase shift in equation (3).[8] The factor K^n in the integral is used to weight the data according to the value of K, for reasons that have been discussed elsewhere.[4, 7] In practice, transforms with $n = 1$ and $n = 3$ have been used routinely. The limits K_{min} and K_{max} of the integral are the minimum and maximum values of K at which experimental data are obtained. The complex transform as shown in equation (6) produces real, imaginary, and magnitude functions. Only the magnitude functions, which are everywhere positive, are used in this review. When a Fourier transform is taken over a finite range of variables, as in equation (6), a termination error is introduced into the function resulting from the transform. This error leads to a set of ripples propagating through the transformed function and contributes significantly to broadening of the peaks.[13] Peak heights are also decreased somewhat. A correction for termination error is made by means of a Hanning window function[14] of the form

$$(\tfrac{1}{2})\{1 - \cos 2\pi[(K - K_{min})/(K_{max} - K_{min})]\} \qquad (7)$$

This function is multiplied by the EXAFS function $\chi(K)$ in the regions of K values corresponding to the first and last 10% of the range investigated. As a result, $\chi(K)$ is forced to assume a half-cosine bell shape terminating at zero at K_{min} and K_{max}. This treatment of the data was very helpful in minimizing the termination ripples that have a period equal to $\sim 2\pi/K$. When necessary, K_{min} and K_{max} were varied in performing the integration to change the period of the ripples to discriminate them from real peaks.

Examination of data at this stage of analysis, i.e., the Fourier transform, is a useful but qualitative intermediate step. Comparison of data to reference materials to evaluate a_j can be used to determine R_j to ± 0.02 Å. A qualitative estimate of the coordination number may also be made from the peak height of the transform but this involves an uncertainty as to the effect of σ_j.

In our analysis of EXAFS data, we next obtain an inverse Fourier transform of the radial structure function $\phi_n(R)$ over a range of R encompassing a single peak corresponding to a particular coordination shell.[4] This procedure determines the contribution to EXAFS arising from that shell. If we consider a range of R from $R'_1 - \Delta R$ to $R'_1 + \Delta R$ encompassing the first coordination shell, the inverse transform yields the EXAFS function $K^n\chi_1(K)$ for the first coordination shell,

$$K^n\chi_1(K) = (2\pi)^{1/2} \int_{R_1' - \Delta R}^{R_1' + \Delta R} \phi_n(R) \exp(-2iKR) \, dR \qquad (8)$$

From equation (3) we have the following theoretical equation for the EXAFS function:

$$K^n\chi(K) = K^n A_1(K) \sin[2KR_1 + 2\delta_1(K)] \qquad (9)$$

where the subscript 1 refers to the first coordination shell. The phase shift $\delta_1(K)$ may be expressed empirically as a function of K.[11b]

$$\delta_1(K) = b + aK + a'K^2 + a''/K^3 \tag{10}$$

The amplitude function $A_1(K)$ is given by the equation

$$A_1(K) = (N_1/KR_1^2)F_1(K)\exp(-2K^2\sigma_1^2) \tag{11}$$

The parameters in equations (9), (10), and (11) are determined by fitting the theoretical EXAFS function of equation (9) to the function $K^n\chi_1(K)$ of equation (8) derived from the EXAFS data. The fitting is accomplished by means of an iterative least-squares procedure.

The use of an iterative least-squares procedure is required because equation (9) is not linear in the various parameters. In this procedure, the EXAFS function $\chi_1(K)$ of equation (9) is first expanded in a Taylor's series about a set of assumed values of the parameters. The expansion expresses $\chi_1(K)$ in terms of variations from the assumed parameter values. Terms higher than first order in the parameter variations are omitted. Application of the least-squares condition then provides a way of determining a set of parameter variations about the assumed values that give the best fit to the EXAFS function of equation (8) derived from the data. The parameter variations may then be taken as adjustments to the original set of assumed parameter values to obtain a new set of parameter values. The least-squares condition is again applied with the new set of parameter values to obtain a new set of parameter adjustments. The procedure is repeated until the parameter adjustments approach zero. The number of iterations required depends on the quality of the initial assumptions of the parameter values.

The least-squares technique that we have applied to EXAFS data is quite general. Problems involving a single shell or a number of shells of neighboring atoms can be handled. Information on all the quantities in equation (9) can be extracted. In practice, parameters derived from experiments on standard materials can be used to limit the number of unknowns in the system of interest. For example, in the analysis of EXAFS data on a dispersed metal catalyst, we begin by analyzing data on the pure metal, the structure of which is known (i.e., R_1 and N_1 are known). The amplitude function $A_1(K)$ in question (9) is derived from the values of the maxima and minima in the EXAFS function of equation (8) determined from the experimental data. For values of K other than those corresponding to maxima or minima, the values of $A_1(K)$ are obtained by interpolation. The phase shift in the form of equation (10) is then determined by using the least-squares fitting procedure. From the amplitude function $A_1(K)$ we determine the quantity $F_1(K)\exp(-2K^2\sigma_1^2)$. In the case of a dispersed metal catalyst the latter quantity becomes $F_1(K)\exp(-2K^2\sigma_1^2)\exp(-2K^2\,\Delta\sigma_1^2)$, where $\Delta\sigma_1^2$ is the difference between σ_1^2 values for the dispersed metal catalyst and the pure metal. Note that $\delta_1(K)$ and $F_1(K)$ are assumed to be the same for the pure metal and the corresponding dispersed metal catalyst. The transference of phase shift information from one system to another involving the same pair of absorbing and scattering atoms has been demonstrated to be a satisfactory calibration technique.[15, 16] The assumption that $F_1(K)$ also is the same for the pure metal and the dispersed metal catalyst seems reasonable on the basis that it is determined primarily by the type of back scattering atom. After the foregoing assumptions are made, application of the iterative least-squares fitting procedure to the first shell EXAFS function derived from experimental data on the dispersed metal catalyst yields values of N_1 and R_1 for the catalyst. Also, we obtain the quantity $\Delta\sigma_1^2$ defined earlier in this paragraph.

In order to evaluate our analysis procedure and gain experience in its application, we have investigated a variety of materials with known structural parameters. The feature in the analysis procedure that was of particular interest was the strong coupling between N_1 and σ_1^2 (or $\Delta\sigma_1^2$) that occurs in the least-squares refinement program. Because of their functional dependence, the parameters N_1 and σ_1^2 can vary in concert to fit both the shape and the magnitude of the EXAFS interference function. To investigate this coupling, experiments were designed in which N_1 and σ_1^2 were varied in an essentially independent manner. In addition, the effects of using various transform ranges and inverse transform ranges have been examined. A sampling of this effort is shown in Figure 1 and Table 1.

To examine the influence of changes in σ_1^2 (or $\Delta\sigma_1^2$), measurements were made on Pt and Ir metal at different temperatures. In each case the metal at 100 K was taken as the reference and the parameters $F_1(K)$ and $\delta_1(K)$ determined at this temperature were used to analyze data at higher temperatures. In the analysis N_1, R_1, and $\Delta\sigma_1^2$ were allowed to vary. The best-fit values for these parameters can be compared with the known values for N_1 and R_1 and the values calculated for $\Delta\sigma_1^2$ from the theory of Beni and Platzman[24] (to be discussed in Section 5). To examine the influence of changes in N_1 measurements were made at 100 K on $Ir_4(CO)_{12}$, in which the central Ir atoms form a tetrahedron. The reference for this analysis was

Table 1. Comparison of Fitted and Known[a] Parameters

Sample	N_1	R_1 (Å)	$\Delta\sigma_1^2$ (Å²)[b]	Remarks[d]
Pt metal at 300 K	12.8	2.774	0.0032	2–20 transform "narrow" inverse
	(12)	(2.775)	(0.0030)	
	11.6	2.773	0.0030	6–20 transform "wide" inverse
	12.2	2.768	0.0031	2–15 transform "wide" inverse
Pt metal at 673 K	12.3	2.756	0.0098	2–20 transform "narrow" inverse
	(12)	(2.784)[e]	(0.0104)	
	13.0	2.756	0.0096	6–20 transform "wide" inverse
	14.4	2.755	0.0098	2–20 transform "wide" inverse
$Ir_4(CO)_{12}$ at 100 K	2.7	2.688	−0.00020	2–20 transform "narrow" inverse
	(3)[c]	(2.686)[c]	(?)[c]	
	3.1	2.685	−0.00019	6–20 transform "wide" inverse
Estimated uncertainty	±20%	±0.010 Å	±10%	

[a] The known value is shown in parentheses.
[b] The known value was estimated using the Beni–Platzman theory discussed later.
[c] G. R. Wilkes, Ph.D. thesis, University of Wisconsin (1965).
[d] "Narrow" and "wide" refer to inverse transforms that included only the primary peak, and the primary peak plus one side lobe (each side), respectively.
[e] Estimated using coefficient of expansion = $8.3 \times 10^{-6}/C°$.

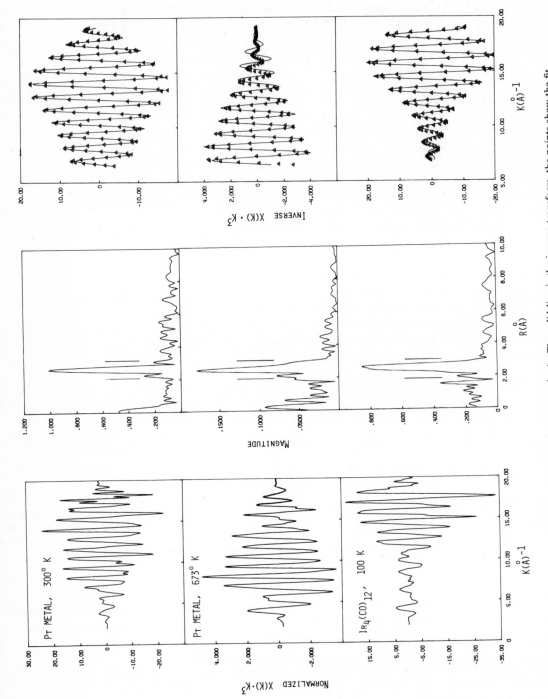

Figure 1. Spectra, transforms, and fits for reference standards. The solid line is the inverse transform, the points show the fit.

the metal at 100 K and again, N_1, R_1, and $\Delta\sigma_1^2$ were allowed to vary. The best-fit values for N_1 and R_1 can be compared to their known values, but it was not possible to make an estimate of $\Delta\sigma_1^2$. In this case we find that $\Delta\sigma_1^2$ has a small negative value, which appears reasonable considering the short Ir–Ir bond length.

The results in Table 1 indicate that our analysis procedure does a reasonably good job in decoupling the highly correlated variables N_1 and $\Delta\sigma_1^2$. We estimate that measurements of N_1 are accurate to $\pm 20\%$ while values for $\Delta\sigma_1^2$ are accurate to $\pm 10\%$. For bond length measurements, R_1, we also find that large increases in $\Delta\sigma_1^2$ result in an apparent shortening of R_1. This effect has been discussed by Eisenberger and Brown.[17] From their estimate and a consideration of our experimental findings (Table 1), we estimate an uncertainty in R_1 of ± 0.010 Å when $\Delta\sigma_1^2 < 0.005$, with no attempt made to correct for the effect (which should be possible). Indeed, relative changes in R_1 for materials with similar $\Delta\sigma_1^2$ may be accurate to ± 0.003 Å. In trials with transforms over different ranges of K and with "narrow" or "wide" (explained in Table 1) inverse transform ranges, similar results were obtained as long as the procedure for reference and sample was identical.

4. Experimental Procedures

The supported metal catalysts to be discussed in the next section consisted of small metal clusters dispersed on a carrier. Either silica or alumina was employed as the carrier. The silica used was Cabosil HS5 (300 m^2/g surface area), obtained from the Cabot Corporation, Boston, Massachusetts. The alumina was prepared by heating beta alumina trihydrate at 600°C for 4 hours, and had a surface area of approximately 200 m^2/g. The concentration of metal in the catalysts was 1.0 wt. %. A simple impregnation procedure[18] was employed in the preparation of the catalysts. Aqueous solutions of appropriate metal salts or acids were used in the impregnations. The amounts of impregnating solution of appropriate concentration employed per gram of silica or alumina carrier were, respectively, 2.2 and 0.65 ml. After impregnation, the preparations were dried at 110°C, except for the osmium preparation, which was dried at 80°C under vacuum. The materials were then reduced in a stream of hydrogen at 500°C, after which they were purged with helium while being cooled to room temperature. They were then stored in air until needed. Prior to EXAFS measurements the catalyst materials were rereduced in the EXAFS apparatus at about 425°C in flowing hydrogen. Chemisorption measurements of metal dispersions were made on samples reduced *in situ*, in apparatus described previously.[18]

All EXAFS measurements were made in the transmission mode using the EXAFS I spectrometer on beam line #1 at SSRL. The spectrometer control software was written by Jon Kirby.[19] Typically, data were collected from about 200 eV on the low-energy side of an absorption edge to about 1600 eV on the high-energy side of the L_3 edge. In the vicinity (± 50 eV) of an edge data were collected at about 1-eV intervals, and 3–4-eV step intervals were used in the regions removed from the edge. In most experiments a one second counting time per point was used, but frequently multiple scans were made and added to improve signal/noise.

Because of the tendency of dispersed metal catalysts to strongly adsorb reactive gases such as oxygen, it was necessary that the catalyst environment be controlled so that measurements on well-defined catalyst states could be achieved. In particular, it was desirable to be able to vary the catalyst environment so that conditions equivalent to those used in catalyst characterization[18] experiments (i.e., chemisorption) could be simulated. To these requirements must be added those needed to optimize the actual EXAFS measurements (minimum beam attenuation by windows, low temperature, etc). To meet these requirements, a furnace assembly was designed that provided a wide usable temperature range along with the ability to control the catalyst environment. The furnace is illustrated in Figure 2.

The furnace assembly consisted of three main components: the housing, the sample cell, and the cell mounting assembly. The housing was comprised of an aluminum box with entrance and exit slots for the x-ray beam and a vacuum connection for coupling to a vacuum pump. Mylar film was used to cover the entrance and exit slots. The interior of the box contained a resistance heater surrounded by ceramic heat reflectors for heating the sample. The ceramic reflectors were slotted on the entrance and exit side for passage of the x-ray beam. Electrical connection to the heating element was made through the wall of the aluminum vessel with vacuum tight connections, and the power level to the heating element was controlled by a variable transformer. Because of the wide range of operating

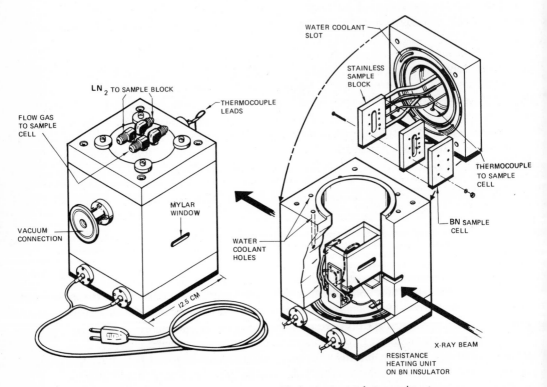

Figure 2. Controlled atmosphere furnace assembly for *in situ* catalyst experiments.

temperatures (100–800 K), it was necessary to thermally isolate the furnace assembly from the x-ray detectors located on either side. This was accomplished by circulating ambient water through the aluminum walls. Additional insulation was provided by evacuating the space around the sample cell assembly interior of the reactor during operation.

The cell mounting assembly consisted of a stainless steel mounting block attached to the top of the vacuum-sealable aluminum housing by four stainless steel tubes. The mounting block was drilled and tapped to allow liquid nitrogen to circulate for cooling purposes, and to provide an inlet and outlet for gas flow through one face of the block. The mounting block also contained an embedded thermocouple for temperature measurement. The cell that contained the catalyst was fabricated from boron nitride to limit the x-ray absorption by the cell itself. It consisted of two parts, and the cell and the mounting block were clamped tightly together with bolts. By scrupulous polishing of the cell mating surfaces, it was possible to obtain a reasonably good seal. At the junction of the two parts of the cell, there was a cavity to hold a catalyst sample of the desired thickness. Machined channels in the boron nitride cell in contact with the mounting block allowed gas to flow from the entrance line attached to the mounting block, through the catalyst bed, and back to the exit line attached to the mounting block. Thus, when the entire unit was assembled a closed gas flow loop existed that could be coupled to appropriate gas lines at the top of the aluminum housing. Since the region around the boron nitride cell was evacuated during operation, any minor leaks that occurred around the boron nitride mating surfaces were outward from the sample rather than into the sample cavity. The depth of the cell cavity that contained the catalyst sample could be varied between about 1 and 5 mm to accommodate a variety of sample types. To further limit x-ray absorption by the cell, the two areas of the cell assembly in the x-ray path and in contact with the sample were machined to a thickness of 1 mm.

Control of the catalyst atmosphere was accomplished by maintaining a constant gas flow through the catalyst bed at atmospheric pressure. To allow a variety of different gases to be used, a multivalved manifold was constructed that would accommodate up to four gas cylinders. The gases typically used in our experiments included hydrogen, helium, a mixture of 1% oxygen in helium, and carbon monoxide. The hydrogen was used for reducing the various catalysts since they were all exposed to air prior to measurements. Helium provided an inert atmosphere, and the oxygen–helium mixture and carbon monoxide were used for chemisorption experiments. The gas manifold that controlled access to the various gases was equipped with a Pd catalyst bed to remove traces of oxygen from the hydrogen and a multicomponent sorption bed to remove traces of water, CO_2, and hydrocarbons from all gases. A flow meter was included in the system, but evidence that a positive gas flow was maintained was furnished by a bubbler attached to the end of the exit line.

For the EXAFS experiments the cell was charged with approximately $\frac{1}{2}$ g of catalyst in the form of 40–60 mesh particles so that gas flow would not be restricted. In a typical experiment the catalyst was initially reduced by flowing hydrogen over the catalyst bed as the temperature was increased from room temperature to 700 ± 25 K. Hydrogen flow was continued at this temperature for 15–60 min to ensure complete reduction. Actually the edge structure discussed later could be used as a sensitive indicator of catalyst condition. Following reduction, the temperature was lowered rapidly to 100 K for the x-ray measurements. For chemisorption studies other than hydrogen, the temperature was lowered from 700 K to about 650 K with flowing hydrogen to eliminate contamination of the catalytic metal with water from the support. At 650 K the hydrogen flow was replaced with helium to remove all hydrogen from the catalyst, and the temperature was lowered to room temperatures. At this point the helium flow was replaced with either $1\% \ O_2$ in He or CO, depending on the experiment, to carry out the chemisorption of O_2 or CO on the catalytic metal. The temperature was subsequently lowered to 100 K for the EXAFS measurement.

5. Structure of Catalysts

5.1. Dispersed Metal Catalysts

The application of EXAFS to highly dispersed metal catalysts appears to have much promise.[20–22] Catalysts of this type consist of small metal clusters dispersed on the surface of a carrier, typically a refractory oxide such as silica or alumina. Recently, we have conducted EXAFS studies on dispersed platinum, iridium, and osmium catalysts with a metal concentration of 1 wt. %.[23] The alumina and silica carriers on which the metals were dispersed had surface areas of approximately 200 and 300 m^2/gm, respectively. The dispersion of the metal, defined as the ratio of surface atoms to total atoms in the metal clusters, was estimated from gas chemisorption measurements (H_2, CO, and O_2) to be 0.8 ± 0.2 for all of the catalysts. Differences in the amounts of a given gas chemisorbed on the different catalysts were small, and it was concluded that the dispersions were not significantly different.

Figure 3 shows EXAFS data at 100 K on the dispersed osmium catalyst and on a sample of metallic osmium used as a reference. Included in the figure are plots of the EXAFS function $\chi(K)K^3$ versus K and Fourier transforms of the EXAFS function over the range $2 \leq K \leq 20$. As discussed previously, such Fourier transforms provide radial structure functions. An inverse transform over a range of R just encompassing the main peak gives an EXAFS function representing the first coordination shell shown by the solid line function on the right, along with demonstrations of the quality of fit to these functions achieved using parameters derived from the data in the manner described earlier. The points represent values calculated with the "best fit" parameters. As can be seen readily from the figure, the EXAFS fluctuations for the dispersed osmium catalyst are markedly smaller than those for the metallic osmium. Correspondingly, the magnitudes of the peaks in the Fourier transforms are also much smaller. (Note that the scales in the figure are not the same for the dispersed osmium catalyst and the metallic osmium.) These features are a consequence of a lower coordination number and/or a higher degree of disorder of the osmium atoms in the dispersed catalyst. The degree of disorder is characterized by

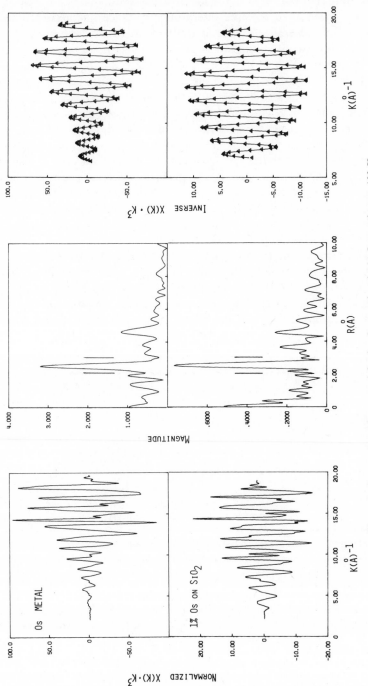

Figure 3. Spectra, transforms, and fits for Os metal and 1% Os catalyst at 100 K.

the root-mean-square deviation of the interatomic distances about the equilibrium values. We note that the peaks in the radial structure functions are located at values of R slightly lower than the true interatomic distances, as a consequence of the phase shift discussed earlier. In the analysis of the EXAFS data, Fourier transforms were taken over varying ranges of the wave vector K, e.g., over the ranges $2 \leq K \leq 15$ and $6 \leq K \leq 20$ in addition to the range $2 \leq K \leq 20$. This procedure is useful in distinguishing true peaks in the transformed EXAFS functions from peaks or ripples arising solely as a consequence of taking transforms over a finite range of K. The different ranges of K employed in taking the transforms, while leading to slight differences in the detailed results of the data analysis, did not significantly alter any conclusions.

A summary of properties derived from the EXAFS data on the dispersed platinum, iridium, and osmium catalysts is given in Table 2. The results in the table were obtained using Fourier transforms of the original EXAFS data over the range $2 \leq K \leq 15$. Included in the table are results on the average coordination number N_1, which refers to the average number of nearest metal atom neighbors (i.e., atoms in the first coordination shell) about an atom in a metal cluster. While the estimated uncertainty in N_1 is high (± 2 atoms), the values are significantly lower than 12, which is the value characteristic of large crystals of platinum, iridium, and osmium. This result is expected, since most of the atoms in the metal clusters are surface atoms with lower coordination numbers than the atoms in the interior of a crystal. Also, atoms at corners and edges have lower coordination numbers than

Table 2. Properties of Dispersed Metal Catalysts Derived from EXAFS Data[f]

Metal[a]	Carrier	N_1[b]	R_1[c] (Å)	$\Delta\sigma_1^2$[d] (Å²)	σ_1[e] (Å) Bulk metal	σ_1[e] (Å) Dispersed metal
Os	SiO$_2$	8.3	2.702 (2.705)	0.0022	0.027	0.054
Ir	SiO$_2$	9.9	2.712 (2.714)	0.0027	0.030	0.060
	Al$_2$O$_3$	9.9	2.704	0.0029	—	0.061
Pt	SiO$_2$	8.0	2.774 (2.775)	0.0018	0.044	0.061
	Al$_2$O$_3$	7.2	2.758	0.0037	—	0.075

[a] The concentration of metal in the catalyst was 1.0 wt. % in all cases.
[b] Average coordination number (nearest neighbors) of the metal atoms in the dispersed metal clusters (estimated uncertainty ± 2 atoms).
[c] Interatomic distances (nearest neighbors) in the metal clusters (estimated uncertainty 0.010 Å). The values in parentheses are the corresponding distances for the bulk metals.
[d] The difference between the mean-square deviation σ_1^2 of interatomic distance about the equilibrium value in the metal clusters and the value of σ_1^2 for the bulk metal at 100 K (estimated uncertainty $\pm 10\%$).
[e] Root-mean-square deviation of interatomic distance about the equilibrium value at 100 K. The values for the bulk metals were calculated using Debye temperatures of 500, 420, and 240 K, respectively, for Os, Ir, and Pt. The values for the dispersed metals were obtained using the calculated values for the bulk metals and the values of $\Delta\sigma_1^2$ determined from the EXAFS data.
[f] From Via, Sinfelt, and Lytle.[23]

the interior atoms in surface planes of crystals and become increasingly important as the size of a metal crystal decreases. The values of N_1 for the iridium catalysts appear to be higher than those for the platinum and osmium catalysts, although chemisorption data do not indicate such a difference. However, it should be noted that the differences indicated by the EXAFS results are about equal to our estimated uncertainty in the determination of N_1.

Also included in Table 2 are values of the interatomic distances R_1. The known values for the bulk metals are shown in parentheses for comparison. For all the catalysts except platinum dispersed on alumina, the interatomic distance in the metal clusters differed from that in the bulk metal by 0.01 Å or less, i.e., within our estimated uncertainty of ± 0.01 Å. In the case of platinum on alumina, the catalyst with the lowest coordination number, the interatomic distance was about 0.017 Å lower than that of bulk platinum. Interestingly, the data suggest an effect of the type of carrier. For both iridium and platinum, the interatomic distance appears to be shorter for alumina supported clusters than for silica supported clusters, suggesting a stronger interaction of the clusters with the alumina. While this suggestion would be consistent with other information on these types of catalysts, more data are needed to establish the effect of the carrier on interatomic distance.

The quantity $\Delta\sigma_1^2$ in Table 2 is the difference between σ_1^2 for the metal clusters and the corresponding quantity for the bulk metals at 100 K, where σ_1^2 is defined as the mean-square deviation of interatomic distance about the equilibrium value. We note that for all the catalysts the value of $\Delta\sigma_1^2$ is positive, i.e., the values of σ_1 are higher for the dispersed metal catalysts than for the corresponding bulk metals. The values of σ_1 for the bulk metals were calculated from values of u_x^2, the mean-square displacement of the atoms about their equilibrium positions, by means of the equation

$$\sigma_1^2 = 2u_x^2 - 2DCF \tag{12}$$

where DCF is the displacement correlation function, values of which can be determined from the paper by Beni and Platzman.[24] The mean-square displacement u_x^2 was in turn calculated from the equation

$$u_x^2 = \frac{3\hbar^2}{km\theta} \left[\frac{\phi(\theta/T)}{\theta/T} + \frac{1}{4} \right] \tag{13}$$

where $\hbar$ is Planck's constant, k is Boltzmann's constant, m is the mass of the atom, T is the absolute temperature, θ is the Debye temperature, and ϕ is the Debye function.[25] The Debye temperatures for platinum, iridium, and osmium were taken from Kittel.[26] The values of σ_1 for the dispersed metals were then determined from the calculated σ_1 values for the bulk metals and the experimentally determined values of $\Delta\sigma_1^2$. The values of σ_1 for the dispersed metals are greater by a factor 1.4–2 than the corresponding values for the bulk metals. This result may in part be an effect of the support. The interaction of the metal atoms with the support, presumably through bonding to the oxygen-or-hydroxide-rich surface[20, 22] may distort the normal structure of the metal lattice. This effect may be significant in catalysis.

Lytle *et al.*[22] have reported a study of the interaction of oxygen with a 1% Ru on silica catalyst. The results given in Figure 4 show the data and transforms for the reference materials, Ru and RuO_2, and for the clean catalyst, catalyst + O_2, and oxidized catalyst. This presents a clear picture of the ability of EXAFS to investigate solid–gas interactions *in situ*. The clean dispersed catalyst shows Ru–Ru bonds as in the pure metal but with diminished magnitude because of a smaller average coordination number due to particle size effects. When the catalyst was exposed to 7.6-mm O_2 *in situ*, Ru–O bonds were clearly resolved from the still extant Ru–Ru bonds. O_2 chemisorption left the catalyst structure essentially intact in the interaction with exposed surface atoms. When the catalyst was heated to 400°C in O_2, bulk oxidation occurred and destroyed the Ru lattice as expected. It was also noted that knowledge of the catalyst surface area from a separate chemisorption measurement and EXAFS measurement of the average number of chemisorption bonds leads to a determination of an average site symmetry, i.e., an average over all of the different sizes and shapes of crystallites containing the catalytic atoms. Paren-

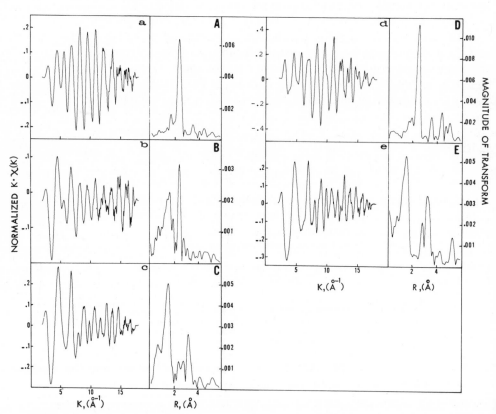

Figure 4. EXAFS data, $K\chi(K)$ versus K and the associated Fourier transforms, magnitude versus radial distance: (a, A) 1% Ru on SiO_2, reduced in H_2; (b, B) after exposure to 7.6 mm O_2 at 25°C; (c, C) after exposure to O_2 at 400°C; (d, D) Ru metal (the factor of two error in the scale for $\chi(K)$ in the original paper has been corrected[22]); (e, E) RuO_2.

thetically, we note that if uniformly sized and shaped metal clusters could be prepared, perhaps by decarbonylation of metal cluster compounds, the EXAFS determination of the geometry of reaction sites would have a more precise meaning.

5.2. Metal Oxide Catalysts

Friedman *et al.*[27] reported a study of the interaction of cupric ions with alumina using a variety of physical techniques, i.e., x-ray absorption spectroscopy, x-ray diffraction, diffuse reflectance optical spectroscopy, ESCA, and EPR, all combined to yield a description of the phases present and the cation site locations. It was found that at low concentrations Cu entered the defect spinel structure of γ-alumina in predominantly a distorted octahedral environment with $\sim 10\%$ Cu in tetrahedral coordination. At temperatures $> 600°C$ there was a slow transformation toward spinel ($CuAl_2O_4$) formation. At metal loadings up to the support capacity (~ 4 wt. $\%$ Cu per 100 m^2/g) the dispersed phase was not present as microcrystals of CuO. This was a definitive finding of the EXAFS experiments. Above the support capacity a crystalline phase of CuO formed on the surface and was detected by both EXAFS and by x-ray diffraction. The EXAFS data analysis proceeded only as far as Fourier transforms, which were then used as fingerprints in a proportional manner to identify the different absorbing atom environments. Careful selection of reference materials and EXAFS data all obtained at 77 K (to minimize the contribution of the temperature factor) were important factors in the success of the experiment.

Kohatsu *et al.*[28] have investigated hydrotreating catalysts, NiMo on alumina, in the calcined and sulfided form. Using well chosen reference materials (NiO, $NiMoO_4$, MoO_3, MoS_2), parameters for phase shift and envelope functions were established and transferred to the unknown catalyst samples. The results are summarized in Table 3. Note that in the sulfided material the Ni remained in oxygen coordination. Consideration of a reduced second-sphere coordination of Mo in the essentially MoS_2 environment of the sulfided sample resulted in a model composed of an open network of $[MoS_6]$ subunits or isolated clusters rather than edge sharing as in MoS_2.

Table 3. Cation Coordination in NiMo Alumina Catalysts

	N_1	R_1 (Å)
Oxide form		
Ni	6 (0)	2.04–2.05
Mo	4 (0)	1.73–1.74
	2 (0)	2.33
Sulfide form		
Ni	6 (0)	2.04–2.05
Mo	6 (S)	2.28–2.35

Table 4. Wilkinson's Catalyst Results

Catalyst and Technique	Bond	N_1	R_1 (Å)
Solid,	Rh–Cl	1	2.376
by x-ray	Rh–P_1	1	2.214
diffraction	Rh–P_2	2	2.326
Solution,	Rh–Cl	1	2.35
by EXAFS	Rh–P_1	1	2.23
	Rh–P_2	2	2.35
Polymer bound,	Rh–Cl	2	2.33
by EXAFS	Rh–P_1	1	2.23
	Rh–P_2	2	2.16
Hydrogenated,	Rh–Cl	1	2.29
by EXAFS	Rh–P_1	1	2.20
	Rh–P_2	1	2.38

5.3. Homogeneous Catalysts

A definitive, ground-breaking study was reported by Reed et al.[29] on an EXAFS investigation of the structure of Wilkinson's catalyst $(Ph_3P)_3$ RhCl, polymer-bound Wilkinson's catalyst, and hydrogenated polymer-bound Wilkinson's catalyst. Their results suggested formation of a dimer in the polymer-bound material, which was then cleaved by hydrogenation of the chloride bridges:

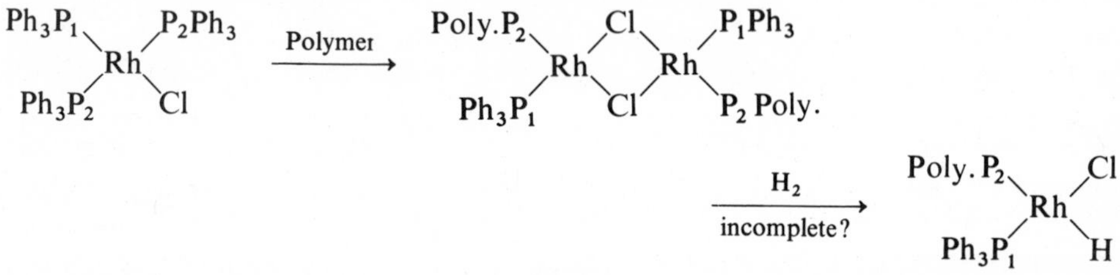

The data analysis proceeded by evaluation of scattering amplitudes and phase shifts from reference compounds containing Rh–P and Rh–Cl bonds and then used them in the analysis of the catalyst data. Fourier-filtered (first shell) fits were made by varying N_1 and R_1. The criterion for the best structure was the minimization of the fit residuals for a particular combination of ligands. Note that although Cl and P are next neighbors in the periodic table, there were sufficient differences in scattering amplitude and phase shift to discriminate between them. The results are summarized in Table 4.

6. Near-Edge Structure

In addition to the edge spectroscopy discussed in Chapter 9, there have been recent studies of the L edges of third period transition metals of catalytic interest. The L_3 or L_2 x-ray absorption threshold resonance (the "white line" of the older literature) has been qualitatively described as a dipole transition from a core hole, $2p_{3/2}$ or $2p_{1/2}$, respectively, to the vacant d states of the absorbing atom.[30, 31] (This and related topics have been reviewed in the book by Azaroff.[32]) With this point in mind Lytle[33] compared L_3 spectra of Ta, Ir, Pt, and Au where the d-band occupancy varies and found a correlation of the "white-line" area with the number of unfilled states estimated from band-structure calculations. The sensitivity of this transition to chemical combination was also noted as well as the effect of the environment on a supported Pt catalyst (which will be discussed in detail later).

Gallezot *et al.*[34] investigated the L_3 edge of Pt clusters supported on zeolites and correlated the area of the L_3 white line and the edge shift with catalytic activity through a quantity defined as electron deficiency. Qualitatively a large white line was evidence of a large electron deficiency with the largest effects occurring when the catalyst was heated to high temperatures in O_2. A much smaller effect was present when the catalyst had been hydrogen reduced.

Lytle *et al.*[35] have proposed a model based on chemical ionicity to account for the strength of the threshold resonance in the presence of different ligands. This model assumes that as bonding d electrons "flow" into chemical bonds and become delocalized from the parent metal atom, the resulting d-electron vacancy will be mapped by the L_3 absorption transition. Combining the concepts of Batsanov[36] and Pauling[37] the amount of electron flow or coordination charge on element A after formation of an A–B bond is

$$\eta_A = m_A I_{AB} \tag{14}$$

where m_A is the valence of the absorbing atom, $X_{A \text{ or } B}$ is the electronegativity, and I_{AB} is the Pauling single-bond ionicity of elements A–B.

$$I_{AB} = 1 - \exp[-\tfrac{1}{4}(X_A - X_B)^2] \tag{15}$$

The concept of coordination charge as used here has intuitive appeal as the fraction of valence electrons transferred from the atom due to chemical bonding. This retains the traditional chemical concept of integral valency (or oxidation state) modified by an easily calculated ionicity using tabulated values of electronegativities. For a pure element $\eta = 0$ and η can vary from m to 0 in compounds. The identification of η with the change in area of the threshold resonance suggests that this quantity can be isolated by the difference spectrum, i.e., spectrum of compound minus spectrum of pure metal. In Figure 5 difference spectra are shown for compounds of Au, Pt, and Ir. Each spectrum was normalized to unit absorption in the region of the edge after the resonance and then shifted relative to each other to obtain the best alignment in the region of the onset of absorption from -20 to 0.0 eV (0.0 defined as the first inflection point of the absorption rise) before the spectrum of the pure metal was subtracted from each of its compounds. The area versus η correlation is made in Figure 6. Considering the simple theory, the correlation is remarkable. It appears that difference spectra can isolate the d-electron vacancy and a simple ionicity theory can achieve a degree of organization even between different elements. This calibration curve can be used to infer coordination charge and, hence, d-electron concentration in unknown materials containing the same elements.

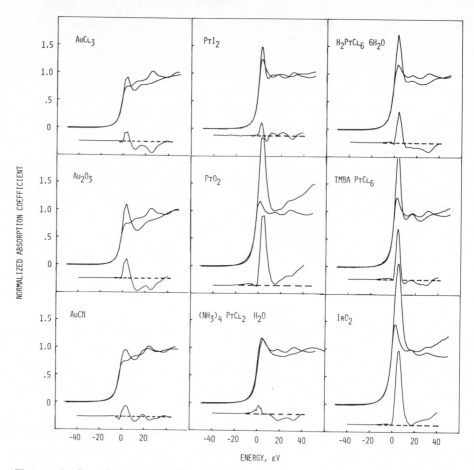

Figure 5. L_3 absorption edges and difference spectra (compound minus pure metal) for compounds of Pt, Au, and Ir. The difference has been shifted downward for clarity.

In Figure 7 difference spectra are shown for 1% Pt and 1% Ir supported on Al_2O_3. The samples were reduced *in situ* in H_2, cooled in He as described previously, and then exposed to 1% O_2 in He at 25°C. The data are summarized in Table 5. Note that each catalyst shows a substantial, positive peak even in the reduced, He-cleaned condition. This effect may be a consequence of the small size of the metal clusters or of bonding of metal atoms of the clusters to oxygen in the alumina surface. If one assumes the latter possibility to be the dominant one, the areas of the peaks can be related through the calibration of Figure 6 to a coordination charge from which one can estimate the number of bonds with the support. The iridium and platinum catalysts have values of η equal to 0.4 and 0.15, respectively. From these values one can estimate roughly that about 70% of the Ir and 40% of the Pt atoms are bonded to the support. This quantity should be

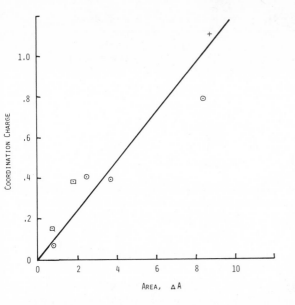

Figure 6. Coordination charge versus area ΔA of the peak in difference spectra for compounds of Pt(⊙), Au(□), and Ir(+).

sensitive to the metal particle size and shape. After exposure of the catalysts to O_2, additional Pt–O and Ir–O bonds form, but the η values indicate the chemisorption bond may be different from the metal–oxygen bond in the metal oxide in that the increase in η is less than expected from the measured (oxygen/metal atom) ratio. One possibility is that the metal atom has less than the normal valence.

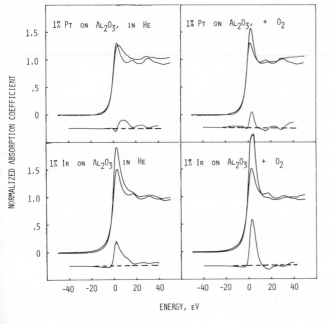

Figure 7. Difference spectra for 1% Pt and Ir on alumina catalysts reduced in H_2, cleaned in He, and exposed to 7.6 mm O_2 at 25°C.

Table 5. L_3 Threshold Resonance of 1% Pt and 1% Ir on Al_2O_3

Catalyst	Condition	Threshold resonance area	Chemisorption[a]: O/metal atom
1% Ir	Reduced in H_2, cooled in He	3.7	—
1% Ir	Exposed to 1% O_2 in He at 25°C	4.9	0.86 ± 0.1
1% Pt	Reduced in H_2, cooled in He	1.1	—
1% Pt	Exposed to 1% O_2 in He at 25°C	1.5	0.72 ± 0.1

[a] Determined by a separate chemisorption experiment.

7. Status and Outlook

We have shown initial applications of x-ray absorption spectroscopy to the determination of the structure of catalysts. Although the work is in an early stage, clearly it adds important new dimensions to catalyst characterization. EXAFS data analysis has reached a maturity where it can be applied to real catalytic problems resulting in accurate models of structure and surface processes. Since many important catalytic systems fall within the happy coincidence of an x-ray soft support containing a heavier catalytic atom, the penetrating and element-specific qualities of x-ray edge spectroscopy are ideally suited for these investigations, i.e., EXAFS determines bond distance, near-neighbor type, coordination number, and the disorder parameter, while near-edge spectroscopy senses changes in bonding electron distribution. The phenomenological description of near-edge structure presented here can only be considered as an interim guide to an understanding of the effect.

Because of the infancy of the subject we have drawn heavily upon our own work and the few publications and preprint reports available to us at the present time (November, 1978). However, many others have begun work. We know of nine active research programs in the subject area. Results from most of these have at least reached the stage of oral presentation and publication should be rapidly forthcoming.

As discussed in Chapter 3, user interest has prompted creation of new dedicated national synchrotron radiation laboratories to be built at Brookhaven and Wisconsin, a parasitic facility at Cornell, plus expansion and 50% dedication at Stanford. The planned improvements of on-site facilities including increased x-ray flux, improved detection systems, data processing, and on-line reaction cells will make user access much more straightforward and should considerably expand the user community. We expect important new developments of catalytic interest related to:

1. the structure of chemisorbed species,
2. the effects of chemisorbed species on the properties of catalysts (e.g., on disorder),

3. studies of very dilute catalysts, i.e., catalysts in which the active component is present at concentrations as low as 0.001%,
4. the nature of catalysts at various stages of preparation and use,
5. the nature of the catalyst system when catalysis is actually occurring, and
6. new approaches to EXAFS investigations, e.g., time resolved EXAFS in a dynamic system and surface EXAFS experiments using electron detection methods with low-atomic-number elements such as C, O, Al and Si.

Successful completion of research in these areas should resolve the structural ambiguities that have been present in models of catalysts and catalysis. These results should be sufficiently detailed and accurate to discuss catalytic structure with the same detail and confidence that has previously been possible only in more structurally regular systems. That knowledge will certainly increase our understanding of catalysts and will perhaps lead to significant improvements.

ACKNOWLEDGMENTS

We are grateful for experimental support at SSRL from the excellent staff and for facility funding by DOE and NSF. The research of F. W. Lytle was partially supported by NSF grants CHE 76-11255 and DMR 77-12919.

References

1. J. H. Sinfelt, *Science* **195**, 641 (1977).
2. J. H. Sinfelt, *Annu. Rev. Nat. Sci.* **2**, 641 (1972).
3. F. W. Lytle, D. E. Sayers, and E. A. Stern, *Phys. Rev. B* **11**, 4825 (1975).
4. E. A. Stern, D. E. Sayers, and F. W. Lytle, *Phys. Rev. B* **11**, 4836 (1975).
5. W. H. McMaster, N. Kerr del Grande, J. H. Mallet, and J. H. Hubbell, Compilation of X-ray Cross Sections, *UCRL-50174* **II**(1), University of California, Livermore (1969).
6. D. E. Sayers, E. A. Stern, and F. W. Lytle, *Phys. Rev. Lett.* **27**, 1204 (1971).
7. E. A. Stern, *Phys. Rev. B* **10**, 3027 (1974).
8. C. A. Ashley and S. Doniach, *Phys. Rev. B* **11**, 1279 (1975).
9. P. A. Lee and J. B. Pendry, *Phys. Rev. B* **11**, 2795 (1975).
10. P. A. Lee and G. Beni, *Phys. Rev. B* **15**, 2862 (1977).
11a. B. K. Teo, P. A. Lee, A. L. Simons, P. Eisenberger, and B. M. Kincaid, *J. Amer. Chem. Soc.* **99**, 3854 (1977).
11b. P. A. Lee, B. K. Teo, and A. L. Simons, *J. Amer. Chem. Soc.* **99**, 3856 (1977).
12. D. E. Sayers, Ph.D. dissertation, University of Washington (1971).
13. J. Waser and V. Schomaker, *Rev. Mod. Phys.* **25**, 671 (1953).
14. G. D. Bergland, *IEEE Spectrum* **6**(7), 41 (1969).
15. P. H. Citrin, P. Eisenberger, and B. M. Kincaid, *Phys. Rev. Litt.* **36**, 1346 (1976).
16. S. P. Cramer, T. K. Eccles, F. Kutzler, K. O. Hodgson, and S. Doniach, *J. Am. Chem. Soc.* **98**, 8059 (1976).
17. P. Eisenberger and G. Brown, *Solid State Commun.* **29**, 481 (1979).
18. J. H. Sinfelt and D. J. C. Yates, *J. Catal.* **8**, 82 (1967).
19. J. A. Kirby, Manual for Data Collection Program, UCID-401 (Feb. 1978).
20. I. W. Bassi, F. W. Lytle, and G. Parravano, *J. Catal.* **42**, 139 (1976).
21. J. H. Sinfelt, G. H. Via, and F. W. Lytle, *J. Chem. Phys.* **68**, 2009 (1978).
22. F. W. Lytle, G. H. Via, and J. H. Sinfelt, *J. Chem. Phys.* **67**, 3831 (1977).

23. G. H. Via, J. H. Sinfelt, and F. W. Lytle, *J. Chem. Phys.* **71**, 690 (1979).
24. G. Beni and P. M. Platzman, *Phys. Rev. B* **14**, 1514 (1976).
25. J. S. Kasper and Kathleen Lonsdale (eds.), *International Tables for X-Ray Crystallography*, Vol. II, p. 241, The Kynoch Press, Birmingham, England (1959).
26. C. Kittel, *Introduction to Solid State Physics* 5th edition, p. 126, Wiley, New York (1976).
27. R. M. Friedman, J. J. Freeman, and F. W. Lytle, *J. Catal.* **55**, 10 (1978).
28. I. Kohatsu, D. W. Blakely, and H. F. Harnsberger (to be published).
29. J. Reed, P. Eisenberger, B. K. Teo, and B. M. Kincaid, *J. Am. Chem. Soc.* **99**, 5217 (1977).
30. N. F. Mott, *Proc. R. Soc. London* **62**, 416 (1949).
31. Y. Cauchois and N. F. Mott, *Philos. Mag.* **40**, 1260 (1949).
32. L. V. Azaroff (ed.), *X-Ray Absorption Spectra*, McGraw-Hill, New York (1974).
33. F. W. Lytle, *J. Catal.* **43**, 376 (1976).
34. P. Gallezot, R. Weber, R. A. Dalla Betta, and M. Boudart, *Z. Naturforsch* **34A**, 40 (1979).
35. F. W. Lytle, P. S. P. Wei, R. B. Greegor, G. H. Via, and J. H. Sinfelt, *J. Chem. Phys.* **70**, 4849 (1979).
36. Cited in I. A. Ovsyannikova, S. S. Batsanov, L. I. Nasonova, L. R. Batsanova, and E. A. Nekrasova, *Bull. Acad. Sci. USSR Phys. Ser.* **31**, 936 (1967) [English translation].
37. L. Pauling, *The Nature of the Chemical Bond*, Cornell University Press, Ithaca, New York, 1st edition (1939) or 2nd edition (1948).

13

X-Ray Absorption Spectroscopy of Biological Molecules

S. DONIACH, P. EISENBERGER, and KEITH O. HODGSON

1. Introduction

The advent of intense, tuneable x-ray sources has made possible a new field of spectroscopy for the study of local molecular structure and chemical properties of specific atoms in biological molecules. Biological applications of synchrotron radiation to date have used x rays emerging from a beryllium window, and hence of relatively short wavelength—varying from about 3 keV (4 Å) up to 25–30 keV (0.5–0.3 Å). Using this radiation, K-shell absorption has been studied for elements with atomic number Z above 19 (potassium) to beyond $Z = 42$ (molybdenum). In the same range, L-shell absorption starts at $Z = 45$ (rhodium) and goes all the way through plutonium ($Z = 94$).

Thus with x-ray beams whose energy range is limited by the Be window, x-ray absorption spectroscopy (XAS) may be used to study biological molecules containing any element with $Z \geq 19$. (Such molecules are often referred to as "bioinorganic.") Future applications will extend the available range of x-ray wavelengths beyond 2 keV (6 Å), so that the K eages of phosphorus (2.15 keV) and sulfur (2.47 keV) will become accessible. XAS has proved particularly fruitful in bioinorganic systems because of the fact that atoms of an absorbing chemical element usually occur in very few chemical bonding sites in a given molecule, and furthermore, the information extracted concerning the local structure and chemical states

S. DONIACH • Department of Applied Physics, Stanford University, Stanford, California 94305.
P. EISENBERGER • Bell Laboratories, Murray Hill, New Jersey 07974.
KEITH O. HODGSON • Department of Chemistry, Stanford University, Stanford, California 94305.

of these binding sites is frequently of considerable biochemical and biophysical interest.

The information contained in x-ray absorption spectra and the fine structure found above the edge called EXAFS—the type, number, and distances of atoms surrounding the central x-ray absorber—is a large part of what one seeks to know about the structure of metal sites in biological molecules. The positions and intensities of absorption edge features contain information about the oxidation state, site symmetry, and electronic structure of the atom under study. Furthermore, because the interesting x-ray absorption features of different elements are well separated in energy, this type of spectroscopy is a unique method for probing the local environment of specific elements in a complex sample.

The limited spatial range of structural information obtainable from x-ray absorption data is often an advantage. In contrast to crystallographic studies, the information present in EXAFS is restricted to the immediate vicinity of the absorber (out to about 3.5 Å), and long-range features such as secondary and tertiary structure are not revealed. This means that one need only refine a small number of variables to obtain the structure around the absorbing atom, whereas single-crystal diffraction results depend for accuracy on refining all of the atomic positions in the crystal. The distances obtained from careful EXAFS studies can be accurate to better than 0.02 Å, while most protein crystallographic results have accuracies almost an order of magnitude worse.

As will be shown below, careful numerical analysis of EXAFS data can also yield information leading to the identification of a bonding element (e.g., sulfur may be distinguished from nitrogen) and can give the number of neighbors in a given shell with accuracies of about $\pm 20\%$.

Thus, although EXAFS can just approach small molecule crystallography in accuracy, and cannot reveal the macromolecular features obtained from protein crystallography, it will often be the method of choice for the study of specific metal binding sites in large molecules.

At this point, some experimental advantages of x-ray absorption spectroscopy should be noted. First, this technique can be applied to any state of matter; therefore, crystals, transparent solutions, or frozen samples are not required. Second, the selection rules are such that an absorption edge and associated fine structure will always exist—paramagnetic or isotopically enriched samples are unnecessary. The lack of isotopic sensitivity makes labeling experiments impossible, but metal and ligand substitution studies are conceivable. Finally, x-ray absorption spectra can be collected in minutes or hours, and once the data are obtained, the results may be available within days or weeks, in contrast to the years often required for high-resolution single-crystal protein diffraction data analysis and refinement.

As will be shown in this chapter, the application of EXAFS (the spectral region 50 to 800 eV above an absorption edge) to bioinorganic problems may be divided into two basic types: (1) refinement (and often correction) of metal coordination sites already known from crystallographically determined structures, and (2) determination of completely unknown coordination spheres or further elucidation of those proposed from spectroscopic data or partially refined crystal structures.

On the other hand, the applications of x-ray absorption edge structure measurement (the spectral region within 30 eV below or above the absorption edge) to bioinorganic compounds are less quantitative because of the greater difficulty in their interpretation (see Chapter 10, Section 2). However, they are also of considerable potential interest on account of the large effects of binding site symmetries, oxidation state, and ligand charge distribution on the details of the spectrum.

2. Discussion of Experimental Techniques Appropriate to Biological Molecules

Two features distinguish XAS studies of biological molecules from that of other systems: the fact that the absorbing element is in a low-Z matrix (composed mainly of C, N, and O with some sulfur and phosphorus), and the fact that the absorbing element is frequently present in a very low concentration relative to the organic matrix. Examples are hemoglobin [1 Fe in about 1000 C, N, O (not including a possible solvent), i.e., about 20 millimolar] and nitrogenase [1 Mo in about 12,000 C, N, O (1.5 millimolar)].

It is therefore necessary to develop measurement techniques of considerable sensitivity. X-ray absorption measurement techniques are discussed in general in Chapter 10. Here, we confine ourselves to special features of the techniques of importance for biological systems.

2.1. Transmission versus Fluorescence

As samples become increasingly dilute, the measurement of cross sections by direct absorption becomes less favorable than its monitoring via secondary x-ray emission, i.e., fluorescence.[1, 2]

Denoting the mass absorbance of the sample by μ_T and the absorbance of the specific element of interest by $\mu_A(\omega)$, assumed to be $\ll \mu_T$ (where ω measures the photon energy), the transmission through a sample of thickness x is given by

$$I_t(\omega) = I_0(\omega) \exp[-(\mu_T - \mu_A)x][1 - \mu_A(\omega)x] \tag{1}$$

where the signal of interest is

$$S(\omega) = I_0(\omega) \exp[-(\mu_T - \mu_A)x]\mu_A(\omega)x \tag{2}$$

while the noise is proportional to $(I_t)^{\frac{1}{2}}$. Thus the signal-to-noise ratio to first order is

$$S/N = (I_0 e)^{-\mu_T x/2} 2\mu_A(\omega)x \tag{3}$$

which is a maximum when

$$d(S/N)/dx = 0, \qquad \text{or } \mu_T x = 2 \tag{4}$$

Thus the optimum sample thickness is given by equation (4) and, under these conditions, the signal-to-noise ratio is given by

$$S/N = (I_0)^{\frac{1}{2}} 2e^{-1} \frac{\mu_A(\omega)}{\mu_T} \tag{5}$$

In the experiment the net absorbance is determined by taking the log of I_0/I. The above analysis assumes that the precision is limited by the measurement of I_t. In practice, various systematic effects such as incomplete cancellation of the fluctuations in the source (because the two ion chambers and electronics are not matched) changes in the beam position together with an inhomogeneous sample, or varying sample thickness can limit the detectable signal more severely than expected from equation (5). To determine the sensitivity of the technique given by equation (5) some estimate is needed for the incident photon flux $I_0(\omega)$. At the SPEAR storage ring, on the order of 3×10^{10} photon s^{-1} are available in 1×20 mm area (depending on electron beam conditions and photon energy).

With the use of a focusing mirror, about 10^{12} photons s^{-1} eV^{-1} are available in an area of 3 mm^2 spanning a region from 3.5 to 9 keV, with considerably fewer photons below 3.5 keV (due to the absorption by the windows) and a rapidly decaying spectrum at higher energies owing to the decreasing reflectivity of the mirror. A conservative estimate of photon number for the purpose of calculating sensitivities is of the order of 10^{11} photon s^{-1}. For example, for a S/N of 3, equation (5) predicts that the fractional change in absorption that can be measured in one second is

$$\frac{\mu_A(\omega)}{\mu_T} \cong 1.3 \times 10^{-5} \tag{6}$$

Considering a dilute iron metalloprotein one has

$$\mu_B \simeq 10 \text{ cm}^{-1}$$

$$\mu_A = N_A \sigma_A \tag{7}$$

where N_A is the concentration of Fe atoms and σ_A is the absorption cross section, which equals 3.85×10^{-20} cm^2. Hence, assuming that the required precision desired for the measurement of μ_A is $\delta\mu_A/\mu_A \sim 1\%$, one finds that $N_A \approx 10^{18}$ cm^{-3} or ≈ 2 mM. The systematic effects mentioned above make the practical limitation closer to about 10 mM.

The fluorescence technique makes use of the fact that an inner-shell vacancy may relax by undergoing a radiative transition from a higher-energy occupied shell. The radiative probability f, or "fluorescent yield," is a monotonically increasing function of atomic number. Selected values of f are given in Table 1.

For simplicity we shall study the geometrical arrangement where the incident and fluorescent radiation make equal 45° angles with the sample normal. The

Table 1. Values of Fluorescent Yield for K-Shell Fluorescence[2]

Element	Z	f
Ca	20	0.163
Fe	26	0.342
Cu	29	0.440
Mo	42	0.764

incident radiation has energy ω, the fluorescent radiation has energy ω_f, and the detector subtends a solid angle Ω. The fluorescence counting rate for a thick, dilute sample is then given by

$$I_f = \frac{I_0 f (\Omega/4\pi)\mu_A(\omega)}{\mu_t(\omega) + \mu_t(\omega_F)} \qquad (8)$$

In the fluorescence case one also has a background due to the elastic and inelastic scattering from the sample as well as from the fluorescence of the other atoms. All these background contributions are at energies that are usually significantly different from the fluorescence energy of the atom of interest but that will contribute to the noise. Detailed calculations and experimental experience show that with broad-bandwidth detectors like scintillation counters the signal and background are roughly equivalent at concentrations of ~ 10 mM. For systems such as hemoglobin where the iron concentration was ~ 25 mM an array of scintillation detectors that does not distinguish the signal from the background but that does have a high counting rate has been used successfully.[1, 3]

Leaving aside the background effect for the moment, the signal-to-noise ratio is given by

$$S/N = \left(\frac{I_0 (f(\Omega)/4\pi)\mu_A(\omega)}{\mu_t(\omega) + \mu_t(\omega_F)} \right) \qquad (9)$$

Comparing equation (9) with equation (6) we see that the S/N of a transmission experiment scales as $\mu_A(\omega)$, but for the fluorescence experiment the S/N scales as

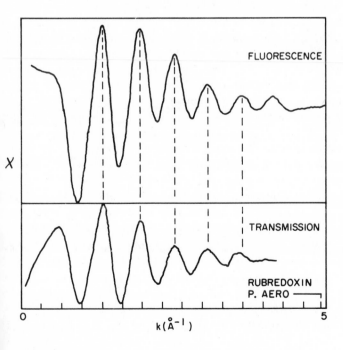

X

$k(\text{Å}^{-1})$

FLUORESCENCE

TRANSMISSION

RUBREDOXIN
P. AERO

0   5

Figure 1. A comparison of the fluorescence and transmission data of two different lyophilized samples of oxidized *Peptococcus aerogenes* rubredoxin showing the better signal-to-noise ratio in the former spectrum. In both cases, the raw absorption data were converted from the energy to the k scale and smoothed. The background noise was removed by a single best-fit cubic. The two spectra have different positions of the peak near $k = 11$ Å^{-1}, presumably reflecting errors introduced into the transmission data by the background removal.

$\mu_A(\omega)^{\frac{1}{2}}$. For dilute, thick samples we may set $\mu_t(\omega) \doteq \mu_t(\omega_f)$, and the two techniques have equal sensitivity when

$$\mu_A = f(e^2/8)(\Omega/4\pi)\mu_t \qquad (10)$$

As a concrete example, we consider a dilute Fe metalloprotein, assuming $(\Omega/4\pi) = 0.01$, $\mu_t = 10$ cm^{-1}, which yields $\mu_A = 1.0 \times 10^{18}$ cm^{-3}. Below this concentration the fluorescence technique has better S/N at the absorption edge. One can easily imagine a twentyfold or more improvement in detector solid angle, thus making the fluorescence technique competitive at concentrations as high as 2×10^{19} cm^{-3}.

Using equation (8), $I_0 = 10^{11}$ photon s^{-1}, a time of 1 s, and the other parameters above, we find $\mu_A(\omega)/2\mu_T = 4 \times 10^{-8}$, which is three orders of magnitude more sensitive than the transmission case. Thus, in principle, one can detect an EXAFS modulation of an absorption cross section in one second with a signal-to-noise ratio of three in a 2×10^{-6} M solution provided background photons can be eliminated. The improved signal-to-noise ratio obtainable with fluorescence is illustrated in Figure 1.

2.2. Fluorescence Detectors with Energy Discrimination

To approach theoretical signal-to-noise ratios in more dilute systems, one must use a fluorescence detector that can selectively count only the photons at the fluorescent energy of the atom of interest. Until recently, solid state Ge or Si detectors have been used either singly or in an array. The limited total data counting rate obtainable before saturation from a single solid state detector ($\sim 20,000$–$50,000$ counts s^{-1}) means one will have a low signal rate for a dilute system where the intensity of photons at the wrong energy is much greater than those at the flourescent energy. For example, at 1 mM concentration, the ratio of wrong to right photons is roughly 10 to 1 and thus each detector can only provide at best 5000 fluorescent counts s^{-1}. Since approximately 0.1% statistical accuracy is required to obtain a signal-to-noise ratio of 10/1 on a 1% modulation, it would take $200/N_d$ per data point where N_d is the number of detectors. Since there are roughly 200 data points per spectrum in an experiment, using this approach would take $40\,000/N_d$ s. Because of the expense of solid state detectors, N_d cannot be too large; of course, for more dilute systems the saturation will even be worse. Also, one must consider questions of radiation damage, and thus, high collection efficiencies should be used.

Recently an ingeneous new fluorescence detector has been developed and successfully used to study 1 mM concentrations.[4] This new detector is shown in Figure 2 and is based on Bragg scattering from an array of doubly bent (in a barrel shape) pyrolitic graphic crystals. The barrel produces a point image at a distance $2L$ from the source point. At the waist of the barrel, the radius B is such that $\lambda = 2d \sin \theta_B = 2dB/L$, where d is the graphite crystal lattice spacing and θ_B is the Bragg angle. The shape of the barrel is chosen such that as one either moves closer or further from the source (i.e., the photon hits the surface of the crystal at a different place) one changes the configuration of the crystal and B so that Bragg's

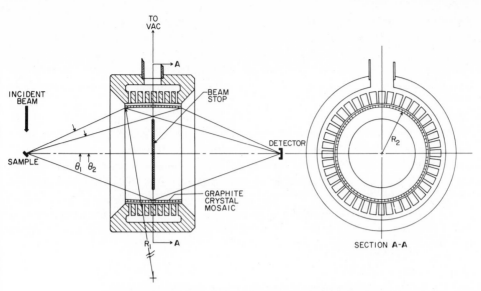

Figure 2. Sketch of a barrel configuration analyzing detector.

law is still satisfied. Thus, since d of the graphite planes is fixed, each different fluorescence energy (wavelength) will require a unique B/L. To avoid having to make a new array for each wavelength (at considerable cost) the graphite crystals are first stuck onto a flexible backing that is then, through the use of a vacuum, pulled down on a metal barrel whose shape is chosen for a particular wavelength. Thus, the same set of crystals can be used for all elements.

The detector resolution is directly related to the size of the regions from which the fluorescent photons are emitted and thus also to the size of the incident beam. Assuming that we are interested in fluorescence at energy $E_F(\lambda_F)$ with an incoming beam of radial size D and that both the incident and scattered beams are at 45° to the specimen surface, one then finds, using Bragg's law, that

$$\Delta E/E = \cot \theta_B \Delta\theta = \cot \theta_B D/L$$

The solid angle collected by the detector is given approximately by

$$\Omega/4\pi = \tfrac{1}{2}Bl/L^2$$

where l is the length of the barrel. Thus the smaller D is, the smaller L can be made, which in turn increases the solid angle by $1/L^2$. For this reason, this type of detector is best used on a line with a focusing mirror (see Chapter 10). At Stanford this means that D can be about 2–3 mm (while still collecting 6 mrad of horizontal flux), while it will be as small as 0.5 mm on future dedicated rings.

In its successful utilization in the study of dilute Fe systems, the parameters were $L = 10$ cm, $B = 2.5$ cm, $l = 5$ cm, and $D = 3$ mm, which give $\Delta E = \pm 950$ eV and $\Omega/4\pi = 0.06$. Since the $K\alpha$ fluorescence is 700 eV away from the absorption edge, this system provided adequate resolution. For a similar concentration of Fe the fluorescence signal to background incident on the crystal was 1/20 while that reflected was 10/1. Thus the detector system provided a 200/1 discrimination and so enabled the study of concentrations as low as 0.1 mM.

A word of caution is appropriate at this point. The 1% absorption modulation assumed in our discussions is representative of the modulation at intermediate energies above the edge. If higher-resolution data is required, one must go to higher energies above threshold where the signal becomes smaller because of Debye–Waller effects; hence, the limitation in sensitivity will be affected accordingly. Secondly, the calculation assumed a somewhat favorable case where the absorbing Fe atoms have an absorption cross section about 100 times that of the other constituents of the molecule that have low Z's. At a given energy above the K shell the rate of change of the absorption cross-section ratio for two atoms goes as $(Z_1/Z_2)^4$. Thus, for Fe $(Z = 26)$ coordinated to oxygen $(Z = 8)$, one gets a factor of about 100, which was used above. However, for Ca^{2+} ions in a membrane with a lot of phosphorus present, the ratio is almost unity. Thus, in general, high-Z atoms in a low-Z medium are easier to measure.

3. EXAFS Data Analysis for Biological Molecules

The reader is referred to Chapter 10 for a detailed discussion of the methods of analysis of EXAFS data. For bioinorganic molecules, there is frequently a close overlap of near-neighbor and next-neighbor shells. For this reason the nonlinear least-squares curve-fitting analysis, combined with Fourier filtering, has proved to be the most practical approach for most biological systems studied to date.[5, 6, 10]

The essence of EXAFS curve-fitting analysis is to use a parameterized function that will model the observed EXAFS, and then to adjust the structure-dependent parameters in this theoretical EXAFS expression until the fit with the experimentally observed EXAFS is optimized. The final values of the optimized parameters will then yield structural information about the compound under study.

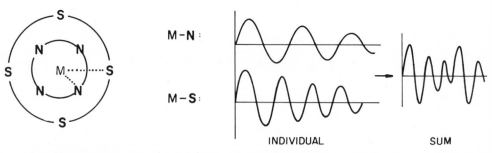

Figure 3. Schematic representation of EXAFS for a hypothetical metal complex with two coordination shells (S and N). Here

$$\text{EXAFS}(M - N) = N_N[\text{amplitude N scat}][\sin(2R_{M-N} + \alpha_{M-N})]$$

$$\text{EXAFS}(M - S) = N_S[\text{amplitude S scat}][\sin(2R_{M-S} + \alpha_{M-S})]$$

where the amplitude $= c_0 e^{-c_1 k^2}/k^{c_2}$, the phase $= \sin(a_0 + a_1 k + a_2 k^2)$, $a_1 = 2R + \alpha$. The parameters c_1, c_2, a_0, a_2 are determined from the model data fits; the constant c_0 gives N_N and N_S, a_1 gives the distance, and a_0 is characteristic of Z.

This procedure is illustrated schematically in Figure 3 for a metal surrounded by a coordination shell of sulfur and of nitrogen atoms. The N and S shells each contribute a sine wave component to the overall EXAFS. Each wave (M–N and M–S) has parameters that determine its amplitude shape and phase behavior. The phase is modeled by the function indicated and the constants a_0 and a_2 are determined for each wave from curve fitting data from model compounds of known structure. The amplitudes of the M–N and M–S waves are parameterized and the constants c_1 and c_2 determined from model data fits. Alternatively, the phase and amplitude parameters can be calculated, and tabulated values are available for most elements.[7, 8] Then the overall multiplier of each wave, N, is directly proportional to the number of atoms in the shell.

Once determined, the constants c_1, c_2, a_0 and a_2 (which describe the EXAFS for the S or N wave) can be used in fits to the EXAFS data of unknowns. From the fit to the data of the unknown structure, the EXAFS analysis provides the distance (from the a_1 term) or number of atoms (from c_0) in each coordination shell. In practice, up to four shells of ligands out to about 4.5 Å from the metal can be analyzed. The approximations and errors in this analysis have been discussed in detail.[5, 6, 9, 10]

3.1. Scatterer Identication

The combined effects of phase shifts and amplitude envelopes on the overall EXAFS are illustrated in Figure 4. For hypothetical Mo–0, Mo–S, and Mo–Mo distances of 2 Å, it is clear that there are phase differences in the EXAFS. These

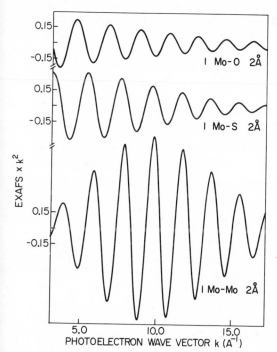

Figure 4. The effects of phase and amplitude on waves from three different Mo–scatterer interactions each at a distance of 2.0 Å. Note that the difference in the phases of the oxygen and sulfur waves allows them to be distinguished on this basis alone (even though the atoms are at the same distance). Likewise, Mo–Mo and Mo–O have different phases. Although Mo–Mo and Mo–S have similar phases, their vastly different amplitude envelopes allow them to be distinguished as well. These plots were made using the empirical phase and amplitude parameters given in Cramer *et al.*[5]

differences are caused by the different scatterer phase shifts. The Mo–Mo wave is seen to be substantially larger than the Mo–S wave and has a different shape. It would thus be impossible to describe Mo–S EXAFS with Mo–Mo parameters. In the case of the Mo–O and Mo–S, the waves have an origin that differs by almost π radians, thus, one could not describe Mo–O EXAFS with Mo–S parameters. It is these effects that allow qualitative structural information to be extracted about the type of atoms in each coordination shell. In the fitting analysis, these differences in phase shift are reflected in the constant part of the phase a_0, For the same absorber, different scatterers will have different values of a_0. The effects, however, are modulo 2π rad and thus the a_0 term for one scatterer could unfortunately have the same value as another. When the a_0 terms do differ by a nonintegral value of 2π rad, it is possible to use the a_0 value to identify the type of scatterer. This effect can be used to "fingerprint" the scatterer type. However, the a_0 values do not vary rapidly enough to distinguish between scatterers adjacent in the same row of the periodic table (for example, between carbon and nitrogen).

3.2. Amplitudes and Numbers of Scatterers

In the EXAFS formula, the EXAFS amplitude depends on the number of scatterers N_s, as well as R, k, $|f(\pi, k)|$, and σ. Since R is obtained from the frequency of the EXAFS and k is known, calculation of scatterer numbers from EXAFS is possible if $|f(\pi, k)|$ and σ is known. Information about the k dependence of $f(k, \pi)$ comes both from theoretical calculations and from the measurement of known model compounds. An important feature of this function is its dependence on the atomic number Z of the scattering atom. Some understanding of this may be obtained from the Born approximation for f:

$$f(\pi, k) = \int d^3r \; V_{\text{scatterer}}(r) e^{ik \cdot r}$$

Using a screened Coulomb potential to measure the inner region of the scattering atoms

$$V_{\text{scatterer}}(r) = Z_{\text{eff}} e^2 \frac{e^{-\mu r}}{r}$$

one has

$$f(\pi, k) = \frac{4\pi e^2 Z_{\text{eff}}}{(\mu^2 + k^2)}$$

so that $f(\pi, k)$ falls off as $1/k^2$ for photoelectron wavelength $2\pi/k$ small compared to an effective inner-shell screening length $2\pi/\mu$. In energy terms, $f(k)$ falls off as $1/E$, where $E = \hbar^2 k^2/2m$ is the photoelectron energy, and E becomes of the order of or larger than the K-shell binding energy of the scattering atom.

Figure 5 shows calculated values of $f(\pi, k)$ for a variety of scattering atoms.[11] It may be seen that while $f(k)$ falls off fairly rapidly for low-Z scatterers such as C, N, and O, it remains large, or even has a maximum, for higher-Z scatterers. This characteristic difference between the form of the EXAFS amplitude for low- and

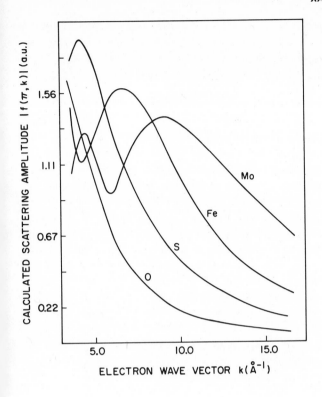

Figure 5. Theoretical back scattering amplitudes.

high-Z atoms can provide a useful characteristic to help identify metal–metal reflections in molecules of unknown structure.

For low-Z scatters a useful empirical approach has been to parameterize the total amplitude envelope using the function form: $c_0 \exp(-c_1 k^2)/k^{c_2}$. For Mo[5] and Fe[11] compounds, these parameters were obtained by curving fitting model compound EXAFS. The amplitudes obtained agreed reasonably well for the same scatterers,[5] in accord with theory, which predicts that EXAFS amplitudes should be independent of the nature of the x-ray absorbing atom. Thus, each shell of scatterers is described in the fit by only two parameters: the overall amplitude (c_0), which gives the number of atoms in the shell, and the linear part of the phase ($2R + a_1$), which provides the distance information. Obtaining accurate scatterer

Table 2. Accuracy of EXAFS Distances and Coordination Numbers

For 18 known structures analyzed	Mean $\lvert R_{XSTAL} - R_{EXAFS} \rvert = 0.017\text{Å}$
	Mean $\dfrac{\text{No.} - \text{No.}_{EXAFS}}{\text{No.}} = 19\%$
For 4 unknown structures (with x-ray structures completed subsequently)	Mean $\lvert R_{XSTAL} - R_{EXAFS} \rvert = 0.007\text{Å}$
	Mean $\dfrac{\text{No.} - \text{No.}_{EXAFS}}{\text{No.}} = 20\%$

numbers from this approach requires that the model compound and the unknown have similar Debye–Waller factors. For a variety of Mo structures it was possible to determine the absolute number of scatterers with an accuracy of about 20%.[5, 10, 11] A summary of the accuracy obtained from EXAFS analysis is given in Table 2.

3.3. An Example of Structure Determination

The above approach to determining structure is illustrated by the EXAFS analysis for tris(2-amidobenzenethiolate)Mo(VI), which at the time of the analysis had yet to be crystallized. Assuming that sulfur would ligand to the Mo, a one-shell fit was first carried out. The results of fitting a Mo–S wave are shown in Figure 6a. It is apparent that a single scatterer shell alone cannot reproduce the presence in the data of the beat that is characteristic of at least two different absorber–scatterer distances that alternately constructively and then destructively interfere. Adding a second shell of atoms (in this case nitrogens), dramatically improves the fit, as seen in Figure 6b, and allows the fit to reproduce the beat. From the four variables in this two-shell fit, 3.2 sulfur atoms at 2.419 Å and 3.1 nitrogen atoms at 1.996 Å were calculated.

The uniqueness of the choice of atomic type can be tested by using phase and amplitude parameters characteristic of different Mo–scatterer pairs. For example, a

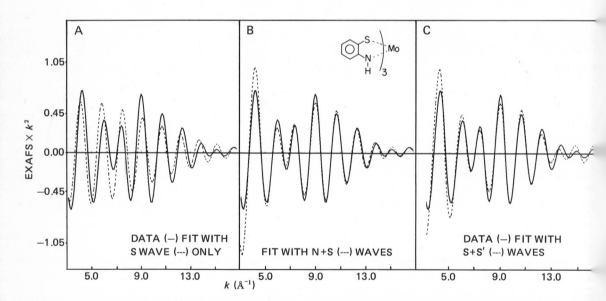

Figure 6. Curve fitting analysis of Mo EXAFS for $Mo(SC_6H_4N)_3$. The data (——) are fit (– – – –) over a range of k (4 to 14 Å^{-1}) with only one sulfur wave in (A), and it can be seen clearly that this is inadequate to describe the phase and amplitude of the observed EXAFS. The fit in (B) includes a second coordination shell (N) and gives an excellent fit. The fit in (C) is obtained by using a second sulfur wave instead of a nitrogen wave (in this fit, the overall amplitude on this sulfur becomes negative).

reasonable fit with a second shell of sulfurs is impossible because the a_0 term for Mo–N and Mo–S waves are so different. The result of a fit with two different sulfur waves is shown in Figure 6c. The observed fit is not substantially worse than that for the correct N and S wave fit shown in the middle of the figure. However, the coefficient on the second sulfur wave of the S, S' fit becomes *negative*, thereby changing the phase of the fit wave for the S' shell by about π rad. This allows the fit to simulate, insofar as possible, an N wave with S-wave parameters. Likewise, the large and characteristic amplitude of a Mo–Mo pair (see Figure 4) makes it impossible to use these parameters to describe the Mo–N wave. Using such logic, in combination with a knowledge of reasonable bond distances, one can generally arrive at a structure. It should be emphasized that it would be quite difficult to distinguish between scattering atoms that are adjacent in the periodic table (such as N and O) and this introduces an ambiguity.

It is also instructive to observe the numerical agreement between the observed and calculated values. This is conveniently done by calculating $\sum_x \chi^2$, the sum of the weighted squares of the residuals. For the S only fit, χ^2 is 1.56. Adding the N shell reduces χ^2 to 0.21 (approximately a 12-fold reduction!), while fitting with the S, S' model results in a χ^2 of 0.32. These results establish the unique nature of the S, N fit to the data for the compound.

Subsequent to the EXAFS analysis, the structure of tris(2-amidobenzenethiolate) Mo(VI) was determined by Yamanouchi and Enemark.[12] It is shown in Figure 4 of Reference 12 along with a comparison to the distances determined by EXAFS. The agreement is seen to be excellent. A summary of the accuracy of distances and coordination numbers calculated by EXAFS on a variety of Mo complexes[13] is given in Table 2. Similar results on other inorganic complexes and proteins containing Cu[9] and Fe[6, 14, 15] have been obtained. As a coordination environment becomes more complex (for example, with a spread in coordination distances produced by Jahn–Teller distortions), theoretical formulation[6] and practical experience[5, 9] reveal that for most data, distances to the same ligand type cannot be resolved unless they differ by more than about 0.15 Å.

4. Selected EXAFS Applications to Problems of Biological Significance

The EXAFS technique has been applied in two basic ways to biological problems: (a) verification (or in some cases revision) of structures know from other techniques (crystallography), and their extensions to noncrystalline situations and to changes in atomic spacing upon substrate binding; and (b) determination of unknown structures. Each type of problem dictates a slightly different analysis procedure. When the basic structure has already been determined from crystallography, one need only determine the distances for the absorber–scatterer interactions that contribute to the EXAFS. In verifying postulated structures, one must also vary the scatterer identifications to ascertain that the proposed atomic types in the coordination sphere fit the data better than any reasonable alternative structure. Finally, in the prediction of unknown structures, one must take into account chemically reasonable bond lengths and coordination numbers, the known chemistry of the system, and the possibility that certain scatterers might not be observed in the EXAFS because of their distances and high relative thermal motion. In the latter case, one often may be limited to determining a set of characteristics that may be fulfilled by a number of different structures.

During the four to five years from the first EXAFS experiments using synchrotron radiation to the time of writing, four structural problems of considerable biological significance have been solved, or advanced in a major way by the use of the EXAFS technique. These are:

1. accurate determination of the Fe–S bond lengths in rubredoxin;
2. determination of changes in Fe–N distances on going from oxy- to deoxyhemoglobin;
3. discovery of a chemical model for the active Mo–Fe site in nitrogenase; and
4. discovery of a model for the binding of Mo in xanthine oxidase and sulfite oxidase.

In addition to these major advances, studies of both EXAFS and absorption-edge structures have been carried out for a number of other systems of biological significance. In this section, a brief summary is given of results obtained on a

Table 3. Summary of Metalloprotein X-Ray Absorption Studies

Protein	Reference	Metal	Conclusions
Rubredoxin	(16)	Fe	Fe–S = 2.24 Å
	(6)		Fe–S = 2.26 Å
	(17)		Fe–S = 2.30 Å
	(14)		Fe–S = 2.27 Å
Hemoglobin	(18)	Fe	$Fe-N_{normal} = Fe-N_{Kemps} \pm 0.02$ Å
Hemoblobin	(19)	Fe	$Fe-N_{porphyrin}^{(oxy)} = 1.986$ Å
			$Fe-N_{porphyrin}^{(deoxy)} = 2.055$ Å
P-450-LM-2 (low-spin ferric)	(20)	Fe	$Fe-N_{porphyrin} = 2.00$ Å $Fe-S_{axial} = 2.19$ Å
Chloroperoxidase (high-spin ferric)	(20)	Fe	$Fe-N_{porphyrin} = 2.05$ Å $Fe-S_{axial} = 2.30$ Å
Ferritin		Fe	Fe–O = 1.95 Å; Fe–Fe = 3.29 Å Probable layer structure
Nitrogenase	(21–23, 56)	Mo	Mo, Fe, S cluster Mo–S = 2.35 Å Mo–Fe = 2.73 Å
Azurin	(9)	Cu	Cu–N = 2.00 Å Cu–S = 2.09 Å
Cytochrome oxidase	(24, 49–52)	Cu	Inequivalent Cu, one Cu is not redox active
Hemocyanin	(25, 54)	Cu	Deoxy form—both Coppers + 1 Oxy form—both Coppers + 2 No S coordination Probable Cu–Cu interaction
Carbonic anhydrase	(26)	Zn	Cu(I) = 2.65 Å in iodide complex
Ca binding proteins	(53)	Ca	Ca–O 2.32 ± 0.05 Å to 2.43 ± 0.05 Å depending on type of protein.

number of systems. The examples given are of work on metalloproteins, though other biological applications, such as the binding of Ca^{2+} to membranes, are also being studied. A summary of results of x-ray absorption studies of metalloproteins is given in Table 3.

4.1. Rubredoxin*

The small protein rubredoxin ≈ 6000 mol. wt, with one iron tetrahedrally coordinated to four cysteinate sulfurs, has been intensively studied by EXAFS[6, 14, 16, 17] and the iron–sulfur distances have been determined. Before the EXAFS experiments the crystal structure of *Clostridium pasteurianum* rubredoxin had been determined by x-ray diffraction with high-quality data that was analyzed with very few assumptions about the structure.[29] Fe–S bonds (R_3) were clustered about 2.30 Å (which is the normal distance) and a fourth bond (R_1) was at 2.05 Å, although in an earlier publication it was reported to be as short as 1.95 Å. The standard deviations (σ) were 0.045.

The original EXAFS study of *Peptococcus aerogenes* rubredoxin was done in transmission with a data set extending out to $k \approx 11$ Å^{-1}.[16] $\phi(k)$ was determined by assuming phase transferability and the average Fe–S distance was determined to be 2.24 ± 0.04 Å. By a least-square fitting of a model with R_3 and R_1 to the complete data, it was found that $|R_3 - R_1| = 0.00 \pm 0.15$ Å. This was a fit to both the amplitudes and phases of the data. Shortly afterwards another transmission EXAFS experiment was reported[17] on *C. pasteurianum* rubredoxin, the same molecule whose x-ray crystal structure had been determined. The results were interpreted as showing that the average Fe–S bond length was 2.30 ± 0.04 Å and that the spread of distances was less than 0.06 Å. The data set extended out only to $k \approx 6.4$ Å^{-1} and the tight limits on the spread of distances were obtained by fitting the amplitude to a Debye–Waller factor and showing that only the very small spread in Fe–S distances mentioned above was required to fit the amplitude. A comparison of these reports naturally suggested that the amplitude fit looked like a good way to fit the data, possibly because of its presumptive nature, assuming as it does that the $f(\pi, k)$ term could be fixed by comparison with model compounds. However, a recent reinvestigation of the *P. aerogenes* rubredoxin EXAFS, in which the spectra were obtained with a fluorescence spectrometer and careful attention was paid to the magnitudes of the errors, has clarified these issues.[6] The spectrum in k space used for this analysis is shown in Figure 1, where it can be seen to extend out to a value of $k_{max} \simeq 13$ Å^{-1}. In the recent analysis it is shown that once the model allows two different distances, which cannot be separated by filtering, then the difference of those distances ΔR, which in this case is defined as $R_3 - R_1$, can be determined to an accuracy just slightly better than $\Delta R \approx \pi/2k_{max}$ or $\approx \pm 0.12$ Å for the present data. This was shown to be true analytically for the case where one makes a least-squares fit to the amplitude and phase of $\chi(k)$. The reason for this limitation upon determining ΔR is that for small values of ΔR it is possible to show

---

* Sections 4.1 and 4.2 extracted from Reference 27.

in closed form that ΔR and the Debye–Waller factor (σ) are formally equivalent. In other words, ΔR and σ can take a range of values with equivalent effects upon the fit, or one can trade phase for amplitude rigorously. Furthermore, it was shown that separating phase and amplitude, as discussed above, and fitting the phase separately gave the same accuracy and limits upon ΔR as did the fit of both phase and amplitude together. By fitting the phase to a one-distance model, the value of k_{max} was determined. Removing the vibrational contributions gave a static spread of distances of $|R_3 - R_1| = 0.05 \pm 0.05$ Å. The comparison between the average distances determined in oxidized and reduced rubredoxin and those obtained by EXAFS and x-ray crystallography in the model compounds of $[Fe(S_2-o-xyl)_2]$ prepared by Holm and his collaborators[30] showed that the protein distances were identical to those in the unstrained model systems. Recent interpretations of the crystallographic data have revised the distances and now indicate that $R_3 - R_1 = 0.10$ Å(L. H. Jensen, to be published).

4.2. Hemoglobin

The possibility of obtaining accurate distances between iron atoms and their ligands in hemoglobin posed two questions about the function of hemoglobin that have been answered by recent EXAFS experiments. Numerous experimental results have been interpreted as being consistent with the simple Monod–Wyman–Changeaux model of hemoglobin wherein two quaternary states corresponded to two different oxygen affinities.[38] The first question was concerned with the structural basis for the lower oxygen affinity in one of these forms. Was there, in the deoxygenated state, strain energy localized in the vicinity of the iron that was contributing appreciably to the 3-kcal difference in oxygen binding energy between the two forms?

The second question concerned the pathway, in both high- and low-affinity forms, by which information about the ligation of the iron was transmitted to the globin. Did this proceed solely by the iron forced out of the plane in deoxyhemoglobin and into the plane in oxyhemoglobin by changes in the iron–nitrogen bond lengths as had been proposed by Hoard[33] and Perutz[37]? Or was this pathway delocalized as had been suggested for the strain energy responsible for the difference between the high- and low-affinity forms[39, 40]?

Typical hemoglobin absorption spectra converted to k space are shown for oxyhemoglobin in Figure 7a. Spectra of similar signal-to-noise ratios that differ in their features were obtained for deoxyhemoglobin. These spectra were taken with a fluorescence spectrometer with exposures of about 20 min and a total x-ray flux of $\sim 10^{13}$ photons. The hemoglobin samples were solutions about 24 mM in heme, with a total volume of ~ 0.15 cm^3, for a total of 2×10^{18} iron atoms. Since about one half of the photons are absorbed by the iron and the other half are absorbed or scattered by the rest of the sample, there is very little chance of two photons being absorbed by the same iron.

The information obtained from analyzing these and similar hemoglobin spectra was at first qualitative,[18] but recently[19] it has been possible to analyze

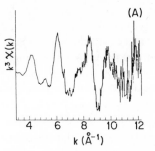 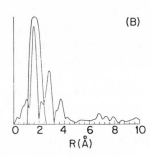 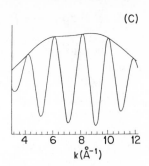

Figure 7. (A) The oscillating part of the absorption spectrum of a concentrated solution of oxyhemo-globin (~25 mM in heme), after conversion to k space, multiplication by k^3, and background removal. Measurements were made by fluorescence. The Fourier transform of (A) is shown in (B), where the large peak contains contributions from the 4Fe–N in-plane ligands and the oxygen and nitrogen axial ligands. The filter window is applied to the transformed data (B) and the inverse transform is computed. The resulting filtered data are shown (C). The Fourier transform technique also yields the amplitude (C) and phase (not shown) of the filtered data. Further analysis, described in the text, removes the contribu-tions of the axial ligands from the phase function, making it possible to determine the in-plane Fe–N distance to ± 0.01 Å.

the hemoglobin x-ray absorption data quantitatively, thereby obtaining distances accurate to ± 0.01 Å of the Fe–N_p bonds (Fe–porphyrin nitrogens) in oxy- (HbO$_2$) and deoxyhemoglobin (Hb), as well as in the oxygenated (PFO$_2$) and deoxygenated (PF) forms of Fe(T$_{piv}$PP), the so-called picket fence porphyrins that have been synthesized by Collman[32] and have been shown to be capable of reversible oxygenation.

The consequences of the observed distances are first that the Fe–N_p distance, which increases by ~0.12 going from high-spin ferric to high-spin ferrous in ionic compounds[33] does not do that in hemes. This has been expected[34] on the basis that electrons, when added to compounds like hemes or iron hexacyanides with extensive π bonding, go to the ligands, and the iron charge does not change upon reduction of the complex.

Second, this rather short Fe–N_p distance in Hb and PF must be considered in conjunction with recent x-ray diffraction studies of five-coordinated metalloporphy-rins, which show that the relaxed porphyrin ring would have a center–N_p distance of 2.045 Å.[35, 36] This center–N_p distance, determined by EXAFS for Hb and PF, means that by the Pythagorean theorem the iron is $0.2^{+0.1}_{-0.2}$ Å above the plane. This implies that the iron is not forced out of the plane by long Fe–N_p bonds as had been proposed by Perutz[37] and Hoard.[33] Rather its position is determined by the uncompensated steric hindrance of the axial ligand.

Third, since there is no difference between Fe–N_p in Hb and PF, it is clear that to within the present accuracies there are no strains induced in these bonds by the globin. Furthermore, since high- and low-affinity deoxyhemoglobins have identical Fe–N_p distances, no strains are induced by the changes of quaternary structure. Hence, it may be seen that the recurring theme in the literature about the globin straining the Fe–N_p bond with the implication that this strain is energetically significant has no support from these experiments.

4.3. Nitrogenase

The molybdenum–iron(MoFe) protein of nitrogenase contains two molybdenums and 24–32 atoms each of iron and acid-labile sulfide per molecular weight of 220 000. Despite many hypotheses that molybdenum is part of the reduction site for dinitrogen, before the advent of x-ray absorption spectroscopy using synchrotron radiation there was no unambiguous means of observing the state of molybdenum in this protein. Recent absorption edge and EXAFS studies[21, 22] thus constitute the first work to provide any information about the molybdenum environment in nitrogenase.

EXAFS studies were initially carried out with lyophilized MoFe component from *Clostridium pasteurianum* and the results of these studies have recently appeared.[22] Furthermore, studies of the MoFe component from *Azotobacter vinelandii*[21, 56] and of the FeMo cofactor isolated therefrom have been carried out.[21] A summary of these results, along with more recent studies of model complexes, will be given below and evidence will be developed for the structure of the Mo site in the resting state of nitrogenase.

X-ray Absorption Studies of the MoFe Component from Clostridium pasteurianum. The first x-ray absorption spectrum recorded on lyophilized MoFe protein revealed a distinct beat pattern (see Figure 8). The presence of such a beat, remin-

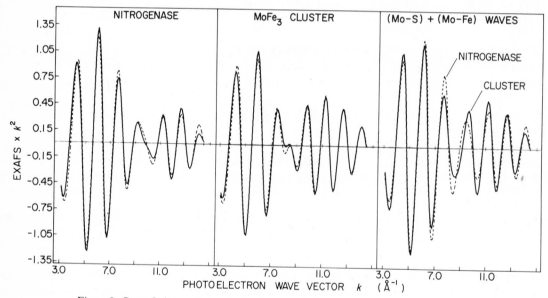

Figure 8. Curve-fitting analysis of nitrogenase and the Mo–Fe cluster compound shown in Figure 9. The nitrogenase data on the left and the cluster compound data in the middle are shown, each fitted to three waves—S, Fe, and S'. A comparison of the sum of the S + Fe waves for each fit is shown on the right. The close agreement between the two sums (and likewise the very close agreement between the parameters determined from the fits) establishes that the Mo_1Fe_3 cluster possesses a Mo structural fragment in common with nitrogenase. Further details are given in Section 4.3 of the text.

iscent of that seen in model complexes, immediately suggested at least two shells of scatterers around the Mo. Careful curve fitting analysis of the EXAFS, using methods identical to those outlined Section 3, identified the lower-frequency component as sulfur as 2.35 Å from Mo. The absolute phase of this wave is in excellent agreement with Mo–S ligation while being significantly different from Mo–O values. The amplitude of the wave indicated about four sulfurs.

A second conclusion is that no Mo–O bonds exist in the semireduced state of nitrogenase. This result is based upon the position and shape of the edge and the absence of a lower-frequency component in the EXAFS, which would be present if there were Mo–O bonds.[22]

The second set of scattering atoms is about 0.4 Å further out from the Mo than are the primary S ligands. This shell of atoms, initially suspected to be Mo, has a phase behavior that is significantly different from Mo–Mo or Mo–S parameters and is close to Mo–Fe parameters. Two to three Fe atoms at a distance of 2.72 Å are calculated from the analysis.

Finally, a third set of neighboring atoms must be included to produce a model that completely fits the data. This shell is best fit by one to two sulfur atoms at 2.47 Å.

These coordination distances and numbers are chemically reasonable and may be interpreted through comparison with known structures as representing three or four bridging sulfides, one or two terminal RS^- groups, and two to three μ-sulfido bridged Fe atoms around Mo. The numbers suggest the presence of a novel Mo, Fe, S cluster in nitrogenase, At least two specific models may be proposed (see Figure 9).

It is also important to observe that these results dictate the absence of Mo–O, Mo–S, or Mo–Mo moieties at the Mo site.

Comparison of the Mo EXAFS in Nitrogenase with That of a $Mo_1Fe_3S_4$ Model Complex. The presence of a Mo, Fe, S cluster in nitrogenase suggested that synthetic attempts might lead to an inorganic analog of the Mo site. Noting the spontaneous self-assembly of the ferredoxin analog clusters $[Fe_4S_4(SR)_4]^{2-}$ from simple reagents, Holm and his co-workers initiated a similar approach to the synthesis of Mo, Fe, S clusters.[23] Recently, the first such complex, $[Mo_2Fe_6S_9(SEt)_8]^{3-}$ has been isolated and structurally characterized.[23] The molecular structure of the complex contains a dimer of two $Mo_1Fe_3S_4$ cubes.

A comparison of the EXAFS of this compound to that of Mo in nitrogenase is shown in Figure 8. It can be seen that the data are quite similar, especially the EXAFS phases. There are small differences in amplitude around $k = 9$ Å^{-1}. Curve-fitting analysis reveals that the Mo–S (sulfide) and Mo–Fe interactions in the

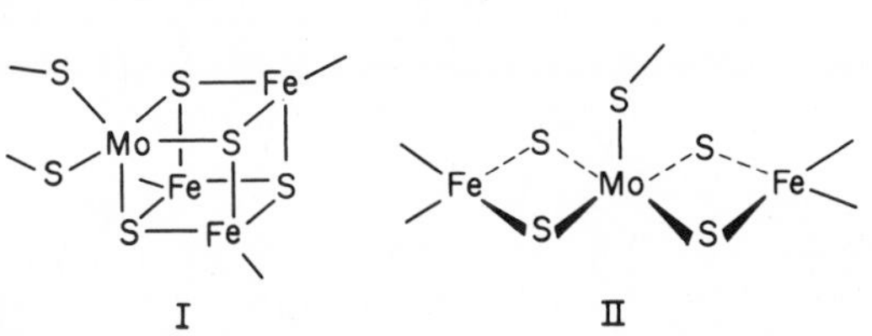

Figure 9. Two possible candidate structures for the Mo–Fe binding site in nitrogenase.

cluster and in nitrogenase are virtually identical. The major difference between
the model and the protein may reside in the nature of the external ligands to Mo. It
must be emphasized that differences in stoichiometry and physical properties, and
the absence of evidence in the EXAFS for the distant fourth cube sulfide, point out
that the $[Mo_2Fe_6S_9(SEt)_8]^{3-}$ complex in its entirety should not be considered an
analog of the Mo sites in nitrogenase.

EXAFS Comparisons between MoFe Components of Different Organisms. All of
the initial studies described above were carried out with MoFe protein isolated
from *Clostridium pasteurianum.* In order to confirm these results and make a comparison between MoFe components from two different species, we have also
studied the MoFe component from *Azotobacter vinelandii.*[21] The EXAFS data
recorded on the *A. vinelandii* MoFe protein are shown (Figure 10) compared with
the data for the lyophilized *C. pasteurianum* protein. Detailed analysis shows that
the conclusions about primary sulfur ligation and the presence of a Mo–Fe interaction are substantiated and that the Mo environment in both proteins is virtually
identical.

EXAFS Study of the FeMo Cofactor. EXAFS studies of the FeMo cofactor
isolated by Shah and Brill have been performed.[21] EXAFS of the FeMo cofactor
isolated from *A. vinelandii* MoFe protein was recorded on a rigorously anaerobic
solution of the material. Visual comparison of the data (Figure 10) with that of the
intact MoFe proteins indicates that the basic features of the Mo environment in the
cofactor are preserved during the extraction process and that the intact protein and
the cofactor share a common Mo site. These data lend support to the idea of a
common Mo site in the nitrogenase MoFe protein.

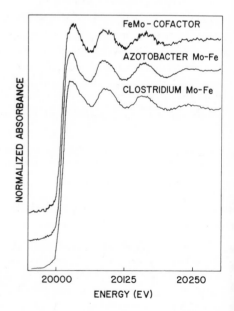

Figure 10. Comparison of EXAFS for cofactor and nitrogenase.

4.4. Cytochrome P-450 and Chloroperoxidase

Cytochrome P-450 and chloroperoxidase are heme proteins with unusual catalytic and spectroscopic properties. A question of fundamental importance to the explanation of these properties is the nature and influence of the iron axial ligands during the various stages of the enzyme catalytic cycles. Through comparison of the physical properties of synthetic porphyrin complexes with various known axial ligands, a strong case has been made for axial cysteinate sulfur ligation in cytochrome P-450. The strong spectral resemblance of chloroperoxidase to P-450 would suggest sulfur ligation in the latter enzyme as well, but attempts to detect a free sulfhydryl group available for iron ligation have been unsuccessful. It would be quite disturbing if all the unique spectral properties of cytochrome P-450, which had been explained by the presence of thiolate ligation, were reproduced in a protein without such ligation. The EXAFS studies[20] on P-450 and chloroperoxidase were designed to determine whether an axial sulfur ligand was really present, and if so, what the Fe–S distance was.

The protein EXAFS data were analyzed using empirically obtained phase-shift and amplitude functions for Fe–N, Fe–O, Fe–S, and Fe–C_α interactions. Studies on a variety of synthetic iron porphyrins showed that by fitting the EXAFS with three waves corresponding to the Fe–N_{porph}, Fe–C_α, and Fe–X_{axial} interactions, distances with an accuracy better than 0.025 Å could be obtained.[11, 20] It is worth noting that a different Fe–C phase shift was required for carbon atoms beyond the first coordination sphere than was relevant for the Fe–C interaction in ferrocene. Applying this procedure to the protein data gave results consistent with thiolate ligation in both cases, with an Fe–S distance of 2.30 ± .03 Å in high-spin chloroperoxidase and 2.19 ± 0.03 Å in low-spin ferric P-450.[20]

4.5. The "Blue" Copper Proteins

The "blue" copper proteins are another group of macromolecules whose unusual spectroscopic and chemical properties have been explained by postulating cysteine thiolate ligation. Recent EXAFS studies of azurin by Tullius et al.[9] have confirmed this assignment and provided the first quantitative details of a blue copper first coordination sphere.

The EXAFS data for the oxidized form of azurin can only be interpreted when nitrogen and short sulfur shells are included in the fit to the data. The Cu–N and Cu–S distances are determined to be 1.97 Å and 2.10 Å, respectively. The presence or absence of a second methionine sulfur, as suggested from the x-ray crystal structure of plastocyanin (a plant blue-copper protein), could not be definitively established over the range of data available.

Similar EXAFS studies on two other blue copper proteins, stellacyanin and plastocyanin, have also been carried out.[41] Both proteins have nitrogen as well as a short sulfur in their primary coordination sphere. For these two proteins, the reduced state [Cu(I)] has also been studied by EXAFS. A dramatic change in the

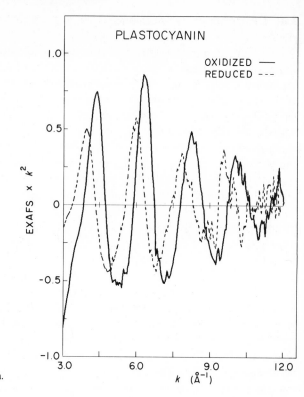

Figure 11. EXAFS for plastocyanin.

EXAFS spectrum (Figure 11) is seen upon reduction, and analysis of the data reveals that the Cu–S bond in the reduced state has lengthened to about 2.20 Å in both proteins. The remarkably short Cu–S distance in the oxidized proteins and the observed geometric changes upon reduction are important components in helping explain the unusual redox and spectroscopic properties of the "blue" copper site.

4.6. Xanthine Oxidase and Sulfite Oxidase

Xanthine oxidase is a molybdenum containing enzyme that consists of two ca. 150 000 dalton subunits, each containing one Mo, two $Fe_2S_2(SR)_4$ clusters, and one FAD (flavine adenine dinucleotide). It typically oxidizes xanthine to uric acid at the Mo site, the reaction responsible for its name but not necessarily for its biological significance, and reduces O_2 to H_2O_2 at the FAD. During the last twenty-five years, xanthine oxidase has been the subject of intensive experimental study. It is now one of the most well-characterized enzymes of bioinorganic significance. Electron paramagnetic resonance (EPR) and spectral analyses have elucidated the kinetics and mechanisms of its reactions with a variety of substrates and have assisted in describing a number of chemically induced and naturally occurring inactive forms. Despite the continuous efforts of several productive research groups, the structure of the Mo site of the resting states of the active enzyme and of the inactivated forms had proved elusive until recent x-ray absorption spectroscopy experiments at Stanford University. These have provided a description of the Mo environment in the oxidized state of the enzyme.[42]

Sulfite oxidase catalyzes the oxidation of sulfite to sulfate. The enzyme contains one molybdenum and one heme per subunit and utilizes water as the source of oxygen. Several different states of sulfite oxidase have been identified by a combination of EPR and optical spectroscopic measurements, but little was known about the molecular details of the Mo site until the EXAFS studies recently reported.[43]

The Mo K absorption edges of oxidized and dithionite-reduced xanthine oxidase are shown in Figure 12. Unlike the smooth edge characteristic of the nitrogenase FeMo component, both xanthine oxidase edges exhibit two distinct inflection points. The presence and intensity of the first inflection at about 20,003 eV (assigned to a $1s \rightarrow 4d$ bound-state transition) argues strongly for at least one and more likely two terminal oxo groups bound to Mo in the oxidized form of the protein.[42] Based on the absorption edges of numerous other Mo complexes with biologically relevant ligands, the position of the second inflection at 20,016 eV also dictates the presence of oxo groups. Upon reduction, the higher-energy inflection is seen to move by about 3.1 eV to lower energy. This shift could result from either a formal two-electron reduction of the Mo [Mo(VI) to Mo(IV)] while maintaining the

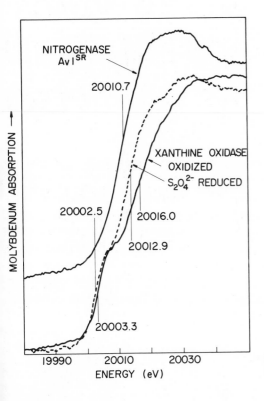

Figure 12. X-ray absorption edges for xanthine oxidase and nitrogenase.

same ligation, or loss of one oxo group, or a combination of both. Only the Mo(IV) or the Mo(VI) oxidation states are consistent with the absence of a Mo EPR signal in the reduced enzyme. Considering the relatively low-energy value for the second inflection, the edge results are then most consistent with reduced xanthine oxidase containing Mo(IV).

The Mo EXAFS of oxidized xanthine oxidase has been analyzed with curve-fitting techniques. The most obvious feature of the EXAFS data (Figure 13) is the presence of a "beat" in the amplitude envelope, which is direct evidence of at least two different Mo–L distances. Detailed curve-fitting analysis suggests that in its oxidized form xanthine oxidase contains the MoO_2^{2+} structural unit. Similar results are also found for sulfite oxidase.[43]

The average Mo–O distances, 1.71 Å, are the same for both proteins. Two sulfur atoms are present, at 2.42 Å and 2.54 Å for sulfite and xanthine oxidase,

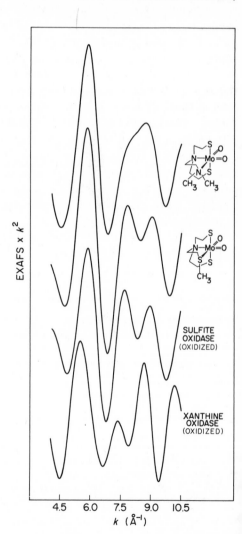

Figure 13. EXAFS data for xanthine oxidase, sulfite oxidase, and model clusters.

respectively. A more distant sulfur is also present at about 2.85 Å in both enzymes. The reason for the similarity of the EXAFS spectra of sulfite oxidase and xanthine oxidase is clear—at least five ligands of the Mo are the same and at similar distances. The different appearance of the beat region is simply a manifestation of the longer distance to the shorter sulfur ligands in xanthine oxidase.

EXAFS analysis has also been performed on two model compounds containing tripod ligands (see Figure 13 for the structures) synthesized by Dr. E. I. Stiefel and his collaborators at the Charles F. Kettering Research Laboratory.[13] The EXAFS spectra of these two are compared in Figure 13 with the spectra of the oxidized forms of xanthine oxidase and sulfite oxidase. There are substantial differences between the two model compound spectra in the beat region between κ of 6.5 and 8.5 Å^{-1}, despite the basic structural similarity of the two compounds. This demonstrates the ability of the EXAFS method to distinguish the substitution of one ligand [$-SCH_3$ versus $-N(CH_3)_2$] in the metal's coordination sphere. The EXAFS spectrum of the model compound containing $-SCH_3$ can be seen to be virtually superimposable on that of resting-state sulfite oxidase.

These results suggest structural models for the resting oxidized state of xanthine oxidase and sulfite oxidase. The Mo–O distance of about 1.7 Å is similar to that found in Mo complexes that have several terminal oxo groups, and is not consistent with a bridging oxo group distance. The intermediate sulfurs into two enzymes are most likely thiolates, although the distances in xanthine oxidase are slightly longer (ca. 0.1 Å) than observed for thiolates in other MoO_2L_4 complexes. This suggests that in xanthine oxidase these sulfurs may be *trans* to the Mo–O groups [which are normally *cis* to each other in Mo(VI) complexes] due to the well known *trans* effect of Mo–O. The longer sulfur ligand in each enzyme could be a thioether; an alternative formulation for this sulfur could be as the second sulfur of a persulfide. Unfortunately, no examples of this type of coordination yet exist in Mo complexes. Finally, for sulfite oxidase, the EXAFS analysis reveals the loss of one terminal oxo group upon reduction of the enzyme by dithionite.

4.7. Hemocyanin

Hemocyanin is an oxygen transport protein that is found in the blood of arthopods and mollusks. The proteins from different species all contain two copper atoms per subunit and x-ray absorption spectroscopy has been used to study the copper environment for both oxy and deoxy forms of hemocyanin. The hemocyanins from *Busycon canaliculatum* and *Megathura crenulata* have been studied by two different groups.[25, 47]

As described in more detail in Section 5.3, a study of the edge features of oxy and deoxy hemocyanins reveals that the edge shape and position is consistent with the presence of Cu(I) in the deoxy and Cu(II) in the oxy form of the protein. Analysis of the EXAFS data has been carried out to determine the features of the coordination environment of the Cu atoms. The phase and amplitude behavior of the atoms contributing to the first-shell scattering strongly suggests that only first-row atoms are present and higher-Z scatterers such as S are excluded. For *B*.

canaliculatum hemocyanin, about five N or O ligands at 1.96 Å and about three N or O ligands at 1.95 Å were found for the oxy and deoxy forms, respectively. A similar result was obtained for the *M. crenulata* hemocyanin where about six atoms of N or O were found at 1.97 Å in the oxy and about four atoms of N or O at 1.99 Å in the deoxy form. These numbers are consistent with other chemical and spectroscopic evidence.

There is clear evidence in the EXAFS data for shells further out from the first. Specifically, in the hemocyanins from both species, there is evidence for a Cu atom at about 3.4 Å in the deoxy form of the proteins. This shell can be fit best by Cu parameters, though because of the unfortuitous congruency of the absolute phase for Cu and N, a large number of low-Z scatterers can be used to fit this shell as well. In the case of the *B. canaliculatum* deoxyhemocyanin, the authors noted that using a low-Z scatterer for this second shell resulted in an unusually high Debye–Waller factor and an unreasonable distance.[47] For the *B. canaliculatum* oxyhemocyanin, the same authors also indicate the presence of an even longer Cu at 3.7 Å.

These results suggest that hemocyanin functions through a Cu–Cu dimeric structure. EXAFS studies, along with other studies, such as resonance Raman and electronic spectroscopy, suggest that each Cu atom is liganded by several imidazoles and that in the oxy state, the Cu atoms are bridged by a bound peroxide.

4.8. Ferritin

Ferritin is a protein, found predominantly in animals, which is primarily responsible for the storage of iron. The protein is known to consist of a protein sheath surrounding an inorganic Fe core of the stoichiometry $(FeOOH)_8$-$(FeO \cdot OPO_3H_2)$. The detailed structure of the core has been the source of considerable speculation. EXAFS studies have recently been applied to delineate the first molecular information about the Fe micellar core.[48]

Analysis of the EXAFS data revealed that each Fe atom was surrounded by about seven low-Z neighbors (O or N) at an average distance of about 1.95 Å. The second shell consists of about seven Fe neighbors at an average distance of 3.29 Å. Utilizing the known stoichiometry and density of ferritin and its core, the authors postulate a model in which the iron core is a layered arrangement with the iron layered between two nearly close-packed layers of oxygen atoms with appropriate sixfold symmetry. The O–Fe–O layers are proposed to only weakly interact with adjacent layers. The phosphorus content is accounted for by terminating the layers into a strip, which is consistent with the known size of the core. These features suggest that the ferritin core is best described as "a strip folded back and forth upon itself in the form of a pleat."[48]

4.9. Cytochrome Oxidase

Cytochrome oxidase is a membrane-bound enzyme involved in the mitochondrial electron transfer system. It contains both two inequivalent heme groups and a pair of copper binding sites.

EXAFS studies have been made from both the iron and copper absorption edges.[49–52] Although much of this data is currently being analyzed, preliminary conclusions, based on the Fe EXAFS data, are listed below:

a. A large structure change occurs on oxidation/reduction; a large signal in the second shell of the oxidized cytochrome oxidase appears to be retained in the formal mixed-valance state (cytochrome a_3 still oxidized) but is not observed in the reduced state (all components reduced).

b. The average bond length for the first shell in the fully reduced oxidase is 1.98 Å, which would correspond to a bisimidazol axial ligation in the first shell of a six-coordinated iron.

c. A deconvolution analysis that assumes a bisimidazole ligation on the heme (with consequently no change on oxidation/reduction) gives the bond length for heme a_3 in the fully reduced and the mixed-valence state of 2.05 Å.

These measurements were made using crystal focusing lenses for x-ray fluorescence developed by J. Hastings, J. Brown, and P. Eisenberger (graphite) and L. Powers, M. Marcus, B. Kincaid, and B. Chance (LiF).

Edge and EXAFS studies of cytochrome oxidase (~ 1 mM in Cu) were made in the oxidized, reduced, and mixed-valence states of the protein at low temperatures ($-70°$ to $-190°$C). No radiation damage was noted in these measurements within experimental error ($\sim 10\%$) and addition of ferricyanide to the oxidized state showed no change in optical absorption, indicating that the protein was in the fully oxidized state.

4.10. Calcium Binding Proteins

Calcium is an important constituent of biological systems where it acts as a "secondary messenger" for communication across cell membranes. Preliminary studies have been made of both calcium binding proteins and calcium modulator proteins[53] indicating that EXAFS measurements can be made on proteins having calcium concentrations of a few μmol/mg or ~ 1 mM. Analysis of a series of model compounds showed the possibility of determining bond lengths with an accuracy of ± 0.02 to ± 0.04 Å and coordination numbers to about 15–20%. The measurement of K absorption in Ca in biological systems suffers from rather high background due to the relatively low x-ray energies (4 keV). The analysis is complicated by the fact that model compounds show a considerable spread (0.1 Å) of near-neighbor distances, but simplified because Ca coordinates exclusively to oxygen.

Bond distances deduced from EXAFS for Ca bound in the proteins concanavolin A, thermolysin, and MCBP (muscle calcium binding parvalbumin) are consistent with those deduced using x-ray crystallography.

In terms of averaged bond lengths observed, the calcium-binding proteins thermolysin and concanavolin A appear to have different binding sites than the calcium-modulator proteins MCBP (muscle binding parvalbumin) and tropin C (TrC), which cyclicly bind and release calcium.

5. X-Ray Absorption Edge Structure for Biological Molecules

As discussed in Chapters 10 and 4, the quantum mechanics determining the features of the x-ray absorption spectrum just below or just above an inner-shell absorption edge is considerably more complex than in the EXAFS region. As may be seen in Figure 14, the general features of the "edge region" are the existence of one or more distinct absorption lines (broadened by the core-hole lifetime) in the range −20 to 0 eV, a discontinuous rise in cross section at the continuum edge, followed by a series of fairly narrow resonant peaks superposed on the continuum cross section in the range 0 to ∼30 eV. The absorption lines in the pre-edge region are interpreted as "bound → bound" transitions, and for an isolated atom would be labeled in terms of angular momentum quantum numbers, e.g., for a Cu(I) ion (d^{10} configuration) the lowest available transitions would be $1s \rightarrow 4s$, $1s \rightarrow 4p$. For the case of a Cu(II) ion (d^9 configuration) there would also be a $1s \rightarrow 3d$ transition. For isolated atoms, the strength of these transitions would also be determined by the dipole selection rule, so that of the above three transitions, only the $1s \rightarrow 4p$ transition would be allowed on this basis.

For an ion in a molecular cluster, on the other hand, angular momentum is no longer a good quantum number, and a correct analysis must be based on the symmetry group (point group) of the cluster in question. Thus for a cluster with inversion symmetry (e.g., a square-planar Cu–nitrogen complex) parity violating transitions such as $1s \rightarrow 4s$ and $1s \rightarrow 3d$ would still be forbidden to zero order in

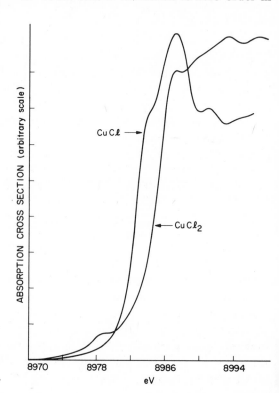

Figure 14. Cu absorption edges.

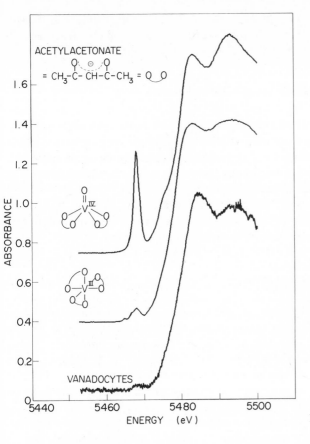

Figure 15. Vanadium edges.

atomic displacements, while for a cluster without inversion symmetry (e.g., a tetrahedral Cu–nitrogen complex) the $4p$ and $4s$ orbitals can become hybridized to a molecular T_2 state so that both $1s \rightarrow 4s$ and $1s \rightarrow 4p$ atomic transitions have nonzero projections for dipole allowed molecular transitions.[15]

Some of these features are illustrated in Figure 14 where it may be seen that the small "$3d$-like" absorption about 8 eV below the edge is absent for Cu(I) and present for Cu(II). At the time of writing, the origin of the oscillator strength for the $3d$ transition is not fully established. Some authors have speculated that it is vibronic in origin (as in an optical transition). However, the observed lack of temperature dependence of this oscillator strength is a bit of a mystery (zero-point motion would lead to a nonzero vibronic contribution at $T = 0$ K, but some additional strength would be expected as the temperature is increased).

Another possibility for which preliminary calculations show reasonable orders of magnitude (C. R. Natoli, D. Misemer, and S. Doniach, private communication) is that there is an appreciable electric quadrupole contribution to this transition, since the wave vector of the absorbed photon is in the region of 4 Å^{-1}.

The dependence of oscillator strength on molecular symmetry is nicely illustrated in Figure 15 (T. Tullius and K. Hodgson, to be published), where a set of

$1s \rightarrow 3d$-like transitions are seen to be very weak for octahedrally coordinated vanadium ions, but become drastically increased when the symmetry is reduced by the replacement of two oxygen bonds by a single V=O bond.

5.1. The Effect of Oxidation State on Edge Position

By simple electrostatics it will cost more energy to remove a core electron as the positive charge on an ion is increased. Thus the absolute *position* of the continuum edge is observed to increase by 3 eV or so for each oxidative electron removal from a given ion. The problem with pinning this down is that the actual position of the edge is hard to determine unambiguously as it frequently overlaps bound → bound and resonance features of the spectrum. Nevertheless, a study by Cramer and Hodgson[44] showed that the edge position correlates reasonably well with other measures of ionicity for a series of molybdenum complexes.

5.2. Continuum Spectral Features

Theoretical studies by Natoli *et al.* (see Chapter 10) suggest that absorption peaks in the near-edge continuum region of the spectrum correspond to "eigenphase" resonances of the outgoing electron wave in the molecular potential.[15] Physically, this corresponds to a kind of temporary "trapping" of the electron by a combination of the centrifugal barrier and the potential due to the coordination cage of the ion. To date, quantitative analysis of these peaks has only been done for two or three model compounds (Natoli, Misemer, Kutzler, Doniach, and Hodgson, to be published) so that the general phenomenology of these "post-edge features" in relation to molecular structure remains to be established. Certain simple feature are clear: the strong resonance for Cu(ı) in Figure 16 correlates with the more closed

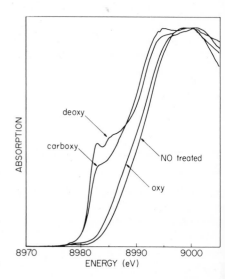

Figure 16. X-ray absorption edge of copper in four forms of hemocyanin.

nature of the tetrahedral molecular cage for this compound—as it deforms to a square-planar geometry for Cu(II), the resonance features weaken, indicating a lower amplitude for "trapping" of the electron wave in the open-geometry cluster. Similarly, the very strong resonances observed for $L_{2,3}$ edges in a wide variety of materials (so-called white lines) probably correspond to a combination of reinforcement of the normal outgoing d-wave centrifugal barrier by the molecular potential and the multiplying effect of the $(2l + 1)$ multiplicity for the d-wave final state.

5.3. Applications to Specific Biological Molecules*

Rubredoxin. Shulman, Yafet, Eisenberger, and Blumberg[45] have studied the edge structure for the K-shell Fe absorption. They observe a sevenfold increase in the $1s \rightarrow 3d$-like transition strength on going from an octrahedrally coordinated to a tetrahedrally coordinated Fe site.

Hemocyanin. Hemocyanin is an oxygen transport protein found in arthropods and mollusks. X-ray absorption edge spectroscopy has been used to elucidate changes in the valency of the copper atoms upon binding of oxygen.[25] In deoxyhemocyanin (see Figure 16), the edge occurs at lower energy with inflection points of the two main transitions falling in the normal range for Cu(I) [and lower than for Cu(II)]. Furthermore, there is no evidence of a pre-edge $1s \rightarrow 3d$ bound-state transition consistent with the formulation as Cu(I) (which has a filled $3d$ shell). Upon binding oxygen, the edge can be seen to shift to higher energy on average by about 3 eV. Also, careful inspection of the region before the main edge reveals the presence of a low-intensity $1s \rightarrow 3d$ transition. These observations are only consistent with a net oxidative change in the valency of the Cu atoms upon O_2 binding. The formulation of the oxy state of hemocyanin as two Cu(II) atoms bound to a peroxide is most consistent with these observations. In the carboxy form of the protein, edge-fitting analysis suggests that one copper remains fully reduced while the other is partly oxidized, thereby giving an observed edge that is approximately a 1 : 1 sum of the oxy plus deoxy edges.

Cytochrome Oxidase and Cytochrome-C Oxidase. Hu et al.[24] have reported measurements of the Cu edge. Their analysis of the data suggests the existence of two distinct Cu-binding sites in the oxidized form of the enzyme. However, the analysis is based on an identification of a $1s \rightarrow 4s$-like transition, which would require a more detailed study of the effects of molecular symmetry for an unambiguous conclusion.

A comparison of radiation effects on cytochrome oxidase at 300 K and at 200 K suggests that radiation at 300 K results in reduction of cytochrome oxidase in a time of 20 min. Thus, Chance and his collaborators suggest that a low-temperature trapping technique combined with on-line optical assay is essential to ensure the redox state of the copper and iron atoms of cytochrome oxidase. They suggest that the measurements of Hu, Chan, and Brown, which cite the presence of reduced copper in otherwise oxidized cytochrome oxidase, may be the result of

---

* See also the summary in Table 3.

radiation damage.[24, 54] A satisfactory resolution of these disparate observations has yet to be obtained.

In measurements carried out at low temperatures (see Section 4.9), the oxidized state and enzymatically reduced (ascorbate, TMPD, cytochrome C) state with carbon monoxide were found to be similar to those observed by Hu, Chan, and Brown[24] in the reproducible features and a shift of $\sim$5 eV is observed with a change in edge shape.[50] Stellacyanin was studied in the oxidized and reduced states as a model protein. The similarity of features as well as of reduced states suggests that stellacyanin may be a good model for one of the coppers present in cytochrome oxidase.

Vanadium in Cells of Ascidia ceratodes (Tunicates). Tunicates are marine organisms of the Cordata family that occur in salt water tide pools and sea beds. It has long been known that they contain high concentrations of vanadium (up to 1% dry weight in certain cases). X-ray absorption edge studies of the whole cells from the blood of these organisms have been used to establish the nature of the vanadium ion *in situ.*[46] A study of the vanadium edges of the intact cells (called vanadocytes) and of cells that have been allowed to lyse, and of a model compound containing the vanadyl ion $[V(IV)=O^{2+}]$ reveals that for the compound with vanadyl ion, the edge clearly has a sharp pre-edge feature that corresponds to a transition from the 1s level into a molecular orbital comprised of ligand p plus metal $3d$ orbitals. This distinctive feature is absent in both the tunicate edges. Thus, the intact cells contain no (or very little) vanadyl and contrary to what would be expected, lysis does not lead to oxidation to vanadyl. There must be some mechanism associated with the cells for keeping the vanadium reduced. Thus, the vanadium ion in tunicates is best described as V(III) in a highly acidic aquo environment (the acid is known to be present from pH measurements on the whole cells).

Calcium Binding in Biological Systems. Sensitivity of x-ray absorption edge features to coordination geometry has been observed for a variety of model compounds by Powers *et al.*[53] and by Bianconi *et al.*[55] Edge spectra studies of the Ca–ATP complex (involved in the process of Ca transport across membranes by Ca–ATPase) show a marked similarity to the edge structure of $CaHPO_4$ supporting the hypothesis that Ca is bound to the oxygen atoms in the $(HPO_4)^{2-}$ group of the ATP.

6. Prospects

Since the start of the field in 1974, unique data of biological significance has already emerged from XAS studies of metalloproteins (see Sections 4.2 and 4.3 on hemoglobin and nitrogenase). However, it is clear that this field of spectroscopy is still in its infancy. Prospects for extensions of its application to biological systems include:

    a. technical improvements—increased x-ray flux and improved detection techniques—leading to

- detailed study of binding site coordination changes as a result of enzyme–substrate interactions using EXAFS, and
- monitoring of chemical kinetics using time-resolved x-ray spectroscopy;

b. theoretical improvements leading to

- improved analysis of the relationship of edge features to the chemical state of the absorber, to provide information about binding site charge distribution and site symmetry as a function of enzyme–substrate interaction.

References

1. J. Jaklevic, J. A. Kirby, M. P. Klein, A. J. Robertson, G. S. Brown, and P. Eisenberger, *Solid State Commun.* **23**, 679–682 (1977).
2. Bambynek *et al.*, *Rev. Mod. Phys.* **44**, 716 (1972).
3. P. Eisenberger, R. G. Shulman, G. S. Brown and S. Ogawa, *Proc. Nat. Acad. Sci. U.S.A.* **73**, 491 (1976).
4. J. Hastings, P. Eisenberger, and J. Brown (to be published).
5. S. P. Cramer, K. O. Hodgson, E. I. Steifel, and W. E. Newton, *J. Am. Chem. Soc.* **100**, 2748 (1978).
6. R. G. Shulman, P. Eisenberger, B. K. Teo, D. M. Kincaid, and G. S. Brown, *J. Mol. Biol.* **124**, 305 (1978).
7. P. A. Lee, B. K. Teo, and A. L. Simons, *J. Am. Chem. Soc.* **99**, 3856 (1977).
8. B. K. Teo, P. A. Lee, A. L. Simons, P. Eisenberger, and B. M. Kincaid, *J. Am. Chem. Soc.* **99**, 3854 (1977).
9. T. Tullius, P. Frank, K. O. Hodgson, *Proc. Nat. Acad. Sci. U.S.A.* **75**, 4069 (1978).
10. S. P. Cramer and K. O. Hodgson, *Prog. Inorg. Chem.* **25**, 1 (1979).
11. S. P. Cramer, Ph.D. thesis, Stanford University (1977).
12. Y. Yamanouchi and J. Enemark, *Inorg. Chem.* **17**, 2911 (1978).
13. J. N. Berg, K. O. Hodgson, S. P. Cramer, J. L. Corbin, A. Elsberry, N. Pariyadath, and E. I. Striefel, *J. Am. Chem. Soc.* **101**, 2274 (1979).
14. B. Bunker and E. A. Stern, *Biophy. J.* **19**, 253 (1977).
15. F. W. Kutzler, C. R. Natoli, D. K. Misemer, S. Doniach, and K. O. Hodgson, *J. Chem. Phys.*, submitted.
16. R. G. Shulman, P. Eisenberger, W. E. Blumberg, and N. A. Stombaugh, *Proc. Natl. Acad. Sci. U.S.A.* **72**, 4003 (1975).
17. D. E. Sayers, E. A. Stern, and J. R. Herriott, *J. Chem. Phys.* **64**, 427 (1976).
18. P. M. Eisenberger, R. G. Shulman, G. S. Brown, and S. Ogawa, *Proc. Natl. Acad. Sci. U.S.A.* **73**, 491 (1976).
19. P. Eisenberger, R. G. Shulman, B. M. Kincaid, G. S. Brown, snd S. Ogawa, *Nature (London)* **274**, 30 (1978).
20. S. P. Cramer, J. H. Dawson, K. O. Hodgson, and L. P. Hager, *J. Am. Chem. Soc.* **100**, 7282 (1978).
21. S. P. Cramer, W. O. Gillum, K. O. Hodgson, L. E. Mortenson, E. I. Steifel, J. R. Chisnell, W. J. Brill, and V. K. Shah, *J. Am. Chem. Soc.* **100**, 3814 (1978).
22. S. P. Cramer, K. O. Hodgson, W. O. Gillum, and L. E. Mortenson, *J. Am. Chem. Soc.* **100**, 2298 (1978).
23. T. E. Wolff, J. M. Berg, C. Warrick, R. H. Holm, K. O. Hodgson, and R. B. Frankel, *J. Am. Chem. Soc.* **100**, 4030 (1978).
24. V. W. Hu, S. I. Chan, and G. S. Brown, *Proc. Natl. Acad. Sci U.S.A.* **74**, 3821 (1977).
25. T. K. Eccles, Ph.D. thesis, Stanford University (1977).
26. G. S. Brown, G. Navon, and R. G. Shulman, *Proc. Natl. Acad. Sci. U.S.A.* **74**, 1974 (1977).
27. R. G. Shulman, P. Eisenberger, and B. M. Kincaid, *Annu. Rev. Biophys. Bioeng.* **7**, 559 (1978).
28. P. H. Citrin, P. Eisenberger, and B. M. Kincaid, *Phys. Rev. Lett.* **36**, 1346 (1976).

29. K. D. Watenpaugh, L. C. Sieker, J. R. Herriott, and L. H. Jensen, *Acta Crystallogr.* **29**, 943 (1973).
30. R. W. Lane, J. A. Ibers, R. B. Frankel, G. C. Papaefthymion, and R. H. Holm, *J. Am. Chem. Soc.* **99**, 84 (1977).
31. B. Kincaid, P. Eisenberger, K. O. Hodgson, and S. Doniach, *Proc. Natl. Acad. Sci. U.S.A.* **72**, 2340 (1975).
32. J. P. Collman, *Acc. Chem. Res.* **10**, 265 (1977).
33. J. L. Hoard, *Science* **174**, 1295 (1971).
34. R. G. Shulman and S. Sugano, *J. Chem. Phys.* **42**, 39 (1965).
35. J. L. Hoard, in *Porphyrins and Metalloporphyrins*, K. M. Smith (ed.), p. 317, Elsevier, Amsterdam (1975).
36. G. B. Jameson, G. A. Rodley, W. T. Robinson, R. R. Gagne, C. A. Reed, and J. P. Collman, *Inorg. Chem.* **17**, 850 (1978).
37. N. F. Peruz, *Nature (London)* **228**, 726 (1970).
38. R. G. Shulman, J. J. Hopfield, and S. Ogawa, *Q. Rev. Biophys.* **8**, 325 (1975).
39. J. J. Hopfield, *J. Mol. Biol.* **77**, 207 (1979).
40. R. G. Shulman, S. Ogawa, K. Wuthrich, T. Yamane, J. Peisach, and W. E. Blumberg, *Science* **165**, 251 (1969).
41. T. Tullius and K. O. Hodgson (submitted for publication).
42. T. D. Tullius, D. M. Kurtz, Jr., S. D. Conradson, and K. O. Hodgson, *J. Am. Chem. Soc.* **101**, 2776 (1979).
43. S. P. Cramer, H. B. Gray, and K. Rajagopalan, *J. Amer. Chem. Soc.* **101**, 2772 (1979).
44. S. P. Cramer, T. K. Eccles, K. O. Hodgson, and L. E. Mortenson, *J. Am. Chem. Soc.* **98**, 1287 (1976).
45. R. G. Shulman, Y. Yafet, P. Eisenberger, and W. E. Blumberg, *Proc. Natl. Acad. Sci. U.S.A.* **73**, 1384 (1976).
46. T. Tullius, W. O. Gillum, R. Carlson, and K. O. Hodgson, *Inorg. Chem.*, in press.
47. J. M. Brown, L. Powers, J. Larabee, B. Kincaid, and T. Spiro, Structural studies of the hemocyanin active site I: EXAFS and analysis, *J. Am. Chem. Soc.* (submitted for publication).
48. S. M. Heald, E. A. Stern, B. Bunker, E. M. Holt, and S. L. Holt, *J. Am. Chem. Soc.* **101**, 67 (1979).
49. B. Chance, L. Powers, and J. S. Leigh, Jr., *Porphorin Chemistry Advances*, Longo *et al.* (eds.), pp. 9–15, Ann Arbor Science Publications, Ann Arbor, Michigan (1979).
50. L. Powers, W. E. Blumberg, B. Chance, C. Barlow, J. S. Leigh, Jr., J. C. Smith, T. Yonetani, S. Vik, and J. Peisach, *Biophys. Biochem. Acta*, **546**, 520 (1979).
51. L. Powers, P. Eisenberger, and J. Stamatoff, *Developments in Biochemistry: Cytochrome Oxidase*, King *et al.* (ed.), 189, Elsevier/North-Holland, New York (1979).
52. L. Powers, P. Eisenberger, and J. Stamatoff, *Frontiers in Biological Energetics–Electrons to Tissues*, Vol. 2, Dutton *et al.* (eds.), 863, Academic Press, New York (1978).
53. L. Powers, P. Eisenberger, and J. Stamatoff, *Ann. N.Y. Acad. Sci.* **307**, 113 (1978).
54. V. Hu, S. Chan, and G. Brown, *FEBS Lett.* **84**, 287 (1977).
55. A. Bianconi, S. Doniach, and D. Lublin, *Chem. Phys. Lett.* **59**, 121 (1978).
56. Joseph Smith, Ph.D. thesis, University of California at Berkeley (1978).

14

X-Ray Fluorescence Microprobe for Chemical Analysis

C. J. SPARKS, Jr.

1. Introduction

X-ray photoionization of elements and the detection of their characteristic fluorescent radiation has long been a basic research tool in atomic physics (see Chapter 4). It is also a widely applied analytical technique for the determination of elemental composition in support of both research and technology. The analytical application is the subject of this chapter. Synchrotron radiation must offer some important advances over present analytical techniques or little justification can be found for the use of a facility remote to most users. The analytical capabilities of the limited synchrotron facilities available will not supply all the routine analytical services required. However, increasing demands on analytical services caused by advanced technologies and growing concern for environmental monitoring is exceeding the performance capabilities of standard analytical methods. Justifications for applying synchrotron radiation to measurements of chemical composition include lowering of the detection limits, reducing heat or damage to the sample, improving the spatial resolution and contrast of microprobe analysis, reducing the time for analysis, improving the means of chemical identification by measuring absorption edge shifts or EXAFS, providing more accurate quantitative analysis, and

* Research sponsored by the Materials Sciences Division, U.S. Department of Energy under contract W-7405-eng-26 with the Union Carbide Corporation and the National Science Foundation through support of the Stanford Synchrotron Radiation Laboratory.

C. J. SPARKS, Jr. • Metals and Ceramics Division, Oak Ridge National Laboratory, Oak Ridge, Tennessee 37830.

extending analytical measurements to samples, configurations, and environments that are impractical, if not impossible, to analyze with present techniques.

X rays are unique in their interaction with atoms as illustrated by the large resonances that exist in the photoelectric cross section. Increases by factors of 5 to 10 occur in the cross section when the energy of the incident x ray just exceeds the binding energy of an inner-shell electron. These high x-ray cross sections for selective photoionization of specific electrons yield better signal-to-background ratios than those for charged particle excitation which include large contributions from loosely bound electrons. Electron microprobes are widely used because of the ease of focusing electron sources of high intensity to small beam diameters. More recently, proton excitation has received attention because the signal-to-noise ratio is superior to that for electrons. We will evaluate synchrotron radiation as an excitation source by comparing it to electron and proton excitation since the techniques are similar. Some comparisons between charged particle and x-ray excitation are made to conventional x-ray-source excitation since few data exist for synchrotron radiation. However, conventional x-ray sources (2 to 60 kilowatts dissipated by electrons impinging on metal targets) have insufficient intensity to compete with synchrotron sources in terms of detectable limits or spatial resolution.

Parameters important to this evaluation are presented in this chapter. They include fluorescence cross sections, backgrounds beneath the fluorescence signals, spatial resolution for microanalysis, energy deposition in the sample, and the incident intensities achievable from synchrotron radiation through x-ray optics. It will be shown that x rays, in comparison to charged particles, have fluorescence cross sections 10 to 10^3 times higher, and fluorescence signal-to-background ratios 10 to 10^5 times larger. We will also show that x rays deposit only 10^{-3} to 10^{-5} as much energy in the sample for the same elemental detectability. In addition, the intensity of x rays available from synchrotron radiation sources will produce a similar spatial resolution as for that available from charged particle microprobes with at least a 10-fold reduction in the amount of element detected for the same beam exposure time. These advantages of synchrotron radiation as an ionization source are sufficient justification to develop, from the intense flux available from storage rings, a unique and powerful analytical microprobe.

2. Quantitative Analysis and Fluorescence Cross Sections

Equations relating measured fluorescence intensities to mass fractions and to the number of exciting projectiles are necessary for quantitative analysis of elemental compositions and for predicting their theoretical limits of detection. The principal parameters for the fluorescent radiation leaving the sample are the absorption cross sections. For exciting radiation entering the sample the principal parameters are the partial photoionization cross sections, fluorescence yields, and absorption. Cross sections are compared with charged particle excitation to emphasize the differences in the number of x rays to the number of charged particles required for the same minimum detectable limit and the same amount of energy deposition.

2.1. Equations for Quantitative X-Ray Fluorescence Analysis

The fluorescence intensity from n_z atoms of element z is derived from the definition of the fluorescence cross section σ_{zij}, given as the ratio of the power radiated per atom to the incident intensity,

$$\sigma_{zij} \equiv P_{zij} n_z^{-1} I_0^{-1} \tag{1}$$

where I_0 is the power per unit area of a monoenergetic incident x-ray beam capable of photoejecting an electron from the ith shell of an atom of element z, and $P_{zij} n_z^{-1}$ is the fluorescence energy s^{-1} atom^{-1} from element z. The fluorescence cross section σ_{zij} is the product of the partial or subshell ionization cross section σ_{zi}, the probability that this vacancy results in a fluorescence event ω_{zi} (fluorescence yield), the fraction of these fluorescence events F_{zij} (fractional radiative rate) belonging to the particular fluorescence line of interest, and the electron-hole transfer factor $T'_{zi,k}$ (≥ 1), which accounts for the transfer of holes from deeper vacancies by Auger, Coster–Kronig, and radiative transitions:

$$\sigma_{zij} = (\sigma_i \omega_i F_{ij} T'_{i,k})_z \tag{2a}$$

The subscript i denotes the subshell ionized, j identifies the final state of an x-ray transition, and k the subshell to which a vacancy has been shifted. Only σ_{zi} is dependent on the kind of excitation projectiles and their energies. For K-shell ionization,

$$\sigma_{zKj} = (\sigma_K \omega_K F_{Kj})_z \tag{2b}$$

as $T'_{K,k}$ must be unity. For a $K\alpha_1$ x-ray fluorescence event,

$$\sigma_{zK\alpha_1} = (\sigma_K \omega_K F_{K\alpha_1})_z \tag{2c}$$

For an x-ray emitted during the filling of an L_2 subshell vacancy,

$$\sigma_{zL2j} = \sigma_{L_2} \omega_2 F_{L2j}[1 + (\sigma_{L_1}/\sigma_{L_2})f_{1,2}]_z \tag{2d}$$

where the square brackets contain the hole transfer factor $T'_{zi,k}$ and where $f_{1,2}$ is the Coster–Kronig yield. The assumption in equation (2d) is that only L_1 and L_2 subshell ionization events occurred. Bambynek et al.[1] present a more complete discussion of these hole transfer factors and fluorescence yields.

The intensity I_z observed at a detector a distance R away is $P_{zij}/4\pi R^2$. Equation (1) becomes

$$I_z = I_0 \sigma_{zij} n_z / 4\pi R^2 \tag{3}$$

To a good approximation, one incident photon produces only one ionization since multiple ionizations and fluorescent radiation produced by secondary processes can be ignored. Thus, we can use units of counts s^{-1} cm^{-2} for the intensities given in equation (3) rather than units of energy.

In most practical cases, the exciting radiation is attenuated in traversing the target and the fluorescent radiation is attenuated on the way out. Attenuation may occur along the path to the detector and the detector efficiency may not be unity.

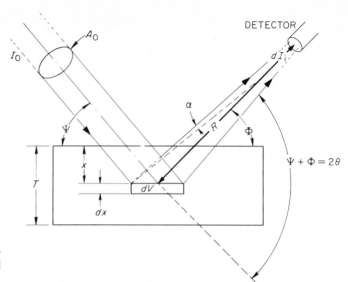

Figure 1. Geometry for calculating the fluorescence intensity from a sample with parallel faces (after Sparks[2]).

With reference to Figure 1, we may express the fluorescence intensity in terms of the number of z atoms in the small volume $dV = dxA_0/\sin \psi$, where the A_0 is the area of the incident beam. If we integrate the expression over the sample thickness (T) excited by the incident radiation, the result is

$$C_z = \frac{I_z 4\pi R^2 (\mu_{s,0} + \mu_{s,i} \sin \psi/\sin \phi)/\rho_s}{P_0 D_z \sigma_{zij} \langle 1 - \exp\{-[(\mu_{s,0} + \mu_{s,i} \sin \psi/\sin \phi)/\rho_s]\rho_s T/\sin \psi\}\rangle} \qquad (4)$$

Definitions of the terms in this equation (with a consistent set of units) are:

C_z Concentration of element z in mass of z per unit mass of sample

R Specimen-to-detector distance in cm

I_z Fluorescence intensity in counts per second (cps) per area of detector receiving slit in cm^2, and is understood to be (power in cps)/A_D, where A_D is the detector area in cm^2

$\mu_{s,0}/\rho_s$,   Mass absorption coefficients of the sample in cm^2/g
$\mu_{s,i}/\rho_s$ for the exciting and the fluorescent radiation, respectively

P_0 Power in cps of the exciting radiation incident on the sample capable of ionizing the i shell or subshell of element z

D_z Unitless absorption factor for the fluorescent radiation from element z transversing the path from sample surface to detector (including the detector efficiency)

$\sigma_{zij} =$ Fluorescence cross section for the exciting energy
$(\sigma_i \omega_i F_{i,j} T'_{i,k})_z$ P_0 in cm^2/g. This cross section must be summed to include the radiative transitions for the various k

shells and subshells that contribute energy to the
measured intensity

$\rho_s T$ Mass per unit area of a sample of uniform thickness
in g/cm^2 determined by dividing its mass by the
area or by knowing its actual density and thickness

For an infinitely thick sample (i.e., one that completely stops the incident beam) the exponential goes to zero and equation (4) becomes

$$C_z = \frac{I_z 4\pi R^2 (\mu_{s,\,0} + \mu_{s,\,i} \sin \psi/\sin \phi)/\rho_s}{P_0 D_z \sigma_{zij}} \tag{5}$$

For transparent samples with negligible absorption for the incident and fluorescent radiation, equation (4) becomes

$$C_z = \frac{I_z 4\pi R^2}{P_0 D_z \sigma_{zij} \rho_s T/\sin \psi} \tag{6}$$

and one need not know the absorption coefficients for the incident and fluorescence energy in the sample, but instead the mass per unit area $(\rho_s T)$ of the sample.

Substitution of a pure element or known stoichiometric compound for the sample determines many of the parameters in equation (4). For an infinitely thick pure element z, $C_z = 1$, and equation (4) may be rearranged to give

$$I_z^{\text{pure}} = P_0 D_z \sigma_{zij}/4\pi R^2 (\mu_{s,\,0}^{\text{pure}} + \mu_{s,\,i}^{\text{pure}} \sin \psi/\sin \phi)/\rho_s^{\text{pure}} \tag{7}$$

The terms common to equations (4) and (7) and identified by the parameter Q_z are

$$Q_z = 4\pi R^2/P_0 D_z \sigma_{zij} \tag{8}$$

where Q_z is determined by

$$Q_z = 1/I_z^{\text{pure}}(\mu_{s,\,0}^{\text{pure}} + \mu_{s,\,i}^{\text{pure}} \sin \psi/\sin \phi)/\rho_s^{\text{pure}} \tag{9}$$

in units of $g\ cm^{-2}\ cps^{-1}$. Many of the most difficult parameters to determine experimentally (area of the detector, detector efficiency, absorption in the path to the detector, power of the exciting radiation, and fluorescence cross section) can be replaced by a measurement of the fluorescence from a pure element and the tabulated values of its mass absorption coefficients for the incident and fluorescent energy. Substitution of Q_z as given in equation (8) for the same parameters contained in equation (4) results in

$$C_z = \frac{Q_z I_z(\mu_{s,\,0} + \mu_{s,\,i} \sin \psi/\sin \phi)/\rho_s}{1 - \exp\{-[(\mu_{s,\,0} + \mu_{s,\,i} \sin \psi/\sin \phi)/\rho_s]\rho_s T/\sin \psi\}} \tag{10}$$

Thus determination of the elemental concentration depends on our knowledge of the mass absorption coefficients of the incident and fluorescent radiation in the pure element standard and in the sample of unknown composition. Methods of determining these coefficients and the mathematical treatment of fluorescence measurements for quantitative analysis and secondary sources of fluorescence have been reviewed by Sparks.[2]

Empirical methods utilize the parameters determined from several standards that bracket the suspected composition of the unknown sample. These methods are reviewed by Heinrich.[3] A review of the effect of sample particle size, surface roughness, and the methodology of x-ray fluorescence analysis has been given by Müller.[4]

In a photoionization event, a photon is annihilated in a single encounter. Charged particles lose energy by straggling energy-loss processes so that the cross sections used in the equations above must be replaced with integrals to account for the changing cross sections as the energy of the charged particle is reduced. The simplified mathematics for x rays contributes to the accuracy of quantitative analysis, which approaches $\pm 1\%$ with similar composition standards[4] and about $\pm 5\%$ with pure element standards.[2] Proton excitation gives relative errors of about ± 5–10% with similar composition standards,[5] and electron microprobe excitation at best gives relative errors of $\pm 6\%$ for concentrations above 20 wt. % and errors of $\pm 10\%$ or greater for lower concentrations depending on the homogeneity of the sample.[6]

A measurement of the absorption edge jump of an element in a sample would readily provide the mass fraction of that element in the sample from the following relationship:

$$\left(\frac{\mu}{\rho}\right)_s = C_1\left(\frac{\mu}{\rho}\right)_1 + C_2\left(\frac{\mu}{\rho}\right)_2 + \cdots + C_z\left(\frac{\mu}{\rho}\right)_z \tag{11}$$

The mass fraction (C_z) is obtained by taking the ratio of the measured absorption edge jump of the element in the sample to the absorption edge jump of the pure element. Detection of the change in the mass absorption coefficient of the edge has a signal-to-noise ratio of only 10 at best (maximum absorption jump for a pure element). Fluorescence intensities will be shown to equal the background beneath the peak even at concentrations as low as 10^{-7}. About 0.1 of the incident x rays responsible for the signal from the edge of interest produce fluorescence, and if 0.01 of the total fluorescence is detected, then the fluorescence signal is 10^{-3} of the edge signal obtained by transmission. From the definition of the minimum detectable limit given in the next section, it can be shown that measurement of fluorescent radiation can be 100 times more sensitive for detecting elements $(Z \geq 14)$ than measurement of the absorption edge jump. In EXAFS measurement, the EXAFS signal is only 0.01 of the absorption edge. Thus, fluorescence detection is 10^3–10^4 times more sensitive for following EXAFS variations than measuring the transmitted beam.

2.2. Minimum Detectable Limit

We adopt the derivations of Currie[7] based on Poisson counting statistics for a definition of the detection limits. For 95% confidence in detection, the signal N_s equals $3.29(\eta N_b)^{1/2}$, where N_b is the background beneath the signal. Depending upon how well the background is known, η has a value between one and two. In this discussion we will set $\eta = 1$ for simplicity. With N_s defined as the number of counts and C_z as the mass fraction, a definition for the minimum detectable limit (MDL) is

$$\text{MDL} = 3.29 C_z (N_b)^{1/2} / N_s = 3.29 (B/t)^{1/2} C_z / A_D I_z \tag{12}$$

where B is the background count rate beneath the fluorescence signal, t the counting time, A_D the detector area, and I_z/C_z the number of fluorescence counts per unit of time and mass fraction as defined previously. To obtain the smallest MDL in terms of the mass fraction, the flux of exciting particles (P_0) and the fluorescence cross section (σ_{zij}) should be as large as possible. If the background doubles, the counting time must double to maintain the same MDL. The minimum detectable mass of a pure element is obtained by substituting the mass of the elemental target irradiated by the incident beam $(A_0 \rho_s T)$ for C_z. For detecting or analyzing the smallest mass, the intensity (I_0) rather than P_0 of the incident beam must be maximized.

It would be useful to have a relative figure of merit that gives the ratio of the number of exciting x rays to the number of charged particles required to produce the same MDL for the same sample composition. The number of fluorescence events (N_s) is proportional to the number of incident x rays (N_x), electrons (N_e), or protons (N_p) times their respective fluorescence cross sections (σ_s). Similarly, the background (N_b) is proportional to the number of incident events times their respective cross sections for background production (σ_b). Thus, equation (12) for the MDL may be written for x-ray excitation as

$$\text{MDL}_x = 3.29 (N_x \sigma_{xb})^{1/2} / N_x \sigma_{xs} \tag{13}$$

and similarly for charged particles. The relative figure of merit is defined as N_x/N_e for x-ray versus electron excitation and N_x/N_p for x rays versus protons. N_x/N_e is obtained by equating the right-hand side of equation (13) for x-ray excitation to that for electron excitation and solving for the ratio,

$$N_x/N_e = (\sigma_{es})^2 \sigma_{xb} / (\sigma_{xs})^2 \sigma_{eb} \tag{14}$$

and likewise for N_x/N_p. The most significant factor favoring x-ray excitation is that the ionization cross sections enter into the figure of merit relationship as the square [equation (14)].

The relationship stated in equation (14) can also be expressed in terms of the number of counts in the signal (N_s) to the number of counts in the background (N_b). To obtain the same MDL, the ratio of the incident x rays to electrons depends on the signal to noise in the following way:

$$N_x/N_e = \sigma_{es}(N_s/N_b)_e / \sigma_{xs}(N_s/N_b)_x \tag{15}$$

and similarly for x rays compared to protons (N_x/N_p).

In the next sections, these relationships will be used to draw comparisons between x-ray and charged-particle excited fluorescence.

2.3. Comparison of X-Ray and Charged-Particle Fluorescence Cross Sections

A comparison of the x-ray fluorescence cross sections and backgrounds permits an evaluation of the relative figure of merit as given in equation (14). These values will be used to calculate the energy deposition for the same MDL. Beam spreading in the sample for x rays will be compared with that for electrons and protons. These comparisons will provide the basis for evaluating synchrotron radiation as an excitation source for microanalytical analysis.

Protons and heavier charged particles are used for fluorescence excitation because they produce less bremsstrahlung than electrons by the ratio of the mass of an electron to the mass of a proton.[8] However, ejected electrons contribute the major bremsstrahlung background. The early work of Johansson et al.[9] focused attention on the application of proton excited fluorescence for trace element analysis. Charged particles heavier than protons can also be used for fluorescence analysis.[9–11] A comparison by Cooper[11] between proton and alpha particle excitation revealed that alpha particles give detectable limits that are about a factor of ten higher than protons for elements $Z \geq 30$.

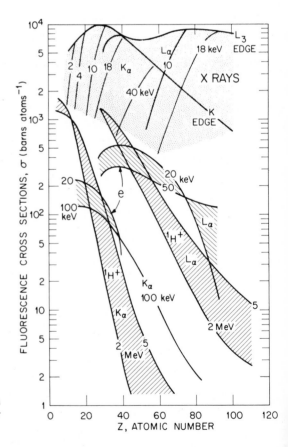

Figure 2. Fluorescence cross sections for the elements when excited by x rays, protons ($^1H^+$), and electrons (e) (after Sparks et al.[50]).

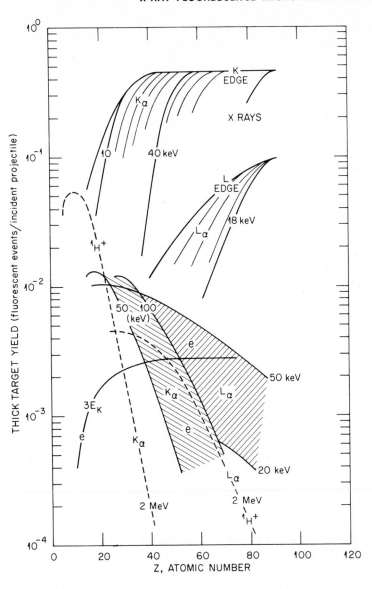

Figure 3. Thick-target fluorescence yields for x rays, protons, and electrons (after Sparks et al.[50]).

A comparison of fluorescence cross sections σ in barns atom^{-1} as defined in equation (2) is given in Figure 2 for the elements. X rays, electrons, and protons are compared at the energies typically used for fluorescence excitation. The ratio of these cross sections is the same as the ratio of the fluorescent intensities when the sample is so thin that absorption is negligible for the incident and fluorescent radiation. Figure 3 shows the thick-target yield $(I_z 4\pi R^2/P_0)$ as defined in equation (7) for pure elements. The number of fluorescence events are those at 180° to the incident radiation, and normal to the target surface. Thick-target yields give the maximum number of fluorescence events possible per incident particle. Data for the x-ray fluorescence cross sections were taken from compilations by Krause *et al.*,[12,13] and thick-target yields were calculated using the same data with absorption coefficients from McMasters *et al.*[14] Electron fluorescence cross sections and thick-target yields were taken from several sources.[15-22] Earlier data[15] for $L\alpha$ electron fluorescence cross sections gave lower values than more recent measurements.[21,22] However, all data was scaled to the higher values for this comparison. Proton ($^1H^+$) fluorescence cross sections and thick-target yields were taken from References 18 and 23–28 and are less uncertain than electron fluorescence cross sections. Nevertheless, the cross sections are sufficiently accurate to conclude that 10^{-1} to 10^{-3} fewer x rays are required to produce the same fluorescence signal as charged particles for commonly used energies. To obtain the maximum fluorescence signal-to-background (S/B) ratio with electron excitation, energies about three times the K- or L-shell binding energies are used. This curve for the K shell is labeled $3E_K$ in Figure 3. In thick targets, secondary processes[17] can contribute significant fluorescence signals when charged particles are the exciting radiation. This accounts for the inversion in the 20 and 50 keV electron data between Figures 2 and 3.

The uppermost curves for the x-ray fluorescence cross sections and thick-target yields (labeled K edge and L_3 edge in Figures 2 and 3) are for exciting x-ray energies 1.007 times the energy of the K or L_3 binding energies.[13] Absorption edge fine structure is neglected. The curves labeled 2, 4, 10, 18, and 40 keV in Figure 2 are the five different x-ray energies needed to excite every element above $Z = 10$ such that the fluorescence cross section is greater than 10^3 barns atom^{-1}. The use of these few energies will maintain x-ray fluorescence cross sections that are 20–200 times larger than those for electrons and 5–1000 times larger than those for protons for $Z \geq 10$. Absorption edge resonances for x-ray ionization are so large that for energies just above an absorption edge about 90% of the ionization events will occur in a specific shell or subshell. With the continuum radiation available from storage rings, an energy can be chosen to increase the sensitivity for the detection of a specific atom. Ionization cross sections for charged particles do not contain such sharp resonances. The means for selecting a specific energy from the synchrotron radiation continuum will be discussed in Section 4.

With a fixed energy of the incident radiation, the fluorescence cross section increases as Z decreases for charged-particle excitation, but the reverse is true for fixed energy x-ray excitation. If simultaneous multielement analysis with emphasis on the lower-Z elements is the primary goal, charged particles could have an advantage in fluorescence cross sections for $Z \leq 15$. Since the natural abundance of the elements decreases with increasing Z, x rays have the desirable property of increased sensitivity for the less-abundant heavy trace elements. Choices for the exciting x-ray energy depend on target composition, elements of major interest, and background.

2.4. Backgrounds Beneath the Fluorescence Signals

Though there are few references to direct measurements of the integrated intensity of fluorescence lines above the background, reliable estimates can be made of the S/B ratios achievable from the literature on x-ray, electron, and proton excited fluorescent spectra. For monochromatic x-ray excitation,[29] fluorescence measurements made with solid state detectors result in detector-generated background clearly dominating over that generated in the sample.[11, 30–34] For charged particles, bremsstrahlung radiation clearly dominates the background.

Cooper[11] compared fluorescence spectra excited with ^{55}Fe, ^{109}Cd, and a Mo transmission target x-ray tube (approximations to a monoenergetic x-ray source) with those excited with 2–4-MeV protons and 30–80-MeV alpha particles. X rays and protons gave similar S/B ratios. However, the solid state detector contributed most of the background in the x-ray excited spectrum. Alpha particles produced from 1 to 0.1 as much S/B as x rays or protons.

Jaklevic[31] has modeled the background for samples composed of 250 ng cm^{-2} each of Al, S, Ca, Fe, Cu, or Br on 25-mg cm^{-2} carbon substrates, which are akin to biological or air filter samples. The elemental concentrations corresponded to 10 ppm by weight if uniformly distributed throughout the substrate. Spectral data were calculated assuming a solid state Si(Li) detector with a resolution of 200 eV (FWHM) independent of energy. Spectra typical of x-ray fluorescence analysis are presented in Figures 4 and 5. Background beneath the fluorescence signals for proton and electron excitation as calculated by Jacklevic[31] are shown in Figures 4a and 4b. The relative heights of the fluorescence peaks above the bremsstrahlung background are about 10^2 greater for 3-MeV proton excitation than for 20–40-keV electron excitation (note change from log to linear scales). Figures 5a and 5b from Jaklevic[31] show the calculated fluorescence spectrum assuming 17- and 4.5-keV x-ray excitation, respectively. The continuum background is mostly generated in and by the solid state detector from the impinging elastically and inelastically scattered radiation. A comparison of Figures 5a and 5b with Figure 4b shows comparable S/B values that slightly favor proton over x-ray excitation for a region around $Z \simeq 30$. Jaklevic[31] assumed a continuous and flat x-ray intensity distribution incident on the same samples to calculate the fluorescence S/B spectrum shown in Figure 5c. In this case, the background is from sample scattering of the continuous radiation used to excite the sample and not from detector generated background. The spectrum would be similar to that excited with a continuous synchrotron radiation spectrum. Fluorescence S/B values are one to ten times larger for continuum x-ray than for electron excitation, but about 10^{-2} times smaller than for proton excitation. Goulding and Jaklevic[30] used the calculations of Folkmann *et al.*[33] on proton-induced backgrounds to conclude that monochromatic x-ray sources would give about a factor of ten improvement over protons in fluorescence signal to bremsstrahlung background. Thus, monoenergetic x rays produce fluorescence signals to background about 10^3 times larger than for electrons and about ten times larger than for protons.

Fluorescence S/B for thick targets excited by monoenergetic x rays as compared to electrons can be estimated by assuming that each incident x ray will create one photoejected electron of about the same energy as the x ray. These photoejected electrons are assumed to have the same cross sections for bremsstrahlung production as those electrons that arrive from outside the sample. Thus, each incident x ray at the very worst would produce the same bremsstrahlung background as each incident electron. Since the fluorescence signal produced by each incident x ray is 10 to 10^3 times larger than that produced by each incident electron, the signal to background ratio will be 10 to 10^3 larger for monoenergetic x-ray excitation. This ratio is similar to Jaklevic's[31] value of 10^2. However, his value would have been even larger if the solid state detector had not contributed the limiting background. Better ratios will be achieved with the abatement of detector background.

The advantages of monoenergetic x rays as an excitation source for fluorescence becomes apparent when the relative S/B ratios and the cross sections of x-rays and charged particles are used in equations (14) and/or (15) to determine the relative ratio of x rays to charged particles required to produce the same MDL. Typically, x-ray fluorescence cross sections are 10 to 10^3 higher (increasing with Z) than those for charged particles. A conservative estimate is that monoenergetic x-rays have the same cross section for background production as electrons and a cross section ten times larger than that for protons. With these estimates of the

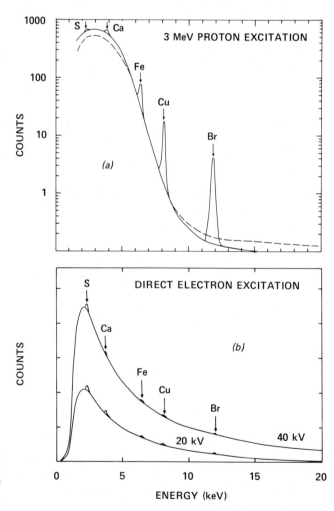

Figure 4. Calculated fluorescence spectrum from thin samples excited by (a) 3-MeV protons and (b) 20- and 40-keV electrons (after Jaklevic[31]).

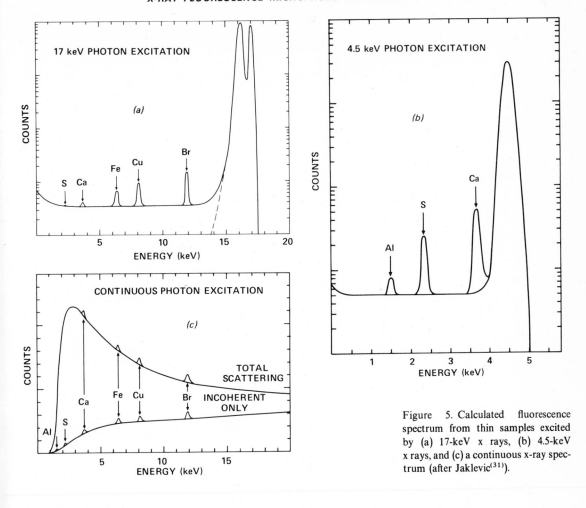

Figure 5. Calculated fluorescence spectrum from thin samples excited by (a) 17-keV x rays, (b) 4.5-keV x rays, and (c) a continuous x-ray spectrum (after Jaklevic[31]).

relative fluorescence and background cross sections, the value of N_x/N_e ranges from 10^{-2} to 10^{-5} and N_x/N_p from 1 to 10^{-4}. The smaller values occur when the x-ray energies are just above the absorption edges. These numbers will be used in the next section as typical values for an evaluation of the energy deposited in a specimen for the same MDL.

A direct comparison of an x-ray- and proton-excited fluorescence spectra in which the solid state detector does not generate the limiting background is made in Section 3.

2.5. Energy Deposited Versus Fluorescence Production

It is important in many applications to limit the amount of power dissipated in the sample. Volatile components may be evaporated, redistribution of elements may occur, chemical bonding is easily affected, and organic materials are especially vulnerable to both heat and radiation damage. Nondestructive analysis of biological materials requires limiting the dose to levels below that usually employed for metals and minerals in electron and proton microanalysis. An estimate of the energy deposited by monochromatic x rays in the energy region 5–40 keV to that deposited by 20–100 keV electrons is readily obtained for thick samples from the ratio of x rays to electrons (N_x/N_e) required to produce the same MDL. A typical value of N_x/N_e is 10^{-3} when the x ray and electron excitation energy is the same. For x-ray energies chosen just above the absorption edge and electrons at energies typically three times that energy, x rays would deposit an energy that is 3×10^{-4} of that deposited by electrons in thick samples for the same MDL. Similar values are found for both K and L fluorescence.

Since typical proton energies of 2 MeV are 10^2 higher than typical x-ray energies and since N_x/N_p averages about 6×10^{-3}, then about 10^{-4} as much energy will be deposited by monoenergetic x rays as protons for the same MDL in thick samples.

For thin targets, the calculation of the energy deposited in the sample is more involved because of the straggling energy-loss processes of charged particles. Increasing ionization cross sections for the less tightly bound electrons and high cross sections for plasmon excitations of the least-bound electrons contributes to the energy-loss processes for charged particles. Generally, the energy lost by the incident charged particle to each electron in the atom is similar since the larger amount of energy removed by inner-shell ionization is compensated for by the higher cross sections but smaller energy loss per event for the less tightly bound electrons.[35] For example, ionization of the K shell in an atom containing 30 electrons will account for about $\frac{1}{15}$ of the total energy loss for charged particles. Consequently, $Z/2$ times as much energy goes into ionizations of no interest as go into those producing fluorescence. For x rays of an energy just above a K absorption edge, about 90% of their energy is lost by ionization to that K shell. An approximation is made that charged particles will deposit $(Z/2)[N_x/(N_e \text{ or } N_p)]$ as much energy as x rays in thin targets. Thus, from 10^{-2} to 10^{-6} as much energy is deposited by x-ray excitation of thin targets as by electron and proton excitation at typical energies for the same MDL.

Calculations of the number of ionizations per keV of energy deposited in pure element samples 1 μm thick are taken from the work of Shaw and Willis[36] and given in Table 1. In this comparison, the x-ray energies were not chosen to be just

Table 1. Number of Ionizations/keV of Energy Deposited in Sample 1 μm Thick[a]

	Photons		Electrons		Protons	
	5 keV	20 keV	20 keV	80 keV	3 MeV	5 MeV
Al K	0.21	—	2.0×10^{-2}	—	5.9×10^{-3}	—
Cu K	—	6.9×10^{-2}	5.3×10^{-4}	1.2×10^{-3}	3.2×10^{-5}	1.0×10^{-4}
Au L_3	—	7.6×10^{-2}	5.9×10^{-4}	1.1×10^{-3}	1.4×10^{-5}	4.8×10^{-5}

[a] Shaw and Willis.[43]

above the absorption edges of the elements. For typical excitation energies, they find that x-ray excitation deposits from 10^{-1} to 10^{-2} as much energy as electron excitation and from 10^{-2} to 10^{-3} as much energy as proton excitation for the same number of ionizations. Their calculations were not based on the MDL criteria and did not consider background beneath the fluorescence signals. If we divide their numbers by the relative gain in cross section for x rays just above the respective absorption edges of Al K, Cu K, and Au L_3 and the factor of 2 to 3 reduction in the energy of the required x ray, then 10^{-2} to 10^{-4} less energy is deposited by x rays for the same fluorescence signal.

Kirz et al.[37] have considered the relative merits of x rays versus electrons and protons in terms of radiation dose for microanalysis of thin biological samples. They calculated that electrons (> 10 keV) and protons (> 5 MeV) deposited the same amount of energy within a factor of three for the same fluorescent signal. For x-ray energies 10% above the absorption edge energies of the elements, they calculated that the energy deposited for each fluorescence event is 5×10^{-2} at $Z = 6$ to 5×10^{-4} at $Z = 20$ less than that deposited by charged particles. If background had been considered by these authors, their values would have shown that even less energy is deposited by monoenergetic x rays for the same MDL. Further discussion of radiation damage limits to biological samples for soft x rays is given in Chapter 8 by Kirz and Sayre.

If heat lost by conduction in a microparticle sample is neglected, the radiative heat loss is proportional to the fourth power of the temperature difference between the sample and its surroundings. Thus, the temperature rise would be ten times larger for a 10^4 times greater power dissipation in the sample. Shaw and Willis[36] calculated the electron beam heating of a 1-μm^3 sample in which the energy loss per charged particle is 2 keV. For 10^{-10} A (6.24×10^8 electrons s^{-1}) incident on the sample, they found the equilibrium temperature to be 1042 K, reaching 894 K in 10^{-2} s after exposure to the beam. This heating of the sample, which can be a problem in charged-particle microprobe analysis, will be markedly reduced by x-ray excitation.

2.6. Summary of the Properties of X Rays and Charged Particles for Fluorescence Excitation

Table 2 gives comparisons of the fluorescence excitation properties of x rays and charged particles as discussed above. The high fluorescence cross section of x rays is the major factor that accounts for an approximately 10^3-fold reduction in number and 10^4-fold reduction in energy deposited by x rays for the same MDL achieved with electrons or protons. Thus, an x-ray beam only 10^{-3} as intense as a charged particle beam would have a comparable MDL. We have yet to compare the number of photons and charged particles available per unit time. These numbers are necessary for our comparison of the MDL achievable for the same time and spatial resolution. This discussion will be presented after the section on the x-ray optics for concentrating the x rays from the synchrotron radiation source onto the sample.

Table 2. Comparison of X Rays to Electrons and Protons for Exciting Characteristic X-Ray Fluorescence for Elementary Analysis

Properties compared[a]	Range of values	Z ranges for elements	
Relative fluorescence cross sections			
$K\alpha$, x rays/electrons, σ_{xs}/σ_{es}	30–200	15–50	
$L\alpha$, x rays/electrons, σ_{xs}/σ_{es}	12–60	40–92	
$K\alpha$, x rays/protons, σ_{xs}/σ_{ps}	6–800	15–50	
$L\alpha$, x rays/protons, σ_{xs}/σ_{ps}	10–350	40–92	
Relative thick-target fluorescence yields			
$K\alpha$, x rays/electrons	5–150	15–50	
$L\alpha$, x rays/electrons	10–50	40–92	
$K\alpha$, x rays/protons	$1–10^3$	10–50	
$L\alpha$, x rays/protons	$3–10^3$	40–92	
Relative signal to background			
x rays/electrons	$10^2–10^3$	≥ 10	
x rays/protons	10^b	$\neq 30 \pm 5$	
Relative background cross sections			
x rays/electrons, σ_{xb}/σ_{eb}	≤ 1	≥ 10	
x rays/protons, σ_{xb}/σ_{pb}	10^c	$\neq 30 \pm 5$	
Relative number of projectiles required to produce same minimum detectable limit			
x rays/electrons, N_x/N_e	$7 \times 10^{-3}–3 \times 10^{-5}$	15–92	
x rays/protons, N_x/N_p	$1–2 \times 10^{-5}$	15–92	
Relative amount of energy deposited for same minimum detectable limit			
x rays/electrons	$2 \times 10^{-3}–10^{-5}$	15–92	
x rays/protons	$10^{-2}–10^{-7}$	15–92	
Accuracy of concentration determinations	X Rays	Electrons	Protons
Similar standards[d]	$\leq 1\%$	6% @ > 20 wt. %	5–10%
Pure element standards	5%	> 10%	> 10%

[a] X-ray energies are 1.007 to 3 times the absorption edge energy; electron energies are 20–100 keV, and proton energies are 2–5 MeV.
[b] Lower values in the $25 < Z < 35$ region.
[c] Higher values in the $25 < Z < 35$ region.
[d] Standards prepared to bracket sample composition.

2.7. Microprobe Spatial Resolutions; Charged-Particle Intensities

To be competitive with electron and proton microprobes in spatial resolution, x-ray beams of similar resolution will need to be produced. Even though electron beam diameters approaching 5 nm can be achieved, spreading of the electron beam by scattering processes in the sample produces about a 1-μm-diam source of x rays in thick samples. For example, a 20-keV electron beam of vanishingly small diameter will produce an interaction volume of ~ 2 μm diam in Al, ~ 1 μm diam in Cu, and ~ 0.4 μm diam in Au.[38] Electron beam spreading in samples that are thin compared to the interaction volume is typically about equal to the sample thickness.[38] Thus, a spatial resolution of about 1 μm for an x-ray microprobe would compete with that obtained in electron microprobes. The intensity available from the brightest cold-field emission sources exceeds 2×10^{12} and can approach 2×10^{15} electrons s^{-1} μm^{-2}.[39, 40] How much of this flux can be safely utilized without affecting the sample by overheating depends upon sample thickness, volatility of constituents, melting temperature, thermal conductivity, means of cooling, etc. In Section 2.5 we found that $\sim 10^9$ electrons s^{-1} depositing 2 keV per electron in a 1-μm-diam sample heats the sample to 1042 K if only radiative cooling is permitted.[36]

Most proton microprobes have beam diameters in excess of 30 μm, but Cookson *et al.*[41] have shown that beam diameters approaching 2 μm are achievable which contain 0.6 nA (3.7×10^9 proton s^{-1}) of 3-MeV protons. The spatial resolution achievable with proton beams is limited to about 1 μm because of the difficulty associated with efficient focusing of heavy charged particles[41] and multiple scattering of the proton beam in the samples.[42] Though several hundred nA of protons can be focused into areas of about 1 cm^2, proton beam diameters approaching 1 μm result in 10^6 less particles per square micrometer than obtainable with electron sources. Willis *et al.*[43] found that selective vaporization of the organic matrix of freeze-dried liver pellets took place with a beam current as low as 40 nA (2.5×10^{11} protons s^{-1}) in a 0.4-cm^2 area.

Inherently, x-ray excitation offers the highest spatial resolution of any fluorescence microprobe because the low scattering cross section limits the small lateral spreading of the x-ray beam. The cross section of x rays for direct photoionization exceeds by orders of magnitude the contribution from the secondary processes arising from photoejected electrons. Though these secondary processes enlarge the interaction volume, their contribution to the total fluorescence is small. Elastic scattering of x rays is strongly peaked in the forward direction ($\theta < 3°$) except where Bragg diffraction occurs.[44] Compton scattering is most intense in the back scattering direction.[44] Thus, the x-ray beam spreads but little in lateral directions since elastic and Compton scattering cross sections are only about 3 to 5% of the photoionizations cross sections at the energies considered here.[45] X-ray beams approaching 1 μm in diameter will have a spatial resolution comparable to those obtained with charged-particle excited microanalysis in thick samples. In samples with thicknesses below 1 μm, x-ray beams below 1 μm diam will be necessary to match the spatial resolution of electron microprobes. The instrumentation required for x-ray probes of this diameter and their intensities will be considered in Section 4.

3. Results of X-Ray Fluorescence Measurements with Synchrotron Radiation

Little use has been made of synchrotron radiation for fluorescence analysis because of its limited availability. A paper by Horowitz and Howell[46] in 1972 described the first microprobe application of synchrotron radiation. Radiation focused by an ellipsoidal mirror from the now dismantled Cambridge Electron Accelerator was used for a scanning x-ray microscope with a 2-μm-diam beam in the 0.5–3.5-keV region (see Chapter 8). Exploratory experiments by Sparks and Hastings[47] at SSRL in 1975 assessed the application of the continuum spectrum of synchrotron radiation for fluorescence excitation. In 1977, Sparks et al.[48] reported the use of monochromatic synchrotron radiation at SSRL for the fluorescence analysis of small (~80-μm-diam) particles. This and subsequent work[49, 50] by the author and collaborators (see Acknowledgments) furnish most of the material for this section. A report from Novosibirsk, USSR,[51] gave a limited survey of fluorescent spectra with MDL's (obtained in 1000 s) of 1.6×10^{-8} and 7.2×10^{-8} g g^{-1} for Zn and Au, respectively, in aqueous solutions. Fluorescence was excited with ~10^{11} photons s^{-1} cm^{-2} diffracted from a flat graphite monochromator at an energy resolution of $\Delta E/E \simeq 10^{-2}$ and detected with a solid state detector. Though this experiment was not optimized for intensity, the MDL is about a factor of 30 lower than obtained with the usual unpolarized and filtered x-ray sources.[30]

There has been no sustained effort to instrument and utilize synchrotron radiation for fluorescence analysis. The data presented in this section will be used to provide design information and to predict the results expected from a nearly optimized experimental facility (described in Section 4).

3.1. Fluorescence Excitation with the Continuum Radiation

A simple fluorescence excitation system would utilize the continuum radiation rather than a monochromatic beam if lower S/B and increased energy deposition could be tolerated. Continuum radiation scattered by the sample would be an appreciable part of the radiation reaching the detector (Figure 5c) and would contribute a large background beneath the fluorescence signals. The obvious advantages of the continuum spectra are (a) capacity for multielement analysis and (b) simplicity of design. Horowitz and Howell[46] used the continuum spectrum focused with a doubly curved mirror set to reflect energies from 1 keV (Be window cutoff) up to 3.5 keV followed by collimation of the beam with 2-μm-diam holes in Au foils. Samples of Al, S on Si, and a Cu electron microscope support grid were scanned in an x-y pattern and the scattered or fluorescent radiation detected with large-area proportional counters. The raster motion covered a 200 μm × 500 μm area of the sample in one to two minutes giving a two-dimensional picture of objects with about a 2-μm resolution. It was proposed that the large depth of field, roughly 1000 μm, possible with synchrotron radiation would permit micrographs of an object for stereoscopic viewing. Horowitz and Howell[46] did not use a solid state detector to separate the energy of the fluorescent radiation from the large amount of scattered radiation. Thus, their work provided only qualitative observations of detection limits and spatial resolution of the system. Our results[47] reported here make use of a solid state detector with 1–5% energy resolution, giving more quantitative information with which to determine the S/B ratios and detection limits possible with the continuum synchrotron spectrum. No focusing of the beam was attempted.

An enhancement of the fluorescence signal over the scattered background can be obtained by making use of the fact that synchrotron radiation is nearly linearly

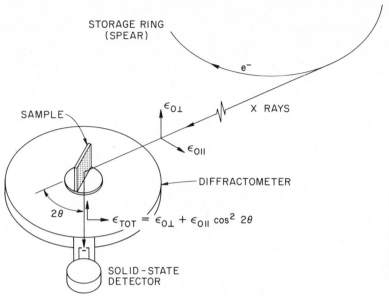

STORAGE RING
(SPEAR)

e^-

$\epsilon_{0\perp}$ X RAYS

SAMPLE

$\epsilon_{0\parallel}$

DIFFRACTOMETER

2θ

$\epsilon_{TOT} = \epsilon_{0\perp} + \epsilon_{0\parallel} \cos^2 2\theta$

SOLID-STATE
DETECTOR

Figure 6. Schematic of the experimental arrangement used at SSRL for continuum excited fluorescence measurements. The scattered radiation is minimized at $2\theta = 90°$ since the radiation is linearly polarized. (After Sparks and Hastings.[47])

polarized with the electric vector in the plane of the orbit (see Chapter 2). No radiation is Compton or elastically scattered into the detector at 90° to the incident radiation for the electric vector that lies in the plane of the scattering if we neglect higher-order scattering processes. Thus, radiation scattered from the sample is minimized by placing the detector in the orbital plane at 90° to the incident beam direction.

The experimental arrangement shown in Figure 6 was installed at SSRL[47] in the direct radiation 14 m downstream from a SPEAR bending magnet so that the detector could intercept radiation from the sample at $2\theta = 90°$. At this angle the horizontal electric vector, parallel to the plane of the storage ring, ε_{011}, lies along the detector axis. Two slits with 3 mm × 3 mm openings separated by 0.4 m and made from 6-mm-thick high-purity Al defined the central portion of the incident beam upstream from the sample. A similar 3 mm × 3 mm slit located 11 mm beyond the sample defined the radiation incident into the Si(Li) detector (resolution of 180 eV FWHM at 5.9 keV). The 2θ scanning range of the diffractometer was 55 to 140°.

Measurements of the intensity scattered from polystyrene (C_8H_8) were made to determine the energy distribution in the direct beam and to obtain an approximate value of the incident beam polarization. Polystyrene has a scattering cross section that is almost constant[52] for $(\sin \theta)/\lambda \geq 0.2$ (λ is the wavelength of the incident radiation in Å) and produces an almost undistorted energy spectrum of the incident radiation at 2θ scattering angles above 60° and energies above 5 keV. Figure 7 shows the measured synchrotron radiation spectrum scattered by polystyrene at $2\theta = 55°$ and 90°. The peak intensity occurs at an energy above the expected energy of 1 keV (for 1.85 GeV) because of absorption by the C heat shields and Be

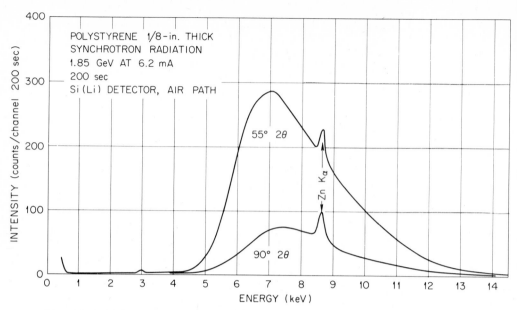

Figure 7. Reduction in scattered intensity from polystyrene at $2\theta = 90°$ compared to 55° because of the linear polarization of the synchrotron radiation (after Sparks and Hastings[47]).

windows upstream. The measured intensity scattered at $2\theta = 90°$ is from the perpendicular component of polarization and double-scattering processes.[53] For this geometry, the former is insensitive to the scattering angle and the latter has a small dependence on the scattering angle. This intensity scattered at $2\theta = 90°$ was taken to be a constant background from $2\theta = 55°$ to 90° and was subtracted from the scattering intensity measured at various angles. The resulting integrated intensities are shown to fit a $\cos^2 2\theta$ dependence in Figure 8. If all the scattering at $2\theta = 90°$ came from the perpendicular component of polarization and none from double-scattering processes, this component would account for $4.7 \pm 0.3\%$ of the incident radiation passing through a 3-mm-high slit 14 m from the source. When scattering at $2\theta = 90°$ is compared with unpolarized radiation from a conventional x-ray source (50% of the intensity in each of the two polarization components), synchrotron radiation reaching the detector is reduced by a factor of 11, i.e., 50/4.7, for the same incident intensity on the sample. For x rays generated at ten times the critical energy, and therefore 97% linearly polarized,[54] a maximum reduction of 16.7 in the amount of scattered radiation reaching the detector is achieved if all the vertical emittance is intercepted by the sample.

The fluorescent and scattered radiation spectra recorded at $2\theta = 90°$ from two NBS standard reference materials (SRM) are shown in Figures 9 and 10. A He beam path was used for the 1.85-GeV runs to enhance the detection of low-Z elements, and a 0.4-m air path upstream from the sample and a 0.15-m air path between the sample and detector were used for the 3.1-GeV runs. A comparison of Figure 9a with 9b and Figure 10a with 10b shows the effect of increasing the electron energies from 1.85 to 3.1 GeV on extending both the fluorescence spectra and background to higher x-ray energies. The numbers on the fluorescence peaks in units of 10^{-6} gg^{-1} (ppm) are either NBS certified concentrations or are taken from Ondov *et al.*[55] The integrated fluorescence peak to background is observed to be about one for concentrations of 5×10^{-4} gg^{-1} with a solid state detector having a resolution of ~ 180 eV (FWHM) at an energy of 5.9 keV. Detection with a crystal spectrometer having a resolution of the natural energy width of the 1.61-eV (FWHM) Fe K_{α_1} line[56] would give an integrated peak to background of one at concentrations of 5×10^{-6} gg^{-1}. The fluorescence signal to background of the continuum synchroton radiation excited fluorescence spectrum is 10^2 times larger than for electron excitation,[19, 57] a factor of ten larger than for excitation with raw radiation from conventional x-ray sources,[4] and 10^{-1} to 10^{-2} times as large as for proton excitation.[11, 30, 31] A direct comparison of the fluorescence spectrum from NBS orchard leaves excited with the continuum synchrotron radiation, Figures 9a and 9b, can be made to that excited with 3-MeV protons, Figure 11.[43, 58, 59] Though a polyethylene absorber was used to attenuate the low-energy x-ray signal for the proton excited spectrum of Figure 11, this does not affect the S/B. The proton excited spectra has a S/B that is larger by 10 at Mn $K\alpha$ to 10^2 at Sr $K\alpha$ than obtained with the continuum synchrotron radiation, Figure 10b. However, the synchrotron continuum produces a larger S/B than 3-MeV protons for Ca $K\alpha$ fluorescence (3.69 keV). This results from the large secondary bremsstrahlung background in the proton excited spectrum at these energies, whereas the large

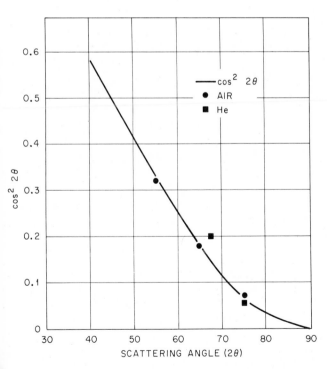

Figure 8. The decrease in scattered intensity follows a $\cos^2 2\theta$ dependence, verifying the linear polarization of synchrotron radiation (after Sparks and Hastings[47]).

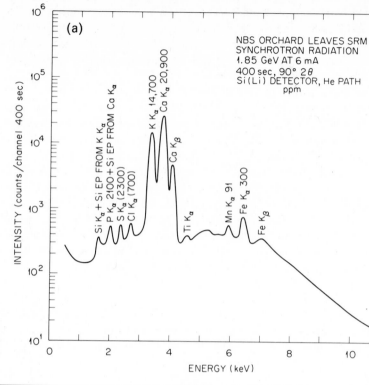

Figure 9. Continuum synchrotron radiation excited fluorescence spectrum from NBS orchard leaves with the elemental concentrations listed in units of 10^{-6} gg^{-1}: (a) 1.85-GeV electron energies and (b) 3.1-GeV energies in SPEAR. Concentrations in parentheses are not certified. (After Sparks and Hastings.[47])

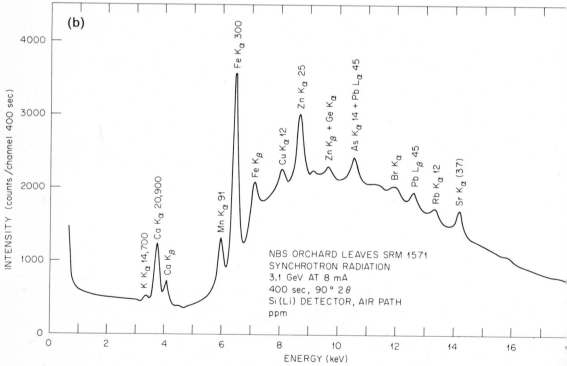

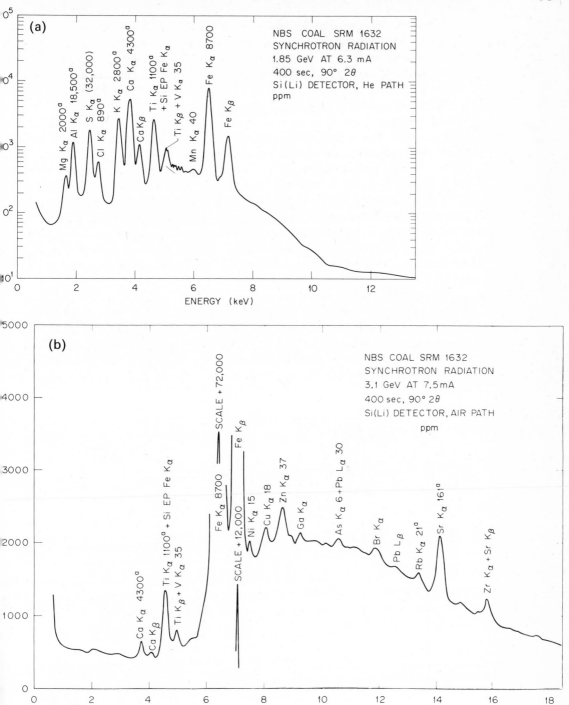

Figure 10. Continuum synchrotron radiation excited fluorescence spectrum from NBS coal with the elemental concentrations listed in units of 10^{-6} gg^{-1}: (a) 1.85-GeV electron energies and (b) 3.1-GeV energies in SPEAR. (After Sparks and Hastings.[47])

absorption in the beam line for synchrotron radiation < 6 keV reduces the amount of this energy scattered into the background. This finding suggests that filters in the incident continuum radiation could be effectively used to reduce the intensity in the lower-energy region of the spectrum that is scattered by the sample and contributes to the background beneath the fluorescence signals of interest. We will explore the use of filters later.

Total count rates of $(2-5) \times 10^3 \text{ s}^{-1}$ were observed, though only 0.13 of the total emittence in one milliradian impinged on the sample and only 6×10^{-5} of the total solid angle of the fluorescent radiation was collected. Count rates greater than 10^6 s^{-1} (which exceed the $5 \times 10^4 \text{ s}^{-1}$ count rate capabilities of single-diode solid state detectors) would be achieved by increasing the solid angle of the detector by 10^2 and intercepting all the vertical emission in 2 mrad of the horizontal synchrotron beam. Multidiode solid state detectors or crystals diffracting a single fluorescence energy into proportional or scintillation detectors would be required to effectively utilize the fluorescence intensity excited by the available synchrotron radiation. For example, see the description of the barrel monochromator in Chapter 13.

A parameter useful to our evaluation of the continuum radiation spectrum for fluorescence excitation is the photoionization cross section as a function of the energy bandwidth above the absorption edge. From the diagram of the absorption edges shown in Figure 12, we find that their slopes are very similar and, therefore, the fractional decrease in cross section above the K edges is similar for an energy that is the same multiple of the edge energy, i.e., the fractional decrease in the

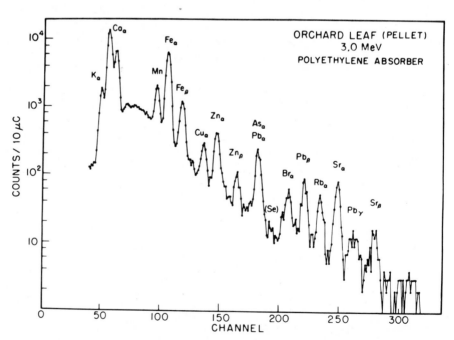

Figure 11. 3-MeV proton excited fluorescence spectrum from NBS orchard leaves (after Willis *et al.*[59]).

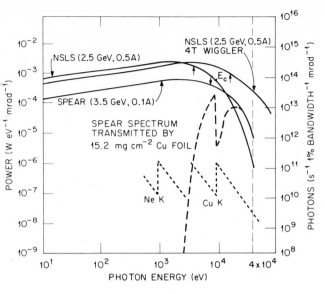

Figure 12. Continuum radiation spectrum from SPEAR and the NSLS in units of watts (left-hand ordinate) or number of photons (right-hand ordinate) and the filtered spectrum transmitted by a 15.2-mg cm^{-2} Cu filter. The K absorption edges of Ne and Cu are schematically represented. Arrows denote the critical energy E_c above which one-half of the power in the spectrum occurs.

photoionization cross section is nearly the same at 20 keV for an absorption edge at 10 keV as at 2 keV for an edge at 1 keV. Thus, wider energy bands of radiation are more effective for ionization of higher-energy absorption edges.

Filters placed in the incident beam will transmit large energy bands from the continuum radiation and reduce the amount of lower-energy radiation that can be scattered as background beneath the fluorescence signals. A spectrum transmitted by a filter inserted in the incident synchrotron beam is shown by the dashed curve in Figure 12. This curve is the calculated spectrum from SPEAR passed by a 15.2-mg cm^{-2} Cu foil that transmits 1% of the SPEAR radiation at the absorption edge maximum. Intensities at energies higher than the absorption edge of Cu could be reduced by adjusting the reflection angle of the mirror to exclude that energy above the minimum at 9 keV.

The efficiency of large energy bandwidths for exciting fluorescence compared to more monochromatic x-ray beams with energies just above the absorption edge can be calculated from the mean value of the partial photoionization cross section. An approximate empirical expression for the partial photoionization cross section is

$$\sigma_i = AE^{-k} \tag{16}$$

where k has a value of about three.[60] At the absorption edge, $\sigma_e = AE_e^{-3}$ and the mean value of σ_i becomes

$$\bar{\sigma}_i = \sigma_e \int_{E_e}^{E_{\max}} \frac{E^{-3}}{E_e^{-3}} \, dE \bigg/ \int_{E_e}^{E_{\max}} dE = \frac{\sigma_e E_e^3}{2 \, \Delta E} \left(\frac{1}{E_e^2} - \frac{1}{E_{\max}^2} \right) \tag{17}$$

where $\Delta E = E_{\max} - E_e$ and the subscript e denotes the absorption edge. The ratio $\bar{\sigma}_i/\sigma_e$ is plotted in Figure 13 (curve A) for various values of $\Delta E/E_e$. For example, a $\Delta E/E_e = 0.01$ (where $E_{\max}$ begins at the absorption edge) is 98.5% as efficient for

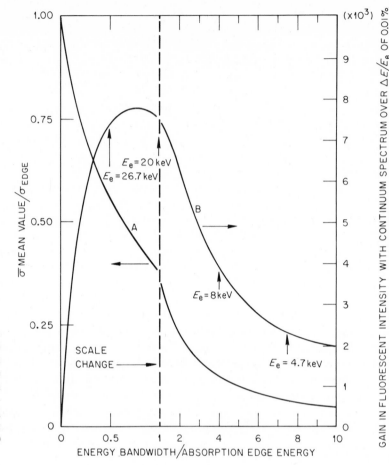

Figure 13. Effectiveness of various energy bandwidths in exciting fluorescence; (A) decrease in absorption cross sections as a function of the energy width above the absorption edge energy and (B) gain in fluorescent intensity over a $\Delta E/E$ of 0.01 % for the SPEAR continuum spectrum at 3.5 GeV, which does not exceed 40 keV.

photoionization as the same number of photons with energies at the absorption edge. $\Delta E/E_e = 1$ is 37.5% as efficient as photons with an energy at the absorption edge and contains 10^4 times as many photons as $\Delta E/E_e = 0.01\%$ (assuming a flat distribution of intensity as a function of energy). Thus, the fluorescence signal could be $37.5\% \times 10^4$ or 38×10^2 times larger than that from a beam where $\Delta E/E_e = 0.01\%$. With the assumption that the intensity of the radiation is constant as a function of $\Delta E/E$ to 40 keV and then decreases to zero, we can calculate the increase in fluorescence intensity obtainable with the continuum spectrum over that from a beam with $\Delta E/E_e = 0.01\%$. This increase in fluorescence signal as a function of $\Delta E/E_e$ is plotted in Figure 13 (curve B). The gain in fluorescence signal obtainable with the continuum synchrotron spectrum over that from a beam with a 0.01% bandwidth reaches a maximum of about 7700 times greater near $\Delta E/E_e = 1$ and $E_e \cong 25$ keV. Little additional intensity is gained by using $\Delta E/E_e > 1$. For the same MDL [see equations (12)–(14)] a 100-fold increase in fluorescence signal permits a 100^2-fold increase in background or a factor of 100 increases in S/B.

In comparison to continuum excitation, the filtered radiation will produce a signal to background about 10 times larger, a fluorescence signal lower by a factor of about 2 to 7, about a factor of 2 smaller MDL, and about a 10- to 40-fold reduction in energy deposited. The synchrotron continuum radiation can be used to produce a more monoenergetic x-ray source by exciting elemental targets. These secondary targets are chosen to have chararacteristic energies just above the absorption edge of the principle elements to be fluoresced. The additional complexity of focusing the secondary radiation to microprobe size after having focused the continuum spectrum may not be worth the anticipated reduction of about 2 in the MDL over that achievable with direct monochromatization of the continuum radiation.

We shall show in Section 3.2 that the S/B is improved by a factor that is $\geq 10^3$ for $Z \leq 40$ when monoenergetic rather than continuum x rays are used for excitation. Thus, a $\sim 10^3$-fold increase in signal with continuum radiation would have to offset a 10^3-fold decrease in S/B to maintain the same MDL. However, the amount of energy deposited in the sample would increase by more than a factor of 10^3 (increase in incident radiation divided by the efficiency for photoionization). We shall find in Section 3.2 that excitation with continuum synchrotron radiation becomes more competitive with monochromatic excitation for lower-Z elements and when simultaneous multielement analysis is important.

Discussions in the next section and in Section 4 on the x-ray optics needed to achieve microprobe beams from synchrotron radiation sources will permit more quantitative evaluation of the detectable limits with different excitation sources for fixed exposure times.

3.2. Fluorescence Excitation with Monochromatic Radiation

The motivation to instrument a fluorescence experiment with a tunable monochromatic beam of synchrotron radiation as an excitation source was provided by reported evidence for the existence of primordial superheavy elements[61] (SHE) between $Z = 114$ and 127. Weak spectra excited by 4.7- and 5.7-MeV protons were interpreted as possibly containing L_α lines for SHE, although only one spectral line was observed for each reported element. In addition, 5th period K lines and proton-induced nuclear reactions[62] are possible sources of radiation that hindered a unique interpretation of the reported spectra. Thus, additional experimental evidence was necessary for confirmation. A tunable monochromatic x-ray excitation source can eliminate the uncertainty in element identification since the energy can be varied to excite the absorption edge of interest, allowing a unique correlation of fluorescence energies with their absorption edge energies.[48] The absorption edges of interest were between 28 and 40 keV. Particular interest in $Z = 126$ with an L_3 edge near 36 keV[63] dictated maximizing the intensity for 37-keV x rays. As the samples were small (30 to 125 μm diam), it was useful to focus the radiation. A curved mosaic graphite crystal was chosen to both focus and monochromatize the radiation. Fluorescence spectra of widely used standard samples were observed

with this system to compare with conventional x-ray and charged-particle fluorescence excitation.[50] With this information, a better assessment can be made of the
advantages of monochromatic synchrotron radiation for fluorescence and diffraction microprobe analysis.

3.2.1. Experimental Arrangement

A schematic of the experimental arrangement for the energy dispersive XRF
spectrometer[48–50] installed at SSRL is shown in Figure 14. Two milliradians (horizontal angle ψ) of radiation impinged on a monochromator made of hot-pressed
pyrolytic graphite curved to a radius of 10 cm and located 17 m from the source.
This produced a focus of 37-keV x rays at the sample located 1 m from the
monochromator. Demagnification of the 3.2-mm (FWHM) horizontal source width
by $\frac{1}{17}$ and broadening by the 0.4° (FWHM) mosaic spread of the graphite monochromator produced a focus ~ 0.8 mm wide at the sample. Three Ta slits before the
sample limited the beam to a cross section 1 mm wide by 0.45 mm high.
The arrangement of the Ta slits to define the incident beam and a vertical scan of
the intensity distribution are shown in Figure 15. Fluorescence intensity from a
long horizontal 25.4-μm-diam W wire is plotted in Figure 15 as a function of its
position in the beam. No attempt was made to align the three slits other than by
visual inspection. The constant intensity distribution over ~ 0.1 mm encourages
the belief that uniform intensity distributions can be obtained for the accurate
quantitative analysis of samples smaller than the beam diameter.

For reliable trace element analysis, extraneous fluorescent radiation must be
minimized. All slits were made from 0.5–1-mm-thick Ta covered with 0.25–0.5-mm-
thick Al (10^{-6} gg^{-1} purity) to reduce Ta L fluorescence. The Cu specimen chamber
walls were covered with 0.25-mm-thick Ta followed by 0.25-mm-thick Al. All other
surfaces that could contribute fluorescence into the beam path were similarly
covered with Ta and/or Cu foil followed by the high purity Al to ensure that K
lines of fifth-period elements would not interfere with possible L lines of SHE.

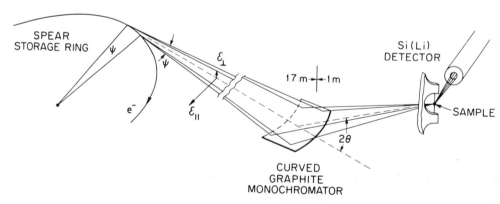

Figure 14. Schematic of the single-curved mosaic crystal installed at SSRL for fluorescence excitation
with monochromatic x rays (after Sparks et al.[48, 49]).

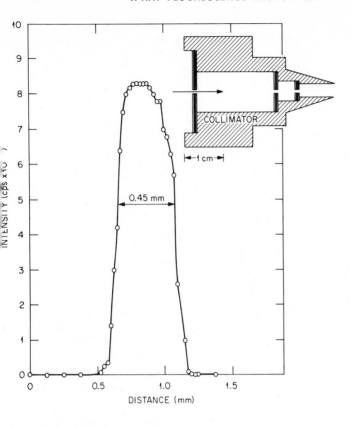

Figure 15. A three-slit collimator used to define a 0.45 mm × 1 mm beam is shown with a vertical scan of the intensity distribution obtained through this collimator.

Slits placed just upstream of the monochromator selected 2 mrad of horizontal beam spread and 1 mm of the approximately 8-mm vertical spread (see Webb *et al.*[64] and Chapter 2 for a discussion of the effects of betatron oscillations on emission angles). The vertical beam height at the monochromator was limited to 1 mm to prevent the energy spread on the sample from overlapping absorption edges of interest. A range of x-ray energies could be chosen by changing the scattering angle of the monochromator. The useful range was confined to between 20 and 45 keV by the permissible vertical travel of the sample chamber. The Bragg scattering angle θ is 2.86° for 37-keV x rays from the (00·2) reflection of graphite, which has a $2d$ spacing of 0.671 nm (6.71 Å). A 0.45-mm vertical slit located 1 m beyond the crystal selected an energy band of 0.4 keV from an energy of 37 keV. Intensities incident on the samples were determined from measurements of the fluorescence from known reference standards which included NBS standards, thin pure metal foils, Cs atoms implanted in Al foils, and Cd and U atoms in equal amounts loaded into Dow cation-exchange resin beads of ~75-μm diam. The number of Cs and U atoms were determined by neutron-activation analysis.

Radiation from the sample was intercepted by a Si(Li) detector with a 5-mm-thick diode internally collimated to 9.8 mm in diam and with a 0.25-mm-thick Be window. The detector was placed at 90° with respect to the incident beam and parallel to the plane of the storage ring to reduce the amount of scattered radiation reaching the detector. Detector resolution was sacrificed (240 eV at 6.4 keV and 450 eV at 39.2 keV) to achieve counting rates in excess of 4×10^4 cps. At the closest distance of the Si diode to the sample (3.0 cm), 6.7×10^{-3} of the total solid angle was intercepted. A He atmosphere was maintained throughout the system. The value of the incident intensity could be found from equation (4) with the following information: the solid angle of radiation intercepted by the detector, the tabulated values of the fluorescence cross sections and mass absorption coefficients, detector efficiency, and the measured fluorescence intensities from the standards. For the geometry of this experiment, the observed intensity was $(3.6 \pm 0.9) \times 10^{10}$ photons s^{-1} (0.45 mm^2)$^{-1}$ at 37 keV ($\Delta E = 0.4$ keV) when SPEAR operated at 3.5 GeV and 30 mA in the colliding beam mode. In the single-beam mode with the same electron energy and stored current, the lower emittance increased the intensity at the sample by 2.3 ± 0.4. At 3.7 GeV and 40 mA in the single-beam mode, an intensity of $(1.9 \pm 0.5) \times 10^{11}$ photons s^{-1} (0.45 mm^2)$^{-1}$ was observed at the sample. Typical SPEAR running conditions were 3.4 GeV and 20 mA for most of the fluorescence spectra shown here.

3.2.2. Fluorescence Spectra and Detectable Limits

A fluorescence spectrum from one of the standards used to determine the intensity in the incident beam is shown in Figure 16. Each data point is plotted as a vertical line with a length equal to twice the square root of the number of counts. The ~ 75-μm-diam Dow cation-exchange resin bead contained $(4.6 \pm 1.7) \times 10^9$ Cd atoms, which is a concentration of 2.2×10^{-6} gg^{-1} in the bead. Fluorescence spectra from leached resin beads with no Cd gave a similarly intense Sn $K\alpha$ fluorescence but no detectable Cd $K\alpha$ fluorescence. If this number of Cd atoms were uniformly distributed over the 0.45-mm^2 incident beam area, they would correspond to 8×10^{-4} of a monolayer of bulk Cd. The MDL for Cd atoms in this spectra is calculated from equation (12) to be 2.7×10^8 atoms for a counting period of 2200 s. The advantages of monochromatic excitation for improving S/B is evident in Figure 16 since the intensity of the elastic and Compton scattered radiation would become the background beneath the fluorescence peaks if continuum radiation were used as the excitation source.

A direct comparison between fluorescence spectra excited with synchrotron radiation[49] and 5.7-MeV protons[61] is shown in Figure 17. The spectrum labeled ORNL was excited with 37-keV synchrotron radiation and plotted every 20 eV. The FSU data excited with 5.7-MeV protons is plotted for a channel width of 61.8 eV and, therefore, appears three times more intense in the comparison in Figure 17 than is actually the case. An analysis of the data shows that a conservative MDL of 4.3×10^8 atoms of element $Z = 126$, if present, could have been detected after eight hours of excitation with the synchrotron radiation. This is 55 ± 10 times fewer atoms than could have been detected in the proton-beam induced fluorescence spectrum (if the $Z = 126$ atoms were uniformly distributed in the inclusion).[49] The S/B of the spectra in Figure 17 was 5 to 50 times larger for x-ray than for proton excitation in the spectral region of 8 to 28 keV.

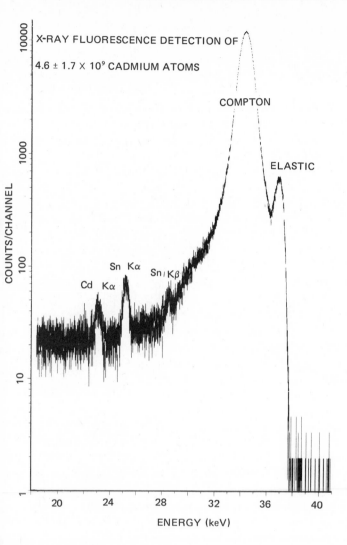

Figure 16. Fluorescence spectrum excited with 37-keV synchrotron radiation from a 75-μm-diam Dow cation-exchange resin bead loaded with 5×10^9 Cd atoms.

Choice of the x-ray excitation energy below 38.5 keV avoided exciting the K edges of the more abundant elements in the inclusions (La and Ce at 38.93 and 40.45 keV, respectively), which would produce intense K lines and larger background in the region between 22 and 32 keV. With excitation energies below the La and Ce K edges, inelastic resonance scattering[65, 66] from these edges dominated the background between 24 and 32 keV. Targets of Ce and La irradiated with 37-keV x rays revealed that the more abundant Ce in inclusion 19-D contributed most of the inelastic resonance scattering to the background. The other major contribution to the background is the bremsstrahlung generated by the slowing down of photoejected electrons. As the inelastically scattered radiation[65] and bremsstrahlung[67] are only weakly polarized, they cannot be separated from the

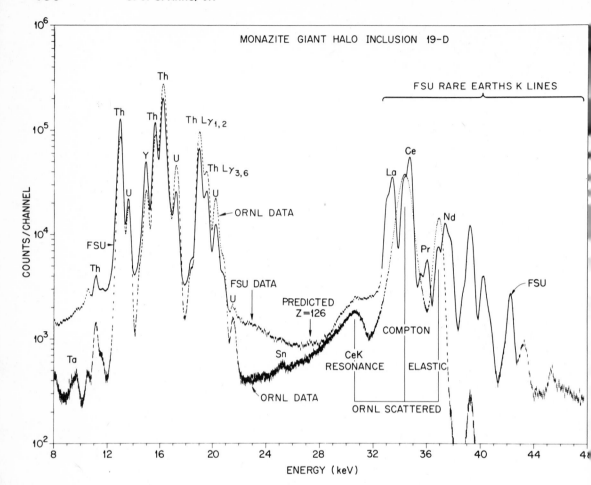

Figure 17. Fluorescence spectrum from a monazite particle excited with 37-keV synchrotron radiation (ORNL DATA) shows an improved signal-to-background ratio over that excited with 5.7-MeV protons (FSU DATA). Data points every 20 eV in ORNL spectrum and every 61.8-eV in FSU spectrum. (After Sparks et al.[49])

fluorescent radiation. Thus, this radiation limits the background achievable beneath monoenergetic x-ray-excited fluorescence spectra.

Fluorescence spectra were also obtained from NBS standards with low-Z matrices. These spectra are shown in Figures 18–20 for orchard leaves (SRM-1571), bovine liver (SRM-1577), and coal (SRM-1632), respectively.[50] Some of the elemental concentrations in units of 10^{-6} gg^{-1} are given for the fluorescence lines. Sensitivity increases with increasing Z, and Cd $K\alpha$ lines for concentrations of $(0.1–0.3) \times 10^{-6}$ gg^{-1} are observable in all three spectra. Most of the count rate, $(17–25) \times 10^3$ cps for these three spectra, was contributed by scattered radiation rather than from fluorescence. (Note in Figure 20 that because of the pulsed nature of the synchrotron radiation the pulse pile-up spectra consists of peaks easily

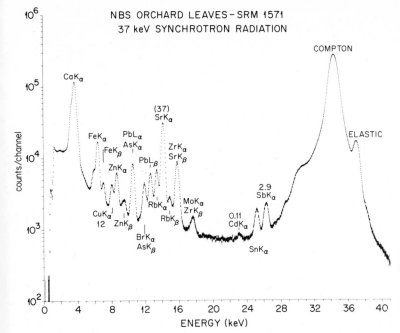

Figure 18. Fluorescence spectrum from NBS orchard leaves excited with 37-keV x rays. Numbers of fluorescent lines gives the concentration of that element in units of 10^{-6} gg^{-1}. (After Sparks *et al.*[50])

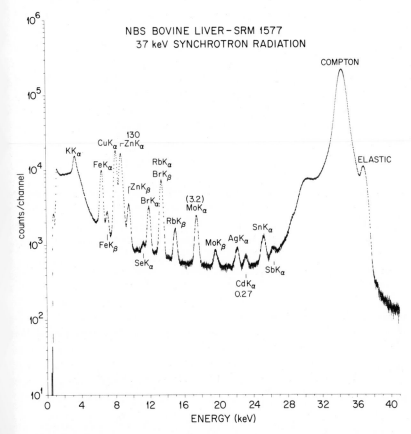

Figure 19. Fluorescence spectrum from NBS bovine liver excited with 37-keV x rays. Numbers on fluorescent lines give the concentration of that element in units of 10^{-6} gg^{-1}. (After Sparks *et al.*[50])

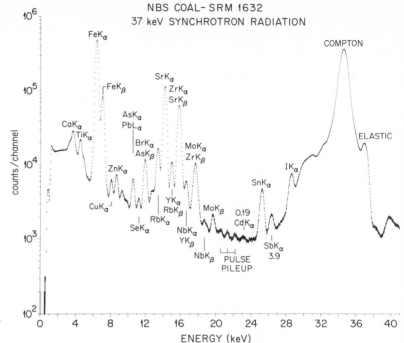

Figure 20. Fluorescence spectrum from NBS coal excited with 37-keV x rays. Numbers on fluorescent lines gives the concentration of that element in units of 10^{-6} gg^{-1}. (After Sparks *et al.*[50])

confused with fluorescence lines and unlike the broad bumps with low-energy tails obtained with sources emitting randomly spaced pulses.)

Fluorescence spectra of two of the same NBS standards excited with negligibly polarized and monochromatic Mo $K\alpha$ radiation (17.44 keV) from conventional x-ray sources[29] are presented in Figures 21 and 22. Comparisons of the results from Figures 18 and 20 with Figures 21 and 22 show that the S/B are slightly larger for the synchrotron-excited spectra. The 37-keV x rays have a fluorescence cross section ten times less than the 17.44-keV x rays for the fluorescence lines common to both spectra. Thus, for the same exciting energies the spectra with synchrotron radiation would give a factor of ten larger S/B. The linear polarization of synchrotron radiation results in a tenfold reduction in scattered (relative to fluoscent) radiation reaching the detector and accounts for this improvement in S/B. The Compton and elastically scattered radiation reaching the detector is degraded in the diode by insufficient charge collection to become the major component of the background, especially in those samples with low-Z matrices. Similar S/B's are observed for orchard leaves excited with 37-keV synchrotron radiation (Figure 18) as with protons (Figure 11). For the same reasons given above, 17-keV synchrotron radiation would produce a factor of ten improvement in S/B and, therefore, result in a factor of ten larger S/B than for proton excitation. In principle, S/B's even greater than ten could be achieved with x rays by eliminating the solid state detector contributions to the background. For proton excitation, the background is mostly from electron-generated bremsstrahlung in the sample and, therefore, not separable from the fluorescence signal.

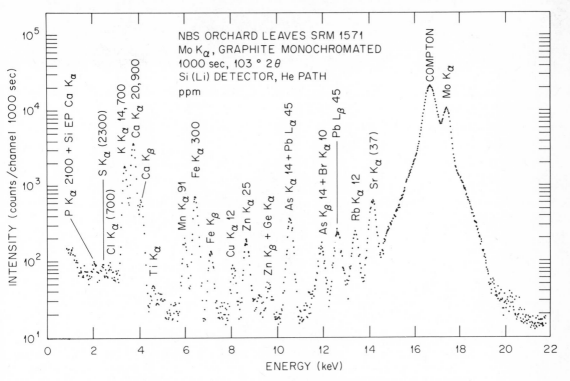

Figure 21. Fluorescence spectrum from NBS orchard leaves excited with unpolarized Mo $K\alpha$ (17.4 keV) x rays. Numbers on fluorescent lines give the concentration of that element in units of 10^{-6} gg^{-1}. (After Sparks *et al.*[29])

Comparisons of S/B's for spectra from orchard leaf and coal show larger values for those excited with 37-keV x rays (Figures 18 and 20) than excited with continuum radiation (Figures 9 and 10). These values range from one at Ca $K\alpha$ to 120 at Sr $K\alpha$. A choice of monochromatic excitation energies closer to the Sr K edge would give a fluorescence cross section larger by a factor of ten than 37-keV x rays and S/B's values 10 to 1200 times larger. As before, the detector contribution dominates the background from these low-Z matrices or S/B's even more favorable than 1200 would have been achieved. It is apparent that monochromatic beams result in much greater sensitivity for those elements with absorption edge energies sufficiently below the excitation energy that their fluorescence energies are also below the Compton scattering, e.g., Cd or Sb K fluorescence excited with 37-keV x rays.

In general, the smallest MDL will be achieved if the incident radiation scattered by the sample (and having the same energy as the fluorescence lines of interest) is reduced in intensity until its contribution to the background is similar to that contributed by intrinsic processes (bremsstrahlung from photoejected electrons and inelastic resonance scattering); if the energy bandwidth of the incident radiation is up to twice as large as the energy of the absorption edge; and if diffracting crystals are used instead of solid state detectors for energy analysis.

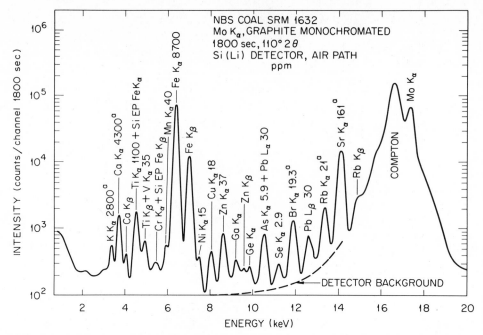

Figure 22. Fluorescence spectrum from NBS coal excited with unpolarized Mo $K\alpha$ (17.4 keV) x rays. Superscript *a* denotes uncertified concentration of that element in units of 10^{-6} gg^{-1}. (After Sparks *et al.*[(29)])

Table 3. Minimum Detectable Limits Obtained in 100 s for 37-keV Synchrotron Radiation and Proton-Excited Fluorescence[a]

| Element | Synchrotron[b] | | PIXE[c, d] | |
	C_D (ng cm^{-2})	W_D $A = 6.4 \times 10^{-3}$ cm^2 (ng)	$C_D{}^d$ (ng cm^{-2})	$W_D{}^e$ $A = 7.1 \times 10^{-6}$ cm^2 (ng)
V	1.6	0.010	8.4	0.0042
Fe	—	—	3.7	0.0018
Ni	0.36	0.0023	—	—
Se	—	—	1.7	0.0008
Sr	—	—	3.1	0.0015
Nb	0.10	0.0006	—	—
Cd	0.05	0.0003	21	0.010

[a] Millipore filter was used as substrate background for all methods. $W_D = AC_D$, where A is the irradiated area of the sample.
[b] SPEAR operating at 3.7 GeV and 40 mA.
[c] Data on C_D compiled by Jaklevic and Walter[(68)] for proton-induced x-ray emission (PIXE).
[d] Proton accelerator at Duke University operating at 3 MeV and 70 nA.
[e] Data of column 3 normalized to a 30-μm-diam proton beam of 3 MeV and 1 nA.

3.2.3. Minimum Detection Limits for 37-keV Radiation

The minimum detectable limits derived from this experiment[50] are shown in Table 3. A listing of the MDL taken from the work of Jaklevic and Walters[68] for proton-excited fluorescence (PIXE) is given for comparison. The basis for comparison is Currie's[7] definition of the MDL given in equation (12) and applied in both cases under the favorable conditions of a single element backed with a Millipore filter as a low-Z substrate. Comparisons are made between the mass per unit area C_D, which emphasizes differences in power incident on the samples, and the mass W_D, which stresses the power per unit area or intensity. Synchrotron radiation of 37-keV x rays is not optimum for exciting low-Z elements, yet produces a lower C_D for all elements listed and a much lower W_D for Cd than proton excitation. From the beam current, energy, and time listed in Table 3, we can calculate the energies incident on the sample. The 37-keV x-ray beam requires 6.7×10^{-6} as much energy as the proton beam to achieve the same C_D for Cd and 6.1×10^{-3} as much energy to achieve the same W_D for Cd. Lower W_D values will be achieved with x-ray optics designed for greater demagnification of the source (see Section 4).

3.2.4. Intrinsic Background for X-Ray-Excited Fluorescence: Summary of Signal-to-Background Ratios

Filters of Mo and Cd were placed in the incident beam between the monochromator and sample and between the sample and detector to determine the source of background in the fluorescence spectra. No absorption edges were observed in the spectrum when the filters were placed between the monochromator and sample. This observation eliminated the possibility that lower energies of an insufficiently monochromated incident beam were being scattered by the sample as a source of background. When the Mo and Cd filters were placed between the sample and detector, increasingly prominent absorption edges were observed with increasing Z for pure element samples from Si through Pb (the highest Z element used). For Pb ($Z = 82$) more than 95% of the background was generated in the sample, and for Si ($Z = 14$) less than 70% came from the sample with the remainder being generated in the solid state detector. For targets with $Z \leq 8$, the solid state detector's contribution to the background was $\geq 90\%$ of the total.

In Figure 16, where the solid state detector contributes most of the background, the Cd $K\alpha$ count rate for a Cd concentration of $\sim 1 \times 10^{-6}$ gg^{-1} is 0.23 cps. The background rate for the natural width of the Cd $K\alpha_1$ line (~ 10 eV) is 0.005 cps. Even this S/B of 46 at a concentration of $\sim 10^{-6}$ gg^{-1} will be exceeded since diffracting crystals will also eliminate solid state detector background. A summary of the S/B's for x-ray- and charged-particle-excited fluorescence derived from our data[47-50] and many other sources[11, 18, 19, 21, 29-32, 53, 56-59, 61] is given in Figure 23. The background used was that in an energy bandwidth equal to the natural width of the $K\alpha_1$ or $L\alpha_1$ fluorescence line (FWHM). Efficient intensity collection with crystal analyzing optics at energy resolutions approaching 1 eV are possible because of the micron-sized source. There are fewer data on proton and monochromatic x-ray excitation of high-Z matrices, which makes it less certain than the other data. X-ray excitation gives factors of 10^3 to 10^5 improvement in

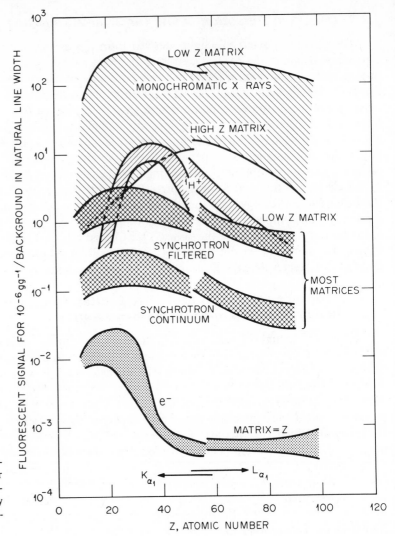

Figure 23. Comparison of the fluorescent signal-to-background ratio for various excitation radiations at a concentration of 10^{-6} gg^{-1} for an x-ray detection system with an energy resolution of the natural linewidth.

S/B's over electron excitation, and a factor of ten improvement over proton excitation for low-Z matrices of thin samples. Thus, x rays would produce high image contrast in a scanning microprobe. The use of a solid state detector to record the fluorescence would lower all the curves in Figure 23 by at least the energy resolution of the detector divided by the natural width of the fluorescence line. This would vary from about 30 for U $L\alpha_1$ to about 150 for Ca $K\alpha_1$. In theory a low S/B can be overcome by good statistics, but in practice the larger the S/B the greater the reduction in the problem associated with background characterization.[7]

Synchrotron radiation intensities obtainable with mirrors, crystals, and combinations will be explored in the next section. With this information we will calculate the MDL that could be achieved with a feasible synchrotron radiation microprobe and compare this with electron and proton microprobes.

4. Optics for an X-Ray Microprobe

In this section, several combinations of mirrors and crystal monochromators will be compared in terms of flux in a 1-μm-diam area. Our intent is not to investigate all possibilities but to illustrate a feasible optical system that demonstrates the advantages of synchrotron radiation excitation for a microprobe. These intensities will provide our final basis for comparing x-ray and charged-particle microprobes in terms of detectable limits and energy deposited. Schemes for detecting the fluorescent radiation will not differ greatly among the various excitation modes except that x rays do not require the gas-free environments with restricted volumes associated with charged-particle excitation. Present microprobe fluorescence detection methods could be adapted with greater flexibility to x-ray systems.

The optical considerations tor focusing x rays can be discussed in terms of rays.[69] Each ray from the synchrotron source is a vector that is assigned a fractional part of the area of the beam cross section with an energy spectrum similar to that of every other ray. (More detailed considerations of the source characteristics given in Chapter 2 do not appreciably affect the conclusions drawn here.) The design of the optics should be such that each ray from the source delivers the desired $\Delta E/E$ to as small an area on the target as possible. For a microprobe, it is advantageous to vary both the $\Delta E/E$ and the energy of the radiation. As fluorescence is emitted isotropically from the atom, the convergence angles at which the exciting radiation enters the target need be no less than those required to keep the beam divergence within the target from reducing the spatial resolution of the probe. Two optical elements will be considered: mirror surfaces from which the incident x-ray energy below the mirror cutoff can be redirected by total reflection, and crystal diffraction which selects only a limited $\Delta E/E$ from each ray for redirection by Bragg scattering.

Because x-ray energies from 2 to 40 keV are desirable to optimize the intensity of the emitted fluorescent radiation, the x-ray microprobe should provide both the incident energy range and the energy detection capability to analyze for all elements above $Z = 10$. To achieve the highest intensities, it is necessary to intercept as much of the available vertical and horizontal divergence of the synchrotron radiation as possible and to demagnify the source size. At the highest excitation energies (40 keV), the angle of incidence for total reflection from mirror surfaces has decreased to 2.1 mrad for a high-density Pt-coated mirror,[70] and to about 50 mrad for crystal diffraction from Si. For such small angles, mirrors of large dimensions are required to intercept most of the available synchrotron radiation. For practical considerations, we will limit the total mirror length to about 1 m, but this length may require more than one segment. Radii of curvature should be greater than about 3 cm because of limitations imposed in the polishing of mirrors.[71] The optical surface of the mirrors will be assumed to be of sufficient quality as to not introduce significant broadening errors in the focus.

The intensities achievable with these optics will be combined with the results on S/B and Figures 2 and 3 to produce our final comparisons between the minimum detectable limits and energy deposition for synchrotron and charged-particles fluorescence microprobes.

4.1. Mirrors for Focusing the Continuum Spectrum

Mirrors will reflect all photon energies incident at angles below the critical angle (θ_c) for total reflection. The angle for total reflection is proportional to the square root of the electron density of the mirror surface. For a Pt surface, the maximum energy E in keV that is totally reflected at θ_c is given by Franks[70] as

$$E^{-1} = 11.85 \sin \theta_c \qquad (18)$$

Pt has nearly the largest θ_c of any usable mirror surface. To reflect 40-keV x rays the value of θ_c is 2.1 mrad for Pt-surfaced and 2 mrad for Au-surfaced mirrors. Though the high-energy cutoff at θ_c increases with increasing electron density of the mirror surface, the sharpness of the cutoff decreases.[70]

An increase in the value of the scattering angle of incidence (θ) reduces the length of the mirrors required to intercept a given divergence. From the geometry shown in Figure 24, we can calculate the various parameters listed in Table 4, which characterize a mirror with a Pt-coated surface. The following geometrical relationships were used to derive these parameters assuming that $\sin \theta = \theta$[69, 70]:

$$\text{Magnification } M = F_2/F_1 \qquad (19a)$$

$$R_s \simeq (2F_1 F_2 \sin \theta)/(F_1 + F_2) \qquad (19b)$$

$$R_m \simeq R_s/\sin^2 \theta \qquad (19c)$$

$$L_m \geq (F_1 \phi_V/\theta) + (R_s/8\theta) = L_1 + L_2 \qquad (19d)$$

$$L_s \simeq F_1 \phi_H \qquad (19e)$$

A study of Table 4 reveals that as the reflected x-ray energy approaches 40 keV increasing longer mirrors (L_1) are required to intercept a vertical divergence (ϕ_V) of 0.2 mrad. Both the opening angle of the synchrotron radiation and the vertical angular acceptance of the mirror increases with decreasing photon energy. Thus, the mirror could be tilted in the vertical plane to accept the total opening angle of the radiation for lower photon energies. At 40 keV, a mirror 0.95 m long is required. If we limit the mirror to 1 m in length, practically no horizontal divergence could be focused, i.e., the mirror can only be singly curved. Therefore, a doubly curved mirror to reflect 40-keV x rays would be limited to a magnification

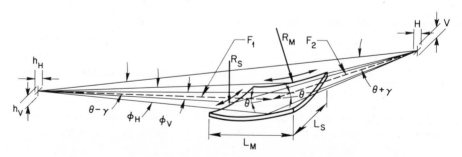

Figure 24. Doubly curved mirror and the parameters used to define the geometry of the mirror.

Table 4. Design Parameters and Divergences Intercepted for a Doubly Curved Platinum-Coated Mirror Located 10 m from the Synchrotron Radiation Source

M	E (keV)	θ (mrad)	R_s (mm)	ϕ_H (mrad)	R_m (m)	ϕ_V (mrad)	$L_s \leq R_s$ (mm)	L_m $L_1{}^a + L_2{}^b$ (m)
$\frac{1}{10}$	40	2.1	3.8	0.38	866	0.2	3.8	0.95 + 0.23
1	40	2.1	21	2.1	4762	0.2	21	0.95 + 1.25
1	40	2.1	21	0.9	4762	0.15	9	0.75† + 0.25
$\frac{1}{10}$	30	2.8	5.1	0.51	649	0.2	5.1	0.71 + 0.23
$\frac{1}{10}$	20	4.2	7.6	0.76	433	0.2	7.6	0.48 + 0.23
$\frac{1}{10}$	10	8.4	15.3	1.53	216	0.2	15.3	0.24 + 0.23
1	10	8.4	84	8.4	1190	0.2	84	0.24 + 1.25
$\frac{1}{10}$	5	16.8	31	3.1	108	0.2	31	0.12 + 0.23
$\frac{1}{10}$	1	84	153	15.3	22	0.2	153	0.024 + 0.23

^{a} Length of mirror needed to intercept a vertical divergence (ϕ_V) of 0.2 mrad for the reflecting angle θ. The entry marked with a dagger is an exception.
^{b} Additional length of mirror needed to intercept ϕ_H with the curvature R_s.

of one and to intercepting less than the total vertical opening angle of synchrotron radiation. The parameters for this mirror are listed as the third entry of Table 4. Mirrors are impractical for focusing, in the sagittal plane (perpendicular to the plane of scattering), x-ray energies ≥ 20 keV because of the small values of R_s required. Mirrors focus in the meridian plane (the plane of scattering) much more strongly with large radii of curvature for R_m.

To compare the synchrotron flux delivered to a 1-μm-diam area by the several mirror systems shown in Figure 25, we calculate the gain in number of photons ΔE^{-1} s^{-1} μm^{-2} for these mirror optics compared to those from an unfocused beam. The Kirkpatrick–Baez mirror system[72] of Figure 25b produces the greatest intensity of the sample of mirrors considered (Table 5). A simple singly curved, elliptically bent mirror is as effective as a doubly curved ellipsoidal mirror in producing intensity. We choose from Table 5 a conservative value of 100 as the gain in intensity achievable with mirrors over the unfocused radiation at 10 m from the source.

Table 5. Design Parameters and Relative Intensity of Mirror Systems for Focusing the Synchrotron Continuum Radiation $\leqslant 40$ keV

Mirror	F_1 (m)	F_2 (m)	F_3 (m)	M_V	M_H	R_s (mm)	ϕ_H (mrad)	R_m (m)	ϕ_V (mrad)	L_s (mm)	L_m L_1 (m)	L_2 (m)	Source size (mm^2)	Relative intensity
None	10	—	—	—	—	—	—	—	—	—	—	—	—	1
Ellipsoidal	10	10	—	1	1	21	0.9	4761	0.15	9	0.75	0.25	0.20	68
Kirkpatrick–Baez														
Mirror 1	10	1	1	—	$\frac{1}{5}$	∞	—	1587	0.21	—	1	—	0.20 }	1046
Mirror 2	10	1	1	$\frac{1}{11}$	—	∞	0.19	873	—	—	1	—	0.20 }	
Elliptical	10	1	—	$\frac{1}{10}$	—	∞	—	865	0.21	—	1	—	—	73

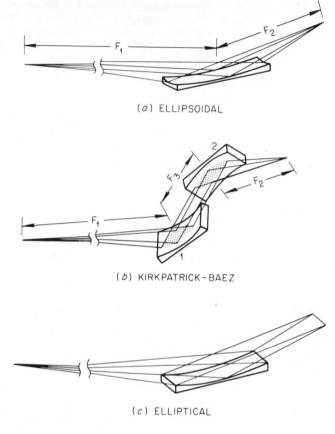

Figure 25. Mirror focusing schemes for the continuum radiation: (a) doubly curved ellipsoidal mirror, (b) Kirkpatrick–Baez system of two singly curved mirrors, and (c) singly curved mirror for meridian focusing only.

The flux incident on a sample at 10 m from the source is taken from the values of the fluences, hereafter labeled P, and given in Figure 12 (right-hand ordinate). These values of P are for the total vertical emission (~ 0.2 to 0.6 mrad) and 1 mrad of horizontal emission. In terms of P, the intensity at 10 m is $5 \times 10^{-8}P$ photons $s^{-1}(\Delta E/E = 1\%)^{-1}\,\mu m^{-2}$. Since the results of Figure 13 (curve B) show that the continuum spectrum produces $(2-7) \times 10^3$ more fluorescence than a $\Delta E/E$ of 0.01%, we must multiply P by 20 to 70 to eliminate the restriction of a $\Delta E/E$ of 1%. Thus, the continuum intensity in equivalent monoenergetic x rays just above the absorption edge (using the factor of 20) is $\sim 10^{-6}P$ photons $s^{-1}\,\mu m^{-2}$. A typical value of P taken from Figure 12 is 10^{14}. This number of photons would produce the same fluorescence as $10^{14} \times 10^{-6} = 10^8$ photons with energies just above the adsorption edge. With a mirror focusing system, the gain of 100 increases this to 10^{10} photons $s^{-1}\,\mu m^{-2}$.

We will now explore the optics for maximizing the intensities of monoenergetic x rays before making final comparisons of the various fluorescence excitation methods.

4.2. *Mirrors and Crystals for Focusing Monoenergetic X Rays*

Diffracting crystals focus sagittal rays more strongly than mirrors because R_s is proportional to the scattering angle, which is typically 20 times larger for crystal diffraction than for mirror reflection for x-rays. Thus, the use of crystals as both monochromators of the continuum radiation and sagittal focusing devices appears attractive. Many possible configurations of crystals, and crystals with mirrors, exist. For selecting the smallest $\Delta E/E$, the doubly curved crystal configurations for point focusing x rays described by Berreman *et al.*[73] could produce the more intense focused beams. However, doubly curved optics with fixed radii produce the minimum focus for only one energy. Single crystals and crystal mosaics with flexible backings that provide variable double radii for continuous energy selection would be advantageous. Only the simpler singly and doubly curved crystals with fixed radii will be considered here.[73]

To determine the intensity delivered to the target by the various optical configurations, we must know the following: source solid angle, $\phi_V \phi_H$, which contributes to the intensity at the sample; the fraction of the horizontal and vertical source dimensions accepted by the optics (F_{h_V} and F_{h_H}); the energy bandwidth (ΔE_v) passed by the optics; the horizontal (H) and vertical (V) dimensions over which the beam is spread at the target; and the efficiencies (ξ) with which each optical element reflects or diffracts the incident radiation within the band pass, including an absorption factor for the beam line and windows. With these parameters the intensity I_0 at the target can be written as

$$I_0 = P\phi_V \phi_H F_{h_V} F_{h_H} \Delta E_V \xi / VH \tag{20}$$

where P is given in Figure 12. Equation (20) is applied to each optical element beginning with that nearest to the source. To illustrate, we will apply the equation to calculate the intensity for two x-ray focusing systems used at SSRL for which the intensity has been measured. In the scattering plane (typically the vertical plane), the optical conditions for calculating the parameter V of equation (20) are shown in Figure 26. Each incident ray contains the continuum spectrum and will undergo diffraction such that the diffracted fan of each ray (A', B', and C') will contain an energy band

$$\Delta E_V = E\omega \cot \theta \tag{21}$$

where ω (FWHM) is the mosaic spread or Darwin width of the crystal. The diffracted radiation is spread in the scattering plane over the dimension V, which in a Gaussian approximation is

$$V = \{(\omega F_2)^2 + [\phi_V(F_1 + F_2)]^2 + (h_V F_2/F_1)^2\}^{1/2} \tag{22}$$

Here the term $\phi_V(F_1 + F_2)$ is the beam spread from the vertical divergence and would be replaced by a focusing error if the optics were made to focus in the vertical plane. The term $h_V F_1/F_2$ is the source height multiplied by the magnification factor (dashed lines, ray BB', Figure 26).

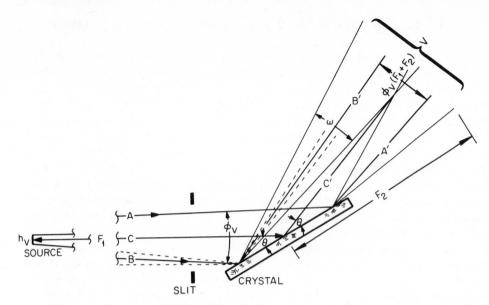

Figure 26. Beam spread V in the scattering (vertical) plane when the incident radiation contains the continuum spectrum. The radiation is spread by the vertical divergence (ϕ_V), the crystal mosaic spread (ω), and the source height (h_V) shown as dashed lines on ray BB'.

In the sagittal plane shown in Figure 27, the diffracted radiation is spread along the dimension

$$H = [(F_2 2\omega \sin \theta)^2 + (h_H F_2/F_1)^2 + (F_e)^2]^{1/2} \tag{23}$$

where F_e is the focusing error. For a nonfocusing optical element, F_e would be replaced by $\phi_H(F_1 + F_2)$. The effect of divergence in the sagittal plane on the energy band passed by the crystal is usually negligible compared to the divergence in the scattering plane.

We compare the observed intensity and that calculated[74] for a singly curved graphite monochromator with a mosaic spread of 0.4° and mounted in the geometry shown in Figure 14. A 1 mm × 34 mm slit was placed just in front of the

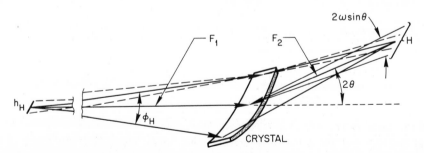

Figure 27. Beam spread H in the sagittal (horizontal) plane caused by the crystal mosaic spread ($2\omega \sin \theta$) and the horizontal source width (h_H) shown as dashed lines.

graphite monochromator and collected a horizontal divergence (ϕ_H) of two milli-rads of radiation. This slit collected only $\frac{1}{8}$ of the FWHM vertical beam divergence since the natural vertical divergence, electron beam oscillations, and source height combined to produce a beam ~ 8 mm high at 17 m in the colliding-beam mode.[64, 75] The total source height contributed to the focus since the convergence angle (1.6/17,000) is small compared to the mosaic spread of the monochromator. The sagittal plane focusing was accomplished by curving the monochromator to a radius R_s of 10 cm. The 3.2-mm horizontal width of the source was magnified by $\frac{1}{17}$ to 0.2 mm but the 0.4° mosaic spread of the monochromator ($\theta = 2.86°$ for 37-keV radiation) added $F_2 2\omega \sin\theta = 0.7$ mm to the width of the focus so that $F_{h_H} = 1$ for a 1-mm-wide receiving slit. Each ray incident on the graphite monochromator contributed a ΔE_V of 5180 eV as calculated from equation (21) for the 37-keV central ray. But this radiation is spread in the scattering plane over a distance V of 7.07 mm as calculated from equation (22). Here the 1-mm height of the slit before the monochromator is substituted for $\phi_V(F_1 + F_2)$. An efficiency of diffraction (ξ) of $\frac{1}{4}$ was used for graphite.[74] For a SPEAR current of 30 mA at 3.5 GeV, the number of 37-keV photons s^{-1} 370-eV^{-1} mrad^{-1} is 2.5×10^{12}. Thus, the intensity through a 1-mm^2 slit at the focus is calculated with equation (20) to be

$$I_0 = \underbrace{\frac{2.5 \times 10^{12}}{s(370\text{ eV})\text{mrad}}}_{P} \times \underbrace{\frac{30\text{ mA}}{100\text{ mA}} \times \frac{1}{8}}_{\phi_V/\phi_{total}}$$

$$\times \underbrace{(2\text{ mrad})}_{\phi_H} \times \underbrace{(5180\text{ eV})}_{\Delta E_V} \times \underbrace{\frac{1}{4}}_{\xi}/\underbrace{(7.07\text{ mm})}_{V} \times \underbrace{(1\text{ mm})}_{H}$$

$$= (9.3 \pm 3) \times 10^{10}\text{ photons s}^{-1}\text{ mm}^{-2}$$

This calculated value agrees well with the measured value of $(8 \pm 2) \times 10^{10}$ photons s^{-1} mm^{-2}.[49]

Another check on the reliability of predicting I_0 is given for 7.1-keV x rays for a focusing system described by Hastings et al.,[76] which consists of a bent Pt-coated cylindrical mirror furnished by Howell and Horowitz[69] (from their original microprobe) followed by a pair of flat Ge (111) crystals. See Figure 18 in Chapter 3 for a drawing of this mirror–monochromator system. The mirror is located 10 m from the source at SPEAR and its double curvature is designed to focus with 1 : 1 magnification 10 m from the mirror. With a length of 60 cm and a reflection angle θ of 10 mrad, the mirror intercepts all of the vertical divergence. The horizontal divergence (ϕ_H) of 10 mrad intercepted by the 10 cm wide mirror is reduced to 4.8 mrad since the beam is smeared to 125 cm by the sagittal curvature (the dimension L_2 given in equation (19d)). A mirror efficiency of 1 is used for the x rays. For a 21″ (FWHM) Darwin width typical of nearly perfect Ge (111) reflections for 7.1-keV x rays ($\theta = 15.47°$), a ΔE_V of 2.62 eV is diffracted from each ray as calculated from equation (21). However, the focus will contain an even larger ΔE because the convergence angle of the radiation from the mirror (0.25 mr) onto the crystal adds an additional 5 eV (FWHM) to the spread. The efficiency with which the Ge (111) crystals diffract 7.1-keV x rays is taken as 0.9. Both F_{h_V} and F_{h_H} are unity because the white radiation from every point on the source will contain an energy that will be at the proper Bragg angle for diffraction by the first crystal, and therefore, diffracted by the second crystal in this nondispersive (parallel) arrangement. At 3.7 GeV and 40 mA, SPEAR produces $\sim 4 \times 10^{11}$ 7.1-keV photons s^{-1} eV^{-1} $mrad^{-1}$ in the total vertical emission (see data contained in Chapter 2). These values are used in equation (20) to calculate the intensity at the focus as

$$I_0 = \underbrace{\frac{4 \times 10^{11}}{s \text{ eV mrad}}}_{P} \times \underbrace{1}_{\phi_V/\phi_{total}} \times \underbrace{(4.8 \text{ mrad})}_{\phi_H} \times \underbrace{(2.62 \text{ eV})}_{\Delta E_V}$$

$$\times \underbrace{1}_{\varepsilon_1} \times \underbrace{0.9}_{\varepsilon_2} \times \underbrace{0.9/(1.6 \text{ mm})}_{\varepsilon_3 \quad V} \times \underbrace{(3.2 \text{ mm})}_{H}$$

$$= (4 \pm 3) \times 10^{12} \text{ photons } s^{-1} \ (1.6 \text{ mm} \times 3.2 \text{ mm})^{-1}$$

Hastings et al.[76] gave the measured intensity as 10^{12} s^{-1} (2 mm × 4 mm)$^{-1}$.

Having shown that the agreement between calculated and observed intensities is acceptable for the purpose of estimating the flux obtainable from synchrotron radiation, we can now proceed with confidence to calculate the intensity available for a 1-μm^2 microprobe.

Several focusing systems for monochromatic x rays were investigated. Of those, the performance of the singly curved sagittal-focusing crystal as shown in Figure 27 and the doubly curved crystal similar to the doubly curved mirror shown in Figure 24 are presented for comparison. The singly curved crystal is simple and has been tested,[48] as has the doubly curved crystal described by Berreman et al.[73] Here we propose to form a doubly curved surface from a single crystal since one of the radii, R_m, is so large that the result is essentially a cylindrically bent crystal. Our intent is to show that feasible x-ray-focusing systems for synchrotron radiation can deliver sufficient x-ray intensity that a microprobe with 1-μm^2 resolution can be constructed with sensitivities for elements that exceed those of charged-particle microprobes and with large reductions in the amount of power deposited in the sample.

The intensity used in the following calculations is from Figure 12 for the NSLS. In the calculation of the focused intensity, we assume a 0.25-mm vertical and 0.8-mm horizontal source size typically proposed for a bending magnet region in the NSLS. A larger source size typical of SPEAR has a negligible effect on the intensity when the crystals have mosaic spreads $\geq 0.4°$ or focus energies below 10 keV. Low-β regions of the storage rings produce even smaller source sizes.[54] Wiggler magnets will generate 10 times more intensity in the energy region from 20 to 40 keV than used in these calculations (see Chapters 2 and 21 for a discussion of source sizes and wiggler magnets). Thus, the intensities calculated are conservative.

All of the necessary parameters and the calculated number of photons s^{-1} μm^{-2} at the focus for the following three focusing systems are given in Table 6: (1) a 0.4° mosaic spread graphite monochromator bent for sagittal focusing, (2) a perfect Ge (111) crystal in the same configuration as the graphite monochromator but ground flat to reduce focusing errors, and (3) a doubly curved but asymmetrical cut Ge (111) crystal similar to the mirror geometry as shown in Figure 24. (The

Table 6. Various Parameters for Three Microprobe Focusing Systems and the Number of Photons per Second in 1 μm^2

E (keV)	ΔE_T (eV)	θ (deg)	ω (arc sec)	F_1 (m)	F_2 (m)	M_V	M_H	ξ	θ_V (mrad)	θ_H (mrad)	V (mm)	H (mm)	L_m (mm)	L_s (mm)	R_s (mm)	R_m (m)	$I \times 10^{-6}$ photons $s^{-1} \mu m^{-2}$
40	864	2.65	1440	10	1.0		0.1	0.25	0.10	8.4	7.3	0.65	22	84	84	∞	0.7
30	788	3.53	1440	10	0.74		0.074	0.25	0.12	8.4	5.5	0.65	22	84	84	∞	6.6
20	673	5.30	1440	10	0.48		0.048	0.25	0.15	8.4	3.9	0.65	22	84	84	∞	31
10	372	10.65	1440	10	0.24		0.024	0.33	0.21	8.4	2.6	0.65	22	84	84	∞	122
5	88	21.69	1440	10	0.12		0.012	0.33	0.34	8.4	2.2	0.65	22	84	84	∞	342
40	15	2.71	3.7	10	1.0		0.1	0.9	0.10	8.4	1.10	0.08	21	84	84	∞	0.3
30	11	3.62	4.8	10	0.74		0.074	0.9	0.12	8.4	1.3	0.06	21	84	84	∞	3.5
20	7	5.43	7.1	10	0.48		0.048	0.9	0.15	8.4	1.6	0.04	21	84	84	∞	21
10	4	10.94	14	10	0.24		0.024	0.9	0.21	8.4	2.3	0.02	21	84	84	∞	131
5	2	22.26	26	10	0.12		0.012	0.9	0.34	8.4	3.7	0.016	21	84	84	∞	511
40	4.8	2.71	1.2	10	1.0	0.1	0.1	0.9	0.10	8.4	0.026	0.08	277	84	86	116	5
30	3.4	3.62	1.5	10	1.0	0.1	0.1	0.9	0.15	8.4	0.027	0.08	158	84	115	87	42
20	2.3	5.43	2.2	10	1.0	0.1	0.1	0.9	0.21	8.4	0.033	0.08	74	84	172	58	193
10	1.1	10.94	4.4	10	1.0	0.1	0.1	0.9	0.34	8.4	0.047	0.08	23	84	172	29	277
5	0.5	22.26	8.2	10	1.0	0.1	0.1	0.9	0.34	8.4	0.047	0.08	10	84	653	15	2010

Section headings within table:
- Graphite Singly Curved to a Fixed Radius
- Germanium (111) Singly Curved to a Fixed Radius and Ground Flat
- Asymmetrically Cut Germanium (111) Crystals Each Doubly Curved to a Magnification of 0.1

[a] NSLS operating at 2.5 GeV and 500 mA with 0.25 mm × 0.8 mm (FWHM) source size.

effects of thermal loads may increase the energy bands passed by the crystals, but are not expected to decrease the diffracted intensity.) The parameter symbols in Table 6 have been defined in Figures 24, 26, 27, and in the text. All three of the focusing systems can intercept the total vertical divergence of the synchrotron radiation.

A comparison of the vertical and horizontal dimensions (V and H) of the focus given in Table 6 shows that the mosaic crystal distributes the x rays over a much larger area but with an intensity comparable to that of a nearly perfect crystal. Thus, the total power diffracted by the mosaic crystal is much larger than for the Ge crystal. Above 15 keV, the graphite monochromator produces a 2 to 3 times higher intensity and about 60 times the energy spread ΔE_T within the μm^2 area than calculated for the Ge crystal. In the third system, focusing is achieved in the scattering plane with a Pt-coated mirror. This reduces the dimension V in the scattering plane and accounts for the increase in intensity over that from the singly bent Ge crystal. Five doubly curved crystals can be chosen to focus the most useful photon energies of 2, 4, 10, 18, and 40 keV as discussed in Section 2.3. As the radii of the curved crystals are fixed, the sample and detection equipment must be moved to follow the focus when changing energy. More complicated two-crystal systems can be envisioned which will keep the focus fixed in space.

As wiggler magnets are only now being evaluated, the values of I listed in Table 6 for the doubly curved Ge crystal are to be multiplied by 0.1 of the gain to be expected from the installation of a $4T$ wiggler magnet at the NSLS for the comparison given in the next section. Thus, the values of I in Table 6 for the doubly-curved Ge crystal are to be multiplied by 9, 7, and 3 for the x-ray energies 40, 30, and 20 keV respectively when the comparison is made with proton and electron sources given in the next and final section. Intensities that approach 10^9 photons s^{-1} μm^{-2} are calculated for this focusing system (Table 6). The sagittal radius R_s of the Ge crystal is varied to keep the focus at a fixed position.

Several possibilities (besides wiggler magnets and smaller source sizes) exist for increasing the x-ray intensity over that used in this comparison. Anisotropic mosaic crystals (discussed by Freund[77]), in which the mosaic spread is smaller in the plane perpendicular to the scattering plane, could decrease the dominant mosaic contribution to the beam spread labeled H in Table 6. In a multilayer monochromator, the d spacing can be varied through the thickness by controlling the layer thickness. Such layered crystals could diffract x rays which are parallel and contain larger $\Delta E/E$ for increased intensities.[78] Additionally, demagnification of the source size by greater than the factor of ten used here is possible.[79]

5. Final Comparative Analysis and Conclusions

From the data presented, the MDL can be calculated from equation (12), which on rearranging becomes

$$\text{MDL} = 3.29 C_z (N_s/N_b)^{-1/2} (N_s)^{-1/2} \qquad (24)$$

where N_s/N_b is the signal to background given in Figure 23 for a concentration C_z of 10^{-6} gg^{-1}. Typical values for the S/B were chosen that matched the elements to similar atomic-numbered matrices. The number of fluorescence counts (N_s) from 10^{-6} gg^{-1} of a second element embedded in a thick target of similar Z can be calculated from equation (5) where $\sigma_{zij}/(\mu_{s,0} + \mu_{s,i} \sin \psi/\sin \phi)/\rho_s$ is the thick-target yield. With the substitution of $N_s = I_z A_D t$, where A_D is the detector area and t the time, equation (5) becomes

$$N_s = \frac{A_D}{4\pi r^2} P_0 t \times \text{thick-target yield} \qquad (25)$$

A reasonable solid angle of 10^{-3} is assumed for a detector with an x-ray efficiency of one. The number of fluorescence counts in one second from a thick target containing only 10^{-6} gg^{-1} of the element when excited with P_0 photons s^{-1} contained in a 1-μm^2 area is

$$N_s = P_0 t \times \text{thick-target yield} \times 10^{-9} \qquad (26)$$

Values of P_0 used in equation (26) were taken from the following sources: continuum and filtered synchrotron radiation from Table 5 and text of Section 4, monochromatic x-ray values, $I = P_0 \mu\text{m}^{-2}$, multiplied by 0.1 of the gain expected from a wiggler magnet from Table 6 for the doubly-curved crystal configuration similar to Figure 24, and values of 10^{13} electrons s^{-1} μm^{-2} and 10^9 protons s^{-1} μm^{-2} from References 39–41. Thick-target yields were taken from Figure 3 for x-ray energies just above the absorption edges, for electron energies three times the absorption edge energy, and for 5-MeV protons scaled by the ratios of the 5–2-MeV data of Figure 2. With these calculated values of N_s and the data from Figure 23, equation (24) was solved for the MDL values plotted in Figure 28. Values for thick targets in units of 10^{-6} gg^{-1} s^{-1} μm^{-2} are given on the left-hand ordinate. Synchrotron radiation produces a lower MDL than electrons or proton in a 1-μm^2 area with the more monoenergetic x rays ($\Delta E/E \leq 3\%$) producing the lowest limit of detection for $K\alpha$ fluorescence. Since the MDL depends on $(N_s)^{-1/2}$, a reduction of 10 in the diameter of the incident synchrotron radiation beam would reduce the incident power by 100 and raise the MDL by 10, making it similar to that for the other probes. Thus, even with spatial resolutions of 0.1 μm, synchrotron radiation would have detection limits as low as proton and electron microprobes.

In thick targets each of the incident microprobe beams deposit the power shown on the right-hand ordinate of Figure 28. This comparison shows that x-ray excitation deposits only 10^{-3} to 10^{-5} as much power for a MDL smaller by a factor of 10. For the same MDL, the power deposited by x rays would be 10^{-5} to 10^{-7} as much. Since thin samples are more transparent to x rays than to charged particles, the relative amount of power per unit area deposited by x rays compared to charged particles would be even less.

X-ray, proton, and electron microprobes will be complimentary with each having certain advantages. The advantages that x-ray microprobes offer over charged-particle microprobes are the following: orders of magnitude less power deposited in the sample with reduction in surface contamination; improved sensitivity levels in both thick and thin samples; improved spatial resolution in thicker samples; reduction in the analysis time; analysis of difficult sample configurations and in difficult environments, improved contrast (S/B ratios), and negligible charge

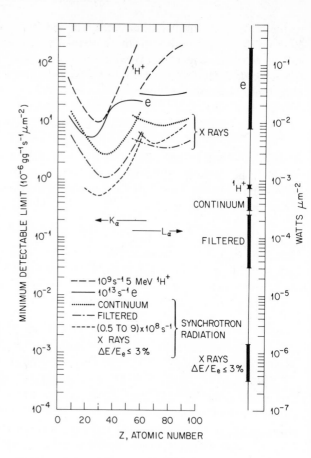

Figure 28. Comparisons of minimum detectable limits in units of 10^{-6} gg^{-1} s^{-1} μm^{-2} for thick samples and power in watts μm^{-2} (right vertical scale) among charged-particle and synchrotron radiation fluorescence microprobes. The μm-diam beams contain 10^9 s^{-1} 5-MeV protons, 10^{13} s^{-1} electrons at three times the energies of the absorption edge, $(0.5-9) \times 10^8$ s^{-1} x rays at energies of the absorption edges, and continuum and filtered synchrotron radiation as described in the text.

collection. The power density achievable with modern electron optics is 10^{11} (typically) to 10^{16} (most advanced) electrons s^{-1} into an area 1 μm in diameter.[39, 40] Such high power in small areas could vaporize the target. Therefore, electron microprobes will excel when targets are very thin, in which case most of the electrons are transmitted and the high spatial resolution of the small electron beam diameter is preserved. Just as with electrons, x-ray microprobes can also be used for diffraction to determine the atomic structure of small particles or volumes with the same incident radiation used for fluorescence excitation. Chemical bonding information could be obtained from measurements of absorption edge energies and fluorescence energies, but intensity would be sacrificed to achieve the necessary energy resolution of about 1 eV. Present and future developments will improve means for extracting electromagnetic radiation from stored electron beams as discussed in Chapter 21. The insertion of these devices into the higher energy storage rings like the 8 GeV ring at Cornell and the 18 GeV ring (PEP) at the Stanford Linear Accelerator Center will produce orders of magnitude greater intensity at the sample than used in the comparisons given here.

Electron-energy-loss spectroscopy provides information on both the elemental concentration and chemical bonding and competes favorably with EXAFS measurements made with synchrotron radiation at energies below 1 keV.[39, 80] However, the small S/B in absorption edge measurements ($\sim 10^{-7}$ of that obtainable with monochromatic x-ray-excited fluorescence) limits detection to concentrations which are 10^2 times higher than for fluorescence detection.

In conclusion, sufficient justification for the instrumentation of a fluorescence and diffraction microprobe that utilizes synchrotron radiation rather than charged particles as an excitation source is provided by the improved fluorescence signal over background, the lower detectable limits, and the reduction in the amount of energy deposited to achieve the same minimum detectable limits.

ACKNOWLEDGMENTS

The author expresses his appreciation to the following colleagues who assisted in obtaining the experimental results presented here (some have not been previously published): E. Ricci, S. Raman, M. O. Krause, H. L. Yakel (also for reviewing the manuscript), A. J. Millet and D. L. Holcomb of the Oak Ridge National Laboratory, R. V. Gentry (visiting scientist from Columbia Union College, Takoma Park, Maryland), and J. B. Hastings (now at Brookhaven National Laboratory), and is grateful to N. J. Zaluzec for charged-particle cross sections, to the staff of SSRL whose assistance made our research possible, and to B. S. Borie and G. Ice for helpful comments on the manuscript.

References

1. W. Bambynek, B. Crasemann, R. W. Fink, H. -U. Freund, H. Mark, C. D. Swift, R. E. Price, and P. V. Rao, X-ray fluorescence yields, Auger and Coster–Kronig transition probabilities, *Rev. Mod. Phys.* **44**, 716–813 (1972).
2. C. J. Sparks, Jr., in *Advances in X-Ray Analysis*, R. W. Gould, C. S. Barrett, J. B. Newkirk, and C. O. Rudd (eds.), Vol. 19, pp. 19–52, Kendall Hunt, Dubuque, Iowa (1976).
3. K. F. J. Heinrich, in *Advances in X-Ray Analysis*, R. W. Gould, C. S. Barrett, J. B. Newkirk, and C. O. Rudd (eds.), Vol. 19, pp. 75–84, Kendall Hunt, Dubuque, Iowa (1976).
4. R. O. Müller, *Spectrochemical Analysis by X-Ray Fluorescence*, Plenum Press, New York (1972).
5. T. B. Johnasson, R. E. Van Grieken, J. W. Nelson, and J. W. Winchester, Elemental trace analysis of small samples by proton induced x-ray emission, *Anal. Chem.* **47**, 855–860 (1975).
6. D. R. Beaman and L. F. Solosky, Accuracy of quantitative electron probe microanalysis with energy dispersive spectrometers, *Anal. Chem.* **44**, 1598–1610 (1972).
7. L. A. Currie, Limits for qualitative detection and quantitative determination, *Anal. Chem.* **40**, 586–93 (1968).
8. A. S. Rubin, T. O. Passell, and L. E. Bailey, Chemical analysis of surfaces by nuclear methods, *Anal. Chem.* **29**, 736–743 (1957).
9. T. B. Johansson, R. Akselsson, and S. A. E. Johansson, X-ray analysis: Elemental trace analysis at the 10^{-12} g level, *Nucl. Instum. Methods* **84**, 141–143 (1970).
10. B. M. Gordon and H. W. Kraner, On the development of a system for trace element analysis in the environment by charged particle x-ray fluorescence, *J. Radioanal. Chem.* **12**, 181–188 (1972).
11. J. A. Cooper, Comparison of particle and photon excited x-ray fluorescence applied to trace element measurements of environmental samples, *Nucl. Instum. Methods* **106**, 525–538 (1973).

12. M. O. Krause, E. Ricci, C. J. Sparks, Jr., and C. W. Nestor, Jr., in *Advances in X-Ray Analysis*, C. S. Barrett, E. C. Leyden, J. B. Newkirk, and C. O. Rudd (eds.), Vol. 21, pp. 119–127, Plenum Press, New York (1978).

13. M. O. Krause, C. W. Nestor, Jr., C. J. Sparks, Jr., and E. Ricci, X-ray fluorescence cross sections for *K* and *L* x rays of the elements, Oak Ridge National Laboratory Report ORNL-5399 (June 1978).

14. W. H. McMasters, N. K. Del Grande, J. H. Mallett, and J. H. Hubbell, Compilation of x-ray cross section, Sec. II, Rev. I, University of California, Lawrence Radiation Laboratory Report UCRL 50172 (May 1969).

15. E. S. H. Burhop, The inner shell ionization of atoms by electron impact, *Proc. Cambridge Philos. Soc.* **36**, 43–49 (1940).

16. J. Green and V. E. Cosslett, The efficiency of production of characteristic x radiation in thick targets of a pure element, *Proc. Phys. Soc. London* **78**, 1206–1214 (1961).

17. M. Green, The angular distribution of characteristic x radiation and its origin within a solid target, *Proc. Soc. London* **83**, 435–451 (1964).

18. L. S. Birks, R. E. Seebold, A. P. Batt, and J. S. Grosso, Excitation of characteristic x rays by protons, electrons, and primary x rays, *J. Appl. Phys.* **35**, 2578–2581 (1964).

19. L. S. Birks, R. E. Seebold, B. K. Grant, and J. S. Grosso, X-ray yield and line/background ratios for electron excitation, *J. Appl. Phys.* **36**, 699–702 (1965).

20. M. Green and V. E. Cosslett, Measurements of *K*, *L*, and *M* shell x-ray production efficiencies, *Br. J. Appl. Phys.* (*J. Phys. D*) **1**, 425–436 (1968).

21. E. Strom, Emission of characteristic *L* and *K* radiation from thick tungsten targets, *J. Appl. Phys.* **43**, 2790–2796 (1972).

22. D. V. Davis, V. D. Mistry, and C. A. Quarles, Inner shell ionization of copper, silver, and gold by electron bombardment, *Phys. Lett. A* **38**, 169–170 (1972).

23. H. W. Lewis, B. E. Simmons, and E. Merzbacher, Production of characteristic x rays by protons of 1.7- to 3-MeV energy, *Phys. Rev.* **91**, 943–946 (1953).

24. E. M. Bernstein and H. W. Lewis, *L*-shell ionization by protons of 1.5- to 4.25-MeV energy, *Phys. Rev.* **95**, 83–86 (1954).

25. W. T. Ogier, G. J. Lucas, J. S. Murray, and T. E. Holzer, Soft x-ray production by 1.5-MeV protons, *Phys. Rev.* **134**, 1070–1072 (1964).

26. J. M. Khan, D. L. Potter, and R. D. Worley, Studies in x-ray production by proton bombardment of C, Mg, Al, Nd, Sm, Gd, Tb, Dy, and Ho, *Phys. Rev.* **139**, 1735–1746 (1965).

27. R. C. Bearse, D. A. Close, J. J. Malanify, and C. J. Umbarger, Production of $K\alpha$ and $L\alpha$ x rays by protons of 1.0–3.7 MeV, *Phys. Rev. A* **7**, 1269–1272 (1973).

28. T. L. Hardt and R. L. Watson, Cross sections for *L*-shell x-ray and Auger-electron production by heavy ions, *Atom. Data Nucl. Data Tables* **17**, 107–125 (1976).

29. C. J. Sparks, Jr., O. B. Cavin, L. A. Harris, and J. C. Ogle, in *Trace Substances in Environmental Health*, Vol. VII, D. D. Hemphill (ed.), pp. 295–304, University of Missouri, Columbia (1974).

30. F. S. Goulding and J. M. Jaklevic, XRF analysis—Some sensitivity comparisons between charged-particle and photon excitation, *Nucl. Instrum. and Methods* **142**, 323–332 (1977).

31. J. M. Jaklevic, Proceedings of the Energy Research and Development Administration X- and Gamma-Ray Symposium (Conference 760539), Ann Arbor, Michigan (May 1976), pp. 1–6.

32. V. Valkovic, Proton-induced x-ray emission: Application in medicine, *Nucl. Instum. and Methods* **142**, 151–158 (1977).

33. F. Folkmann, C. Gaarde, T. Huus, and K. Kemp, Photon induced x-ray emission as a tool for trace element analysis, *Nucl. Instum. and Methods* **116**, 487–499 (1974).

34. P. Kirkpatrick and L. Wiedmann, Theoretical continuous x-ray energy and production, *Phys. Rev.* **67**, 321–339 (1945).

35. D. H. Madison and E. Merzbacher, in *Atomic Inner Shell Processes*, B. Crasemann (ed.), Vol. 1, pp. 1–72, Academic Press, New York (1975).

36. R. W. Shaw, Jr., and R. D. Willis, in *Electron Microscopy and X-Ray Applications to Environmental and Occupational Health*, P. A. Russell and A. E. Hutchings (eds.), pp. 51–64, Ann Arbor Press, Michigan (1978).

37. J. Kirz, D. Sayre, and J. Dilger, Comparative analysis of x-ray emission microscopies for biological specimens, *Ann. N.Y. Acad. Sci.* **306**, 291–305 (1978).

38. D. E. Newbury and H. Yakowitz, in *Use of Monte Carlo Calculations in Electron Probe Microanalysis and Scanning Electron Microscopy*, K. F. J. Heinrich, D. E. Newbury, and H. Yakowitz (eds.), Natl. Bur. Stand. U.S. Spec. Publ. **460**, 15–44 (1976).

39. M. Isaacson and M. Utlaut, A comparison of electron and photon beams for determining microchemical environment, *Optik (Stuttgart)* **50**, 213–234 (1978).

40. P. Bovey, I. Wardell, and P. M. Williams, in *8th International Conference on X-Ray Optics and Microanalysis and 12th Annual Conference of the Microbeam Analysis Society, August 18–24, 1977*, p. 117A (available from K. F. J. Heinrich, NBS, Washington, DC).

41. J. A. Cookson, A. T. G. Ferguson, and F. D. Pilling, Proton microbeams, their production and use, *J. Radioanal. Chem.* **12**, 39–52 (1972).

42. P. Horowitz and L. Grodzins, Scanning proton-induced x-ray microspectrometry in an atmospheric environment, *Science* **189**, 795–797 (1975).

43. R. D. Willis, R. L. Walter, R. W. Shaw, Jr., and W. F. Gutknecht, Proton induced x-ray emission analysis of thick and thin targets, *Nucl. Instum. and Methods* **142**, 67–77 (1977).

44. B. E. Warren, *X-Ray Diffraction*, Addison-Wesley Publishing Co., Reading, Massachusetts (1969).

45. E. Storm and H. I. Israel, Photon Cross Sections from 0.001 to 100 MeV for Elements 1 through 100, Los Alamos Scientific Laboratory Report LA-3753; UC-34, Physics; TID-4500 (1967).

46. P. Horowitz and J. Howell, A scanning x-ray microscope using synchrotron radiation, *Science* **178**, 608–611 (1972).

47. C. J. Sparks, Jr., and J. B. Hastings, X-Ray Diffraction and Fluorescence at the Stanford Synchrotron Radiation Project, Oak Ridge National Laboratory Report ORNL-5089 (June 1975), p. 8.

48. C. J. Sparks, Jr., S. Raman, H. L. Yakel, R. V. Gentry, and M. O. Krause, Search with synchrotron radiation for superheavy elements in giant-halo inclusions, *Phys. Rev. Lett.* **38**, 205–208 (1977).

49. C. J. Sparks, Jr., S. Raman, E. Ricci, R. V. Gentry, and M. O. Krause, Evidence against superheavy elements in giant-halo inclusions re-examined with synchrotron radiation, *Phys. Rev. Lett.* **40**, 507–511 (1978).

50. C. J. Sparks, Jr., E. Ricci, S. Raman, M. O. Krause, R. V. Gentry, H. L. Yakel, and J. B. Hastings, X-ray fluorescence analysis with synchrotron radiation, *Anal. Chem.* (1980).

51. V. E. Il'in, G. M. Kazakevich, G. N. Kulipanov, L. N. Mazalov, A. M. Matyushin, A. N. Skrinski, and M. A. Sheromov, X-Ray Fluorescence Element Analysis with the Use of Synchrotron Radiation, The Institute of Nuclear Physics, SOAN, USSR, Reprint IYAF 77–57 (1977).

52. C. J. Sparks and B. Borie, in *Local Atomic Arrangements Studied by X-Ray Diffraction*, J. B. Cohen and J. E. Hilliard (eds.), Vol. 36, pp. 5–46, Metallurgical Society Conference, Gordon and Breach, New York (1965).

53. B. E. Warren and R. L. Mozzi, Multiple scattering of x-rays by amorphous samples, *Acta Crystallogr.* **21**, 459–461 (1966).

54. K. Green, Proposal for a national synchrotron light source, J. Blewett (ed.), Brookhaven National Laboratory Report BNL 50595, Vol. II (Feb. 1977).

55. J. M. Ondov, W. H. Zoller, I. Olmez, N. K. Aras, G. E. Gordon, L. A. Rancitelli, K. H. Abel, R. H. Filby, K. R. Shah, and R. C. Ragaini, Elemental concentrations in the National Bureau of Standards environmental coal and fly ash standard reference materials, *Anal. Chem.* **47**, 1102–1109 (1975).

56. M. O. Krause and J. H. Oliver, Natural widths of atomic *K* and *L* levels, *Kα* x-ray lines, and several *KLL* Auger lines, *J. Phys. Chem. Ref. Data* **8**, 329–338 (1979).

57. J. V. Gilfrich and L. S. Birks, Spectral distributions of x-ray tubes for quantitative x-ray fluorescence analysis, *Anal. Chem.* **40**, 1077–1080 (1968).

58. R. L. Walters, R. D. Willis, W. F. Gutknecht, and R. W. Shaw, Jr., The application of proton-induced x-ray emission to bioenvironmental analysis, *Nucl. Instrum. and Methods* **142**, 181–197 (1977).

59. R. D. Willis, R. L. Walter, B. L. Doyle, and S. M. Shafroth, Wavelength-dispersion analysis of PIXE spectra, *Nucl. Instrum. and Methods* **142**, 317–321 (1977).

60. J. Victoreen, The absorption of incident quanta by atoms as defined by the mass photoelectric absorption coefficient and the mass scattering coefficients, *J. Appl. Phys.* **19**, 855–860 (1948).

61. R. V. Gentry, T. A. Cahill, N. R. Fletcher, H. C. Kaufman, L. R. Medsker, J. W. Nelson, and R. G. Flocchini, Evidence for primordial superheavy elements, *Phys. Rev. Lett.* **37**, 11–15 (1976).

62. J. D. Fox, W. J. Courtney, K. W. Kemper, A. H. Lumpkin, N. R. Fletcher, and L. R. Medsker, Comment on evidence for primordial superheavy elements, *Phys. Rev. Lett.* **37**, 629–631 (1976).

63. C. C. Lu, T. A. Carlson, F. B. Malik, T. C. Tucker, and C. W. Nestor, Jr., Relativistic Hartree–Fock–Slater eigenvalues, radial expectation values, and potentials for atoms, $2 \le Z \le 126$, *Atom. Data* **3** (1971).

64. N. G. Webb, S. Samson, R. M. Stroud, R. C. Gamble, and J. D. Baldeschwieler, A focussing monochromator for small-angle diffraction studies with synchrotron radiation, *J. Appl. Crystallogr.* **10**, 104–110 (1977).

65. C. J. Sparks, Jr., Inelastic resonance emission of x rays: Anomalous scattering associated with anomalous dispersion, *Phys. Rev. Lett.* **33**, 262–65 (1974).

66. C. J. Sparks, Jr., in *Anomalous Scattering*, S. Ramaseshan and S. C. Abrahams (eds.), pp. 175–191, International Union of Crystallography, Munksgaand, Copenhagen (1975).

67. H. W. Koch and J. W. Motz, Bremsstrahlung cross section formulas and related data, *Rev. Mod. Phys.* **31**, 920–955 (1959).

68. J. M. Jaklevic and R. L. Walter, in *X-Ray Fluorescence Analysis of Environmental Samples*, T. G. Dzubay (ed.), pp. 63–75, Ann Arbor Science, Michigan (1977).

69. J. A. Howell and P. Horowitz, Ellipsoidal and bent cylindrical condensing mirrors for synchrotron radiation, *Nucl. Instum. Methods* **125**, 225–230 (1975).

70. A. Franks, X-ray optics, *Sci. Prog.* (*London*) **64**, 371–422 (1977).

71. V. Rehn (personal communication).

72. P. Kirkpatrick and A. V. Baez, Formation of optical images by x rays, *J. Opt. Soc. Am.* **38**, 766–744 (1948).

73. C. W. Berreman, J. Stamatoff, and S. J. Kennedy, Doubly curved crystal point focussing x-ray monochromators: Geometrical and practical optics, *Appl. Opt.* **16**, 2081–2085 (1977).

74. C. J. Sparks, Jr., in *Workshop on X-Ray Instrumentation for Synchrotron Radiation Research*, H. Winick and G. Brown (eds.), SSRL Report No. 78/04, pp. III 35–46 (May 1978).

75. P. Pianetta and I. Lindau, High resolution x-ray spectroscopy using synchrotron radiation: Source characteristics and optical systems, *J. Electron Spectrosc. Relat. Phenom.* **11**, 13–38 (1977).

76. J. B. Hastings, B. M. Kincaid, and P. Eisenberger, A separate function focussing monochromator system for synchrotron radiation, *Nucl. Instum. Methods* **152**, 167–171 (1978).

77. A. Freund, A neutron monochromator system consisting of deformed crystals with anisotropic mosaic structure, *Nucl. Instum. Methods* **124**, 93–99 (1975).

78. A. M. Saxena and B. P. Schoenborn, Multilayer neutron monochromators, *Acta Crystallogr. Sect. A* **33**, 805–813 (1977). Also, T. W. Barbee, Jr., and D. C. Keith, in Workshop on X-Ray Instrumentation for Synchrotron Radiation Research, H. Winick and G. Brown (eds.), SSRL Report No. 78/04, pp. III-26–34 (May 1978).

79. W. H. Boettinger, H. E. Burdette, and M. Kuriyama, X-ray magnifier, *Rev. Sci. Instrum.*, **50**(1), 26–30 (1979).

80. B. M. Kincaid, A. E. Meixner, and P. M. Platzman, Carbon *K* edge in graphite measured using electron-energy-loss spectroscopy, *Phys. Rev. Lett.* **40**, 1296–1299 (1978).

Small-Angle X-Ray Scattering of Macromolecules in Solution

H. B. STUHRMANN

1. Introduction

The use of synchrotron radiation for small angle scattering is an obvious choice. The radiation emerging from multi-GeV electrons (or positrons) in circular motion is concentrated into a narrow cone of about 10^{-4}-rad aperture about the instantaneous flight direction. This provides optimal conditions for the technique of small-angle scattering and simplifies the design of instruments. Further, the available intensity is, in principle, very large. Do these favorable factors lead to new developments, and what are the results? The answer which I shall try to give to this question is necessarily incomplete, as it is based on somewhat less than one year's experience with small-angle scattering using synchrotron radiation.

A situation similar to the present one existed in the early days of neutron small-angle scattering. The properties of high-flux reactors, which became available ten years ago, initiated the development of new instruments of gigantic size.

Though there are fundamental differences between the processes of neutron scattering and x-ray scattering, there is an important similarity: synchrotron radiation x-ray beams, like neutrons beams, are produced at only a few places in the world. However, unlike neutron high-flux reactors, the x-ray synchrotron radiation sources are still used largely in a parasitic mode.

Both x-ray synchrotron radiation and thermal neutrons have a broad energy spectrum (Figure 1). However, in almost every experiment, selection of a narrow energy band from the continuum has to be made, which stresses the importance of appropriate mirrors and monochromators in both cases.[1]

H. B. STUHRMANN • European Molecular Biology Laboratory, Outstation at the Deutsches Elektronen-Synchrotron (DESY), Hamburg, Germany.

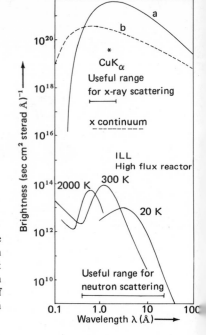

Figure 1. Brightness of synchrotron radiation of the storage ring DORIS (a) and of the synchrotron DESY (b). For comparison the emission of an x-ray tube (Cu K radiation and bremsstrahlung from a 60-kW tube) and 'of the high-flux reactor of the Institut Max von Laue–Paul Langevin (ILL) is given. The maximum of the neutron wavelength distribution is shifted to longer wavelengths at lower temperatures of the moderator (hot source: graphite at 2000 K; thermal source: D_2O at room temperature; and cold source: liquid deuterium at 20 K).

Furthermore, position-sensitive counters have greatly facilitated neutron small-angle scattering experiments, and they have also proven very useful in ordinary x-ray small-angle scattering. Their practical impact on experiments with synchrotron radiation has however become only apparent during the last months.

2. The Small-Angle Scattering Intensity

Small-angle scattering from a dilute solution of monodisperse systems provides low-resolution models of the dissolved particles. The wealth of information that can be obtained from the small-angle scattering intensity as a function of momentum transfer k $[k = (4\pi/\lambda) \sin \theta, \lambda = \text{wavelength}, 2\theta = \text{scattering angle}]$ is best described in a quantitative way by introducing spherical harmonics for the description of the scatterer. It appears that with increasing resolution the number of structural models that can be associated with a measured scattering intensity $I(k)$ increases rapidly.[2]

But experimental techniques are available that can be used to reduce the ambiguity of structure determination. These are based on the variation of the solvent density, which leads to a different appearance for the same macromolecular structure.[2-5] Though this technique was first developed for x-ray small-angle scattering,[6] its full power could only be demonstrated with neutron scattering, where tremendous changes of contrast (including inversion of the sign of the contrast) can be achieved in H_2O–D_2O mixtures. The measuring time for one neutron small-angle scattering curve has been reduced to about half an hour, including the measurement of both solution and solvent.[7]

From the point of view of intensity, x-ray synchrotron radiation from modern storage rings is even more powerful (Figure 1). It appears that small-angle scattering measurements could be done in milliseconds with good precision. Apart from the flux of the primary beam, many other factors have to be considered too. These are the scattering probability of the particles in solution and the crucial parts of the small-angle instruments, the monochromators and detectors.

2.1. The Probability of Small-Angle Scattering from a Dilute Solution

Small-angle scattering of dilute solutions of large macromolecules is usually performed on an absolute scale. This is due to the fact that the dissolved particles scatter the incident beam nearly independently, at least in the useful range of scattering angles and at low particle concentration. The extrapolated zero-angle scattering $J(0)$ is proportional to the square of the excess scattering density ρ and the volume V of the particle.

$$J(0) \sim (\rho V)^2 \tag{1}$$

It is also proportional to the incident photon flux P and to the number of scatterers in the irradiated sample volume. In a more quantitative way one obtains[8]

$$I(0) = \frac{P}{a^2} \frac{v^2 \rho^2}{N_L} M \, dc \tag{2}$$

where P = incident flux of photons

a = distance between sample and detector (cm)

M = molecular weight of particle

$N_L = 6 \times 10^{23}$

v = specific volume of solute in cm^3/g

ρ = contrast excess scattering density in cm/cm^3

c = concentration of solute in g/cm^3

d = thickness of sample in cm

At small angles the scattering curve is described by

$$J(k) \sim 1 - \tfrac{1}{3} R^2 k^2 \tag{3}$$

where $k = (4\pi/\lambda) \sin \theta$, λ is the wavelength, 2θ is the scattering angle, and R is the radius of gyration of the particle. The decrease in small-angle scattering can be approximated by a Gaussian with half-width k_H given by

$$k_H = \frac{1.4}{R} \tag{4}$$

Though formula (2) is very important for the determination of molecular weights, a more general formula that describes the probability W of an incident photon being scattered into a cone of small divergence is sometimes preferable:

$$W = \frac{2\pi \int J(Z) Z \, dZ}{P} \tag{5}$$

where Z is the distance of a point in the scattering pattern from the primary beam. Assuming a two-phase system in electron density as a convenient model for solutions of macromolecules, we obtain

$$W = 2\rho^2 cv\lambda^2 d \int_0^\infty H(x)\, dx \tag{6}$$

The characteristic function $H(x)$ resembles the Patterson function in crystallography.[9] For spherical particles with radius r we obtain

$$W = \rho^2 cv\lambda^2 \, dr \tag{7}$$

The optimal thickness d of a sample is given by the reciprocal of the absorption coefficient. If we take water as a simulation of biological material, an optimal thickness d (cm) $= 0.3/\lambda^3(\text{Å}^3)$ is obtained. The large increase of W with the square of the wavelength is compensated by an even stronger decrease of the optimal thickness of the sample:

$$W = 0.3\rho^2 cvr/\lambda \tag{8}$$

In the case of neutrons, the absorption of water depends only slightly on the wavelength, leading to a scattering probability of

$$W = 0.3\rho^2 cvr\lambda^{3/2} \tag{9}$$

For heavy water D_2O the optimal thickness increases by nearly one order of magnitude. Correction for absorption decreases all following values of W by $1/e = 0.37$.

As an example we calculate the contrast ρ of a protein in solution. The electron density of the protein $(0.45\ \text{Å}^{-3})$ and that of water $(0.33\ \text{Å}^{-3})$ differ by $0.1\ \text{Å}^{-3}$. Multiplication of the excess electron density $(0.1\ \text{Å}^{-3} = 0.1 \times 10^{24}\ \text{cm}^{-3})$ by the scattering length of one electron $(0.28 \times 10^{-12}\ \text{cm})$ yields the contrast $(2.8 \times 10^{10}\ \text{cm}^{-2})$. Nearly the same contrast is encountered with proteins in neutron scattering. The scattering probability W for spherical proteins therefore becomes

$$W = 2.3 \times 10^{-4} cvr/\lambda \qquad \text{(x rays)} \tag{10a}$$

$$W = 2.3 \times 10^{-4} cvr\lambda^{3/2} \qquad \text{(neutrons and proteins in } H_2O) \tag{10b}$$

$$W = 3 \times 10^{-3} cvr\lambda^2 \qquad \text{(neutrons and proteins in } D_2O) \tag{10c}$$

These equations are useful for a rough estimation of the integral small-angle scattering of globular particles. For myoglobin $(c = 10$ mg/ml, i.e., $c = 0.01$ g/cm^3, $v = 0.74$ cm^3/g) we obtain, with $r = 20$ Å and $\lambda = 1.6$ Å,

$$W = (2.3 \times 10^{-4})(0.01)(0.74)(20/1.6) = 2.1 \times 10^{-5}$$

With neutrons $(\lambda = 10$ Å) W may increase considerably:

$$W = (2.3 \times 10^{-4})(0.01)(0.74)(20)(10)^{3/2} = 1.1 \times 10^{-3} \qquad \text{(myoglobin in } H_2O)$$

$$W = (3.0 \times 10^{-3})(0.01)(0.74)(20)(10)^2 = 0.044 \qquad \text{(myoglobin in } D_2O)$$

Ribosomes are about five times larger in diameter than myoglobin. The contrast is also increased by a factor of two due to large RNA content. The probability for a photon to be scattered by a 1% solution in the case of x rays is therefore

$$W = 2.3 \times 10^{-4}(2)^2(0.01)(0.55)(100/1.6) = 3 \times 10^{-4}$$

Ribosomes from *E. coli* are matched in a H_2O–D_2O mixture containing about 60% D_2O. This means an increase of contrast by a factor 1.5 compared to proteins in H_2O. The scattering probability of ribosomes in H_2O for 10 Å neutrons therefore is

$$W = (2.3 \times 10^{-4})(1.5)^2(0.01)(0.55)(100)(10)^{1.5} = 9 \times 10^{-3}$$

In heavy water the contrast for ribosomes is lower. Nevertheless, in a sample with an optimal thickness of about 10 mm, one out of eight neutrons are scattered into small angles. It is clear that with even larger particles (viruses) and longer wavelengths, multiple scattering will become predominant for neutrons.

Multiple scattering does not appear to be of any importance with x rays as long as aqueous solutions of biological macromolecules are considered. The use of long wavelengths limits the sample thickness severely (Figure 1), and x rays with wavelengths greater than 2 Å become increasingly impractical (Figure 2) for scattering experiments.

As the relevant variable in a scattering experiment is the momentum transfer $k = (4\pi/\lambda) \sin \theta$, the means of achieving low momentum transfer measurements are different for x rays and neutrons. With neutrons the choice of long wavelengths is very advantageous, as both the resolution of momentum transfer and the integral

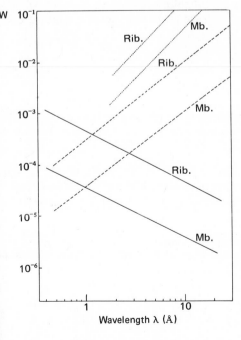

Figure 2. The probability W of photons and neutrons being scattered into small angles: x rays (——); neutrons (H_2O as solvent) (– – –); neutrons (D_2O as solvent) (. . . .); Rib = 50s subunit of *E. coli* ribosomes (mol. wt. = 1.6×10^6); Mb = Myoglobin (mol. wt. = 18,600).

small-angle intensity are increased. Limitations result from the decreasing spectral density at higher wavelengths, and multiple scattering, which may not always be desirable.

With x rays, the only way to resolve low momentum transfer is through high angular resolution. The increasing small-angle scattering intensity at low angles suggests the choice of short wavelengths, which deteriorate the resolution of momentum transfer. A compromise has to be made. This is easily done with synchrotron radiation, as the spectrum of radiation from storage rings covers the whole useful range of x-ray wavelengths (Figure 1).

2.2. The Energy Spectrum of the Source

The intensity of the primary beam is another important factor that influences the expected counting rate of scattered photons or neutrons. For both neutron reactors and electron storage rings, the wavelength spectrum is a continuum with one maximum. The most probable wavelength is a function of the particle energy and the radius of deflection (Figure 3) in the case of synchrotron radiation. The mean velocity ($\sim 1/\lambda$) of neutrons depends on the temperature of the moderator (water at room temperature, liquid deuterium as a cold source at 20 K, and heated graphite at 2000 K) (Figure 1).

The cutoff to short wavelengths is around 0.3 to 0.5 Å for both synchrotron radiation and thermal neutrons. At longer wavelengths the brightness of synchrotron radiation remains relatively high, whereas the spectrum of neutrons drops rather fast. Soft x rays can only be handled in vacuum in order to avoid absorption, whereas cold neutrons offer no technical difficulties. Although the flux of cold neutrons ($\lambda > 10$ Å) is relatively low, the increase in the total small-angle scattering intensity compensates to a large extent for the weak primary beam (Figure 2).

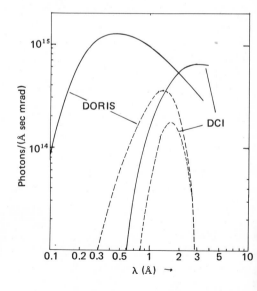

Figure 3. The wavelength distribution in an aperture of 1 mrad width and infinite height. Full lines refer to the values of the storage ring DORIS (4.8 GeV, 10 mA) and DCI at Orsay (1.8 GeV, 100 mA). The broken lines represent the measurable radiation after absorption by air and losses due to detector efficiency: DORIS (absorption by 1 m air, detector: 10 mm Ar–CO_2) DCI (absorption by 30 cm air, detector 6 mm Xe).

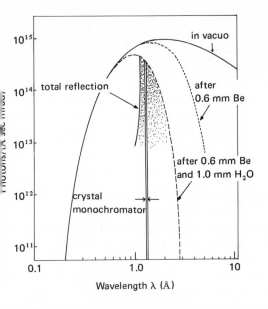

Figure 4. The wavelength spectrum of synchrotron radiation as emitted from DORIS (3 GeV, 50 mA) *in vacuo* (——), after absorption by a Be window (. . .), and after additional absorption by a water sample (---). The cutoff due to total reflection from a mirror is shown on the left side of the figure. The crystal monochromator would yield a very narrow wavelength distribution of $\Delta\lambda/\lambda = 10^{-4}$.

In small-angle experiments the size of the primary beam is no longer negligible relative to the half-width of the central peak of small-angle scattering. The angular resolution $\Delta\theta/\theta$ can be around 0.01 to 0.1. From the definition of momentum transfer it is clear that the wavelength band taken out of the spectrum should have a width $\Delta\lambda/\lambda$ that is comparable to the angular resolution. A universal way of monochromatization is offered by Bragg reflection from a single crystal, which results in a fractional wavelength spread of the order 10^{-4}. This is unnecessarily small for small angle scattering. For neutrons a very special technique of monochromatization is given by mechanical velocity selection. A fractional wavelength spread of about 0.1, which meets the requirements of small-angle scattering ideally, is achieved.

A similar monochromator does not exist for x rays and one has to admit a still broader wavelength distribution of $\Delta\lambda/\lambda = 0.3$ or more. A method of rough monochromatization is offered by the use of filters. The long x-ray wavelengths are absorbed by the sample and the short wavelengths can be eliminated by total refraction from mirrors at grazing incidence (Figure 4). At a 2.5-mrad glancing angle, quartz mirrors will reflect only wavelengths beyond 1 Å. The critical angle is proportional to the wavelength and to the square root of the scattering density of the mirror.

The gain in intensity at $\Delta\lambda/\lambda = 0.3$, compared to 10^{-4} from monochromator crystals, is tremendous. As the distortion of the central peak of small-angle scattering due to the wavelength spread is still small, information about particle size and shape in solution can be obtained in a few milliseconds. High-flux neutron beams are achieved in a similar way. The bent neutron guides (rectangular glass profiles) cut off short wavelengths. At the long wavelength side the spectral density decreases rapidly, which leads to a $\Delta\lambda/\lambda$ of about 0.5 to 1. No monochromatization at all is needed with energy dispersive methods. On the contrary, a wavelength spectrum of uniform density would be ideal. Thermal neutrons and synchrotron radiation meet this requirement fairly well. In the case of neutrons, the energy analysis is done with time-of-flight methods, whereas semiconductor detectors can be used in the case of x-ray scattering.

The measurement is done at a fixed (small) angle. The intensity as a function of the wavelength corresponds to the small-angle scattering curve as can be seen from the definition of momentum transfer. The wavelength resolution of 0.01 compares well to the angular resolution.

2.3. The Efficiency of Gas-Filled Chambers

Though considerable progress has been made in the development of fast position-sensitive counters (PSC) in this decade, the bottle-neck in scattering experiments is still the detector with its signal processing and data acquisition system. So far the maximum counting rate of a PSC hardly exceeds 10^5 cps. This is too low for future small-angle experiments as can be deduced from the Figures 1–3. The rate determining step very often is given by the time-to-amplitude conversion and the digitation of the analog signal (ADC), which takes a few microseconds. The cycle time of multichannel analyzers is shorter (about 1 μs) and the lifetime of a signal at the amplifier stage is considerably below one microsecond.

The probability of a photon (or neutron) being seen by the detector (its efficiency) depends on the nature of the counter gas. For neutrons, boron fluoride is used to create charged particles from the reaction of neutrons with boron nuclei, whereas argon or xenon together with carbon dioxide is used for x-ray counters. The efficiency depends on the absorption cross section of the gas. It increases with the depth of the chamber, the pressure of the gas, and with the wavelength of both photons and neutrons.

As a typical example we mention the position-sensitive counters of A. Gabriel. The length of a one-dimensional counter is about 60 to 80 mm, where the two-dimensional counters have a sensitive area of 200×200 mm. The wire spacing for the anode and cathode is 2 mm. The resolution is somewhat better than 2 mm in both x and y axes of the coordinate plane of the detector. The precision of the position (linearity) of the image points is not worse than 1 mm. The sensitivity pattern over 90% of the area is practically homogeneous (less than 5% deviation). At the edges of the area the local sensitivity may deviate from the mean value by 10% to 20%. The dead time of the detector system is about 8 μs. The depth of the counter, i.e., the distance between the anode and cathode plane, is 10 mm. So far we used a mixture of 7 parts argon and 3 parts carbon dioxide, which absorbs about 15% of 1.5-Å photons.

The decreasing efficiency of the detector at shorter wavelengths introduces another limit to the useful range of the x-ray spectrum. If one takes into account the absorption of x rays by beryllium windows, some air and helium paths, and the efficiency of the detector, the resulting spectrum of synchrotron radiation shows a relatively narrow region of nearly maximum intensity (Figure 5). These curves are

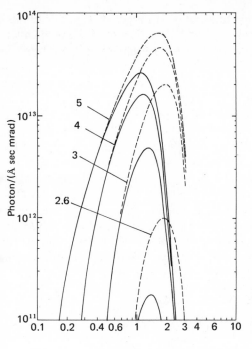

Figure 5. The measurable wavelength distribution in an aperture of 1 mrad width and infinite height at various particle energies (2.6, 3, 4, and 5 GeV) of the storage ring DORIS. The detector consisted of a 10-mm $Ar/CO_2(7:3)$ mixture and was 10 mm deep. The absorber consisted of 1-m air + 0.9-mm Be (——) or 0.1-m air + 0.9-mm Be (– – –).

of immediate use for the estimation of small-angle scattering. As the wavelength width of photons emerging from the monochromator is nearly constant, the spectrum should be multiplied by the wavelength. On the other hand the integral intensity decreases as $1/\lambda$. This means that the two operations cancel. After multiplication by the wavelength, the resolution of the crystal, and the solid angle of radiation accepted by the sample, one obtains the reduction factor for the flux of incoming photons. This does not change the optimum conditions for small-angle scattering, which can be taken from Figure 5.

3. Small-Angle Scattering Instruments

Any scattering experiment has to define the primary beam in direction and energy (or wavelength), i.e., the initial momentum of the photon or neutron has to be known. The direction of the beam can be defined by slits, which simply cut out a narrow solid angle of the radiation emerging from the source. This property is also shared by mirrors and monochromators, which, because of their finite size, accept only a small part of the radiation. In a further step, only those photons with wavelengths in a given small range are allowed to travel to the small-angle instrument. A poor monochromatization is achieved with mirrors, which sometimes may be acceptable for the experiment. In most cases a subsequent crystal monochromator will diffract a monochromatic beam into the small-angle instrument. It is also proposed to extract the monochromatic radiation with a crystal first and to eliminate higher orders of the Bragg reflection by total reflection from a mirror afterwards. This appears to be advantageous as the monochromator crystals seem to suffer less damage from radiation than mirrors.

In most cases both mirrors and monochromators are bent. As the reflecting planes of the monochromator are orthogonal to that of the mirror, a drastic demagnification of the beam cross section can be achieved. An accepted beam cross section of 6 × 30 mm might be decreased to 0.8 × 1.0 mm. The angle of convergence is small enough in order to obtain up to 1000-Å resolution, because of the long distances (several meters) between the focusing elements and the focus.

If only one monochromator crystal is used, the direction of the emerging monochromatic beam depends on the wavelength according to Bragg's law. The instrument has to rotate around the monochromator whenever a change in wavelength is desired. Though the mechanical problems have been solved for medium-size optical benches and instruments, the rotation of a very long instrument causes many problems. Therefore a double-crystal monochromator system has been developed, which leads to constant beam position at all wavelengths.

3.1. Instruments with One Monochromator Crystal

Small-angle instruments using one crystal are run at LURE, Orsay (near Paris) and SSRL, Stanford (California). The apparatus designed for small-angle experiments at LURE has been improved to match the requirements of studies of biological systems in solution.[10] Monochromatization and horizontal focusing is obtained by a bent germanium crystal. The low vertical divergence of the beam allows one to obtain a point-focus geometry by using slits in this direction, as represented in Figure 6. The scattering curves are recorded with a linear position-sensitive counter (with conductive wire) made by A. Gabriel. The window is about 8 cm long. The samples are contained in quartz capillary tubes (1-mm diameter) horizontally positioned in a thermostated sample holder. Absolute scale measurements are obtained by the use of a black carbon sample as a reference. Emphasis has been placed on optimizing the signal-to-noise ratio thanks to carefully adjusted slits and maximum path under vacuum.

The small-angle instrument used at SSRL has a silicon crystal as a monochromator. The beam is diffracted at a Bragg angle of 45°, thus eliminating all but a series of discrete bands. Only the band of longest wavelength (1.48 Å), resulting from the (333) reflection, is used for scattering data. Consequently, this band is isolated from shorter wavelength bands ($\lambda = 1.08$ Å, 0.88 Å, and shorter). Electronic discrimination alone suffices to eliminate bands with wavelengths of 0.88 Å and shorter, and is almost adequate to eliminate the band at 1.08 Å. However, to ensure that the latter is excluded, a copper foil (25 mm thick) is used as a filter.

The beam is shaped by a single defining slit. A guard slit with dimensions slightly larger than those of the defining slit is present so that most of the radiation scattered from the edges of the defining slit can not reach the detector. Aluminium slit edges, bevelled at 5°, were shaped to attenuate the 1.48 Å radiation with mini-

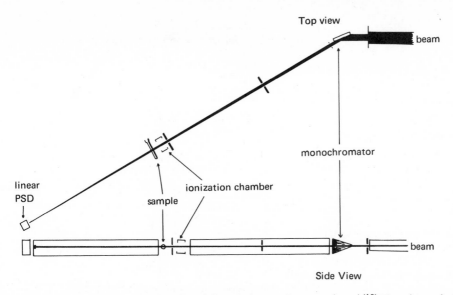

Figure 6. Small-angle instrument with a bent (triangular) monochromator (LURE).[10] The tubes on both sides of the sample cell are evacuated. The distance between sample and detector can be varied as required by the experiment.

num scattering. Lead, required to exclude x rays of higher energies, generated undesirable scattered radiation at about 1.4 Å. Therefore, the lead edges were inset.

The sample cell is made of two parallel 8-μm-thick polyimide sheets (Kapton, made by Dupont). In order to minimize absorption the x rays passed through helium at atmospheric pressure. A position sensitive counter (PSD 100, made by Tenelec) with a carbon coated quartz fiber of high resistance was used. The count rates were 1000/s or less.[11]

3.2. The Mirror–Monochromator System

The gain in intensity from a single monochromator setup compared with a similar arrangement using a GX6 Elliott rotating anode is presently of the order of 50.[10] A further improvement is expected from the introduction of focusing mirrors.

The EMBL outstation at Hamburg runs two x-ray small-angle instruments, where vertical and horizontal demagnification of the beam cross section is achieved by mirrors and a bent germanium crystal, respectively (Figure 7). The quartz mirror accepts the radiation at grazing incidence. At a glancing angle of 3 mrad, wavelengths shorter than 1 Å are eliminated. At a distance of 20 m from the source (storage ring DORIS) the beam height is approximately 10 mm. Eight quartz mirrors 20 cm in length, 5 cm in width, and 2 cm in thickness are aligned in such a way that each flat surface is a tangent to an ellipse with focal points in the source (20 m) and in the sample region (4–6 m). Horizontal focusing is achieved by bending the Ge monochromator (111 reflection). All movements of mirrors and monochromators are remote controlled. The focal spot size is smaller than 1 mm^2.

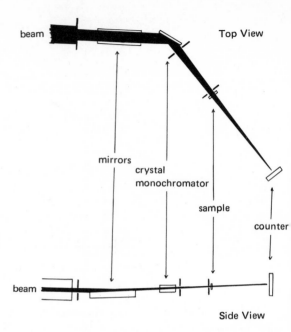

Figure 7. Schematic top and side view of the low-angle diffraction cameras at the storage ring DORIS. Eight quartz mirrors, each 20 cm in length, 5 cm in width, and 2 cm in thickness, are aligned in such a way that their surfaces are tangents of an ellipse. This leads to a focal point at a distance of about 3 to 5 m. Vertical focusing is achieved by a bent Ge crystal (111 reflection).

One of the optical benches is movable, whereas the other has a fixed takeoff angle of $2\theta = 26°$. The movable optical bench can accept wavelengths between 1 and 2 Å using the 111 reflection from a germanium crystal. The gain factor compared to classical equipment is of the order of some hundred at the present conditions of parasitic use of synchrotron radiation. Several kinds of position-sensitive counters were used. A two-dimensional PSC based on an image intensifier television system has been tested successfully during the last months. This detector is very efficient at high intensities.[12] Linear position sensitive counters (conducting wire) from A. Gabriel were preferred when the scattered intensity was weak.

3.3. The Double-Monochromator System

The inherent advantage of the double-monochromator system is the invariance of the emerging beam direction at different wavelengths and its rapid tunability. A further advantage lies in the fact that it easily explores the vertical plane of the hall for the installation of further instruments.

As synchrotron radiation is almost fully linearly polarized, the polarization factor is nearly one at all angles for upward reflection. This allows the use of long wavelengths at high angles without appreciable losses due to the polarization factor.

In order to maintain the same position of the beam leaving the second monochromator, the crystals were mounted on separate goniometer heads, i.e., the principle of one double-monochromator crystal was abandoned. The second crystal has to be displaced whenever the glancing angle of the first crystal is changed. As

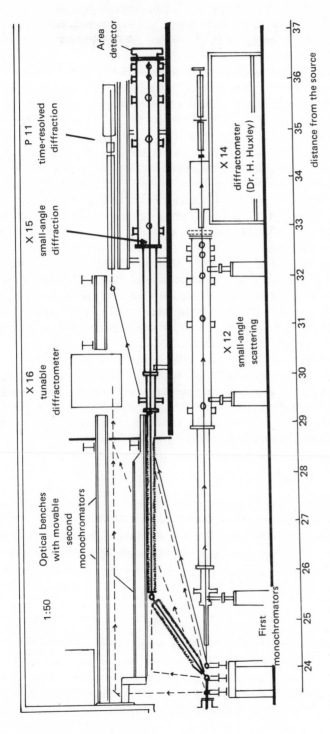

Figure 8. Small-angle instrument (X 15) with a double-monochromator system. The first crystal at a distance of 23.8 m from the source deflects the beam upwards. A second crystal hanging beneath an iron girder 125 cm above the main beam deflects the beam into the horizontal direction. The second monochromator can be moved continuously along a distance of 4 m. The steel tubes (from 29 to 36.5 m distance from the source) can be evacuated. At present the whole instrument is filled with helium.

the two crystals had to be separated, the distance between them hardly affects the quality of the radiation. There might even be an improvement at greater distances between the monochromators because less background radiation created by the first monochromator will be accepted by the second. We have mounted the second monochromator 125 cm above the axis of the main beam (Figure 8). With a displacement of the second monochromator along an H-shaped iron girder 6 m in length, Bragg angles between 8° and 45° can be achieved. The wavelength range extends from 1 to 4 Å when the 111 reflection of germanium crystals is used. Both crystals were cut asymmetrically at 7° with respect to the 111 plane of the crystal and they were aligned in such a way that the vertical height of the beam was decreased twice. A line-shaped horizontal beam of typically 1 × 10 mm results. This beam geometry is quite convenient for the use of capillary tubes as sample holders. But it is also an unnecessary waste of beam intensity that could be avoided by sample cells with a greater vertical height and a different monochromator setting. As the resolution of the detector is about 2 mm, a sample cell several millimeters in height would not only be tolerable, but also desirable in order to accept the full vertical divergence of the beam.

The small-angle instrument consists of three connected vacuum tubes with diameters between 150 and 350 mm. A position-sensitive area detector from A. Gabriel collects the scattered photons and stores the scattering pattern in the memory of a multichannel analyzer (IN 90, made by Intertechnique). There are 8192 picture elements stored with 20-bit resolution. The number of channels along the x and y axis can be 2^m2^n, where $m + n$ may not exceed 13.

Though small-angle scattering depends only on one variable, a two-dimensional counter is preferred because it accepts the total small-angle scattering. It offers unique advantages for scattering experiments and it is also very convenient for the alignment of monochromators and slits. It should also be mentioned that this double-monochromator system is easily transformed into a Bonse–Hart small-angle camera by putting the sample between the monochromators.

3.4. The Energy Dispersive Method

Small-angle scattering is usually represented as a function of the momentum transfer $k = (4\pi/\lambda) \sin \theta$. The definition suggests two ways of measuring $J(k)$: Either $J(k)$ is recorded as a function of the scattering angle 2θ at constant wavelength λ, or the scattering angle is kept constant and $J(k)$ is measured as a function of the energy $E = hc/\lambda$,

$$ k = \frac{4\pi \sin \theta}{\lambda} = \frac{4\pi E \sin \theta}{12.398} $$

where E is measured in KeV and λ in Å. The general layout of the apparatus is shown in Figure 9. It is very simple because the role of x-ray optics and the detection of photons is combined in an energy analyzing semiconductor detector. The distance between sample and detector was 6 m. The slit system, which consisted of blocks of high-purity polished lead with a thickness of 12.5 mm, was found

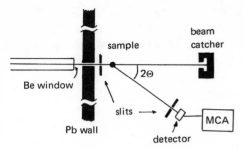

Figure 9. Principle of the energy dispersive x-ray method using synchrotron radiation. 2θ is a fixed angle. A semiconductor detector is used. MCA is the multichannel pulse height analyzer.

adequate to completely stop the unused portions of the main beam and also to avoid fluorescence lines at the 5-GeV particle energy of the DESY synchrotron. However, recent experience with synchrotron radiation from 7-GeV electrons seems to indicate that thicker slits are necessary.

The semiconductor detectors producing the energy discrimination were placed behind the secondary slits: A lithium-doped silicon detector has a range of detection of approximately 4–40 KeV. Consequently, one has a k range extending over about one order of magnitude. Two or three settings of the detector (or better, 2 or 3 detectors at different angles) will cover the interesting range of small-angle scattering.[13]

4. Experimental Results

Though small-angle scattering with synchrotron radiation will have to be judged finally from its use for time-resolved experiments, so far only test experiments are being done in different synchrotron laboratories in order to find optimal conditions for the experiment.

At LURE (Orsay) the performance of the apparatus has been tested on solutions of aldolase and ferritin.[10] With a sample-to-detector distance of one meter (resolution 850 Å) and a dilute solution (3.5 mg/ml aldolase, mol. wt. 150,000) a good determination of the radius of gyration is obtained in 30 min. In such an unfavourable case—great distance, very dilute solution—the protein scattering at low angles accounts for 70% of the total scattering. A series of experiments on the *E. coli* 50 s ribosomal subunit in solvents of variable density has been carried out in the following conditions: 30 mg/ml, sample–detector distance = 380 mm, and angular range corresponding to 230–15 Å. The ribosome scattering accounted for 98–99% of the total scattering at 200 Å and 30% at 15 Å. Kinetic studies of some phase transitions in concentrated lipid–water systems have also been started, these involve changes in the conformation of the hydrocarbon chains of the lipids. The first results are promising.

Solutions of the myosin subfragment $-1(S1)$ (mol. wt. $= 1.15 \times 10^5$) have been studied by Kretzschmar, Mendelson and Morales[11] at the Stanford Synchrotron Radiation Laboratory (SSRL). Two tests were made for radiation damage of S1. First, samples of S1 were subjected to large doses of radiation by irradiating the whole area of a sample cell (75 mm²) for about one hour. The ATPase activities of samples tested after the exposure were indistinguishable from those of nonirradiated samples taken from the stock solution. Second, the scattering measured during the first 30-min exposure was compared to the scattering measured in a 30-min period following a 140-min irradiation. The two curves were indistinguishable. Thus, neither test indicated any radiation damage.

The ratio between the scatter from the solute (7.9 mg/ml) and that from the solvent was 90% of the total scatter at small angles, whereas at 30 mrad it was only 20% of that due to the solvent.

At the EMBL outstation in Hamburg small-angle scattering for solutions have been done with the instrument using two monochromators (Section 2.3), whereas small-angle diffraction projects on fibrous structures were carried out at the double-focusing beam lines (Section 2.2). The results from the latter will be described elsewhere.

Test experiments have been made with solutions of different biological macromolecules (ferritin, ATPase from bacteria, bovine serum, albumin, myglobin). The results from ATPase solution (mol. wt. = 330,000) can be compared readily with those from LURE and SSRL. A 1% solution of ATPase gives rise to scattering which is 70% of the total scattering at very small angles. The background is independent of the angle. It appears to be higher as no precautions were taken to suppress the fluorescence emitted from the lead slits. With a very dilute solution of 50s ribosomal particles (10.4 mg/ml) a peak to background of 1 : 1 was found. After a measuring time of 10 min the radius of gyration from the ATPase and 50s ribosomal subunit were determined from the solutions mentioned above with good and satisfactory precision, respectively (Figure 10).

A concentrated ferritin solution ($C = 100$ mg/ml) has been measured to greater angles in order to test the spatial resolution of the detector. Any distortion of the linearity would deform the spherical pattern and, on integration over the detector area at constant scattering angle, details of the outer part of the scattering curve

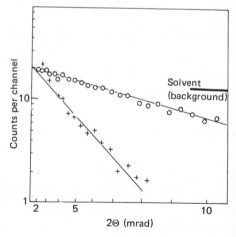

Figure 10. Small-angle scattering (Guinier plot) of ATPase (O) and the 50s subunit of *E. coli* ribosomes (+). The concentrations are 10 and 0.4 mg/ml, respectively. The measurements were made with the double-monochromator instrument in 10 min. The wavelength was 1.54 and 1.23 Å for ATPase and the 50s subunit, respectively. The distance between sample and detector was 3.77 m.

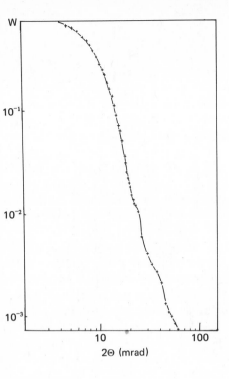

Figure 11. Small-angle scattering of ferritin ($C = 100$ mg/ml). Measuring time was 10 min.

might be lost. This is not the case with the present area detector, as there is no difference between our results (Figure 11) and those reported earlier.

A more convincing test for the spatial resolution can be obtained from the measurement of diffraction patterns. Figure 12 shows the pattern obtained from collagen fibers. This collagen sample is used for rapid calibration of the beam intensity. 50 counts per second in the third-order reflection correspond to 10^7 incident photons per second.[15]

As there were 80 cps in the third order, the effective intensity was 1.5×10^7 cps. The correction for the efficiency of the detector at 1.23 Å leads to 10^8 incident photons per second. The estimations from small-angle scattering of protein solutions lead to similar values. The measured primary beam at the sample is about one order of magnitude weaker than the expected one (positron energy at DORIS = 4.7 GeV, current = 10 mA). This may be due to incomplete exchange of air by helium in the 14-m path between the first monochromator and the detector (Figure 8).

5. Conclusions

The development of diffuse small-angle scattering techniques using synchrotron radiation has started only very recently. Most of the results were obtained during the last months. They are very promising, as a considerable reduction of the

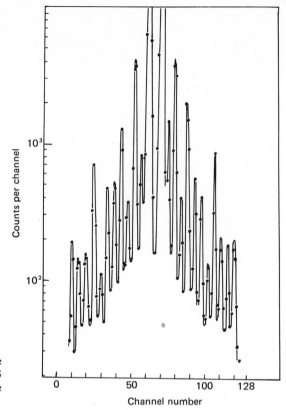

Figure 12. Diffraction pattern of collagen. Measuring time was 200 s. The spot size of a diffraction peak covers about 5 picture elements of the two-dimensional position-sensitive counter.

measuring time compared to that with classical equipment could be obtained almost immediately. The small-angle instruments are now rapidly being improved. A major problem is the low counting rate of position-sensitive counters, which is mainly due to the associated data acquisition system. From the available small-angle scattering data produced at synchrotron radiation facilities, it can be seen that so far the counting rates from the experiment did not lead to electronic pileup. However, this will change very soon when the instruments are improved and when reasonable compromises are made in the wavelength resolution. It is also expected that the x-ray flux from storage rings will increase in the next years.

X-ray small-angle scattering is a low-resolution method. The experimental data are not amenable to unequivocal interpretation in terms of a structural model. The combination of x-ray scattering with neutron scattering experiments will certainly be very useful because the latter can provide more detailed information about the particle structure in solution (contrast variation, specific deuteration of parts of molecules). Once the correlation between neutron and x-ray small-angle scattering of a given biological system is established, the analysis of time-dependent scattering patterns obtained at powerful synchrotron radiation sources becomes meaningful.

References

1. H. B. Stuhrmann, *Q. Rev. Biophys.* **11**, 71–98 (1978).
2. H. B. Stuhrmann, *Acta Crystallogr. Sect. A* **26**, 297–306 (1970).
3. H. B. Stuhrmann, *J. Appl. Crystallogr.* **7**, 173–178 (1974).
4. H. B. Stuhrmann, in *Neutron Scattering for the Analysis of Biological Structures, Brookhaven Symp. Biol.* **27**, 3 (1975).
5. H. B. Stuhrmann and A. Miller, *J. Appl. Crystallogr.* **77**, 325–345 (1978).
6. H. B. Stuhrmann and R. G. Kirste, *Z. Phys. Chem.* (*Frankfurt on Main*) **56**, 333–337 (1967).
7. K. Ibel, *J. Appl. Crystallogr.* **9**, 296–309 (1976).
8. O. Kratky and I. Pilz, *Q. Rev. Biophys.* **11**, 39–70 (1978).
9. G. Porod, *Kolloidz.* **124**, 84–114 (1951).
10. A. Tardieu (private communication).
11. K. M. Kretzschmar, R. A. Mendelson, and M. F. Morales, *Biochemistry* (in press).
12. G. T. Reynolds, J. R. Milch, and S. M. Gruner, *Rev. Sci. Instr.* **49**, 1241–1249 (1978).
13. J. Bordas and J. Randall, *J. Appl. Crystallogr.* **77**, 434–441 (1978).
14. F. A. Fishbach and J. W. Anderegg, *J. Mol. Biol.* **13**, 569–576 (1965).
15. W. Farugi (private communication).

16

Small-Angle Diffraction of X Rays and the Study of Biological Structures

G. ROSENBAUM and K. C. HOLMES

1. Design Criteria for Small-Angle Diffraction

1.1. Conditions Imposed by the Specimen and Experiment

1.1.1. Specimens

Specimens fall broadly into three categories, *crystals*, *fibers*, or *solutions*, depending on the degree of order. The categories *crystals* and *solutions* are the subject of Chapters 17 and 15, respectively, so that they will not be considered in detail here. However, the experimental conditions necessary for the registration of diffraction patterns from crystals and fibers with large unit cells (ca. 50 nm) are similar and thus the two will be considered together in the following discussion. In the main, similar considerations apply to solution studies. However, as they need additionally very low parasitic scattering and as the size of specimen is not generally limiting, design strategy is considerably different. For crystals and fibers the specimens are usually very small (100–500 μm). A limited amount of parasitic scattering in defined directions is tolerable. In the following analysis we do not

G. ROSENBAUM and **K. C. HOLMES** ● European Molecular Biology Laboratory, and Max Planck Institut für medizinische Forschung, Heidelberg, Germany.

explicitly consider the parasitic scattering. In critical situations the rules for solution scattering apply.

Crystalline specimens are characterized by the dimensions of the crystal (t_x, t_y, t_z) and by the mosaic spread of the crystal (m_x, m_y). Specimens with all degrees of coherence have been studied. Sometimes it is necessary to take the intrinsic width (c_x, c_y) of a Bragg reflection into account.

Fibrous specimens consist of an array of microcrystals randomly orientated about one axis so that the Bragg reflecting condition is always obeyed. Such specimens are characterized by their linear dimensions (t_x, t_y, t_z). The concept of mosaic spread for such specimens is not very meaningful. Such samples generally have broad Bragg reflections (e.g., $c_x = c_y = 0.05$ nm^{-1} for muscle fibers). Often there is so little spatial coherence between neighboring molecules that the fibrous molecules scatter independently $(c_{x, y} \to \infty)$, in which case the fiber diffraction pattern consists of a set of layer lines along each of which the intensity is continuous. Such samples are known as noncrystalline fibers. They make special demands on the experimental arrangement since the background must be low in order to observe the relatively weak layer-line streaks.

An important boundary condition imposed by biological specimens is the need to maximize signal while minimizing dose. This generally makes the use of monochromatic beams mandatory.

1.1.2. Experimental Requirements

A certain angular resolving power (a_x, a_y) is required at the detector. a_x is generally the angular separation of two neighboring Bragg reflections,

$$a_{x, y} = 2[\theta(h_2) - \theta(h_1)] \tag{1}$$

It is necessary to distinguish between two kinds of resolution: the closest angular approach to the direct beam and the order-to-order resolution. In the following analysis we consider only the order-to-order resolution because the criterion for the order-to-order resolution is also a *necessary* condition for the resolution from the direct beam. The additional conditions imposed by the resolution from the direct beam are the same as those pertaining to small-angle scattering (Chapter 15).

In most biological experiments involving synchrotron radiation one wishes to perform the experiment in as short a time as possible, either because the specimen is unstable or because one is carrying out time-resolved studies of structure (see Section 2.1). This provides a clear design criterion, namely the maximalization of flux through a certain set of Bragg reflections or the maximalization of intensity at the detector plane. The second criterion is appropriate for situations where the signal-to-noise ratio is important (e.g., high background). The well-defined optical properties of the beam can be important in improving the signal-to-noise ratio.

On account of its limited spatial resolution the detector can impose strong constraints on experimental design (see Section 1.4). In particular the detector resolution can determine the physical size of the optical system.

1.2. X-Ray Optics and Optimalization

1.2.1. Definitions

1.2.1.1. Coordinate System. The x–z plane is taken to be in the plane of the synchrotron with the z axis pointing outwards along the beam. The y axis points vertically upwards.

1.2.1.2. Definition of Variables. Variables are defined in Table 1 and illustrated in Figures 1, 4, and 5. The subscripts x and y are used for equations pertaining to the x–z plane or y–z plane, respectively. In cases where either an inequality in the x–z plane or in the y–z plane is limiting, the equation in the x–z plane will be used as representative of the limiting condition.

1.2.2. Optical Principles of Curved Mirrors and Curved Crystal Monochromators*

1.2.2.1. The Optical Elements. The most readily obtainable x-ray focusing elements of sufficient perfection to be used with synchrotron radiation are cylindrical lenses: curved mirrors used at grazing incidence and curved crystal monochromators used at small Bragg angles. A review of curved mirrors and curved crystal monochromators for conventional sources has been given by Witz.[1] Curved crystal monochromators behave as mirrors when used with white radiation so that the two classes of focusing elements can be treated in a unified way. Monochromators have the extra property that a variation in the glancing angle θ (the Bragg angle) leads to a variation in the reflected wavelength (see Section 1.2.2.4).

The combination of a curved mirror at right angles to a curved monochromator was introduced for conventional x-ray sources by Huxley and Holmes (see Huxley and Brown[2]) in their investigations of diffraction from muscle fibers and was introduced for synchrotron radiation by Barrington Leigh and Rosenbaum.[3] For many applications this turns out to be the optimal combination.

Most of the electrons in atoms behave as if they were free when illuminated with x radiation. As a result the refractive index of solids for x rays is slightly less than unity and total external reflection can be observed from polished surfaces at grazing angles of incidence. Typically the critical grazing angle for total external reflection (θ_c) lies between 3 and 10 mrad. Electron dense materials have a larger θ_c but have also higher absorption, which leads to the critical reflection phenomenon becoming unsharp. The surface of the mirror must be very well polished.[4] A very good optical finish on fused quartz or glass is necessary. Polished metals are not sufficiently flat. However, metal surfaces can be prepared by vacuum deposition on polished quartz or glass.

Perfect single crystals that show total primary extinction are used as monochromators. The importance of using perfect single crystals rather than mosaic crystals is that they give *specular* reflection essentially without dispersion within the narrow range of wavelengths allowed by dynamical theory. Mosaic crystals degrade

---

* See also Chapters 3, 13, and 14.

Table 1. Definition of Variables

θ	Bragg angle—glancing angle of incidence or reflection
θ_c	Critical angle of total external reflection
x, y, z	Orthogonal coordinates in the experimental area
t_x, t_y, t_z	Dimensions of the specimen
m_x, m_y	Mosaic spread of specimen
u	Distance from source to mirror (monochromator)
v	Distance from mirror to focus
R	Instantaneous radius of curvature of mirror (monochromator)
ψ	Azimuthal coordinate of general point A on mirror surface
σ	Angle between Bragg planes and surface of monochromator
l	Length of mirror
λ	Wavelength of radiation
λ_c	Critical wavelength
SA	Spherical aberration
B	Luminosity
S_x, S_y	Dimensions of (virtual) source
s_x, s_y	Dimensions of image
D	Specimen–detector distance
a_x, a_y	Required angular resolution at detector plane
d_x, d_y	Width of a Bragg reflection at detector plane
c_x, c_y	Width of a Bragg reflection in reciprocal space
F	Photon flux in a given Bragg reflection
I	Intensity in a given Bragg reflection
P_h	Integrated reflectivity of Bragg reflection
P_0	Flux in primary beam
q	$= D/v$
r	$= v/u$

the luminosity of synchrotron radiation because they are dispersive and only in special circumstances can a mosaic monochromator such as graphite be successfully employed (see Section 2.4).

1.2.2.2. The Focusing Condition—Radius of Curvature. Rays emanating from a source S are focused by a curved mirror or monochromator $MO'M$ (Figure 1) to a focus I. For a mirror the glancing angle of incidence (θ) must be equal to the glancing angle of reflection. For a monochromator it is usual to cut the surface of the crystal at an angle σ to the Bragg planes (Guinier[5]) so that the glancing angle of incidence is $\theta + \sigma$ and the glancing angle of reflection is $\theta - \sigma$.

The focusing condition (for $\psi \ll 1$) requires that OAS and OAI be triangles. Using in addition the condition of specular reflection the *change* in θ in going from the origin O to the general point A is

$$\Delta\theta = \theta_A - \theta_0 = \psi - z \sin(\theta_0 + \sigma)/u_0 = z \sin(\theta_0 - \sigma)/v_0 - \psi \qquad (2)$$

Moreover

$$\psi = dy/dz = l/R = z/R_0 \qquad (3)$$

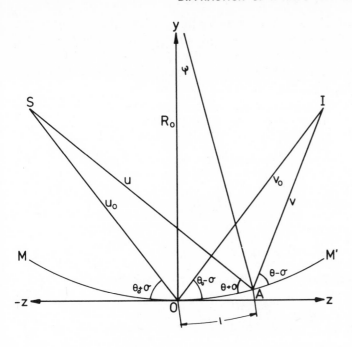

Figure 1. Focusing geometry: rays emanating from S are focused to I by the mirror or monochromator MOM'. The radius of curvature of the mirror (monochromator) is R_0. The angle of incidence on the mirror is $\theta + \sigma$, the angle of reflection is $\theta - \sigma$. For a mirror $\sigma = 0$.

From (2) and (3) one obtains

$$2/R = \sin(\theta - \sigma)/v + \sin(\theta + \sigma)/u \qquad (4)$$

and

$$\Delta\theta = z/2[\sin(\theta - \sigma)/v + \sin(\theta + \sigma)/u] \qquad (5)$$

If $\sigma = 0$ (mirror or symmetrical monochromator), (4) is the lens equation

$$2/R \sin \theta = 1/u + 1/v \qquad (6)$$

For symmetrical curved crystal monochromators used with monochromatic radiation the Bragg law imposes the additional constraint $u = v$, whereupon (6) becomes

$$u = v = R \sin \theta \qquad (7)$$

Equations (7) are the Johann focusing conditions.[6] Equation (4) is the lens equation for an asymmetrically cut curved crystal.

Equations (4) and (6) demonstrate that the *local focusing* depends only upon the instantaneous radius of curvature and the glancing angle of incidence. It is shown below that the radius of curvature is the reciprocal of the coefficient of the second-order component in the power expansion of the mirror surface.

For a long mirror u and v are different for the two ends of the mirror so that the mirror should be appropriately shaped in order to bring the ray bundle to a focus at I. For exact on-axis focusing we require the general point A to lie on an ellipse (with foci S and I) generated by

$$u + v = u_0 + v_0 \qquad (8)$$

For any curve not satisfying (8) the elements of the mirror will not focus to the same point but will produce an image caustic. By analogy with normal optics we refer to this effect as the spherical aberration.

　　1.2.2.3. Spherical Aberration. We calculate the spherical aberration for a cylinder and for a logarithmic spiral. Let the form of the mirror be

$$y = A_2 z^2/2! + A_3 z^3/3! + \cdots \tag{9}$$

The radius of curvature is given by

$$R = [1 + (dy/dz)^2]^{3/2}/(d^2y/dz^2) \tag{10}$$

For $(dy/dz)^2 \ll 1$ we may expand (10) to give

$$d^2y/dz^2 \approx 1/R(z) + (3/2R_0)(dy/dz)^2 \tag{11}$$

Substituting (3) in (11) we obtain

$$d^2y/dz^2 = 1/R(z) + 3z^2/2R_0^3 \tag{12}$$

Comparing the second derivative of (9) with (12) as $z \to 0$ one obtains

$$A_2 = 1/R_0 \tag{13}$$

Furthermore, neglecting the second-order term in (12), we obtain

$$1/R(z) = 1/R_0 + A_3 z \tag{14}$$

for the first-order dependence of $1/R$ on z. From (9) we obtain

$$\psi = dy/dz = z/R_0 + A_3 z^2/2 \tag{15}$$

The way R and ψ vary with z determine the quality of the focus: the value of A_3 determines whether or not the mirror can focus to a point. The condition for a perfect focus is equation (8), the generating equation of an ellipse. Combining (8) with (4) and (14) we obtain for an ellipse

$$A_3 = \tfrac{3}{8}[\sin 2(\theta_0 - \sigma)/v_0^2 - \sin 2(\theta_0 + \sigma)/u_0^2] - \sin 2\sigma/4u_0 v_0 \tag{16}$$

Therefore to produce an aberration-free focus the form of the mirror must be

$$y = (1/R_0)z^2/2! + A_3 z^3/3! \tag{17}$$

where R_0 and A_3 are given by (4) and (16).

　　The generating equation of a cylinder is

$$R = R_0 \tag{18}$$

Therefore, for a cylinder, $A_3 = 0$ [equation (14)]. The spherical aberration (SA_{cyl}) may be found by calculating how much $\psi_{cylinder}$ deviates from $\psi_{ellipse}$ and by multiplying this by $2v$. Thus from (15)

$$SA_{cyl} = vl^2\{\tfrac{3}{8}[\sin 2(\theta_0 - \sigma)/v_0^2 - \sin 2(\theta_0 + \sigma)/u_0^2] - \sin 2\sigma/4v_0 u_0\} \tag{19}$$

where l has been substituted for z. Equation (19) gives the distance from the focus of the rays coming from the extremities of the mirror and represents an overestimate of the spherical aberration.

Logarithmic spirals are often used as monochromators (de Wolff[7]). For a logarithmic spiral,

$$\theta = \theta_0 \tag{20}$$

measured along radial lines emanating from S. We find

$$A_3 = -\sin 2(\theta + \sigma)/2u_0^2 \tag{21}$$

Therefore, proceeding as for the cylinder, we obtain

$$SA_{log} = vl^2[\tfrac{3}{8} \sin 2(\theta_0 - \sigma)/v_0^2 - \tfrac{7}{8} \sin 2(\theta_0 + \sigma)/u_0^2 - \sin 2\sigma/4v_0 u_0] \tag{22}$$

Note that the cylinder and logarithmic spiral having the same focal properties (same u, v, and σ values) are indistinguishable to the second order since in each case the coefficient A_2 in (9) is $1/R_0$ where R_0 is given by the lens equation (4) or (6). The two forms differ in A_3 which alters the spherical aberration in the two cases.

1.2.2.4. Wavelength Inhomogeneity. If the form of the monochromator is a logarithmic spiral with S as origin, the angle $\theta + \sigma$ is constant and there is no wavelength inhomogeneity.

If the form of the monochromator is a cylinder, the variation in θ is found by putting $\theta = \theta_0$ and $z = l$ in equation (5), viz.,

$$\Delta\theta = \tfrac{1}{2}l[\sin(\theta_0 - \sigma)/v + \sin(\theta_0 + \sigma)/u] \tag{23}$$

$\Delta\theta$ may be expressed as a wavelength variation by differentiating Bragg's law,

$$\Delta\lambda/\lambda = \cot\theta\Delta\theta \tag{24}$$

If the object and image distances are chosen so that

$$u/v = \sin(\theta_0 + \sigma)/\sin(\theta_0 - \sigma) \tag{25}$$

$\Delta\theta$ is zero. Equations (25) together with (4) are the Guinier focusing conditions,[5] which are *necessary* conditions when the monochromator is used with monochromatic radiation. For an obliquely cut crystal ($\sigma \neq 0$) they replace the Johann focusing conditions. Conversely, the Guinier focusing conditions produce no wavelength inhomogeneity for a cylindrical monochromator used with white radiation. Note that the Guinier focusing conditions also apply to the logarithmic spiral. Moreover, a logarithmic spiral not obeying (25) is not monochromatic and has the same wavelength inhomogeneity as the cylindrical crystal having the same radius of curvature and cut angle.

1.2.3. Optical Phase Space

In Chapter 2 the general properties of the source have been described. A special property of the synchrotron source is the correlation between the direction of emission and position in the plane of the source. This is due to the fact that the trajectories of the electrons in the ring are themselves part of an optical system. Moreover, because of the colinearity of the electrons and the emitted radiation the electron trajectories are essentially part of the external optical system. Therefore, the electron paths must be included in a consideration of the optical properties of the source.[8–10] We may do this by a consideration of the luminosity B expressed as a function of position and angle in an optical phase space (Green[8] and Hastings[9]). If B is plotted on a graph whose axes are the y axis and the angle y' between the direction of observation and the x–z plane, contours of constant luminosity are elliptical (Figure 2a). In Figure 2 the width of the cone of emission of the radiation from each electron, which is approximately a Gaussian distribution with standard deviation 0.16 mrad, has been taken into account. At some value of $z(z_y)$ the major axes of the ellipse are parallel to y and y' (Figure 2b). This is the virtual source for the y direction. At this point the luminosity function may be expressed as the product of a function depending only on y and a function depending only on y'. The factorization of B at the virtual source allows one to apply normal geometrical optics to the beam issuing from this point. The effects of the defining slits of the x-ray camera can be represented on the same diagram. In the plane of the slits (which will be some meters from the source) the slits are represented on an optical phase space diagram as a pair of vertical lines. These are shown (broken lines) in Figure 2a. The trace of the slits (i.e., the parallel lines delineating that region of the ray bundle that can be accepted by the defining slits) rotates as one moves along z. The slits are set to accept the whole of the vertical divergence of the beam (Figure 2b). However it can transpire that the size of the specimen forces one to reduce the aperture so that it cuts into the luminosity function in Figure 2b.

The source for the x axis is the point at which the line through the x-ray camera is tangent to the electron trajectory. The phase-space diagram for the x axis is shown in Figure 3. x' is the angle between the direction of observation and the y–z plane. The source is unbounded in the x' direction, the actual bounds being provided by the horizontal entry slits of the x-ray camera. The projections of the entry slits of the x-ray camera on the source plane are shown by solid lines in Figure 3.

The apparent dimensions of the x-ray source (S_x, S_y) will be the width of the luminosity functions at the virtual source points. In practice one has to establish empirically where the source points are. Compared with conventional sources the

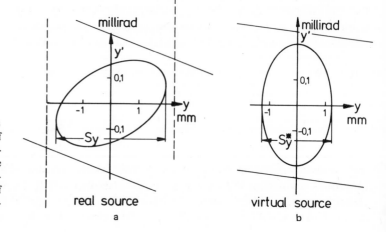

Figure 2. Optical phase space: contours of constant luminosity as a function of the height above the plane of the synchrotron (y) and the angle out of the plane (y') at a general point (a) and at the virtual source (b) are elliptical. The trace of the entrance slits of the first optical element are also shown.

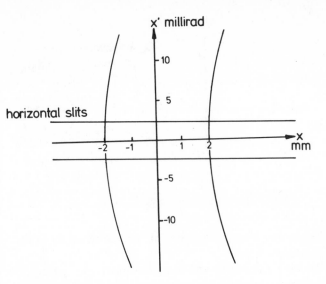

Figure 3. Optical phase space: in the plane of the synchrotron the effective size of the source is the width of the electron beam at the tangent point. The angular range accepted by the optical system is shown by the trace of the horizontal slits.

unusual property of synchrotron radiation is that the source is virtual and astigmatic. The use of a pair of cylindrical lenses for the x-ray optical system allows the correction of the astigmatism of the source. None of the peculiar effects encountered in high-resolving-power spectrometers[11] are encountered in a small-angle diffraction camera except that the height of the virtual source as determined from its image may be considerably smaller than the height of the electron beam at the tangent point.

At the virtual source the luminosity in the x direction is the product of a Gaussian distribution in x with a constant function of x'. In the y direction the luminosity is the product of a Gaussian distribution in y with an essentially Gaussian distribution in y'.

The average luminosity ($\bar{B}$ photons per unit solid angle per unit area per wavelength interval) is an invariant of the system: all nonlossy optical elements leave this quantity unaltered—it can only be degraded. This result can be obtained by applying Liouville's theorem to the function B in the optical phase space. The average luminosity is calculated by taking the total number of photons emitted per unit wavelength interval per radian in the x–y plane and by dividing this number by the area of the *virtual* source ($S_x S_y$) and by the apparent angle of vertical divergence.

1.2.4. The Optimum Condition

In the following we evaluate the photon flux for a focusing camera and for a slit (pinhole) camera using geometrical optics. For a given size of specimen and for a required angular resolution at the detector we arrive at an optimum camera geometry. In the following equations the x coordinate is considered; a similar set of equations exist for the y coordinate. We derive inequalities for the x and y axes,

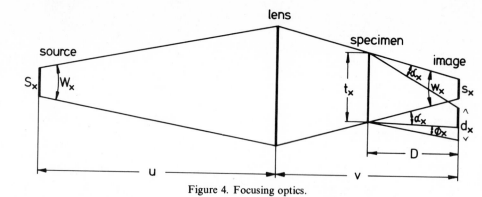

Figure 4. Focusing optics.

one of which will be limiting in any given situation. This will be taken to be the x axis for the following development.

We may represent a focusing camera by the diagram shown in Figure 4. The width of a diffraction spot d in the detector plane is governed by the demagnified source size (s_x, s_y), the intrinsic width of the reflection, and a dispersion term to account for the wavelength spread (because the profiles are Gaussian or approximately Gaussian we add the half-widths vectorially),

$$d_x^2 = s_x^2 + \lambda^2 c_x^2 D^2 + [(\Delta\lambda/\lambda)\theta(h)D]^2 \tag{26}$$

where the variables are explained in Table 1 and Figure 4; $\Delta\lambda/\lambda$ is the bandpass of the monochromator or the wavelength inhomogeneity (see Section 1.2.2.4), whichever is greater, and $\sin\theta$ has been replaced by θ. The mosaic spread (m_x, m_y) does not appear explicitly in equation (26) if $m_x > (\Delta\lambda/\lambda)\theta$: if the incident beam is monochromatic only a small proportion of the crystallites can diffract. On the other hand, if w_x (the angle of convergence of the beam at the specimen) should be greater than m_x, only a small proportion of the available incident beam can be diffracted, leading to an unfavorable signal-to-dose and signal-to-noise ratio. This situation should be avoided. For fibers m_x is essentially 2π so that there is no restriction on w_x except that given by the size of the specimen.

We will refer to the quantity $[\lambda^2 c_x^2 + (\Delta\lambda/\lambda)^2\theta^2]^{1/2}$ as the *intrinsic broadening* ϕ.

For resolving two reflections we assume that the distance between reflections should be twice the width (s_x) of a reflection (see Figure 4) so that

$$d_x < (a_x/2)D \tag{27}$$

Replacing $\lambda^2 c_x^2 + (\Delta\lambda/\lambda)^2\theta^2$ by ϕ^2 in (26) and substituting (26) in (27) we obtain

$$s_x < [(a_x/2)^2 - \phi_x^2]^{1/2}D \tag{28}$$

The relationship between source and image dimensions is given by

$$s_x = v/uS_x \tag{29}$$

The specimen dimensions (t_x, t_y) determine w_x, w_y,

$$w_x = t_x/D \tag{30}$$

as long as the condition $w_x < m_x$ is obeyed (for crystals if $w_x > m_x$, then w_x is *determined* by the mosaicity of the crystal).

The total flux F into a Bragg reflection h is

$$F = \bar{B}S_x S_y W_x W_y P_h \, \Delta\lambda = \bar{B}s_x s_y t_x t_y P_h \, \Delta\lambda/D^2 \tag{31}$$

where P_h is the integrated reflectivity of the reflection h (as P_h is specimen dependent it will not be considered further). The optimum condition is obtained by making D as small as possible, viz.,

$$s_x = \{(a_x/2)^2 - \phi_x^2\}^{1/2}D \tag{32}$$

Using the optimum condition (32) we obtain for the optimum flux

$$F_{\text{opt}} = \bar{B}[(a_x/2)^2 - \phi_x^2](S_x/S_y)t_x t_y \, \Delta\lambda \tag{33}$$

The optimum intensity is obtained by dividing F_{opt} by $s_x s_y$.

If ϕ is small compared with a_x (i.e., the intrinsic broadening is small compared with the required resolving power), the optimum condition is

$$s_x = (a_x/2)D \tag{34}$$

and the opimum flux is

$$F_{\text{opt}} = \bar{B}(a_x/2)^2(S_x/S_y)t_x t_y \, \Delta\lambda \tag{35}$$

Commonly with synchrotron radiation sources the experimentalist has no control over the source size S and little control over the distance u between the optical system and the source. Introducing the dimensionless parameter q, which is the ratio of the specimen–detector distance D to the image distance v the inequality (28) becomes

$$S_x/u < [(a_x/2)^2 - \phi_x^2]^{1/2}q \tag{36}$$

and the flux

$$F = \bar{B}S_x S_y t_x t_y/q^2 u^2 \, \Delta\lambda \tag{37}$$

The optimum condition is

$$q = S_x/\{u[(a_x/2)^2 - \phi_x^2]^{1/2}\} \tag{38}$$

D may be chosen to match the spatial resolution in the focal plane ($Da_x = qa_x v$) to the spatial resolution of the detector. In addition q may be chosen to optimize flux. One is not completely free in ones choice of q: if q is larger than 2/3 the parasitic scattering in the small-angle region of the image plane becomes too strong. Therefore the best attainable angular resolution is

$$a_x \approx 3(S_x/u) \tag{39}$$

Equation (39) shows that the best resolution is dictated by the parameters of the source (u often being fixed by architectural considerations).

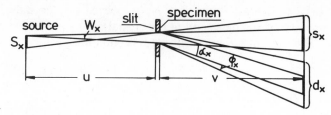

Figure 5. Slit collimation.

For a slit camera (Figure 5), setting $v = D$ and neglecting the intrinsic broadening, we have as the resolution condition in the detector plane

$$s_x < a_x/2v \qquad (40)$$

where

$$s_x = t_x(1 + v/u) + S_x v/u \qquad (41)$$

Introducing the dimensionless parameter r

$$r = v/u \qquad (42)$$

we obtain as the condition for resolving two reflections

$$t_x < (a_x/2u - S_x)r/(r + 1) \qquad (43)$$

where the flux is given by

$$F = \bar{B}S_x S_y t_x t_y/u^2 \, \Delta\lambda \qquad (44)$$

The intensity is obtained by dividing (44) by $s_x s_y$. u is generally fixed by the shielding of the storage ring so that the only free parameters that may be used to satisfy the angular resolution condition are r and t_x. To increase flux we wish to maximize the right-hand side of (43). The largest value that r can realistically take is ca. 1.5 so that (43) becomes

$$t_x < 0.6(a_x/2u - S_x) \qquad (45)$$

To satisfy (45) it will usually be necessary to choose a value of t_x that is less than the actual width of the specimen. The use of focusing optics allows one to avoid this situation since one has one more degree of freedom q.

Rosenbaum[12] has evaluated the gain to be expected by using focusing optics rather than slit collimation and arrives at a factor 10 *in flux* for a typical muscle experiment if in each case the optimum arrangement is used. If, as is often the case, the properties of the detector and specimen make the *intensity* rather than the *flux* the important factor, then the gain to be expected from focusing optics compared with slit collimation for a muscle experiment is about 3000!

In situations where a high background pertains, arising either from the specimen or detector, it is necessary to maximize intensity rather than flux in order to improve the signal-to-noise ratio. We consider only the focusing case. Dividing (35) by $s_x s_y$ we find

$$I_{opt} = \tfrac{1}{4}\bar{B}t_x t_y(a_x/S_x)^2(S_x/S_y)(u/v)^2 \, \Delta\lambda \qquad (46)$$

Equation (46) shows that for a given angular resolution and source size the correct procedure is to make (u/v) as large as possible. The limit is set by the resolution from the direct beam and the spatial resolution of the detector.

1.2.5. Choice of Wavelength

The wavelength range 0.1–0.3 nm is easily obtainable from synchrotron radiation sources. The boundaries are set by λ_c and the transmission properties of vacuum windows.

The integrated reflectivity of a mosaic crystal (see Zachariasen,[13] pp. 166–167) is

$$P_h/P_0 \propto \lambda^2 V \exp(-\mu t_z) \tag{47}$$

where the volume V is $t_x t_y t_z$. The linear absorption coefficient is approximately proportional to λ^3. For a crystal of given area $t_x t_y$ we have

$$P_h/P_0 \propto \lambda^2 t_z \exp(-a\lambda^3 t_z) \tag{48}$$

where a is a constant.

If we seek the maximum of (48), treating t_z as a variable and holding λ constant, we find

$$t_z(\text{optimum}) = 1/(a\lambda^3) \tag{49}$$

and at this maximum

$$P_h/P_0(\text{optimum}) \propto 1/(a\lambda) \tag{50}$$

Equation (50) applies only if the width of the specimen remains constant. If the specimen is isodimensional we find

$$P_h/P_0 \propto \lambda^{-7} \tag{51}$$

but note that the optimum wavelength is different from that given by (49)!

Frequently, the specimen size is a factor over which we have little control, in which case we seek the maximum of (48). Treating λ as a variable we find

$$\lambda(\text{optimum}) = (2/3at_z)^{1/3} \tag{52}$$

showing that even for a variation of specimen thickness over a range of 10:1 the optimum wavelength does not vary by more than 2.2:1. At the maximum

$$P_h/P_0 \propto a^{-2/3} t_z^{1/3} = (2/3)^{1/3}/a\lambda_{\text{opt}} \tag{53}$$

The actual choice of λ is often imposed by other requirements (e.g., sensitivity of the detector).

For semicrystalline and noncrystalline fibers, where we need to measure the intensity in reciprocal space as a continuous function of the reciprocal space radial coordinate R, if we vary the wavelength at constant volume the *scale* of the diffraction pattern will alter but not its intensity. Every detector has a given minimum resolution element (a pixel). The average intensity falling on each pixel will be the same but at longer wavelengths more pixels will be illuminated. For a given interval of the reciprocal space radius R we therefore collect more counts by going to a longer wavelength ($\alpha\lambda^2$), but the limit is determined by absorption just as in equations (52) and (53) above.

1.3. Small-Angle Diffraction Cameras for Synchrotron Radiation

1.3.1. Remote Control

In the case of synchrotrons the high background radiation levels make it impossible to work in the neighborhood of the direct or reflected beam. The apparatus must stand in a shielded area that is not available to the experimenter when the beam shutter is open. Thus all normal manipulations such as aligning cameras must be performed by remote control. For storage rings the situation is somewhat better and it is possible to work within the shielding walls. Even so the direct beam is never available to the experimentalist so that the primary optical elements that sit in the direct beam must be remotely controlled. Moreover, the air scatter from the reflected, monochromatic beam is so strong that by working for a few minutes a day in the neighborhood of the beam one would exceed the proscribed dose in most countries. The easiest solution to most of these problems lies in a convenient remote control system. Remote control offers a number of advantages once it has been installed. The most attractive of these is the possibility of computer control and semiautomatic alignment.

1.3.2. Vacuum Window

The vacuum system of the synchrotron must be separated from the optical system of the small-angle camera. This is generally achieved by means of a beryllium window. The heat loading on such windows is not such a problem as was first thought. Calculations of the distribution of heat due to absorbed radiation yield a maximum temperature increase of 90°C for the worst conceivable case (DORIS at 2.3 GeV and 6 A). No special problems have been experienced but note that the window acts as a filter for soft x radiation and makes it impossible to extract radiation softer than about 0.3-nm wavelength from x-ray beam lines so equipped. See also Chapter 3 for a discussion of beryllium windows.

1.3.3. Curved Mirrors

In a number of experiments fused quartz (Spectrosil) has been employed as a mirror (Franks and Breakwell[4]). The general experience with synchrotrons and storage rings has been that deposited metals are unstable when exposed to the primary beam and they have not been widely used. However, Hastings et al.[14] have successfully employed platinum deposited on a fused-quartz substrate.

The relationship between the critical grazing angle (θ_c) and the wavelength for fused quartz is[15]

$$\theta_c(\text{mrad}) = 26.3 \ \lambda(\text{nm}) \tag{54}$$

so that θ_c is typically 3–4 mrad. Only wavelengths longer than those given by (54) can be reflected. In this way the mirror acts as a high-frequency filter, which is particularly valuable for eliminating higher harmonics (see Section 2.3).

To accept the full vertical divergence of the synchrotron radiation at 20 m using a grazing angle of three milliradians one needs a mirror 1.5 m in length. Typically the radius of curvature of the mirror is between 2 and 4 km. Maintaining such a radius of curvature over such a mirror length presents problems: if the mirror is thick enough to be stable it is too thick to bend. Horowitz[16] allowed his mirror to bend under gravity. However, leaving a glass under strain, even its own weight, leads to flow. Webb *et al.*[10] used a relatively thin mirror and provided closely spaced adjustable supports. The European schools[3, 4, 15, 17, 18] have favored segmented mirrors, each segment being thick enough to be stable against flow but thin enough to be bent. For Spetrosil a thickness of 15–20 mm seems to be the right compromise between stability and flexibility for a 20-cm mirror segment. Longer segments would need to be thicker to retain enough stability against flow and would thereby become too inflexible to be bent. 20-cm mirror lengths can be relatively easily manufactured and polished; longer mirrors are very expensive.

Segmented mirrors need well-machined mechanical holders to allow for their bending and mutual alignment[3, 4, 15, 18]. The bending of each segment is traditionally achieved by pressure on pairs of steel pins separated by 15–20 mm sitting on the back and front surfaces at each end of the mirror.[4] The pins on the front face are cut away in the middle to allow the x-ray beam through. This method is not well-suited for application to synchrotron radiation since it necessitates placing a steel pin in the direct beam where it fluoresces brightly. Harmsen and Rosenbaum[18] have used metal plates glued to the ends of the mirror segments to apply the couple. This method of bending has the advantage that both concave and convex bending is possible. In this way possible permanent distortions of the mirrors caused by flow can be compensated. The radius of curvature of each segment is so large that the detailed nature of the bending achieved (e.g., circle or ellipse) is not important. The heights of the mirrors must be adjustable to allow for their mutual alignment. The precision of this alignment needs to be about 3 μm (Rosenbaum[12]), in order to overlap the images produced by each mirror segment with a precision of 1/3 of the image diameter. The heights are adjusted to bring the images of the segments into coincidence at the focus. The overall form of the segmented mirror is, therefore, operationally defined and will be elliptical. The mutual alignment of 8×20 cm segments[15, 18] in practice turns out to be straightforward if an in-line television monitor is used to view the focus.[3, 19]

1.3.4. Curved Crystal Monochromators

The following crystals have been used as monochromators for synchrotron radiation: quartz (10.1 planes)[3, 19, 20]; silicon (111 planes)[10]; germanium (111 planes)[15, 17, 18]; and graphite (001 planes).[21] The first three are perfect single crystals with very narrow rocking curves that give specular reflection without appreciable dispersion. Of the three, germanium has the widest bandpass and has zero reflectivity for second-order components. Germanium is at present the most widely used material. Graphite is a mosaic monochromator with a mosaic spread of about 0.5°. A mosaic monochromator can only be used if the distance between the monochromator and the detector is very short so that the dispersion introduced by the monochromator does not lead to a spreading out of the beam. Graphite has been used for photographing the fiber diffraction diagram of DNA[21] (Figure 9).

The use of bent single crystals to focus x rays has a history nearly as long as that of x rays themselves (see Witz[1]). The bending is traditionally achieved by applying couples to both ends of the crystal slab (see, for example, de Wolff[7]). When used with synchrotron radiation it is important that no material should be present on the front surface of the monochromator on account of fluorescence. In the DESY camera[3, 19] steel pins were glued to the back surface of the monochromator. Symmetrical couples at both ends of the crystal were generated by pressure on these pins. In theory it is possible with such an arrangement to generate good approximations to a number of different bending profiles (e.g., cylinder or logarithmic spiral) merely by altering the ratio of the two couples,[7, 22] but in practice, on account of the large image and object distances, alterations of the detailed form of the monochromator surface do not produce significant changes in the performance of the monochromator. A novel method of achieving a cylindrical form has been used by Lemonnier et al.[23] Lemonnier and collaborators make use of the fact that if a plate with the shape of an isosceles triangle is held firmly at the base and a force is applied to the apex, the plate takes up the form of a cylinder. The main disadvantage of this system of bending is that the center of the monochromator is displaced sideways during bending.

In the classical form introduced by Johann,[6] curved crystal monochromators have a cylindrical shape and the source, crystal, and focus are symmetrically arranged on the circumference of a circle. The crystal is elastically bent to a radius of curvature equal to twice that of the focusing circle. The simple Johann geometry has the restriction, for a monochromatic source, that the object and image distance must be equal. This condition can be relaxed by cutting the crystal at an angle σ to the Bragg planes. The focusing conditions are now the Guinier conditions [equations (4) and (25)]. However, with synchrotron radiation it is no longer necessary to obey (25)—the penalty incurred for departing from (25) is an increase in the wavelength inhomogeneity—but it is always necessary to obey (4) since this is an expression of the condition of specular reflection. For synchrotron radiation the monochromator is employed near the Guinier conditions [equations (4) and (25)] but usually far enough from the Guinier conditions to produce a wavelength inhomogeneity which exceeds the bandpass of the monochromator by a factor 10 or more (regardless of the exact form of the bending).

On account of the asymmetric cut (σ) the beam leaving the monochromator will be narrower or wider than the beam entering the monochromator in the ratio

$$N_u/N_v = \sin(\theta + \sigma)/\sin(\theta - \sigma) \tag{55}$$

where N_u and N_v are the widths of the beam on the input and output sides of the monochromator. For a monochromatic source the Guinier conditions pertain so that the condition

$$N_u/N_v = u/v \tag{56}$$

must hold. Equation (56) shows that the angle of divergence before the monochromator is equal to the angle of convergence after the monochromator, a necessary condition for monochromatic radiation, but a condition that no longer holds for synchrotron radiation.

A further effect of the asymmetric cut is to alter the bandpass of the monochromator. Dynamical theory (see Zachariasen,[13] pp. 123–126) shows that the integrated reflectivity (P_h/P_0), for Laue diffraction (white source) from a perfect crystal without absorption, is

$$P_h/P_0 \propto (N_v/N_u)^{1/2} \tag{57}$$

$(P_h/P_0$ is the effective bandwidth of the monochromator).

Using synchrotron radiation, equation (56) can be disregarded and we may choose a convenient value of σ. In practice a value of four for N_u/N_v can be achieved without incurring significant absorption losses. The bandwidth is thereby decreased by a factor of two but this is more than compensated for by the increase in intensity resulting from the compression of the beam. Since a passive optical element cannot raise the luminosity, some compensating effect must occur: the outgoing beam has more crossfire than the incoming beam so that the size of the image is enlarged to the value it would have from a consideration of the ratio v/u, thereby maintaining a constant luminosity.

A further effect is that the width of the illuminated monochromator crystal as seen by the detector through the guard slits is reduced. This is important when considering the effects of parasitic scattering on the small-angle resolution about the direct beam.

Webb[22] has proposed that a logarithmic spiral is preferable to a cylindrically bent crystal for use with synchrotron radiation. For the reasons given in 1.2.2.3, the difference between a logarithmic spiral and a cylinder only shows up in the spherical aberration of the monochromator. Since the effect is very small (of the order of a few microns) it can be neglected in all practical situations.

1.3.5. Mirror–Monochromator Camera at DESY

The early work on the use of synchrotron radiation as a source for x-ray diffraction was done on the DESY synchrotron (Rosenbaum et al.[20]; Barrington Leigh and Rosenbaum[3, 19]). A description of this apparatus serves as an introduction to the ensuing discussion since it was used as a model for a number of other cameras. Barrington Leigh and Rosenbaum employed a grazing-incidence curved mirror to reflect and focus the beam in the vertical $(y–z)$ plane and a curved monochromator to focus and monochromatize the beam reflected in the horizontal $(x–z)$ plane. This is not the best direction with respect to the polarization of the source, but the intensity losses invoked are small (10–20%) and are adequately compensated by the convenience of having the x-ray beam approximately parallel to the floor of the laboratory. Typically specimen–film or specimen–detector distances between 40 and 120 cm were employed. The x-ray mirror consisted of two 20-cm sections 10 mm thick, each of which might be bent to focus the beam. Adjustments for mutual alignment of the two mirrors were provided. Fused quartz

(Spectrosil) x-ray mirrors of suitable flatness were supplied by A. Franks.[4] The mirrors were bent by applying equal couples to the two ends. No deterioration of the surface of these mirrors was observed after years of exposure to the direct beam. However, they became permanently deformed after a few months bending, showing that they were too thin.

A bent quartz monochromator cut at 7° to the 10.1 planes focused the beam in the horizontal plane. The beam was also compressed fourfold by the asymmetric cut of the crystal. Subsequently, a germanium crystal cut at 7° to the 111 planes was substituted for quartz. The beam after the monochromator was about 6 mm × 1.5 mm and focused to a point less than 200 μm in diameter. The distance between the monochromator and the focus could be varied between 1 and 3 m. A diagram of the small angle diffraction camera is given in Figure 6.

The mechanical design criteria employed in setting up this apparatus were that components should be light and that all movements should be remotely controllable. The mirrors and monochromator were mounted on the end of an optical bench along which the reflected beam traveled. All the components of the x-ray camera (slits, specimen holder, film holder, counter holder) were mounted on the optical bench. The angle between the bench and the direct beam (2θ) was fixed at 26° ($\lambda \approx 0.15$ nm for quartz or germanium). Two sets of slits were used: a set of primary slits in front of the mirror box and a set of guard slits in front of the specimen. The apparatus was located in an experimental hall that could not be entered when the beam shutter was open. Remote control was achieved by the use of small dc motors fitted with reduction gears. Helipots were used as position sensors. Helipots or slide resistors were mounted on a table in the control area in a form that mimicked the actual physical layout. Power amplifiers mounted in racks *close* to the experiment (in order to avoid cross-talk) were used to drive the various motors until the helipot output voltage was the same as the voltage given by the control table. Forty movements could be remotely controlled. Three remotely con-

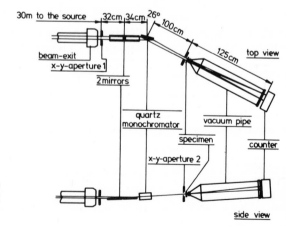

Figure 6. Plan and elevation of the mirror–monochromator camera at DESY.

trollable television cameras were used for monitoring: an overview camera mounted near the roof of the experimental hall, a camera that ran along a rail parallel to the optical bench and was capable of dropping a fluorescent screen into the beam, and a sensitive image-intensifier camera mounted in line with the x-ray beam on the optical bench. This camera viewed the primary beam falling on a CsI (Tl) scintillator crystal through magnifying zoom optics that could be adjusted between ×20 and ×100.

The mirrors were mounted in a helium-filled plexiglass case fitted with thin mylar windows. The monochromator housing was also helium filled. During experiments all other x-ray paths were enclosed in evacuated plexiglass tubes fitted with mylar windows. Using a germanium monochromator, a flux of 7×10^8 photon/s ($\lambda = 0.15$ nm) was routinely obtained from this camera when the synchrotron was operating at 7.2 GeV and 10 mA. The gain in speed over a small-angle camera of similar performance used with a fine-focus rotating anode tube (Elliott Gx13) was about 40.

1.3.6. Mirror–Monochromator Camera at NINA

Haselgrove et al.[17] used a heavier construction method and the remote control was carried out with stepping motors. The number of control circuits was kept small and the individual drive motors could be addressed individually through a multiplexer. A 2×20 cm segmented cylindrically bent mirror (Franks and Breakwell[4]) was used with a cylindrically bent germanium monochromator cut at 7° to the 111 planes. The mirrors and monochromator were housed in a helium-filled enclosure.

1.3.7. Mirror–Monochromator Camera at SPEAR

Instead of the segmented mirrors favored by the DESY and NINA groups, Webb et al.[10] used a single 1.2 m length of float glass. They used a silicon monochromator cut at 8.5° to the 111 planes bent to a logarithmic spiral.[22] A flux of 6×10^8 photons/s was recorded at the focus with SPEAR running at 3.7 GeV and 20 mA. The focus was about 0.5×0.5 mm². The wavelength could be varied. The radiation safety system at SPEAR deserves attention. A number of experiments are arranged along each beam line, each taking a fraction of the beam. Each experiment is situated inside a radiation hutch that has its own beam shutter and its own set of radiation interlocks. Adjustments within a hutch must be carried out by remote control when the beam shutter is open.

1.3.8. Monochromator–Mirror Camera at VEPP-3

A variation on the mirror–monochromator theme has been used on the storage ring VEPP-3 for studies on muscle (Vasina et al.[24, 25]). In this camera the monochromator (quartz) was *before* the mirror and sat in the direct beam of the storage ring. This appeared to give less scattering around the direct beam than was experienced with the alternative arrangement. No special problems with heating the monochromator were experienced. Since the beam dimensions in VEPP-3 were small, Vasina and her collaborators were able to work close to the source point $(u_x = 3 \text{ m})$ and were able to achieve a high intensity with relatively small optical elements.

1.3.9. Mirror–Monochromator Cameras at DORIS

The second generation of small-angle diffraction cameras is now appearing. These cameras are incorporated into more complex beam lines and the optical elements are inside massive vacuum chambers (Chapter 15). The first of these on the storage ring DORIS (Harmsen and Rosenbaum[18]) employs 8×20 cm mirror segments and a cylindrically bent germanium monochromator (Figure 7). Focusing is achieved by overlapping the reflected strips of radiation from each of the mirrors giving a focus 500–500 μm in height (the value depends upon the wavelength used since this affects the grazing angle on the mirror). In principle by bending individual mirrors it will be possible to reduce this value by a factor of two or three but at present the mirror bending mechanism has not been commissioned. The camera

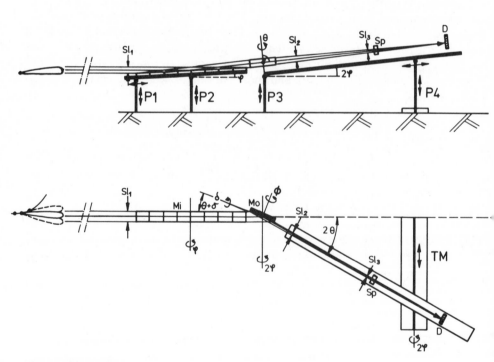

Figure 7. Plan and elevation of the mirror–monochromator camera at DORIS. Each of the posts P1, P2, P3, and P4 may be adjusted in height. In addition P4 may traverse to allow alteration of the wavelength. An eulerian cradle allows fine positioning of the monochromator (Mo) relative to the mirrors (Mi). The monochromator can be focused. The eight mirror segments may be separately aligned and focused. The mirrors and monochromator (Ge) are housed in a vacuum chamber. All movements are remotely controllable.

is a revised, enlarged, and improved version of the DESY camera. The main innovation is that the wavelength may be continuously varied. The whole optical bench, which is 4.2 m long, may be rotated about an axis through the monochromator. The monochromator axis must stay accurately at right angles to the plane of the mirror or one obtains a horseshoe-shaped focus. Since this setting has to be rather accurate, movements that allow the monochromator axis to be adjusted over a small range of angles are also provided.

The revised remote control is smoother and more reliable than that of the DESY camera. The electronics are designed and built to minimize failure in the 60 separate control circuits. All adjustments are independent. Control is normally available from a control desk fitted with potentiometers as for the DESY camera. In addition, each control circuit can be directly controlled by a computer that can either function as a laboratory notebook by recording and restoring positions or can be used to optimize settings. This part of the development awaits implementation.

The optical elements are designed so that the direct beam, which is 250 mm wide, may be divided into sections without the optics of the first section casting a shadow on the following sections. This requirement imposes strong constraints on the design of the mirror bender and mirror supports. The optical system (eight mirrors and a germanium monochromator) are housed in an evacuated chamber that also contains another mirror–monochromator system and also provides the beam path for the direct beam (see Chapter 15). In a typical muscle experiment on this camera, 10^{10} photon/s have been obtained at 3.7 GeV and 10 mA.

A second camera of a similar kind has been set up on DORIS by Hendrix, Koch, and Bordas.[15] Hendrix *et al.* have used the same design of mirror block as Harmsen and Rosenbaum but have used a triangular monochromator.[23] The optical bench is constructed for a fixed wavelength. The design philosophy employed in constructing the bench was that it should be solid and as far as possible be constructed from commercially available optical bench parts. The flux measured with a calibrated ionization chamber was 5×10^{11} photons/s at 0.15 nm (4.6 GeV and 20 mA).

The remote control system is built round an address bus. To position components in the same way as has been used by Harmsen and Rosenbaum,[18] dc drive motors are used. However, each motor driver carries an address. When the bus address lines signal that a given motor is required the desired motor is connected to the analog control lines. This kind of remote control system offers a simple design without the need for massive cable runs. The disadvantage is that only one or a very few motors (depending on the number of channels implemented) can be adjusted at the same time. Experience is needed to judge which of the two systems now being used at DORIS will be the easiest to use and maintain.

1.3.10. The Separated Function Focusing Monochromator at SPEAR

Hastings, Kincaid and Eisenberger[14] have used a 60-cm-long platinum-coated toroidal mirror[16] in the symmetrical mode ($u = v$) to collect six milliradians of synchrotron radiation and focus it to a point. Shortly before the focus a double-crystal monochromator (germanium 111) is inserted. The main advantage of the system is that the wavelength may be altered without disturbing the focus. The disadvantage of the present setup is the equality of v and u, which produces an image the same size as the source. For most applications involving small-angle diffraction it is essential to make the ratio v/u as large as possible. Therefore the apparatus is unsuitable for most experiments needing high resolution at the detector plane from small samples. See Chapter 3 for a description of this instrument.

1.4. Conditions Imposed by Detector Technology

Four kinds of detectors have been used with synchrotron radiation: x-ray film, T.V. detectors, one-dimensional position-sensitive proportional counters (PSD's), and multiwire proportional chambers (MWPC's) as area detectors. The characteristics of the detector are important in determining experimental design. In all cases where the detector or the environment of the detector imposes a high background it is necessary to maximize intensity rather than flux. Detectors have typically between 250 and 1000 resolution elements along an axis.

X-ray film is the simplest detector to use: it has very good spatial resolution, quantum efficiency is high, and data storage is easy. Measuring the optical density (OD) is relatively simple and straightforward. The OD has virtually the same statistics as grain counting,[26] which means that it is as good a measure of the number of quanta falling on a given area as photon counting. The disadvantages of film include the fact that it is difficult to use for time-resolved studies and that it has an intrinsic fog level equivalent to about 200,000 counts/mm^2. This high fog level drives one to seek the finest possible focused beam so as to attain a reasonable signal-to-noise level. In order that the fog level should not be higher than the signal for a signal of 10,000 counts the image in the focal plane (d_x) may not be bigger than about 250 μm. Therefore 250 μm is the upper limit for the efficient use of film.

X-ray phosphors [ZnS(Ag)] in conjunction with high-sensitivity television tubes have been employed for the registration of x-ray diffraction diagrams.[27, 28] Milch[29] has used a silicon intensified target vidicon (SIT-vidicon) coupled to the phosphor by a fiber-optics face plate to study the diffraction from muscle. A raster of 256×256 picture elements (pixels) may be read out from an area of 25×25 mm. On account of the small size of the detector and the relatively high spatial resolution of the system one is required to use a small focal spot. Moreover, on account of the relatively high dark current of this device one is forced to work with a high intensity at the focal plane (i.e., high v/u value, small D). On account of its composition the phosphor has a well-defined maximum in its sensitivity at 0.13 nm, which fixes the wavelength requirement.

A number of authors have used PSD's for measurements on muscle or muscle fibers.[24, 25, 30–33, 42] Generally such counters have a low background and good spatial resolution (<200 μm). Moreover, the size of the counter is not critical, making such counters easy to employ. One major drawback has been the saturation count rate. If the counter is employed with an analog readout system (time–voltage converter + voltage–digital converter) the data rate, which is determined by the electronics, cannot exceed about 50 000/s *over the whole length* of the counter.

For intact muscle samples, which have a relatively large volume and can present 10^6 counts/s on the zero-layer line when used with a storage ring, this is a real limitation on the type of experiment that can be carried out. Gebhardt and Rosenbaum have developed a fast counter that can be used to effect a time–digital conversion in less than a microsecond. For samples consisting of isolated bundles of muscle fibers the count rate rarely exceeds 50,000 so that the conventional PSD is adequate.

MWPC's are just beginning to find application in small-angle diffraction.[34-36] Such detectors have two main drawbacks: a relatively course spatial resolution (1–2 mm) and the total count rate is limited to 50,000 (or with improved electronics to 10^6) over the whole area of the counter. To use such counters it is necessary to use a relatively large specimen–detector distance ($D > 2$ m) and to screen out the center of the diffraction pattern when measuring the weaker parts. See Chapter 3 for a discussion of MWPC's.

1.5. A Comparison of the Theoretical and Actual Performance of Mirror–Monochromator Systems

The synchrotron radiation emitted from an accelerator or storage ring may be accurately calculated, so much so that synchrotron radiation may be used as a standard for calibration. This has been demonstrated for different spectral regions by a number of investigators.[37, 38] The first crude investigations of the application of synchrotron radiation in small-angle diffraction in biology[20] using a curved quartz-crystal monochromator demonstrated an agreement within 50% between the observed and calculated flux. Nevertheless a number of reports have been made of an apparent shortfall in the performance of the mirror–monochromator system.[10, 39] We suspect these come about because inadequate attention has been paid to the actual limiting aperature and to absorption in the specimen and monochromator. Often the flux has been calculated for fully open apertures, which in practice have not been attainable.

Using the Harmsen–Rosenbaum camera at DORIS (3.87 GeV and 15 mA) we have measured a flux of 1.5×10^{10} photons/s at $\lambda = 0.125$ nm. The calculated flux may be evaluated from the following data:

horizontal aperture $= 21$ mm
vertical aperture $= 3$ mm
distance of aperture from source $= 21.3$ m
mirror reflectivity (estimated) $= 50\%$
total absorption of windows and air paths (calculated) $= 30\%$
polarization factor $= 0.93$
bandpass of monochromator (germanium 111, $\sigma = 7°$) (calculated[40])
   $\Delta\lambda/\lambda = 1.2 \times 10^{-4}$

Using the computed values for the storage ring DORIS at 0.125 nm under the operating conditions cited above we find

$$F_0 = 1.94 \times 10^{10} \text{ photons/s}$$

The discrepancy of 30% between the two values is probably to be sought in the fact that the beam is actually about 2 mm high at the source, resulting in penumbra at the slits that have not been taken into account.

A similar observation and calculation has been made by Hendrix *et al.*[15]

There is apparently no great discrepancy between the observed and calculated flux for a mirror–monochromator camera.

2. Applications

2.1. Muscle Fibers; Real Time Experiments on Muscle

2.1.1. Frog Muscle

Contraction and cell movement are central properties of all eukariotic cells. Moreover, muscular contraction as a very efficient chemomechanical process is interesting in its own right. Small-angle x-ray diffraction from intact muscle, such as frog sartorius, offers a unique method for the study of the molecular structure and function of muscle. Muscle consists of a matrix of two interdigitating protein filaments containing actin and myosin (see, e.g., H. E. Huxley[41]). When the muscle is stimulated the two sets of filaments slide past each other. This apparently comes about through the action of "cross bridges" on the myosin filaments, which attach cyclically to specific binding sites on the actin filament and undergo a conformational change during attachment to generate the sliding motion.

The myosin filaments are arranged on a hexagonal lattice with (in vertebrate muscle) the actin filaments located on the trigonal positions. The cross bridges form a regular three-start helix around the thick filament with a pitch of 42.9 nm. The diffraction pattern from the myosin cross bridges consists of a set of layer lines spaced at 42.9 nm of which the central or equatorial layer line is of special importance. During contraction neither the spacings of the reflections nor the layer-line spacing alters but the intensities of the low-order equatorial reflections, in particular the $10\bar{1}0$ and $11\bar{2}0$, alter dramatically.[2] These changes reflect a large movement of mass, presumably the cross bridges, from the neighborhood of the myosin filaments to the neighborhood of the actin filaments on activation. By measuring the equatorial reflections, therefore, one may follow the time course of the conformational change taking place in the cross bridge during activation and, furthermore, one may hope to follow the conformational change taking place during contraction. This is most readily done by perturbing the steady state established during contraction by means of a quick length jump. It is known from physiological experiments of this kind that the time scale of molecular events is about one millisecond so that in order to carry out a temporal examination of the equatorial reflections from muscle one needs to be able to collect data in 10 ms or less. With the best fine-focus x-ray tubes now available and using PSD's the strongest part of the diffraction pattern can be obtained in about 10 s so that the kind of experiment envisaged can be performed by repeated stimulation of the muscle for about 1000 stimulations.[2, 42] There are considerable difficulties in repeating steady-state physiological experiments so many times and the perturbation experiments are even more difficult. Moreover, the conformational changes that are presumed to take place in the cross-bridge during the contraction cycle would best be registered in the weak high-order layer lines, which are very difficult to record. Two groups working with intact vertebrate muscle (frog sartorius) have turned to synchrotron radiation. H. E. Huxley and collaborators[17] started at NINA but have now shifted to DORIS. In some preliminary and unpublished experiments recently carried out with a SIT-vidicon and the mirror-monochromator focusing camera,[18] H. E. Huxley, J. R. Milch, and R. Goody were able to record the equatorial reflections in 100 ms under conditions that were not optimal. Using VEPP-3 Vasina *et al.*[24, 25] have been able to record the equatorial pattern from frog muscle in 1000 ms using a PSD.[32] Clearly the required time resolution can now be reached and experimental design bifurcates: systems reliability must be improved to the point where it is conceivable to carry out complicated physiological experiments on a synchrotron radiation beam (a careful optimization of the optical system is required in order to produce the highest possible signal-to-noise ratio) and the experimental design on the physiological side must be optimized—this is closely linked to a better understanding of the diffraction pattern from the muscle.

2.1.2. Insect Flight Muscle

Whereas frog muscle is an excellent system for studying the dynamics and physiology, muscles from other phyla have advantages when it comes to studying structure. The most beloved of these is the flight muscle from the tropical water bug *Lethocerus*. This muscle is highly crystalline and gives comparatively well-ordered fiber diagrams (Figure 8). Samples are normally stored for some time in a mixture of glycerine and water, a treatment that breaks up the membranes and makes the crystalline actin–myosin matrix permeable to reagents or metabolites (e.g., ATP). The disadvantage of the glycerinated preparation compared with an intact muscle

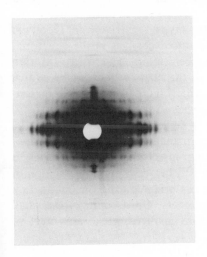

Figure 8. A small-angle x-ray diffraction diagram obtained from a fiber bundle of *Lethocerus* flight muscle in rigor using the DESY camera with a quartz monochromator (6 GeV and 10 mA, exposure 24 h). The strong layer-line repeat corresponds to a Bragg spacing of 38.5 nm.

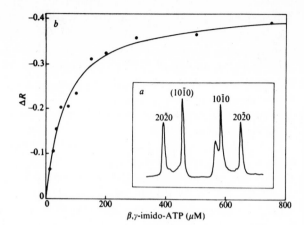

Figure 9. A titration curve obtained by adding β,γ-imido-ATP to insect flight muscle and measuring the ratio of the 10.0 and 20.0 reflections.[30] This is a novel use of synchrotron radiation made possible by the high data rate.

is that when the muscle matrix is doing work and consuming ATP the processes involved are diffusion-limited so that for experiments involving active muscle the fiber bundle may not be thicker than about 100 μm. In contrast, frog muscle can be relatively large (5 × 3 × 1 mm³) since the ATP is generated internally.

The first application of synchrotron radiation in the field of low-angle diffraction was an intensive study of the structure of the flight muscle from the water bug *Lethocerus* in the absence of ATP (the so-called *rigor* state) (Holmes, Tregear, and Barrington Leigh[43]). Data were collected with the mirror–monochromator camera at DESY[3, 19] (Figure 8). Samples consisted of glycerinated fiber bundles about 300–500 μm in diameter.

The general aim of such studies is to characterize as many equilibrium states of muscle as possible as a preparation for the problem of interpreting the diffraction patterns from dynamical experiments. One of the states that may be investigated may be induced by adding an analog of ATP (β, γ-imido-ATP). In order to determine whether or not this analog was binding to the actin–myosin matrix or not Goody *et al.*[30] used synchrotron radiation to follow the induced changes in the equatorial reflections produced by altering the analog concentration. This x-ray titration showed that the binding constant of the analog to the fibers was the same as that measured when the analog bound to isolated myosin (Figure 9) so that it is reasonable to assume that one is observing the same process in the intact fibers.

In spite of the small sample volume, Goody, Milch, and Tregear, using the mirror–monochromator camera on DORIS,[18] have managed to collect data extending to 10-nm resolving power from oscillating insect flight muscle using a SIT-vidicon[29] as a detector. Sufficient data to allow a low-resolution description of the changes in conformation undergone by the cross bridge during the contraction cycle should become available from such experiments. On account of the small sample volume and the weakness of the reflections involved, great care must be given to improving the signal-to-noise ratio.

2.2. Collagen

Collagen is the most important connective tissue in the body and an under-standing of its structure and the molecular basis of its elasticity is of some medical significance. Moreover, collagen samples are a useful test material for small-angle diffraction cameras! The fundamental repeat of collagen fibers is 67 nm. Bowitz *et al.*[44] have addressed themselves to the problem of explaining the elasticity of collagen. For small strain ($<3\%$) the small-angle diffraction does not alter in spacing whereas for large strain (7%) the small-angle spacing increases appreciably (68.5 nm) showing that the component molecules have slipped relative to each other. Although these changes are reversible, under such strain the collagen fibers tear after a few minutes so that all measurements must be made quickly. Moreover, the samples must be small (200–300 μm) in order to be sure that the whole of the sample is subjected to the same strain. Bowitz *et al.* were able to collect data in 100 s using the DESY camera. Using an energy dispersive counter, however, Bordas, Munro, and Glazer[45] were able to collect data in a similar time on NINA, which was a weaker source than DESY. Moreover, Vasina *et al.*[24, 25] were able to obtain data from collagen in 100 ms by the same method. In general, the energy dispersive method is not suitable for biological samples because of the very unfavorable signal-to-dose ratio. In the present case, however, it could probably be used to advantage.

The mirror–monochromator camera, with its high angular resolution and well collimated beam can be used to record the two-dimensional fiber diagram from collagen (A. Miller, see Barrington Leigh and Rosenbaum[19]). Moreover White *et al.*[46] have been able to study the orientation of collagen in bone by the use of the DESY camera. The very high intensity allows the use of large specimen–film dist-ances, which reduces the diffuse scatter from the bone relative to the Bragg reflections from the collagen whilst still retaining the ability to record the weak high-order reflections from collagen. The DORIS camera has recently been used to record the diffraction from the collagen matrix in cornea (G. Elliott and A. Harm-sen, unpublished result), a task which proved very difficult with conventional sources.[47] The much better signal-to-noise properties of synchrotron radiation compared with conventional sources are advantageous in such situations.

2.3. DNA Fibers

Orientated bundles of DNA fibers give detailed fiber diffraction patterns ex-tending out to 0.3-nm resolution (Figure 10). The diffraction from DNA fibers is not normally a situation that would call for synchrotron radiation. However, Mokul-'skii and collaborators[21] have found it very advantageous to use synchrotron radiation to photograph the cesium salt of DNA. By analysis of the cesium salt one can locate the counter ions in the crystal lattice, a task that has not proved possible with light counter ions. However, the absorption of the cesium salt is so high that no scattering can be obtained with conventional x-ray sources giving 0.15-nm radiation. The intensity available from x-ray tubes with silver or molybdenum

Figure 10. A fiber diffraction pattern from DNA fibers obtained by Mokul'skii and collaborators at VEPP-3. The optical system consisted of a single graphite monochromator that gives a very high second-harmonic component. Thus the diffraction pattern consists of two diffraction patterns superimposed (courtesy of Dr. M. A. Mokul'skii).

targets is much smaller than can be obtained from copper targets. Moreover, the small-focus high-brilliance x-ray tubes necessary for such experiments are not available for hard radiation. Therefore the intense hard radiation available from a storage ring proved ideal for obtaining data from the cesium-DNA fibers. The diffraction pattern of DNA is not small-angle scattering and most of the optical criteria discussed above are not relevant. For example, it was possible for Mokul'skii and collaborators to work with a mosaic monochromator and no mirror because the mosaic spread of the specimen is even greater than that of the monochromator and because only a very limited angular resolution was required at the detector plane. Thus the specimen–film distance could be kept small.

Figure 10 was obtained using a graphite monochromator (without a mirror) and shows the diffraction pattern twice: once with λ and once with $\lambda/2$ (subsequent studies were made with a quartz monochromator). This unique photograph demonstrates the difficulties of working with a monochromator with a high second-order structure factor if one does not use reflection from a mirror as a high-frequency filter.

3. Extensions of the Methodology

3.1. Wide-Band Monochromators—A Technological Challenge

The bandwidths ($\Delta\lambda$) of perfect single-crystal monochromators are ca. 0.0004 nm at wavelengths around 0.15 nm (for references see Boeff et al.[40]). In order to record a diffraction pattern with adequate angular resolution we require that the contribution of the intrinsic broadening to the width of a reflection [equation (26)] should be small. In a practical situation a bandwidth of 0.1 nm would be quite adequate so that existing monochromators appear to have an unnecessarily small bandwidth. In theory it is possible to increase the bandwidth by distorting the crystal so that the distance between Bragg planes varies as the x-ray beam penetrates the crystal. This may be done either by mechanical stress (i.e., bending) or by diffusing in foreign atoms. In a detailed theoretical study, Boeff et al. have shown that, for x rays, not more than a factor of four could be obtained by such methods. The bending radii normally used for monochromators (10–20 m) are not sufficient to cause any appreciable increase in the bandwidth. In an unpublished study Rek and Harmsen have investigated the effect of implanting foreign ions. By implanting boron in a perfect single crystal of silicon they have increased the integrated reflectivity by a factor of three. The topographic studies necessary to determine the extent of damage to the surface are in hand but the results are not yet available.

An alternative approach is the use of specially fabricated multilayer monochromators. This approach has been successfully used for soft x rays.[48] For 0.15-nm x rays the thickness of the layers may not be greater than 5 nm. With such small distances, diffusion across the layers becomes a major problem.

A completely different kind of approach has been suggested by Stuhrmann[49] based on the use of mirrors and filters to define a wavelength interval. The difficulty with this method for low-angle diffraction is that, although the short wavelength cutoff can be well defined by critical external reflection from a suitable mirror, the long wavelength cutoff can only be defined by absorption, which is not sharp. In principle it would be possible to devise a long wavelength filter by using a *transmission* mirror set for total external reflection. Such a mirror could be constructed, for example, by depositing a thin layer of nickel on a taut mylar foil. Such an arrangement has the advantage of being nondispersive.

Wide-band monochromators, if they can be manufactured, are not without their problems. One of the benefits accruing from the use of narrow-bandpass monochromators such as quartz or germanium is that the *dispersion* introduced by the monochromator is usually negligible. However, with a wider bandpass this will not be the case. It will then be necessary to use two monochromators as a non-dispersive pair in the classical Bonse–Hart manner. This makes it very difficult to use the monochromators at the same time as a focusing element.

3.3. Towards an Optimum Design

The pioneer experiments on the use of synchrotron radiation as a source for x-ray diffraction[20] are now eight years old. A considerable mass of experience, some of it obtained under very difficult conditions, has been accumulated. The x-ray optical systems must be simplified since, at present, they are too difficult to set up and maintain and take too long to alter between experiments. Furthermore, the dictates of the experiment itself may well be such that the experimental group has neither the resources nor the emotional strength to cope with a complex optical system. This is generally the case in physiological experiments. On the other hand, the optical requirements set out in Section 1 are very important. By moving away from the optical maximum as dictated by the requirements of angular resolution and specimen size it is rather easy to reach a situation where the synchrotron radiation source cannot match the performance of a conventional source used under optimum conditions. One cannot simply abandon x-ray optics.

The new dedicated sources have a high luminosity and small beam size. It is, therefore, often possible to operate without a mirror, as has been done successfully at Novosibirsk (VEPP-3) and at Orsay (DCI). One relies on the small vertical divergence of the beam and its small size to produce adequate collimation in the vertical plane. This may be augmented, as was done by Vasina et al.,[24, 25] by a single-segment mirror *after* the monochromator. The mirror should be nickel plated to produce an adequate aperture and may be manually controlled since it does not sit in the primary beam.

An alternative approach is to use a toroidal mirror (see Section 1.3.10). However, the manufacture of asymmetric toroidal mirrors presents formidable problems. Alternatively, a symmetrical mirror could be employed if the x-ray source could be reduced to a suitable size (ca. 200 μm) by limiting the synchrotron and betatron oscillations.

References

1. J. Witz, Focusing monochromators, *Acta Crystallogr. Sect. A* **25**, 30–41 (1969).
2. H. E. Huxley and W. Brown, The low angle x-ray diagram of vertebrate striated muscle and its behavior during contraction and rigor, *J. Mol. Biol.* **30**, 383–434 (1967).
3. J. Barrington Leigh and G. Rosenbaum, A report on the application of synchrotron radiation to low-angle scattering, *J. Appl. Crystallogr.* **7**, 117–121 (1974).
4. A. Franks and P. R. Breakwell, Developments in optically focusing reflectors for small-angle x-ray scattering cameras, *J. Appl. Crystallogr.* **7**, 122–125 (1974).
5. A. Guinier, Rayon X—Sur les monochromateurs a cristal courbe, *C. R. Acad. Sci.* **223**, 31–32 (1946).
6. H. H. Johann, Die Erzeugung lichtstarker Röntgenspektren mit Hilfe von Konkavkristallen, *Z. Phys.* **69**, 185–206 (1931).
7. P. M. de Wolff, in *Selected Topics on Crystallography*, J. Bouman (ed.), North-Holland Publishing Co., Amsterdam (1951).

8. G. K. Green, Spectra and Optics of Synchrotron Radiation, B.N.L. Report 50522, Brookhaven National Laboratory, Upton, N.Y. (1976).

9. J. B. Hastings, X-ray optics and monochromators for synchrotron radiation, *J. Appl. Phys.* **48**, 1576–1584 (1977).

10. N. G. Webb, S. Samson, R. M. Stroud, R. C. Gamble, and J. D. Baldeschwieler, A focusing monochromator for small angle diffraction studies with synchrotron radiation, *J. Appl. Crystallogr.* **10**, 104–110 (1977).

11. P. Pianetta and I. Lindau, Phase space analysis applied to x-ray optics, *Nucl. Instrum. Methods* **152**, 155–159 (1978).

12. G. Rosenbaum, Die Anwendung von Synchrotron Strahlung in der Biologie, Ph.D. thesis, University of Heidelberg (1980).

13. W. H. Zachariasen, *Theory of X-Ray Diffraction in Crystals*, Dover Publications, New York (1967).

14. J. B. Hastings, B. M. Kincaid, and P. Eisenberger, A separated function focusing monochromator system for use with synchrotron radiation, *Nucl. Instrum. Methods*, **152**, 167–171 (1978).

15. J. Hendrix, M. H. J. Koch, and J. Bordas, A double focusing x-ray camera for use with synchrotron radiation, *J. Appl. Crystallogr.* **12**, 467–472 (1979).

16. P. Horowitz and J. A. Howell, A scanning x-ray microscope using synchrotron radiation, *Science* **178**, 608–611 (1972).

17. J. C. Haselgrove, A. R. Faruqi, H. E. Huxley, and U. W. Arndt, The design and use of a camera for low-angle x-ray diffraction experiments with synchrotron radiation, *J. Phys. E.* **10**, 1035–1044 (1977).

18. A. Harmsen and G. Rosenbaum, A remote controlled optical bench for small-angle diffraction with synchrotron radiation, *J. Appl. Crystallogr.*, in preparation.

19. J. Barrington Leigh and G. Rosenbaum, Synchrotron x-ray sources: a new tool in biological structural and kinetic analysis, *Annu. Rev. Biophys. Bioeng.* **5**, 239–270 (1976).

20. G. Rosenbaum, K. C. Holmes, and J. Witz, Synchrotron radiation as a source for x-ray diffraction, *Nature (London)* **230**, 129–131 (1971).

21. T. D. Molkul'skaya, M. A. Mokul'skii, A. A. Nikitin, I. Ya. Skuratovskii, C. E. Baru, G. N. Kulipanov, V. A. Sidorov, A. N. Skrinskii, and A. G. Khabakhphashev, X-ray structural research of biopolymers using synchrotron radiation and a multi-channel detector (in Russian), *Crystallography* **22**, 744–752 (1977).

22. N. G. Webb, Logarithmic spiral focusing monochromator, *Rev. Sci. Instrum.* **47**, 545–547 (1976).

23. M. Lemonnier, R. Fourme, F. Rousseaux, and R. Kahn, X-ray curved-crystal monochromator system at the storage ring DCI, *Nucl. Instrum. Methods* **152**, 173–177 (1978).

24. A. A. Vasina, V. S. Gerasimov, L. A. Zheleznaya, A. M. Matyushin, B. Sorikin, L. N. Skrebnitskaya, V. N. Shelestov, G. M. Frank, Sh. M. Avakyan, and A. I. Alikhanyan, Experience with the use of synchrotron radiation for x-ray diffraction studies of biopolymers, *Biophysics (USSR)* **20**, 813–820 (1975).

25. A. A. Vasina, The use of synchrotron radiation for structure research on biopolymers (in Russian), *Mol. Biol.* (USSR) **8**, 242–307 (1976).

26. U. W. Arndt, X-ray film, in *The Rotation Method*, U. W. Arndt and A. J. Woncott (eds.) pp. 207–218, North-Holland Publishing Co., Amsterdam (1977).

27. U. W. Arndt and D. J. Gilmore, X-ray television area detectors for macromolecular structural studies with synchrotron radiation sources, *J. Appl. Crystallogr.* **12**, 1–9 (1979).

28. Geo. T. Reynolds, J. R. Milch, and S. M. Gruner, High sensitivity image intensifier–TV detector for x-ray diffraction studies, *Rev. Sci. Instrum.* **49**, 1241–1249 (1978).

29. J. R. Milch, Slow scan SIT detector for x-ray diffraction studies using synchrotron radiation, *I.E.E.E. Trans. Nucl. Sci.* **NS-26**, 338–345 (1979).

30. R. S. Goody, J. Barrington Leigh, H. G. Mannherz, R. T. Tregear, and G. Rosenbaum, X-ray titration of binding of β,γ-imido-ATP to myosin in insect flight muscle, *Nature (London)* **262**, 613–615 (1976).

31. R. J. Podolsky, R. St. Onge, L. Yu, and R. W. Lymn, X-ray diffraction of actively shortening muscle, *Proc. Natl. Acad. Sci. U.S.A.* **73**, 813–817 (1976).

32. S. E. Baru, G. I. Provitz, G. A. Savinov, V. A. Sidorov, I. G. Feldman, and A. G. Khabakhpashev, One-coordinate detector for rapid multisnap recording of x-ray pictures, *Nucl. Instrum. Methods* **152**, 195–197 (1978).

33. H. Hashizume, K. Mase, Y. Amemiya, and K. Kohra, A system for kinetic x-ray diffraction using a position sensitive counter, *Nucl. Instrum. Methods* **152**, 199–203 (1978).

34. A. Gabriel, F. Dauvergne, and G. Rosenbaum, Linear, circular, and two dimensional position sensitive detectors, *Nucl. Instrum. Methods* **152**, 191–194 (1978).

35. S. E. Baru, G. I. Provitz, G. A. Savinov, V. A. Sidorov, A. G. Khabakhpashev, B. N. Shuvalov, and V. A. Yakovlev, Two-coordinate x-ray detector, *Nucl. Instrum. Methods* **152**, 209–212 (1978).

36. G. E. Schulz and G. Rosenbaum, The multi-wire proportional chamber as an area detector for protein crystallography in comparison with photographic film; guidelines for future development of area detectors, *Nucl. Instrum Methods* **152**, 205–208 (1978).

37. G. Bathow, E. Freytag, and R. Haensel, Measurement of the synchrotron radiation in the x-ray region, *J. Appl. Phys.* **37**, 3449–3458 (1966).

38. E. Pitz, Absolute calibration of light sources in the vacuum ultraviolet by means of the synchrotron radiation of DESY, *Appl. Opt.* **8**, 255–263 (1969).

39. J. Schelten and R. W. Hendricks, Recent developments in x-ray and neutron small-angle scattering instrumentation and data analysis, *J. Appl. Crystallogr.* **11**, 297–324 (1978).

40. A. Boeff, S. Lagomarsino, S. Mazkedian, S. Melone, P. Puliti, and F. Rustichelli, X-ray diffraction characteristics of curved monochromators for synchrotron radiation, *J. Appl. Crystallogr.* **11**, 442–449 (1978).

41. H. E. Huxley, The structural basis of muscular contraction, *Proc. R. Soc. London Ser. B* **178**, 131–149 (1971).

42. A. R. Faruqi and H. E. Huxley, Time-resolved studies on contracting muscle using small-angle x-ray diffraction. 1. Design of data collection system, *J. Appl. Crystallogr.* **11**, 449–454 (1978).

43. K. C. Holmes, R. T. Tregear, and J. Barrington Leigh, Interpretation of the low angle x-ray diffraction from insect flight muscle in rigor, *Proc. R. Soc. London Ser. B* **207**, 1–12 (1980).

44. R. Bowitz, R. Jonak, H. Nemetschek-Gansler, Th. Nemetschek, H. Riedl, and G. Rosenbaum, The elasticity of the collagen triple helix, *Naturwissenschaften* **63**, 580 (1976).

45. J. Bordas, I. H. Munro, and A. M. Glazer, Small-angle scattering experiments on biological materials using synchrotron radiation, *Nature (London)* **262**, 541–545 (1976).

46. S. W. White, D. J. S. Hulmes, and A. Miller, Collagen-mineral axial relationship in calcified turkey leg tendon by x-ray and neutron diffraction, *Nature (London)* **266**, 421–425 (1977).

47. J. M. Goodfellow, G. F. Elliott, and A. E. Woolgar, X-ray diffraction studies of the corneal stroma, *J. Mol. Biol.* **119**, 237–252 (1978).

48. R. P. Haelbich, A. Segmüller, and E. Spiller, Smooth multilayer films suitable for x-ray mirrors *Appl. Phys. Lett.* **34**, 184–187 (1979).

49. H. B. Stuhrmann, The use of x-ray synchrotron radiation for structural research in biology, *Q. Rev. Biophys.* **11**, 71–98 (1978).

17

Single-Crystal X-Ray Diffraction and Anomalous Scattering Using Synchrotron Radiation

JAMES C. PHILLIPS and KEITH O. HODGSON

1. Introduction

1.1. Uses of Synchrotron Radiation in Macromolecular Crystallography

It has long been apparent that the extreme brightness and broad spectral distribution of synchrotron radiation could be of importance in carrying out several classes of x-ray diffraction experiments. One of the earliest uses of synchrotron radiation for diffraction experiments was at the DESY synchrotron in Hamburg, where a low-angle diffraction camera, which focused and monochromatized the radiation, was set up for work on muscle.[1]

Concurrent with these experimental efforts, there were discussions about the potential uses of synchrotron radiation for protein crystallography. Some of these discussions may be found summarized in papers by Harrison[2] and by Wyckoff.[3] At Stanford, we began to consider this question from an experimental viewpoint in 1974.[4]

JAMES C. PHILLIPS and *KEITH O. HODGSON* ● Department of Chemistry, Stanford University, Stanford, California 94305. J. C. Phillips' current address: European Molecular Biology Laboratory, Hamburg Outstation, c/o DESY, 52 Notkestieg 1, 2000 Hamburg, Germany.

1.1.1. High Intensity

There are many problems that can occur during the course of a protein struc-ture determination by x-ray diffraction. Use of a synchrotron radiation source could help in the solution of some of them. After a protein is isolated, the labor-ious task of growing crystals can begin. If crystals are obtained, they are often small and diffract weakly. It can take months or years to obtain crystals suitable for diffraction studies. Protein crystals have large unit cells, therefore there are many Bragg reflections to be recorded. Moreover, they suffer from radiation damage, so there is only a limited time to collect this large amount of data. The diffracted intensities fall off rapidly with increasing Bragg angle so high-resolution data, which gives a clearer image of the molecule, is difficult to obtain. A high-intensity x-ray beam, such as that provided by synchrotron radiation, would help to alleviate all of the above problems, if it did not produce rapid and extreme radiation damage. Smaller crystals could be used. Data could, in principle, be collected faster, to higher resolution and with greater statistical accuracy.

1.1.2. The Phase Problem

Obtaining diffracted intensities from a crystal is not the only step in the crystal structure determination. The "phase problem" must also be solved. As is well known, the diffraction pattern of a crystal is the Fourier transform of the electron density within the unit cell. The transform has an amplitude and a phase compon-ent (i.e., it is a complex number) but only the amplitude can be directly measured. Normally, the phase is obtained by preparing a second crystal that is identical to the first except for the inclusion of an extra atom, preferably a heavy atom such as platinum or mercury. This is known as an isomorphous derivative of the native crystal. Comparison of the two diffraction patterns gives the needed phase information.[5] Actually, several derivatives are usually necessary.

The same information can, however, be obtained by varying the photon energy of the x-ray beam. For each element there are characteristic photon energies called absorption edges. Below the edge there is not sufficient energy to excite electrons out of an atomic shell, but above the edge there is. Thus, by simply tuning the beam across the absorption edge of the heavy atom, an effect similar to making an isomorphous derivative can be obtained. Above the edge, the atom absorbs and appears not to be in the structure and below it scatters x rays into the diffraction pattern instead of absorbing and is thus "in" the structure. This phenomenon is called "anomalous scattering."

In the case of some proteins, no one has yet succeeded in making heavy atom derivatives. It can often take months or years to obtain good derivatives. Thus, the alternative technique of using tunable synchrotron radiation is of great interest. Synchrotron radiation is essential for the technique as laboratory sources have only a weak continuum (about 10^5 times weaker than synchrotron radiation). Of course laboratory sources have high intensity at the x-ray energies of the characteristic emission lines. These intense emission lines of conventional x-ray tubes never fall on the absorption edges of the elements.

1.1.3. Large Unit Cells

There is an important need to solve the structure of ever larger molecules. This means that crystals with unit-cell sizes of 200 Å or larger must be examined. In order to resolve individual diffraction spots, a highly parallel x-ray beam is required. Collimation of conventional sources produces intensity losses. Synchrotron radiation is naturally well collimated and this feature can be preserved with a suitable focusing monochromator system. Thus, for crystallographic studies on extremely large unit cells synchrotron radiation offers further intensity gains over conventional sources.

1.2. Early Protein Crystallography Results at SPEAR

In early 1974, a synchrotron radiation beam line became available at the (then) Stanford Synchrotron Radiation Project (SSRP). It was logical that some of the possibilities outlined above could be explored. Some preliminary experiments were performed with monochromatized, but unfocused, radiation from a single crystal monochromator built by the authors and the SSRP staff. They produced one precession photograph of a protein crystal taken with radiation before a fire in the SPEAR storage ring halted the work. The photograph showed little intensity gain over a conventional x-ray source but proved that there was no intrinsic reason why single-crystal diffraction experiments could not be performed with synchrotron radiation.

During the repair of the storage ring, a focusing monochromator built by a group from Cal Tech was installed at SSRP.[6-8] This gave much higher flux and was used in further exploratory experiments. In the first series of experiments using this monochromator system, a series of precession photographs of various protein crystals were obtained. Exposure times were reduced dramatically. With SPEAR operating at 3.7 GeV and 40 mA, and with separated beams, the intensity was estimated to be 60 times higher than a fine focus copper x-ray tube operated at 40 KeV and 30 mA. Anomalous scattering effects were found in diffraction data from an iron-containing protein, rubredoxin. This data was obtained with incident beam wavelengths around the iron K edge. Good spot resolution was observed along a 180-Å unit-cell axis in a photograph of nerve growth factor protein. A 2.7-Å-resolution data set of the copper-containing protein azurin was collected using an energy just above the copper K edge. Radiation damage to this crystal was much lower than for an equivalent exposure (in the sense of amount of diffraction data collected) from a lower-power conventional source.[4] The data from the azurin crystal proved of help in locating the copper atoms because of enhancement of the anomalous scattering.[9]

Thus, the early experiments showed that the intensity, tunability, and collimation of synchrotron radiation could all be successfully utilized. In a later experiment, more data on rubredoxin was obtained at a series of wavelengths through the iron K edge. The phases of the Bragg reflections were extracted from this data and differed by 60° on the average from the known phases.[10] This was also an encouraging sign that phasing by anomalous scattering could be a generally applicable method.

1.3. Crystallography at Other Synchrotron Radiation Laboratories*

Experiments carried out at the DESY synchrotron in Hamburg, Germany and the DCI storage ring in Orsay, France soon confirmed the early SPEAR results. At DESY, Harmsen *et al.*[11] used rotation photography to obtain data from protein and virus crystals. They used a doubly focusing mirror–monochromator system. It was concluded that the intensity available from the synchrotron was only advantageous for larger-unit-cell crystals, when the brightness of the source becomes important, and then only by a factor of 16 over conventional sources for unit cells of 100 Å and above. It was thought that the storage ring DORIS would provide 30 times more flux, making the source advantageous for all unit-cell sizes. When diffraction apparatus was set up at DORIS, this proved to be correct.[12]

At DCI, a simple focusing monochromator was set up.[13] Rotation photographs of a wide variety of protein crystals have since been obtained. Intensities are 25–50 times higher than for rotating anode sources. Complete data sets on tyrosyl-t-RNA synthetase[14] and phosphorylase were collected to higher resolution than proved possible with rotating anode sources. These results were reviewed by Fourme.[15]

Some preliminary protein crystallography experiments were also performed at the NINA synchrotron at Daresbury, England,[16] before it was shut down. There are plans to continue the work on the dedicated storage ring SRS now nearing completion there. The characteristics of this new source should be similar to DCI.

1.4. Scope of the Remainder of the Chapter

With the knowledge gained from the early SPEAR experiments, it was possible for us to set requirements for a focusing monochromator and x-ray crystallography equipment that could be used to routinely perform diffraction experiments utilizing the special properties of synchrotron radiation. A high-flux, rapidly tunable, focusing monochromator system well suited to these requirements was set up at the Stanford Synchrotron Radiation Laboratory.[17] In Section 2, a diffractometer system operating on this beam line will be described. This apparatus has been used to accurately measure anomalous scattering terms for cesium and cobalt, and these experiments are described in Section 3. Section 4 is a more detailed description of how these anomalous scattering effects can be used to obtain the phases of a diffraction pattern.

With the knowledge of the strength of the anomalous scattering terms for cesium, it was possible to plan an experiment to solve the crystal structure of a cesium-containing protein that so far has not been solved by conventional techniques. This experiment is described in Section 5 as a case study in the use of synchrotron radiation for structure determination.

Finally, in Section 6, we will attempt to draw conclusions from these results as to what diffraction studies with synchrotron radiation are now possible and by what means the scope of these studies can be widened.

---

* See also Chapter 16 of this volume.

2. A Four-Circle Diffractometer Used with Synchrotron Radiation

2.1. Introduction

In late 1976, a focused monochromatized beam port at the Stanford Synchrotron Radiation Laboratory using synchrotron radiation from the storage ring SPEAR became available to be, in part, a source for single-crystal diffraction experiments.[17] In early 1977, an Enraf-Nonius CAD-4 diffractometer was mounted at this port in order to perform such experiments.

The system was first tested in June, 1977, during low-energy operation of the SPEAR storage ring. Since then, as beam time has been scheduled, further tests and experiments have been carried out.

Experiments to measure the anomalous scattering terms for cesium at the cesium L_3 edge have been conducted and reported.[18] Also, data from protein crystals have been collected.

The use of a diffractometer with a synchrotron radiation source requires specialized equipment and controlling software. We will first outline the relevant characteristics of the focusing monochromator system. Then we will describe the modifications made to the diffractometer, and the design considerations and construction of the alignment carriage on which it was mounted. The software and computer control system will also be discussed. Results of the system tests will be summarized.

2.2. Description of the System

2.2.1. Focusing Mirror–Monochromator System

Hastings *et al.*[17] have described the separated function focusing monochromator system that they constructed and tested at SSRL. This system is described in Chapter 3. It consists of a toroidal mirror, for focusing the x-rays, and two flat germanium (111) crystals in a "parallel" double-reflection arrangement that monochromatizes the white beam. The intensity and angular distributions across the x-ray beam focus are essentially the same as those of the electron beam in the storage ring at the source point. With the beam line in its present configuration, the electron beam source size to x-ray focus size is 1 : 1. The beam is brought into focus by tilting the mirror upwards into the beam. The focused beam therefore travels at a slight upward angle (about 20 mrad) to the horizontal. With a two-crystal monochromator, the monochromatic x-ray beam moves up and down slightly as the wavelength is changed. Over the energy range from 3 to 9 keV, the beam height will change by 4 mm. It is easy to arrange for the diffractometer to track this small motion. Thus this arrangement is well suited for use in x-ray diffraction experiments that require rapid and frequent wavelength changes.

2.2.2. Diffractometer Modifications

Synchrotron radiation has a high degree of polarization in the horizontal plane. To avoid intensity losses due to polarization, the diffractometer was mounted sideways, i.e., with the θ and ω axes horizontal. Enraf-Nonius fitted counterweights to the θ and ω motions. This relieves the load imbalance on the drive motors when the detector arm or ω turret are moving in the vertical plane. An automated oiling system, consisting of reservoir, electric pump, pipelines, and wipers, was installed to ensure that the drive gears are continually lubricated.

2.2.3. Alignment Carriage

For accurate measurements in single-crystal diffractometry, it is essential that the incident x-ray beam pass through the centerline of the instrument. To align the diffractometer to the synchrotron radiation beam, a carriage with four motions (horizontal and vertical translations and pivots about horizontal and vertical axes that pass through the opening of the incident beam collimator) was built. These motions (shown schematically in Figure 1) are both necessary and sufficient to ensure that the most intense portion of the focused beam passes through the fixed centerline of the CAD-4 diffractometer. They are also "decoupled" in that one motion does not substantially affect the adjustment of any other. Thus, alignment is a stepwise rather than an iterative procedure. Figure 2 is a photograph of the diffractometer mounted on the carriage.

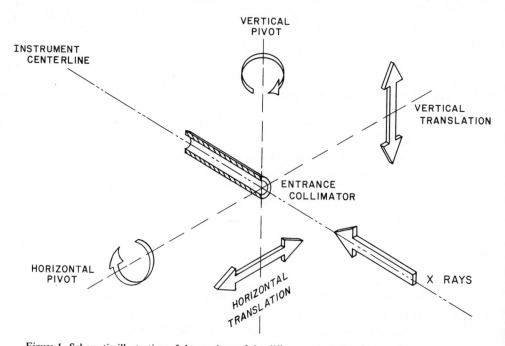

Figure 1. Schematic illustration of the motions of the diffractometer alignment carriage.

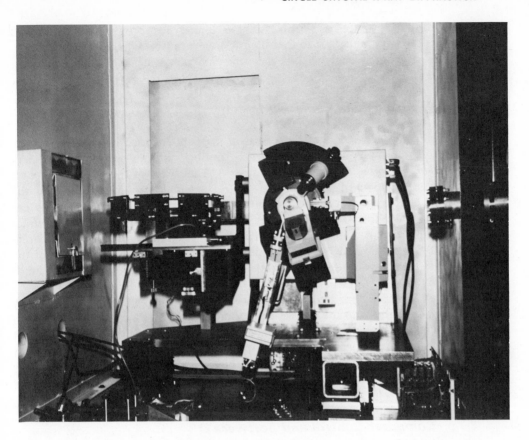

Figure 2. Photograph of the CAD-4 diffractometer mounted on the alignment carriage inside the radiation safety hutch. The horizontal slides may be seen at the bottom corners. The vertical jack is slightly lower and just to the right of center. At the top to the right is the diffractometer and to the left, the rotation camera carousel. The micrometer pivot and vertical guiding slide are hidden. When the θ arm detector is used, the rotation carousel is removed.

A micrometer pushing on a lever arm provides a horizontal pivot. A vertical translation is affected by a jack pushing from under the diffractometer, the motion being guided by a vertically mounted slide. The remaining two motions are provided by the horizontal slides, arranged in a triangle under the base of the carriage and transverse to the incident beam direction, two at the upstream end and one at the downstream end. The rear slide and one of the front slides are driven through lead screws. The third slide is free. If the slides move together, the diffractometer moves horizontally. If only the rear slide is driven, the diffractometer pivots about a vertical axis. All four motions are driven by stepping motors.

Also in the photograph is shown a film holder carousel for rotation photography using the ϕ or ω circle of the diffractometer as a rotation axis. No results using this feature are presented in the present work. But it is pointed out that the carriage does not interfere with the operation of the carousel and the full range of crystal-to-film distance is possible. Enraf-Nonius has provided software and modified the CAD-4 interface so that the rotation carousel–diffractometer combination can be remotely controlled by the same computer system.

2.2.4. Transportable Radiation Hutch

For radiation safety, all experiments at SSRL are enclosed in a sheet-steel box (called a "hutch") with a large lead beam stop behind it in the path of the beam. The hutch system is described in Chapter 3. The doors of the hutch and beam shutters are interlocked so that they cannot be simultaneously opened. For the diffractometer, a large hutch with sliding doors was constructed. The sides are removable (when the x-ray shutters are disabled) to allow easy access to the diffractometer and carriage for maintenance. The roof is also removable. Thus, the diffractometer can be lowered onto the carriage or removed with a crane. One of the doors has lead glass windows in it so that the motions of the diffractometer can be observed.

The hutch is carried on a three-point kinematic mount that is used to crudely adjust the apparatus to the correct height and angle of the beam line. The hutch and kinematic mount may be jacked up onto wheels and removed from the beam port so that other experiments can be set up. So far, the diffraction system has only been used at the beam port described above, but it can also be used at any end-of-line x-ray beam port, existing or planned, at SSRL.

2.2.5. Computer Control and Software

A PDP11/34 computer with 32K word memory and two RK05 disk packs is used to control the diffraction system. There is also a tape drive for transferring data and a Versatec line printer. The standard Enraf-Nonius CAD-4 interface (which includes a PDP8 computer) links the 11/34 computer and diffractometer. A CAMAC interface is used in controlling the stepping motors that drive the four motions of the alignment carriage and the monochromator position. The ion chamber monitoring is also controlled by the CAMAC system. Figure 3 is a block diagram of the entire apparatus and illustrates the function and control chain of each component.

The 11/34 computer operates under the DEC RSX11-M real-time system. The programs supplied by Enraf-Nonius to operate the diffractometer time share with other programs for automatic alignment, main beam monitoring, and wavelength calibration. These latter programs were all developed from a software package for x-ray absorption spectroscopy data collection written in our laboratories.[19]

The alignment program adjusts each motion of the carriage in turn until the intensity of the beam passing through the diffractometer collimator is maximized, this being measured by the detector of the diffractometer when set to $\theta = 0°$.

The electron beam current in a storage ring decays with time and therefore so does the x-ray beam intensity. To normalize the diffracted intensities to the same

SSRL 4 CIRCLE GONIOMETER SYSTEM

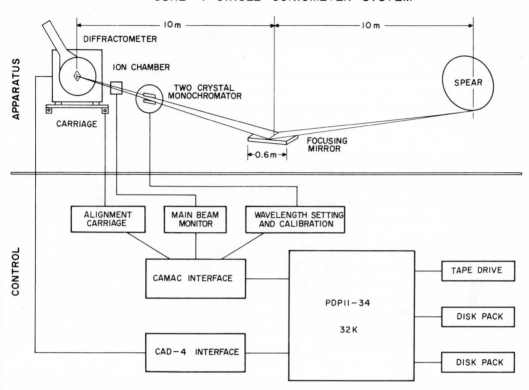

Figure 3. Block diagram of the complete diffraction system showing the purpose and control of the various components.

incident beam intensity, the reading of an ion chamber placed in front of the collimator was constantly recorded in a disk file along with the computer system time. The diffractometer data collection program was modified to include the system time in the data file along with all the other information about the measurement of a reflection. During data reduction, a further program combines the information in the two files.

The monochromator is calibrated by the method, well known in x-ray absorption spectroscopy, of relating an absorption edge of an element to a monochromator setting. The absorption spectrum is taken using the ion chamber and the CAD-4 detector to measure the incident flux and that transmitted through a standard sample. Graphs of such spectra are used to plan anomalous scattering experiments.

The CAD-4 interface has an input that is normally used to sense the high voltage on the x-ray tube and interrupt data collection if there is no voltage. This was connected to signals indicating the status of the storage ring ("beam," "dump," "inject," or "fault") so as to allow data collection only in the "beam" status.

2.3. Performance of the System

2.3.1. Diffractometer Positioning Accuracy

Software provided by Enraf-Nonius checks the accuracy with which the goniometer may be positioned. Since being mounted sideways, this software has been run periodically. There has been no deterioration in the positioning accuracy of any of the motions. It remains within the manufacturer's specification of $\pm 0.01°$ for all motions.

2.3.2. Alignment

In the initial stages of testing of the apparatus, the diffractometer was approximately aligned by maximizing the flux through the entrance collimator. Then a standard alignment procedure involving measurements on a test crystal was carried out.

It was found that the diffractometer was extremely well aligned purely by the method of maximizing the flux through the collimator. In all further testing and experimentation, this simple alignment procedure was the only one carried out. It proved to be extremely important as spatial instabilities in the synchrotron radiation beam make frequent alignment checks necessary in some modes of operation of the storage ring.

In SPEAR there is a beam position monitor that keeps the vertical height of the electron beam, and therefore the synchrotron radiation beam, constant.[20] However, whenever there is a major orbit change in the storage ring, the horizontal source point changes and the focusing mirror has to be realigned. This problem was noticed when the mirror was first used at the Cambridge Electron Accelerator and the solution, which is to provide a pivot motion on the mirror so that it may be brought back into alignment, has been described by Howell.[21] Hastings et al.[17] incorporated this feature into the setup at SSRL. The mirror may be realigned in a few seconds by maximizing the reading of the main beam ion chamber, but then the diffractometer is misaligned. The frequency of the orbit changes varies with the mode in which the storage ring is operating. Electrons can be injected into SPEAR at energies of 2.6 GeV or lower. If high-energy physics experiments are being performed at lower energies, the stored current can be replenished without dumping the beam. The orbits may be stable for several days and the diffractometer need not be realigned. When the operating energy is higher than the maximum injection energy, the beams are periodically dumped and the ring is refilled at 2.6 GeV, then the energy is increased to the operating energy. In this mode of operation, orbit changes can occur at every fill and thus the mirror and diffractometer must be realigned. It is thus fortunate that alignment is a simple procedure taking a few minutes, which is performed fully under the control of the computer. If this had not been the case, at each fill, the sample crystal would have had to be removed, a test crystal mounted, its orientation refined, and then alignment adjustments to the diffractometer made. Then the sample would have had to be replaced and its orientation refined before data collection could proceed. This procedure would have taken most of the beam time available before the next fill so data collection would have been impossible during high-energy operation.

If periodic checks of the alignment are carried out, then there is no difficulty in using the diffractometer for accurate measurements of diffracted intensities. Orientation matrices with correct cell parameters may be obtained, and during data collection reflections occur at the predicted goniometer settings.

2.3.3. *Parameters of the Focused, Monochromatized Beam*

Intensity profiles of the main beam are shown in Figure 4. The beam profile has a FWHM of 2.4 mm in the vertical and 4.2 mm in the horizontal. Across the central 1-mm-diam portion of the beam, which is that used for the diffraction experiments, there is a smooth 6% intensity variation. The measurements were made at 1.8 GeV and 10 mA with colliding beams.

Reflection widths were measured for a crystal of ammonium tartrate and a crystal of hen egg-white lysozyme. A typical ω scan was 0.012° and 0.015° FWHM, respectively. Figure 5 shows a comparison of the diffraction profile of a weak reflection from the lysozyme crystal when measured with both synchrotron radiation and a conventional source. This source was a fine-focus x-ray tube operated at 40 keV and 30 mA. It provided Ni-filtered Cu Kα radiation. The same reflection from the same crystal gave an ω peak with 0.066 FWHM.

An estimate of the beam intensity was obtained by comparing integrated intensities of reflection peaks from a crystal of lysozyme measured using the conventional source described above with those obtained using the focused synchrotron radiation. At SPEAR conditions of 3.7 GeV and 20 mA with colliding beams the integrated intensities were 40 ± 5 times higher using the synchrotron radiation beam at 1.54 Å. The peak intensities were 240 times higher, corresponding to the narrower reflection widths. The true measure of the relative photon flux in the two beams is the comparison of integrated reflection intensities, not peak intensities.

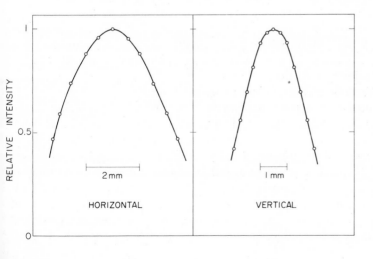

Figure 4. Horizontal and vertical intensity distributions across the focus of the synchrotron radiation beam from SPEAR; 1.8 GeV, 10 mA.

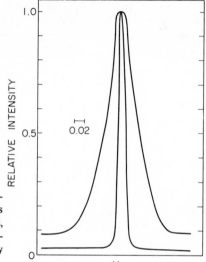

Figure 5. The lysozyme (014) reflection. A comparison of ω scans using a fine-focus tube and focused synchrotron radiation as x-ray sources. The scan widths are 0.066° and 0.015° for the fine-focus tube and synchrotron radiation sources, respectively. The peak to background ratio is three times higher with synchrotron radiation. Both scans are normalized to the same peak height, but actually the synchrotron radiation peak is far higher.

However, the high, narrow reflection peaks can be advantageous and their use, in certain circumstances, will be discussed later.

For the planning of anomalous scattering experiments, it is important to know what kind of count rates can be expected at other wavelengths where atomic scattering factors and absorption by the crystal (and the capillary in which protein crystals are mounted) are different. Count rates at 1.7 Å were similar to those at 1.54 Å. At 2.1 Å the gain over a sealed tube was 7 times and at 2.5 Å the gain factor was 2. This data was obtained using a crystal of gramicidin A complexed with cesium.[22] The cesium may be contributing towards the absorption and may distort these figures somewhat.

The beam size agrees well with the figures of 2 mm × 4 mm estimated by Hastings et al.[17] For a beam of this size, the brightness within the central portion used in the diffraction experiments should be 4×10^{10} photons sec^{-1} mm^{-2} at 3.7 GeV and 20 mA, using the figure of 10^{12} photons sec^{-1} at 3.7 GeV and 40 mA given by Hastings et al.[17] for the whole beam. This should lead to an expected flux gain of 75 over a fine-focus x-ray tube rather than the 40-fold increase measured.

The width of a Bragg reflection is a convolution of four factors—crystal size, crystal mosaicity, beam divergence, and beam spectral bandwidth.[23] The reflection peak widths measured from the ammonium tartrate give an upper bound to the angular divergence and bandpass factors of the central portion of the focused beam of $0.012 \pm 0.002°$. For focused white radiation of a given angular convergence incident upon a monochromatizing crystal with a much smaller angular acceptance, the convergence of the beam governs the overall bandpass and there is a correlation between a photon's direction of propagation and its energy. From the value of 5 eV at 7.1 keV for the bandpass given by Hastings et al.,[17] a beam convergence of 0.011° can be inferred. Thus the upper limit imposed by the diffraction experiment is probably close to the actual value.

2.3.4. Accuracy of Measurement of Diffracted Intensities

In a paper on a diffractometer rotating anode system, Massey and Manor[24] performed a test in which repeated measurements of diffracted intensities were made. A quantity R, the "machine error coefficient," was defined as the increase in the rms error of the measurements over that expected from counting statistics alone,

$$\sigma_{tot} = (\sigma_{counts}^2 + R^2)^{1/2} \tag{1}$$

where all quantities are expressed as fractions. They found a value of 1.26% for R, which is typical of conventional diffractometer systems. A similar test was conducted on the diffractometer system at SSRL. For normal data collection techniques with ω–2θ scans of 2° and scan speeds that are automatically adjusted so that appropriate numbers of counts for 2% statistics are accumulated, the value of R was found to be 6.5%. If 0.3° scans were used at a speed of 1° per minute, R was found to be 2.4% with a σ_{tot} of 2.7%. This was for 18 reflections measured 13 times. The scan width of 0.3° was found to be the minimum necessary to ensure that the reflections were intercepted although an accurate orientation matrix was obtained. This test was carried out during one storage ring fill. A reproducibility test was carried out using the same reflections from the same crystal on a CAD-4 diffractometer with a fine-focus x-ray tube source. An R value of 1.0% was obtained.

Using the above narrow, slow scan technique, a shell of data from the ammonium tartrate crystal was measured. Two hundred reflections were measured in symmetry related pairs, and the quantity R_{sym} was calculated, where

$$R_{sym} = \sum |I_{hkl} - I_{\bar{h}\bar{k}\bar{l}}| / \sum (I_{hkl} + I_{\bar{h}\bar{k}\bar{l}}) \tag{2}$$

and was found to be 3.3%. For the same data from the crystal using a sealed tube source R_{sym} was 2.6%. The data from the two sources were scaled and compared. The quantity R', where

$$R' = \sum |F_{SR} - F_{ST}| / \sum (F_{SR} + F_{ST}) \tag{3}$$

(F_{SR} is the value of the structure factor measured with synchrotron radiation and F_{ST} is that measured with a sealed tube source) was computed and found to be 1.8%. The synchrotron radiation data collection was carried out over 8 hours and 5 storage ring fills.

All of the above described system tests were performed while SPEAR was running at 2.2 GeV, with intensities approximately 5 times *less* than from a fine-focus x-ray tube.

2.3.5. Tests Using a Graphite Monochromator

For comparison purposes some tests were done in the same beam line with graphite (002) crystals of 1° FWHM mosaic spread (nominally) as monochromatizing crystals. Sparks[25] gives predictions of what the beam size, bandpass, and flux should be under these conditions. The beam focus was 6 mm × 16 mm, the bandpass (estimated from reflection widths) was 260 eV at 7.1 keV, and the relative flux increase of the whole beam, over that obtained with the germanium crystals, was 40-fold. At comparable SPEAR operating conditions, reflection peak count rates were 1.6 times higher and integrated intensities were 120 times higher with the graphite monochromator. Data collection rates are not so high, as wide scans were necessary. The integrated intensities were 7.6 times lower than those from a fine-focus x-ray tube. Reflection widths were 0.8° to 1.0° wide (depending upon θ). These values agree well with Sparks' predictions. In a reproducibility test, a value of R of 1.77% was found. The value of σ_{tot} was 2.1%. This test involved measuring 18 reflections 24 times and was carried out over a period of 30 hours and during 11 fills of the storage ring. SPEAR was being operated at 1.84 GeV and 8 mA average current. No checks of alignment were carried out during the data collection.

2.4. Discussion of System Tests and Possible Improvements

The problem of keeping a diffractometer aligned to a spatially unstable beam was readily solved by the use of the computer automated alignment carriage. Alignment by the simple procedure that has been described is sufficient, probably because the focused synchrotron radiation beam is intrinsically well collimated. This contrasts with the tedious alignment required with a conventional source.

The intensity measurements reported here were made 20 months later than those of Hastings *et al.*[17] Deposits of carbon have been forming on the beryllium windows of the beam transport system in this period and there may have been deterioration in the reflectivity of the mirror or crystal monochromators. Also, it is suspected that the helium transport system is contaminated with heavier gases. This being the case, there is no serious disagreement between the two intensity measurements. At a recent workshop on synchrotron radiation beamline design, such degradative processes were discussed.[26]

The results of the system tests show that diffracted intensities can be measured slightly less reliably than from a conventional source (assuming that the sample is such that extremely accurate data can be obtained by conventional means). Diffraction data using the synchrotron source can be collected and correlated over a long time period. There is no difficulty in obtaining true structure factors from the synchrotron radiation data, only trivial modifications to conventional data reduction procedures are necessary. From the values of R obtained with germanium crystal-monochromatized radiation, it is clear that care must be taken in measuring the integrated intensities of extremely narrow reflections. As the positional accuracy of the instrument is nominally only a factor of two better than the reflection width, this is not surprising. [That measurements of 2.7% accuracy were found for slow scan speeds indicates that the CAD-4 probably tracks better than its rated positional accuracy during a scan]. The lower value of R with the wide reflections from graphite monochromatized radiation tends to confirm this interpretation. It also appears that there is some error due to other causes perhaps associated with the decay of the incident beam and the Gaussian distribution of intensity across the focus. Nevertheless, it is clear that data of sufficient quality for a typical structure determination is readily obtainable.

The present method of monitoring the main beam decay has two drawbacks. First, the entire beam is monitored, not just the central portion after collimation. The intensity of the whole beam and the collimated beam are proportional to a good approximation (measurements of the ratio of the two over an extended period showed that it remained constant within experimental error). The method of relating the time at which the reflection is recorded to an intensity/time file has slight errors. To reduce these errors, the CAD-4 interface has recently been modified to record ion-chamber readings as well as diffracted intensities. Software to record the main beam intensity at precisely the same time as the diffraction spot is being implemented. A detector to monitor the collimated beam is being developed. These changes should improve the quality of the data slightly.

At present, the system is not fully automatic. Adjustments at the start of each fill require an operator to be present. However, it is a simple matter to modify the software so that the mirror and diffractometer are aligned in turn before data collection is restarted upon the receipt of the "beam" signal from SPEAR. For an anomalous scattering experiment, the precise monochromator setting could be checked by taking a spectrum of a sample placed where the beam stop would normally go. This spectrum could be compared to a standard by a curve-fitting routine and corrections made if necessary. With such software, the system could be safely left unattended during data collection.

2.5. Possible Uses of the Systems

2.5.1. Macromolecular Crystallography

This apparatus can and has been used to perform anomalous scattering experiments in which a Bragg reflection is measured as a function of incident beam photon energy as this energy is varied across an absorption edge of an element in the crystal. It is well known that such an experiment can provide phase information for protein structure determination when the protein contains a naturally occurring heavy atom when only nonisomorphous derivatives have been prepared, or as a supplement to isomorphous derivative data.

The high intensity makes it possible to collect data on weakly diffracting samples such as those often encountered in protein crystallography. In fact, at the limit of resolution of the diffraction pattern all protein crystals diffract weakly. It is possible to use the increased intensity to collect more accurate data with better counting statistics, to collect more data per crystal, and to extend the resolution of the data collected. A data collection strategy using all or some of these features is possible. Massey and Manor[24] also obtained increased flux with their apparatus and suggested these uses. However, the flux gains with synchrotron radiation are higher. The results described herein also represent the lower limit on possible intensity gains. When SPEAR is operated as a dedicated source of synchrotron radiation 150 mA of stored current (averaged over a fill at 3.4 GeV) are possible.[27] Moreover, the electron beam source size is reduced when there are no collisions with the positron beam.[28] Therefore, flux gains more than directly proportional to the increased current are expected. Orbits are more stable during dedicated operation. It is planned that SPEAR will be operated as a dedicated storage ring for 50% of the time starting in early 1980. The flux available at the National Synchrotron Radiation Light Source under construction at Brookhaven National Laboratory should be comparable to SPEAR and possibly even higher.

The brief experience with a highly mosaic graphite monochromatizing crystal indicates that for experiments where a narrow bandpass incident beam is not required, additional flux gains are possible. The relationship between flux gain and increased data rates is, however, complex. For a wide bandpass beam, a large scan, and thus more time, is required. After a point, the increased flux is matched by the increased time necessary to scan the wider reflection. Also background noise is raised without raising the peak height so the signal-to-noise ratio decreases. It is probably best to adopt a compromise strategy. For a sample of mosaic spread less than about 0.1°, a monochromatizing crystal of this mosaic spread is best. The reflection width will be within the accuracy capability of the instrument, and the scan width will not be increased beyond the minimum necessary, while the integrated counts will be raised. For a sample of higher mosaicity, the best solution is a monochromatizing crystal of a somewhat lower mosaicity so that integrated intensities are raised but the necessary scan width is not. The monochromator at SSRL is set up so that crystals may be easily changed, thereby allowing an initial survey preceding a data collection run. With a wide bandpass beam, ω–2θ scans are necessary.

An opposite approach can also be taken in some circumstances. In the case of an extremely small protein crystal mounted in a capillary tube, there is some problem in distinguishing the peak of the weak diffraction spot over the noise from scattering from the capillary and air. If the crystal is of low mosaicity, then there will be a signal-to-noise gain as well as an intensity gain due to the narrow reflections standing higher above background. Figure 5 shows this effect for a weak reflection from lysozyme of low mosaicity. In practice, the actual signal-to-noise gain is a function of the necessary detector aperture widths as well as relative reflection widths. In this regard, the monochromatic nature of the synchrotron radiation beam helps as it is not necessary to increase the aperture width to account for $K\alpha_1$–$K\alpha_2$ splitting at higher Bragg angles. For experiments of this nature, a high quality germanium or similar crystal monochromator is required. To further increase data collection rates peak-top counting may help. Software that searches for the Bragg peak and then positions the crystal at the peak could be developed. How fast a peak search could be done compared to the increased data rate during counting determines whether overall data rates are improved. Massey and Manor[24] reported narrow reflection widths from their apparatus also. The potential signal-to-noise advantage was less than that noted here however.

The narrow reflection widths also mean that crystals with large unit cells can be investigated with this apparatus. Arndt and Willis[29] gave a general expression for the maximum lattice constant for which diffraction peaks can be resolved with terms for the source and crystal sizes, the crystal mosaicity, and the source wavelength spread. The source and crystal size determine the angular divergence of the beam incident upon the sample for the usual case of a conventional x-ray source, which is isotropic across the focal spot. For the focused synchrotron radiation source, the vertical convergence replaces these terms. As discussed above, the vertical convergence determines the wavelength spread in the beam. When these factors are taken into account appropriately, spots along a 1000-Å unit-cell axis could be resolved with 1.5-A radiation for a crystal of similar mosaic spread to the monochromator. For a typical macromolecular crystal the mosaic spread would be the dominating factor in determining the resolution that can be achieved.

2.5.2. Other Types of Experiments

The above uses are primarily concerned with macromolecular crystallography, and reflect the main research interests of the authors. There are other classes of experiments possible with four-circle goniometers using a synchrotron radiation source. With the knowledge that accurate data can be readily obtained, it is certain that some of the experiments described below will be pursued.

When synchrotron radiation is focused with suitable optics, an extremely high flux of x rays can be directed onto a sample of typical size for a diffraction experiment. Thus, data could be collected rapidly on samples that diffract too weakly from a normal source or from rapidly decaying samples, for example, radioactive samples with short half-lives. Kinetic studies in which a sample is subjected to a stimulus, perhaps a shock wave or a flash of a laser beam, and then the time course of the diffraction pattern is studied are also experiments that would benefit from higher flux. The higher the flux then, the shorter the time interval in which statistically significant data can be obtained.

The high brightness of a synchrotron radiation source enables one to produce a highly parallel x-ray beam with narrow spectral bandwidth and reasonable flux. A four-circle goniometer coupled with a beam line of these characteristics could be used in studies where a scattering pattern is investigated with high-angular and/or high-energy resolution. For example, diffuse scattering could be measured with a very precise definition of the momentum transfer.

The broad spectrum of synchrotron radiation is vital in a wide variety of diffraction experiments, mostly concerned with measuring or utilizing anomalous scattering effects. The proceedings of a recent conference contain many examples of the use of these effects in diffraction studies.[30] The effects are strongest and most interesting at absorption edges. Only a synchrotron radiation source can provide a powerful flux of radiation at wavelengths close to absorption edges and thereby provide for optimal use of anomalous scattering to solve the "phase problem" of crystallography.

There are other types of experiment in geology and materials science. By correct choice of wavelengths around the absorption edge of an element, that element's contribution to the diffraction pattern can be extracted from the total pattern and its sites in the crystal structure can be found. Thus, in the study of naturally occurring minerals where different elements substitute for each other in the crystal lattice, questions about which site is being substituted can be answered. Also, binary alloys can be studied in this way. This technique is especially useful when the two elements are close together in the periodic table and normal scatter-ing is very similar for the two elements so that a diffraction study at a single general wavelength cannot distinguish between the two. Again, these studies can only be carried out with a tunable synchrotron radiation source and by far the best measur-ing apparatus would be a four-circle goniometer.

Another type of diffraction experiment in which the broad spectrum of synch-rotron radiation is of great significance is the energy dispersive method. If a quite complex crystal with a large number of diffraction spots were examined, it would be useful to have the sample mounted on a four-circle instrument in order that all reflections may be examined without remounting the sample.

The high degree of polarization, while not a unique property of synchrotron radiation, as radiation from a conventional source can also be polarized, can be used in combination with other properties that are unique. For example, if fluorescence from a sample is being measured as a function of sample orientation, then a measurement with the detector at $2\theta = 90°$ would minimize the error from Bragg and diffuse scattering. For such a measurement, the diffractometer must have the 2θ axis vertical. This is discussed in more detail by Sparks in Chapter 14.

Clearly, unique experiments are possible with this diffractometer–synchrotron radiation source combination. More experiments are planned by us in order to further investigate and utilize this apparatus.

3. Measurement of the Anomalous Scattering Terms for Cesium and Cobalt at Noncharacteristic Wavelengths

In order to further understand the phenomenon of anomalous scattering, we made measurements of the effects at wavelengths unavailable from conventional sources, using the apparatus described in Section 2. These experiments were per-formed in collaboration with Professor D. H. Templeton and Dr. L. K. Templeton of the Lawrence Berkeley Laboratory.

Correlations between large anomalous scattering effects for cesium and distinc-tive features in the cesium x-ray absorption spectrum were found. Absorption spectra of many elements were obtained to see if the large anomalous scattering effects for cesium are general. A brief discussion of anomalous scattering, a descrip-tion of the principles of the measurement, experimental details, and a discussion of the results follow.

3.1. The Phenomenon of Anomalous Scattering

James[31] has given a good account of the fundamentals of anomalous scatter-ing. A summary of this is presented below.

In structure analysis by x-ray diffraction, the x-ray scattering factors of ele-ments are important quantities. They are a measure of how strongly and with what phase shift x rays are elastically scattered into the diffraction pattern. For a given atom, the atomic scattering factor f is defined as

$$f = f_0 + f' + f'' \tag{4}$$

The normal scattering amplitude (f_0) is that value appropriate for very short wavelengths, and decreases as the scattering angle increases. The wavelength dependence of f is, by convention, described by the correction terms f' and f'', the real and imaginary parts of the "anomalous" scattering. They account for the absorption of the x rays by the electrons of finite binding energy in the atom. The f' term represents in-phase scattering and the f'' term represents scattering shifted by 90° in phase. As for many types of dispersion phenomena f' and f'' are linked by a Kramers–Kronig relationship,

$$f'(\omega) = \sum_i \int_0^\infty \frac{\omega'^2 (dg/d\omega')_i \, d\omega'}{\omega^2 - \omega'^2} \tag{5}$$

$$f''(\omega) = \tfrac{1}{2}\pi\omega \sum_i (dg/d\omega)_i \tag{6}$$

where $(dg/d\omega)_i$ is the oscillator density for oscillators of type i and the sum runs over all absorption edges. The oscillator density is related to the atomic absorption coefficient for the ith oscillator $[\mu(\omega)_i]$ by

$$(dg/d\omega)_i = \frac{mc}{2\pi^2 e^2} \mu(\omega)_i \tag{7}$$

Close to an absorption edge f' and f'' vary rapidly with wavelength. There is a sharp jump in f'' from below to above the edge (in energy) while f' dips to a large negative value through the edge.

There are many reasons why crystallographers concern themselves with anomalous scattering. For accurate structure determination, proper account must be taken of these effects in least-squares procedures in which observed and predicted structure factors are compared in order to best fit the coordinates of the structure. Also, in some cases these effects can be useful in determining structures where other methods fail. The availability of accurate values of the anomalous scattering factors is of great benefit in both problems.

3.2. Previous Determinations of f' and f"

There have been both experimental measurements and theoretical determinations of the anomalous scattering factors. Calculations were made by James,[31] Parratt and Hempstead,[32] Dauben and Templeton,[33] Templeton,[34] and Cromer.[35] Cromer and Liberman[36] used relativistic wave functions to calculate f' and f'' for various wavelengths characteristic of conventional x-ray tubes. These values are commonly accepted and widely used by crystallographers in structure determination. There seems to be little grounds for doubting the calculations at general wavelengths, but close to absorption edges there are effects that are not included in the calculations. Recent experimental work has therefore concentrated on absorption edges. Fukamachi and Hosoya[37] have measured f' and f'' for gallium near the gallium K absorption edge by energy dispersive diffractometry. Bonse and Materlik[38, 39] have reported values for nickel at the nickel K edge. Their method used an x-ray interferometer. Values for copper[40] have been reported for the K edge. In all the above work there were definite but not large deviations from theory. The maximum values of f' and f'' reported were -10 and 6 electrons, respectively, for gallium. There has been only one study close to an L edge, that of tungsten.[41] They reported a value of f' of -10.1 electrons at a wavelength that differs by 6% from the L_3-edge wavelength. Our experiments were the first measurements made of f' and f'' using wavelengths that differ by less than 1% from that of an L absorption edge.

3.3. Principles of the Present Method of Determining the Anomalous Scattering Factors

The method involves measuring a set of Bragg reflections from a crystal of known structure that contains a suitable heavy atom. Then a least-squares refinement is performed in which the measured values of the structure factors are compared with those calculated from the known structure. The parameters of the structure (atomic sites, occupancy, and thermal motion factors) are held constant. An overall scaling factor, f', and f'' are varied until the best fit of measured and calculated values is obtained.

The value of any structure factor is a function of the three variables and the known structure of the crystal, so in effect, a set of simultaneous equations with three unknowns is being solved. Each Bragg reflection measured adds one more equation. In principle, only three reflections need be measured, but in practice more are obtained to greatly improve the reliability of the results. It is best to measure a set of reflections some of which have little heavy atom contribution, and others with large heavy atom contributions where the heavy and light atom contributions are exactly in phase or 180° out of phase, and still others where the heavy and light atom contributions are out of phase by close to 90°. These three subsets ensure that the scale factor, f', and f'' are independently well determined.

3.4. Experimental Details

The crystal structure of cesium tartrate was refined using data collected with Mo $K\alpha$ radiation to an R factor of 1.6% by Templeton and Templeton.[42] Anomalous scattering terms were varied during the structure refinement. The details of the least-squares procedure are discussed in this paper also.

The measurements of diffracted intensities as a function of wavelength were done using the diffractometer–focusing monochromator system at SSRL. After aligning the diffractometer to the x-ray beam, an absorption spectrum of a sample of cesium tartrate was measured as described in Section 2 and was used to decide at what monochromator settings data was to be taken. Then at each selected setting the intensities of 16 to 19 Bragg reflections from a cesium tartrate crystal were measured. At some settings the measurements were repeated several times, at others the measurements were done only twice. The measurements were made over a period of many months during different operating conditions of the storage ring. The beam intensity and count rates thus varied a great deal and, on different occasions, different numbers of measurements were necessary to obtain structure factors to sufficient statistical accuracy.

The crystal structure of tris(ethylenediamine) cobalt (III) chloride tartrate pentahydrate has been solved to high accuracy.[43] A crystal of this substance was used in an experiment, similar to the one on cesium, in order to obtain the anomalous scattering terms for cobalt. Using the synchrotron radiation source, a set of reflections was measured at several wavelengths around the cobalt K edge. The edge was calibrated by taking an absorption spectrum of cobalt tartrate.

Structure factors were obtained from the raw data by the usual procedure in crystallography except that a main beam decay correction and a polarization correction (assuming completely polarized radiation) were applied as described in Section 2. A spherical absorption correction was applied to the cesium tartrate data. An analytical correction was made to the cobalt tartrate data. Then a least-squares refinement was performed on each of the data sets in turn to obtain the values of f' and f''.

3.5. Discussion and Correlation of Results with Absorption Spectra

The complete details of the measurements have been published[18, 44] and some rather interesting points have been raised.

Figure 6 is a plot of the results for cobalt with a wavelength scale relative to that of the edge. The values obtained by Bonse and Materlik[39] for nickel and by Freund[40] for copper are also plotted relative to their respective K edges and the theoretical values of Cromer and Liberman[36] for f'' are also shown. There is a good agreement with theory for the values of f'' and with the other experiments for f'. This is a strong indication that the method employed to measure f' and f'' is valid.

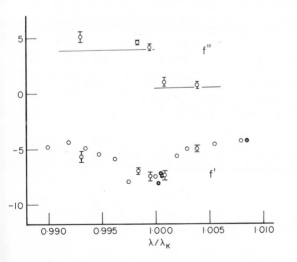

⚯   Templeton, Templeton, Phillips & Hodgson, Co
●   Freund, Cu
○   Bonse & Materlik, Ni
—   Cromer & Liberman

Figure 6. Anomalous scattering terms near K edges on a relative wavelength scale. Experimental values of f' and f'' are shown along with a theoretical calculation of f''. The values of f' for cobalt reported here agree well with measurements made on copper and nickel. The theoretical and experimental values of f'' are not in serious disagreement. The method of measuring f' and f'' used here thus appears to give sensible results in simple cases.

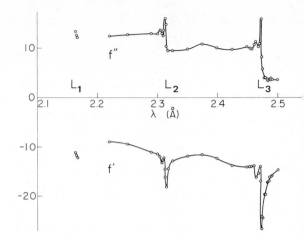

Figure 7. Anomalous scattering terms for cesium through the L edges. The oscillations at the L_3 and L_2 edges are similar in shape but are twice the scale for the former over the latter.

Figure 7 shows the values for cesium. Figures 8 and 9 are expansions of the regions close to the L_3 and L_2 edges, respectively. There is a marked similarity in the shapes of the two curves but the L_2 edge shows variations on a scale half that of the L_3 edge. As there are four electrons involved in the L_3 edge jump and two in the L_2 edge this is not surprising. For the L_3 edge, Cromer[45] has calculated the values of f' and f'' and they are also shown in Figure 8. The edge jump of f'' is reproduced but the large feature at the edge and the smaller oscillation just above the edge in energy are not present in the theoretical calculation. To account for these oscilla-

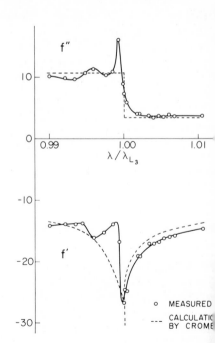

Figure 8. An expansion of Figure 7 near the L_3 edge. A calculation by Cromer (1978)[45] is also shown. The calculation reproduces the edge jump of f'' well but not the large spike at the edge. The oscillations in f' and f'' are not reproduced in the calculation.

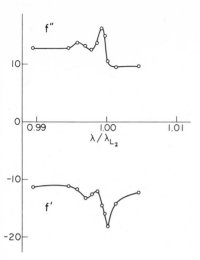

Figure 9. An expansion of Figure 7 near the L_2 edge showing the oscillations more clearly.

tions one must refer to the ideas of x-ray spectroscopy. According to equations (6) and (7), the atomic absorption coefficient and f'' are closely related.

There has been a great increase in understanding of x-ray absorption edges in recent years due to the data that has been collected at synchrotron radiation laboratories utilizing spectrometers that can be tuned rapidly through the high intensity continuum. Atomic absorption coefficients are influenced by bound-state transitions close to the nominal edge, there can be shifts in the edge position due to the chemical state of the atom, and there can be oscillations in the absorption coefficient extending far above the edge due to the environment of the atom (EXAFS). In principle f'' should show these effects also. An absorption spectrum of cesium tartrate through the Cs L_3 edge is shown in Figure 10, minus a linear background. The values of f''/ω are also plotted. As the absorption spectrum is only a relative measure of the absorption coefficient μ, the f'' values were scaled suitably and set to an arbitrary origin. The absorption spectrum and f'' accurately coincide. The edge peak and post-edge bump are present in both measurements. It thus seems that all features present in absorption spectra will be present in f'' and indirectly in f' via the Kramers–Konig relationship. The large values of f'' found at the edge of cesium appear to be due to a bound-state transition of cesium. Such peaks in absorption spectra have been known for a long time and are called "white lines." To see if the results for the anomalous scattering of cesium might be general, L-edge absorption spectra were taken of simple compounds of some elements with L-edge wavelengths suitable for x-ray diffraction. The L_3 spectra of lanthanum, cerium, praseodymium, europium, gadolinium, dysprosium, erbium, ytterbium, rhenium, osmuim, and iridium all show large peaks at the edge, as do the L_2-edge spectra. The L_1 edges examined did not show such large features. Figures 11 and 12 show the spectra of several L_3 edges on an energy scale with the peak shifted to zero in each case. It thus seems reasonable to expect anomalous scattering effects greater than current theoretical values close to the L_2 and L_3 edges of many

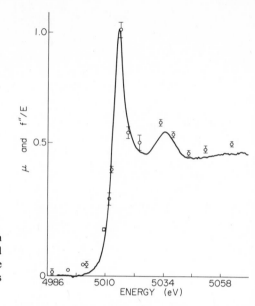

Figure 10. Absorption spectrum of cesium tartrate and f''/ω for cesium plotted on an energy scale. Both curves have been normalized and set to an arbitrary origin as the absorption spectrum is only a relative measurement of the variations in μ through the edge. The two curves coincide quite well.

elements. Standard crystallography is currently carried out using the theoretical values of Cromer and Liberman[36] on compounds that contain elements with absorption edges close to the wavelength of the radiation used. Effects of bound-state transitions or EXAFS in the values of f' and f'' are not taken into account during structure refinement.

For phase determination by use of anomalous scattering, the larger the effect the better. The results for cesium indicate that macromolecular crystal structures could be phased using anomalous scattering only. If the indications of the absorption spectra prove true, then a quite general method would become available for structure determination. Moreover, there exists much expertise in binding com-

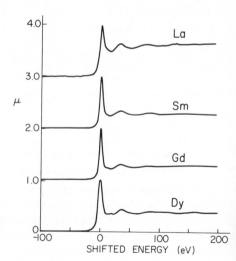

Figure 11. L_3-edge spectra of simple compounds of lanthanum ($LaCl_3$), samarium ($SmCl_3$), gadolinium ($GdCl_3$), and dysprosium (Dy_2O_3). All show a large spike at the edge and a smaller bump approximately 30 eV above the edge, as does cesium tartrate.

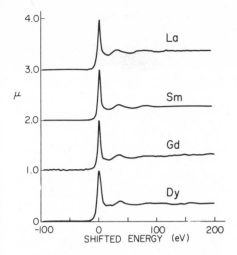

Figure 12. L_2-edge spectra of the same compounds as for Figure 11. The spectra of the two edges are remarkably similar.

pounds containing some of the heavy elements listed above (particularly the lanthanides) to proteins as NMR shift reagents.

The method used here to determine f' and f'' makes certain assumptions that are not perfectly valid. It is assumed that there is no variation of the anomalous scattering terms with scattering angle or orientation of the crystal with respect to the incident beam polarization vector. As the measurements are made at different wavelengths, the scattering angle and crystal orientation change from measurement to measurement. That results at the L_2 and L_3 edges were so consistent indicates that f' and f'' are not very sensitive to scattering angle or polarization. If the full sphere of reflections were measured such effects could be looked for. This involves vastly more data time than has so far been available. It is planned to do this in the future.

In conclusion, the results of the measurements of the anomalous scattering terms for cesium show effects larger than those calculated *ab initio*. The large effects are correlated with features in the absorption spectrum of cesium. Since such features are common in *L*-edge spectra, it is anticipated that large anomalous scattering factors may also be found for many other elements. It is also possible that they might be successfully used to obtain phase information from crystals of macromolecules. Such uses are considered below.

4. Use of Anomalous Scattering Effects to Phase Diffraction Patterns from Macromolecules

As mentioned in Section 1, anomalous scattering can be used to solve the "phase problem" in crystallography. The experiment is best performed using a synchrotron radiation source, since the results discussed in the previous section show that large effects can then be obtained. It seems appropriate to discuss the method in detail with particular reference to the results on the *L*-edge anomalous scattering of cesium.

4.1. Previous Work on Phasing Macromolecular Structures

To solve the phase problem for protein structure determination, one normally prepares a series of heavy-atom-containing crystals of the protein. They should all be identical to the crystals of protein with no heavy atom except for the inclusion of the extra atom. Such a crystal pair are called the native and the heavy atom isomorphous derivative. The difference between the native and the derivative can be thought of as the diffraction pattern from a single atom. This can be solved for the position of the atom within the unit cell. Then the phase of the native structure may be estimated. Knowing the position of a heavy atom, the amplitude and phase of its contribution to the diffraction pattern can be calculated. Then the phase difference between the native and the derivative can be calculated as one then knows the amplitudes of two complex numbers and the amplitude and phase of the difference between them. There remains an ambiguity of phase that is resolved by using data from a second derivative. This method was originally demonstrated for protein structures by Perutz and co-workers.[5] In practice, a phase probability function is calculated and a Fourier transform with weighted coefficients is used.[46] Use of many derivatives improves the quality of the map.

In an experiment in which only one x-ray wavelength is used, anomalous scattering effects, as manifested in the breaking of Friedel's law, can provide phase information. Measurement of Friedel pairs can resolve the phase ambiguity when only a single isomorphous derivative is available, and is used in general to improve the accuracy of phase information.[47, 48] Use of Friedel pair information in combination with isomorphous derivative information in the location of heavy atoms in protein crystals has been discussed by Kartha and Parthasarathy[49] and Matthews.[50] The way in which Friedel pair information is currently used in phase determination in protein crystallography was originally formulated by North[51] and Matthews.[52]

It has long been realized that anomalous scattering effects can be used to completely determine the phases of the diffraction pattern. A detailed explanation of the principles has been given by Raman.[53] Through an absorption edge of the heavy atom the real (f') and imaginary (f'') parts of the anomalous scattering vary drastically. Changes in the real part are mathematically analogous to isomorphous derivative changes. Thus, measurements using several different wavelengths can also be used to obtain the phase.

There has been a theoretical estimate of how well anomalous scattering from sulfur and copper can be used to solve the structure of organic molecules, using molybdenum and tungsten L emission radiation.[54] It was concluded that a structure containing 1000 light atoms could be solved.

A phase determination of the diffraction pattern from crystals of an iron-containing protein was made using data obtained with nickel and cobalt $K\alpha$ radiations whose wavelengths occur on either side of the iron K absorption edge.[55] The phases obtained solely from the use of anomalous scattering differed by an average of 50° from those obtained by multiple isomorphous replacement. As mentioned in Section 1, phases of rubredoxin were obtained to 60° accuracy using multiple wavelengths around the iron K edge.[10] Arndt[56] has discussed multiple-wavelength phasing using synchrotron radiation with the emphasis on data collection rates.

Anomalous scattering effects have also been used in neutron diffraction. Singh and Ramaseshan[57] have discussed the method. It is analogous to that used for x rays. Schoenborn[58] obtained phases from a cadium derivative of myoglobin using two wavelengths of neutrons. Koetzle and Hamilton[59] used neutrons at three wavelengths to solve the phase problem for $Na[Sm(EDTA)8H_2O]$.

4.2. A Quantitative Assessment of Anomalous Scattering Phasing

We will first give the mathematical basis of multiple-wavelength phasing in a form analogous to that used for the multiple isomorphous replacement method. Then, utilizing the recently acquired data on anomalous scattering effects at L edges discussed in the previous section, we will examine how well a diffraction pattern could be phased using these phenomena and what is the best data collection strategy. Specifically, we will address the following questions: (1) What and how many wavelengths should be used? (2) Need Friedel pairs be measured? (3) What kind of accuracy can be expected for a given ratio of light to heavy atoms?

4.2.1. The Effect of Anomalous Scattering on the Diffraction Pattern

Figure 13 shows the contributions to a Bragg reflection from various scatterers within the unit cell; the addition of each contribution of different amplitude and phase is represented as a sum of vectors on the complex plane. The total scattering vector **F** is given by

$$\mathbf{F} = \mathbf{F}_p + \mathbf{F}_H + \mathbf{f}' + \mathbf{f}'' \qquad (8)$$

where

$\mathbf{F}_p$ = sum of scattering from the light atoms
$\mathbf{F}_H$ = normal scattering from the heavy atom
$\mathbf{f}', \mathbf{f}''$ = real and imaginary parts of the anomalous scattering of the heavy atom, which are respectively 180° and 90° out of phase with $\mathbf{F}_H$

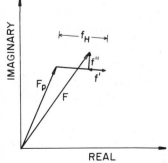

Figure 13. The amplitude and phase of a Bragg reflection from a protein crystal containing a heavy atom is represented as a vector on the complex plane (**F**). F is the vector sum of the protein contribution $\mathbf{F}_p$, the contribution from the normal scattering of the heavy atom $\mathbf{f}_H$, and that from the real, $\mathbf{f}'$, and imaginary, $\mathbf{f}''$, parts of the heavy atom anomalous scattering. $\mathbf{f}'$ and $\mathbf{f}''$ are respectively 180° and 90° out of phase with $\mathbf{f}_H$.

Figure 14. The contributions to a Bragg reflection at three wavelengths through an absorption edge. f′ and f″ vary rapidly with wavelength but F_p and f_H are essentially constant. Also shown is the Friedel related reflection, reflected through the real axis. f″ adds differently to the Friedel pair reflections and thus their amplitudes are different. Total vectors have been omitted for clarity.

$|F|$ is the only measurable quantity. When the wavelength of the incident x-ray beam is varied through an absorption edge of the heavy atom, f_H'' and f_H'' will vary in magnitude while F_p and F_H stay essentially constant. Figure 14 shows the summation of vectors for three different wavelengths. The Friedel related reflection $\bar{F}$ is also shown for three wavelengths, inverted through the real axis. The total scattering vectors are omitted for clarity. It is clear that F is altered slightly in both amplitude and phase as f′ and f″ vary.

4.2.2. Criteria Governing the Choice of Wavelengths

A measurement of the amplitude of any three of the resultant F's is sufficient in principle to obtain the phase of the Bragg reflection unambiguously. A graphical solution, the Harker construction,[60] for the case of a measurement of a reflection at three wavelengths is shown in Figure 15. Circles are drawn with radii equal to the measured $|F|$ and centered on the appropriate heavy atom plus anomalous scattering vector. The common intersection of all three circles determines the correct phase. Any two measurements give two possible phases where the two circles intersect. Each pair of intersections is symmetric about the line joining the two

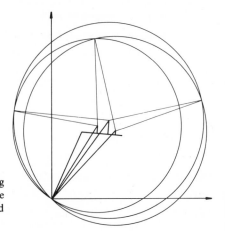

Figure 15. Harker construction showing how the measurement of a Bragg reflection at three wavelengths can lead to an unambiguous solution for the phase. Any pair of measurements give two possible solutions. The third measurement resolves this ambiguity.

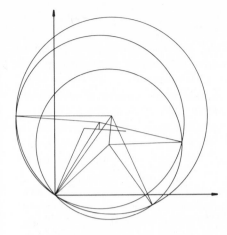

Figure 16. As for Figure 15 but now a measurement of Friedel related reflections plus a measurement of one of the pair at a second wavelength is assumed. Again an unambiguous phase determination is possible.

heavy atom plus anomalous scattering vectors. Figure 16 is the solution for the case of the measurement of Friedel pairs at one wavelength and one of the pair at a second wavelength. Measurement of the reflection at two wavelengths plus the Friedel-related reflection at a third wavelength would also lead to an unambiguous value for the phase.

In deciding what measurements to make it is clear that the largest possible changes in $\mathbf{F}$ should be induced so that they can be accurately measured. Also, the ambiguity should be broken by the best choice of measurements such that the line of symmetry of any pair is in a different direction to that of the other pairs. Hoppe and Jakubowski[55] suggested that measurements be made such that two of the lines of symmetry be as close as possible to right angles with each other. This is correct if only two of the three possible pairings of three measurements are used in phase determination, but is modified if more use of the measurements is made.

In the multiple isomorphous replacement (MIR) method, phases are obtained from the native and derivative data by calculating a phase probability function $P(\phi)$ according to the formula

$$P(\phi) = \exp[-x^2(\phi)/2\sigma^2] \tag{9}$$

where $x(\phi)$ (the "lack of closure") is the difference between the derivative amplitude as measured and as calculated from the native amplitude and the heavy atom parameters assuming a phase ϕ. Sigma (σ) is the estimated average error in the measurements for that particular derivative. The total probability function is the product of the functions for each derivative. The phase used to calculate electron density maps is then the weighted mean of this probability function. This method was introduced by Blow and Crick.[46]

When this method is used with normal MIR data, there is generally no attempt to relate the measurements between two derivatives as there can be worse errors due to imperfect isomorphism between two derivatives than between a native and a derivative. Also the introduction of a heavy atom into the crystal often causes problems. Derivative crystals may crack more easily, diffract less strongly, and suffer more from radiation damage than the native crystal. Thus the error in a measurement between two derivatives can be higher than that between a native and a derivative. A third factor is the error due to the imperfect estimation of the heavy atom parameters, which are again compounded if two derivative measurements are compared. In a phase determination by multiple-wavelength measurements, the above considerations are greatly modified. There are no isomorphism errors. Measurements would be made on the same crystal that would contain a heavy atom. Therefore, although the data may still be poor, it is all of similar quality (except that Friedel-related reflections differ by worse absorption errors). Errors in the estimation of heavy atom parameters would effect the correlation of any pair of measurements in exactly the same way. Therefore in each phase analysis the total phase probability function can be taken as the product of those between all possible pairs of the measurements assumed. If data from several measurements are correlated in this way, it is possible that the correct choice of the minimum three measurements is not that of Hoppe and Jakubowski.[55] Rather than pick measurements such that two of the symmetry lines are at right angles, it is conceivable that the three symmetry lines are 60° apart from each other.

Figure 17 shows some values of the anomalous scattering terms for cesium at the Cs L_3 edge plotted with f' as abscissa and $\pm f''$ as ordinate to show what vector differences between measurements at different wavelengths and Friedel pairs are possible with a small change of wavelength (the point $f' = 13$, $f'' = 13$ is an interpolation). It is quite possible to obtain large vector changes closely perpendicular to each other. Particularly note that a measurement of a Friedel pair (points 1 and 2 on Figure 17) coupled with measurements at two other wavelengths (points 3 and 9 or 5) are close to the supposed ideal case with the largest variation in one component of the anomalous scattering associated with little change in the other. Moreover, the wavelengths involved are within $\sim 28\%$ of each other. The approxi-

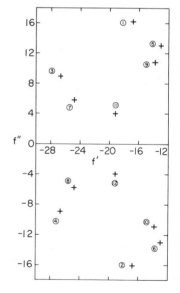

Figure 17. Plot of f' versus $\pm f''$ for cesium through the L_3 edge. This plot shows the vector changes in the total **F** that are made possible by altering the wavelength or measuring Friedel-related reflections. The points are referred to by their numbers in the text.

mation that other scattering factors are not changing with wavelength will be valid in such a small range. Points 1, 4, and 6 are closer to the 60° condition while still being quite widely separated.

4.2.3. How Many Measurements Should be Made?

As described above, each measurement makes possible the calculation of more phase probability functions; the Jth measurement gives $(J - 1)$ more. In some senses this is similar to using several isomorphous derivatives. However, as data from more wavelengths are taken, the magnitudes of the intensity changes will decline. Also, there is little possibility of phasing reflections for which the structure factor of the anomalous scatterer is low, whereas this becomes less true as more derivatives with different sites are used in the MIR method. Collecting more data also uses more synchrotron radiation beam time, at present a precious commodity. There is also the question of the availability of a sufficient quantity of crystals.

4.2.4. A Method for Assessing Various Data Collection Strategies

When the anomalous scattering terms are known, as with cesium, it is possible to consider the above questions in a quantitative way. To do this we wrote a FORTRAN computer program that generates artificial data at various wavelengths for an arbitrary phase and analyzes it assuming different combinations of measurements.

The program steps are as follows:

(1) Random number generators are used to obtain an arbitrary phase from 0–360° and heavy atom structure factors.

(2) The heavy atom anomalous scattering vectors for various wavelengths are added to a constant protein vector using the above obtained phase and structure factor. The relative scattering amplitudes of the protein and the anomalous scattering are set using the following formula derived by Crick and Magdoff[61]

$$\frac{2\Delta F}{F} = \frac{\Delta I}{I} = \left(\frac{2N_H}{N_L}\right)^{1/2} \frac{f_A}{f_L} \tag{10}$$

where $\Delta F/F$, $\Delta I/I$ = relative amplitude and intensity change due to the isomorphous addition of the anomalously scattering electrons

N_H = the number of heavy atoms in the unit cell
N_L = number of light atoms (taken to be molecular weight/14.7).
f_A = anomalous scattering factors
f_L = average normal scattering factor for the protein (taken to be seven electrons, an average of C, N, and O)

(3) Gaussian distributed noise with an rms error of σ is added to the resultant F's to obtain the artificial data.

(4) The phase is recovered from the data assuming various different measurements have been made.

The lack of closure $x(\phi)_{12}$ between two measured amplitudes M_1 and M_2 at wavelengths i and j is calculated as follows: From equation (8) and assuming no error in the measurement,

$$M_1 = F_p + F_H + f'_i \pm f''_i \tag{11}$$

where the sign of f''_i is set according to the Friedel pair to which M_1 corresponds. Similarly

$$M_2 = F_p + F_H + f'_j \pm f''_j \tag{12}$$

Therefore,

$$M_1 = M_2 + f_i - f_j \pm f''_i \mp f''_j \tag{13}$$

By squaring equation (6) the lack of closure due to error is found to be:

$$x(\phi)_{12} = M_1 - [M_2^2 + f'^2_i + f'^2_j + f''^2_i + f''^2_j + 2M_2 \cos \phi(f'_i - f'_j) \\ + 2M_2 \sin \phi(\pm f''_j \mp f''_j)]^{1/2} \tag{14}$$

Phase probability functions for all possible pairings of the measurements are assumed and they are multiplied together to obtain the final phase.

(5) The differences between the correct phase and those obtained by the different phasing methods are accumulated.

(6) The procedure is repeated many times; then average phase differences and other information are printed.

The estimates of the average phase error expected, as obtained by the above calculation, will be lower than that to be expected in an actual experiment as several effects have not been taken into account. The heavy atom parameters and the anomalous scattering terms are assumed to be known precisely, which would not be true for a real experiment. The formula of Crick and Magdoff used to set the scale between the heavy atom and protein contributions assumes average F's based on Wilson statistics, which are known not to be obeyed by protein crystals. The structure factor of the protein was not varied, although varying this would probably not affect the results so much, and what is required is an estimate of the ability to phase average reflections rather than very weak or very strong ones. Also the weakest and strongest reflections are measured the least and most accurately, respectively, so keeping a constant protein vector with a constant error is a good approximation. Thus the estimates presented should be taken only as upper limits on the phasing power of the method. However, they can certainly be used as a guide in the planning of an experiment, indicating the relative advantages of one data collection strategy over another.

4.2.5. An Example of the Use of the Multiple-Wavelength Phasing Method

The effects of making different measurements and more measurements were investigated using the above described computer program as a function of molecular weight of the protein and error in the measurements. Molecular weights of 12,000, 25,000, 50,000, and 100,000 per anomalously scattering atom were used. The range of σ was 0.5–3% of F, corresponding to a 1–6% error in the intensity measurement. The mean phase errors calculated by the program are given in Reference 62 for all of the ten methods described below.

In method I, the minimum of three measurements (points 1, 2, and 3 of Figure 17) is assumed. The points are chosen to maximize the Friedel pair difference, then to obtain the largest possible difference of f'. In method II, the same measurements plus the Friedel pair at the second wavelength (point 4) are assumed to be measured. In method III the minimum of three measurements (points 3, 5, and 6) are again assumed. There is a trade-off of a smaller f'' for a larger difference in f' between this and method I. Method IV again assumes that the Friedel pair reflection at the second wavelength has also been measured (point 4). Method V assumes all six measurements of methods I–IV have been made (points 1, 2, 3, 4, 5, and 6). Method VI, involving measurements at points 2, 3, and 5, is a compromise between methods I and III.

To investigate whether adding in information at other wavelengths improves phasing ability, methods VII–IX were tested. In VII, points 7 and 8 are added to those of method V. In VIII, points 9 and 10 are added to those of method V. Method IX assumes all ten measurements at the five wavelengths have been made.

If single-counter diffractometry is the data collection method, it can be inefficient in measuring Friedel pairs. Data collection is halted while driving between the pairs, which takes much longer than positioning on a reflection with one index higher, and this must be done every few measurements if all phasing measurements of one Bragg reflection are to be made close together in time on the same crystal. Method X assumes measurements of all positive index reflections only at three wavelengths (points 1, 3, and 11).

Of the four methods that assume that only the minimum of three measurements have been made, method VI appears to be slightly better than method III. Method I, utilizing the maximum Friedel pair splitting is less favorable than methods III and VI, where somewhat more f' changes occur. Method X is clearly unfavorable and would only be used in specialized circumstances, where the requirements for measuring Friedel pairs causes great inefficiencies in data collection, and there is a high ratio of heavy to light atoms.

Of the two methods that assume that all measurements at two wavelengths have been made, method IV appears better than method II, just as method III was better than method I assuming three of the four measurements. When data from all three measurements is added, there is a further improvement in the phase determination. Adding in a fourth or fifth wavelength again reduces the phase error. However, it would be more efficient to spend data collection time in improving accuracy of the best measurements rather than collecting data at lots of wavelengths, for the most part. A two or three wavelength method with the emphasis on accuracy of the data collection appears best. However, there is a limit to the accuracy with which measurements can be made. If it turns out that there are no limitations other than instrument accuracy, that is if a plentiful supply of samples and data collection time are available, then measurements at many wavelengths will further improve the phase determination.

The above conclusions as to possible data collection methods apply only where the anomalous scattering curves are similar to those for cesium. For each anomalous scatterer, a similar analysis should be performed in order to decide on the best data collection method, once the anomalous scattering curves are known.

4.3. Other Data Reduction Considerations

Data taken with multiple wavelengths would in practice be analyzed in a similar fashion to MIR data. After heavy atom sites are found, there would be a refinement of the phases by adjustment of the heavy atom parameters as with MIR methods.[63] Schoenborn[58] did this during the anomalous scattering phasing of the neutron diffraction data from myoglobin. It is a simple matter to adapt existing computer programs to calculate the above described phase probability functions. Estimates of the mean lack of closure error (σ) can be obtained from comparison of the observed and calculated lack of closure on centric reflections, exactly as with MIR methods.[46]

It should be noted that phase refinement of multiple-wavelength data may prove simpler than that of MIR data. For either method, each measurement contributes a constraining equation to the refinement procedure but each derivative also adds new variables (the coordinates, occupation, and thermal motion of each site). There are a constant number of variables, however many measurements are included, for multiple-wavelength data, as the heavy atom site is constant independent of the changes in its anomalous scattering.

4.4. Conclusions

By simple analogy with concepts used in MIR phase determinations, a method for estimating the phasing power of multiple-wavelength techniques has been developed. The estimates are only lower limits on the expected accuracy so no absolute comparison with an MIR phase determination is possible. However, knowing the anomalous scattering terms, the best possible data collection strategy can be planned, taking into consideration the availability of samples and data collection time.

Multiple-wavelength data can be analyzed quite simply. Only minor adaptions of existing MIR computer software is necessary. No special problems in data analysis should occur; in fact, there should be rapid convergence of phase refinement procedures since the ratio of observations to variables is larger than with MIR data.

5. Progress on the Structure Determination of Gramicidin A: A Case Study of the Use of Synchrotron Radiation in X-Ray Diffraction

A group at Stanford has been conducting an x-ray diffraction study of various forms of gramicidin A, a transmembrane channel protein, for some time and has published some preliminary conclusions.[22] It was not possible to solve the crystal structure as no isomorphous derivatives could be prepared. A derivative containing cesium was prepared but it was not isomorphous to the native crystals so no phase information was obtained. Direct methods and model building techniques were tried without success. The cesium derivative is thought to be of biological significance, so it was planned to solve the structure using the multiple-wavelength techniques described in Section 4. Data collection was carried out using the synchrotron radiation diffraction facility at SSRL described in Section 2. However, during the testing of the apparatus, a potassium derivative was found. After x-ray diffraction data was collected, it was found to be isomorphous with the cesium derivative, the cesium and potassium atoms occupying the same sites. An unambiguous phase determination was not, however, possible.

5.1. The Crystallographic Problem and the Possibility of Its Solution Using Synchrotron Radiation

The cesium and potassium crystals are isomorphous, with Cs^+ and K^+ bound at the same sites. They are therefore suitable for an isomorphous heavy atom phase analysis in which the effective heavy atom is the difference between cesium and potassium (36 electrons). Two full sites 5 Å apart were found in each dimeric channel of gramicidin A. With effectively only one derivative, it is possible to only obtain the phase with an ambiguity. In such a case, it is possible to estimate phases by computing the weighted average of the two possible phases. However, the potassium/cesium sites form a centrosymmetric array so this method (called SIR, single isomorphous replacement) gives centrosymmetric phases that lead to a map of the molecule with its mirror image superimposed. It proved impossible to interpret this map. Friedel pair differences are required to obtain true phases. Potassium shows weak anomalous scattering. At the Cu $K\alpha$ wavelength cesium has an f'' of about eight electrons. However, significant Friedel pair differences were not observed. At low resolution, the gramicidin A molecule is a tube and the cesium atoms lie inside the tube giving rise to a nearly centrosymmetric structure. Also the crystals diffract weakly and it was possible to collect data of only about 5% accuracy. The partially centrosymmetric structure gives relatively weak Friedel pair splitting that does not show over the noise in the data. By collecting data with an incident beam of energy 2.47 Å at the peak of absorption of the Cs L_3 edge, f'' is at least doubled, making it possible that statistically significant Friedel pair differences would be observed. Only by using synchrotron radiation can an intense x-ray beam at this wavelength be obtained.

5.2. Preliminary Results of Structural Studies to 3.8-Å Resolution Using Synchrotron Radiation

5.2.1. The Experiment

During a brief period of high energy running of SPEAR in June, 1978, the first stages of experiments to obtain unambiguous phases for ion-bound gramicidin A were performed. Dr. Roger Koeppe, Mr. Jeremy Berg, and Professor Lubert Stryer joined in a data collection effort using the SSRL diffractometer system.

Data were collected at 2.47 Å at the peak of absorption of the cesium L_3 edge where f'' is maximum. Most reflections with d spacings of 3.8 Å or more were measured (along with Friedel pairs) during the time allocated to the experiment.

Despite the high intensity of the incident beam and the potential high damaging power of 5-keV radiation to biological systems, there was no detectable falloff in the diffracted intensities for the one crystal used for the data collection.

5.2.2. Data Analysis

The Friedel pair differences were scaled to the Cu $K\alpha$ data by the following formula:

$$\Delta = \frac{2(F_{2.47} - \bar{F}_{2.47})}{F_{2.47} + \bar{F}_{2.47}} F_{1.54} \qquad (15)$$

where $F_{1.54}$ is the value for the Cu $K\alpha$ data. As the Friedel pairs were measured consecutively their ratio suffers little error from long term instability in the synchrotron radiation intensity. By this method of scaling the accurate relative Friedel pair difference with the accurate structure amplitudes available from the Cu $K\alpha$ data, an accurate measurement of the Friedel pairs throughout the diffraction pattern was possible without relying on the long term reproducibility of the storage ring beam.

Friedel pair differences averaged out to 15% of the observed average intensity. This is similar to that of the isomorphous differences between the cesium and potassium derivatives. This result is to be expected from the measurements as the normal scattering powers of cesium and potassium differ by 36 electrons and Friedel differences are proportional to $2f'' = 32$ electrons at the Cs L_3 edge.

Phases were calculated using the isomorphous and Friedel pair data. The effect of including Friedel pair information into the phase calculation is shown graphically in Figure 18. Phase probability functions computed according to the method of Blow and Crick[46] are plotted. Shown in Figure 18 for the (1, 10, 1) reflection are the individual isomorphous and anomalous probability function

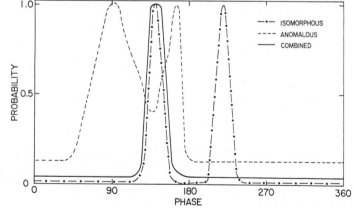

Figure 18. Phase probability functions for the gramicidin A (1, 10, 1) reflection based on isomorphous and anomalous differences. The two separate curves, and their product, the combined curve, are shown. The combined curve shows a single maximum at a general phase.

curves and their renormalized product (the combined curve). Isomorphous or ano-malous data alone each give two equally likely phases. A combination of the two allows the correct phase to be chosen because there will be only one peak common to both the isomorphous and anomalous probability distributions. This result is general for all reflections when isomorphous and anomalous data for a given derivative are combined. Note also that the isomorphous phase probability function is symmetric about 180°, as the heavy atom is in an approximately centrosymmetric position. Thus, using a weighted mean of the two phases (the SIR method) results in a centrosymmetric phase.

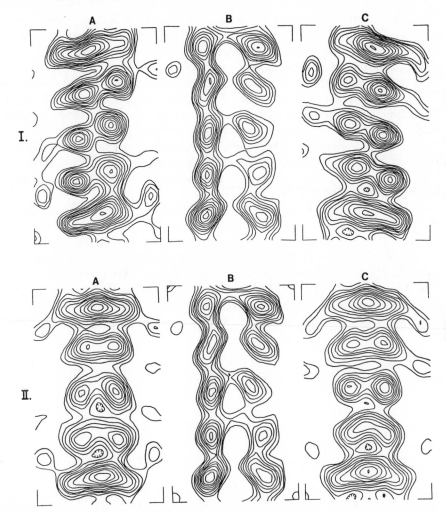

Figure 19. (I) Sections of the 3.8-Å resolution electron density map parallel to the channel axis, computed using the "combined" phases. (A) Section 3.2 Å in front of the center of the channel. (B) Section through the center of the channel. (C) Section 3.2 Å behind the center of the channel. There is clear evidence for a helical structure. (II) Same as I but for isomorphous phases only. No upward progression of the helix can be seen.

The Friedel pair differences were arbitrarily divided by three. This was done so that the phase probability function based on Friedel pair differences is not sharply peaked. The effect is to make the Friedel pair differences pick out one of the two possible phases based on the isomorphous data, but no other reliance is placed on the Friedel pair data. In Figure 18 it is clear that the peaks in the anomalous curve are broader and that the final phase is extremely close to one of the two possible isomorphous phases.

Using the combined isomorphous and anomalous phases, a 3.8-Å Fourier map (Figure 19) was computed from which a tentative interpretation on the structure of cation-bound gramicidin can be made. The 3.8-Å Fourier map shows that both of the independent dimers in the asymmetric unit are channels centered around the heavy atom positions. Both dimers are decidedly helical and many of the side chains are visible, although an unambiguous tracing of the amino acid sequence may not be possible at this resolution. A significant problem in the interpretation of the map is the tight packing in the crystal and the apparent interleaving of trypto- phan residues (numbers 9, 11, 13, and 15 in the sequence) from neighboring chan- nels, which make the molecular boundaries difficult to distinguish. Higher- resolution anomalous scattering data should result in a much improved map.

There was a dramatic improvement in the electron density upon inclusion of the Friedel pair phase information. Figure 20 shows a possible model, a double helix. Figure 19 shows contoured sections of electron density corresponding to planes through A–A, B–B, and C–C of Figure 20.

Figure 19(II) was computed using SIR (centrosymmetric) phases and shows *no evidence of a helical tilt or progression from one turn to the next.* Figure 19(I), on the other hand, was computed using the combined SIR and anomalous data and *clearly shows the helical nature* of the polypeptide backbone. Figure 19(IA) is a section through the channel, parallel to but off its axis. Figure 19(IC) is a similar section but on the opposite side of the channel. Both sections are viewed from the same direction. The upward progression of the strands is in a different direction on the two sections. Only a helical structure of a single handedness could give rise to such features.

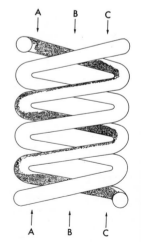

Figure 20. Schematic drawing of a double stranded helix. It is not yet known whether the gramicidin A dimer consists of an intertwined double helix, or of two single-stranded helices. The approximate locations of the planes used to calculate the maps shown in Figure 19 are indicated.

6. Summary and Discussion

Our early crystallography experiments[4, 10] indicated that the intensity and tunability of synchrotron radiation could be of use in macromolecular structure determination. These results have stimulated a series of experiments in which structure determinations are being carried further than was possible by use of conventional sources. The case of ion-bound gramicidin A is an example. The high-resolution data collection at DCI is also an important advance.[15]

Instrumentation for the technique is also becoming more sophisticated, as represented by the diffractometer system at SSRL. Groups at SSRL and DCI are planning to go a step further and use electronic area detectors for data collection. New facilities for synchrotron radiation, at present in construction around the world, will be coming into operation in the next few years. It is likely that, with the greater availability of beam time, many diffraction studies will be carried out along the lines of the original ones. These experiments will also benefit from the experience gained already. This is especially true for the anomalous scattering experiments.

It is evident from the gramicidin study that a detailed knowledge of the anomalous scattering curves greatly aids in the planning of an experiment. It is clearly desirable that measurements be made on more elements and also that a more fundamental understanding of anomalous scattering at absorption edges be reached. It is also evident from the results of the anomalous scattering measurements that the "white lines" could result in much larger anomalous scattering effects than had hitherto been thought. This would greatly increase the molecular weight range for which anomalous scattering experiments might be considered feasible in obtaining phase information.

If the "white lines" are to be used, then it is necessary to take care that the optics and monochromator system produce a narrow bandpass x-ray beam. This in turn implies a lower flux than could be achieved in an optical system designed for purely high data rates at a general wavelength. This could lead to data being collected in two separate runs. A high-flux beam line would be used to collect data rapidly to provide an "amplitude" data set, much in the same way as the DCI experiments. Then an accurate measurement of a Bragg reflection at many wavelengths would be made using a narrow bandpass, rapidly tunable beam line. This data collection would proceed more slowly and may use many crystals. There is no problem in relating phases from one crystal to another, unlike amplitudes, which are best measured from one crystal. In the gramicidin experiment, this method has been used to a certain extent, with the synchrotron radiation experiment being primarily designed to measure the Friedel pair ratio, the ratio from one reflection to the other then being scaled with an "amplitude" data set collected by conventional techniques.

In the future, these and possibly other techniques will be applied in macromolecular crystallography using synchrotron radiation sources. The knowledge that accurate data can be collected using synchrotron radiation sources by diffractometry will encourage the planning of other kinds of diffraction experiments also. The new synchrotron radiation laboratories will certainly be used by crystallographers and diffractionists with many different research interests.

ACKNOWLEDGMENTS

We wish to thank many of our colleagues for their interest in and support of this work during the past four years. Lyle Jensen provided initial insight into the usefulness of synchrotron radiation in protein crystallography and his inspiration helped get this project started. Much of the early work was done with Alex Wlodawer and Marguerite Yevitz Bernheim. Julia Goodfellow contributed to the NGF studies. The diffractometer modifications were made by Enraf-Nonius and John Cerino of SSRL was most helpful in the design of the diffractometer carriage. The measurements of the anomalous scattering terms for cesium and cobalt were carried out in collaboration with David and Lilo Templeton of the University of California at Berkeley and the Lawrence Berkeley Laboratory. Seb Doniach and Frank Kutzler contributed useful ideas on the nature of x-ray absorption edges. David Templeton and Paul Phizackerley made many useful suggestions concerning improvements to the diffractometer system. The work on gramicidin is being done in collaboration with Roger Koeppe, Jeremy Berg, and Lubert Stryer. Fellowship support was provided by the National Institutes of Health to Alex Wlodawer and by NATO to Julia Goodfellow. This work would not have been possible without the grant support of the National Institutes of Health (CA 16748) and the availability of the resources of the Stanford Synchrotron Radiation Laboratory (supported by the NSF in cooperation with the Stanford Linear Accelerator Center and the U.S. Department of Energy).

References

1. G. Rosenbaum, K. C. Holmes, and J. Witz, *Nature (London)* **230**, 434–437 (1971).
2. S. C. Harrison, in *Research Applications of Synchrotron Radiation*, R. W. Watson and M. L. Perlman (eds.), 105–108, Brookhaven National Laboratory Report BNL 50381 (1973).
3. H. W. Wyckoff, in *Research Applications of Synchrotron Radiation*, R. W. Watson and M. L. Perlman (eds.), 133–138, Brookhaven National Laboratory Report BNL 50381 (1973).
4. J. C. Phillips, A. Wlodawer, M. M. Yevitz, and K. O. Hodgson, *Proc. Natl. Acad. Sci. U.S.A.* **73**, 128–132 (1976).
5. D. W. Green, V. M. Ingram, and M. F. Perutz, *Proc. R. Soc. London* **225**, 287–307 (1954).
6. N. G. Webb, *Rev. Sci. Instrum.* **47**, 545–547 (1976).
7. N. G. Webb, S. Samson, R. M. Stroud, R. C. Gamble, and J. D. Baldeschwieler, *Rev. Sci. Instrum.* **47**, 836–839 (1976).
8. N. G. Webb, S. Samson, R. M. Stroud, R. C. Gamble, and J. D. Baldeschwieler, *J. Appl. Crystallogr.* **10**, 104–110 (1977).
9. E. T. Adman, R. E. Stenkamp, L. C. Sieker, and C. H. Jensen, *J. Mol. Biol.* **123**, 35–47 (1978).
10. J. C. Phillips, A. Wlodawer, J. M. Goodfellow, K. D. Watenpaugh, L. C. Sieker, L. H. Jensen, and K. O. Hodgson, *Acta Crystallogr. Sect. A* **33**, 445–455 (1977).
11. A. Harmsen, R. Leberman, and G. E. Schulz, *J. Mol. Biol.* **104**, 311–314 (1976).
12. H. B. Stuhrmann (private communication of results to be published).
13. M. Lemonnier, R. Fourme, F. Rousseaux, and R. Kahn, *Nucl. Instrum. Methods* **152**, 173–177 (1978).
14. C. Monteilhet, R. Fourme, and D. M. Blow, European Physical Society Fourth General Conference, York, United Kingdom (1978).
15. R. Fourme (to be published in *Trends in Physics*).
16. J. R. Helliwell, Ph.D. thesis, Oxford University, United Kingdom (1977).
17. J. B. Hastings, B. M. Kincaid, and P. Eisenberger, *Nucl. Instrum. Methods* **152**, 167–171 (1978).
18. J. C. Phillips, L. K. Templeton, D. H. Templeton, and K. O. Hodgson, *Science* **201**, 257–259 (1978).

19. T. K. Eccles, Ph.D. thesis, Stanford University, Stanford, California (1977).
20. H. Winick, in *Synchrotron Radiation Research*, A. N. Mancini and I. F. Quercia (eds.), 43–44 (1976). See also Chapter 3 of this volume.
21. J. A. Howell Ph.D. thesis, Harvard University, Cambridge, Massachusetts (1974).
22. R. E. Koeppe II, K. O. Hodgson, and L. Stryer, *J. Mol. Biol.* **121**, 41–54 (1978).
23. L. E. Alexander and G. S. Smith, *Acta Crystallogr.* **15**, 983–1004 (1962).
24. W. R. Massey and P. C. Manor, *J. Appl. Crystallogr.* **9**, 119–125 (1976).
25. C. J. Sparks, in *Proceedings of the Workshop on X-Ray Instrumentation for Synchrotron Radiation Research*, H. Winick and G. Brown (eds.), SSRL Report 78/04, III-35–46 (1978). See also Chapter 14.
26. V. Rehn, in *Proceedings of the Workshop on X-Ray Instrumentation*, H. Winick and G. Brown (eds.), SSRL Report 78/104, VII-11–13 (1978).
27. G. E. Fischer, in *Synchrotron Radiation Research*, K. O. Hodgson, H. Winick, and G. Chu (eds.), SSRL Report 76/100, Appendix G (1976).
28. A. Garren, M. Lee, and P. Morton, in *Synchrotron Radiation Research*, K. O. Hodgson, H. Winick, and G. Chu (eds.), SSRL Report 76/100, Appendix H (1977).
29. U. W. Arndt and B. T. M. Willis, *Single Crystal Diffractometry*, 169–179, Cambridge University Press (1966).
30. S. Rameseshan and S. C. Abrahams (eds.), *Anomalous Scattering*, Munkssaard, Copenhagen (1975).
31. R. W. James, *The Optical Principals of the Diffraction of X-Rays*, Chapter 4, Cornell University Press, Ithaca, New York (1948).
32. L. G. Parratt and C. F. Hempstead, *Phys. Rev.* **94**, 1593–1600 (1954).
33. C. H. Dauben and D. H. Templeton, *Acta Crystallogr.* **8**, 841 (1955).
34. D. H. Templeton, *Acta Crystollogr.* **8**, 842 (1955).
35. D. T. Cromer, *Acta Crystallogr.* **18**, 17–23 (1965).
36. D. T. Cromer and D. Liberman, *J. Chem. Phys.* **53**, 1891–1898 (1970).
37. T. Fukamachi and S. Hosoya, *Acta Crystallogr. Sect. A* **31**, 215–220 ((1975).
38. U. Bonse and G. Materlik, *Z. Phys.* **253**, 232–239 (1972).
39. U. Bonse and G. Materlik, *Z. Phys. B* **24**, 189–191 (1976).
40. A. Freund, in *Anomalous Scattering*, S. Rameseshan and S. C. Abrahams (eds.), 69–84, Munksgaard, Copenhagen (1975).
41. J. C. M. Brentano and A. Baxter, *Z. Phys.* **89**, 720–735 (1934).
42. L. K. Templeton and D. H. Templeton, *Acta Crystallogr. Sect. A* **34**, 368–371 (1978).
43. D. H. Templeton, A. Zalkin, H. Ruben, and L. K. Templeton (manuscript in preparation).
44. D. H. Templeton, L. K. Templeton, J. C. Phillips and K. O. Hodgson, *Acta Crystallogr. A.*, in press.
45. D. T. Cromer (private communication).
46. D. M. Blow and F. H. C. Crick, *Acta Crystallogr.* **12**, 794–799 (1959).
47. D. M. Blow, *Proc. R. Soc. London Sect. A* **247**, 302–336 (1958).
48. D. M. Blow and M. G. Rossmann, *Acta Crystallogr.* **14**, 1195–1202 (1961).
49. G. Kartha and R. Parthasarathy, *Acta Crystallogr.* **18**, 745–749 (1965).
50. B. W. Matthews, *Acta Crystallogr.* **20**, 230–239 (1966).
51. A. C. T. North, *Acta Crystallogr* **18**, 212–216 (1965).
52. B. W. Matthews, *Acta Crystallogr.* **20**, 82–86 (1966).
53. S. Raman, *Proc. Indian Acad. Sci. Sect. A* **50**, 95–107 (1959).
54. A Herzenberg and H. S. M. Lau, *Acta Crystallogr.* **22**, 24–28 (1967).
55. W. Hoppe and U. Jakubowski, in *Anomalous Scattering*, S. Rameseshan and S. C. Abrahams (eds.), 437–461, Munksgaard, Copenhagen (1975).
56. U. W. Arndt, *Nucl. Instrum. Methods* **152**, 307–311 (1978).
57. A. K. Singh and S. Ramaseshan, *Acta Crystallogr. Sect. B* **24**, 35–39 (1968).
58. B. P. Schoenborn, in *Anomalous Scattering* S. Ramaseshan and S. C. Abrahams (eds.), 407–416, Munksgaard, Copenhagen (1975).
59. T. F. Koetzle and W. C. Hamilton, in *Anomalous Scattering*, S. Ramaseshan and S. C. Abrahams (eds.), 489–502, Munksgaard, Copenhagen (1975).
60. D. Harker, *Acta Crystallogr.* **9**, 1–9 (1956).
61. F. H. C. Crick and B. S. Magdoff, *Acta Crystallogr.* **9**, 901–908 (1956).
62. J. C. Phillips, Ph.D. thesis, Stanford University, Stanford, California (1978).
63. R. E. Dickerson, J. C. Kendrew, and B. E. Strandberg, *Acta Crystallogr.* **14**, 1188–1195 (1961).

Application of Synchrotron Radiation to X-Ray Topography

M. SAUVAGE and J. F. PETROFF

1. Introduction

During the last two decades, single-crystal materials have acquired a major importance in advanced technology, particularly in the fields of electronics and optics. Moreover, single-crystal studies have allowed a better understanding of material properties such as plastic deformation or mechanical strengths of metals and alloys. Accordingly, a nondestructive technique like x-ray topography, which is able to provide a map of the defect distribution in crystals, has found numerous applications. One limitation of the method lies in the rather long exposure time, but this drawback has been overcome by the advent of highly intense synchrotron radiation x-ray sources.

X-ray topography is an imaging technique based on the difference in reflecting power between perfect and distorted parts of a crystal. Since it is only sensitive to strain fields extending over more than several microns, x-ray topography is mainly used for the study of dislocations, planar defects (stacking faults, domain walls in ferroelectric and magnetic materials, growth defects, etc., or large precipitates. The geometrical resolution lies in the micron range. Several experimental techniques have been developed to deliver a two-dimensional image. The most widely used is the Lang technique[1] whose principle will be briefly described.

M. SAUVAGE and **J. F. PETROFF** ● Laboratoire de Mineralogie-Cristallographie, associe au CNRS, Université P. et M. Curie, 4 Place Jussieu, 75230 Paris Cedex 05, France and LURE (CNRS, Université Paris-Sud) Bât. 209c, 91405 Orsay, France.

1.1. Features of X-Ray Topography

A single crystal is adjusted for Bragg reflection of a characteristic radiation line (usually $K\alpha_1$) issued from a point focus F (Figure 1) and collimated by a narrow slit S_1 to prevent reflection of other emission lines. If the crystal is perfect enough, the propagation of x rays follows the predictions of the dynamical theory applied to a spherical incident wave (N. Kato[2]). Energy is propagated in the so-called Borrmann triangle bounded by the incident and reflected directions. At the exit surface the wave fields are split into a reflected and a transmitted beam. Only the reflected beam is allowed to reach a fine-grained photographic plate P through the slit S_2. The width of this instantaneous image is governed by the slit S_1, which is of the order of 100 μm. In order to obtain a two-dimensional projection, the crystal and the photographic plate are simultaneously traversed in front of the beam. The divergence of the incident beam is usually much larger than the intrinsic reflection width of the sample and due to the scanning process, the photographic plate records an integrated intensity.

Other techniques delivering an integrated intensity pattern have been introduced; for example, the Berg–Barrett method, optimized by J. B. Newkirk[3] and designed for reflection topography.

In such settings, the defect images are comprised of three parts: the direct or kinematical image built with that fraction of the incident beam that does not belong to the reflection domain, and the intermediary and dynamical images resulting from strain-induced perturbations among the wave fields propagating within the Borrmann fan. The relative importance of the three types of images depends on the absorption of the sample, which is adequately described by μt, the product of the linear absorption coefficient μ and the crystal thickness t. A detailed analysis of dislocation contrasts may be found in articles by A. Authier[4] and J. E. A. Miltat and D. K. Bowen[5]; the details of stacking fault images are fully understood in the case of nonintegrated spherical wave patterns, as obtained by the Lang technique without scanning the sample. (For a review see A. Authier.[6])

However, in some settings, the divergence of the incident beam can be made significantly smaller than the reflection width of the sample. These are the two-crystal arrangements either in the $(+, -)$ parallel setting (U. Bonse[7]) shown in

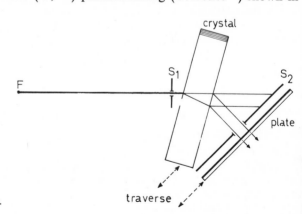

Figure 1. Lang topography setting.

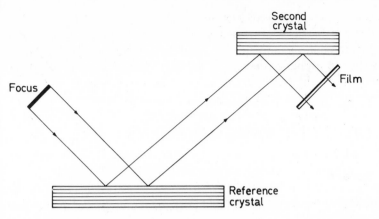

Figure 2. Double-crystal arrangement in the $(+, -)$ parallel setting.

Figure 2, or with a highly asymmetric first crystal as was largely developed in Japan by K. Kohra and his co-workers (K. Nakayama *et al.*[8]). Narrow-passband multireflection monochromators may also be used to deliver a pseudo-plane wave (K. Kohra *et al.*,[9] S. Takagi *et al.*[10]). These methods are extremely sensitive to minute strains in crystals.

The geometrical resolution is governed by the local divergence on the sample and lies usually in the micron range but the actual width of dislocation images depends in each case on the diffraction conditions and on the sensitivity of the technique: 5 to 100 μm in the Lang method and several hundred microns with two-crystal settings. Typically the maximum defect density that can be accommodated is 10^3–10^4 per cm^2, curved crystals or samples with a subgrain structure cannot be homogeneously imaged without use of an additional compensating device degrading the resolution and leading to a decrease in the average intensity. With the most powerful laboratory generators exposure times range from a few minutes in favorable cases to several hours.

1.2. Relevant Characteristics of Synchrotron Radiation Sources for X-Ray Topography

As can be inferred from the previous section, the important parameters for a topography setting are the wavelength and angular spread of the incident beam together with its lateral extension. Synchrotron radiation sources will now be examined with respect to these properties. See also Chapter 3.

1.2.1. Wavelength Spread

The spectrum is continuous with a broad maximum in the number of emitted photons close to a wavelength given by

$$\lambda_{max}(\text{Å}) = \frac{4.2R}{E^3}$$

where R is the bending radius in the magnet expressed in meters and E the kinetic energy of the accelerated particles expressed in GeV. However, most of the x-ray beam lines are conventional vacuum lines isolated both from the ring ultrahigh vacuum and from the atmospheric pressure of the experimental area by beryllium windows. Absorption shifts the maximum towards shorter wavelengths and introduces a long wavelength cutoff around 6 Å. In white beam settings the sample sees the whole spectrum but tunable monochromatic topography can be performed at any selected wavelength as will be described in the next sections.

1.2.2. Angular Spread

One has to distinguish between vertical and horizontal divergence: the vertical divergence is governed by the vertical source size, the emission cone aperture classically denoted by γ^{-1}, and by the trajectory angular distribution in the electron bunch characterized by a full width at half-maximum which we call β. However the local divergence experienced by a point on the sample depends only on the angle δ subtended by the source Sv at the sample location provided δ is smaller than β or γ^{-1}, which is usually the case (Figure 3a). The local horizontal divergence is evaluated in the same way although in some facilities with broad horizontal sources it might be controlled by β or γ^{-1}. But, in addition, the average incidence angle varies

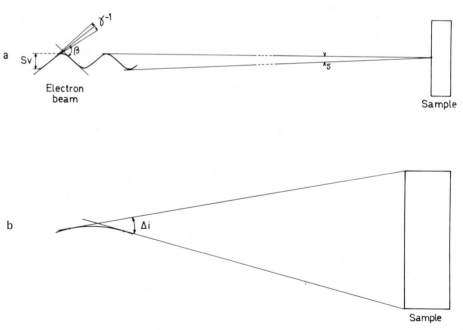

Figure 3. (a) Unscaled diagram of the three components of the vertical divergence: γ^{-1}, β, δ, and (b) variation Δi of the average incidence angle along the horizontal dimension of the sample.

Table 1. Relevant Features of Synchrotron Radiation Facilities[a]

Synchrotron radiation facility	S_h (mm)	S_v (mm)	D (m)	δ_h	δ_v	γ^{-1}	r_v (μm/cm)
NINA	0.5	0.5	47	10^{-5}	10^{-5}	10^{-4}	0.1
DESY	8	4	40	2×10^{-4}	10^{-4}	$<10^{-4}$	1
DCI	6	1.5	20	3×10^{-4}	0.7×10^{-4}	3×10^{-4}	0.7
SPEAR	3.2	1.6	17	1.8×10^{-4}	0.9×10^{-4}	1.6×10^{-4}	0.9
VEPP 3	1.5	0.2	5	3×10^{-4}	0.4×10^{-4}	2×10^{-4}	0.4

[a] S_h and S_v are the horizontal and vertical source size; D is the distance between the source and the topography beam port; δ_h and δ_v are the angles subtended by the source in the horizontal and vertical planes; r_v is the loss of vertical resolution per centimeter from the sample.

between both ends of the irradiated area proportionally to the orbit arc intercepted by the sample (Figure 3b). In white beam techniques, the geometrical resolution depends only on the local divergences; the loss of resolution per centimeter of distance between sample and detector for the existing and future facilities is given in Table 1.

1.2.3. Lateral Extension

Typically, the beam delivered at a topography station is one cm high and a few cm wide. However, in the vertical direction, this beam is truly homogeneous only over a height equal to the vertical size of the electron bunch. This limitation is more pronounced for short wavelengths and has to be taken into account in multiple-crystal arrangements.

1.2.4. Intensity and Polarization Properties

These properties have been extensively described in Chapter 2. The high intensity serves mainly to bring the exposure time on photographic detectors down into the second range for standard white beam topographs and in the minute range for highly absorbing materials and to allow a more comfortable use of direct viewing detectors. Moreover, monochromatic experiments with extremely narrow bandwidth are made possible. The radiation emitted by a nondivergent point source would be 100% plane polarized in the orbit plane but the finite source size and the betatron oscillations lead to a smearing of the polarization purity and the beam is around 95% linearly polarized in the horizontal plane. This property compels most of the multiple-crystal arrangements to confine the successive beams to the vertical plane to prevent an unwanted intensity loss. On the other hand, it also opens a new field of investigation; namely, polarized x-ray optics.

1.3. Available and Future Facilities

X-ray topography studies have been performed in six different synchrotron radiation facilities:

Synchrotron NINA,	Daresbury (United Kingdom)
Synchrotron DESY,	Hamburg (West Germany)
Storage ring DCI,	Orsay (France)
Storage ring SPEAR,	Stanford (USA)
Synchrotron ARUS,	Yerevan (USSR)
Storage ring VEPP 3,	Novosibirsk (USSR)

The relevant characteristics of these machines are listed in Table 1. In the early eighties, three novel facilities dedicated to synchrotron radiation research will be completed at Brookhaven (USA), Daresbury (UK), and Tokyo (Japan), and at least two of the existing ones will be extended and partly dedicated (SPEAR and DORIS). Important programs of topography and x-ray optics in nearly perfect crystals are planned at the Daresbury synchrotron radiation facility where four beam ports are dedicated to these topics and at the Japanese Photon Factory as well.

2. Description of Experimental Techniques

The techniques described in this section are not only imaging techniques such as white beam or monochromatic topography, but, in addition, another method relying also on dynamical theory predictions is included; the so-called x-ray interferometry technique (U. Bonse and M. Hart[11]) which permits accurate measurements of the dispersion correction f'.

2.1. White Beam Topography

Up to now this is the most widely used technique, which is not surprising when one considers the simplicity of the experimental arrangement.

When a single crystal is immersed in a white x-ray beam (Figure 4) a number of lattice planes (hkl) select out of the continuous spectrum the proper wavelengths to be reflected according to Bragg's law

$$2d_{hkl} \sin \theta = n\lambda$$

where d_{hkl} is the interplanar spacing of (hkl) lattice planes, θ is half the scattering angle and n is an integer. This is the well-known Laue method, which has been already applied as a low-resolution imaging technique to obtain information on the strain distribution in samples by A. Guinier and J. Tennevin[12] and by L. G. Schulz.[13] However, due to the low divergence of the synchrotron radiation beam, each spot of this particular Laue pattern is a high-resolution topograph that

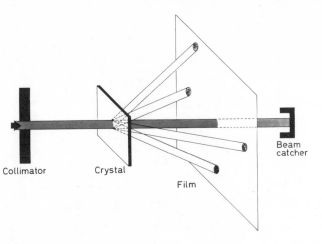

Figure 4. Synchrotron radiation Laue pattern.

compares favorably with the best results obtained in the laboratory. The possibilities of this technique have been thoroughly investigated by T. Tuomi et al.[14] and by M. Hart.[15] The advantages are obvious:

1. There is no need for accurate positioning of the sample; for most studies, a fraction of a degree is sufficient compared to a few seconds of arc in the Lang camera. Perfect stability is required only during the exposure, which lasts a few tens of seconds instead of hours.

2. There is no need for scanning since the instantaneous image is two-dimensional. However, a compromise has to be found between the image area and the resolution: for a large area the photographic plate should be located far from the sample to avoid spot overlapping, which leads to a degradation of the resolution (see Table 1).

3. Several spots may be simultaneously recorded and in favorable cases, an instantaneous Burgers vector assignment is possible with a judicious choice of the crystal orientation (T. Tuomi et al.[14]).

Thus in evaluating the reduction of exposure time provided by synchrotron radiation white beam topography, one has to take into account three factors: first the higher intensity, second the absence of scanning (within certain limits the exposure time does not depend on the sample size), and third the simultaneous recording of several *hkl* images. Another advantage is the possibility of obtaining homogeneous images even with warped samples or crystals presenting a subgrain structure. (Being irradiated by a continuous spectrum, each part of the sample actually finds the proper wavelength matching its local parameter and orientation.) In extreme cases like polycrystalline materials, reconstructed topographs may still be obtained.[16] Monotonic warpage only results in a wavelength and Bragg angle shift over the sample width; adjacent misoriented areas reflect nonparallel beams, which leads to a black or white orientation contrast in the boundary image. A quantitative study of subgrain boundaries in LiF has been performed by M. Hart[15] while

Figure 5. Cryogenic equipment for magnetic studies. The height of the cryostat in about 60 cm (after Tanner et al.[19])

evidence for the contribution of orientation contrast to dislocation images has been produced by B. K. Tanner et al.[17] Additional information may be found in a review article by B. K. Tanner.[18]

A unique aspect of synchrotron radiation topography is the opportunity it affords to select out of the incident spectrum a wavelength on the low-energy side of the absorption edge of a given element by a suitable orientation of the investigated sample (see Section 3.4).

Finally when one considers the resolution loss per centimeter between sample and detector (Table 1), it appears that for a number of facilities, distances up to 10 or even 20 cm can be tolerated. As a consequence, experiments where the sample is surrounded by cumbersome equipment are made feasible. For example, a helium

Figure 6. Magnetic domain structure in a terbium single crystal revealed by low-temperature white beam reflection topography (after Tanner et al.[19]).

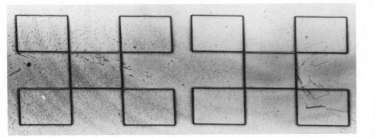

Figure 7. *n, n, n* white beam reflection topograph of a silicon wafer partially coated with an oxide film (after Hart[15]).

cryostat coupled to an electromagnet (Figure 5) has been used by B. K. Tanner and his co-workers[19] in their study of antiferromagnetic perovskites (see Section 3.2.1) and also for rare-earth metals (Figure 6).

However, the white beam topography technique also presents a few drawbacks. For all but very special cases, the images result from a superposition of several harmonics whose relative importance may be estimated through knowledge of the emission spectrum of the source, the successive structure factors, the absorption by the beryllium window, air, and sample, and the detector response. As an example, Figure 7 is an *n, n, n* silicon reflection topograph whose "composition" is given in Table 2 (M. Hart[15]). It is easy to understand that a quantitative interpretation of contrast is made more intricate especially for planar defects. Long-range strain fields are not detected. This is the negative aspect of curvature compensation. Moreover, misoriented areas superimposed in the beam path cannot be imaged separately, which prevents a detailed investigation of heteroepitaxial systems like III–V heterojunctions or garnet materials. Then, besides the very simple white beam setting, there was a need for a more sophisticated instrument delivering a monochromatic beam.

Table 2. Relative Intensities in Silicon *n,n,n* Diffraction Topographs

hkl	λ (Å)	%
111	4.434	5.3
222	2.217	5.8
333	1.478	62.0
444	1.019	21.9
555	0.887	4.5
666	0.739	0.02
777	0.633	0.5

a After Hart.[15]

2.2. *Monochromatic Topography and Specialized Monochromators*

2.2.1. *Two-Axis Spectrometers*

Two-axis spectrometers have been designed to perform monochromatic topography both at SSRL (Stanford) and at LURE-DCI (Orsay). Figure 8 is a block diagram of the Orsay unit (Figure 9): to preserve the polarization properties and to take advantage of the smaller vertical divergence, successive diffracted beams are confined in the vertical plane. The necessary crystal adjustments together with the various slit and detector positioning are remotely controlled. Fine rotations of both axes are driven by piezoelectric translators providing an accuracy of 1″ for the first axis where the monochromator is located and better than 0.1″ for the second one. The detector is an argon filled ionization chamber, while an air ionization chamber is inserted between the two crystals to monitor the monochromatized beam. Somewhat different solutions have been adopted by W. Parrish and his co-workers at SSRL; their unit is more compact and fully computer controlled and the detector is a scintillation counter working in the current mode.

2.2.2. *Specialized Monochromators*

The definition of a "monochromatic" beam is critically dependent on its application. As far as imaging techniques are concerned, the important criterion is the ratio of the incident divergence $\Delta\theta$ to the intrinsic width of the sample reflection profile δ:

$\Delta\theta \gg \delta$ As in the case of the Lang and Berg–Barrett techniques described in Section 1.1, the images are interpreted in terms of integrated intensity

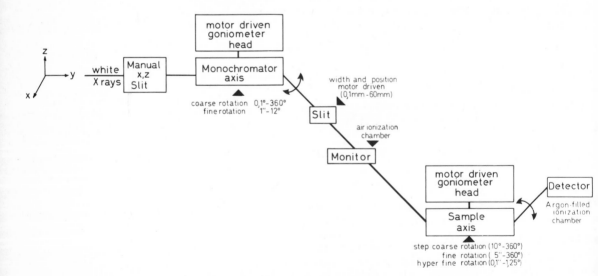

Figure 8. Block diagram of the LURE–DCI two-axis spectrometer. The figures between brackets indicate the accuracy and the range of the various movements.

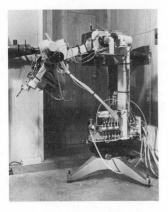

Figure 9. The LURE–DCI two-axis spectrometer; the separation between the axes is 50 cm.

calculations. The kinematical image dominates in low-absorbing wafers.[4, 5]

$\Delta\theta \ll \delta$ The incident wave can then be considered a plane wave; the image contrast is similar to the one observed with two-crystal spectrometers in the parallel setting.[7]

$\Delta\theta \sim \delta$ No universal theory is available in this case.

A review of the various designs making best use of the angular divergence of synchrotron radiation may be found in an article by K. Kohra and his co-workers.[20] Perfect crystals should be used to avoid parasitic images and to ensure a rigorously homogeneous lattice constant. In the present state of the art of crystal growth technology, only silicon and germanium single crystals are available. In most cases, silicon prevails over germanium except when a high-intensity output is required, even with a broad bandwidth. The monochromator crystal will be tailored in order to meet the user's requirements. Figure 10 gives some examples of solutions from the simplest type to the most sophisticated and the corresponding Du Mond diagram. The various devices ought to be compared in terms of tunability, λ and θ passband, harmonic rejection, and suppression of tails.

As shown by J. Beaumont and M. Hart,[21] multiple parallel reflections, although not reducing the full width at half-maximum, ensure tailless output profiles with a negligible loss of peak intensity in low-absorbing materials. Thus channel-cut multiple-reflection crystals are superior to single-reflection monochromators.

Harmonic elimination may be achieved in various ways. A review of the principle methods has been presented by U. Bonse and his co-workers.[22] The critical parameter is the angular difference $\Delta\beta_{\pm}^{(m, n)}$ in the peak positions for the mth and nth harmonics at two successive reflection sites labeled I and II. According to reference,[22] $\Delta\beta_{\pm}^{(m, n)}$ takes on the general form

$$\Delta\beta_{\pm}^{(m, n)} \propto (m^{-2} - n^{-2})\left[\frac{N_{\mathrm{I}}(1 - 1/b_{\mathrm{I}})}{\sin 2\theta_{B_{\mathrm{I}}}} \pm \frac{N_{\mathrm{II}}(1 - 1/b_{\mathrm{II}})}{\sin 2\theta_{B_{\mathrm{II}}}}\right]$$

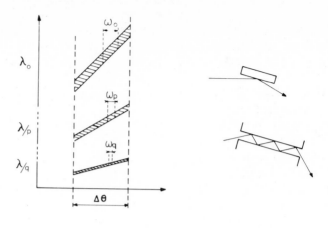

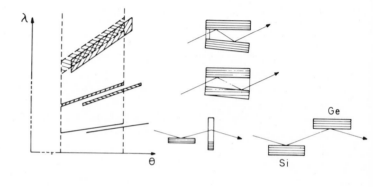

Figure 10. Possible arrangements for semibroad-band monochromators ($\Delta\theta \sim \Delta\lambda/\lambda \sim 10^{-4}$) and corresponding Du Mond diagrams. (Top) Simple devices without harmonic rejection. (Bottom) Harmonic rejection is achieved by tilting, symmetric–asymmetric, Bragg–Laue, or silicon–germanium combinations.

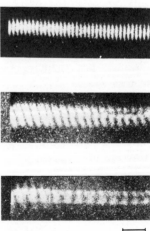

Figure 11. Successive display of the interferometer wedge fringe pattern corresponding to the fundamental (top), first (middle), and second (bottom) harmonics (after Bonse et al.[22]).

where $N_{I(II)}$ is the electron number per unit volume of the reflecting material at site I(II), $b_{I(II)}$ is the algebric asymmetry parameter at site I(II) (b is negative in the Bragg case), and $\pm$ signs refer to $(+, +)$ or $(+, -)$ configurations for reflections I and II.

Harmonic rejection will be effective as soon as $\Delta\beta_{\pm}^{(m,n)}$ is significantly larger than the intrinsic reflection width of each individual harmonic. Experimental data concerning the combinations of a silicon 220 (site I) with a germanium 220 (site II) in the $(+, -)$ Bragg setting and also of a silicon 220 in the Bragg setting (site I) with a silicon 220 in the Laue setting (site II) are presented in Reference 22. Figure 11 shows the output of an LLL interferometer (i.e., an interferometer working in the triple Laue setting) irradiated by a beam first monochromatized by a 220 reflection on a silicon crystal in the Bragg setting. Wedge fringe patterns for the fundamental (220), first harmonic (440), and second harmonic (660) of the monochromator output are successively displayed by rotating the interferometer over a few seconds of arc.

A different solution thoroughly investigated by M. Hart and A. R. D. Rodrigues[23] relies on an elastically tilted channel-cut crystal. Here also the efficiency of multiple reflections for harmonic elimination (the fundamental is 10^5 times more intense than the 1st harmonic) and for the suppression of unwanted Bragg peaks is demonstrated.

All the arrangements sketched in Figure 10 are tunable but one has to take into account the wavelength dependence of the harmonic rejection conditions.

The passband in θ and λ is mainly governed by the divergence $\Delta\theta$ of the synchrotron radiation beam according to

$$\frac{\Delta\lambda}{\lambda} = \Delta\theta(\tan\theta)^{-1}$$

which for most facilities has a value close to 10^{-4} if diffraction takes place in the vertical plane. If a higher resolution is required, $(+, +)$ configurations should be used as shown in the Du Mond diagram of Figure 12a. Then, tunability can only be preserved if the two crystals are independent. Harmonic rejection is achieved by coupling with one of the procedures described above. Monolithic versions deliver a fixed wavelength but are more simple to use. The principle of a monochromator designed by H. Hashizume et al.[24] is represented in Figure 12b. Its performance characteristics are listed in Table 3.

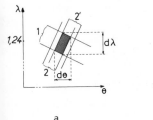

a

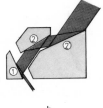

b

Figure 12. Harmonic free narrow-band monochromator. (a) Du Mond diagram showing the reduction in λ and θ spread. (b) Schematic of the monolithic monochromator.

Table 3. Performance of a Narrow-Band Harmonic-Free Monochromator[a]

Output wavelength	1.2378 Å
Wavelength band $\Delta\lambda/\lambda$	7×10^{-6}
Divergence	0.3″
Photon flux (DCI, 1.72 GeV, 100 mA)	2×10^7 Photons/sec/cm^2
Beam size	15×15 mm^2

[a] After Hashizume et al.[24]

An important property of these monolithic monochromators is the high accuracy in the wavelength definition and not only in the bandwidth $\Delta\lambda/\lambda$. By properly selecting the couple of reflections 1 and 2 among the variety provided by silicon (or germanium) crystals, the output wavelength can be brought very close to absorption edges, in the angstrom range. For example, a multiple-reflection monochromator, whose output wavelength is only 30 eV above the bromine K edge has been designed for polarization experiments at LURE–DCI.

2.2.3. Monochromatic Topography

As pointed out before, a major purpose of monochromatic topography is the separate imaging of epilayers and substrate in heteroepitaxial systems. As an example, Figure 13 shows 444 reflection topographs of a magnetic garnet epilayer on a GGG substrate taken at different locations on the reflection profiles by W. Parrish and his co-workers[25] at SSRL. The first crystal was a (110) oriented silicon. For the selected reflection both crystals were nearly in a $(+, -)$ parallel configuration and the wavelength was 2.25 Å.

By use of a narrow bandpass monochromator such as the one mentioned in Section 2.2.2, a pseudo-plane-wave topograph of any single-crystal material may be obtained provided that its intrinsic reflection width is markedly larger than 0.3″. For instance, misfit dislocation images recorded at a variety of positions on the epilayer and substrate reflection profiles in a quaternary heterojunction (Ga, Al) (As, P)/GaAs are presented in Figure 14. A detailed interpretation of the contrast is in progress[26] but a quick inspection of the images reveals obvious differences in the deformation fields. A contrast reversal of the dislocation image is observed between both wings of the epilayer profile (Figures 14a and 14c) implying that the lattice plane rotation induced by the dislocation in the epilayer has a unique sense (right-handed *or* left-handed). On the other hand, a black contribution appears in the dislocation image on both wings of the substrate profile (Figures 14d and 14f), which indicates that, in the substrate, lattice planes are rotated in two opposite directions at the dislocation line. Dislocation images may be as narrow as 1 μm when recorded in what might be designated as a weak-beam mode (see Figure 14g).

However, it should be remarked that the horizontal divergence of the beam is kept unchanged by the monochromator, which broadens the horizontal component of the defect images. In the present example, this effect was not observed since the dislocation lines had been intentionally oriented horizontally.

(a)

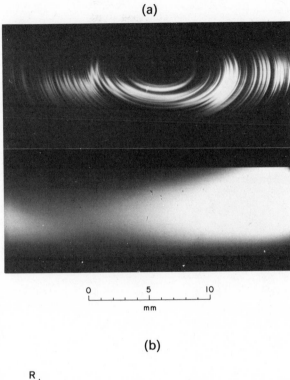

(b)

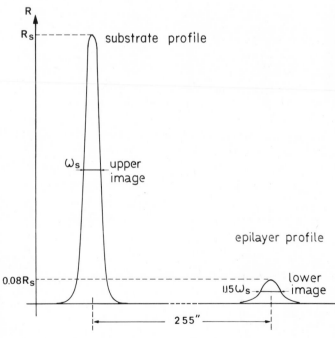

Figure 13. (a) Garnet heterostructure observed with a two-crystal spectrometer. Upper image: swirl pattern in the Czochralsky-grown GGG substrate. Lower image: defect-free magnetic garnet epilayer, 2 μm thick; (b) Schematic of the reflection profiles: image locations (courtesy of W. Parrish et al.[25]).

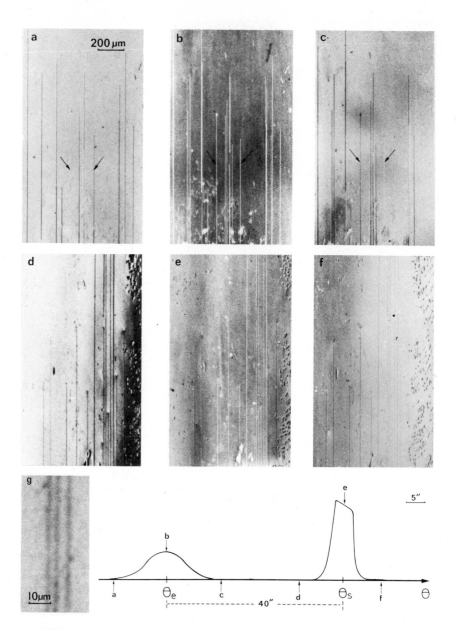

Figure 14. Pseudo-plane-wave images of misfit strain fields in quaternary heterojunctions (Ga, Al) (As, P)/GaAs. In the epilayer: black or white images with contrast reversal between both wings (a) and (c), white images at peak position (b). In the substrate: black images on both wings (d) and (f) and white images at peak position. (g) Extremely narrow images in the "weak-beam" region.

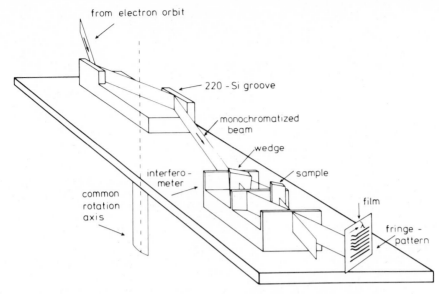

Figure 15. Interferometry setup at DESY (after Bonse and Materlik[27]).

2.3. Interferometry

X-ray interferometry has been introduced by U. Bonse and M. Hart.[11] A schematic view of the experimental setup installed at DESY by U. Bonse and G. Materlik[27] is represented in Figure 15. A standing wave field is produced by beam splitting and recombination. The introduction of a wedge-shaped phase object in one of the beam paths generates a fringe pattern as is shown in Figure 11. The insertion of the phase object under investigation leads to an additional shift of the fringe pattern whose measurement provides an accurate determination of the refractive index of the material. The knowledge of the wavelength dependence of the anomalous forward scattering amplitude in the vicinity of absorption edges is of a particular interest. Such experiments take maximum advantage of the continuous spectrum delivered by synchrotron radiation sources and an important research program is being developed in that field at DESY.

2.4. New Developments in Direct Viewing Detectors

The particular features of white beam topography make this technique a most suitable tool for kinetic experiments. A modification of the sample environment may lead to significant changes in its orientation that classical topography would not be able to accomodate. However, as long as detection is achieved by nuclear plates, only slowly changing systems can be investigated. As an example, by using an automatic plate exchanger, I. M. Buckley-Golder et al.[28] could reduce the time loss between successive exposures down to 1.5 s, which indicates clearly the present

Figure 16. Si; comparison between an x-ray-sensing TV camera image (top) and the same topograph recorded on Kodak film type M (bottom).

limitations of the technique. A step forward will only be provided by direct viewing detectors. Several attempts have been made both with high-power laboratory generators (for a review see Reference 29) and with synchrotron radiation. Direct and indirect x-ray imaging devices have been tested. The latter, where an intermediate image is produced in an x-ray phosphor layer, appears to be of more flexible application since optical systems may be used to magnify or demagnify the image at will. For example, comparison between the instantaneous display of an indirect x-ray-sensing TV camera and the same area recorded on an x-ray film[30] can be made in Figure 16. Individual misfit dislocations in epitaxial silicon are clearly resolved. The resolution of this standard equipment, supplied by Thomson CSF, is around 25 μm. Due to the high photon fluxes delivered by synchrotron radiation sources, a resolution better than 10 μm should be expected from specially designed cameras. Only then will synchrotron radiation topography reach its full potential in the field of real-time kinetic experiments.

3. Selected Examples of Applications of Synchrotron Radiation Topography

As was pointed out in the previous section, kinetic studies where the system changes slowly over a few minutes can be uniquely performed by use of synchrotron radiation topography in the present state of the art. In the following examples controlled modifications of the sample environment are produced and their incidence on the crystal defect organization is continuously observed by rapid sequences of topographs.

3.1. Recrystallization Experiments

In this type of work, the starting material is a polycrystalline sample submitted to an appropriate mechanical and thermal treatment prior to the *in situ* recrystallization annealing. The unique advantage of white beam topography is that the growing grains are imaged as soon as they start developing, whatever their orientation might be. Figure 17 gives an example of the initial stage of growth observed by J. Gastaldi and C. Jourdan[31] during the *in situ* recrystallization of aluminum by the strain–anneal technique around 300°C. As is demonstrated in Figure 17, the law stating that a nucleus should be bounded by faces parallel to low index planes is also valid for solid state growth, a result that could not be obtained by other techniques. After nucleation is completed, the recrystallization process is governed by grain boundary migration. During the course of its progression through

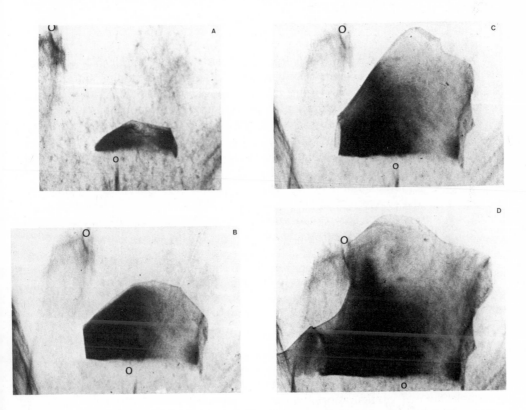

Figure 17. Al; initial stage of polyhedral growth of a recrystallized grain (after Gastaldi and Jourdan[31]).

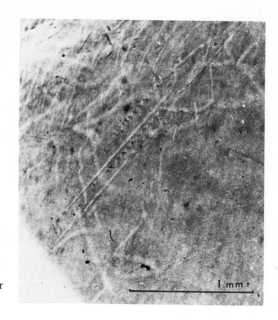

Figure 18. Al; screw dislocations climbing into helices (after Gastaldi and Jourdan[32]).

the strained matrix, the growth front of the recrystallized grain hits all sorts of obstacles that either reduce its velocity or even completely stop its motion. In the latter case, the growth front splits into two wings progressing on both sides of the pinning point. After incorporation of the obstacle, a certain mismatch may occur between the two wings and the closure stresses are plastically relaxed by generation of long straight dislocations,[32] mainly of screw character as shown in Figure 18. As soon as they are created, these dislocations climb into helices through a vacancy diffusion process and further growth leads to the elimination of some segments through the outer surfaces of the sample. The only tracks of the defect generation process available after cooling, prior to the advent of synchrotron radiation topography, were more or less regular arrays of edge dislocation segments. It is worth noting that the assumptions made at that time on the defect origin[33] have been fully confirmed by the present experiment.

In the example of aluminum recrystallization, large single crystals of approximately 1 cm^2 are obtained. However, statistical information on material recrystallizing with a smaller grain size, such as Fe–3%Si, may be also obtained by use of white beam topography.[34]

3.2. Domain Wall Motion in Magnetic Materials

In magnetic crystals, a major effort has been devoted to the interpretation of stable domain configurations appearing under a given applied field. Moreover, information on the magnetostrictive properties may also be obtained by coupling an applied stress to the magnetic field.

3.2.1. Antiferromagnetic Perovskites KNiF₃ and KCoF₃

A quantitative study of domain wall motion in antiferromagnetic perovskites $KNiF_3$ and $KCoF_3$, extending the previous data,[35] has been performed by M. Safa and B. K. Tanner.[36] As was mentioned in Section 2.1, cumbersome equipment is required for this experiment (Figure 5), implying a minimum distance of 20 cm between sample and nuclear plate. However, at NINA, this means only a loss of 2 μm in the geometrical vertical resolution (see Table 1). In addition, thanks to the use of white radiation, the orientation of the crystal in the cryostat need not be very accurate and the thermal stability should only be ensured over small periods. Sequences of topographs have been taken under a varying applied field parallel to different crystallographic directions with the help of an automatic plate exchanger.

In the antiferromagnetic phase the spins may take either of the cube axes directions, [100], [010], [001]. Thus three types of domains, respectively labeled d_x, d_y, and d_z, may be found. A specimen cut parallel to the (100) plane exhibits the four types of domain walls sketched in Figure 19: A walls containing [010] between d_x and d_z, B walls containing [001] between d_x and d_y, C walls containing [0$\bar{1}$1], and D walls containing [011] between d_y and d_z.

Simple thermodynamic considerations show that under an applied field **H**, a minimum in the stored free energy is reached when the spins align normal to the applied field. Consequently, domain wall rearrangements aiming to minimize the areas where the condition $\mathbf{H} \cdot \mathbf{s} = 0$ is not fulfilled should be observed. These trends may be expressed phenomenologically as forces **F** on walls separating two domains with spins $\mathbf{s}_1$ and $\mathbf{s}_2$ taking on the form

$$|\mathbf{F}| \propto (\mathbf{H} \cdot \mathbf{s}_1)^2 - (\mathbf{H} \cdot \mathbf{s}_2)^2$$

The sequence of topographs presented in Figure 20 shows the domain wall behavior under a field parallel to [001], which first increases (Figure 20a to e) and then decreases (Figure 20f) in strength. Domains d_x and d_y are in a minimum energy state. Hence, the observed displacements should reduce the area occupied by domains d_z. Indeed, in the well-resolved upper part of the sample, image A, C, and D gradually disappear while the B walls remain unchanged except for an extension into the region previously occupied by the A walls. In a second sequence the field was applied parallel to [011], and both d_y and d_z are unstable. In this case, the A and B walls are submitted to a displacement force leading to an extension of the d_x area. However, there is no force on the C and D walls, since the products $\mathbf{H} \cdot \mathbf{s}$,

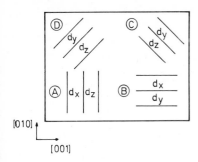

[010]

[001]

Figure 19. Schematic domain configuration in a (100) perovskite wafer.

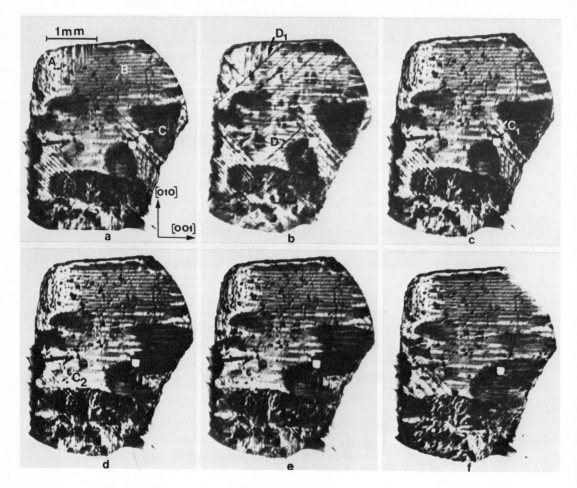

Figure 20. KNiF₃; domain wall motion under an applied magnetic field parallel to [001]. Increasing the field from 0.09 T (a) to 0.42 T (e) leads to the disappearance of all domain walls but the B walls. Decreasing the field down to 0.01 T (f) brings back some of the C walls (after Safa and Tanner[36]).

although nonvanishing, are equal on both sides of the walls. During the course of these experiments, domain wall motion took place only between orientation-dependent threshold and maximum values of the applied field. The displacement of an isolated domain wall under a controlled field was quite satisfactorily accounted for by simple theoretical considerations and a certain amount of reversible motion could be observed.

3.2.2. Ferromagnetic Alloy Fe–3%Si

This body-centered cubic alloy is ferromagnetic at room temperature and its domain configuration has been thoroughly investigated by x-ray topography.

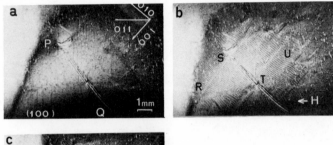

Figure 21. Fe–3%Si; changes in the domain pattern under an applied field parallel to [110] (after Stephenson *et al.*[37]).

However, dynamic experiments on domain wall motion have only been made possible by the availability of synchrotron radiation x-ray sources.

By use of white beam topography at DESY, J. D. Stephenson and his co-workers[37] have observed domain wall displacements under a step increase of the magnetic field. As expected, the sample showed a different behavior whether the applied field was parallel to the easy $\langle 100 \rangle$ or to the hard $\langle 110 \rangle$ magnetization direction. Figure 21 is a sequence of topographs recorded during a cycle parallel to $\langle 110 \rangle$. The influence of an applied stress on the domain configuration was also investigated.

A truly dynamic experiment has been performed at LURE-DCI, by J. Miltat.[38] Domain wall motion under an alternating magnetic field has been observed by stroboscopic topography where a beam chopper synchronized with the field had been inserted in the path of the monochromatic beam on the two-axis spectrometer. Figure 22 is an example of such topographs. In this experiment, the high intensity of the incident beam enables only 1/70th of the reflected beam to be used, with exposure time around 15 min. This new concept of stroboscopic topography should find other applications to all kinds of reversible displacements.

3.3. Plastic Deformation

Quite naturally, besides changing the temperature or the applied field around the sample crystal, one might think of changing the applied stress to study plastic deformation. The aim of the experiment presented in this section was to estimate the influence of surface orientation on the activated slip systems, at room temperature, in a body-centered cubic material Fe–3%Si. It has been performed at LURE–DCI by J. Miltat and D. K. Bowen[39] and a preliminary account of the experimental results may be found in a paper by J. Miltat.[40]

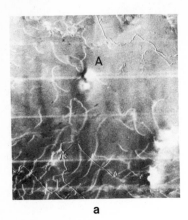
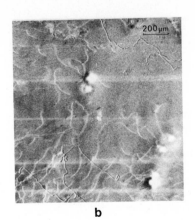

a   b

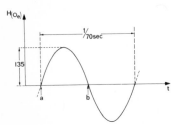

Figure 22. Fe–3%Si; behavior of 90° {110} magnetic domain walls under a sinusoidal applied field. Stroboscopic illuminations for two phase settings (a) and (b). (a) The precipitate at A is seen to interact with the magnetic structure since wall image doubling on the right side indicates that the wall occupies mainly two positions. (b) The broadening of the wall image means that the wall may occupy a continuous range of positions. Two-crystal arrangement: Ge 220, Fe–3%Si 110 in the $(+, +)$ setting, $\lambda \simeq 0.9$ Å. (Courtesy of J. Miltat.)

Figure 23. Fe–3%Si; Sample A: (a) Initiation of a single slip system on surface defects (101 reflection, $\sigma = 208$ MPa). (b) Extinction of glide bands by use of 101 reflection (scale mark 200 μm). (After Miltat.[40])

Figure 24. Fe–3%Si; Sample *B*: (a) Zero stress topograph: all subgrains are simultaneously imaged and individual dislocations are revealed. (b) Black–white contrast of slip bands populated with dislocations exhibiting a strong edge component under a stress $\sigma = 248$ MPa. (After Miltat and Bowen.[39])

Two samples *A* and *B* with the same tensile axis but different surface orientations have been successively deformed *in situ* and sequences of topographs were taken during the step increase of the applied stress. Exposure times were short enough, about 20 s, to prevent any significant contamination by a parasitic creep. Although the Schmid factors for all slip systems are identical in *A* and *B*, both samples showed a markedly different behavior, which however could not always be interpreted in terms of Vesely's law. For example, in sample *A*, slip bands clearly visible in Figure 23a (diffraction vector 101) are out of contrast in Figure 23b (diffraction vector 101). This means that a single slip system, namely (101) [111], has been activated, which is quite satisfactory since this system presents the largest Schmid factor and the smallest angle between Burgers vector and surface. Conversely, in sample *B*, since no extinction was ever observed, it is believed that at least two systems have been activated, although one of them exhibits a very low Schmidt factor. The black and white contrast displayed by the slip band images in sample *B* is typical of active dislocation sources with a strong edge component (Figure 24b).

One should be aware that results of plastic deformation experiments are never easy to unravel. The major contribution of white beam topography to the field is to facilitate not the interpretation but the collection of a larger amount of useful data. This is accomplished in three ways: first, by providing homogeneous images of samples (such as most metallic materials) with a subgrain structure (Figure 24a), second, by being insensitive to possible changes introduced in the sample orientation by the plastic strain (resulting in fairly good images even at very high strain

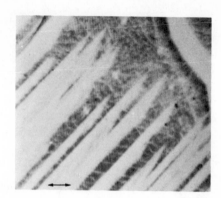

Figure 25. Fe–3%Si; Sample A: 101 reflection at a high stress level ($\sigma = 260$ MPa). Good resolution is still preserved. (After Miltat.[40])

levels, as shown in Figure 25), and third, by reducing the exposure time to durations too short for parasitic creep to occur (as was already mentioned at the beginning of this section).

3.4. Tunable Wavelength Topography in Absorbing Materials

For compounds with at least one element whose atomic number is approximately between 25 and 45, the selected wavelength can be tuned to the K absorption edge of this particular atom, thus minimizing the photoelectric absorption in the sample. The study of quaternary epilayers (Ga, Al) (As, P) over GaAs substrates performed by J. F. Petroff and M. Sauvage[41] demonstrates the capabilities of this method. For each hkl reflection, the crystal was oriented in order to select out of the incident white spectrum a wavelength close to 1.2 Å, just above the Ga K edge and still near the As K edge. In these conditions, the absorption coefficient μ was the same as for the characteristic line Ag $K\alpha$ (0.559 Å), but the resolution of the images and the reflecting power were significantly enhanced since both quantities depend on λ^n with n between 1 and 2.

Heterojunctions on GaAs are a basic element of infrared laser diodes. It has been shown that the defects present in the active layer have a deleterious effect on device performance. The present work was devoted to a thorough characterization of misfit dislocations in a single heterostructure grown by liquid phase epitaxy at the CGE laboratory (Marcoussis, France). The growth conditions of an epilayer on a GaAs substrate depend critically on the composition. As shown in Figure 26a, a ternary layer $Ga_{1-x}Al_xAs$ grows in a strain-free state at 850°C since the lattice parameters of layer and substrate are matched for any x value. However, due to a difference in thermal expansion coefficients, a mismatch develops on cooling and relaxes through an elastic curvature of the heterostructure (Figure 26a). It was shown by G. Rozgonyi et al.[42] that by adding a small amount of phosphorus to the melt, the misfit was transferred at the epitaxy temperature. Then, if the high temperature stress is kept below the critical value for plastic relaxation, the room temperature heterostructure may be flat and defect free (Figure 26b), except for the

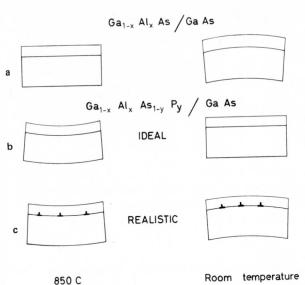

Ga$_{1-x}$ Al$_x$ As / Ga As

a

Ga$_{1-x}$ Al$_x$ As$_{1-y}$ P$_y$ / Ga As

IDEAL

b

REALISTIC

c

850 C   Room temperature

Figure 26. (a) Elastic relaxation of the misfit stress in the heterostructure Ga$_{1-x}$Al$_x$As/GaAs. (b) Defect-free heterostructure Ga$_{1-x}$Al$_x$As$_{1-y}$P$_y$/GaAs. (c) Elastic and plastic relaxation of the misfit stress in the heterostructure Ga$_{1-x}$Al$_x$As$_{1-y'}$P$_{y'}$/GaAs.

Figure 27. Ga$_{0.7}$Al$_{0.3}$As$_{1-y}$P$_y$/GaAs (reflection 113, $\lambda \simeq 1.2$ Å): (a) Severely warped heterostructure at room temperature (radius of curvature 1.5 m). (b) Same sample at 400°C: the parameters of the epilayer and substrate are matched; the heterostructure is flat.

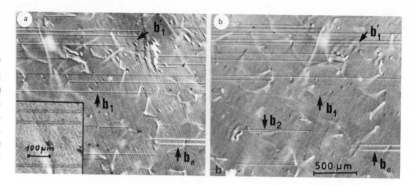

Figure 28. $Ga_{0.7}Al_{0.3}As_{1-y}P_y/$ GaAs; Burgers vector assignment by use of 422-type reflections. (a) $\overline{4}2\overline{2}$; dislocations with Burgers vector $b_1 = (a/2)[01\overline{1}]$ are out of contrast. (b) $2\overline{4}2$; dislocations with Burgers vector $b_1 = (a/2)[\overline{1}01]$ are out of contrast. Insert: enlargement of Figure 28a.

substrate-grown imperfections. However, a small misadjustment of phosphorus and a high-temperature misfit stress larger than the critical value may lead to the situation sketched in Figure 26c. Here, plastic and elastic stress relaxation takes place at high temperature; then, misfit dislocations together with a residual curvature are observed at room temperature. Such imperfect heterostructures are the object of the present study.

Being insensitive to warpage, white beam topography was indeed the most suitable technique for this purpose. As an example, in the sample shown in Figure 27a, the rotation of the reflecting planes per millimeter was larger than two minutes. Then, due to the continuous variation of Bragg angle, the dislocations, though actually straight, appear as curved lines. By heating this crystal *in situ* up to 400°C, the dislocation images are observed to straighten progressively (Figure 27b). This experiment provides clear evidence for the bimetal strip behavior classically attributed to such heterostructures.

During the course of defect characterization, primary misfit dislocations originating either from substrate threading dislocations or from surface or edge defects have been unambiguously identified as 60° dislocations by use of the proper 422 reflection bringing each family to a minimum contrast (Figure 28). Such reflections are highly asymmetric and almost impossible to obtain satisfactorily without synchrotron radiation.

Secondary misfit dislocations of pure edge character arise obviously from reactions as is illustrated in Figure 29. The high resolution of the topographs may be estimated on the enlarged view (boxed area in Figure 28) where double black dislocation images are observed with a peak to peak distance of about 8 μm.

Tunable white beam topography is thus a most powerful technique to identify the defects in any epitaxial structure involving III–V substrates like GaAs, GaP, InAs, GaSb, etc. Considering the technological importance of these compounds, there is no doubt that systematic experiments will be performed in the near future at the various x-ray synchrotron radiation facilities. However, as pointed out in Section 2.2, a separate analysis of dislocation strain fields in layer and substrate requires an incident beam whose wavelength and angular spread are narrow enough to prevent a simultaneous reflection by the epilayer and the substrate. In the present example this meant that $\Delta\lambda/\lambda$ or $\Delta\theta$ was smaller than 10^{-4}.

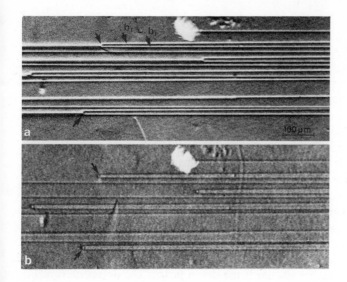

Figure 29. $Ga_{0.7}Al_{0.3}As_{1-y}P_y$/GaAs: Lomer-type dislocations generated by reactions (arrows). (a) $\bar{2}20$ reflection and (b) 220 reflection: the pure edge segment is totally out of contrast.

3.5. Miscellaneous Applications

Synchrotron radiation white beam topography should provide invaluable information on nondestructive phase transitions. The feasibility of such experiments has been demonstrated by J. Bordas et al.[43] in the case of $BaTiO_3$ and experiments on different ferroelectric materials are in preparation at LURE–DCI.

Another field of application is polytype analysis. In their study on ZnS, performed at NINA, I. T. Steinberger and his co-workers[44] took advantage of the simultaneous imaging of all polytypes. Then by careful indexing and skillful application of "jig-saw puzzles," the different forms coexisting in the sample could be identified and their respective behavior under thermal treatment investigated.

To conclude this section it is worth pointing out that after a period of daydreaming about the potentialities of synchrotron radiation topography, arousing some skepticism in the scientific community, a sufficient amount of serious work has now been performed to convince most of the people working in the field that this new source actually means a definite step forward for x-ray topography.

4. Future Developments

With the present generation of storage rings, progress in kinetic experiments may be anticipated as soon as real-time detectors of good quality are available. Then, evolving phenomena with velocities compatible to TV frequencies will be investigated. As an example, high-temperature plastic deformation in semiconductor single crystals and bicrystals will be performed at LURE–DCI by A. Mathiot, A. George, and their co-workers. As was mentioned above, stroboscopic topography, successfully introduced by J. Miltat, should be developed.

In addition, ultrasensitive detectors, able to record an image with a few pulses, will provide "snapshots" of rapid evolution in the sample.

Progress may also be expected in the field of polarized x-ray optics. Up to now, only preliminary pendellösung experiments have been performed both at LURE–DCI[45] and at Yerevan.[46, 47] Due to interference, the intensity of Bragg reflections by a perfect crystal, in the transmission geometry, is an oscillating function of the sample thickness. As a consequence, a fringe pattern is observed with wedge-shaped crystals. The quoted experiment gave photographic evidence of the splitting of a plane-polarized incident wave, inclined at 45° to the ray plane, along the two allowed vibration directions associated with a Bragg reflection process, namely, the normal (σ case) and parallel (π case) to the ray plane. The splitting is revealed through a periodic fading of the fringe contrast due to incoherent addition of the σ and π patterns whose periodicities are slightly different. As mentioned in Section 1.2, synchrotron radiation is only 95% plane polarized in the orbit plane, which is not sufficient for high accuracy measurements. By use of a polarizing monochromator successfully tested at LURE–DCI, the polarization ratio has been raised to 10^6. Such beam quality should bring down further the detection limits of anisotropy effects in crystals in the x-ray frequency range.

Finally, the insertion of wigglers and undulators (see Chapter 21) and the next generation of storage rings will supply photon fluxes two or three orders of magnitude larger than the present facilities. However, these new sources will not be fully exploited if a major effort is not directed towards an improvement of the detector capabilities. Transformations in single crystals occuring during a few milliseconds or diffraction effects based on very weak interactions might then be investigated.

ACKNOWLEDGMENTS

The authors wish to express their thanks to the various investigators whose results have been used in this review chapter and who kindly provided original topographs. We are pleased to acknowledge the help of A. Jeanne-Michaud in making the drawings and producing the photographic material.

References

1. A. R. Lang, The projection topograph: a new method in x-ray diffraction microradiography, *Acta Crystallogr.* **12**, 249–250 (1959).
2. N. Kato, A theoretical study of Pendellösung fringes, *Acta Crystallogr.* **14**, 526–532 and 627–636 (1961).
3. J. B. Newkirk, Observation of dislocations and other imperfections by x-ray extinction contrast, *Trans. AIME* **215**, 483 (1959).
4. A. Authier, Contrast of dislocations images in x-ray transmission topography, in *Advances in X-Ray Analysis*, J. B. Newkirk and G. R. Mallet (eds.), Vol. 10, p. 9–31, Plenum Press, New York (1967).
5. J. E. A. Miltat and D. K. Bowen, On the widths of dislocation images in x-ray topography under low absorption conditions, *J. Appl. Crystallogr.* **8**, 657–669 (1975).
6. A. Authier, Section topography, in *Topics in Applied Physics*, H. J. Queisser (ed.), Vol. 22, p. 145–189, Springer-Verlag, Berlin (1977).

7. U. Bonse, X-ray picture of the field of lattice distortions around single dislocations, in *Direct Observations of Imperfections in Crystals*, J. B. Newkirk and J. H. Wernick (eds.), p. 431–460, Interscience, New York (1962).

8. K. Nakayama, H. Hashizume, A. Miyoshi, S. Kikuta, and K. Kohra, Use of asymmetric dynamical diffraction of x-rays for multiple crystal arrangements of the $(n_1, + n_2)$ setting, *Z. Naturforsch. Teil A* **28**, 632–638 (1973).

9. K. Kohra, T. Matsushita, H. Ishida, and T. Ishikawa, Plane-wave x-ray topography, Eleventh International Congress of Crystallography, Abstract 11.2-26, p. S257, Warsaw (August 1978).

10. S. Takagi, K. Ishida, and A. Ohtsuka, Dislocation image by plane-wave topography, Eleventh International Congress of Crystallography, Abstract 11.2-18, p. S254, Warsaw (August 1978).

11. U. Bonse and M. Hart, An x-ray interferometer, *Appl. Phys. Lett.* **6**, 155–160 (1965).

12. A. Guinier and J. Tennevin, Sur deux variantes de la méthode de Laue et leurs applications, *Acta Crystallogr.* **2**, 133–138 (1949).

13. L. G. Schulz, *Trans AIME* **200**, 1082 (1954).

14. T. Tuomi, K. Naukkarinen, and P. Rabe, Use of synchrotron radiation in x-ray topography, *Phys. Status Solidi A* **25**, 93–106 (1974).

15. M. Hart, Synchrotron radiation—Its application to high speed, high resolution x-ray diffraction topography, *J. Appl. Crystallogr.* **8**, 436–444 (1975).

16. J. D. Stephenson, V. Kelha, M. Tilli, and T. Tuomi, Reconstructed topographs of polycrystalline Fe-3%Si crystals and observation of their magnetic domains using synchrotron radiation, *Nucl. Instum. Methods* **152**, 319–322 (1978).

17. B. K. Tanner, D. Midgley, and M. Safa, Dislocation contrast in x-ray synchrotron topographs, *J. Appl. Crystallogr.* **10**, 281–286 (1977).

18. B. K. Tanner, Crystal assessment by x-ray topography using synchrotron radiation, *Prog. Cryst. Growth* **1**, 23–55 (1977).

19. B. K. Tanner, M. Safa, and D. Midgley, Cryogenic x-ray topography using synchrotron radiation, *J. Appl. Crystallogr.* **10**, 91–99 (1977).

20. K. Kohra, M. Ando, T. Matsushita, and H. Hashizume, Design of high resolution x-ray optical systems using dynamical diffraction for synchrotron radiation, *Nucl. Instrum. Methods* **152**, 161–166 (1978).

21. J. H. Beaumont and M. Hart, Multiple Bragg reflection monochromators for synchrotron x-radiation, *J. Phys. E* **7**, 823–829 (1974).

22. U. Bonse, G. Materlik, and W. Schröder, Perfect crystal monochromators for synchrotron radiation, *J. Appl. Crystallogr.* **9**, 223–230 (1976).

23. M. Hart and A. R. D. Rodrigues, Harmonic-free single-crystal monochromators for neutrons and x-rays, *J. Appl. Crystallogr.* **11**, 183–189 (1978).

24. H. Hashizume, M. Sauvage, J. F. Petroff, B. Capelle, and P. Riglet, Harmonic-free monochromators with narrow band pass, European Physical Society Fourth General Conference, Abstract S.R.20, York (September 1978).

25. W. Parrish and C. G. Erickson, Diffraction topography at Stanford Synchrotron Radiation Laboratory, Eleventh International Congress of Crystallography, Abstract 14.2-3, p. S331, Warsaw (August 1978).

26. J. F. Petroff, M. Sauvage, P. Riglet, and H. Hashizume, Synchrotron radiation plane-wave topography. I. Application to misfit dislocation imaging in III–V heterojunctions, *Phil. Mag.* (accepted for publication); P. Riglet, M. Sauvage, J. F. Petroff, and Y. Epelboim, Synchrotron radiation plane-wave topography. II. Comparison between experiments and computer simulations for misfit dislocation images in III–V heterojunctions, *Phil. Mag.* (accepted for publication).

27. U. Bonse and G. Materlik, Precise interferometric measurement of the Ni K edge forward scattering amplitude with synchrotron x-rays, *Z. Phys. B* **24**, 189–191 (1976).

28. I. M. Buckley-Golder, B. K. Tanner, and G. F. Clark, A simple automatic cassette for x-ray synchrotron topography, *J. Appl. Crystallogr.* **10**, 502 (1977).

29. R. E. Green, Jr., Direct display of x-ray topographic images, in *Advances in X-ray Analysis*, H. F. McMurdee, C. S. Barrett, J. B. Newkirk, and C. O. Rund (eds.), Vol. 20, p. 221–235 (1977).

30. M. Sauvage, X-ray topography settings: white beam topography and direct viewing detectors, two-axis spectrometers, *Nucl. Instrum. Methods* **152**, 313–317 (1978).

31. J. Gastaldi and C. Jourdan, New possibilities for recrystallization study afforded by the x-ray synchrotron radiation topography advent, *Phys. Status Solidi A* **49**, 529–537 (1978).

32. C. Jourdan and J. Gastaldi, Kinematic studies of misfit recrystallization dislocation formation in aluminium single crystals during recrystallization, *Scr. Metall.* **13**, 55–59 (1979).

33. J. Gastaldi and C. Jourdan, Dislocations et croissance des cristaux d'aluminium par recristallisation secondaire, *J. Cryst. Growth* **35**, 17–27 (1976).

34. I. B. McCormack and B. K. Tanner, Application of x-ray synchrotron topography to in-situ studies of recrystallization, *J. Appl. Crystallogr.* **11**, 40–43 (1978).

35. B. K. Tanner, M. Safa, D. Midgley, and J. Bordas, Observation of magnetic domain wall movements by x-ray topography using synchrotron radiation, *J. Magn. Magn. Mater.* **7**, 337–341 (1976).

36. M. Safa and B. K. Tanner, Antiferromagnetic domain wall motion in $KNiF_3$ and $KCoF_3$ observed by x-ray synchrotron topography, *Philos. Mag. B* **37**, 739–750 (1978).

37. J. D. Stephenson, V. Kelhä, M. Tilli, and T. Tuomi, Quasi life-time topography of magnetic domain movements in (100) [001] Fe-3%Si polycrystals using white synchrotron radiation, Eleventh International Congress of Crystallography, Abstract, 14.2-1, p. S330, Warsaw (August 1978).

38. J. Miltat and M. Kleman, Interaction of moving {110} 90° walls in Fe–Si single crystals with lattice imperfections, *J. Appl. Phys.* **50**, 7695–7697 (1979).

39. J. Miltat and D. K. Bowen, Slip initiation in iron–silicon single crystals: study by conventional and synchrotron radiation x-ray topography, *J. Phys. (Paris)* **60**, 389–401 (1979).

40. J. Miltat, X-ray topography and dynamics: description of two experiments performed at LURE–DCI, *Nucl. Instrum. Methods* **152**, 323–329 (1978).

41. J. F. Petroff and M. Sauvage, Misfit dislocation characteristics in quaternary heterojunctions $Ga_{1-x}Al_xAs_{1-y}P_y$/GaAs analysed by synchrotron radiation white-beam topography, *J. Cryst. Growth* **43**, 628–636 (1978).

42. G. A. Rozgonyi, P. M. Petroff, and M. B. Panish, Control of lattice parameter and dislocations in the system $Ga_{1-x}Al_xAs_{1-y}P_y$/GaAs, *J. Cryst. Growth* **27**, 106–117 (1974).

43. J. Bordas, A. M. Glazer, and H. Hauser, The use of synchrotron radiation for x-ray topography of phase transitions, *Philos. Mag.* **32**, 471–489 (1975).

44. I. T. Steinberger, J. Bordas, and Z. H. Kalman, Microscopic structure studies of ZnS crystals using synchrotron radiation, *Philos. Mag.* **35**, 1257–1267 (1977).

45. M. Sauvage, J. F. Petroff, and P. Skalicky, Synchrotron radiation topographic evidence for the interaction of plane-polarized x-ray photons with perfect crystals, *Phys. Status Solidi* **43**, 473–477 (1977).

46. I. P. Karabekov, D. I. Egikian, R. A. Mikaelyan, J. G. Bagdasarian, and L. I. Datsenko. The investigation of the polarization fading of waves in wedge-shaped silicon crystal by means of synchrotron radiation, Yerevan Institute of Physics, Preprint 258(51)-77.

47. L. I. Datsenko, V. G. Bagdasaryan, D. L. Yegikyaj, I. P. Karabekov, and R. A. Mikaelyan. Investigation of the interference pendulum bands of V-shape silicon crystal in (0.33–0.80 Å) of synchrotron radiation, Yerevan Institute of Physics, Preprint 240(33)-77.

19

Inelastic Scattering

P. EISENBERGER

1. Introduction

Historically the initial utilization of synchrotron radiation has been closely tied to various absorption spectroscopies. Successful photoemission and EXAFS studies have also been responsible for the rapid growth in synchrotron radiation research activity. However, as useful and important as those techniques are, it is highly likely that the second generation of dedicated facilities and synchrotron users will be remembered for the beginning of a successful utilization of scattering techniques. The delay in the application of synchrotron radiation inelastic scattering techniques is directly related to the fact that for most inelastic scattering experiments momentum resolution is not critical. Thus one can afford to collect a relatively large solid angle (typically $2° \times 2°$) of the characteristic emission of a conventional source for the experiment. In the above solid angle a 10-kW x-ray tube would provide 10^{12} photons/s in the $K\alpha$ 2.5-eV-wide emission line. This is slightly *more* than has been available in the most intense beam line at the Stanford Synchrotron Radiation Laboratory. Even for that flux inelastic scattering experiments typically produced counting rates of only 1–10/s. However, dedicated sources hold the promise of 10^{13}–10^{14} photons/s in a 1-eV bandwidth. When these sources are in operation, a whole new generation of experiments will become possible. The breath of new studies is made possible not only because of the high flux, but also because the high collimation and tunability of the synchrotron radiation enables one to perform high-energy resolution and/or resonant types of experiments. Those experiments are virtually impossible with a conventional source.

In Section 2 we will review briefly the general scattering formalism with an emphasis on resonant processes and the possibility of directly measuring phonon

---

P. EISENBERGER ● Bell Laboratories, Murray Hill, New Jersey 07974.

excitations. The possible experimental techniques that will be utilized will be described in Section 3. Finally, in Section 4, the previous experiments and future applications will be discussed.

2. Theory

It is beyond the scope of this review to provide all the details of the theory for resonant and nonresonant Raman, Compton, and thermal diffuse scattering. We will only briefly review them here. More details can be obtained from other reviews.[1-4] In general the nonrelativistic inelastic cross section can be written as

$$\frac{d\sigma}{dr\,d\omega} = r_0^2 \frac{\omega_2^2}{\omega_1^2} S(\mathbf{k}, \omega) \tag{1}$$

$$\frac{d\sigma}{dr\,d\omega} = r_0^2 \frac{\omega_2}{\omega_1} \sum_f \sum_i |M_{fi}|^2 (E_f - E_i - \omega) \tag{2}$$

where $r_0 = e^2/m_0 c^2$ is the classical radius of the electron, ω_1 and ω_2 are the incident and scattered frequencies, $\omega = \omega_1 - \omega_2$, i and f are the initial and final states, and M_{fi} is the matrix element that connects them. In general equation (2) can be used to describe transitions from the initial state of the system to an electron excited to a bound or collective excited state (Raman), or an excited state in the continuum (Compton), or one on which the electron remains in the ground state but the lattice has been excited (thermal diffuse scattering). For each of the above classes of excitations there exists two regimes in which the matrix element M_{fi} can be calculated. These two regimes arise from the fact that x rays couple to the electronic system via

$$H_c = \frac{e^2 A^2}{2mc^2} + \frac{e\mathbf{p} \cdot \mathbf{A}}{mc}$$

where $\mathbf{A}$ is the vector potential of the photon and $\mathbf{p}$ is the momentum of the electron to which one is coupling. Thus H_c involves two terms, the A^2 or nonresonant part and the $\mathbf{p} \cdot \mathbf{A}$ or resonant part. Collecting terms to order A^2 one finds for M_{fi} that

$$M_{fi} = \langle f|\exp(i\mathbf{k} \cdot \mathbf{r})|i\rangle(\varepsilon_1 \cdot \varepsilon_2)$$
$$+ \sum_n 1/m \frac{\langle f|\varepsilon_2 \cdot \mathbf{p}\exp(-i\mathbf{k}_2 \cdot \mathbf{r})|n\rangle\langle n|\mathbf{p} \cdot \varepsilon_1 \exp(i\mathbf{k} \cdot \mathbf{r})|i\rangle}{E_N - E_0 - \omega_1 - i\Gamma_0}$$
$$+ \sum_n 1/m \frac{\langle f|\varepsilon_1 \cdot \mathbf{p}\exp(i\mathbf{k}_1 \cdot \mathbf{r})|n\rangle\langle n|(\mathbf{p} \cdot \varepsilon_2)\exp(-i\mathbf{k}_2 \cdot \mathbf{r})|i\rangle}{E_N - E_0 + \omega_2 + i\Gamma_0} \tag{3}$$

where $\mathbf{k}_1$ and $\mathbf{k}_2$ and the wave vectors of the incident and scattered photons, $\mathbf{k} = \mathbf{k}_1 - \mathbf{k}_2$, ε_1 and ε_2 are the polarizations of the incident and scattered photons, and Γ_0 is the lifetime of the hole state created in the excitation. In the x-ray regime the momentum transfer k is given, for small energy loss ($\omega_1 \approx \omega_2$), by $k = 2k_1 \sin \theta/2$, where θ is the angle between the incident and scattered photon. If ω_1 is far away from $E_N - E_0$, then the first term will dominate, while when $\omega_1 \approx E_N - E_0$, the second term will be the largest. Near resonance a more detailed calculation would replace the delta function in equation (2) with a Lorentzian with a width characteristic of the lifetime of the final state into which the electron is excited.[5]

The nonresonant Compton and Raman cross sections follow naturally by specifying the appropriate initial and final state wave functions. Formally only the first term in M_{fi} is associated with $S(\mathbf{k}, w)$, which is the dynamic structure factor.[6] At low-momentum transfer the transitions to the bound or collective states dominate, while at large k the transitions to the free-like continuum states dominate.[6] The resonant Compton and Raman cross sections also follow naturally from equation (3). However, to obtain the thermal diffuse scattering (TDS) cross sections an extra coupling must be included. One must invoke the Born–Oppenheimer approximation, which asserts that the electrons bound to the atoms follow the motion of the nuclei. One can derive the TDS cross section by letting r become $\mathbf{r} + u(r, t)$ in the first term for M_{fi}. Here $\mathbf{r}$ is the average position and $u(r, t)$ represents the thermal displacements.[1] In this way the low-frequency motion of the nuclei (phonons) produces low-frequency electron charge density variations that can then be directly measured by inelastic x-ray scattering. Again both resonant and nonresonant coupling is possible depending upon the frequency of the x rays and the specific energy levels of the atoms being studied. For the resonant case one must go to third order perturbation theory and specifically include an electron–phonon coupling term (e.g., deformation potential).

The cross section for nonresonant phonon scattering that relates the scattered photon rate to the incident rate can be written in the general form for a monatomic lattice,

$$n_s = n_{in} e^{-2m} \frac{r_0^2}{2\mu} N_A f_A^2(k) \sum_{pj}{}' \left\{ \begin{matrix} n_{pj} \\ n_{pj} + 1 \end{matrix} \right\} \left(\frac{\hbar^2 k^2 / 2M_A}{\hbar\omega_{pj}} \right) (\varepsilon \cdot \varepsilon_{pj})^2 \, \partial\Omega \qquad (4)$$

where $\sum'_{pj}$ includes the modes accepted by the energy resolution and angular resolution $\partial\Omega$ with the two obviously connected by dispersion. In the above μ is the absorption length, N_A is the number of scatterers with scattering factor $f_A(k)$ and mass M_A per cubic centimeter, and n_{pj} is the occupation probability of the mode with wave vector p, polarization state ε_{pj}, and frequency $\hbar\omega_{pj}$. For copper metal at $k = 4$ Å^{-1}, $2\mu = 350$ cm^{-1}, $f_A(k) = 20$, $\hbar\omega_{pj} = 0.1$ eV, $n_{pj} = 0$, $\varepsilon_1 \cdot \varepsilon_{pj} = 1$, $e^{-2m} = 1$, and $N_A = 10^{23}$ atoms/cm^2, one finds

$$n_s = n_{in} \times 10^{-5} \, \partial\Omega \qquad (5)$$

where n_{in} is the number of incident photons/s meV.

3. Experimental Techniques

The fundamental spectrometer for performing inelastic x-ray scattering experiments will be the triple-axis spectrometer, where the nomenclature comes from neutron scattering. In a typical experimental apparatus there are three "axes," monochromator, sample, and analyzer. The details of each axis will, of course, depend upon the desired energy and momentum resolution. Before discussing any details of the low-resolution ("eV") and the high-resolution ("meV") forms, a general description of the function of each axis would be helpful.

The monochromator axis in the case of synchrotron radiation will very likely consist of three components. The first mirror would collect the radiation emitted from the source and reimage it (usually with 1:1 magnification) at some convenient distance from the ring (5–30 m). A pair of flat crystals for the resonant experiments or a curved crystal for fixed energy or high-resolution experiments will monochromatize the white synchrotron radiation by Bragg scattering. Finally a second mirror or asymmetric flat crystal may be used to determine the momentum and spatial extent of the incident beam by either magnifying (increasing size, decreasing momentum spread) or demagnifying (decreasing size, increasing momentum spread). Thus the first "axis" creates a beam of size s, angular spread $\delta\theta_H$ and $\delta\theta_V$, and energy spread ΔE, with an intensity in photons/s given by

$$I_{in} = LB(\text{photons/s eV mrad mm}^2)s\ \delta\theta_H\delta\theta_V\Delta E \tag{6}$$

where B is the brightness of the synchrotron radiation, which, apart from losses L due to the finite reflectivity of the mirrors and crystals, will be unaffected by the various mirrors and crystals.

The second "axis" will consist of the sample in its experimental environment (dewar, pressure cell, etc.) mounted on a goniometer. In an ideal setup this would be a four-circle goniometer that is capable of orienting the crystal axis with respect to the incident beam as well as orienting the analyzer arm so as to define the momentum transfer $\mathbf{k}$.

The third axis would be mounted on the detector arm and again would either be a pair of flat crystals or curved crystals. One would rotate the crystals with respect to the scattered beam to analyze the energy distribution of the radiation scattered with momentum transfer $\mathbf{k}$. In this way $S(\mathbf{k}, \omega)$ can be experimentally measured. Of course one can take various types of scans by suitable choice of motion of the axis. One that may prove useful in resonant studies is one in which $\mathbf{k}$ and ω are fixed but the incident frequency is changed.

The use of a double crystal Bragg spectrometer in the antiparallel mode to obtain resolutions on the order of 1 eV is fairly straightforward and has been used in many experiments with conventional sources. The main problem with that approach is caused by the relationship between ΔE, the resolution of the apparatus, and $\Delta\theta$, the angular acceptance, which is given by

$$\Delta E/E = \cot\theta_B\Delta\theta \tag{7}$$

The signal strength in a scattering experiment with a resolution better than the width of the spectrum being studied is given by

$$S = I_{in}S(\mathbf{k}, w)\left|\frac{dw}{d\theta}\right|\left|\frac{dk}{d\theta}\right|d\theta_V^2 \tag{3}$$

Thus S will vary as ΔE^2 on the output and, of course, the input Bragg crystal spectrometer has the same relationship. Thus from equation (6), I_{in} also varies as ΔE^2 and therefore S varies as ΔE^4.

By using a curved crystal one can remove the coupling between ΔE and angular range $\Delta \theta$ of collection. The crystal is bent so that over the whole range $\Delta \theta$ the input rays make the same Bragg angle with the crystal surface. In this case S will vary only as ΔE^2. Thus in going from 1-eV to 1-meV resolution there is a difference of 10^6 in intensity in the two approaches. For the case of a bent perfect crystal the resolution is given by

$$\Delta E/E = \cot \theta_B \Delta \theta_D$$

where $\Delta \theta_D$ is the Darwin width of the particular reflection being used. A curious fact is that for a beam polarized perpendicular to the plane of scattering one finds that

$$\Delta \theta_D = CNF_{hkl} d_{hkl}^2 \tan \theta_B$$

where N is the number of unit cells per cubic centimeter and F_{hkl} and d_{hkl} are the structure factor and d spacing of the hkl reflection respectively. Therefore one finds that

$$\Delta E/E = CNF_{hkl} d_{hkl}^2$$

This implies that the fractional resolution is a constant for a given scattering plane *hkl*. However, there are many secondary reasons that still make it very desirable to choose an *hkl* so that $\theta_B \approx 90°$. For example, a simple spherical shape for the curved crystal is appropriate in the backward direction. Also abberrations due to finite source size and the effect of the imperfect geometry of the scattering planes are the smallest at 90° and thus, in general, one has greater tolerance to the effects of heating and crystal imperfections. It is beyond the scope of this review to discuss the various technological problems that remain to be solved in order to successfully construct a 1–10-meV spectrometer. At this point in time one can only be sure that it will be attempted. If one used curved crystals on the input that collected the total vertical emission and 0.3° in the horizontal, n_{in} in equation (5) for a dedicated storage ring (SPEAR) operating at 150 mA and 3.5 GeV would be approximately (allowing for some losses) 10^{10} photons/meV. With a $\partial \Omega$ of 1/40 000 on the output (1° × 1°), one would have a scattering rate of 2/s meV. The small sample sizes required (10^{-4} to 10^{-6} cm^3) would give an x-ray scattering technique a distinct advantage over neutron scattering where samples of approximately 1 cm^3 are required.

In addition to the above special detection system, another type of spectrometer system may turn out to be useful for studying Compton scattering. Following the successful utilization of solid state detectors and a gamma-ray source, a pilot study[7] was attempted for utilizing synchrotron radiation rather than a gamma-ray source. The gamma-ray detector, which is described in various publications,[8] would be mounted on the third axis. The other special apparatus required for such a study would be a beam line with no mirror (the grazing angle of incidence required to reflect 60–100-keV radiation is too small to be useful) and a monochromator that could provide the required resolution and efficient angular acceptance of the source. Since in these studies the resolution requirements are not severe because the solid state detector on the output only has 300–500-eV resolution, to first order one need only be concerned with obtaining an efficient crystal that will accept the full angular range of the incident beam (3×10^{-4} radians in the vertical). To obtain 300-eV resolution at 100 keV with a 3×10^{-4}-rad acceptance would require a crystal with a spacing of $2d_{hkl} \simeq 1.2$ Å. If the mosaic spread of such a crystal matched the input divergence of the beam then one would have a very efficient monochromator. The SPEAR storage ring at Stanford, operating at 4 GeV and 100 mA would provide a flux of 2.4×10^{10} photons/s with such a spectrometer. At the 8-GeV facility in construction at Cornell such a monochromator system could provide a flux of 2×10^{13} photons/s for a 100-mA current. PEP (at Stanford) operating at 15 GeV and 55 mA would also provide 2×10^{13} photons/s. All of these are superior to that readily obtainable from a laboratory-based radioactive source, which typically provides only about 10^6–10^7 photons/s incident on the sample. Of course, one can also perform high-resolution Compton scattering experiments by utilizing lower-energy photons. For example, with 4-eV resolution at 40 keV one would have approximately a momentum resolution of 0.03 a.u. This would be sufficient to see temperature-dependent effects. In energy terms the Doppler shift associated with the Compton process[3, 6] is given by

$$\frac{\mathbf{k} \cdot \mathbf{p}}{m} \approx 2(E_R KT)^{1/2}$$

where $E_R = k^2/2m$ is the recoil energy given the electron in the Compton scattering process. At 40 keV near the backward direction, $E_R \approx 6$ keV and thus $\mathbf{k} \cdot \mathbf{p}/m \approx 12$ eV.

4. Previous Experiments

As stated previously in the Introduction, there have been very few experimental results on inelastic scattering utilizing synchrotron radiation because the cross section is small and, for poor-resolution experiments, the total photon flux of existing parasitic storage rings is weak compared to high-powered x-ray tubes. Two types of work have been attempted, each of which utilized a special feature of synchrotron radiation. The first was resonant Raman studies[5, 8] that utilized the tunability of the radiation to resonantly enhance the cross section, thus making the study impossible. The second utilized the high-energy tail of the synchrotron spectrum to perform Compton scattering studies.[7] While there are not very many photons at 60–100 keV compared to the flux in the 10–20-keV region, because of its high brightness and the relative weakness of a one curie radioactive source, even nonideal synchrotron radiation provides a distinct advantage. The pilot study[7] was conducted at the NINA synchrotron operating at 4.6 GeV and 20 mA (10^{10} photons/s eV at 1 Å). They found a factor of 100 enhancement in signal rate (half the resolution) over gamma-ray experiments. With fluxes like 10^{13} photons/s eV, which will become available at PEP, Cornell, or Brookhaven and SPEAR with wigglers, enhancements of 10^5 can be expected.

In two papers,[5, 8] the details of the fundamental aspects of resonant Raman scattering were verified. The first of those studies used a solid state detector that had a Gaussian resolution function with a σ of 104 eV. The study confirmed in copper metal the expected Raman dispersion and intensity enhancement of the L-shell electron-to-continuum transition as the incident energy approached the K absorption edge of copper. Above the edge a large and fixed energy fluorescence signal was observed. A clear demonstration of the usefulness of synchrotron radiation is provided by comparing this study with the earlier work of Sparks et al.,[9] which utilized a conventional tube. The other synchrotron-radiation-based study of resonant Raman scattering used a high-resolution triple-axis spectrometer in which the 111 silicon Bragg crystal analyzer was fixed to detect radiation scattered by roughly 90°. The input channel-cut monochromator had a resolution of 0.9 eV FWHM while the output analyzer was 0.8 eV at FWHM. This was the first high-resolution study of resonant Raman scattering in the x-ray regime. It revealed that in the transition region from resonant scattering to fluorescence, the linewidth of scattered radiation is narrower than the lifetime-limited width of the fluorescent radiation. It also provided a fairly clear method for identifying the exact position of the absorption edge. Even in the resonance regime where one had an intensity enhancement of about a 1000 over normal A^2 scattering, these experiments resulted in a signal of only 2 counts/s. However, the combined use of a mirror (factor of 10 increase), a dedicated storage ring (factor of 50 increase), and a monochromator better matched to the input beam (factor of 5 increase) should result in comparable signals in the A^2 regime.

5. Future Prospects

We have already explicitly discussed the possibility of performing high-resolution phonon spectroscopic studies and high- and low-resolution Compton scattering studies with the future dedicated storage ring facilities. It was briefly indicated that normal A^2 Raman scattering with 1-eV resolution should also be possible. This should enable one to investigate plasmon dispersion curves and other electronic excitations with improved resolution.[10] In short, generalized studies of the electronic $S(\mathbf{k}, \omega)$ for various systems can be performed utilizing synchrotron radiation.

The previous investigations of the basis of resonant scattering have also been described. In the future resonant scattering should become a very powerful tool for elucidating new phenomena as well as for sorting out the properties of a system composed of many types of atoms. In the former category one can include resonant resolution studies of the Fermi electrons with the possibility of gaining more detailed insight into the x-ray threshold problem. In addition, one can, by working near resonance, investigate interferences between the resonant and nonresonant parts of the cross section [i.e., the first and second terms in equation (3)].[11] The magnitude and sign of those interferences can contain interesting information about the electronic wave functions.

EXAFS has been useful in decomposing a complicated structural problem into a simpler one; resonant scattering can be used to decompose a complicated electronic and vibronic excitation spectrum from a complex material into simpler ones. By either straight resonance spectroscopy or differential spectroscopy near an absorption edge of a specific atom one should be able to isolate the contribution of a specific atom to the electronic or vibrational spectrum of a complex material. Investigations of impurities in solids, metalloproteins in biology, and complex molecules in chemistry can all potentially benefit from such an approach.

Two conclusions can be drawn from this review of inelastic scattering. The more immediate and conservative evaluation is that with the improved flux available from dedicated storage rings and with the improved development of x-ray spectrometers, one should be able to greatly expand the information that will be obtainable by inelastic scattering by utilizing synchrotron radiation. The experiments will be difficult, and the technology fairly exacting. It will not all occur in a big push, because inelastic scattering has not traditionally been a very active area.

A more long-term and fairly optimistic view is that if and when wiggler devices are developed and fluxes of 10^{15} photons/s eV can be produced, then just as in high-energy physics, a generalized scattering approach will become the dominant experimental technique for the evaluation of electronic and vibronic properties of systems of interest to the physicist, chemist, and biologist.

References

1. R. W. James, *Optical Principles of the Diffraction of X-rays*, Cornell University Press, Ithaca, New York (1965).
2. J. M. Jauch and F. Rohrlich, *The Theory of Photons and Electrons*, Addison-Wesley, Cambridge, Massachusetts (1955).
3. P. M. Platzman and N. Tzoar, *Phys. Rev.* **139**, 410 (1965).
4. T. B. Bannett and I. Freund, *Phys. Rev. Lett.* **34**, 273 (1975).
5. P. Eisenberger, P. M. Platzman and H. Winick, *Phys. Rev. Lett.* **36**, 623 (1976).
6. P. Eisenberger and P. M. Platzman, *Phys. Rev. A* **2**, 415 (1970).
7. M. Cooper, R. Holt, P. Pattison and K. R. Lea, *Commun. Phys.* **1**, 159 (1976).
8. P. Eisenberger, P. M. Platzman and H. Winick, *Phys. Rev. B* **13**, 2377 (1976).
9. C. J. Sparks, Jr., *Phys. Rev. Lett.* **33**, 262 (1974).
10. P. Eisenberger, P. M. Platzman and K. C. Pandry, *Phys. Rev. Lett.* **31**, 311 (1973).
11. P. M. Platzman (private communication).

20

Nuclear Resonance Experiments Using Synchrotron Radiation Sources

R. L. COHEN

1. Introduction

Synchrotron radiation (SR) experiments using nuclear resonance can be divided into two categories: those in which the nuclei are acting as individual scattering or fluorescence centers, and those in which interference from scattering due to many cooperating nuclei is dominant. It is convenient to call the former class nuclear fluorescence or absorption experiments, and the latter nuclear Bragg-scattering experiments. We will discuss here first the nuclear fluorescence experiments, which are relatively straightforward in their approach and goals, and then discuss the possibilities of the Bragg-scattering experiments. At the time of this writing (August 1978), more than a dozen theoretical papers have been written analyzing various aspects of prospective nuclear Bragg-scattering experiments, and three experimental groups have mounted programs to implement such experiments. However, the nuclear Bragg scattering of SR has not yet been observed. Nuclear resonance absorption of SR has been experimentally observed in one recent experiment[1] using Fe^{57}.

It is interesting to note that despite the enthusiasm with which physicists are pursuing these "new" experiments, both the Bragg scattering and the fluorescence experiments were realized almost twenty years ago using conventional excitation sources. A conventional x-ray-tube source was used[2] to excite and observe nuclear resonance fluorescence in F^{19} and Mn^{55}. That experiment was preceded by

R. L. COHEN ● Bell Laboratories, Murray Hill, New Jersey 07974.

others[2] using bremsstrahlung from betatron-accelerated electrons, and gamma rays from nuclear reactions.[3] At about the same time, nuclear Bragg scattering was being observed[4] using Fe^{57}, with radiation coming from a source of $Co^{57} \rightarrow Fe^{57}$. All of these early experiments were both extremely difficult and very limited in the information they supplied. They did not have great impact on the mainstream of nuclear and solid state physics, and did not lead to extensive further work. Why, then, this sudden enthusiasm for similar experiments with SR sources? The answer, basically, is that the unique characteristics of synchrotron radiation—the collimation, the polarization, the pulsed nature, and the intensity—may make possible experiments that are qualitatively different from those previously carried out. Only time will tell whether the severe experimental difficulties involved in this new series of experiments are justified by the new insights gained.

Many of the problems faced in the implementation of nuclear SR experiments can be seen by examining the formula for the cross section for the absorption of a photon of energy E by a nucleus:

$$\sigma(E) = \frac{1}{2\pi} \frac{h^2 c^2}{E_0^2} \frac{2I_e + 1}{2I_g + 1} \frac{1}{1 + \alpha} \left[1 + \frac{4(E - E_0)^2}{\Gamma^2} \right]^{-1} \qquad (1)$$

where E_0 is the transition energy; α is the internal conversion coefficient of the transition; $\Gamma = 1/\tau$ is the uncertainty principle width of the state, derived from $\tau = 1.44 t_{1/2}$; and I_g and I_e are the spins of the nuclear ground and excited states.

Some of these values are given in Table 1 for some states that might be used for SR experiments. It is important to realize that the cross section that determines the strength of nuclear resonance scattering from broad-band sources is the energy-integrated cross section, and is thus proportional to Γ, so that the best states for such experiments would appear to be short-lived. The time resolution of detectors is limited, however, so that for states of half-life shorter than about 20 ns, it is not possible to discriminate between the scattering from electrons and the scattering

Table 1. Some Nuclear Transitions That Are Promising for Synchrotron Radiation Experiments

Isotope	Transition energy (keV)	Half-life[a] (ps)	Peak resonance cross section (10^{-20} cm^2)	Natural abundance (%)
F^{19}	110	600	20.0	100
Fe^{57}	14.4	97800	256.0	2.2
Kr^{83}	9.4	0.14×10^6	107.0	11.6
Gd^{157}	64	0.46×10^6	23.0	15.7
Tb^{159}	58	105	10.5	100
Ho^{165}	95	22	8.3	100
Ta^{181}	6.2	6.8×10^6	167.0	100
Ta^{181}	136	40	6.0	100
Re^{187}	134	10	5.4	63
Hg^{201}	32.2	200	0.95	13

[a] A half-life of 1000 ps corresponds to a resonance of 4.6×10^{-7} eV.

from nuclei by their different time dependences. As will be seen below, this discrimination is a vital part of all of the experimental arrangements that have been proposed to date.

Significant interference arises from the direct electronic scattering of the SR by the atomic electrons, and from x rays following core-level photoelectric absorption. Although cross sections for these processes are much smaller than the peak cross sections for the nuclear resonances, the electronic processes are broadband, rather than being sharply tuned. If the synchrotron radiation is filtered to $\sim$1-eV bandwidth before carrying out the nuclear scattering experiment, the electronic scattering processes will act on *all* the x rays, while the nuclear scattering will be effective only for a narrow ($\sim 10^{-8}$ eV wide) energy slice of the 1-eV-wide radiation. The electronically scattered radiation can easily be many orders of magnitude larger than the nuclear scattering we want to investigate. This interference has been a persistent problem in nuclear resonant scattering experiments; already in 1962 Seppi and Boehm[2] pointed out:

> This problem might be solved by taking advantage of the instantaneous character of the atomic scattering as compared to the relatively long lifetime of the nuclear excited states. Through the use of a pulsed x-ray beam and a properly gated detector it should be possible to observe only the nuclear excitation events in the sample.

With the inherently short pulses of synchrotron radiation, it is relatively easy to implement this approach, by detecting only delayed photons emitted by the relatively long-lived nuclear excitations. To use this discrimination, however, one must study nuclear states with lifetimes of at least 20 ns or so, to have an adequate time separation between the electronic and nuclear scattered radiations. Thus, very short-lived states, which have the largest resonance widths, cannot be studied this way. In fact, all of the currently planned experiments use the 14.4-keV transition in Fe^{57}, with a 100 ns half-life, although the energy-integrated cross section is much smaller than that of many other candidates.

Another approach taken to increase the effective ratio of nuclear to electronic scattering is the use of nuclear Bragg scattering. Experiments of this kind are discussed in Section 3.

2. Single Nucleus Excitations

2.1. Mössbauer Effect

The Mössbauer effect, more formally known as "recoil-free gamma-ray resonance absorption," was first reported in 1958. Since then, experimental research with the use of the Mössbauer effect has diffused rapidly into the diverse fields of solid state physics, metallurgy, chemistry, and biochemistry.[5]

In conventional Mössbauer spectroscopy, the experimental arrangement may be described as follows: An atomic nucleus makes a transition from an excited state to its ground state, emitting a gamma ray. This gamma ray has approximately the right energy to be resonantly absorbed by a nucleus of the same kind in its ground state. Small perturbations in the energy of nuclear levels in the absorber can be measured by observing the change in gamma ray energy required for the gamma ray to be resonantly absorbed. The nuclear hyperfine (hf) interaction and isomer shifts measured in this way provide useful information about the environment of the atom under study.

Although a number of claims have been made that SR could replace conventional sources for Mössbauer spectroscopy experiments, it is my feeling that this is extremely unlikely. The convenience and low cost of radioactive sources will make them difficult to displace. Although there are a very few possible Mössbauer levels that cannot easily be reached from radioactive parents, most of those nuclear states are undesirable or uninteresting for some other reason in any case. It is my feeling that Mössbauer experiments using SR will be worthwhile only where the unique characteristics of SR provide basically new possibilities.

Ruby[6] has already proposed the most straightforward approach to synchrotron Mössbauer experiments. His proposal is to look at fluorescent de-excitation radiation (x rays, gamma rays, or conversion electrons) from the Mössbauer absorber, in a scattering geometry. By gating the counters *off* during the radiation pulse, ($\sim 10^{-10}$ s) all of the Rayleigh, Compton, and x-radiation (following photoeffect) should be eliminated, as they are prompt (on the scale of 10^{-10} s). Thus, provided that the Mössbauer absorbing state had a lifetime of ~ 1 ns or longer, all of the observed radiations should arise from SR x-ray absorption exciting the nuclear state. This allows one to detect the occurrence of SR x-ray resonance absorption, but does not allow Mössbauer spectroscopy, because the 1-eV bandwidth of the monochromator is much greater than the nuclear absorption width, $\sim 10^{-8}$ eV. An immediate useful result of this configuration is that one has a simple, absolutely stable monochromatic detector with a linewidth orders of magnitude smaller than that of the crystal monochromator. Since count rates from 1 to 10^4 counts/s could be obtained, the nuclear resonance detector would be an excellent tool for investigating and improving the characteristics of the monochromator.

To perform Mössbauer resonance experiments despite the broad monochromator output spectrum, Ruby[6] proposed (see Figure 1) putting a "notch filter" in the beam, upstream of the fluorescing absorber, as was done in the old gamma-ray resonance fluorescence experiments.[3] This filter would be another Mössbauer absorber, whose absorption energy could be varied by Doppler modulation, as in a conventional Mössbauer experiment. When the energy of the line in the filter was the same as that of a resonance absorption line in the absorber, a dip in fluorescence counting rate would occur. By varying the filter energy, the absorption cross section of the absorber could be plotted out as a function of energy.

A quick calculation will demonstrate that for conventional Mössbauer experiments, even neglecting certain restrictions that will be discussed below, and assuming that detectors could be perfectly gated and would not be saturated by the electronically scattered synchrotron radiation, this technique will not replace the use of radioactive sources. At present, under good SSRL operating conditions, up to 10^{10} photons/s eV at 10 keV are available after a channel-cut crystal monochroma-

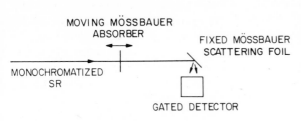

MOVING MÖSSBAUER
ABSORBER

FIXED MÖSSBAUER
SCATTERING FOIL

MONOCHROMATIZED
SR

GATED DETECTOR

Figure 1. Basic configuration to do Mössbauer experiments with SR. X-rays in the incoming beam within the resonance width of the nuclear transition are absorbed by the nuclei in the fixed foil, and secondary radiations from the excited nuclei are observed by the gated detector. The detector is turned on after the SR flash to make it insensitive to the (prompt) electronically scattered SR. The moving absorber acts as a sharp adjustable filter to absorb narrow regions of the incoming beam energy. (After Ruby[6].)

tor that subtends 1 mrad. In the normal 10^{-8}-eV linewidth of an Fe^{57} Mössbauer experiment, we would then have about 10^2 photons/s, which is much less than what a conventional source would produce. (A 50 mCi Co^{57} source, costing $\sim$$2000, gives $\sim 10^5$ 14-keV photons/s into a solid angle of 0.01 sr.) The increased flux that will be available with new monochromators, wigglers, and undulators may result in a thousandfold increase in counting rate (see Table 2). This intensity is still only comparable to that from the radioactive source. The situation may be changed if undulators provide two orders of magnitude further increase.

For the present, then, the question is whether the other characteristics of synchrotron radiation make it possible to do experiments that cannot be done with conventional sources. There are three characteristics of synchrotron radiation that suggest interesting possibilities: the polarization, the pulsed nature, and the good collimation.

Polarization. Polarized Mössbauer gamma rays have been proposed and, in a few cases, used[7] for determining the direction of magnetization and sublattice magnetizations. These experiments can be performed with conventional sources by filtering out some of the hyperfine lines of a split source; the radiation remaining is partly polarized. This is accomplished at a considerable loss in intensity. The availability of "free" polarization would undoubtedly make such experiments relatively more attractive and competitive with neutron scattering.

Table 2. Comparison of Radioactive Sources and Storage Rings for Nuclear Resonance Experiments

Trait	Radioactive source Co^{57}, 30 mCi, $1 K	Storage ring SSRL, measured 1 mrad, 33 mA, 3.7 GeV	Storage ring, BNL, calculated 2.5 GeV, 500 mA
14-keV photons/s, solid angle			
2 eV, 10^{-2} sr	10^5	10^{10}	10^{13}
10^{-8} eV, 10^{-2} sr	10^5	50	5×10^4
10^{-8} eV, 10^{-7} sr	1	50	5×10^4
Availability	100%	1%	10%
Signal/Background	1	10^{-8}	10^{-8}
Polarization	Possible	Ideal	Ideal
Timing	Very limited	Ideal	Possible

Pulsed Beam. There has occasionally been discussion of Mössbauer experiments to measure the hyperfine structure of ions in metastable, low-lying, optically excited states; typically of lifetime 10^{-3} s. These experiments have not been feasible in the past. Even though it is possible to put a large fraction of the ions into the optically excited state with a strong light pulse, they quickly decay to the electronic ground state. The power required to keep a large fraction of the ions in the excited state continuously is too large to supply easily. If, however, the light were pulsed to do the optical excitation in synchronization with the synchrotron radiation pulses, the Mössbauer experiment could take place before the excited ions had decayed. Other experiments could be based on the electronic excitation of the solid by the synchrotron radiation pulse itself. For example, Fe or Sn could be doped into Si, and the effects on the Mössbauer spectrum of electrons excited into conduction states could be observed. This could provide direct measurements of trapping times. Using time-of-flight delay techniques for the Mössbauer synchrotron radiation, it would also be possible to make Mössbauer measurements as a function of time after an initial exciting synchrotron radiation pulse. In insulators and semiconductors, a rich variety of effects would be expected to arise. All these experiments would be impossible with radioactive sources.

Collimation. The very good collimation of the synchrotron radiation suggests that certain Mössbauer experiments, which are severely limited by solid angle considerations, would be greatly eased by the use of synchrotron radiation. One such experiment is the gravitational red shift measurement, where the size of the effect is proportional to the distance between the source (at the base of a tower) and the absorber (at the top); some compromise must be reached between the increasing effect and decreasing solid angle as the tower height increases. For current synchrotron intensities and a 100-m tower, the synchrotron radiation beam is equivalent to that from a source of 0.25 Ci, which is realizable without enormous expense. Anticipated improvements in SR intensity using wigglers would make a synchrotron source practical for these experiments. The equivalence between gravitational and inertial mass, now established for gamma rays to $\sim 10\%$ by conventional Mössbauer experiments,[8] could probably be improved by two orders of magnitude by such experiments.[9]

The inherently good collimation of SR would also facilitate measurements of the interference effects between nuclear and electronic Bragg scattering that allow direct determination of the x-ray scattering phase factors.[10–12] A few such experiments have been carried out using Mössbauer radioactive sources,[10, 11] but they have been extremely difficult. The main limitation is that because of the small acceptance angle of a Bragg reflection, the count rate into a Bragg peak using a radioactive source is very small, typically 1/s to 1/min. This problem would be ameliorated by the use of SR sources, which have angular divergence comparable to the acceptance angle of the crystal.

Mössbauer Experiments in which Radioactive Sources Cannot be Used. In a few cases, there is no easily available radioactive parent for the level to be studied, or the obvious source produces severe interfering lines. Examples of the former are F^{19}, K^{40}, Ni^{61}, Sb^{121}, Gd^{157}, and Hg^{201}; of the latter, Ge^{73}, Ta^{181}, and Tm^{169}. Thus, Mössbauer studies with a large number of isotopes could be carried out more easily by using synchrotron radiation than by using conventional sources. Research using radioactive Mössbauer absorbers (i.e., long-lived radioactive species) would also be easier with SR since gated counters and a small solid angle could be used to reduce the background from the decay of the absorber nuclei.

2.2. Nuclear Excitation without the Mössbauer Effect

One of the major difficulties to be expected in the "Mössbauer experiments without radioactive sources" is that nuclear resonant excitations will arise not only from zero-phonon processes (Mössbauer effect), but also from one- and multi-phonon excitations, because the exciting radiation bandwidth is much wider than the phonon spectrum. Since these processes will result in nuclear excitation indistinguishable from that of the Mössbauer absorption, only f (the recoil-free fraction, or Debye–Waller factor) of the emitted secondary radiation will arise from zero-phonon absorptions. Thus, the resonance intensity is diluted, just as in conventional experiments, by f.

Since the absorption line for non-recoil-free transitions is inhomogeneously broadened by the phonon spectrum no useful hyperfine information can be obtained from that absorption. It is, however, possible to take advantage of this effect and derive hyperfine information from the precession rate of the excited nuclear state.[13] This would essentially be a cross between a perturbed angular correlations (PAC) experiment, where the precessing state is fed by a gamma decay, and implantation perturbed angular correlation technique (IMPACT) experiments, where a charged particle impact provides both the excitation and the initial orientation axis of the system. SR excitation experiments would have the advantages of both these older techniques: the recoil impact would be small, so the nucleus under study would not be displaced, and the size of the gamma fluorescence anisotropy would be very large, since the exciting synchrotron radiation is polarized. The precession could be detected as a time-integral or time-differential effect. The 0.78-μs spacing between SR pulses at SSRL would be ideally matched to the properties of the nuclei of interest.

Another means of detection that has recently been used with great success at the Hahn–Meitner Institute is the "stroboscopic" technique.[14] In that approach, in which the incoming beam is pulsed at a fixed frequency, the applied magnetic field (which provides the hyperfine field) is varied to make the precession frequency of the nuclear excited states equal to the repetition rate of the beam pulses, so that all the nuclei are precessing in phase. This produces an extremely high sensitivity to small hyperfine field perturbations, and it is easy to measure, for example, the Knight shift in a metal.

Excited state double resonance can be carried out by impressing an RF field on the excited nuclei, and inducing transitions between the nuclear hyperfine levels.[15] The change in substate populations would be observable via the altered anisotropy of the emitted radiation. These experiments are also difficult to carry out with radioactive sources,[15] but could become much more attractive with strong SR sources.

The study of hyperfine interactions via the perturbed angular correlation between the absorbed and emitted photons would be a very powerful tool because it would be freed from the limitations of the recoil-free fraction. Thus, medium-energy transitions (50 to 100 keV) could be studied at high temperatures, and paramagnetism and nuclear relaxation times could be studied in liquids. The only requirement would be a nuclear state at a low enough energy to be excited by the synchrotron radiation, and a long enough lifetime to precess noticeably before the decay.

If the incident SR has been monochromatized to a 1-eV bandwidth, the ratio of nuclear scattering to electronic scattering for most of the transitions of interest for perturbed angular correlation work is only about 10^{-6}. Thus, realization of the experiments discussed above will require a radical improvement either in detector design (particularly in time resolution and reduced afterpulsing) or in monochromator design (to provide a narrower energy bandwidth for excitation). Both these areas are discussed below.

2.3. Experimental Results

In contrast to the many possibilities and hopes for SR in exciting nuclear resonances, the experimental results to date have been very limited. The only observation of such an effect reported as of this writing has been the experiment of Cohen, Miller, and West[1] (Figures 2 and 3). In that work, a gated detector was developed to observe the conversion electrons produced from the 14-keV state of Fe^{57}, excited by SR. In principle it should be easy to observe this effect, since even currently available SR beams can put ~ 1 photon/s within the linewidth of the Fe^{57} resonance, and the nuclear cross section is large enough to absorb all of these photons. The principal difficulty arises from the scattering by the atomic electrons, which interact with the entire width (1–5 eV) of the monochromatized incident SR, and thus produce scattering that is several orders of magnitude stronger than that from the nuclei. An obvious answer to this would be to take advantage of the time separation between the (prompt) electronic scattering from the electrons and the delayed reradiation ($t_{1/2} = 100$ ns) from the excited nuclei, as was discussed above. This procedure, however, has two severe problems associated with it: First, the lifetime of the nuclear state must be at least comparable to the recovery time of the detector (e.g., 10–50 ns for fast scintillators with photomultipliers or semiconductor detectors), so that only states with long lifetimes and therefore small integrated cross sections are accessible. Second, radiation detectors are not perfect; afterpulsing in electron multiplier devices and late collection of trapped charge in semiconductor devices are long standing problems of radiation detectors. In resonant scattering of SR, each radiation pulse would give several photons scattered into the detector by electronic processes, whereas only a few (delayed) nuclear scattered pulses per second would be expected. Thus the electronically scattered pulses are perhaps 10^6 times as frequent as the delayed nuclear pulses, and if the detector produces a spurious afterpulse for each 10^2 detected photons, these afterpulses will overwhelm the nuclear resonance scattering signal.

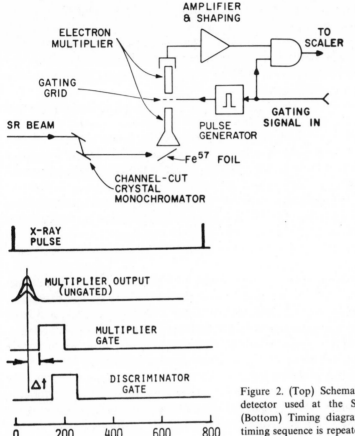

Figure 2. (Top) Schematic drawing of the gated conversion-electron detector used at the Stanford Synchrotron Radiation Laboratory. (Bottom) Timing diagram for the gating circuits shown above. The timing sequence is repeated at the 1.28-MHz rotational frequency of the storage ring.

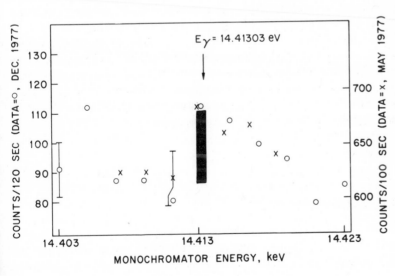

Figure 3. Resonance curves observed by scanning the x-ray monochromator over the nuclear excitation energy. The hyperfine-broadened nuclear resonance is 5×10^{-7} eV wide, so the shape of the observed resonance actually represents the monochromator pass function, about 3 eV wide. Actual numbers of counts are shown, for 100-s ($\times$) or 120-s ($\bigcirc$) counting times, for two different runs. The vertical scales have been selected to merge the data points. Changes to the detector between the two runs account for the reduced background and slightly lower efficiency in the second run. The vertical bar shows the position and calculated size of the resonance under these operating conditions.

In the light of these problems, the detector design shown in Figure 2 incorporates the following features:

1. Conversion electrons arising directly from the nuclear state are detected. This provides increased intensity over the detection of reradiated photons, and also avoids the use of scintillation materials, which have long "tails" on the fluorescence.

2. The electron multiplier is gated off during the SR pulse, so that the prompt electronically scattered radiation (photoelectrons, in this case) is not amplified. Thus, the afterpulsing is virtually eliminated, because the primary pulses are not propagated down the multiplier. It is impractical to gate the electron multiplier input by retarding potential or grid techniques, because the electron energy spectrum extends to 14 keV.

3. The channeltron multiplier used has an inherently low afterpulsing ratio because of its small volume and (in the second stage) helical structure. This property is further enhanced by using a low-noise charge-sensitive preamplifier to detect the amplified electron pulse, so that only a relatively low electron multiplier gain was needed.

The detector design shown in Figure 2 has been used at SSRL on a standard experimental beam line, and the resonance curve shown in Figure 3 was produced. With SPEAR operating at 3.7 GeV and with a stored current of 30 mA, a flux of 2×10^9 14-keV photons/s was incident on the detector foil, with pulses occurring at a 1.28-MHz repetition rate. Approximately 20 nuclear excitations per second were produced, leading to a count rate of ~ 0.2 counts/s after accounting for the solid angle, detector efficiency, and gating fraction. About five prompt photoelectrons per beam pulse were incident on the electron detector. The multiplier and discriminator gating reduced the prompt feedthrough and afterpulsing from these photoelectrons by a factor of approximately 10^6, allowing the relatively weak nuclear resonance absorption signal to be observed as the monochromator was tuned through the nuclear resonance energy. The hyperfine-broadened nuclear resonance and phonon sidebands are very narrow compared to the monochromator passband, and the observed resonance curve is actually a profile of the monochromator function.

Although the detector discussed here could be used for studying monochromator profiles and establishing absolute energy calibrations, it will probably be more useful as a low-background detector in conjunction with the nuclear Bragg-scattering experiments discussed in the next section. Since the nuclear resonance cross section is ~ 500 times larger than the photoelectric cross section, a substantial increase in discrimination against the electronic Bragg-scattered radiation can be obtained.

3. Nuclear Bragg Scattering

3.1. Proposed Experiments

When the outgoing scattered wave fronts from individual nuclei in a crystal constructively interfere, the scattering becomes a cooperative phenomenon, and the scattered intensity is greatly enhanced. This effect is exactly analogous to electronic

Bragg scattering, and thus has been called nuclear Bragg scattering. The analogy is close enough that many of the concepts well developed to understand x-ray scattering in crystals can be used. However, a true analysis of the nuclear scattering case is considerably complicated by the resonant nature of the nuclear scattering cross section,[11, 12, 16, 17] and analysis of a realistic case must include the nuclear hyperfine structure, which divides the nuclear levels into a number of close-lying substates. Many articles have been published recently analyzing the results to be expected in SR-excited nuclear Bragg-scattering experiments.[18–24] The main aim of this section will be to discuss the proposed experiments, emphasizing the compromises and problems that make the proposals so difficult to realize.

Directly comparing the nuclear and electronic cases, the following general statements can be made. All values cited below are for the 14-keV level of Fe^{57}, which has so far been proposed for all nuclear Bragg-scattering experiments.

1. The nuclear scattering cross section is about twenty times larger than the electronic cross section, so that the penetration depth of the radiation is much smaller in the nuclear case, and the "nuclear Darwin width" is much greater.

2. The energy transmission function of a nuclear Bragg "monochromator" is not determined by the nuclear Darwin width, but by the very narrow inherent width of the nuclear resonance ($\Delta E/E = 3 \times 10^{-13} \sim 10^{-8}$ eV for Fe^{57}).

3. Since the nuclear excited states are relatively long lived, the time dependence of the nuclear Bragg-scattered radiation is not simply that of the input radiation, but is much more complex. It is dependent not only on the collective nature of the excited state inside the scattering crystal, but on the nuclear hyperfine interactions.[18, 24]

In standard (radioactive source) nuclear Bragg-scattering experiments, the largest single problem is intensity—a radioactive source emits relatively few photons/s into the solid angle accepted by a crystal Bragg reflection. In SR experiments, as pointed out above, adequate flux is readily obtained from the high intensity (SPEAR, DORIS, VEPP-4) storage rings. In SR nuclear Bragg-scattering experiments, however, the main problem is, as in the fluorescence experiments discussed above, that monochromatization of the SR by electronic Bragg scattering leaves a beam that is still very wide in energy in comparison with that of the nuclear resonance linewidth. Thus, only a very small fraction (typically, 10^{-8}) of the incoming beam will be subject to nuclear Bragg scattering. A number of ingenious schemes have been

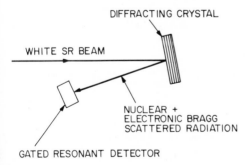

Figure 4. Nuclear Bragg-scattering experiment using a timed resonant detector to provide both energy and timing discrimination between nuclear and electronic Bragg-scattered radiation.

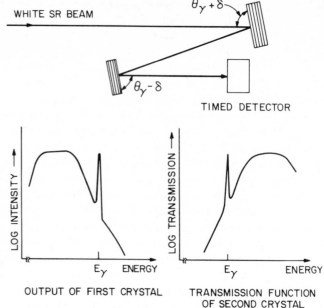

Figure 5. Double nuclear Bragg-scattering experiment to increase discrimination against electronic Bragg scattering. The crystals are misaligned by δ from the scattering angle $\theta\gamma$, corresponding to the nuclear transition energy. If δ is larger than the electronic Darwin width, but smaller than the nuclear Darwin width, the electronic reflection will be greatly attenuated, but the nuclear reflection will have almost the full intensity (after Reference 20).

suggested to enhance the fraction of nuclear Bragg-scattered radiation in the emergent beam. The main aim of these proposals has been to find ways to suppress the electronic Bragg scattering, which inherently occurs at the same angles as the nuclear scattering. It is possible that a number of these techniques may eventually be combined to provide increased discrimination, at the cost of increased complexity. It also appears that timed detectors (as discussed above for the nuclear fluorescence experiments) may be used to allow discrimination between the nuclear and electronic scatterings on the basis of their different time dependences.

Here are brief descriptions of some of these schemes:

1. Single-nuclear Bragg scattering, with gated (timed) detector to discriminate against electronic Bragg scattering: (Figure 4). Use of a Bragg angle near 90° provides high dispersion and reduces the effective energy width of the electronically scattered radiation.[20]

2. Two-crystal nuclear Bragg scattering, with misalignment of the electronic Bragg component[20]: Since the nuclear Darwin width is so much greater than the electronic Darwin width, it should be possible to slightly misalign the crystals from the ideal Bragg angle, and still transmit the nuclear reflected radiation but attenuate the electronic scattering (Figure 5).

3. Use of a polyatomic crystal in which the electronic Bragg-scattering factor is very small for a particular reflection: If only one set of lattice sites in a crystal is occupied by nuclear scattering centers, it may be possible to find reflections for which the electronic scattering factor is close to zero. If the nuclear Bragg reflection does not suffer from the same cancellation, it should thus be possible to get a strong nuclear scattering factor and weak electronic one at the same Bragg angle. The

following three approaches have been proposed; of these, the second and third have already been demonstrated using Mössbauer source (Co^{57}) excitation.

For FeTi, a CsCl-structure ordered intermetallic, odd-order reflections have almost zero intensity because of the near cancellation of the scattering from the Fe and the Ti atoms.[20] For Fe^{57} nuclei on the Fe superlattice, of course, this cancellation does not occur. Thus, the electronic Bragg scattering is much weaker than the nuclear Bragg scattering. This "enrichment" of the nuclear scattered part of the beam can be cascaded by additional reflections (Figure 5).

In $K_4Fe(CN)_6 \cdot 3H_2O$, Black and Duerdoth[10] have shown that for the (080) reflection, the electronic Bragg scattering is negligible due to accidental cancellation of the electronic scattering from the iron ions and all the other ions in the crystal. The nuclear Bragg scattering was readily observed over the very weak electronic Bragg scattering. In some cases it might be possible to improve this accidental cancellation, either by varying the crystal temperature to alter the relative Debye–Waller factors of the ions whose scattering is cancelling, or by partial replacement of one ion (e.g., K^+) by a chemically similar ion (e.g., Na^+) with a different scattering cross section, to adjust the net scattering from one sublattice.

The Soviet nuclear Bragg-scattering research group is planning[21] to use the (777) reflection from α- Fe_2O_3, a reflection that is symmetry forbidden for the electronic Bragg scattering, but allowed for the nuclear Bragg scattering. This apparent contradiction can be explained as follows: The nuclear levels of Fe_2O_3 are split by the hyperfine interaction, because the crystal is magnetically ordered, being a weak ferrimagnet at room temperature. The magnetic structure gives rise to four magnetic sublattices for the iron ions, with different axes for the hyperfine interaction. Thus, the scattering from the nuclear hyperfine levels sees a lower symmetry than that of the crystal and becomes an allowed reflection,[25] Put in a slightly different way, the magnetic structure provides a "hyperfine superlattice" so that the nuclear Bragg reflection would correspond to the superlattice lines of a lower symmetry structure. A suppression of the electronic Bragg scattering by a factor $>10^5$ has been claimed for this approach.[21]

4. Use of a thin crystal[19]: Since the nuclear cross section is so much larger than the electronic, a crystal containing only a few hundred layers of atoms provides good nuclear Bragg-scattering intensity, while the electronic Bragg scattering is still relatively weak.

5. Use of the different multipolarities of the electronic and nuclear scattering[19] (see Figure 6): Since the electronic scattering is E1, while the nuclear scattering is M1, the angular dependences of the differential scattering cross sections are different because of the polarization of the SR. In particular, if the incoming radiation is fully polarized, and the beam is scattered at 90° in the plane of polarization by a Bragg scattering at a 45° Bragg angle, the electronic Bragg scattering is greatly reduced, but the nuclear scattering will still be strong. To make this approach effective, the SR must have its polarization increased (for example, as shown in Figure 6) by one or more Bragg scatterings.

It should be emphasized that many of these ideas can be combined to increase the discrimination against the electronic scattering. However, all of the proposals depend strongly on the perfection of the scattering crystals to eliminate the elec-

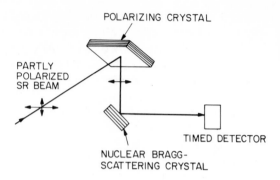

Figure 6. Use of the polarization dependence of the electronic scattering to reduce the electronic scattering. If the second scattering is at a Bragg angle near 45°, the electronic scattering is very weak. The first scattering is primarily to increase the polarization of the SR (after Reference 19).

tronic scattering. Thus, an approach that may appear best in principle may be inferior in practice, because the particular crystal required may be unavailable. Because of the uncertainty arising from these materials science problems, it is difficult to anticipate which of these options will be the most effective, and how many layers of complexity will be required to obtain a useful result.

3.2. Experimental Problems

The basic problem of SR nuclear Bragg-scattering experiments is that there are $\sim 10^8$ times more photons that can be electronically Bragg scattered. The wealth of ingenious ideas outlined above suggests that this basic problem can be overcome, and that the success of these experiments is assured. However, there are a number of experimental problems that impede the realization of the approaches mentioned above. Most of these problems arise from the fact that the Fe^{57} resonance is the one used. Unfortunately (see Table 1) the only other isotope having a long enough half-life to do time-resolved measurements, and still having a reasonable energy-integrated cross section, is Kr^{83}. The obvious materials science problems of making and using diffracting crystals of this isotope seem to have discouraged serious consideration; it remains to be seen whether the obvious problems with krypton are more difficult to overcome than the less obvious problems with iron.

Using Fe^{57}, the most obvious problem is that virtually all high-iron materials are magnetic at room temperatures, so that the nuclear resonance is split by the hyperfine interaction into (normally) 6 lines. Thus the effective resonance scattering cross section is reduced by a significant factor (~ 4), and the Bragg-scattered intensity is reduced by 4^2, or 16 times. Another obvious problem is the low isotopic abundance of Fe^{57}, only 2.2%. Thus, not only must special crystals be grown for these experiments, but they must be grown from exceedingly scarce and expensive ($10/mg) raw material.

Although use of pure iron crystals would appear to be the most direct path, there are additional barriers there. Although the normal (bcc) phase, α-Fe, is stable just below the melting point ($\sim 1530°C$) an additional phase, γ-Fe, is the stable one from $\sim 1380°$ to $\sim 910°C$. Thus α-Fe crystals grown from the melt recrystallize during cooling, and the most direct method for growing single crystals is inaccessible. One widely used remedy for this is to alloy 3–5 atomic percent Si with the Fe;

this addition stabilizes the α-Fe phase up to the melting point, and single crystals can be grown from the alloy melt. However, the crystals obtained by this approach are not of high quality, and the presence of the Si (substitutional for the Fe) decreases the effectiveness of schemes depending on exact cancellation of the electronic scattering. Single crystals of pure iron can be grown by strain-anneal and chemical vapor decomposition techniques, but the crystals obtained are small and of large mosaic spread ($\sim 1°$). A logical solution to these problems would be to epitaxially grow a thin film of pure Fe^{57} on a substrate with a small electronic Bragg scattering. It would obviously require a substantial research project to grow high-quality crystals in this way, although there is extensive literature on the epitaxial growth of Fe crystals.[26]

An alternative approach has been used by the Soviet group,[21] which has chosen to use $\alpha\text{-}Fe_2^{57}O_3$ as the scattering crystal. Good single crystals of Fe_2O_3 are readily grown from borate fluxes, without requiring extremely large quantities of the separated isotope. A mosaic of less than 1' is reported,[21] and a very good reduction of the electronic Bragg scattering on the (777) reflection to be used for the nuclear resonance was achieved.

The preliminary experiments carried out so far suggest that the following features would be very desirable in an experiment to detect the nuclear Bragg scattering:

1. A grazing-incidence mirror, to cut off the SR flux at ~ 15 keV. This will reduce the background due to scattered radiation, and eliminate the presence of λ/n harmonics that might be transmitted by the diffracting crystal.

2. A "premonochromator" will reduce the heating of the diffracting crystal by the white radiation beam (now ~ 1 W; due to increase greatly with high-intensity SR sources), reduce the shielding, ozone, and safety problems inherent in having the high-intensity white radiation beam in the experimental area, and will reduce the contamination and erosion of the crystal from beam-produced ionization.

3. A timed detector, to enhance the observability of the "slow" nuclear Bragg component over the prompt electronic scattering. Use of a resonant-timed detector[1] would provide significant additional discrimination, at some cost in counting rate.

Additionally, the complexities and difficulty of the experiment make it necessary to dedicate a beam line for this purpose, rather than using a "general purpose" line.

4. Conclusions

Although nuclear Bragg scattering from SR sources has not yet been observed, it is likely that it will be within the next year, and will rapidly be developed to a stage where useful beams of highly monochromatic radiation are being generated. For the first few years, the primary interest will be in characterizing the details of the nuclear scattering process and the time dependence of the scattered radiation. Eventually the radiation should be useful for nuclear hyperfine studies, and will probably be intense enough to make long baseline interferometers[19] and determination of x-ray scattering phase shifts possible. The anticipated increase in intensity of SR sources makes schemes that now appear to be of marginal utility likely to be extremely practical in the long run.

Note added in proof. An important innovation, using an impedance-matched grazing incidence reflecting film to pass the nuclear resonant energies and reject the nonresonant radiation, has just been proposed.[27]

ACKNOWLEDGMENT

I thank Dr. Raju Raghavan for suggestions resulting from a reading of the manuscript.

References

1. R. L. Cohen, G. L. Miller, and K. W. West, *Phys. Rev. Lett.* **41**, 381 (1978).
2. E. J. Seppi and F. Boehm, *Phys. Rev.* **128**, 2334 (1962), and references therein.
3. F. R. Metzger, *Prog. Nucl. Phys.* **7**, 53 (1959).
4. P. J. Black and P. B. Moon, *Nature (London)*, **188**, 481 (1960).
5. *Applications of Mössbauer Spectroscopy*, R. L. Cohen (ed.), Academic Press, New York (1976).
6. S. L. Ruby, *J. Phys. (Paris), Colloq.* **6**, 209 (1974).
7. *Mössbauer Spectroscopy*, U. Gonser (ed.), pp. 43–47, Springer-Verlag, New York (1975).
8. R. V. Pound and G. A. Rebka, Jr., *Phys. Rev. Lett.* **4**, 337 (1960).
9. This equivalence has recently been established for microwave frequencies to ~200 ppm by R. C. Vessot, *Gen. Relat. Gravit.* **10**, 181 (1979).
10. P. J. Black and I. P. Duerdoth, *Proc. Phys. Soc. (London)* **84**, 169 (1964).
11. F. Parak, R. L. Mössbauer, U. Biebl, H. Formanek, and W. Hoppe, *Z. Phys.* **244**, 456 (1971), and F. Parak, R. L. Mössbauer, W. Hoppe, U. F. Thomanek, and D. Bade, *J. Phys. (Paris) Colloq.* **6**, 703 (1976).
12. P. J. Black, G. Longworth, and D. A. O'Connor, *Proc. Phys. Soc. (London)* **83**, 925 (1964).
13. This technique has been used for nuclear resonance fluorescence, see F. R. Metzger, *Nucl. Phys.* **27**, 612 (1961).
14. J. Christiansen, H.-E. Mahnke, E. Recknagel, D. Riegel, G. Weyer, and W. Witthuhn, *Phys. Rev. Lett.* **21**, 554 (1968).
15. E. Matthias, in *Hyperfine Structure and Nuclear Radiations*, E. Matthias and D. A. Shirley (eds.), p. 815 ff, North Holland, Amsterdam (1968).
16. Yu. Kagan, A. M. Afanasev, and I. P. Perstenev, *Sov. Phys. JETP* **27**, 819 (1968).
17. J. P. Hannon and G. T. Trammell, *Phys. Rev.* **186**, 306 (1969).
18. S. L. Ruby, in A.I.P. Conf. Proceedings No. 38, p. 50.
19. G. T. Trammell, J. P. Hannon, S. L. Ruby, P. Flinn, R. L. Mössbauer, and F. Parak, *AIP Conf. Proc.* **38**, 46.
20. S. L. Ruby and P. Flinn (unpublished).
21. A. N. Artemyev, V. A. Kabannik, Yu. N. Kazakov, G. N. Kulipanov, E. A. Meleshko, V. V. Sklyarevskii, A. N. Skrinsky, E. P. Stepanov, V. B. Khlestov, and A. I. Chechin, *Nucl. Instrum. Methods* **152**, 235 (1978).
22. R. L. Cohen, P. A. Flinn, E. Gerdau, J. P. Hannon, S. L. Ruby, and G. T. Trammell, discussion panel in *AIP Conf. Proc.* **38**, 140.
23. Yu. Kagan, A. M. Afanasev, and V. G. Kohn, *Phys. Lett.* **68**, 339 (1978).
24. G. T. Trammel and J. P. Hannon, *Phys. Rev.* **B18**, 165 (1978).
25. E. P. Stepanov, A. N. Artemyev, I. P. Perstenev, V. V. Sklyarevskii, and V. I. Smirnov, *Sov. Phys. JETP* **39**, 562 (1974).
26. For a bibliography, see E. Grünbaum, in *Epitaxial Growth, Part B*, p. 611 ff, Academic Press, New York (1975).
27. J. P. Hannon, G. T. Trammell, M. Mueller, E. Gerdau, H. Winkler, and R. Rüffer, *Phys. Rev. Lett.* **43**, 636 (1979).

21

Wiggler Systems as Sources of Electromagnetic Radiation

JAMES E. SPENCER and HERMAN WINICK

1. Introduction

Wiggler systems offer exciting prospects for providing synchrotron radiation (SR) that is both more intense and different in spectral distribution than can be obtained from the normal bending magnets of a synchrotron or storage ring. If one defines a wiggler as any array of electromagnetic fields whose effect on a beam is to produce transverse accelerations with no overall deflection or displacement, then there are many conceivable systems of interest. Because some of these may provide significant improvements and in some cases unique capabilities, one expects to see such devices play an increasingly important role in applications ranging from basic research to industrial technology to extending machine capabilities for high-energy physics research. This introduction discusses the more important concepts for the general reader.

Although wigglers have been used in circular machines for a variety of purposes (e.g., as damping magnets for high-energy physics machines and as models for radiation sources), only recently has a wiggler magnet been designed and used as an intense radiation source for experiments. Since March, 1979, a wiggler magnet has been in routine use in the storage ring SPEAR for synchrotron radiation experiments at the Stanford Synchrotron Radiation Laboratory (SSRL). This magnet (described in detail later in this chapter) produces the most intense beam of synchrotron radiation at SSRL or anywhere in the world and has significantly extended research capability at SSRL. Clearly, more powerful wigglers and specialized undulators can be built that will further expand research possibilities.

JAMES E. SPENCER ● Stanford Linear Accelerator Center, PEP Group, Stanford University, Stanford, California 94305. HERMAN WINICK ● Stanford Synchrotron Radiation Laboratory, Stanford Linear Accelerator Center, P.O. Box 4349, Bin 69, Stanford, California 94305.

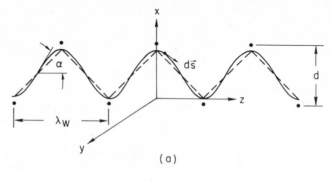

(a)

Figure 1. Schematic trajectories for (a) a sinusoidal planar wiggler and (b) a helical wiggler. If the dots are lattice sites in a monatomic cubic crystal with plane spacing d, the planar trajectory might also represent positive charged particle channeling and the axial trajectory the channeling of negative particles but greatly exaggerated because $\lambda_w \gg d$ (see Reference 25). The angle the trajectory makes with the longitudinal axis is θ and θ_{max} is designated as α. The dashed line represents a sawtooth approximation to the solid trajectory resulting from the equivalent transversely directed impulse function.

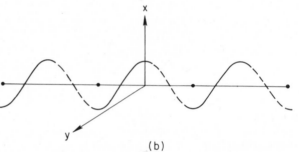

(b)

Although we discuss several rather esoteric possibilities based on this general definition of a wiggler, the main concern will be the special case of pure magnetic fields (in the lab system) that are transverse to the primary beam direction (TM) as produced by either a series of alternating polarity, dipole magnets or a bifilar, helically wound, air-core magnet. These are the two cases of most interest here because they can be inserted directly into the straight section of an existing synchrotron or storage ring and thereby offer increased capabilities in a cost effective manner compared to the usual means of producing SR photons in the ring bending magnets.[1] Figure 1 shows the corresponding particle trajectories, which might have also come about through particle channeling or bremsstrahlung in a crystal. One of the earliest and best discussions of the characteristics of conventional SR produced in uniform magnetic fields is given by Schwinger[2] and some tests of the theoretical predictions are cited in Reference 1. See also Chapter 2 of this book.

In principle, any wiggler has two extremes of operation. With sufficient field range, the same device can be operated as a standard wiggler or as an undulator (also called interference wiggler). Because practical considerations often constrain operation to one mode or the other, it is common to speak of wigglers as being either standard wigglers, which will produce essentially smooth spectral distributions similar to normal SR, or as being undulators, which produce much lower-energy spectra with more structure. Both modes provide control of end-point energy without recourse to changing the stored beam energy, although the range is more restricted for the undulator. Such characteristics should simplify subsequent monochromatization of the radiation and should generally improve the signal-to-noise ratio compared to present SR sources.

The critical energy of SR produced in a ring bending magnet or standard magnetic wiggler is given by[2]

$$\varepsilon_c(\text{keV}) = 2.218E(\text{GeV})^3/\rho(\text{m}) = 0.0665B(\text{kG})E(\text{GeV})^2 \qquad (1)$$

For a variety of technical and economic reasons, the field in the ring magnets is usually limited to $\lesssim 10$–12 kG, whereas wiggler fields of 20 kG or so are possible using conventional iron-core magnets and fields of 50 kG or more can be obtained with superconducting magnets. Thus, standard wigglers can shift the spectrum of relatively low-energy machines into the hard x-ray region ($\varepsilon_c > 10$ keV). This would expand the research possibilities for such machines yielding storage rings with fairly low electron energies ($E < 1$ GeV) in which the normal SR would be used for UV and soft x rays ($\varepsilon_c < 1$ keV) while wigglers could be tuned to produce harder x rays. For instance, in an 0.8-GeV storage ring equipped with a 50-kG superconducting wiggler, the critical energy is 2.1 keV [from equation (1)]. Using the guideline that good flux is available out to five times the critical energy, such a machine would be capable of x-ray diffraction work plus EXAFS studies of all elements up to copper. Furthermore, the signal to background in the experimental areas would generally be better, the operating costs would be lower, and the initial capital equipment costs would be lower compared to higher-energy machines. To achieve the same maximum photon energy with an undulator (i.e., wavelengths on the order of one angstrom) would require machine energies of 5 GeV or higher. While this would improve spectral brightness by providing more directed and monochromatic radiation, such a high-energy dedicated machine would be quite costly, particularly since a high-energy injector would also be needed.

The simplest standard wiggler produces one full oscillation or wiggle of the beam. We will call this a 1λ standard wiggler. Such devices with superconducting coils, which will produce 50-kG fields, four times higher than those typically used in the ring, are being planned for the SRS at Daresbury, the Photon Factory at KEK, and other rings. At SPEAR, with a ring bending radius $\rho = 12.7$ m, it follows from equation (1) that $\varepsilon_c^W/\varepsilon_c^B = 0.38B(\text{kG})/E(\text{GeV})$, i.e., a nearly 13-fold increase in critical energy at 1.5 GeV for such a wiggler. This is equivalent to operating SPEAR at 3.5 GeV, which is near its upper limit.

Increasing the number of oscillations N, for constant wiggler wavelength λ_W, can increase the photon flux by N. Multiperiod magnetic wigglers are being built for several rings. SPEAR now has a conventional (i.e., nonsuperconducting) magnet operating up to 18 kG that produces three oscillations of the beam ($3\lambda_W$), enhancing the intensity and shifting the spectrum to higher photon energy. ADONE is also building a conventional 3λ wiggler. A superconducting system operating up to 35 kG and producing 10 oscillations is under construction at the VEPP-3 ring in Novosibirsk. Another 10λ system has recently been installed in the Tomsk synchrotron and a 24λ system is now under construction for ACO in Orsay. The Tomsk and ACO wigglers will operate up to about 4 kG and run primarily as undulators. The ADONE, SPEAR, and VEPP-3 devices will be operated primarily as standard wigglers.

The first discussions of undulators[3, 4] were contemporary with the development of the maser (*microwave amplification by stimulated emission of radiation*).

The only experimental study until recently was in the submicrowave and millimeter range.[5] With the development of infrared and visible lasers in the early 1960s, there was a loss of interest in the use of undulators in this spectral range. However, the rapid development of storage ring technology and the acceptance of SR as a research tool, together with the inability to obtain significantly shorter wavelengths [i.e., to take the next significant step and produce an x- or gamma-ray laser (graser)], has rekindled interest in wigglers.

The distinction between standard wigglers and undulators is important since one expects a very different response when the field level is varied sufficiently. The wavelength or pitch of the wiggler field λ_w determines the photon wavelengths λ at which structure might appear in an otherwise smooth SR spectrum (see Figure 2). For undulators, peaks are predicted[6] at wavelengths given by

$$\lambda_n \simeq (\lambda_w/2\gamma^2)[1 + cK^2 + (\gamma\theta)^2]/n \equiv (\lambda_w/2\gamma^2)/\xi \qquad c = \begin{cases} \tfrac{1}{2} \text{ (planar)} \\ 1 \text{ (helical)} \end{cases} \qquad (2)$$

where γ is the particle energy in units of rest mass and K is called the field index because the wavelength of the wiggler is not considered variable. At $\theta = 0$ only odd n are allowed for planar devices. For helical undulators (at $\theta = 0$) only the fundamental is allowed. The radiation is linearly polarized for planar wigglers and circularly polarized for helical wigglers. K is defined (and evaluated for electrons) as

$$K = (q/\beta m_0 c^2)\int_{z=0}^{\lambda_w/4} B(s)\,ds \simeq (e/2m_0 c^2)B_1 l_{\text{eff}} = [B_1(\text{G})/3409]l_{\text{eff}}(\text{cm}) \qquad (3)$$

where B_1 is the peak field on the axis and l_{eff} is the effective magnetic length of a pole or half period. For simple sinusoidal fields $l_{\text{eff}} = \lambda_w/\pi = 2/k_w$ and the power radiated in the fundamental is a maximum when $cK^2 \simeq 0.5$.[6] This is at fairly low fields as seen by setting $\theta = 0$, giving $\lambda = 0.75\lambda_w/\gamma^2$ and $B_1(\text{T}) \simeq 1/\lambda_w$ (cm). For presently attainable magnetic systems it will be seen that $\lambda_w > 1$ cm, i.e., $B_1 < 10$ kG, so that some simple and elegant possibilities such as wigglers formed of permanent magnetic materials become feasible. In the weak-field limit ($K \ll 1$), the "critical energy" is independent of field strength and is given by

$$\varepsilon_1(\text{keV}) = 0.9496E(\text{GeV})^2/\lambda_w(\text{cm}) \qquad (4)$$

This defines the undulator regime in which harmonics are minimal and $\lambda_w B_1 \ll 10$ kG cm, whereas $K \gg 1$ or $\lambda_w B_1 \gg 10$ kG cm implies operation as a standard wiggler and can be called the strong-field limit.

The half-width of the peaks for an $N\lambda$ undulator can be as small as $\delta\lambda/\lambda_n \sim 1/nN$ and for maximal power in the angle integrated fundamental ($K \sim 1$ for planar wigglers and $1/\sqrt{2}$ for helical wigglers) the half-width is $<25\%$.[6] Increasing the field level enhances and compresses the higher harmonics so that the observed spectrum eventually becomes white and structureless for typical beam variations and instrument resolutions. Such characteristics clearly offer interesting opportunities and challenges for producing tunable, quasimonochromatic radiation of very high brightness. Because of this, the term undulator radiation (UR) has been used to distinguish it from SR—so named because it was first observed in an electron synchrotron. SR has also been called betatron and cyclotron radiation as well as magnetic bremsstrahlung and Blewett radiation.[1]

Robinson[7] appears to have been the first to suggest the use of standard wigglers in the straight sections of synchrotrons to produce SR for experiments. Recently, Winick and Helm[8] have reviewed standard wigglers in the context of storage rings. The first experiments on undulator radiation were done by Motz and collaborators[5] in the early 1950s. Although several other groups[9–15] have studied wigglers in a variety of situations, only recently (starting in March, 1979) has there been experience with their routine use as radiation sources for experimental purposes. Except for Motz work on planar undulators, equation (2) has only recently been verified experimentally for helical wigglers.[12]

One reported use of standard wigglers was to redistribute damping rates in the Cambridge Electron Accelerator.[10] In this case wigglers with gradients (called damping magnets) enabled electron beams to be stored in the alternating gradient synchrotron. Normally radial betatron oscillations are antidamping in these machines. More recently a similar system has been employed to enable electron beams to be stored in VEPP-4 in Novosibirsk. This ring was originally designed and built as a proton–antiproton ring so that radiation damping was not important in the original design. Such examples clearly show the importance of wigglers for high-energy physics applications—especially when we realize that a fundamental limitation on achieving higher energies in machines like the electron synchrotron or storage ring is radiative energy loss. In such applications the wiggler is neither a harmonic frequency generator nor intensifier nor amplifier of SR but an ion optical device that provides local control of beam energy loss and affects damping, tunes, emittance, and other aspects of machine operation.

For some experiments employing SR (e.g., certain EXAFS or topography experiments), it is usually the transverse source size that is of most concern, i.e., the irradiance (photons/mm^2/s). This can be improved by increasing the number of periods N of the wiggler, the proper selection of its location in the storage ring, and by varying its field or wavelength. For other experiments (e.g., protein crystallography or photoemission) it is the radiance or brightness (photons/mm^2/sr/s) that SR users want to optimize. In these cases, the gain due to increasing the number of periods, the field, or the wavelength may be partially offset by the increase in effective source length and emittance. Brightness is first optimized by properly choosing the location in the ring lattice where the wiggler is to be installed, e.g., where it causes the least increase in beam emittance or even decreases this quantity. This choice, as well as others associated with the application of wigglers to the problems of machine physics, is based on the calculation of synchrotron radiation integrals.[16, 17] We discuss this because machine physics applications seldom optimize SR for experimental uses.

For instance, wigglers have been suggested[18] and are being implemented on the 18-GeV colliding-beam storage ring PEP at SLAC as a means for improving luminosity as well as decreasing damping times to improve injection rates with decreasing beam energy. Because luminosity may be increased by purposely increasing the beam emittance in the beam–beam current limit, the best location for these functions is clearly not optimal for SR brightness.

It will be shown that both the energy spread and the emittance of a stored beam can either be increased or decreased by wigglers. In terms of the subsequent photon beam this implies the possibility of improved brightness or spectral brightness or both depending on the conditions. Undulators, which maximize the spectral radiance (radiance per unit wavelength), should also be useful for the measurement of particle beam energy, current, and emittance.

Motz[4] proposed using undulators to monitor electron beam energies and it has also been proposed[4, 19, 20] to use wigglers to detect or discriminate individual high-energy electrons from heavier particles by the presence or absence of photons. With the increasing energy γ of proton beams at CERN and the Fermi Laboratory (400 GeV) there is also a growing interest in applying the same techniques in use for electron beams[21] to monitor proton beams.[22, 23] Although the number of photons per unit time and solid angle goes as $(\gamma/m)^2$ (see Section 3), the wavelength is in the visible range according to equations (2) and (3) for 400-GeV protons and $\lambda_w = 20$ cm.

A related application concerns the use of electron beams having the same speed as the proton beam (or antiproton beam) to "cool" it or reduce its phase space.[24] The lepton beam would then become "hot" and rather than waste this beam having energies up to 200 MeV by dumping it, it would in turn be cooled in a low-energy electron (or positron) storage ring utilizing wigglers. Because there are no substitutes for quantities such as the radiance or spectral radiance in SR experiments and beam size or emittance in storage rings, we discuss this area in detail. We also discuss some practical design considerations and methods for calculating the effects on particle beams by using the SPEAR wiggler as an example.

A general approach to understanding the radiation expected from magnetic wigglers helps us consider other types of periodic electromagnetic fields. The special case of pure electric fields (in the lab) that are transverse to the primary beam direction (TE) are particularly interesting. The main problem lies in generating sufficient field strengths. This can be solved by using particles channeling along an atomic chain or between the planes of a crystal.[25] The field strengths and number of periods (N) can then be very high, but not easily variable. The subject can be developed in analogy with the TM case except for possible quantum effects arising from shorter wiggler wavelengths ($\lambda_w \gtrsim 1$ Å) and the fact that the atomic potential responsible for the field is not static but can absorb and reemit energy so that effects such as stimulated emission should be possible.[26] As with the TM case, helical trajectories are possible[27] that can either increase or decrease the bremsstrahlung background (from close collisions with individual atoms). Coherent bremsstrahlung by electron beams in crystals has also been observed.[28] This phenomenon is analogous to radiation in an undulator with short wavelength. Because there is ambiguity in the use of certain words like "coherence" and "interference," we discuss this subject in more detail below, including the conditions under which such radiation may be produced in storage rings.

In the limit that the particle velocity approaches that of light ($\beta \to 1/n$), the only differences between TM and TE fields is the direction of polarization of the radiation, which will be rotated by 90° relative to one another when the field directions are the same. Thus, if one could use both TE and TM fields that were independently variable, it would be possible to rotate the plane of polarization of the radiation. This would extend the experimental possibilities considerably. We discuss various ways to perform this rotation with pure TE and TM fields (in the lab system). Other interesting and closely related areas where experiments have been done include the free-electron laser,[12] Compton back scattering of laser photons,[29] and measurement of particle beam polarizations[30] with laser back scattering.

The various methods discussed here all have the potential of providing polarized beams of quasimonochromatic photons over a significant range of energies. Many of them, although still essentially undeveloped, may provide higher energy, intensity, directivity, monochromaticity, tunability, and polarization than obtainable from present SR sources or, in some cases, any source. We should note that although they do not all require a storage ring or synchrotron, these machines may prove most practical for many applications requiring particle energies of appreciable magnitude ($\gamma \gg 1$), e.g., those requiring high natural collimation, intensity, or photon energy. Exceptions would be particle channeling and coherent bremsstrahlung in crystals because of the large number of atoms along a linear chain required to make a single wavelength period.

In addition to serving as a low-energy radiation source, there are also many other potential applications in high-energy physics beyond those already mentioned such as the radiative self-polarization of particle beams via the SR emitted in a wiggler,[30] as well as induction of particle beam polarizations and associated tagging using an external radiation source like the laser. Perhaps the most important of all is the study of the fundamental properties of particles through coherent scattering of high-energy monochromatic radiation from the underlying basic constituents of the particles (quarks) when both they and the radiation fields are polarized (e.g., Bragg scattering from chromodynamic crystals).

Section 2 discusses typical characteristics for wigglers and compares them to lasers. Because there are so many different physical processes involved in the approaches discussed so far, we give some explicit examples to show the underlying physics. Perhaps the best insight into undulators is obtained by considering the Thomson scattering ($K \ll 1$), in the average rest frame of the charged particles, of the virtual photons from the externally applied field. We also establish operational criteria based on the dependence of the radiation spectra on parameters such as the type and specific form of the fields, the number of periods, the wavelength, and the field strengths.

2. General Characteristics of Wiggler Radiation

In this section and the next we discuss some general principles to elucidate the basic processes that control the characteristics of the radiation we are trying to produce. We can relate the usual SR spectrum (Chapter 2) to arbitrary field variations and configurations by an example in which a charge is given an arbitrary acceleration that results in a radiation field. This is related to normal SR, bremsstrahlung, and many of the processes discussed above. We also discuss some desirable characteristics of radiation and compare wigglers and lasers as sources. The most important characteristics are the time structure, energy and tunability, intensity, resolution, and degree of polarization and collimation because our basic goal is to maximize the radiant energy hitting a target area or volume with the desired energy, spectral bandwidth, and polarization. A normal SR source does this with considerable loss of efficiency because it is often necessary to monochromatize or eliminate those frequencies that are not in the desired spectral bandwidth or else to collimate it to eliminate or at least limit its divergence or size. Ideally, a source is tunable without such intermediate corrections.

First, let us suppose that a charge undergoes a collision of some form in which a transfer of momentum occurs. We assume that this transfer is rapid compared to any "macroscopic" phenomena or wave motion that is occurring, such as betatron motion or phase oscillation of the beam, so that it can be regarded as instantaneous. Furthermore, when such momentum transfers occur spontaneously and at random relative to the phase of the various particle motions, such as they may when they are due to normal SR, their overall effect will be to disrupt the particle motions. However, whether they result in beam blowup or damping depends on several factors such as where they occur in the ring lattice.

The situation we consider first is shown schematically in Figure 1 by the dashed line. Except for enhancement of higher harmonics, normal SR might correspond to one such spike with a repeat period T_0, much longer than shown in the figure, corresponding to the fundamental period of the ring and not the wiggler. For instance, at the SPEAR storage ring, the rf accelerates the beam at 358 MHz, the 280th harmonic of the 1.28-MHz orbital frequency. In one-bunch operation the pulse separation is 780 ns. The duration and amplitude of the pulse are correlated with the bunch length, energy, and population of the electron (or positron) bunch, which are themselves correlated, i.e., for any given electron energy there can be significant variations in the pulse duration with electron bunch current via density effects and beam–beam effects, not to mention the machine operating point (nearness to instabilities and the like). One also expects the pulse duration and amplitude to be modulated about their time averaged values by synchrotron oscillations of the bunch with a frequency of about 25 kHz. Such variations can easily be a factor of 2 or more and are one reason why SR users prefer dedicated machines operated at fixed energies.

Wigglers clearly help this situation, e.g., in the case of SPEAR and the SSRL wiggler there is virtually no redistribution of damping rates but there is increased damping so that the ability to "top up" or maintain a more nearly constant beam current is improved. Nonetheless, there are many potential problems that can introduce jitter into the SR source and these must be shown to be negligible, acceptable, or correctable. That this can be done to provide a source that will in any way be comparable to a laser is not at all obvious.

2.1. Directionality

Directionality is a particularly important characteristic of any radiation source. For example, antenna design usually seeks to achieve high directed power. One of the significant advances realized with lasers was their ability to amplify at previously unattainable wavelengths with high directionality (sometimes called space coherence) that was limited primarily by diffraction at their mirrors

$$\delta\theta_L \sim \lambda/D \sim 10^{-4} \text{ rad} \tag{5}$$

where D is the diameter of the mirror (typically 1 cm) and λ is assumed to be 1 μ, which is typical of gas lasers. These cover the widest wavelength range (0.1 μm $\lesssim$ $\lambda \lesssim 300$ μm), i.e., down into the VUV. If you compare this to a microwave antenna you get

$$\delta\theta_M \sim \lambda/D \sim 1 \text{ rad} \tag{6}$$

Because microwaves have 10 000 times greater wavelength it is necessary to increase the antenna size to 100 m to get the same directionality or degree of collimation for 1-cm radiation.

We have already seen from Chapter 2 that for normal SR, the radiation is concentrated in a narrow cone with opening angle

$$\delta\theta_w = \langle \theta^2 \rangle^{1/2} = 1/\gamma = m_0 c^2/E = 2 \times 10^{-4} \text{ rad} \tag{7}$$

for a 2.5-GeV electron beam—ignoring the divergence of the stored beam, which varies with the location in the ring. Atomic bremsstrahlung and Compton scattering of laser photons by relativistic, charged particles have the same characteristic directionality or rms beam divergence (but not the same angular distribution or frequency spectrum), which is a general characteristic that results from using high-energy particle beams, i.e., $\gamma \gg 1$. Ironically, the use of high stored beam energies leads to higher radiation and thicker shield walls resulting in experimental areas that are far from the effective source location. This directionality property is then often necessary just to take full advantage of the desired radiation. Furthermore, if the optics conserves brightness, this implies that the effective source emittance must also be reasonably small to obtain small spot sizes on the target without wasting radiation by using collimators.

In general, a primary difference between standard wigglers and undulators is the amount of instrumental divergence they induce compared to the instantaneous, natural divergence of SR $(1/\gamma)$. If the maximum angular divergence of the reference trajectory due to the wiggler is α, then we can make the following characterization for simple TM and TE wigglers:

$$(\alpha/\delta\theta_w)_{\text{Mag}} = \alpha\gamma = \tfrac{1}{2}(q/\beta m_0 c^2) \int B \, ds \simeq (q/m_0 c^2)\bar{B}\lambda_w/4$$

$$(\alpha/\delta\theta_w)_{\text{Elect}} = \tfrac{1}{2}(q/\beta c) \int (E/p_z) \, ds \simeq (q/m_0 c^2)\bar{E}\lambda_w/4 \tag{8}$$

where $\bar{B}$ and $\bar{E}$ are the average fields experienced by the reference trajectory while passing through half a period of the field. Note that this expression is no different from equation (3), which essentially defines l_{eff} as opposed to $\bar{B}$. It should also be noted that $\alpha\gamma$ is independent of γ and depends only on the average field and wavelength for any particle of charge q and mass m. One can then consider two cases:

$$\alpha\gamma \gg 1 \Rightarrow \text{standard wiggler}$$

$$\alpha\gamma \ll 1 \Rightarrow \text{undulator} \tag{9}$$

i.e., undulator radiation is much more directional than radiation from standard wigglers or normal SR. This has interesting implications for the time and frequency spectrum to be expected from these systems as well as the intensities. Furthermore, $K \ll 1$ for protons and other heavy particles in most wigglers of practical interest.

2.2. Time and Frequency Structure

The shortest laser pulses obtained without collimation are about 1/4 ps, which is equivalent to a pulse length of 0.075 mm. The corresponding situation for SR can be computed from the bunch length of the particle beam. Although this depends on a number of factors, the shortest pulses produced by synchrotrons or storage rings are about 50 ps (see Chapter 9). Further reduction may be possible by using higher radio frequencies (or voltages) but other considerations such as higher-order mode losses due to vacuum chamber discontinuities may limit the current attainable as bunches get shorter. This is an important area of ongoing research that will hopefully provide scaling relations useful for extrapolating from the present generation of machines to the next, with improved characteristics in this area. In SPEAR, the bunch length $(2\sigma_z)$ typically varies from ~ 1 to 6 cm. Bunch lengthening in SPEAR and other rings has been observed for some time,[31] although the causes are still not well enough understood to be predictable or controllable. Considerable effort has gone into this area for the new PEP storage ring since the bunch charge is larger and the length is comparable but smaller than in SPEAR.

Lasers cannot only produce picosecond pulses but these can have repetition rates as high as several hundred MHz. Typical rates for normal SR are on the order of a MHz but multibunch operation or operation with standard wigglers can increase this rate by roughly the ratio of half their wavelength to the ring circumference $2L/\lambda_W$, i.e., by several orders of magnitude. Thus, it appears that the most important area where improvements would be useful is in bunch length and bunch current, although it is unlikely that SR or UR will ever match lasers in the picosecond domain at wavelengths in the near ultraviolet and above. Although there is a good deal of laser research using wavelengths in the vacuum ultraviolet and below, lasers are not presently competitive with synchrotron radiation at these wavelengths, except for specific applications requiring ultrafast timing.

2.3. Monochromaticity

The monochromaticity of a source or the resolution of a detector (sometimes referred to as time coherence) is possibly the single most important feature and is defined as

$$\delta = \left| \frac{\delta\lambda}{\lambda} \right| = \left| \frac{\delta v}{v} \right| = \left| \frac{\delta\varepsilon}{\varepsilon} \right| \tag{10}$$

δ should be as small as possible and the quality factor Q or resolving power, which is the inverse of this quantity, should be as large as practical. In circuit design, a

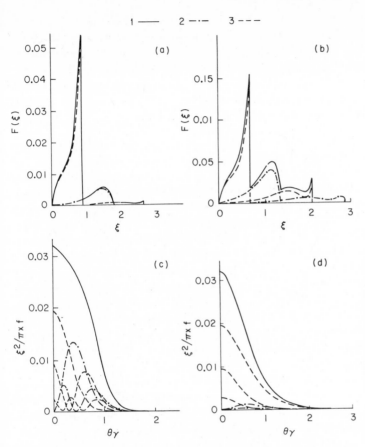

Figure 2. Wiggler angular and frequency spectra as a function of field index K and harmonic number n. Solid lines (1) represent the summed radiation intensity, dot–dashed lines (2) represent the intensity of the even harmonics, and dashed lines (3) represent the intensity of the odd harmonics. Figure 2a has $K = 1/2$, Figure 2b has $K = 1$, i.e., optimal power in the fundamental. Figure 2c has $K = 1$ and gives the angular distribution in the bending (median) plane of the undulator, and Figure 2d is in the vertical plane orthogonal to that of Figure 2c. θ is the angle the radiation makes with the longitudinal axis of the undulator. [See equation (2).]

small Q or large bandwidth is often the goal because this is understood as the acceptance of the device or its ability to transmit an arbitrary input free of distortion. The corresponding quantity here is called energy range (Section 2.4). Whether one wants a high Q ultimately depends on the experiment, i.e., whether one is more concerned about the time or frequency spread, which are reciprocally related. A normal SR source is monochromatized using gratings or crystals to obtain $\delta \sim 10^{-3}$ but with considerable loss of intensity since one essentially loses intensity in direct proportion to the improved resolution by eliminating everything not in the desired bandwidth.

In an undulator operated to maximize the power in the fundamental ($K \simeq 1$), $\delta_{N=\infty} \lesssim 0.25$. This resolution improves with decreasing K and/or collimation[6] as shown in Figure 2. However, the energy and emittance of the particle beam must be folded into this as well as imperfections and fluctuations in the wiggler accelerating field, not to mention possible fluctuations in the ring, which act to increase the natural energy spread and emittance predicted from typical lattice calculations. Although specific calculations should be done for the particular lattice and wiggler involved, a rough guide is to limit the beam divergence to $< \alpha$ and the beam energy spread to <0.1 based on the usual definitions. Finally, it should be pointed out that the number of periods N should generally be as large as possible or this will also result in broadening. In a laser, it generally becomes more and more difficult to achieve a high Q with increasing energy because the boundary conditions for amplification in a laser are $n\lambda/2 = L$, where L is the cavity length or separation of the mirrors, so that the system becomes increasingly intolerant to imperfections with the result that higher modes and instability generally result.

2.4. Tunability and Energy Range

From the standpoint of tunability, we again use the laser as a reference. In that case there are only a finite number of atomic and molecular systems capable of stimulated emission and their frequencies are not easily varied by external parameters such as the field strength as in the case of the wiggler. Small variations in λ have been achieved with changes in temperature and up to $3\lambda_{min}$ has been obtained with pressure variations from 1 to 14,000 atm. The most promising variable wavelength lasers, however, employ liquids (organic dyes) as the active medium and use what is known as a parametric effect. Two frequencies ω_1 and ω_2 are employed to achieve their sum $\omega_1 + \omega_2$ or difference. By varying both with temperature, one then gets tunability but only over a rather limited range.

Frequency doubling[32] is the effect of most interest to us because it can be done with a single intense beam and helps illustrate how wigglers work. In wigglers, the fundamental is more easily varied but its natural resolution is nowhere as good. On the other hand, second-harmonic generation appears to be about all that is currently practical with lasers whereas wigglers can generate several harmonics with reasonable intensity before the spectrum becomes significantly washed out. For example, for $K = 1$, the second harmonic is about 40% of the first as shown in Figure 2. Of course, the best way to vary photon energy is to vary the particle beam energy or possibly the wavelength as discussed in Section 4.3, consistent with equations (2) and (3). Assuming a fixed field, permanent magnet undulator in SPEAR (energy range 1.4–4.0 GeV) implies an energy range of $8\varepsilon_{1\,min}$.

Spontaneous emission in the laser contributes noise and also sets a threshold limit. Since this threshold is proportional to ω^3 and to the linewidth of the spectral transition, it is clear that it places a significant limitation on the highest frequencies achievable with laser-type devices.[33] The so-called free electron laser,[12] which employs mirrors with a wiggler and an electron beam, is a hybrid device whose primary goal is to provide high-power, tunable, coherent radiation over the energy range $(\gamma_{max}/\gamma_{min})^2$ of the machine in which it is operated, i.e., its range is restricted by that of the machine (see Chapter 22).

2.5. Polarization

SR is, in general, elliptically polarized (see Chapter 2), whereas UR can be fully linearly or circularly polarized. Lasers can also produce polarized light that is both linearly and circularly polarized. Furthermore, the same laser can be made to emit linearly polarized light that can be rotated. We see comparatively little difference in this regard since wiggler systems could be designed[34] that could arbitrarily rotate the plane of polarization. The two methods are somewhat analogous. In a gas laser, two optical flats can be placed at an angle to the optical axis (Brewster angle) to protect the mirrors from discharges. These transmit the component polarized normal to their surface. Thus by rotating the Brewster windows one rotates the polarization vector of the plane waves. This technique is called polarization by reflection. Because of alignment problems and the frequencies involved, one also sees use made of the Faraday effect to rotate the polarization vector. In the wiggler scheme, one rotates the primary field direction and thereby achieves the same thing. Notice that mechanical rotation is significantly more difficult for the wiggler because of its size and use in an accelerator. This is discussed in Section 4.1.3.

3. Theoretical Considerations

3.1. Macroscopic Approach–Classical Field Equations

Of the many ways of approaching the overall subject, the most common one is to use Maxwell's equations or more specifically the Lienard–Wiechert potentials. The radiation fields are then related directly to the motion of the individual radiating particles and by inference to the accelerating fields. The response expected from any particular device embedded in a machine results from folding the phase space of the beam, determined by the combined system of storage ring and wiggler, over the measured or modeled field of the wiggler.

There are therefore essentially two steps. First, we ascertain the beam phase space of the machine with the wiggler and then we compute the actual radiation from the device by computing the radiation from a sample of particles averaged over the expected phase space. There are several shortcomings to this method that are obvious but unless we are especially interested in quantum effects, it is most useful for developing a good intuitive feeling for how a radiation field comes about. For instance, we did not say that a charge must undergo acceleration for field lines to "break away." Cherenkov radiation is a good example of how removal of the vacuum invalidates the rule that a uniformly moving charge does not radiate. Furthermore, it is possible to begin the problem with a radiation field present such as for a FEL. The allowed simplifying assumptions then depend on such things as the intensity of the fields, the energy of the particles, and their density.

The electric and magnetic fields E and B associated with an arbitrarily moving charge q in cgs Gaussian units are[35]:

$$\mathbf{E}(\zeta, t) = \frac{(q/R)}{(1 - \hat{\mathbf{n}} \cdot \boldsymbol{\beta})^3} \left\{ \frac{(\hat{\mathbf{n}} - \boldsymbol{\beta})}{R\gamma^2} + \hat{\mathbf{n}} \times [(\hat{\mathbf{n}} - \boldsymbol{\beta}) \times \dot{\boldsymbol{\beta}}]/c \right\}$$

$$\mathbf{B}(\zeta, t) = \hat{\mathbf{n}} \times \mathbf{E} = \frac{(q/R)}{(1 - \hat{\mathbf{n}} \cdot \boldsymbol{\beta})^3} \left\{ -\frac{\hat{\mathbf{n}} \times \boldsymbol{\beta}}{R\gamma^2} + \hat{\mathbf{n}} \times \hat{\mathbf{n}} \times [(\hat{\mathbf{n}} - \boldsymbol{\beta}) \times \dot{\boldsymbol{\beta}}]/c \right\} \quad (11)$$

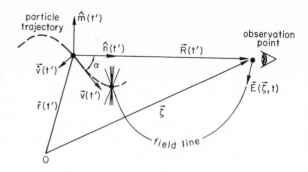

Figure 3. Definition of variables for the radiation process in terms of the retarded time t' and observation time t. Field line schematic shows Lorentz contraction along the direction of the velocity at time t and the electric field at the observer—including the velocity or near field. $\hat{m}$ and $\hat{n}$ are instantaneous orthogonal unit vectors defined so that $\hat{n} \times \hat{m}$ is in the direction of z.

where all the variables are shown in Figure 3 and are to be evaluated at the actual emission time $t' = t - R(t')/c$, which is retarded with respect to the observation time t by R/c. The charge q is in esu. We have included the so-called "velocity" field for completeness because there are some associated effects of potential interest such as higher-order mode losses that can limit the achievable currents in storage rings. If we project the equations for E and B onto $\hat{n}$ and $\hat{m}$ we have the instantaneous longitudinal and transverse fields, i.e., polarizations at (ζ, t):

$$E_n = (q/R^2\gamma^2)/(1 - \hat{n} \cdot \boldsymbol{\beta})^2$$

$$E_m = \frac{(q/R)}{(1 - \hat{n} \cdot \boldsymbol{\beta})^3} \left[\frac{\beta \sin \alpha}{R\gamma^2} + \frac{1}{c} (\beta \sin \alpha \, \hat{n} \cdot \boldsymbol{\beta}) - \frac{1}{c} (1 - \beta \cos \alpha)(\hat{m} \cdot \dot{\boldsymbol{\beta}}) \right] \quad (12)$$

For low values of β the field is dominated by E_n near the charge (Coulomb field) but eventually by the transverse field further out. If the motion is roughly along the direction of observation, then as $\beta \to 1$, E_m will come to dominate E_n regardless of $|R|$ via the extra factor of $1 - \hat{n} \cdot \boldsymbol{\beta}$, which is just $dt/dt' = 1 - \hat{n} \cdot \boldsymbol{\beta}$. The energy radiated into the solid angle $d\Omega$ is given by computing the Poynting vector

$$\mathbf{S}(\zeta, t) = \frac{c}{4\pi} \mathbf{E} \times \mathbf{B} = \frac{c}{4\pi} E^2 \hat{n} - (\mathbf{E} \times \hat{n})\mathbf{E} \quad (13)$$

which is the instantaneous energy flux at the point (ζ, t).

Since the power radiated by the charge is independent of the observation frame, one can transform to the instantaneous proper frame of the particle to compute it:

$$P = \int |\mathbf{S}| R^2 \, d\Omega = P' \quad (14)$$

Lienards's expression[35] for the final result is relevant here,

$$P = \tfrac{2}{3}(q^2/c)\gamma^6[\dot{\boldsymbol{\beta}}^2 - (\boldsymbol{\beta} \times \dot{\boldsymbol{\beta}})^2] = \tfrac{2}{3}(q^2/c)\gamma^6(\dot{\beta}_\parallel^2 + \dot{\beta}_\perp^2/\gamma^2) \quad (15)$$

but requires care in interpreting. Using $\dot{p}_\perp = \gamma m \dot{v}_\perp$ and

$$\dot{p}_\parallel = \dot{\varepsilon}/\beta c = mc\dot{\gamma}/\beta = mc\gamma^3 \dot{\beta}_\parallel \quad (16)$$

we get

$$P = \tfrac{2}{3}(r_e c/\varepsilon_0)(\dot{p}_\parallel^2 + \gamma^2 \dot{p}_\perp^2) \quad (17)$$

where r_e is the classical electron radius and ε_0 is the particle rest energy mc^2. For equal forces, there will be γ^2 more power radiated for the transverse acceleration than when the acceleration is parallel to the velocity. For purely circular motion, which produces SR in the uniform magnetic field B of the ring bending magnets $\dot{\beta}_\parallel = 0$ and $\dot{p}_\perp = q\beta B$, the power is

$$P_\perp = \tfrac{2}{3} r_e^2 c \beta^2 \gamma^2 B^2 = \tfrac{2}{3} q^2 c \beta^4 \gamma^4 / \rho^2 \tag{18}$$

where the last expression is the one given in Chapter 2. Also, the energy lost per turn per charge to SR is

$$U = \oint P_\perp \, ds / \beta c = \tfrac{2}{3} r_e^2 \beta \gamma^2 \oint B^2 \, ds \tag{19}$$

$$U \simeq \tfrac{2}{3} r_e^2 \beta \gamma^2 \sum B_{1i}^2 l_{\text{eff}}^i \propto I_2$$

where l_{eff} is the effective length of the dipole and B_1 is its central field. I_2 is a synchrotron integral defined in Section 4.2.2. To obtain the expression given in Chapter 2 we assume a single type of sharp-cutoff dipole having B, ρ, and l such that the total bend angle per magnet is $\psi = l/\rho$

$$U = \tfrac{2}{3} r_e^2 \beta \gamma^2 B^2 \rho \sum \psi_i = \frac{4\pi}{3} r_e \varepsilon_0 \beta^3 \gamma^4 / \rho. \tag{20}$$

One can use the same expressions for wigglers. If we want to maximize the SR power from a wiggler, we would first make B as large as practicable and next increase the length l. However, if we want to use the SR we may want to restrict the pulse train or divergence at the experimental area to its natural size of $\theta = 1/\gamma$, i.e., $l = \rho/\gamma$. It will be shown that this is possible for undulators but not for standard wigglers.

With these equations and a specification (or measurement) of the externally applied fields one can solve the Lorentz force equations for differing initial conditions on the particle motions to determine the expected radiation fields. Computer ray-tracing codes exist that can solve these equations for a complete storage ring or single device such as a wiggler with an arbitrary number of multipole components in the transverse and/or longitudinal directions. A simple but relevant example illustrating the classical approach would be Thomson scattering. The next section outlines another approach followed by some examples for which analytic solutions exist. The Hamiltonian approach is illustrated by again using the example of Thomson scattering.

3.2. Microscopic Approach–Quantization of the Field

To quantize the electromagnetic field,[36] we first write the relativistic equation of motion of a charged particle in an electromagnetic field in Hamiltonian form, i.e., the total energy as a function of any canonical variables (p, q) such as momenta and coordinates:

$$H\Psi = ih\dot{\Psi} = (H_{rad} + H_{part} + H_{int})\Psi$$

$$H_{rad} = \frac{1}{8\pi}(E^2 + B^2)\,dv, \qquad H_{part} + H_{int} = \sum_i H_i \qquad (21)$$

$$H_i(p, q) = e_i\Phi(\mathbf{r}_i) + \{c^2[\mathbf{p}_i - (e_i/c)\mathbf{A}(\mathbf{r}_i)]^2 + \mu_i^2\}^{1/2}$$

where $\mu_i = m_i c^2$ and $A_\mu = (\mathbf{A}, i\Phi/c)$ is the external field from the magnets, atoms, or lasers as well as the fields produced by the charges themselves. If the momentum is small compared to μ_i then

$$H_{0i} - \mu_i = [\mathbf{p}_{0i} - (e_i/c)\mathbf{A}_0(\mathbf{r}_{0i})]^2/(2m_i) + e_i\Phi(\mathbf{r}_{0i}) + V \qquad (22)$$

Notice that we have used the subscript denoting the Hamiltonian in the "rest" frame because this assumption is valid for an undulator. For a pure electrostatic field ($\mathbf{A} = 0$) this gives the familiar nonrelativistic expression for the energy. Neither H nor H_i includes interaction between particles so far unless we either add another term such as V with subscripts ij, ijk, etc., which then gives coupled equations. However, if we are interested in beam dynamics, such as coherent effects within a beam bunch or various excitation modes in a laser medium, crystal lattice, atom, or even an "elementary" particle, we must consider such terms. It would also include such things as the beam–beam interactions in storage rings. A covariant description (eg., choice of gauge) is clearly important here, since freedom to choose certain reference systems can significantly simplify things. It is useful to separate the external fields from those produced by the particles themselves since these are independent of the positions of the particles.

The quantization of the field is now the only thing remaining to be done before solving the resulting coupled equations. This process can be simple or complex depending on the fields. In most instances, Coulomb gauge (div $\mathbf{A} = 0$) is sufficient (noncovariant) since we are dealing with transverse fields in the average rest frame of the particles,

$$\mathbf{A}(\mathbf{r}, t) = \frac{(2\pi hc)^{1/2}}{L^{3/2}} \sum_k \sum_{s=1}^{2} k^{-1/2}[\hat{e}_s(\mathbf{k})a_{ks}e^{i(\mathbf{k}\cdot\mathbf{r} - \omega t)} + \hat{e}_s^*(\mathbf{k})a_{ks}^\dagger e^{-i(\mathbf{k}\cdot\mathbf{r} - \omega t)}] \qquad (23)$$

where the allowed values of $k_n = 2\pi n/L$ and $L(= \lambda_w)$ is the length of the Born periodicity cube. a_{ks} is an annihilation operator for a photon of momentum $\hbar k$ and polarization s, and $a_{ks}^\dagger$ is the corresponding creation operator. The polarization directions are transverse to the photon direction and to one another. For a bending magnet or planar magnetic wiggler, a natural direction is along the primary field direction, i.e., perpendicular to the dispersion plane (called the π component). The other direction is transverse to this (called the σ component) and it will be the stronger component of polarization.

Remembering that equation (23) can have a number of components due to fields other than the static wiggler field in the laboratory frame (e.g., external pump fields in the FEL) we can insert equation (23) into (22) and (21), which leads to essentially three terms. H_0 is taken to be the total energy of particles and fields when in isolation, H_1 is the interaction between particles and fields ($\mathbf{p}_{0i} \cdot \hat{e}_{ks}$), and

H_2 is the coupling between modes resulting from that interaction $(\hat{\mathbf{e}} \cdot \hat{\mathbf{e}}')$. This last term is relevant for Thomson and Compton scattering, which are important here. Although we do not go into detail, the reader can distinguish the difference between spontaneous and stimulated emission, which occur via the term H_1.

4. Fundamentals of Operation

We have discussed various characteristics of radiation from accelerated charges or, what is essentially the same thing, the scattering of radiation from charged particles. Hamiltonians and a quantum-mechanical description were discussed because they allow one to go beyond the simpler scattering problems to better understand different processes by which radiation can be produced. All of this should help one assess the characteristics of different instruments as well as tailor them to specific needs. In this section, we survey some systems for which analytic solutions have been given (e.g., see References 6, 37, 38, and 39). As we have already seen, because TM and TE fields produce accelerations that are transverse to the direction of motion of the beam centroid, they are more efficient for producing radiation than longitudinal accelerations of the same magnitude (Section 3.1). This difference is one reason why solenoidal fields have not been considered for wigglers, although they have been used in the cyclotron maser. Consequently we consider only "transverse" fields although, as shown below, this is an idealization disallowed by Laplace's equation.

4.1. The Infinite Transverse Wiggler

We begin by assuming a single, ultrarelativistic bunch of particles traveling along the axis of an infinite wiggler $(\hat{\mathbf{z}})$, i.e., $N = \infty$. Figure 1 shows the coordinate conventions. It is useful to transform to a coordinate system in which the reference trajectory for the beam bunch (centroid) is, on the average, at rest in the wiggler. If the centroid of the bunch of particles (charge e and mass m) has velocity βc, then some of the important quantities in this "rest" frame (subscript 0) are related to those in the lab by the Lorentz transformations:

$$\mathbf{E}_0 = \gamma(\mathbf{E} + \boldsymbol{\beta} \times \mathbf{B}) - (\gamma - 1)\hat{\boldsymbol{\beta}}\hat{\boldsymbol{\beta}} \cdot \mathbf{E}$$

$$\mathbf{B}_0 = \gamma(\mathbf{B} - \boldsymbol{\beta} \times \mathbf{E}) - (\gamma - 1)\hat{\boldsymbol{\beta}}\hat{\boldsymbol{\beta}} \cdot \mathbf{B}$$

$$\mathbf{A}_0 = \gamma(\mathbf{A} - \boldsymbol{\beta}\phi) + (\gamma - 1)\hat{\boldsymbol{\beta}} \times (\hat{\boldsymbol{\beta}} \times \mathbf{A}) \qquad (24)$$

$$\phi_0 = \gamma(\phi - \boldsymbol{\beta} \cdot \mathbf{A})$$

$$\mathbf{r}_0 = \gamma(\mathbf{r} - \boldsymbol{\beta}ct) + (\gamma - 1)\hat{\boldsymbol{\beta}} \times (\hat{\boldsymbol{\beta}} \times \mathbf{r})$$

$$t_0 = \gamma(t - \boldsymbol{\beta} \cdot \mathbf{r}/c)$$

The inverse transformations are obtained by reversing the sign of β, $\beta\gamma = \gamma v/c = (\gamma^2 - 1) \sim \gamma[1 - 1/(2\gamma^2)]$, and interchanging subscripts. For any particle i in the bunch, the projection of position and velocity on the z axis is

$z_i = vt + \delta z_i$ and $v_{zi} = v + \delta v_{zi}$ so that its position and velocity in the rest frame will be $z_{0i} = \gamma(z_i - \beta ct)$ and $v_{0zi} = dz_{0i}/dt_0$, i.e., we no longer have time as an independent variable. However, we can define a Lorentz invariant proper time by $d\tau = dt/\gamma$, which is the corresponding interval for an observer moving with speed β along the positive z axis. This is not the proper time for any of the particles until we go to the undulator limit. Then, if the energy spread in the bunch is small, and if the emittance is also small, $\beta_{0zi} = v_{0zi}/c$ will be small so that the simple nonrelativistic Larmor expression of Chapter 2 applies. This also implies weak fields and low harmonics. However, we are interested in a more general development that applies to all wigglers.

In terms of the proper time, the covariant Lorentz force, neglecting radiation reaction, will be

$$\frac{d\mathbf{p}_i}{d\tau} = q\gamma(\mathbf{E} + \boldsymbol{\beta}_i \times \mathbf{B})$$

$$\frac{dp_4}{d\tau} = (iq/mc)\mathbf{p}_i \cdot \mathbf{E} \tag{25}$$

Using

$$\mathbf{B} = \nabla \times \mathbf{A}(\mathbf{r}, t)$$

$$\mathbf{E} = -(1/c)\frac{\partial \mathbf{A}}{\partial t}(\mathbf{r}, t) \tag{26}$$

one can then write $\mathbf{A}(t - r_i/c) = \hat{\mathbf{x}}A_{xi} + \hat{\mathbf{y}}A_{yi}$ for transverse fields and

$$\frac{dp_{1,2i}}{d\tau} = (q/c)\frac{dA_{1,2i}}{d\tau} + F_{1,2i}^{\text{rad}}$$

$$\frac{dp_{3i}}{d\tau} = -(q/mc)\left(p_{1i}\frac{dA_1}{d\tau} + p_{2i}\frac{dA_2}{d\tau}\right) + F_{3i}^{\text{rad}} \tag{27}$$

Ignoring radiation reaction implies that

$$p_{1i}(\tau) = p_{1i}(0) + (q/c)[A_1(\tau) - A_1(0)] \tag{28}$$

i.e., knowing the initial momenta and vector potential at any time implies the momentum at a subsequent time τ'. Since we know the fields in the wiggler, we only need the initial conditions for the particles (i.e., the six-dimensional beam phase space for the machine at the location of the wiggler) to determine the motion throughout the wiggler at any point and thereby the radiation amplitude.

4.1.1. The Flat or Planar Wiggler

For this example we assume a TM field given in the laboratory system of the device by

$$\mathbf{B}(\mathbf{r}) \leftarrow \mathbf{B}(x, 0, z) = \hat{\mathbf{y}} \sum_{n=1}^{\infty} B_n \sin\left(\frac{2\pi nz}{\lambda_{\text{w}}}\right) \xrightarrow{(N=\infty)} \hat{\mathbf{y}}B_1 \sin\left(\frac{2\pi z}{\lambda_{\text{w}}}\right) \tag{29}$$

$$\mathbf{E}(\mathbf{r}) = 0.$$

i.e., a pure sine wave in the median plane of the device ($y \equiv 0$). For reasonably small distances off the median plane, we can then write

$$\mathbf{B}(\mathbf{r}) \simeq \hat{\mathbf{y}}B_1 \sin\left(\frac{2\pi z}{\lambda_\mathrm{w}}\right) - \hat{\mathbf{z}}2\pi(y/\lambda_\mathrm{w})B_1 \cos\left(\frac{2\pi z}{\lambda_\mathrm{w}}\right) \xrightarrow{\lambda_\mathrm{w} \gg 2\pi y_{max}} \hat{\mathbf{y}}B_1 \sin\left(\frac{2\pi z}{\lambda_\mathrm{w}}\right) \quad (30)$$

This satisfies Maxwell's equations and shows that for reasonably large values of λ_w compared to the maximum beam excursion off the median plane (y_{max}) we can ignore the component along the beam direction except possibly when looking at the optical effects of the wiggler on the beam. Thus, it is possible to obtain a rather good transverse field configuration although it can never be purely transverse.

This field produces linearly polarized light whose electric vector is confined mostly along the x direction because the beam oscillations take place primarily in the plane normal to the principal field direction. Notice that even if the initial conditions place the particle in the median plane of the wiggler, quantum deexcitations (emissions) can give it vertical kicks as long as they conserve momentum and energy. Consequently, we call this a flat or planar wiggler simply because the oscillations are confined predominantly to the x–z plane. Notice, however, that when this occurs there will be a linear restoring force tending to bring the particles back to the median plane due to the cosine term in equation (30).

The wiggler should be mirror symmetric about the x–y plane at its center point, since we will eventually make it finite and require (*i*) that there be no deflection of the beam over the length of the wiggler, i.e.,

$$\int B \, ds = 0 \quad (31)$$

and (*ii*) that there be no displacement. Mirror symmetry about the center of the wiggler insures requirement (*ii*) and implies only even terms in equation (29) if we place the z origin there.

4.1.2. The Helical or Axial Wiggler

A helical wiggler, the other main type of magnetic wiggler, can be described in similar fashion:

$$\mathbf{B}(\mathbf{r}) \leftarrow \mathbf{B}(0, 0, z) \xrightarrow{(N=\infty)} B_1\left[\hat{\mathbf{y}} \sin\left(\frac{2\pi z}{\lambda_\mathrm{w}}\right) \pm \hat{\mathbf{x}} \cos\left(\frac{2\pi z}{\lambda_\mathrm{w}}\right)\right] \quad (32)$$

$$\mathbf{E}(\mathbf{r}) = 0$$

It will cause the particle to travel in a helical trajectory and produce predominantly circularly polarized radiation. Except for this difference in polarization, it does nothing more (or less) than the flat wiggler. Practical limitations on the fields achievable with this device for apertures acceptable in a storage ring imply significantly higher comparative costs, which may not be warranted except for high-resolution applications such as the free-electron laser.[12] The larger horizontal aperture required for injection into a storage ring is one reason why most planar wigglers have their bending planes horizontal, i.e., coincident with that of the ring bending magnets. It is useful, however, to be able to rotate the polarization of the radiation. One way to accomplish this is described in Section 4.1.3.

4.1.3. The Rotatable Planar Wiggler

One can conceive of taking a flat wiggler and simply rotating it mechanically about the beam axis. This would suffer the same aperture problems just discussed for the helical wiggler. One could also design a vertical wiggler that could be inserted into the path of the reference trajectory following injection, such as the devices being developed for ACO at Orsay and the Photon Factory at KEK. The question of the necessity or value of vertical versus horizontal polarization could be debated but the ability to *arbitrarily* vary the plane of polarization hardly needs discussion. One aspect of such a system that does require discussion, however, is the rate at which one would like to vary the polarization, e.g., what are the requirements of fast spin-flip experiments. One such method has been discussed by one of the authors[34] based on the use of conventional iron-dominated magnets. The corresponding TM field expression for this device is

$$\mathbf{B}(\mathbf{r}) \xrightarrow{N=\infty} B_1 \left[\psi \hat{y} \sin\left(\frac{2\pi z}{\lambda_w}\right) + (1 - \psi^2)^{1/2} \, \hat{x} \sin\left(\frac{2\pi z}{\lambda_w}\right) \right] \tag{33}$$

where ψ is related to the field angle ($\psi = \cos \phi$) that B makes with the y direction.

4.1.4. The Free-Electron Laser (FEL)

We discuss the FEL because it provides a good transition or stepping stone with which to consider a whole range of new possibilities that should be mentioned within the present context because they are extensions of the basic undulator concept. Thus, we want at least to understand the characteristics of the radiation to be expected from such devices. We refer the reader to Chapter 22 on the FEL for a more complete discussion.[39, 40]

The FEL employs a transverse wiggler (helical or planar) and a cavity terminated by mirrors at either end[12] as in a conventional laser. However, rather than using an active medium, the cavity is evacuated so that the main difference between the FEL and the wigglers just discussed is the addition of a standing wave. If the resulting cavity mode that is excited can be maintained, without seriously influencing the charged particle beam on repeated traversals through the device, or vice versa, then high intensity, coherent radiation should be produced. Coherence is particularly important in this application if narrow linewidths characteristic of conventional lasers are to be obtained. The shortest wavelengths obtained so far are on the order of microns.

4.2. Wiggler Optics and Influence on Stored Beams

Although the wiggler produces no net deflection or displacement of a charged particle beam when properly excited, this should not be understood as being equivalent to a simple drift space for its ion optical transport function. Depending on how it is constructed and excited it can provide either deflection or displacement. Furthermore, it is a dissipative element that does not conserve phase-space volume or its various projections such as the transverse emittance. For instance, it will be seen that wigglers can both increase or decrease important performance parameters of storage ring beams such as rms energy spread and transverse emittance. Thus wigglers are important for both high-energy physics and SR applications and should be carefully considered—especially in machines such as SPEAR that are used for both purposes.

4.2.1. Optics

To good approximation, the wigglers discussed here behave as a drift space in the bend plane and as a converging lens in the direction normal to this plane. These effects are illustrated by Figure 4, which is a schematic layout of the SPEAR wiggler. Because it has 3λ, it is mirror symmetric about the x–y and x–z planes. It provides a maximum deflection angle $\alpha \lesssim 1°[(dx/dz)_{max}]$. In the dispersion plane, the optical transfer characteristics follow from the overall constraints of zero deflection ($\int B\, ds = 0$) and zero displacement (mirror symmetry about x–y) which imply that the angular and spatial dispersions are zero, i.e., the terms m_{16} and m_{26} of the transfer matrix are zero.[17] If by design, $\partial_x B_y = 0$, i.e., the poles are rectangular and sufficiently long in the x direction so that there is no variation over the beam envelope, then $m_{21} = 0$ so there will also be no overall focusing of the beam. This is not the case in the other dimension perpendicular to the dispersion plane. In this case, the primary effects result from "edge focusing" and can be considered as

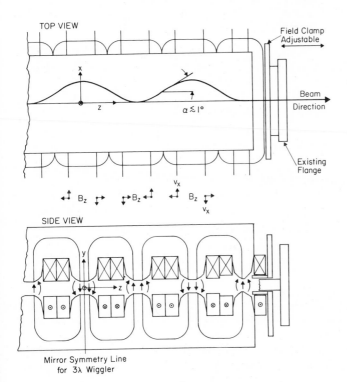

Figure 4. Schematic layout of the 3λ planar wiggler for SPEAR showing coordinate conventions and an exaggerated trajectory illustrating edge focusing for a particle moving above the median plane.

geometrical rather than due to the field distribution that we would have expected to disallow it (analogy with a quadrupole). In this direction there will be a cumulative focusing effect for particles off the median plane due to the longitudinal field component B_z in the vicinity of the entrance and exit of each pole. The effect is cumulative because the field changes sign in synchronism with the transverse velocity $v_x(\simeq c\ dx/dz)$ as shown in Figure 4 (also see Figure 12 for some predicted fields).

Mirror symmetry about the median plane $(y = 0)$ implies $B_z(z, x, 0) = 0$. As one goes off the median plane, the longitudinal field component (B_z) along the primary beam direction begins to grow in such a way that there will be a net force tending to drive a particle that is not moving in the median plane back toward it. This force is a linear function of the displacement y for small excursions and changes sign on going through the pole center or wherever $\partial_z B_y(z) = 0$. This may be seen from the Taylor expansion about the median plane

$$B_y(z, x, y) \equiv B_y(z, y) = \left(1 + \frac{y^2}{2!}\frac{\partial^2}{\partial y^2} + \cdots\right)B_y(z, 0)$$

$$\simeq \left(1 - \frac{y^2}{2}\frac{\partial^2}{\partial z^2}\right)B_y(z, 0) \tag{34}$$

$$B_z(z, x, y) \equiv B_z(z, y) = \left(y\frac{\partial}{\partial y} + \cdots\right)B_z(z, 0)$$

$$\simeq y\frac{\partial}{\partial z}B_y(z, 0)$$

where we have assumed (and made) the magnet long enough in the x direction to ignore any variations in this coordinate over the beam envelope in the magnet. Integrating $B_y(z, x, y)$ gives

$$\int_a^b B_y\ dz \equiv -\frac{y^2}{2}\left[\frac{\partial B_y}{\partial z}(z, 0)\right]_a^b = \frac{y}{2}B_z(z, x, y)\bigg|_b^a \tag{35}$$

since by the constraint, equation (31), $\int_a^b B_y(z, 0)\ dz$ will be zero for any (a, b) of interest. As a result, the off-median plane field integral will be zero between successive pole centers as well as for the wiggler as a whole so that beam steering or perturbations during filling should not result in a lost beam. These expressions can be used in computer codes to predict effects on the particle beam as well as the radiation spectrum expected from a wiggler whose field has been measured only in the median plane.

4.2.2. Effects on Stored Beams

Many important properties of stored beams in rings can be computed from integrals, taken around the ring, of various characteristic functions associated with the guide field.[16, 17] In terms of the azimuthal coordinate s, these functions include $\rho(s)$, the radius of the design orbit: $\beta_x(s)$ the radial betatron amplitude; and $\eta(s)$, the off-energy amplitude. Since most magnets are long compared to the lengths of their entrance and exit fringing fields, the integrals can be approximated by [16]

$$I_1 = \oint (\eta/\rho)\, ds = \sum_i \frac{l_i}{\rho_i} \langle \eta \rangle_i$$

$$I_2 = \oint (1/\rho^2)\, ds = \sum_i \frac{l_i}{\rho_i^2}$$

$$I_3 = \oint |1/\rho|^3\, ds = \sum_i \frac{l_i}{|\rho_i|^3} \tag{36}$$

$$I_4 = \oint \frac{(1 - 2n)\eta}{\rho^3}\, ds = \sum_i \left[\frac{l_i}{\rho_i^3} \langle \eta \rangle_i - 2l_i \left\langle \frac{n\eta}{\rho^3} \right\rangle_i \right]$$

$$I_5 = \oint \frac{H}{|\rho|^3}\, ds = \sum_i \frac{l_i}{|\rho_i|^3} \langle H \rangle_i$$

where $\langle f \rangle_i$ is the mean value of f in the ith element whose effective length is 1_i. The integrals are taken completely around the ring and $H(s)$ is

$$H(s) = \gamma \eta^2 + 2\alpha \eta \eta'^2 + \beta \eta'^2 \tag{37}$$

where $\eta' = d\eta/ds$ and γ, α, and β are machine functions defined by Courant and Snyder[41] for the horizontal plane containing the x–z axes. This is also the median or dispersion plane of the ring bending magnets. Notice that only the ring bending magnets contribute to any of the integrals.

The relationship between these integrals and the important machine parameters is as follows. First, the energy loss per turn per electron from SR is

$$U = [\tfrac{2}{3} r_e E^4 / (mc^2)^3] I_2 \tag{38}$$

with r_e the classical electron radius. We specialize to electrons because of the factor $r = e^2/mc^2$. Adding a wiggler into a ring increases U because $\delta I_2 > 0$, but this perturbation is variable, regardless of whether the particle beam energy remains fixed or not. Thus, whereas

$$I_i(\gamma) \simeq \text{constant} \tag{39}$$

because ρ is constrained to be constant for the ring bends, the perturbation due to the wiggler is

$$\delta I_i(B_1, \gamma) = C(B_{10}, \gamma_0)(B_1/\gamma)^i \tag{40}$$

and depends only on the effective peak field in the wiggler B_1 and the beam energy. The constant (C) depends on details of the wiggler field distribution as well as the ring and should be computed at least once for definite values of the field B_{10} and beam energy E_0 after which one can take advantage of such scaling relations over the range of wiggler excitations for which the field distributions and ring configurations scale ($I_i(\gamma) \simeq$ constant). For the SPEAR wiggler, this covers the full range of fields that were checked, i.e., an order of magnitude in field strengths. We note that it is often a good approximation to assume that η and H do not change through the wiggler so that δI_5 scales like δI_3.

In the absence of other constraints, one would presumably design wigglers to make δI_2 as large as possible, i.e., one would first try to make very high fields with superconducting systems or secondly make $\lambda_w \gg g$, the gap height or solenoid diameter, to achieve as long a uniform field region as permitted by the experiments that employ the radiation. The latter alternative is less desirable because I_2 depends only linearly on the length as well as the fact that this may also limit the brightness achievable with a standard wiggler.

The distribution of energies in a stored beam under equilibrium conditions is a balance between damping (I_2, I_4) and excitation (I_3). The rms energy spread in the stored beam is

$$\left(\frac{\sigma_E}{E}\right)^2 = \left(\frac{55}{32(3)^{1/2}} \frac{\hbar}{mc} \gamma^2\right) \frac{I_3}{2I_2 + I_4} = C(\gamma) \frac{I_3}{2I_2 + I_4} \tag{41}$$

where $\hbar/mc$ is the reduced Compton wavelength of the particles. If we designate σ_E^0 to be the width without the wiggler, the width obtained by energizing it can be written

$$\left(\frac{\sigma_E}{\sigma_E^0}\right)^2 = \frac{(1 + \delta_3)}{(1 + \delta_2)} \xrightarrow{\;\delta_2 \ll 1\;} 1 - \delta_2 + \delta_3 - \delta_2 \delta_3 \tag{42}$$

where $\delta_i = \delta I_i / I_i$ and we have ignored δI_4, which is generally small and ideally zero for SR applications as shown below. The condition for reducing the energy spread of stored beams is $\delta_2 > \delta_3$ but for standard wigglers $\delta_3 > \delta_2$. However, for an undulator we may have

$$\delta_2 - \delta_3 + \delta_2 \delta_3 > 0 \quad \text{or} \quad \delta_2 > \delta_3/(1 + \delta_3) \tag{43}$$

Using a square field approximation implies

$$\rho^i I_i / \rho^j I_j \equiv 1 \tag{44}$$

$$\rho_w^i \, \delta I_i / \rho_w^i \, \delta I_j \equiv 1 \tag{45}$$

Remembering that $\delta I_i > 0$, one then has

$$\rho_w > \rho \quad \text{or} \quad B_w < B \tag{46}$$

To reduce the energy spread in the stored beam, the wiggler has to be operated at fields below the ring bends, i.e., an undulator can be expected to generally improve the energy spread, unless it is operated as a standard wiggler, in which case it would generally worsen it. Notice that this also improves the energy spread expected of the radiation from the undulator so long as there is no significant energy shift because

$$\lambda_1 \propto \lambda_w/2\gamma^2 \to \delta\lambda_1/\lambda_1 \propto \delta\gamma/\gamma$$

The square field approximation is rather good for sufficiently long magnets but for short-wavelength wigglers it is not particularly accurate.[17] We give a practical, numerical example in Section 5 to illustrate how useful these effects might be.

The beam size and emittance, ignoring dispersion, is given by

$$\varepsilon_x = \sigma_x^2(s)/\beta_x(s) = \left(\frac{55}{32(3)^{1/2}} \frac{\hbar}{mc} \gamma^2\right) \frac{I_5}{I_2 - I_4} = C(\gamma) \frac{I_5}{I_2 - I_4} \tag{47}$$

Assuming the wigglers and ring bending magnets have the same dispersion plane implies that

$$\frac{\varepsilon_x}{\varepsilon_x^0} = \frac{(1 + \delta_5)}{(1 + \delta_2)} \xrightarrow{\delta_2 \ll 1} 1 - \delta_2 + \delta_5 - \delta_2 \delta_5 \tag{48}$$

where we have again ignored δI_4 (which can easily be included by redefining δ_2). Specifying δ_5 requires computation of $H(s)$ through the wiggler, which results in a number of terms[16, 17] of which $H_0 \delta I_3 / I_5$ is typically the most important. In the case where $\eta = \eta' = 0$ such as near interaction regions (see Figure 9), the emittance can decrease significantly with increasing wiggler excitations even though δ_2 is generally much less than 1 (see Figure 17). Wigglers at other locations such as between standard cells or in the symmetry straight sections are expected to have the opposite effect.[18, 46] It is instructive to look at the ratio

$$\frac{(\varepsilon_x/\varepsilon_x^0)}{(\sigma_E/\sigma_E^0)^2} = \frac{1 + \delta_5}{1 + \delta_3} \approx \frac{1 + \left(\dfrac{L_w}{L_B}\right)\left(\dfrac{\rho_B}{\rho_w}\right)^3 \dfrac{\langle H \rangle_w}{\langle H \rangle_B}}{1 + \left(\dfrac{L_w}{L_B}\right)\left(\dfrac{\rho_B}{\rho_w}\right)^3} \tag{49}$$

where L_w is the integrated wiggler length and L_B is the integrated length of an isomagnetic guide field having a bend radius ρ_B. If $H(s)$ at the wiggler is approximately zero then the emittance will decrease (and the corresponding SR brightness will be optimal) because of minimal quantum excitation of betatron oscillations by the SR produced by the wiggler as well as increased damping from the wiggler. To keep the ratio constant one would want $\langle H \rangle_w = \langle H \rangle_B \neq 0$ or variable L_w, i.e., a variable number of wigglers or wigglers with a variable number of periods or wavelength per period. This provides control on the contribution of quantum fluctuations to the energy spreading. The range required is reduced by placing the operating range of the wigglers near that of the bends. This is discussed more in Section 5.

Finally, the damping rates in x, y, and E can be written as[16]:

$$\alpha_x = K(I_2 - I_4)$$
$$\alpha_y = KI_2$$
$$\alpha_E = K(2I_2 + I_4) \tag{50}$$
$$\alpha_x + \alpha_y + \alpha_E = 4KI_2 > 0$$

where L is the ring circumference, c is the speed of light and $K = \frac{1}{3} r_e \gamma^3 (c/L)$. Thus damping is always increased by turning a wiggler on and this is typically true in each of the individual modes. Redistribution occurs via the term δI_4, which is a signed quantity in contrast to I_2 or I_3. An example with some values typical for SPEAR is given in Section 5 to illustrate how significant these effects can be. With sufficient rf power and field strength, wigglers would appear to be justified solely on the improved damping they provide for injection and storage capability.[10, 18, 46]

4.3. Practical Design Considerations

Some practical questions that naturally arise in the course of designing wigglers are how best to achieve the complex field patterns of the undulator and the large induction fields of a standard wiggler and how to obtain a net zero field integral over a range of excitation in either type of wiggler. Other questions concern the construction tolerances and how to specify them in a way that achieves the basic design goals most economically. Finally, one must ask what constraints on the design are imposed by the machine in which the wiggler is to be embedded, e.g., is it necessary to maintain the basic superperiod, what is the minimum gap height allowed for beam lifetimes, are higher-mode losses acceptable, and what are the tune changes and increased energy spreads induced by the wiggler? If such changes are not acceptable one must then ask how to compensate or minimize them.

Design of wigglers follows well-established practices and many tools are already available. For instance, design of magnetic wigglers can benefit from use of computer codes like POISSON[42] for realistic simulation of the magnetic circuit over the full range of fields. For standard wigglers one may want to vary the field (or critical energy) from zero to full field without perturbing the particle beam noticeably. This generally implies that one must carefully design the circuit to minimize saturation effects without undue waste of money (overdesign). Notice that even very high field (or very small λ_w) superconducting wigglers may also utilize iron to enhance the field, help shape it, and provide structural strength against forces tending to deform the coils and distort the fields. However, at the end of this section we discuss a design using only permanent magnetic materials. A code[42] also exists to help design permanent magnetic systems, although for our example it was not required.

Even with a good field simulation it is wise to use two adjustable currents (or preferably one current and one shunt or bypass current associated with a single power supply) that can be varied relative to one another to compensate small differences between internal and external parts of the magnet due to relative saturation and/or other end effects over the range of currents desired. These two currents can be considered as tuning controls for ε_c (or I_2) and $\int B\,ds = 0$ that are essentially independent. If two independent current sources are used, it is important to reduce the effect of current ripple and the instantaneous differences in current that exist because the ripples are not in phase. Techniques available are filtering of the power supply output, the use of solid-core iron poles, which allow eddy currents, and the use of sufficiently thick vacuum chamber walls (typically 5 mm or more), which also provide eddy current shielding. Design of the magnetic circuit that couples the fields resulting from the two currents also helps. When properly dealt with, the use of two currents should not be a problem. One can also use adjacent steering magnets to correct for nonzero field integrals if the wiggler imbalance is not too great, so that only one wiggler current is necessary, although two trim currents are needed to eliminate orbit distortions around the ring.

There will be strong coupling between individual poles in a wiggler when the return flux path is through successive poles, i.e., along the beam direction rather

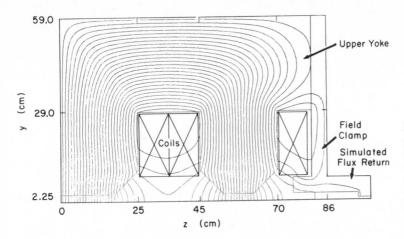

Figure 5. Computer generated flux plot obtained with POISSON for the upper quadrant of the 1λ standard wiggler being built for PEP. This system is mirror symmetric about $y = 0$ and $z = 0$. The flux return associated with the field clamp in the x dimension is shown at lower right.

than transverse to it as in conventional *H*- or *C*-type designs. Such a planar wiggler could be called a multiple *H* or Open *H* design (*OH*). Such a design should be more efficient for very high field wigglers since all the flux (except for leakage) is returned in the adjacent poles rather than through outside paths. In addition to reducing costs, it makes the problem two-dimensional (2D) along the all important beam direction because the poles are rectangular and sufficiently wide to make the 2D simulation optimal. It also reduces beam fluctuations due to variations in the two currents, since an increase in the strength of one pole is accompanied by corresponding increases in adjacent poles of opposite polarity, with the result that $\int B\,ds$ does not change. On the other hand, it implies that care is required to properly terminate the magnet at beam entrance and exit in order to simplify its subsequent use as well as minimize the effects of stray fields. This is the only region in the magnet where a portion of the flux path necessarily goes outside of a plane—to allow the beam to enter and exit the wiggler. One way of simulating the effect of this in a 2D calculation is shown in Figure 5. Thus, one might expect these calculations to be as accurate as any calculations are likely to be, although a limitation is the computer core size required to obtain good resolution of the pole contour. To our knowledge, the first magnet of this type was described in Reference 43 in a somewhat different context. However, in that instance, the problem was completely 2D.

As we have already indicated in equation (2) it is difficult to vary the photon wavelength for a constant beam energy γ without increasing the relative harmonic content of the beam. To remain at the undulator limit ($K \ll 1$) and still vary ε_1 it is desirable to vary only γ or λ_w. One of the more interesting aspects in the design of an undulator is the question of the minimum wavelengths λ_w that are achievable. This also relates to the problem of designing variable wavelength wigglers. This point and some related questions of concern in the design of conventional electromagnetic wigglers were addressed in Reference 43.

Figure 6a shows the field falloff along the beam direction z as a function of pole length in this direction normalized to the full gap height g. To obtain a

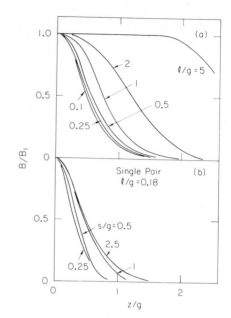

Figure 6. Field falloff in the longitudinal direction for different pole lengths l(a) and pole separations s(b) relative to the central field value B_1. Definition of variables is shown in Figure 4 except for B_1—the field in the vertical direction at the center of the pole $x = y = z = 0$.

uniform field region inside the pole edge requires on the order of a gap or more depending on the degree of uniformity required. For those wigglers for which it is required to maximize $\int B^2 \, dz$ (see Section 4.2.2) or obtain a field that is reasonably linear with excitation current up to the level required for $K = 1$, the pole length l [see equation (3)] should generally be $\gtrsim g/2$ with $s > g$ as verified from Figures 7 and 8 to avoid too much inefficiency from saturation of the steel or loss of amplitude from feeding adjacent poles. If the gap height is limited to $g > 5$ mm by the accelerator, synchrotron, or storage ring it is used in, then $s = \lambda_W/2 > g$ or $\lambda_W > 1$ cm. A similar argument can also be made for helical undulators[38] having a bore diameter > 5 mm by demanding on-axis fields as high as 10 kG, which also implies $\lambda_W > 1$ cm. This is a problem that leads to interesting questions concerning the ultimate limits on J_C and B_C in filamentary superconductors. A comparable limit on λ_W can also be argued for permanent magnet undulators.

Notice that scaling the longitudinal dimensions s and l will neither maintain the field distribution nor scale I_2 nor allow the same ε_C except when $0.5 s/g \sim l/g > 2$, which one can call the intermediate- to long-wavelength regime. From the discussion of equation (2) the problem is that as one decreases s, the field increases become more difficult to achieve because of the onset of saturation effects that were not included in the considerations above. Practicality would seem to dictate wigglers having gaps > 1 cm to avoid radiation and beam clearance problems, not to mention a plethora of other exotic possibilities such as higher-order mode losses. Depending on the machine energies available and the needs of the SR (or UR) users, one can then specify the number of poles and λ_W to be compatible with the constraints of the ring. If λ_W is relatively short, e.g., $\lambda_W < 10$ cm and high fields are required, one should consider a superconducting system to minimize the coil cross sections required to achieve sufficient ampere turns. However, if normal SR is not

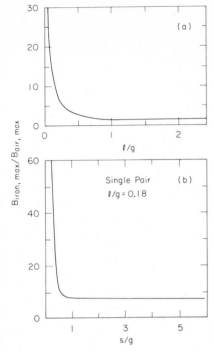

Figure 7. Ratio of maximum field in the iron to the maximum field in the midplane as a function of l/g (a) and s/g (b) for fields well below any appreciable saturation in the steel and the same excitation in ampere turns.

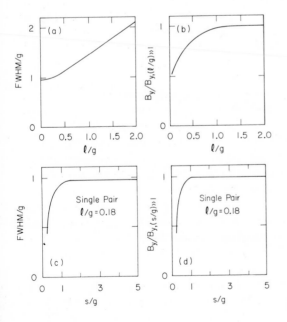

Figure 8. Variation of the peak of the field profile and its FWHM with l/g and s/g for poles that are wide in the x direction.

required, i.e., for undulators, one should consider using permanent magnetic materials since the required fields should not be particularly difficult to obtain except for the very shortest wavelengths, which would be difficult regardless of the approach adopted.

A computer code has recently been developed[42] that can simulate circuits that include permanent magnetic materials such as Alnico, barium ferrite ceramic, and samarium cobalt (as in HICOREX) in very much the same way POISSON does for conventional magnets. For dedicated applications, such systems should find many uses and should be simple to use since they require no power or water cooling. They can be designed to have variable wavelength and also variable field, although this increases the complications and loses some of the elegant simplicity. Several basic designs suggest themselves but perhaps the simplest is one discussed by the authors at the National Conference on Synchrotron Radiation Instrumentation (proceedings to be published in *Nuclear Instruments and Methods*). This consists of two linear arrays of rectangular blocks above and below the median plane. The magnetization direction of successive blocks along the beam direction is rotated by 90° so that each wavelength has four blocks above and below the median plane. A more versatile design is one in which each half-wavelength unit has a pole pair made of permanent-magnet materials with an iron return yoke connecting the poles as in a conventional H-type dipole or C-type if one wants to move it in and out of the beam. The iron would be primarily for structural support and control of stray fields. Such systems could be fully stabilized and subsequently assembled into multiperiod structures that are movable as a whole, but also individually movable to obtain a variable λ_w, except for restrictions such as mentioned above. If a variable induction field were needed, it could be obtained using mechanical screws associated with each $\lambda_w/2$ to control the gap height or possibly to shunt flux away from the gap through a pseudo-return iron path.

5. Applications with Examples

The development of wigglers is comparatively new and can be expected to grow rapidly so that forecasts will probably be quite inaccurate. However, from the properties of standard wigglers and undulators discussed so far, it is hoped there will be new phenomena based on new experimental capabilities. With standard wigglers, the primary benefits are the extension of the spectral range, independent adjustment of the endpoint energy, and a general enhancement of the intensity proportional to the number of poles. From this, it is clear that progress in photon transport and counting systems will be needed to take advantage of the higher photon energies and power densities that will become available. Higher resolution experiments will become more common as well as studies of very dilute systems, which are not feasible now. In fact, there are many practical applications that depend critically on one or more parameters such as power density, irradiance, radiance, spectral power density, etc.

Very little advantage appears to have been taken of the economic advantages of using wigglers, e.g., their equivalence (from the standpoint of net photon flux) to operating a storage ring in single-beam, multibunch mode. Furthermore, as indicated in the Introduction, considerable savings are possible by incorporating wigglers in the initial storage ring design. As an example, consider an 800-MeV storage ring with a bending magnet field of 12 kG with 18 kG and 50 kG wigglers. The critical energy from the bending magnets is 0.51 keV, from the 18-kG wiggler it is 0.77 keV, and from the 50-kG wiggler it is 2.13 keV. It is generally agreed that, for most experiments, useful flux is available out to about five times the critical energy, where the intensity is down by a factor of 25 compared to the critical energy. Using this criterion, the upper limit of useful flux from the 800-MeV machine would be about 2.6 keV from the 12-kG bending magnets, 3.8 keV from an 18-kG wiggler, and 10.7 keV from a 50-kG wiggler.

Thus a medium-sized storage ring (800 MeV) could provide an excellent VUV and soft x-ray flux to a large number of users with many beam channels from the many ring bending magnets serving experiments in surface physics (photoemission), EXAFS [elements up to $Z = 16$ (K-edge EXAFS), and up to $Z = 40$ (L-edge EXAFS)], soft x-ray lithography, and microscopy. An 18-kG wiggler would extend the EXAFS capabilities to $Z = 19$ (K edge) and $Z = 46$ (L edge). Some long-wavelength diffraction work should also be possible but difficult. With a 50-kG wiggler the EXAFS capabilities are further extended up to $Z = 30$ (K edge) and $Z = 70$ (L edge), which includes most elements of current biological interest. Furthermore, excellent diffraction work at the desirable photon energies of 6–10 keV can be performed. Of course there are important experimental applications that require photon energies beyond the 10 keV that an 800-MeV machine with a 50-kG wiggler could provide. Also a 50-kG wiggler magnet is a fairly expensive device and only a relatively small number could be used on most machines. Still it is impressive to see the broad range of science that is accessible to a medium-energy storage ring of only 800 MeV, when equipped with a 50-kG standard wiggler. This appears to be a truly cost-effective approach to a general purpose synchrotron radiation research facility.

The primary benefit of undulators is their high intensity in a limited bandwidth compared to normal SR or even standard wigglers. EXAFS measurements could be extended to more dilute systems. Time-dependent experiments could be performed on a shorter time scale and a variety of higher-order photon processes could be studied such as nonlinear resonance fluorescence or Bragg scattering that require very narrow bandwidths and are quite marginal with present intensities. The interested reader is referred to other chapters for more details on the need for higher intensities. Since there are many applications in the atomic, molecular, and solid state areas, we limit our remaining comments to potential applications in the nuclear and subnuclear areas. Many of these have direct analogies to atomic and molecular processes such as photoionization or photodissociation (CO, H_2O, N_2O, etc.) with corresponding implications for energy applications.

For instance, separation of U^{235} from U^{238} using radiation requires tunable, monochromatic radiation with high power to compete effectively, e.g., > 100 kW is a desirable goal. Undulators or free-electron lasers may prove to be the best alternatives here if they can achieve sufficient power. It appears that > 1 kW is presently achievable[39, 40] with a sufficiently high-intensity storage ring (> 1 A) specifically designed for the task. Certainly, the ratio of energy gain to energy cost would be high in this application so the benefits are clear. This follows directly from the curve of nuclear binding energy per nucleon versus atomic mass number and the basic fission process. The fact that "undesirable" isotopes are generated in the spent fuel can be dealt with in an ecologically sound way using the same techniques used initially to enrich the uranium fuel with U^{235}, i.e., separation of the various elements and their respective isotopes with tunable radiation or other techniques for subsequent biomedical and other applications. For instance, coherent nuclear x-ray scattering has been observed using radioactive sources but the rates were so low as to be of little practical value. Although the possibility of pumping a gamma-ray laser formed of a crystal of the daughter products of such separated radioactive isotopes is extremely remote, it would appear that basic research in the area of element and isotope separation appears to be fully warranted. However, for this to be feasible it would require higher photon energies, small bandwidths, good tuning range, and a number of developments in peripheral areas.

Fusion follows from the same curve of nuclear binding energy as fission. It has been demonstrated using electromagnetic radiation from lasers and active research is now underway to achieve shorter wavelengths and higher power. If one were to again consider using the existing high-power lasers with some of the high-power electron linacs available, it would simplify the basic multibeam laser system and also significantly increase its power. In contrast to the fusion process there have also been proposals for making neutron generators based on the idea of employing the breakup of nuclides such as Li^7 and Be^9 into alphas and the valence neutron by direct nuclear photodissociation.

5.1. The SPEAR Wiggler—A Detailed Example

In this section, we give a detailed discussion of a wiggler[15, 45] that is now operating at the Stanford Synchrotron Radiation Laboratory (SSRL). Although it was developed primarily for the production of high-intensity, tunable SR it can also be operated at low fields as a broadband undulator. It is a 3λ planar wiggler and is located in one of the straight sections of the Stanford Positron-Electron Accelerating Ring (SPEAR) located at the Stanford Linear Accelerator Center (SLAC). Because it can operate up to 18 kG it is also used to improve luminosity during colliding beam experiments, which places further restrictions on its design and operating characteristics. First, we describe the SPEAR ring and then the wiggler, which is the effective source for Beam Line IV at SSRL. Next we describe the beam line that transports the radiation into a wiggler spectrum diagnostic station plus one or more experimental end stations for EXAFS and other experiments. It is interesting to note that the total cost of the beam transport and experimental equipment far exceeds the cost of the wiggler magnet itself. Had the wiggler been made superconducting, the costs would have been more comparable between source and transport system. Finally, we describe some measurements of the effects of the wiggler on stored as well as colliding beams as discussed in Section 4.2.

Since the SSRL wiggler is one of the first of its kind operating as a SR source in a storage ring, there are many detailed questions that arise such as how to operate it in the simplest, most flexible way, how compatible is it with colliding beam operation, etc? It is hoped that experience gained with this instrument and our discussion of such questions will aid others in planning wigglers at other SR facilities as well as subsequent wigglers at SSRL.

5.1.1. Description of the SPEAR Lattice

SPEAR is an oval-shaped ring[44] designed in such a way that counter-rotating electron–positron (e^+, e^-) beams collide at two interaction regions (IR) separated by 180°. The lattice is therefore mirror symmetric about these two points, i.e., it has two superperiods. One-half of one such superperiod is shown in Figure 9. Each superperiod consists of a so-called insertion and a series of standard or unit cells. The primary requirement for colliding beam studies is high luminosity,[16] i.e., a high reaction rate per unit cross section, which is proportional to

$$L = R/\sigma = N_b N_e^2 f/(4\pi\sigma_r^*\sigma_v^*)$$

where N_b is the number of bunches in either beam, N_e the number of electrons (or positrons) in a bunch, f is the revolution frequency, and σ_r^* and σ_v^* are the effective beam width and height in the interaction region. The insertion takes a different form from the repetitive pattern of cells in the rest of the ring in order to reduce the horizontal and vertical beam sizes down to their limiting values in the IR where the beams interact. The reaction rate will then be limited by the critical current density as determined by the beam–beam interaction. The critical current density is sometimes expressed[16] in terms of a limitation on the allowable tune shifts ($\Delta v_m \simeq 0.025$) resulting from one beam passing through the other. This is a positive or

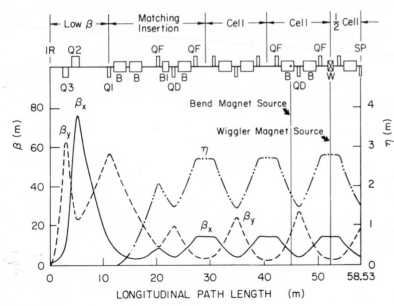

Figure 9. Layout of one-half of a SPEAR superperiod extending from the interaction region (IR) to the symmetry point (SP). The horizontal and vertical betatron amplitude and off-energy function η are shown for tunes used for colliding beam experiments. The wiggler location is labeled W and an adjacent SR source point for beam line III at SSRL is also shown. The vertical η is negligible.

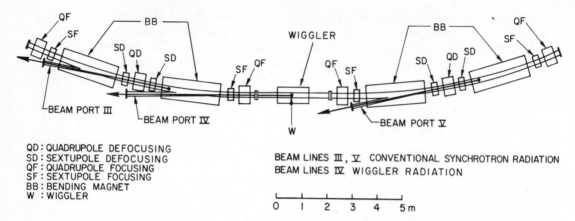

QD : QUADRUPOLE DEFOCUSING
SD : SEXTUPOLE DEFOCUSING
QF : QUADRUPOLE FOCUSING
SF : SEXTUPOLE FOCUSING
BB : BENDING MAGNET
W : WIGGLER

BEAM LINES Ⅲ, Ⅴ CONVENTIONAL SYNCHROTRON RADIATION
BEAM LINES Ⅳ WIGGLER RADIATION

Figure 10. A more detailed layout in the vicinity of the wiggler showing its location in a straight section of SPEAR and some adjacent SR sources in the ring bending magnets (BB).

focusing effect in the vertical direction that sets an upper bound on the allowable tune shift from the wiggler, which must be kept considerably below this or compensated in some way.

Because the wiggler is located in one of the symmetry straight sections of the ring where β_x is large, the particle beam size is comparatively large but the divergence is low. In fact, $\sigma_x \simeq 1.3$ mm at 1.5 GeV, which is comparable to the increased spreading of the photon beam due to the oscillatory motion of the particle beam in the wiggler. Because the dispersion is also high at this point in the lattice, the SR created here more easily excites radial betatron oscillations. The interplay between the increased damping and excitation then results in a change in the emittance whose direction and magnitude depends on wiggler field.[15] As discussed above, this would not be the case for a wiggler located in the IR.

Figure 10 shows the specific layout of the ring in the vicinity of the wiggler in more detail. In this location, the wiggler transport line will see radiation from the fringing fields of both adjacent bending magnets. Thus, when the wiggler is operated as an undulator so that its radiation is well collimated, one expects to see three bumps dispersed in the horizontal direction for the correct wiggler field and sensitivity range of detector.

5.1.2. Description of the Wiggler and Its Operation

The wiggler shown in Figure 11 was fabricated for SSRL by Industrial Coils, Inc. Its length was constrained to be less than 1.22 m by the space available in a straight section of SPEAR. Its gap height was set by several factors, including the vertical beam stay clear of 2.8 cm at its location, the total thickness of the vacuum chamber (1 cm) required by the size of the horizontal beam stay clear, and 10% of the vertical beam stay clear as a safety margin against construction variations along its length. Table 1 gives additional information on the wiggler.

Figure 11. Pictures of the SPEAR Wiggler. The top and middle pictures, taken during magnetic measurements, show the burning and cooling arrangement, and the lower picture was taken during installation into the ring.

Figure 12 shows a partial cross section of the wiggler together with the associated two-dimensional fields predicted with the computer code POISSON. The poles are long in the x direction (compared to the gap height g) so we would not expect to see any variation with this coordinate. For a particle whose motion is constrained to the median plane, mirror symmetry implies that it should experience

Table 1. SPEAR Wiggler (3λ) Summary

Number of poles	7
Maximum central induction B_0 (kG)	20
Pole width w (mm)—coordinate direction ($\hat{x}$)	304.8
Gap height G (mm)—coordinate direction ($\hat{y}$)	41.3
Mechanical pole length—inner (mm) ($\hat{z}$)	88.9
Mechanical pole length—outer (mm) ($\hat{z}$)	66.7
$\dfrac{1}{2I}\int B_y(x, 0, z)\, dz$ for $B_0 = 20$ kG (kG m/kA)	0.576
Effective magnetic pole length—Inner (mm)	98.3
Magnetic wavelength λ_w (mm)	342.9
Total magnet length L (m)—clamp-to-clamp	1.219
Amperes for 20 kG	1708
Turns per pole	28
Power (kW) for 20 kG	250
Total flow rate (gal/min)	42
Maximum temperature rise (°C)	30°
Total magnet weight (kg)	1100

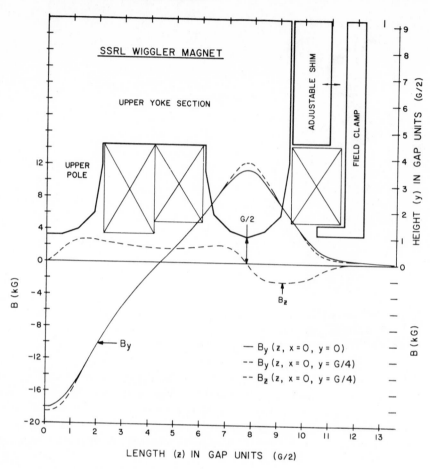

Figure 12. Detailed layout of the end section of the SPEAR wiggler together with the POISSON field predictions.

only the vertical field component $B_y(z, x, 0) = B_y(z)$. This is the major field component that produces the SR and since

$$\int_{-\infty}^{\infty} B_y \, dz \equiv 0$$

the areas above and below the zero-field level in Figure 12 should be equal. To insure this, the adjustable shim and field clamp combination is used at one field, e.g., the highest field likely to be used, after which a current shunt such as described in Reference 46 is used at lower fields (and voltages across the coils). However, it should be pointed out that the final computer calibration table that relates the two currents for operation in the storage ring was determined by observing and minimizing the horizontal orbit distortions as a function of current near the minimum energy at which SPEAR normally operates.

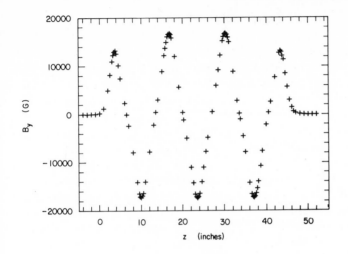

Figure 13. Measured field distribution at 17 kG for the SPEAR wiggler. Similar results at 1 and 10 kG exactly overlay these results when scaled according to peak field values.

The distribution of the vertical field component along the longitudinal direction as well as field integrals were measured for inductions ranging from 0 to 17 kG. Figure 13 shows some typical data taken with a Hall probe at a field of 17 kG. Similar measurements at 1 kG taken on the way up to full field and 10-kG data taken on the way down overlay perfectly when the peaks are scaled to be the same. Consequently, hysteresis effects do not appear to be a limitation on the operation of the wiggler so long as it is properly operated, i.e., it is set with a fixed procedure at a fixed rate. Significant deviations were observed between the measured magnetic field distribution at the highest field levels and predictions made by the computer program POISSON that are attributed to the permeability table used in the design calculations. An interesting characteristic observed during the measurements shows a fixed offset between the peak fields and field integrals associated with polarity for all excitations. The uncompensated results are shown in Table 2. Regardless of whether the net uncompensated field integral for the wiggler as a whole is positive or negative, poles having the same polarity tend to cluster at slightly different central field values and integrals than poles of the opposite polarity. POISSON predicts such effects for the SPEAR wiggler as well as for longer wavelength wigglers such as PEP when simulations with 3, 5, and 7 poles are done.

Table 2. Peak Field (kG)/Integral (kG cm)

Current (A)	Pole 1	− Pole 2	Pole 3	− Pole 4	Pole 5
100	0.94/7.00	1.33/12.5	1.25/11.5	1.34/12.4	1.25/11.5
1000	9.60/74.8	13.2/124.6	12.5/116.4	13.2/124.4	12.5/116.8
1500	13.4/102.0	17.5/164.9	17.1/157.4	17.6/164.9	17.1/158.0

5.1.3. Description of the Wiggler Transport Line

The transport line associated with the wiggler is designated Beam Line IV at SSRL. Its primary purpose is to provide high-intensity, hard x-rays, i.e., tunable SR so that x-ray experiments are possible even when SPEAR operates at low stored beam energies. It was also desired to provide a test station to measure and characterize wiggler radiation as well as provide the capability for additional experimental stations, should experience indicate their feasibility and need. Thus, the wiggler will be operated primarily as a standard wiggler to supply radiation for a single, end of line experimental station that is essentially an EXAFS setup with a rapidly tunable two-crystal monochromator. The advantages that this station will provide over other stations at SSRL such as EXAFS I or II are more beam time (SPEAR is not a fully dedicated SR source), higher intensity, and an easily variable ε_c. Thus, when SPEAR operates at 2 GeV and the wiggler at 18 kG, we expect a critical energy $\varepsilon_c = 4.7$ keV with a nearly sixfold increase in flux compared to the other lines at SSRL using the ring bends as sources. For comparison, one would have to run SPEAR at 3 GeV to achieve the same ε_c on other lines.

Figure 14 shows the layout of the line that exists at this time. Provision has been made for future installation of a curved mirror at $z \simeq 11$ m from the source that would provide focused x rays. Because the initial EXAFS station takes only 2–3 mrad of the total 16 mrad available, there are 6.5 mrad available on either side of the central portion that will be used initially for wiggler diagnostics and ultimately for large-angle UV or other branch lines. At present, an offset fixed mask at 13.2 m allows only 9.5 mrad of beam downstream with 8 mrad on one side of the beam centerline and 1.5 mrad on the other. This is followed by a Varian 15-cm gate valve before the shielding wall that separates SPEAR from the south arc experimental hall. The wiggler diagnostic chamber is at 16 m, followed by a double Be window, an x-ray monochromator at 19 m, and finally the experimental hutch.

5.1.4. Effects of Wiggler on SPEAR Operation

Based on the previous discussion we would expect to see a number of effects whose importance will depend primarily on beam energy. Thus, with increasing stiffness of the beam, the wiggler perturbs the ring less from its normal operating conditions so that one would eventually be able to operate (and vary) the wiggler during colliding-beam operation without noticeable interference with other uses of the ring. An exception to this is below the ψ resonance ($E = 1.55$ GeV), which is near the lower end of the SPEAR operating range. Nevertheless, even here, the wiggler should actually improve luminosity for high-energy physics experiments and thereby provide savings in beam time. In fact, one expects to extend the practical operating range of SPEAR with wigglers by moderating the natural decrease of beam size as well as by decreasing damping times. The first tests of the wiggler at 2.4 GeV demonstrated that beams could be collided with the wiggler at full field and, furthermore, that there was no measurable loss of luminosity while running the wiggler down to zero field under computer control.

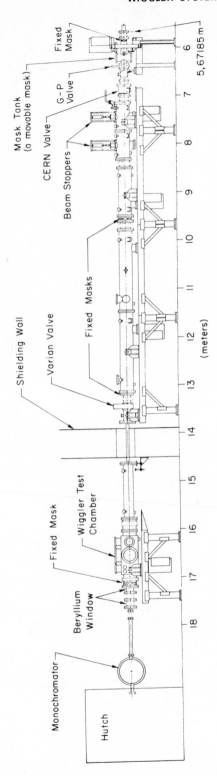

Figure 14. The transport system begins at the exit of the vacuum plumbing, which is part of the SPEAR ring, at 5.67 m from the center of the last pole of the wiggler. A bellows mechanically isolates this system and is followed by a fixed mask with an aperture of 16 mrad horizontally and 2.5 mrad vertically. This is followed by a water cooled movable mask that can be inserted to block the beam from an isolation valve (G–P) or fast acting valve (CERN). The two stoppers are movable stainless steel boxes filled with lead having an effective thickness of 45 cm each. Additional bellows, masks, pumps, valves, and spool pieces follow up to the shield wall that separates the ring from the experimental area.

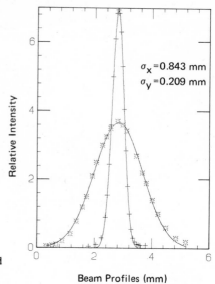

Figure 15. Transverse beam profiles (x, y) in SPEAR $(*, +)$ at 1.8 GeV and Gaussian fits to $I = I_0 \exp\left[-\frac{1}{2}(x - \bar{x})^2/\sigma_x^2\right]$.

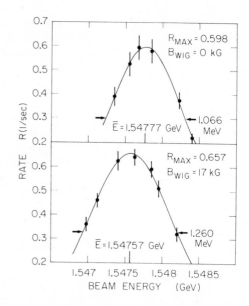

Figure 16. Reaction rate of events having at least three hadrons in the vicinity of the ψ resonance, which has a natural width that is much less than that of the electron beam. The apparent shift in energy of the resonance results partly from an increase in path length due to the wiggler, which has not been taken into account in determining the beam energy.

Table 3. Typical Data

Energy (GeV)	Field (kG)	Measured	Predicted
1.55	10.8	$\Delta v_y = 0.006$	0.004
		$\Delta v_x < 0.001$	0.0
	17.2	$\Delta v_y = 0.012$	0.010
		$\Delta v_x < 0.001$	0.0
		$(\sigma_E/\sigma_E^0) = 1.18$	1.11
1.88	10.8	$\Delta v_y = 0.002$	0.0028
		$\Delta \sigma_x/\sigma_x = 0.9\%$	-0.5%
	17.2	$\Delta v_y = 0.006$	0.0072
		$\Delta \sigma_x/\sigma_x = 3.5\%$	3.1%
(Luminosity improvement here by $>20\%$)			
2.4	16.0	$\Delta v_y = 0.003$	0.0036
		$\Delta \sigma_x/\sigma_x = 1.0\%$	-0.57%

Beam profiles in both transverse dimensions were measured using SR and fit with Gaussian profiles to determine orbit shifts and size changes with wiggler excitation and SPEAR energy. Figure 15 shows some typical profiles and the quality of fits obtained using simple Gaussians. Energy spreading was more difficult to determine. It was measured by looking at the ψ resonance with and without the wiggler. Table 3 compares some typical results to predictions and Figure 16 shows the results of the ψ experiment. Since the two ψ runs were done at different times, the apparent downshift in energy should not be compared directly to predictions although it is in the right direction. In general, it appears that the effects of the wiggler on both storage ring and colliding-beam operation are in quite reasonable agreement with expectations. The calculations[17] were carried out at 1.5 GeV using the detailed field distributions predicted with POISSON. Predictions for other ener-gies and/or wiggler fields can be scaled from 1.5 GeV as discussed above and in References 17 and 46.

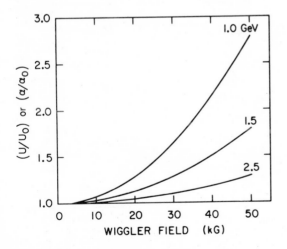

Figure 17. Plots of the increased energy going into SR due to exciting the wiggler as well as the increased damping rates (α_x, α_y, and $\alpha_E/2$) for energies of 1.0–2.5 GeV. Note that this plot is also $1 + \delta_2$ of Section 4.2.2 and that the SPEAR wiggler is limited to 20 kG.

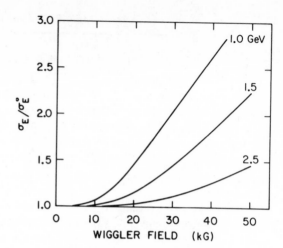

Figure 18. Increased energy spreading in the electron beam
due to the wiggler.

To show some of the expected effects at different energies and wiggler fields we
have calculated the increased SR energy produced by exciting the wiggler in SPEAR
(U) compared to that without the wiggler (U_0) in Figure 17. Since δI_4 is still
comparatively small up to 50 kG, it follows from Section 4.2.2 that this is also the
relative variation of the damping rate to be expected with and without the wiggler.
Figure 18 shows the relative change in the energy spread of the electron beam and
Figure 19 shows the relative change in the radial emittance. At low fields (i.e.,
wiggler fields less than the ring bends) both the energy spread and emittance
decrease somewhat, although this is not evident from the scale required for fields as
large as 50 kG. The difference in signs between predicted and measured beam sizes
in Table 3 may be understood by including η at the profile measurement location (s)
in the ring using $\sigma_{xr}^2(s) = \varepsilon_x \beta_x(s) + \eta^2(s)(\sigma_E/E)^2$.

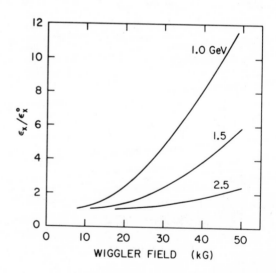

Figure 19. Increased radial emittance due to exciting a wiggler
whose dispersion plane is the same as the ring bends.

5.2. Characteristics of Other Planned Wiggler Installations

Although at the time of this writing (early 1979) there are no wigglers in routine use for SR (except the SSRL wiggler), several magnets have been completed and are being tested, several others are being constructed, and some more are in the design phase. Here we present a review of these systems based on information we have been able to collect from the various laboratories. Table 4 provides a summary comparison of different facilities, arranged in order of decreasing critical energy ε_c of the SR available from the ring bends, beginning first with storage rings and followed by synchrotrons. Certain of these are discussed in more detail below because they represent a class or interesting characteristic or usage. The various systems divide almost equally between superconducting, coil-dominated designs and more conventional, iron-dominated systems. Permanent magnet undulators are being considered at Novosibirsk and Stanford. Also, only two of the wigglers provide horizontal fields (Photon Factory and LURE–ACO) and vertically polarized radiation. In both of these cases the transverse beam contour at the wiggler is approximately circular with $\sigma_x = \sigma_y = 0.5$ mm. Some advantages and disadvantages of horizontal versus vertical configurations are discussed.

All of the "macroscopic" wigglers in Table 4 have $\lambda_w > 1$ cm, which is the limiting value based on the size of apertures typically required in the current generation of machines. Although planar wigglers generally achieve higher fields, this limitation is true for them as well as for helical wigglers, which we have not included because they are discussed elsewhere. The Stanford FEL helical wiggler[12] had $N = 160$ and $\lambda_w = 3.2$ cm with on-axis fields of 2.4 kG.

5.2.1. Photon Factory at KEK, Japan

A 1λ superconducting wiggler magnet[47] is planned for use at the Photon Factory 2.5-GeV storage ring now being constructed at the KEK laboratory in Tsukuba, Japan. This magnet is unusual among present designs in that it will have a horizontal magnetic field providing vertical deflection, thus producing vertically polarized radiation considered suitable for certain diffraction work such as studies of crystal growth.

The maximum field will be 60 kG providing a maximum critical energy of 24.9 keV at 2.5 GeV compared to 4.2 keV from the 10 kG bending magnets. The magnet consists of three superconducting coil pairs. The central full effective pole length is 10 cm in the beam direction and the two end poles are each 5 cm long. Because of the relatively small gap height (10 mm) in the horizontal direction, the magnet is to be moved into place only after the beam is stored, thus providing a larger aperture during injection. The orbit displacement is 10 mm in the wiggler.

Some prototype dipoles and their superconducting housings have been constructed and a full prototype has been fabricated. It will be tested by installing the magnet into the INS 400-MeV storage ring to study its effects on beam dynamics.

Table 4. Status of Some Planned or Existing Wiggler Sources Circa 1978

Facility–machine[a] location	E^b (GeV)	I (mA)	R_B (m)	ε_0^B (keV)	Primary[c] usage	ε_c^B (keV)	Periods (N)	λ_w^d (cm)	ε_1^e (keV)	Gap (mm)	Field (kG)	Status[f]	Reference
PEP (R)													
SLAC, USA	4–12.7	50	165.5	78.1	L/D	215	1	160	1.92	45.0	20.0	C	(45, 46)
SSRL–SPEAR (R)													
SLAC, USA	1.3–4	200	12.8	11.1	SR	19.2	3	34.3	0.44	44.5	18.0	O	(15)
SLAC, USA	1.3–4	200	12.8	11.1	UR	0.5–3	29	6	2.5	27	3.0	D	(15)
Photon Factory (R)													
KEK, Japan	2.5	500	8.3	4.2	SR	24.9	1			10.0	60.0	P/D	(47)
SRS (R)													
Daresbury, UK	1–2	350	5.6	3.2	SR	13.3	1	69.0	0.055	42.0	50.0	C	(48)
NSLS (R)													
BNL, USA	0.7	1000	1.9	0.4	UR	0.13	80	2.54	0.18	12.7	4.0	C	(49)
	2.5	500	8.2	4.2		1.66			2.34				
X-ray ring	2.5	500	8.2	4.2	SR	20.8	1–3	14.0	0.42	35.0	50.0	D	(50)
VUV ring	0.7	1000	1.9	0.4	SR	0.4	1–5	16.0	0.029	50.0	12.0	D	(50)
PULS–Adone (R)													
Frascati, Italy	1.5	60	5.0	1.5	SR	2.8	3	65.4	0.033	40.0	18.5	C	(51, 52)
VEPP-3 (R)													
Novosibirsk, USSR	1.5–2.2	100	6.2	3.8	SR	11.3	10	9.0	0.51	8.0	35.0	C	(53)
LURE-ACO (R)													
Orsay, France	0.1–0.54	180	1.1	0.33	UR	0.078	24	4.0	0.069	22.0	4.0	C	(54)
ARUS (S)													
Yerevan, USSR	4.5	5	24.6	8.2	SR	24.2	1			20.0	18.0	O	(55)
SIRIUS (S)													
Tomsk, USSR	≤1.4	100	4.2	1.4	UR	0.39	10	7.0	0.27	85.0	3.0	O	(14)
Pakhra (S)													
Moscow, USSR	≤1.2	30	4.0	1.0	UR	0.26	20	2.0	0.68	n.a.	2.7	O	(13, 56)

[a] S = Synchrotron and R = Storage Ring.

[b] E is the particle energy for which the wiggler is expected to operate and not the maximum particle beam energy.

[c] L = Luminosity, D = Damping rates, SR = Synchrotron Radiation, UR = Undulator Radiation.

[d] λ_w, the wiggler wavelength, was derived from the separation between adjacent poles when specific predictions were not available.

[e] ε_1 is the fundamental's energy in the weak field limit ($K \ll 1$) computed for the highest possible particle energy, e.g., $E = 18$ GeV for PEP.

[f] P = Prototype, C = Construction, D = Design, and O = Operational.

5.2.2. NSLS at Brookhaven, USA

Several wiggler magnets are planned for the two storage rings now in construction at the National Synchrotron Light Source (NSLS) at Brookhaven. For the x-ray ring (2.5 GeV) or the VUV ring (700 MeV) a superconducting undulator with iron is planned[49] that will have the highest number of periods of any planar wiggler. The gap height is made to be 1.27 cm, which is also the length of each iron pole, to achieve a short wavelength. The total length of the magnet is 2.2 m and the peak field is 4 kG. The system parameters are comparable to the Orsay undulator.

For the x-ray ring, a 50-kG superconducting standard wiggler with up to 3λ is planned. For the VUV ring a 12-kG standard wiggler with up to 5λ is planned that has essentially the same field capability as the ring but would presumably be tunable and provide more intensity than the bends. Both of these are projected for early 1982, i.e., not for initial operation of the ring.[50]

5.2.3. PULS-Adone at Frascati, Italy

A seven pole, 3λ standard wiggler magnet[51] with a peak field of 18.5 kG is in construction for the 1.5-GeV storage ring Adone at Frascati, Italy. The gap height is 40 mm and the total length is 2.1 m. As one of the first wigglers to be initiated with more than three poles, a number of two- and three-dimensional field calculations were carried out for it that led to a rather interesting profile for the yoke, which they describe as a "Toblerone" design.[52] This is intended to reduce the coupling from far away poles. The limit of this design would be a series of separate H-type dipoles except for the design of the flux return circuit. It would not be expected to be as efficient or to have as high a field capacity. It will be interesting to compare the results of field measurements for this structure with those for the SPEAR wiggler.

5.2.4. Pakhra at Moscow, USSR

A 20λ undulator was first operated in May 1977[13, 56] on the 1.2-GeV synchrotron at Pakhra. The magnet consists of a single line of poles above the beam with no return poles below. The magnet is arranged to allow sufficient vertical aperture during injection. After injection the beam is moved vertically close to the poles and the magnet is pulsed on. Typically the magnet was pulsed on at the 100–200-MeV level with the synchrotron operating with a peak energy of about 500 MeV. By turning the rf off shortly after the wiggler is turned on, measurements were made of the visible light produced by the magnet. The magnet can produce a peak field of 2.7 kG but a lower field (0.5 kG) was used to produce undulator radiation in the visible part of the spectrum.

It is also reported[56] that a "universal" helical undulator (see Section 4.1.2) with a 9.6-cm period, 96-cm length, 9-cm inner diameter, and 1-kG peak field along the axis is being built.

5.2.5. LURE–ACO at Orsay, France

A 24λ superconducting undulator magnet[54] is planned for operation in the ACO 540-MeV storage ring. The magnetic field is in the horizontal direction and has a peak value of 4 kG that is obtained with iron poles and superconducting coils to obtain the small wavelength ($\lambda_W = 4$ cm). The prototype of the pole and coil configuration is shown in Figure 20. Such a device should produce intense, quasi-monochromatic peaks based on a fundamental wavelength as short as 180 Å. Figure 21 shows this undulator and the "inverted T" vacuum chamber used to obtain vertically polarized photon beams. The horizontally large aperture is needed for injection. Following injection and damping of the radial betatron oscillations, the beam must either be steered into the wiggler gap or the wiggler moved into the beam.

Two important advantages of vertical polarization are (*i*) the possibility of more uniformly distributing the source brightness over transverse phase space and (*ii*) the ability to distribute secondary beam lines associated with the wiggler source in the horizontal plane rather than vertically without loss of intensity as is typical for horizontal polarizations. These factors can simplify not only the design of the beam lines and experimental areas but also certain individual experiments.

5.2.6. SSRL–SPEAR Undulator

A 29λ permanent magnet undulator is planned for SPEAR. It is 1.95 m long, has a period, λ_W, of 6 cm and is made of $SmCo_5$ blocks with a remanent field, B_r, of about 8.5 kG. Figure 22 shows the design, which is due to K. Halbach of LBL. The peak field on the orbit is given by

$$B_0 = 2B_r \frac{\sin(\pi/M)}{\pi/M} [1 - \exp(-2\pi h/\lambda_W)]\exp(-\pi g/\lambda_W)$$

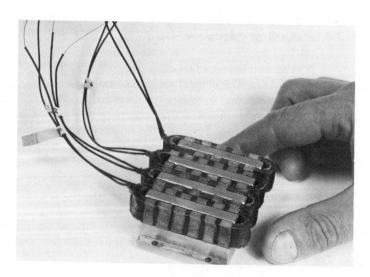

Figure 20. Full-scale prototype of the iron pole and superconducting coils for a 4-cm wavelength undulator for ACO (courtesy of J. Perot, Y. Farge, M. Lemonnier, and Y. Petroff; Orsay).[54]

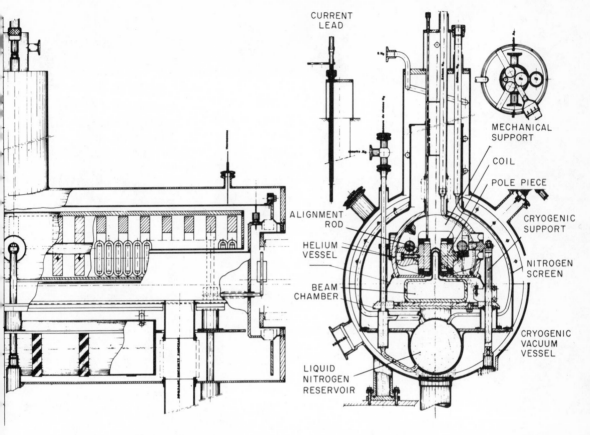

Figure 21. Cross section of the superconducting undulator for LURE–ACO (courtesy of J. Perot, Y. Farge, M. Lemonnier, and Y. Petroff; Orsay).[54]

where M is the number of blocks per period (4 in this case), h is the block height (1.5 cm), and g is the gap height (2.7 to 6.0 cm, remotely variable). The peak field on the orbit varies between 3 and 0.5 kG. With SPEAR operating at 3.5 GeV the fundamental wavelength [see equation (2)] varies between 15 and 6.6 Å. To maintain a zero magnetic field integral over the length of the magnet, end blocks may be changed in their angular orientation as the gap is varied. See also Reference 57.

5.2.7. Sirius at Tomsk, USSR

Detailed study of UR at synchrotrons is complicated by low intensities due to the limited time duration of the UR compared to the SR background coming from the adjacent ring magnets and the attendant difficulties due to beam broadening effects. Nonetheless, some intriguing measurements have recently been made at Tomsk[14, 58] that appear to verify several differences between SR and UR beyond their basic spectral differences. These come mainly in the angular polarization

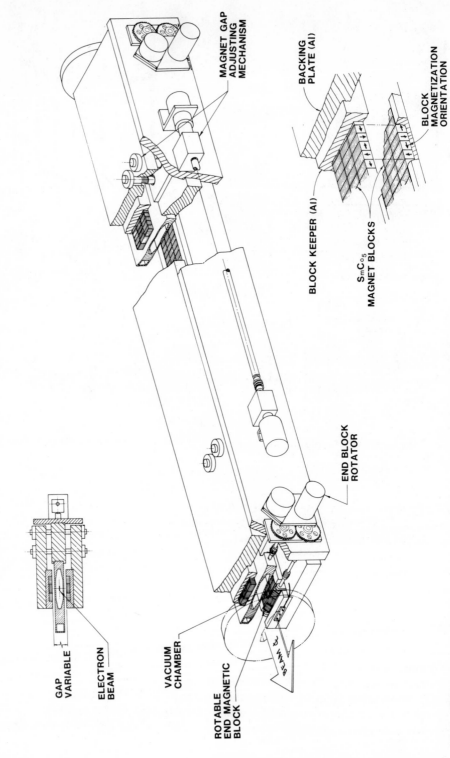

Figure 22. Schematic drawing of a permanent magnet undulator for SPEAR. See text for details.

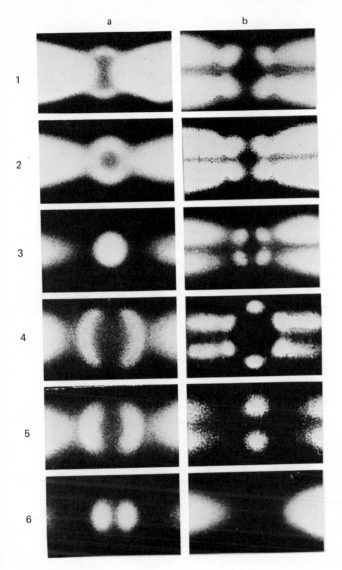

Figure 23. Observations of SR and UR at Tomsk with a 10λ wiggler set for $K = 0.7$ for different electron energies and photon polarizations. The first column gives the σ component and the second gives the π component. Different rows correspond to decreasing beam energy going from top to bottom (1–6). The relative frequency or harmonic number for each spectrum is $n = 0.6(1), 0.8(2), 1.0(3), 1.4(4), 1.8(5),$ and $2.0(6)$ (courtesy of M. Nikitin; Tomsk).

characteristics. For instance, whereas SR is generally elliptically polarized, there are frequencies of UR that are completely linearly polarized regardless of the angle of observation. Figure 23 shows some data for $K \simeq 0.7$ and $B = 268$ G for different electron energies and photon polarizations. The first column corresponds to the σ component of polarization (**E** is parallel to the orbital plane of the electron) and the second column to the π component (**E** perpendicular to the orbital plane). The different rows correspond to decreasing electron energy from top to bottom with the peak of the first harmonic being the third from the top. The peak of the second harmonic for the σ component is in row 6 and shows no intensity along the undulator axis. The SR spectrum without the undulator is shown in the lower-right-hand picture at about twice the frequency of the peak in the fundamental, i.e., near the frequency of the second harmonic. The angular variations of each polarization with electron energy are remarkable over such a small variation of energy (~ 2).

6. Future Directions, Possibilities, and Conclusions

Based on the various considerations and comparisons just discussed, it would appear that the subject of wigglers will become an increasingly important area for future research and development. It is now clear that they do in fact represent the next significant step in the development of high-power, monochromatic sources of radiation. All of the various criteria by which one judges a radiation source imply that considerably more effort should go into this area. While there are many inevitable problems likely to arise, we believe their effects will mostly influence the time scale of developments. For instance, it is clear where certain technical problems are now limiting the field. Improvements will be needed in optical elements and in beam transport systems. The optical elements needed to transport high-power x-ray beams having small emittance are simply not available nor do windows exist that are sufficiently reliable under such conditions for the very high vacuum required in storage rings.

In the same vein, it is clear that any developments in storage rings that allow higher current and lower emittance beams will be important—especially in colliding-beam operation since this would extend undulator radiation into the x-ray region by running parasitically on higher-energy machines intended primarily for high-energy physics research such as PEP or PETRA. However, economic considerations may also justify multiple usage for some proposed lower-energy rings such as ALA (Anello a Luminosita Alta) at Frascati (0.5–1.2 GeV) if the luminosities represent significant improvements over existing machines. Since many practical applications already exist for rings with even lower energies than ALA, wigglers should offer significant economic advantages. Although wigglers have been explored on Tantalus, the 240-MeV ring at Wisconsin, they are not presently in use there.[59]

The fact that we can compare the radiation expected from wigglers to that from lasers is another indication of the potential. While wigglers are certainly no replacement for lasers, one can foresee many applications of the principles of laser physics being incorporated and taken advantage of in this area. The so-called free-electron laser and cyclotron maser are realities. One can expect to see studies of coherent effects within and external to the particle beam bunch as well as exploration of coherent, stimulated channeling effects in crystals. The construction of undulators with permanent magnetic materials appears to be most cost effective.

Looking beyond these possibilities, one might ask where the next step will come that might provide the next order of magnitude improvement in intensity, energy, or resolution. Since we have already admitted that nature was better at providing high wiggler fields than man seems able to devise (e.g., crystal channeling), it may well be that the next development will require a new concept in accelerator technology, e.g., in more controlled or even radiationless acceleration and containment of particle beams. Whereas the major advantages of wigglers is that they provide effectively higher beam currents and energies, literal improvements in these areas would allow wigglers to be operated increasingly as undulators. Such improvements would be "ecological," i.e., more efficient, but will probably only come about from a more unified consideration of the complete cycle of acceleration, control, and usage of particle beams.

ACKNOWLEDGMENTS

The authors have benefitted from many discussions on these subjects and are most indebted for the knowledge and enthusiasm generated. One of the authors (JES) would like to thank Darragh Nagle and Louis Rosen of the Los Alamos Scientific Laboratory for allowing him to pursue this subject during a sabbatical leave from that laboratory. During this period there were many valuable discussions about wigglers and SR with Gosta Brogren who was a visiting scientist at SSRL. The authors wish to thank Bill Brunk for his collaboration on a number of wiggler development projects at SLAC. They also thank Roberto Coisson, Gerry Fischer, Klaus Halbach, Dick Helm, and Phil Morton with whom they had valuable discussions as well as Dick Early for several POISSON calculations. Finally, they would like to thank Linda Burtness for her help with the art work and drawings as well as Mary Lou Beldner, Eleanor Horning, and Joanne Marchetti for their help with the manuscript. This work was supported by the Department of Energy under contract number DE-AC03-765F00515 and by the National Science Foundation under contract number DMR77-27489 with the cooperation of the Department of Energy.

References

1. Synchrotron radiation was first observed by F. B. Elder, A. M. Gurewitsch, R. V. Langmuir, and H. C. Pollock, Radiation from electrons in a synchrotron, *Phys. Rev.* **71**, 829–830 (1947); *Phys. Rev.* **74**, 52–60 (1948); also see note of G. C. Baldwin, *Phys. Today* **28**, 9–10 (1975).
2. J. Schwinger, On the classical radiation of accelerated electrons, *Phys. Rev.* **75**, 1912–1925 (1949); *Phys. Rev.* **70**, 798–799 (1946).
3. V. L. Ginsburg, *Bull. Acad. Sci. USSR Phys. Ser.* **92**, 165–182 (1947).
4. H. Motz, Applications of the radiation from fast electron beams, *J. Appl. Phys.* **22**, 527–535 (1951).
5. H. Motz, W. Thon, and R. N. Whitehurst, Experiments on radiation by fast electron beams, *J. Appl. Phys.* **24**, 826–833 (1953).
6. D. F. Alferov, Yu. A. Bashmakov, and E. G. Bessonov, Undulator radiation, *Sov. Phys. Tech. Phys.* **18**, 1336–1339 (1974); also see A. Hofmann, Quasimonochromatic synchrotron radiation from undulators, *Nucl. Instrum. Methods* **152**, 17–21 (1978).
7. K. W. Robinson, Electron radiation at 6 BeV, CEA Report No. 14 (1956).
8. H. Winick and R. H. Helm, Standard wiggler magnets, *Nucl. Instrum. Methods* **152**, 9–15 (1978).
9. I. A. Grisheev, V. I. Myakota, V. I. Kolosov, V. I. Beloglad, and B. V. Yakimov, *Sov. Phys. Dokl.* **5**, 272–275 (1960).
10. A. Hofmann, R. Little, J. M. Paterson, K. W. Robinson, G. A. Voss, and H. Winick, Design and performance of the damping system for beam storage in the Cambridge Electron Accelerator, Proceedings of the Sixth International Conference on High Energy Accelerators, CEAL-2000 (September 1967), pp. 123–129.

11. A. I. Alikhanyan, S. K. Esin, K. A. Ispiryan, S. A. Kankanyan, N. A. Korkhmazyan, A. G. Oganesyan, and A. G. Tamanyan, Experimental investigation of x radiation of ultrarelativistic electrons in magnetic undulators, *JETP Lett.* **15**, 98–100 (1972).

12. L. R. Elias, W. M. Fairbanks, J. M. J. Madey, H. A. Schwettman, and T. I. Smith, Observation of stimulated emission of radiation by relativistic electrons in a spatially periodic transverse magnetic field, *Phys. Rev. Lett.* **36**, 717–720 (1976); also see D. A. G. Deacon, L. R. Elias, J. M. J. Madey, G. J. Ramian, H. A. Schwettman, and T. I. Smith, First operation of a free electron laser, *Phys. Rev. Lett.* **38**, 892–894 (1977).

13. D. F. Alferov, Yu. A. Bashmakov, K. A. Belovintsev, E. G. Bessonov, and P. A. Cherenkov, Observation of undulating radiation with the Pakhra synchrotron *JETP Lett.* **26**, 385–388 (1977); Part. Accel. **9**, 223–236 (1979). D. F. Alferov, Yu. A. Bashmakov, K. A. Belovintsev, E. G. Bessonov, A. M. Livshits, V. V. Mikhailin, and P. A. Cherankov *Pis'ma Zh. Tekh. Fiz.* **4**, 625–629 (1978); Eng. Trans.: *Sov. Tech. Phys. Lett.* **4**, 251–2 (1978). D.F.

14. A. N. Didenko, A. V. Kozhevnikov, A. F. Medvedev, and M. M. Nikitin, Experimental study of the spectral, angular, and polarization properties of undulator radiation, *Pis'ma Zh. Tekh. Fiz.* **4**, 689–693 (1978). Eng. Trans.: *Sov. Tech. Phys. Lett.* **4**, 277–8 (1978).

15. M. Berndt, W. Brunk, R. Cronin, D. Jensen, A. King, J. Spencer, T. Taylor, and H. Winick, Initial operation of SSRL wiggler in SPEAR, Proceedings of the 1979 Particle Accelerator Conference, *IEEE Trans. Nucl. Sci.* **26**, 3812–3815 (1979). See also H. Winick and J. Spencer in *Nucl. Instrum. Met.* (to be published).

16. M. Sands, *The Physics of Electron Storage Rings. An Introduction*, Proceedings of the International School of Physics "Enrico Fermi," Course XLVI, B. Touschek (ed.), Academic Press (1971); also see R. H. Helm, M. J. Lee, P. L. Morton, and M. Sands, Evaluation of synchrotron radiation integrals, Proceedings of the 1973 Particle Accelerator Conference, *IEEE Trans. Nucl. Sci.* **20**, 900–903 (1973).

17. R. H. Helm, Modeling the Effects of a Flat Wiggler on a Storage Ring Beam, Stanford Linear Accelerator Center, Report PEP-272 (1978); also see R. H. Helm, Effects of the SSRL Wiggler on the SPEAR Beam, SLAC, PEP Report-273 (1978).

18. J. M. Paterson, J. R. Rees, and H. Weidemann, Control of Beam Size and Polarization Time in PEP, PEP Report 125 (1975); also in Wiggler Magnets, H. Winick and T. Knight (eds.), pp. 99–116, SSRP Report No. 77/05 (1977).

19. N. A. Korkhmazyan and S. S. Elbakyan, One Possibility for Detection of Fast Charged Particles, *Sov. Phys. Dokl.* **17**, 345–347 (1972).

20. D. F. Alferov, Yu A. Bashmakov, E. G. Bessenov, and B. B. Govorkov, On using helical wigglers with high energy electron beams from proton synchrotrons for particle separation and quasimonochromatic photon production, Proceedings of the Tenth International Conference on High Energy Accelerators (September 1977), pp. 124–132.

21. A. P. Sabersky, Monitoring the beams in SPEAR with synchrotron light, Proceedings of the 1973 Particle Accelerator Conference, *IEEE Trans. Nucl. Sci.* **20**, 638–641 (1973).

22. R. Coisson, Monitoring high energy proton beams by narrow-band synchrotron radiation, Proceedings of the 1977 Particle Accelerator Conference, *IEEE Trans. Nucl. Sci.* **24**, 1681–1682 (1977). The first observation has recently been reported at CERN by R. Bossart, L. Burnod, R. Coisson, E. D'Amico, A. Hofmann, and J. Mann (submitted to *Nucl. Instrum. Methods*).

23. D. F. Alferov and E. G. Bessenov, Measurement of proton beam parameters by electromagnetic radiation of protons in wiggler magnets, Proceedings of the Tenth International Conference on High Energy Accelerators (September 1977), pp. 387–390.

24. A. Garren (private communication).

25. D. S. Gemmell, Channeling and related effects in the motion of charged particles through crystals, *Rev. Mod. Phys.* **46**, 129–227 (1974).

26. M. A. Kumakhov, Theory of radiation from charged particles in a crystal, *Sov. Phys. JETP* **45**, 781–789 (1977); see also S. Kheifets, The Radiation of the Electrons Channeled Between Planes of a Crystal, Stanford Linear Accelerator, Report PEP-270 (1978).

27. H. J. Kreiner, F. Bell, R. Sizmann, D. Harder, and W. Huttle, Rosette motion in negative particle channeling, *Phys. Lett. A* **33**, 135–136 (1970).

28. G. Bologna, G. Diambrini Palazzi, and G. P. Murtas, High-Energy bremsstrahlung from a silicon single crystal, *Phys. Rev. Lett.* **4**, 572–574 (1960); also see W. K. H. Panofsky and A. N. Saxena, *Phys. Rev. Lett.* **2**, 219–221 (1959).

29. R. H. Milburn, Electron scattering by an intense polarized photon field, *Phys. Rev. Lett.* **10**, 75–77 (1963).
30. R. F. Schwitters, Experimental Review of Beam Polarization in High Energy e^+, e^- Storage Rings, SLAC Report 2258 (1979).
31. P. Wilson, R. Servranckx, A. P. Sabersky, J. Gareyte, G. E. Fischer, A. W. Chao, and M. H. R. Donald, Bunch lengthening and related effects in SPEAR II, Proceedings of the 1977 Particle Accelerator Conference, *IEEE Trans. Nucl. Sci.* **24**, 1211–1214 (1977).
32. P. A. Franken, A. E. Hill, C. W. Peters, and G. Weinreich, Generation of optical harmonics, *Phys. Rev. Lett.* **7**, 118–119 (1961).
33. A. L. Schawlow and C. H. Townes, Infrared and optical masers, *Phys. Rev.* **112**, 1940–1949 (1959).
34. J. E. Spencer, Alternatives for wigglers with iron-dominated magnets, in Wiggler Magnets, H. Winick and T. Knight (eds.), pp. 134–149, SSRP Report 77/05 (1977); also see H. D. Ferguson, J. E. Spencer, and Klaus Halbach, A general ion-optical correction element, *Nucl. Instrum. Methods* **134**, 409–420 (1976).
35. J. D. Jackson, *Classical Electrodynamics*, Wiley, New York (1962).
36. W. Heitler, *The Quantum Theory of Radiation*, Oxford Press, London (1960).
37. B. M. Kincaid, A short period helical wiggler as an improved source of synchrotron radiation, *J. Appl. Phys.* **38**, 2684–2691 (1977).
38. J. P. Blewett and R. Chasman, Orbits and fields in the helical wiggler, *J. Appl. Phys.* **48**, 2692–2698 (1977).
39. P. M. Morton (private communication).
40. C. Pelligrini, The free-electron laser and its possible developments, this volume, Chapter 22.
41. E. D. Courant and H. S. Snyder, Theory of the alternating gradient synchrotron, *Ann. Phys.* (*N.Y.*) **3**, 1–48 (1958).
42. POISSON is an improved version of TRIM and was developed by J. R. Spoerl, R. F. Holsinger, and K. Halbach. PANDIRA is a new code developed recently by K. Halbach and R. F. Holsinger for the design of magnetic circuits that include permanent magnetic materials.
43. G. Blanpied, R. Liljestrand, G. W. Hoffman, and J. E. Spencer, Corrective element for high resolution magnetic optics, *Nucl. Instrum. Methods* **134**, 421–435 (1976).
44. SPEAR Design Report, Stanford Linear Accelerator Center (August 1969); also see A. Garren, M. Lee, and P. Morton, SPEAR Lattice Modifications to Increase Synchrotron Light Brightness, SLAC Internal Report SPEAR-193 (1976).
45. J. E. Spencer, Current status of wigglers at SLAC, Proceedings of the Wiggler Conference, A. Luccio, A. Reales, and S. Stipcich (eds.), Frascati, Italy (June 1978).
46. W. Brunk, G. Fischer, and J. Spencer, Single wavelength standard wiggler for PEP, Proceedings of the 1979 Particle Accelerator Conference, *IEEE Nucl. Sci.* **26**, 3860–3862 (1979).
47. T. Yamakawa, H. Kitamura, and S. Sato, A model of a superconducting wiggler for the Photon Factory, Proceedings of the Second Symposium on Accelerator Science and Technology, March 23–25, 1978, Institute of Nuclear Science, Tokyo, pp. 91–92.
48. D. E. Baynham, P. T. M. Clee, and D. J. Thompson, A 5-tesla wiggler magnet for the SRS, *Nucl. Instrum. Methods* **152**, 31–35 (1978); also V. P. Suller (private communication).
49. W. Sampson, Wiggler Magnets, SSRP Report 77/05, H. Winick and T. Knight (eds.), pp. 51–53, Stanford Synchrotron Radiation Project (1977).
50. S. Krinsky (private communication).
51. M. Bassetti, A. Cattoni, A. Luccio, M. Preger, S. Tazzaari, A transverse wiggler magnet for Adone, Proceedings of the Tenth Internation Conference on High Energy Accelerators, Protvino, USSR (July 1977), Vol. II, pp. 391–399.
52. A. U. Luccio, Status report on the Adone wiggler and x-ray line, Proceedings of the Wiggler Conference, Frascati (June 1978); A. U. Luccio, G. Paroti, and M. Ricci, Computer design of the magnetic field of a transverse wiggler, Proceedings of the Conference on Computation of Magnetic Fields, Grenoble (September 1978).
53. L. M. Barkov, V. B. Baryshev, G. N. Kulipanov, N. A. Mezentsev, V. F. Pindyurin, A. N. Skrinsky, and V. M. Khorev, A proposal to install a superconductive wiggler magnet on the storage ring VEPP-3 for generation of synchrotron radiation, *Nucl. Instrum. Methods* **152**, 23–29 (1978).
54. Y. Farges and Y. Petroff (private communication); see also J. Perot, Design of an undulator with short wavelength for ACO, Proceedings of the Wiggler Conference, Frascati, Italy (June 1978).

55. M. Petrossian (private communication).
56. K. A. Belovintsev (private communication).
57. H. Winick and J. Spencer, *Nucl. Instr. Methods* (to be published).
58. M. M. Nikitin (private communication).
59. W. Trzeciak, A wavelength shifter for the Wisconsin Electron Storage Ring, *IEEE Trans. Nucl. Sci.* **18**, 213–215 (1971).

22

The Free-Electron Laser and Its Possible Developments

C. PELLEGRINI

1. Introduction

The two experiments[1, 2] made at Stanford University in 1975 and 1977, which have shown that it is possible to operate a free-electron laser, have been a turning point in the research on stimulated electromagnetic radiation by relativistic electrons. The interest in this subject and the number of papers discussing it have been increasing very rapidly since the results of these experiments have been known. As a result of all this work we have now a good understanding of the basic principles underlying a free-electron laser and of the feasibility and characteristics of this device.

It is, however, possible that new ideas might be introduced in the near future that will change some parts of our picture of the free-electron laser, perhaps making it a more powerful source of radiation than we think of today.

In this paper I will mainly discuss our understanding of the operation of the free-electron laser and of its capabilities and limitations. After this I will discuss some of the new work being done and what results might be expected from this work.

A large part of this paper will be concerned with the possibility of operating a free-electron laser in an electron storage ring. It seems possible to obtain a high-power tunable laser, with an output power of the order of one kilowatt or more and tunable over a wavelength region of the order of 0.1 to 10 μm, from a free-electron laser operating in an electron storage ring of intermediate energy, 500 or 600 MeV. Such a ring is also a good source of synchrotron radiation in the vuv region. The combination of the synchrotron radiation and of the tunable free-electron laser

C. PELLEGRINI ● Brookhaven National Laboratory, Upton, New York 11973.

would make this ring a unique research tool in many fields, such as, for example, solid state physics, photochemistry, surface physics, and biology. If it should become possible to increase the laser power to the level of 10 kw or more, then the system might become of interest also for advanced industrial applications.

Before discussing the free-electron laser, we will briefly discuss in Section 2 the spontaneous radiation from an electron beam in a wiggler magnet, to introduce some basic quantities and the notations that will be used in the remainder of the paper. Section 3 will cover the basic theory of the free-electron laser. This is followed, in Section 4, by a discussion of the Stanford experiments.

The operation of a free-electron laser in an electron storage ring is discussed in Section 5. A possible choice of parameters for the storage ring and for the laser is described in the same section.

In the final part of the paper the possible new lines of development are discussed. This includes what might be done to increase the free-electron laser power output and the wavelength range over which the system is tunable.

2. Spontaneous Radiation of Relativistic Electrons in Wiggler Magnets

In this section we review briefly some properties of spontaneous radiation of relativistic electrons in a wiggler magnet. A full discussion of this subject can be found in Chapter 21. We will consider only the case of a helical wiggler magnet. However, all the results can be applied, with only minor modifications, also to the case of a transverse wiggler.

Consider one electron moving near the axis of a helical magnet, having N periods of length λ_{w}. The magnet is made by two helical conductors with the current flowing in opposite directions, so that the magnetic field $\mathbf{B}_{\mathrm{w}}$ near the axis is given by[3]

$$\mathbf{B}_{\mathrm{w}} = B_{\mathrm{w}}(\hat{\mathbf{x}} \cos k_{\mathrm{w}} z + \hat{\mathbf{y}} \sin k_{\mathrm{w}} z) \tag{1}$$

where $\hat{\mathbf{x}}$ and $\hat{\mathbf{y}}$ are unit vectors in the x and y directions (see Figure 1), and

$$k_{\mathrm{w}} = 2\pi/\lambda_{\mathrm{w}} \tag{2}$$

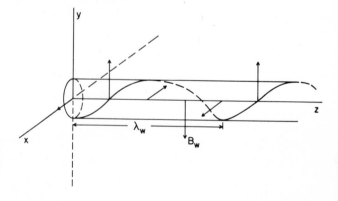

Figure 1. The electron trajectory in a helical magnet; the arrow shows the magnetic field vector at some points along the trajectory; the length of one period of the magnet is λ_{w} and the number of periods is N.

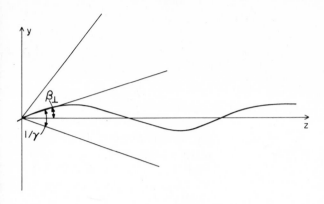

Figure 2. Projection of the electron trajectory on the y–z plane, showing the pitch angle $\beta_\perp$ and the aperture of the radiation cone.

Let

$$\boldsymbol{\beta} = \beta_\parallel \,\hat{\mathbf{z}} + \boldsymbol{\beta}_\perp \tag{3}$$

be the electron velocity in units of the light velocity c. In what follows we will use the assumption that

$$\beta_\parallel \approx 1, \qquad \beta_\perp \ll 1 \tag{4}$$

so that the electron energy, measured in rest mass units $m_0 c^2$, is large

$$\gamma = (1 - \beta_\parallel^2 - \beta_\perp^2)^{-1/2} \gg 1 \tag{5}$$

However, the transverse velocity $\beta_\perp$ is nonrelativistic.

The electron trajectory can be calculated analytically to order $\beta_\perp/\gamma$ using assumption (4). Within this approximation the electron describes a helical trajectory[4, 5] with constant energy (neglecting the energy lost as radiation) such that

$$\boldsymbol{\beta}_\perp = (K/\gamma)(\hat{\mathbf{x}} \cos k_\mathrm{w} z + \hat{\mathbf{y}} \sin k_\mathrm{w} z) \tag{6}$$

where

$$K = \gamma\beta_\perp = \frac{eB_\mathrm{w}\lambda_\mathrm{w}}{2\pi m_0 c^2} \tag{7}$$

The pitch δ of the helix is

$$\delta = \beta_\perp \tag{8}$$

and the radius ρ is

$$\rho = \frac{\beta_\perp \lambda_\mathrm{w}}{2\pi} \tag{9}$$

The electrons moving on the helical trajectory will emit circularly polarized radiation within an angle of the order of $1/\gamma$[4] (Figure 2). Assuming that the pitch angle is smaller than $1/\gamma$ or that

$$K = \beta_\perp\gamma \ll 1 \tag{10}$$

most of this radiation is emitted in a single line of wavelength

$$\lambda_s = \frac{\lambda_w}{2\gamma^2}(1 + K^2 + \gamma^2\theta^2) \tag{11}$$

where θ is the observation angle. The emitted radiation contains also the higher harmonics of this line. These harmonics become stronger when K is increased and can be neglected if $K \ll 1$.

Let us consider only the radiation emitted in the fundamental harmonic and in the forward direction ($\theta = 0$). The shape of the emission line, as a function of the distance from the frequency $\omega_s = 2\pi c/\lambda_s$, is given by[4]

$$I(\omega) \approx \left(\frac{\sin[\pi N(\omega - \omega_s)/\omega_s]}{\pi N(\omega - \omega_s)/\omega_s}\right)^2 \tag{12}$$

where $I(\omega)$ is the radiation intensity per unit frequency interval. The homogenous width $\Delta\omega/\omega$ of the emission line is obtained from (12) and is

$$\frac{\Delta\omega}{\omega} = \frac{1}{2N} \tag{13}$$

Because of the dependence of ω_s on γ, K, and θ, this linewidth can be increased by the energy spread and the angular divergence of the electron beam and by a change in B_w with the distance from the magnet axis.

One way to look at spontaneous radiation emission is to consider the interaction of the electrons with the wiggler magnetic field in the reference frame moving with velocity $\beta^* = \beta_\parallel$ in the z direction. In this frame the electrons execute a circular motion, which, when the parameter K is less than one, is nonrelativistic. In this same reference frame the wiggler magnetic field is transformed to an electromagnetic field

$$B_w^* = \gamma^* B_w \tag{14}$$

$$E_w^* = \beta^* B_w^* \tag{15}$$

where

$$\gamma^* = (1 - \beta^{*2})^{-1/2} = \frac{\gamma}{(1 + K^2)^{1/2}} \tag{16}$$

In the limit $\beta^* = 1$ these fields are identical to those of a plane electromagnetic wave traveling along the z axis, with wavelength

$$\lambda^* = \lambda_w/\gamma^* \tag{17}$$

When $\beta^* \approx 1$, we can, to order $1/\gamma^{*2}$, replace the wiggler field with the field of the equivalent plane electromagnetic wave. This approximation is the same as the Weizsäcker–Williams method of virtual quanta.[6] The interaction between the electrons and the plane wave (Figure 3) gives rise to Thomson scattering,[7] if the condition $\lambda_w^* \gg \hbar/m_0 c$ is satisfied. In the moving frame the back scattered photons

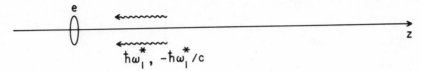

Figure 3. The electrons and the equivalent photons of the wiggler magnetic field as seen in the reference frame moving with velocity β^*.

have the same wavelength λ_W^*. Transforming to the laboratory frame their wavelength is

$$\lambda = \lambda_W^* / \gamma^* (1 + \beta^*) \tag{18}$$

or using (16) and $1 + \beta^* \approx 2$,

$$\lambda = \frac{\lambda_W}{2\gamma^2} (1 + K^2) \tag{19}$$

This is the same relationship as obtained from (11) for $\theta = 0$.

The power radiated per electron over the entire solid angle can be easily calculated in the moving frame.[7] Transforming back to the laboratory frame one obtains

$$P_W = \frac{2}{3} \frac{r_e c}{R_W^2} \frac{\gamma^4 m_0 c^2}{(1 + K^2)^2} \tag{20}$$

where $R_W = \gamma m c^2 / e B_W$ is the radius of curvature of an electron trajectory in a uniform field B_W and r_e is the classical electron radius. The total power radiated by an average electron current i_{av}, in a wiggler of length $\lambda_W N$, can be written as

$$P_T = \frac{2}{3} \frac{\lambda_W Nc}{R_W^2} \frac{\gamma^4 m_0 c^2}{(1 + K^2)^2} \frac{i_{av}}{i_A} \tag{21}$$

where i_A is the Alfvén current ($i_A = 1.7 \times 10^4$ A). Assuming $\gamma = 10^3$, $\lambda = 5$ cm, $N = 100$, $R_W = 10$ m, or $B_W = 1.7$ kG, one has

$$P_T = 18 i_{av} \text{ W}$$

with i_{av} measured in amperes. The radiation wavelength is 2.5×10^{-8} m.

3. Elementary Theory of the Free-Electron Laser

3.1. Stimulated Radiation by Relativistic Electrons

The problem of how to enhance the spontaneous radiation, increasing the radiated power and decreasing the linewidth, has been considered for a long time. Modulation of the electron beam density on a scale length of the same order as the radiated wavelength has been proposed by Motz[8, 9] as a way to produce such an

C. PELLEGRINI

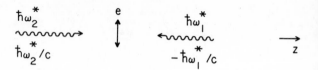

Figure 4. The two sets of photons and the electrons as seen in the moving reference frame.

enhancement. In a different approach, proposed by Madey *et al.*,[10, 11] one wants to use the stimulated radiation by relativistic electrons. This method of stimulated radiation in a wiggler magnetic field is similar to that of stimulated inverse Compton scattering, which was studied by Dreicer,[12] Pantell *et al.*,[13] and Sukhatme and Wolff.[14]

Let us consider again the electron beam and the wiggler magnetic field in the frame of reference moving with velocity $\beta^* = \beta_{\parallel}$. As in Section 2, our system can be described as an ensemble of N_e electrons interacting with n_1 photons of frequency $\omega_1^* = 2\pi c\gamma^*/\lambda_w$ and momentum $-\hbar\omega_1^*/c$ along the z axis. Let us add another set of n_2 photons of frequency ω_2^* and momentum $\hbar\omega_2^*/c$ (Figure 4), such that $\omega_2^* \approx \omega_1^*$. It is now possible to produce stimulated Thomson scattering, transferring photons from state 2 to state 1 or from state 1 to state 2 of the electromagnetic field. If we use a quantum-mechanical description we can say that the transition probability of one photon from state 2 to 1 or from 1 to 2, is proportional to[6]

$$T_{1\to2} \approx n_1(n_2 + 1) \tag{22}$$

$$T_{2\to1} \approx (n_1 + 1)n_2 \tag{23}$$

The two curves describing $T_{1\to2}$ and $T_{2\to1}$ as a function of ω^* are equal apart from a small shift $\Delta\omega/\omega = 4\hbar\omega_1^*/m_0 c^2$. The increase in the number of photons in state 2 is given by the difference $T_{1\to2} - T_{2\to1}$ and, to a good approximation, can also be written as the derivative of $T_{1\to2}$ multiplied by $\Delta\omega$. Defining the gain as

$$G = \Delta n_2/n_2 \tag{24}$$

one obtains for G a curve that is the derivative of $T_{1\to2}$ and is shown in Figure 5.

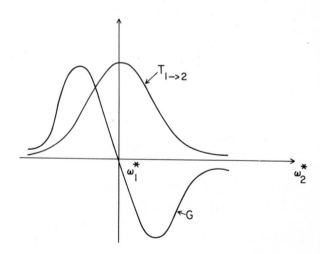

Figure 5. The one photon transition probability $T_{1\to2}$ from state 1 to state 2 of the electromagnetic field and the gain curve G for photons in state 2.

Quantum-mechanical calculations of the gain and other characteristics of the system have been made by Madey,[10] Colson,[15, 16] and Bambini and Stenholm.[17] It is also possible to use a classical description. It has been shown by Madey and Deacon[18] that the classical description is valid as long as two conditions are satisfied. The first is that the photon wavelength should be larger than the electron Compton wavelength, or

$$\lambda \gg \hbar/m_0 c \tag{25}$$

The second is that the power flux of the electromagnetic field S be large, or

$$S \gg \pi^{3/2}\hbar c^2/\lambda^3 \lambda_w \tag{26}$$

This is equivalent to the requirement that there should be many photons in a volume $\lambda^2 \lambda_w$, or

$$n\lambda^2 \lambda_w \gg 1 \tag{27}$$

where n is the number of photons per unit volume. When this condition is satisfied the fluctuations in the number of photons emitted or absorbed are small and the classical description is appropriate. In the following we will assume that (25) and (26) are always satisfied and use a classical description of the laser.

Many classical theories of stimulated radiation by relativistic electrons in a wiggler magnet have been developed. They can be divided into two main groups. In one group we find the calculations of stimulated radiation from a single electron.[15, 19–23] In the second group are the calculations based on the Boltzmann–Maxwell equations or similar techniques that allow not only the calculation of the radiation but also the changes in the electron beam energy distribution and density distribution.[24–29] In the next section we will use the single-electron approach to give a simple description of the free-electron laser. We will also use results derived from theories of the second group in the discussion of free-electron laser operation.

3.2. Classical Theory of Stimulated Radiation in Wiggler Magnets

In this section we will describe a simple single-electron model of the free-electron laser, following the approach developed by Colson.[16, 19] With the help of this model we will obtain the small signal gain. We will then describe in a semiqualitative way other effects including saturation, linewidth, and changes in the electron distribution function, referring the reader to the literature for a complete and rigorous derivation.

Let us consider again a relativistic electron moving in a helical wiggler magnetic field, with N periods each of length λ_w. A circularly polarized plane electromagnetic field is also propagating along the wiggler axis (Figure 6). We will call this system a free-electron laser amplifier. Let the wiggler magnetic field $\mathbf{B}_w$ and the electron velocity be given by equations (1) and (3). The electric and magnetic field of the circularly polarized plane wave are assumed to be

$$\mathbf{E} = E_0[\hat{x} \sin(kz - \omega t + \phi_0) + \hat{y} \cos(kz - \omega t + \phi_0)] \tag{28}$$

$$\mathbf{B} = \hat{z} \times \mathbf{E} \tag{29}$$

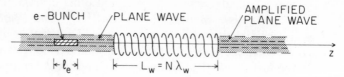

Figure 6. A schematic representation of the electron bunch moving in the wiggler magnet and amplifying a plane wave.

The equations of motion, neglecting spontaneous radiation, are

$$\dot{\mathbf{p}} = e\mathbf{E} + e\boldsymbol{\beta} \times (\mathbf{B} + \mathbf{B}_w) \tag{30}$$

$$\dot{\gamma} = (e/m_0 c)\mathbf{E} \cdot \boldsymbol{\beta} \tag{31}$$

where $\mathbf{p}$ is the electron momentum. These equations can be approximated, to order $1/\gamma^2$, by

$$\dot{\boldsymbol{\beta}} = (e/m_0 \gamma c)\boldsymbol{\beta} \times \mathbf{B}_w \tag{32}$$

$$\dot{\gamma} = (e/m_0 c)\mathbf{E} \cdot \boldsymbol{\beta} \tag{33}$$

showing that the electron trajectory is determined by $\mathbf{B}_w$, while the energy exchange with the electromagnetic field is determined by $\mathbf{E}$.

Equation (32) gives us the same results already discussed in Section 2, i.e., the electron trajectory is a helix with pitch $\beta_\perp = K/\gamma$ and radius $\rho = \beta_\perp \lambda_w/2\pi$, with K given by (7). Using (28) and expression (6) for $\beta_\perp$ we can now write (33) as

$$\dot{\gamma} = -\frac{eE_0 \beta_\perp}{m_0 c} \sin \phi \tag{34}$$

where the phase ϕ is

$$\phi = \left(\frac{2\pi}{\lambda_w} + \frac{2\pi}{\lambda}\right)z - \omega t + \phi_0 \tag{35}$$

An effective energy exchange can take place only when the variation in ϕ, in traversing the wiggler, is small compared to 2π. If $T_w \approx N\lambda_w/c$ is the time it takes one electron to traverse the wiggler, this condition can be written as

$$T_w \dot{\phi} \ll 1 \tag{36}$$

If we ask that $\dot{\phi} = 0$ and use the equations

$$z = \beta_\parallel ct \tag{37}$$

$$\beta_\parallel = \left(1 - \frac{1 + K^2}{\gamma^2}\right)^{1/2} \approx 1 - \frac{1 + K^2}{2\gamma^2} \tag{38}$$

we find that we can satisfy this request for a value γ_r of the electron energy, related to λ and λ_w by

$$\lambda = \frac{\lambda_w}{2\gamma_r^2}(1 + K^2) \tag{39}$$

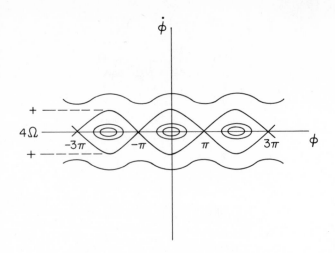

Figure 7. Electron trajectories in phase-space plane ϕ-$\dot{\phi}$.

This is the same relationship as (19) defining the spontaneous radiation wavelength. We call γ_r the resonant energy.

For $\gamma \neq \gamma_r$ and $(\gamma - \gamma_r)/\gamma \ll 1$, one can write

$$\dot{\phi} = \frac{4\pi c}{\lambda_w} \frac{\gamma - \gamma_r}{\gamma_r} \tag{40}$$

From (34) and (40) one can obtain the pendulumlike equation

$$\ddot{\phi} + \Omega^2 \sin \phi = 0 \tag{41}$$

$$\Omega^2 = \frac{2e^2 E_0 B_w}{m_0^2 c^2 \gamma_r^2} \tag{42}$$

We now assume that the change in E_0, produced by the electron in the time T_w, is small and consider $E_0 = $ constant. Then from (41) we obtain a first integral

$$\dot{\phi}^2 = \dot{\phi}_0^2 + 2\Omega^2(\cos \phi - \cos \phi_0) \tag{43}$$

A phase-space representation of (43) is given in Figure 7. For given initial values of ϕ_0 and $\dot{\phi}_0$ at the entrance of the wiggler, the electron will follow a well-defined trajectory in phase space. The energy exchange along this trajectory will then be given by (34).

The electron energy change gives the change in the intensity I of the radiation field. Hence from (41) or (43) and (34) we can calculate the gain G in the electromagnetic field intensity produced by an electron beam of peak current i_p,

$$G = \Delta I/I \tag{44}$$

To evaluate G we assume that the electron beam is bunched, that it has cylindrical symmetry with a transverse cross section area Σ_e, and that the transverse area of the electromagnetic wave is Σ_p. We will also assume, for simplicity, that

$$\Sigma_p = \Sigma_e \tag{45}$$

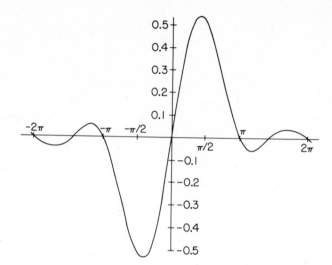

Figure 8. The function $C(x)$, describing the gain as a function of the electron energy.

Since the electron energy loss depends on the initial value of the electron energy and phase, to evaluate the gain we must specify the distribution of these two quantities in the electron beam. Since the scale of the phase change is given by the radiation wavelength λ, which we assume to be in the range between the ultraviolet and the infrared, and the electron bunch length l_e is much longer than λ, it is reasonable to assume a uniform phase distribution. To simplify the calculation we also consider the case of a monochromatic electron beam. We call G_0 the gain for this case.

Under the conditions discussed we obtain

$$G_0 = 2^{5/2}\pi^2\lambda^{3/2}\lambda_{\rm w}^{1/2}N^3\,\frac{K^2}{(1+K^2)^{3/2}}\,\frac{i_p}{i_{\rm A}\Sigma_p}\,C\!\left(2\pi N\frac{\gamma-\gamma_r}{\gamma_r}\right) \qquad (46)$$

where $i_{\rm A}$ is the Alfvén current ($i_{\rm A} = 1.7 \times 10^4$ A). The function $C(x)$ is given by

$$C(x) = -\frac{d}{dx}\left(\frac{\sin x}{x}\right)^2 \qquad (47)$$

and is plotted in Figure 8. One notices that the function $(\sin x/x)^2$ describes the shape of the spontaneous radiation line. Hence the gain is proportional to the derivative of the spontaneous radiation line.[30]

From the behavior of the gain versus the electron energy, described by the function $C[2\pi N(\gamma-\gamma_r)/\gamma_r]$ and shown in Figure 8, one can see that:

    a. the electrons will lose energy if $\gamma > \gamma_r$ and gain energy if $\gamma < \gamma_r$; no energy change occurs if $\gamma = \gamma_r$;

    b. the maximum gain is obtained for $(\gamma - \gamma_r)/\gamma_r \approx 1/4N$;

    c. the gain becomes small and goes to zero when $(\gamma - \gamma_r)/\gamma_r > 1/2N$;

    d. the maximum energy that an electron can transfer to the radiation field is of the order of

$$(\Delta\gamma)_{\max} \approx \gamma_r(1/2N) \qquad (48)$$

Thus the efficiency η of the system has an upper limit $1/2N$, i.e.,

$$\eta \leq 1/2N \tag{49}$$

For a nonmonochromatic electron beam, the effective gain can be obtained by multiplying (46) by the energy distribution function and integrating. It is clear from Figure 8 that it is possible to obtain a large gain only if the width of the energy distribution function $\sigma_E = \langle(\Delta\gamma/\gamma)^2\rangle^{1/2}$, is small compared to $1/2N$. If the opposite is true, i.e., if $\sigma_E > 1/2N$, then the gain will average to a value near to zero.

We can now summarize the characteristics that the electron beam must have to obtain a good energy transfer to the radiation field:

$$\sigma_E < 1/2N \tag{50}$$

$$\Sigma_e \leq \Sigma_p \tag{51}$$

$$\langle\theta^2\rangle^{1/2} < 1/\gamma(2N)^{1/2} \tag{52}$$

The first condition determines the electron energy spread, the second the transverse beam dimension. The last condition (52) determines the electron beam angular divergence θ and is introduced because of the variation of the radiation wavelength with θ [see also (11)].

We want now to estimate the order of magnitude of G_0. To do this we assume that $C(x) \approx 0.5$, $K = 1$, and that Σ_p is determined by the diffraction limit, $\Sigma_p \approx \lambda(N\lambda_W)$. Then one has

$$G_0 = \pi^2 N^2 \left(\frac{\lambda}{\lambda_W}\right)^{1/2} \frac{i_p}{i_A}$$

and for $\lambda = 1$ μm, $\lambda_W = 5$ cm, and $N = 100$,

$$G_0 \approx 2.6 \times 10^{-2} i_p$$

where i_p is in amperes. This shows that the electron beam peak current must be at least of the order of several ampere to yield a gain of several percent.

We can try to increase the gain by varying some wiggler parameters, and not the electron current. For instance, the gain depends strongly on the number of wiggler periods, and to increase N is an easy way to increase G. If we increase G in this way we reach a saturation condition determined by energy conservation. The change in radiation field energy, ΔE_r, is related to the gain and to the electron energy losses, $\Delta E = m_0 c^2 \Delta\gamma$, by

$$\Delta E_r = GE_r = N_e \Delta E \tag{53}$$

where N_e is the electron number. For a given electron number, since $\Delta E \leq E/2N$, the product GE_r has an upper limit. Because of the limitations on ΔE and G, the maximum peak power that can be transferred from the electron beam to the radiation field is

$$P = \eta(Ei_p/e) \tag{54}$$

where η is given by (49) and (Ei_p/e) is the electron beam peak power. A similar equation can be written for the average power P_{av}, using the average current i_{av},

$$P_{av} = \eta(Ei_{av})/e \tag{55}$$

Since N must be of the order of 10 to 100, it follows from (54) and (49) that only a few percent of the electron beam power can be transferred to the radiation field. As an example, if we assume that $N = 50$, $i_p = 100$ A, $E = 50$ MeV, and $\eta = 1\%$, we get a peak power in the radiation field $P = 50$ MW.

When the energy extracted from the electrons is large, the electron energy spread will also increase. This is due to the fact that the gain, as given by (46), is only an average value, obtained by averaging over the initial electron phase. The real energy loss of any given electron depends on the initial value of the phase and the energy, and is different for each electron. It has been shown[30] that the variation in the energy loss is related to average energy loss and that the two are comparable in magnitude. Hence, when the maximum energy is extracted from the electrons, an energy spread as large as the width of the gain curve is produced. Because of this the effective gain decreases toward zero and the system saturates.

Another consequence of the variation of the electron energy loss, depending on the initial phase and energy, is that an initially smooth electron density distribution will become modulated. As it can be seen from (38), a change in energy leads to a change in the parallel velocity and hence to a change in the density distribution on a scale length given by the radiation wavelength. This bunching effect is intimately connected to the energy transfer from the electron beam to the radiation field and to the increase in energy spread. It is indeed impossible to have gain without bunching and energy spread.[24, 26]

3.3. Principles of Operation of a Free-Electron Laser

When the gain G is large enough, it is possible to use the electron beam–wiggler system not only as an amplifier of electromagnetic radiation, but also as a self-sustained oscillator. This is done by adding an optical cavity to the system (Figure 9). We will call this system a free-electron laser oscillator. One of two mirrors forming the optical cavity can be assumed to be perfectly reflecting, while the other will transmit a certain fraction α_0 of the electromagnetic field intensity. We assume, for simplicity, that other loss mechanisms in the optical cavity are small compared to α_0.

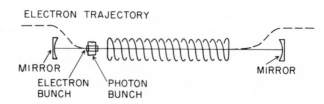

Figure 9. Schematic representation of a free-electron laser oscillator.

For the system to oscillate we must satisfy the condition

$$G \geq \alpha_0 \tag{56}$$

When the oscillator operates in a steady state one must have $G = \alpha_0$, G being the effective gain corresponding to the electron energy spread as determined also by the electron–radiation field interaction. It follows that if we want to operate the system near to its maximum efficiency, $\eta = 1/2N$, we must require that the initial gain G_0, calculated for the initial energy distribution, be much larger than α_0. In most cases this condition can be written as $G_0 \approx 5\alpha_0$. Since α_0 can be of the order of one to two percent, G_0 must be of the order of 10%. This requires a peak electron beam current of the order of a few ampere or larger. When the condition on G_0 is satisfied, the steady-state laser output power is given again by (55).

Since the amplification of the radiation field occurs only where the electron density is nonzero, both the length l_p and the linewidth $\delta\omega$ of the laser radiation pulse are determined by the length l_e of the electron bunch. The laser pulse length l_p is in fact nearly equal or smaller than the electron bunch length $l_p \approx l_e$. This determines the minimum linewidth,

$$\delta\omega = 2\pi c/l_e \tag{57}$$

or, equivalently,

$$\frac{\delta\omega}{\omega} = \frac{\lambda}{l_e} \tag{58}$$

Because of the bunched structure of the electron and photon beams it is also important to "synchronize" the two beams so that the electrons and photons traverse together the wiggler magnet. If we call L_{op} the length of the optical cavity and L_e the distance between the electron bunches, then L_e must be a multiple of L_{op}. We assume for simplicity that

$$L_e = 2L_{op} \tag{59}$$

From the results of this section we can summarize the characteristics that an electron beam must have to operate a free-electron laser oscillator. This is done in Table 1. Not many existing electron accelerators produce a beam with these characteristics. Among them is the Stanford Superconducting Linac, which has been used to make all experiments carried out up to now, and which will be described in Section 4. Electron storage rings and linear electron accelerators, which might be either properly modified existing machines, or specially designed ones, will be used in the future to operate free-electron lasers.

The radiation power output that can be obtained from a laser oscillator operated by a linear accelerator is given by (55), provided that the condition (56) is satisfied. For an electron beam power of one megawatt it is possible to obtain a laser power output of the order of a few tens of kilowatts. The efficiency can be a few percent times the efficiency for producing the electron beam itself. The radiation wavelength can range between a few tenths of a micron and a few hundred microns for electron energies between a few MeV and a few hundred MeV and for $\lambda_w \approx 10$ cm.

Table 1. Electron Beam Parameters Required to Operate a Free-Electron Laser Oscillator near the Saturation Condition. [a]

Parameter	Analytical relationship	Order of magnitude estimate
Electron energy	$\gamma \approx (\lambda_{\mathrm{w}}/2\lambda)^{1/2}$	$E \approx 5\text{–}500$ MeV
Energy spread	$\left\langle \left(\dfrac{\Delta E}{E}\right)^2 \right\rangle^{1/2} < \dfrac{1}{2N}$	$\left\langle \left(\dfrac{\Delta E}{E}\right)^2 \right\rangle^{1/2} \approx 0.1\%$
Angular divergence	$\langle \theta^2 \rangle^{1/2} < \left(\dfrac{1+K^2}{2N\gamma^2}\right)^{1/2}$	$\langle \theta^2 \rangle^{1/2} \approx 10^{-3}$ rad
Transverse bunch area	$\Sigma_e \leq \Sigma_p$	$\Sigma_e \approx 1\text{–}10$ mm
Peak current	$G_0 \approx 5\alpha_0$	$i_p \approx 1\text{–}100$ A

[a] The laser wavelength range is assumed to be between 200 and 0.02 μm, the wiggler parameter K to be less than one, the wiggler period 5 cm, and the number of periods to be 100.

The case of a free-electron laser oscillator operated by an electron storage ring requires special discussion, because of the peculiar fact that in this case a single electron bunch is used to interact repetitively with the laser radiation field. This case will be discussed in Section 5.

4. Free-Electron Laser Experiments

The first experiments demonstrating the possibility of obtaining stimulated emission of radiation by relativistic electrons in a wiggler magnetic field was performed in 1975–76[1, 31–33] by L. R. Elias, W. M. Fairbank, J. M. J. Madey, H. A. Schwettman, and T. I. Smith at Stanford University. The experimental system can be represented schematically as in Figure 6. The incoming radiation field has a wavelength of 10.6 μm and is provided by a CO_2 laser.

The electron beam used in the experiment has an energy of 24 MeV. The beam is bunched and the bunch length $l_e = 1.3$ mm. The distance between bunches $L_e = 24$ cm, corresponding to a repetition frequency of 1300 GHz. The peak electron current can be as high as 70 mA. The energy spread is $\sigma_E = 5 \times 10^{-4}$ and the angular divergence and transverse area are well within the limits required by (51) and (52).

The helical wiggler magnet has a period length $\lambda_{\mathrm{w}} = 3.2$ cm, a total length of 5.2 m, and $N = 162$ periods. The magnetic field has a value $B_{\mathrm{w}} = 2.4$ kG, corresponding to a value of $K \approx 0.7$.

Measurements of the radiation power at the wiggler exit exhibit a modulation at 1.3 GHz, corresponding to the electron beam modulation, showing that the CO_2 laser beam has been amplified. A plot of the spontaneous radiation power and of the gain in the CO_2 beam as a function of the electron energy gives curves in accordance with the theory and similar to those shown in Figure 5.

The peak value of the gain, for a peak electron current of 70 mA, observed in the experiment is 7%. Since the laser beam transverse area in the wiggler magnet is somewhat uncertain (within a factor of 2) it is difficult to make a quantitative comparison between the measured gain and the gain calculated from (46). However, within the limitations of the observation, there is agreement between theory and experiments.

The energy transfer from the electrons to the radiation field is 2×10^{-3}. This is in agreement with the value expected from (49).

Operation of a free-electron laser oscillator was achieved for the first time in a second experiment.[2, 31-34] A schematic representation of the experimental apparatus used is given in Figure 9. In this experiment the optical cavity and the electron energy are adjusted for operation at 3.4 μm. Also, the electron beam characteristics are modified, with respect to the first experiment, to increase the peak current. This is obtained by pulsing the electron gun and increasing the distance between electron bunches to 25.4 m. In this way the peak current is raised up to an estimated value of 2.6 A. The bunch length is assumed to be the same as in the previous experiment. The wiggler characteristics are also unchanged. The electron energy, corresponding to $\lambda = 3.4$ μm, is 43.5 MeV. The optical cavity has a length of 12.7 m, equal to half the electron bunch distance [see (59)], and one of the mirrors has a transmission of 1.5%.

With the free-electron laser switched off (below the threshold for self-oscillations) the spontaneous radiation average power is 10^{-8} W and the linewidth $(\delta\omega/\omega) \approx 1\%$. When the laser gain is raised above threshold the average power increases by almost eight orders of magnitude, to 0.36 W, corresponding to an estimated peak power of 7 kW. The linewidth is also reduced to $(\delta\omega/\omega) \approx 2.3 \times 10^{-3}$, in agreement with (58) and the estimated bunch length.

In the experiment it was also possible to measure the electron beam energy spread at the exit of the wiggler. The value obtained is $\sigma_E = 9 \times 10^{-4}$, compared with an initial value of 5×10^{-4}.

The electron energy is converted into energy of the radiation field with an efficiency of the order of 10^{-4}, lower by an order of magnitude than that obtained in the amplifier experiment, and expected from (49). Some possible reasons for this low value have been given.[32-34] They might be either the system used to measure the total laser energy output, or the synchronization of the electron and photon bunches, or a shortening of the laser pulse. It has been observed in the experiments that a small change in the resonator length, on the order of a few microns, has a large effect on the laser output power. These observations have been discussed in Reference 31 and the effect has been studied by Hopf et al.[26] who have shown that the relative timing of the electron and photon bunch is an important parameter, which can strongly influence the laser gain. However, the maximum gain and output power obtainable are not changed by this effect when the laser is run in proper operating conditions.

The two experiments discussed in this section demonstrated the feasibility of a free-electron laser. Also we have a good understanding of the basic principles governing the stimulated emission of radiation by relativistic electrons in a wiggler magnetic field. There are some particular problems, such as the details of the time and space structure of the radiation pulse or of the electron energy distribution at the wiggler exit, on which more experimental and theoretical work is needed. However there are no reasons to expect important differences between the theoretical predictions of the main laser parameters and what will be observed in future experiments.

5. The Free-Electron Laser Operation in an Electron Storage Ring

5.1. Principal Characteristics of an Electron Storage Ring

This section is dedicated to a discussion of the possibility of operating a free-electron laser in an electron storage ring. A schematic design of such a system is shown in Figure 10.

An electron beam circulating in a storage ring can have very desirable properties for free-electron laser operation. The energy spread, the bunch dimensions, the angular divergence, and the peak current obtainable in these machines can satisfy all the requirements of a free-electron laser over a wide range of electron energies.

However, in a storage ring the same electron bunch is recirculated in the ring and in the laser for a very large number of revolutions. In existing rings the beam lifetime is of the order of several hours, while the revolution period is less than a microsecond, leading to a total number of revolutions of the order of 10^{10}. The periodic interaction of the electrons with the laser radiation field can modify the electron beam characteristics, such as the energy spread. How much these characteristics are modified and what is the effect of these modifications on the laser radiation is the main problem to be studied to predict the performance of a free-electron laser in a storage ring.

The characteristics of an electron beam in a storage ring are well known[35, 36] and some of them will be briefly described here. A discussion of how they are modified by the interaction with the laser will follow in Section 5.2.

An electron circulating in a storage ring moves near a closed equilibrium orbit determined by the electron momentum and by the ring magnetic field. This field provides also the focusing force that keeps the electron near the equilibrium orbit.

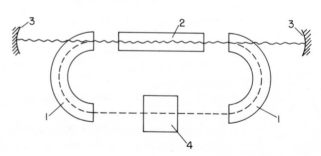

Figure 10. Schematic design of an electron storage ring with the insertion of a free-electron laser oscillator. 1, Bending magnet; 2, wiggler magnet; 3, laser cavity mirrors; 4, radio-frequency cavity.

During its motion the electron loses energy by the emission of synchrotron radiation. This energy is restored to the electrons by the field of one or more radio-frequency cavities. This radio-frequency field also produces a bunching of the electron beam.

The electron bunch dimensions and energy spread are determined by the magnetic structure, the radio-frequency system, and the synchrotron radiation emitted. Of particular importance is the fact the emission of synchrotron radiation provides a damping mechanism for the electron motion around the equilibrium orbit. This damping can balance the diffusion around the equilibrium orbit due to noiselike effects, as the scattering on the residual gas in the vacuum chamber or the fluctuations in synchrotron radiation emission. The balance between the damping and the diffusion determines the steady-state bunch dimensions and energy spread.

As an example we can estimate the energy spread σ_E. Under the most usual conditions, this is determined mainly by the fluctuations in synchrotron radiation emission and by the synchrotron radiation damping and can be written as

$$\sigma_E = \frac{55}{24\sqrt{3}} \frac{r_e \lambda_C \gamma^5}{R^3} \frac{1}{2\tau} \tag{60}$$

where r_e and λ_C are the classical radius and the Compton wavelength of the electron, R is the radius of curvature of the electron trajectory in the ring, and τ is the damping constant.

The damping time constant τ is related to the total energy lost as synchrotron radiation in one revolution, U_0. If T is the revolution period and E the electron energy, one has

$$\frac{1}{\tau} = \frac{1}{T} \frac{U_0}{E} \tag{61}$$

and

$$U_0 = \frac{2}{3} \frac{r_e}{R^2} \gamma^4 m_0 c^2 \mathscr{L} \tag{62}$$

$\mathscr{L}$ being the length of the trajectory in the magnetic field.

Expressions similar to (60) can be obtained also for the beam dimensions and angular divergence, using the appropriate expression for the damping time constant and for the diffusion terms.

5.2. Electron Beam–Free-Electron Laser Interaction

When the free-electron laser is inserted in a storage ring the electrons will also interact with the radiation field in the laser optical cavity. In a single passage of the cavity the electrons lose energy and this energy loss is compensated by the storage ring radio-frequency field. In addition, also the electron beam energy spread increases, as discussed in Section 3.

To discuss the effect on a single electron of the periodic interaction with the laser cavity, it is important to notice that the electron phase at the wiggler entrance with respect to the radiation field, as defined by (35), changes by a quantity much larger than 2π from one revolution to the next. This large phase variation is due to the fluctuation in the electron momentum, due for instance to fluctuation in the emission of synchrotron radiation. A momentum change of the order of a few times 10^{-7} is enough, for a typical storage ring, to change the length of the electron trajectory by one radiation wavelength, at $\lambda = 1 \ \mu\text{m}$, and hence to change the phase by 2π. Because of this random change in the initial phase, which introduces a random term in the electron–laser energy exchange, the effect of the electron–laser interaction is equivalent to a diffusion process.

It has been shown by Renieri[37, 38] that, if we neglect synchrotron radiation damping, this diffusion process causes the electron energy spread to grow and hence the laser gain to decrease, so that in a small number of revolutions the laser would switch itself off. Synchrotron radiation damping now becomes very important, because it can balance the growth term in the energy spread, allowing the system to reach a steady-state condition in which there is a net energy transfer from the electrons to the radiation field. The laser output power is then proportional to the amount of damping provided by synchrotron radiation. An estimate of the laser power output P_L can be obtained simply by multiplying (55) by the ratio of the revolution period to the damping time constant, T/τ, i.e.,

$$P_\text{L} = \eta (E i_\text{av}/e) \frac{T}{\tau} \tag{63}$$

Using (61) one can also write P_L as

$$P_\text{L} = \eta U_0 i_\text{av}/e \tag{64}$$

which shows that P_L is proportional to the total power radiated by the electron as synchrotron radiation, $U_0 i_\text{av}/e$.

The relationship (63), which has been derived in Reference 37, can be interpreted in a simple way. Assume that the electron bunch goes through the laser cavity and that in this passage the maximum energy is extracted from the electron, so that, according to (48) and using $E = m_0 c^2 \gamma$, $\Delta E = \eta_M E$, with $\eta_M = 1/2N$. In doing so the electron energy spread increases and the effective gain is reduced to a small value. After this single passage we "switch off" the laser and wait for a time τ until radiation damping brings the beam back to its initial state. The average laser power that can be obtained in this mode of operation is given by

$$P_\text{L} = N_e \, \Delta E/\tau$$

N_e being the number of electrons. Introducing the average current

$$i_\text{av} = e N_e/T$$

and using (48) we obtain (63).

In the same paper by Renieri[37, 38] it has also been shown that to obtain the maximum laser power, given by (64) with $\eta = 1/2N$, we must also satisfy the condition that the small signal gain G_0, evaluated for the unperturbed energy spread, must be larger than the cavity loss α_0, by a factor of the order of five, or

$$G_0 = 5\alpha_0 \tag{65}$$

This condition is like the one that we discussed in Section 3.3.

The efficiency of the free-electron laser storage ring system, defined as the ratio of the laser power to the power lost by the electrons as synchrotron radiation, is seen from (64) to be given by $\eta \leq 1/2N$, and is the same as in the case of the Stanford experiments, which used an electron beam from a linear accelerator. We must notice that in both cases this is the efficiency of energy transfer from the electrons to the radiation field. The real efficiency must be evaluated by taking into account the total power requirement of the storage ring system, including the power required to operate the magnets and the radio-frequency cavity.

5.3. The Free-Electron Laser Operation in a Storage Ring

To make an estimate of the characteristics of a free-electron laser operating in a storage ring, we need equations (39), (64), and (65). We make the assumption that the unperturbed (i.e., laser off) electron bunch dimensions and energy spread satisfy all the conditions of Table 1.

It is convenient to rewrite the small signal gain formula (46) assuming $C(x) \approx 0.5$ and $\Sigma_p \approx \lambda(N\lambda_{\rm w})$, using (58) and (59) to introduce the synchronization condition and the linewidth, and introducing the average electron current as

$$G_0 = 8\pi^2 N^3 \left(\frac{\delta\omega}{\omega}\right) \frac{K^2}{(1 + K^2)^2} \frac{\gamma i_{\rm av}}{i_{\rm A}} \tag{66}$$

We can now summarize the main equations describing the free-electron laser characteristics as

$$\lambda = \frac{\lambda_{\rm w}}{2\gamma^2}(1 + K^2) \tag{67a}$$

$$8\pi^2 N^3 \left(\frac{\delta\omega}{\omega}\right) \frac{K^2}{(1 + K^2)^2} \frac{\gamma i_{\rm av}}{i_{\rm A}} = 5\alpha_0 \tag{67b}$$

$$P_{\rm L} = \frac{1}{2N} U_0 \left(\frac{i_{\rm av}}{e}\right) \tag{67c}$$

To these we must add the expression (62) for U_0, which we repeat here for convenience

$$U_0 = \frac{2}{3} \frac{r_e}{R^2} \gamma^4 m_0 c^2 \mathscr{L} \tag{62}$$

We want to use now (67) and (62) to make some order of magnitude estimates. The gain formula (67b) has a strong dependence on the number of wiggler periods N. The total wiggler length, $N\lambda_{\rm w}$, is however limited because the wiggler itself is a part of the storage ring system. It is possible to design storage rings where a long wiggler, of the order of ten meters, can be inserted. Here we make the conservative assumption that this length is five meters. Then, for $\lambda_{\rm w}$ of the order of 5 to 25 cm, one has a value of N between 100 and 20.

The linewidth is given by the ratio λ/l_e [see (58)]. The electron bunch length l_e can vary between a few centimeters and one meter. Assuming λ to be between 0.1 and 100 μm, the linewidth can vary in the range 10^{-6} to 10^{-3}. The small signal gain G_0 is proportional to $\delta\omega/\omega$, so that it decreases with λ, thus limiting the possibility of producing laser light in the short wavelength region, 0.1 μm or less.

To increase the output power we want to increase U_0. This can be done by increasing the electron energy to a value as high a possible. An upper limit for γ occurs due to the decrease in laser wavelength with increasing γ. This will reduce $\delta\omega/\omega$ and G_0. To reduce somewhat the dependence of the wavelength on γ, we can attempt to increase K to values higher than one. The wiggler parameter K given by (7) can be increased up to values of the order of ten. As an example, for $\lambda_W = 25$ cm and $B_W = 1.3$ kG, one has $K = 3.2$. It should be pointed out that, while for spontaneous radiation increasing K will increase the intensity of the higher harmonics, a laser oscillator is not influenced by a change in K because of the overall narrow amplification line of the laser cavity–electron beam system.

To obtain a large value of U_0 we can choose a small radius of curvature R. Considering the practical mechanical limitations in a storage ring, we can assume $R = 1$ m. Then for a bending magnetic field of 20 kG we have an electron energy

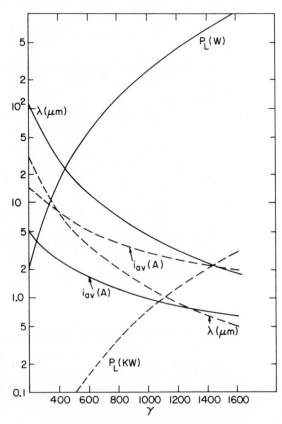

Figure 11. Laser wavelength λ, laser power P_L, and average electron current i_{av} versus electron energy in rest mass units γ. The wiggler parameters used are $\lambda_W = 25$ cm and $N = 20$. The synchrotron radiation energy loss per electron is $U_0 = 20$ keV. The dashed line is for $K = 3$ and $\delta\omega/\omega = 10^{-5}$, and the continuous line is for $K = 6$ and $\delta\omega/\omega = 10^{-4}$.

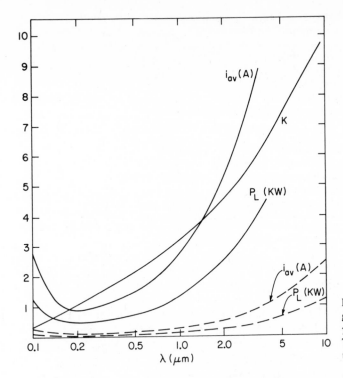

Figure 12. Laser power, electron current, and wiggler parameter k versus λ for fixed electron energy $\gamma = 1200$, $\lambda_{\mathrm{w}} = 25$ cm, $N = 20$, and $U_0 = 20$ KeV. The dashed line is for $\delta\omega/\omega = 10^{-4}$ and the continuous line is for $\delta\omega/\omega = 10^{-5}$.

$E \approx 600$ MeV or $\gamma = 1200$. From (62) we then obtain $U_0 = 2 \times 10^3 \mathscr{L}$ eV/revolution. For a standard storage ring, neglecting the spontaneous radiation in the laser wiggler, one has $\mathscr{L} = 2\pi R$. However it is possible to increase $\mathscr{L}$ for a given R by adding a special high-field wiggler to the ring lattice. In the following we assume $\mathscr{L} = 10$ m and $U_0 = 2 \times 10^4$ eV/revolution. The possible laser wavelength and average power P_{L}, which results from (67), taking also into account the limitation discussed previously, are given in Figures 11 and 12. In Figure 11 we give λ, P_{L}, and the average electron current needed to obtain the required small signal gain, as a function of the electron energy for given values of λ_{w}, N, and U_0, and for two sets of values of K and $\delta\omega/\omega$. In Figure 12 we give the average electron current, laser power, and wiggler parameter K as a function of λ, for given λ_{w}, N, U_0, γ, and for two values of $\delta\omega/\omega$. The values used are $\lambda_{\mathrm{w}} = 25$ cm, $N = 20$, and $U_0 = 2 \times 10^4$ eV/revolution.

The curves given in Figures 11 and 12 show that it is possible to cover a broad range of laser wavelength and power. However, it is very inportant to notice that the electron current that has been obtained in existing storage rings, or that might be obtained in the future ones, is limited by various intensity-dependent effects. A general introductory discussion of these effects can be found in Reference 39. Considering these limitations it seems reasonable to assume an average electron current of the order of a few amperes. A higher current would require extending the technology of electron storage rings beyond their present limits. It must also be noticed that the electron current given in Figures 11 and 12 is the value needed for maximum energy transfer from the electron to the radiation field, as defined by

**Table 2. Indicative Parameters of a
Storage Ring–Free-Electron Laser System**

Laser wavelength	0.1–10 μm
Average laser power	0.1–1 kW
Linewidth	10^{-4}–10^{-5}
Electron beam–laser efficiency	0.5–2.5%
Wiggler magnetic field	0.15–4.5 kG
Wiggler period	5–25 cm
Total wiggler length	5 m
Electron energy	600 MeV
Average electron current	1–2 A
Radio-frequency power	100 kW
Synchrotron radiation power	40 kW
Ring circumference	40 m

(65). At lower currents the laser would still operate, although with a decreasing power output, until a value of $G_0 = \alpha_0$ is reached. At this current the laser would be at the threshold for self-sustained oscillations.

If we use the additional constraint that $i_{av} \leq 2$ A, we can see from Figure 11 that for the case $K = 6$ and $\delta\omega/\omega = 10^{-4}$ one can cover the wavelength region $2 < \lambda < 20$ μm with a power $20 < P_L < 1000$ W. Similarly from Figure 12 one has, for the case $\delta\omega/\omega = 10^{-4}$, that $0.1 < \lambda < 10$ μm and $100 < P_L < 1000$ W, for K smaller than ten; when $\delta\omega/\omega = 10^{-5}$, then $0.1 < \lambda < 0.7$ μm and $500 < P_L < 1000$ W with K less than two.

A list of possible storage ring and laser parameters is given in Table 2. It must be emphasized that all these numerical estimates are only given to illustrate the order of magnitude of the quantities involved in the system and are not based on an optimized design of a storage ring—free-electron laser. Work to design an optimized system is now being carried out at several laboratories and will permit more precise estimates in the future.

6. Conclusions

The results of the two Stanford experiments and the theoretical work done by several authors have given us a good understanding of the physics and of the possible operation modes of a free-electron laser.

When a free-electron laser is operated with an electron beam provided by a linear accelerator, one is limited mainly by the characteristics of the electron beam. Specially designed linear accelerators, producing a high current with small energy spread, might however be used to operate a high-power free-electron laser over a large wavelength region.

A free-electron laser–storage ring system is also feasible. However, to obtain good performance it will be necessary to design high-current rings, using all our knowledge of the storage ring physics and technology. There are still problems to be solved to design an optimized storage ring. Particular attention must be given to the insertion in the ring of the long laser wiggler magnet without impairing the electron beam characteristics. Similar problems arise for the auxiliary wiggler radiator that might be used to obtain a shorter synchrotron radiation damping time.

Other possibilities to improve the characteristics of the system are being explored. One such possibility, proposed by the Stanford group, is to increase the value of the maximum efficiency using special wiggler magnets in which the value of K is matched to the particle energy so that the dependence of the wavelength on the energy is reduced.[30] These special magnets might increase the maximum efficiency or, equivalently, the width of the gain curve.

A possible way to reduce the threshold current needed for self-sustained oscillations has been proposed by S. Krinsky and Vinokurov.[40, 41] This would extend the range of electron currents over which a laser oscillator can be operated. These authors propose splitting the wiggler magnet in two parts. The electron beam energy modulation produced in the first part is transformed into a density modulation by using a dispersive system. In the second part of the wiggler the density modulation enhances the energy exchange with the radiation field. However, the maximum laser power, or the maximum efficiency of the electron–radiation field energy transfer, is not increased by this technique.

Some proposals have also been advanced to overcome the limitation imposed by the proportionality between the laser power and synchrotron radiation damping in a storage ring. One such proposal is based on the idea of compensating for the energy loss of each particle so as to reduce the energy spread at the exit of the laser.[26, 42] This system requires the use of two wigglers, the second one being used to compensate the energy losses, and thus needs more available free space in the storage ring.

We want also to remark that in all the models and the calculations presented in this paper we have neglected the collective radiation effects. By this we mean that the intensity of radiation has been assumed to be proportional to the number of electrons N_e in a bunch of length $l_e \gg \lambda$. However, if it is possible to obtain a bunching of the electron beam on the scale of the optical wavelength λ, then collective radiation from many electrons might be produced leading to a radiation intensity term proportional to $(N_e \lambda^*/l_e^*)^2$, where λ^* and l_e^* are the wavelength and bunch length in the frame of reference moving with velocity β^* (see Section 2). If the condition $(N_e \lambda^*/l_e^*) > N_e$ is satisfied, this collective term becomes the dominant one and the radiation power is increased. It is interesting to notice that bunching on the scale of the optical wavelength will occur as a result of the interaction between the electrons and the radiation field in the wiggler magnet.

As a result of work along the lines discussed here or of the introduction of new ideas, the capabilities of the free-electron laser might be greatly extended. It is, however, important to remark also that without these improvements the electron storage ring–free-electron laser system, as discussed in Section 5, is a unique source of electromagnetic radiation, capable of giving both synchrotron radiation and a high-power continuously tunable laser beam over a wide wavelength region. As such it can be used advantageously in many research fields, such as solid state physics, photochemistry, molecular physics, and biology.

References

1. L. R. Elias, W. M. Fairbank, J. M. J. Madey, H. A. Schwettman, and T. I. Smith, Observation of stimulated emission of radiation by relativistic electrons in a spatially periodic transverse magnetic field, *Phys. Rev. Lett.* **36**, 717–720 (1976).
2. D. A. G. Deacon, L. R. Elias, J. M. J. Madey, G. J. Ramian, H. A. Schwettman, and T. I. Smith, First operation of a free-electron laser, *Phys. Rev. Lett.* **38**, 892–894 (1977).
3. W. R. Smythe, *Static and Dynamic Electricity*, McGraw-Hill, New York (1950).
4. B. M. Kincaid, A short period helical wiggler as an improved source of synchrotron radiation, *J. Appl. Phys.* **48**, 2684–2691 (1977).
5. J. P. Blewett and R. Chasman, Orbits and fields in helical wigglers, *J. Appl. Phys.* **48**, 2692–2698 (1977).
6. W. Heitler, *The Quantum Theory of Radiation*, Oxford University Press, Oxford (1954).
7. J. D. Jackson, *Classical Electrodynamics*, John Wiley and Sons, New York (1962).
8. H. Motz, Applications of the Radiation from Fast Electron Beams, *J. Appl. Phys.* **22**, 527–535 (1951).
9. H. Motz, W. Thon, and R. N. Whitehurst, Experiments on radiation by fast electrons beams, *J. Appl. Phys.* **24**, 826–833 (1953).
10. J. M. J. Madey, Stimulated emission of bremsstrahlung in a periodic magnetic field, *J. Appl. Phys.* **42**, 1906–1913 (1974).
11. J. M. J. Madey, H. A. Schwettman, and W. M. Fairbank, A free electron laser, *IEEE Trans. Nucl. Sci.* **20**, 980–983 (1973).
12. H. Dreicer, Kinetic theory of an electron–photon gas, *Phys. Fluids* **7**, 735–753 (1964).
13. R. H. Pantell, G. Soncini, and H. E. Puthoff, Stimulated photon–electron scattering, *IEEE J. Quantum Electron.* **4**, 905–907 (1968).
14. V. P. Sukhatme and P. A. Wolff, Stimulated Compton scattering as a radiation source—theoretical limitations, *J. Appl. Phys.* **44**, 2331–2334 (1973).
15. W. B. Colson, Theory of a free electron laser, *Phys. Lett. A* **59**, 187–190 (1976).
16. W. B. Colson, Free electron laser theory, High Energy Physics Laboratory, Stanford University, Report HEPL 820 (1977).
17. A. Bambini and S. Stenholm, The free-electron laser: I, Institute of Theoretical Physics, University of Helsinki, Report HU-TFT-78-1 (1978).
18. J. M. J. Madey and D. A. G. Deacon, in *Cooperative Effects in Matter and Radiation*, C. M. Bowden, D. W. Howgate and H. R. Robl (eds.), pp. 313–320, Plenum Press, New York (1977).
19. W. B. Colson, One-Body Electron Dynamics in a Free-Electron Laser, Institute of Theoretical Physics, Stanford University, Report ITP 574 (1977).
20. A. A. Kolomenskii and A. N. Lebedev, Stimulated radiation of relativistic electrons and physical processes in the free electron laser, P. N. Lebedev Physical Institute, Moscow, Preprint 127 (1977).
21. A. Bambini and A. Renieri, The free electron laser: A single-particle classical model, *Lett. Nuovo Cimento* **21**, 399–404 (1978).
22. A. Bambini, A. Renieri, and S. Stenholm, A classical theory of the free-electron laser in a moving frame, *Phys. Rev. A*, **19**, 2013–2025 (1979).
23. V. N. Baier and A. I. Milstein, To the theory of a free electron laser, *Phys. Lett. A* **65**, 319–322 (1978).
24. F. A. Hopf, P. Meystre, M. O. Scully, and W. H. Louisell, Classical theory of a free electron laser, *Phys. Rev. Lett.* **37**, 1215–1218 (1976).
25. H. Al-Abawi, F. A. Hopf, and P. Meystre, Electron dynamics in a free electron laser, *Phys. Rev. A* **16**, 666–671 (1977).
26. F. A. Hopf, P. Meystre, G. R. Moore, and M. O. Scully, in High Power Lasers and Their Applications, C. Pellegrini (ed.), *Proc. Int. Sch. Phys. "Enrico Fermi,"* Academic Press, New York (to be published).
27. F. A. Hopf, P. Meystre, G. T. Moore, and M. O. Scully, in *Novel Sources of Coherent Radiation*, S. F. Jacobs, M. Sargent III, and M. O. Scully (eds.), pp. 41–114, Addison-Wesley, Reading, Massachusetts (1978).
28. A. Bambini and S. Stenholm, The free-electron laser: II, Institute of Theoretical Physics, University of Helsinki, Report HU-TFT-78-2 (1978).
29. A. Bambini and S. Stenholm, The free-electron laser: III, Institute of Theoretical Physics, University of Helsinki, Report HU-TFT-78-8 (1978).

30. J. M. J. Madey, Relationship Between Mean Radiated Energy, Mean Square Radiated Energy, and Spontaneous Power Spectrum in a Power Series Expansion of the Equation of Motion in a Free Electron Laser, High Energy Physics Laboratory, Stanford University, Report HEPL 823 (1978).

31. D. A. G. Deacon, L. R. Elias, J. M. J. Madey, H. A. Schwettman, and T. I. Smith, in *Proceedings of the Third International Conference on Laser Spectroscopy.* pp. 402–409, Springer-Verlag, Berlin (1977).

32. D. A. G. Deacon, L. R. Elias, J. M. J. Madey, H. A. Schwettman, and T. I. Smith, in *Proceedings of the Society of Photo-Optical Instrumental Engineers*, Optics in Adverse Environments **121**, 89–95 (1977).

33. J. M. J. Madey and J. N. Eckstein, Experimental aspects of the free-electron lasers, in *Proceedings of the Workshop on Free Electron Laser*, University of Trento, Trento (1979).

34. H. A. Schwettman and J. M. J. Madey, Final Technical Report to ERDA, Contracts EY76-S-03-0326, PA49 (1977).

35. M. Sands, in Physics with Intersecting Storage Rings, B. Touschek (ed.), *Proc. Int. Sch. Phys. "Enrico Fermi,"* Vol. XLVI, pp. 257–409, Academic Press, New York (1971).

36. C. Bernardini and C. Pellegrini, Linear theory of motion in electron storage rings, *Ann. Phys. (N.Y.)* **46**, 174–199 (1968).

37. A. Renieri, The free electron laser: the storage ring operation, Centro di Frascati del CNEN, Frascati (Roma), Report No. 77/33 (1977).

38. A. Renieri, in High Power Laser and Their Applications, C. Pellegrini (ed.), *Proc. Int. Sch. Phys. "Enrico Fermi,"* Academic Press, New York (to be published).

39. C. Pellegrini, Colliding beams accelerators, *Ann. Rev. Nucl. Sci.* **22**, 1–24 (1972).

40. N. A. Vinokurov and A. N. Skrinsky, Institute of Nuclear Physics, Novosibirsk, Preprint 77/59.

41. N. A. Vinokurov and A. N. Skrinsky, Institute of Nuclear Physics, Novosibirsk, Preprint 77/67.

42. P. Meystre, G. T. Moore, M. O. Scully, and F. A. Hopf, Toward a resolution of the velocity spread problem in the free electron laser using electron echo techniques (to be published).

Index

743